Bioinspired Legged Locomotion

Bioinspired Legged Locomotion

Models, Concepts, Control and Applications

Maziar A. Sharbafi
André Seyfarth

Butterworth-Heinemann
An imprint of Elsevier

Butterworth-Heinemann is an imprint of Elsevier
The Boulevard, Langford Lane, Kidlington, Oxford OX5 1GB, United Kingdom
50 Hampshire Street, 5th Floor, Cambridge, MA 02139, United States

Notices

Knowledge and best practice in this field are constantly changing. As new research and experience
broaden our understanding, changes in research methods, professional practices, or medical treatment
may become necessary.

Practitioners and researchers must always rely on their own experience and knowledge in evaluating and
using any information, methods, compounds, or experiments described herein. In using such information
or methods they should be mindful of their own safety and the safety of others, including parties for
whom they have a professional responsibility.

To the fullest extent of the law, neither the Publisher nor the authors, contributors, or editors, assume any
liability for any injury and/or damage to persons or property as a matter of products liability, negligence
or otherwise, or from any use or operation of any methods, products, instructions, or ideas contained in
the material herein.

Library of Congress Cataloging-in-Publication Data
A catalog record for this book is available from the Library of Congress

British Library Cataloguing-in-Publication Data
A catalogue record for this book is available from the British Library

ISBN: 978-0-12-803766-9

For information on all Butterworth-Heinemann publications
visit our website at https://www.elsevier.com/books-and-journals

Working together
to grow libraries in
developing countries

www.elsevier.com • www.bookaid.org

Publisher: Joe Hayton
Acquisition Editor: Sonnini R. Yura
Editorial Project Manager: Mariana Kuhl
Production Project Manager: Kiruthika Govindaraju
Designer: Vitoria Pearson

Typeset by VTeX

Maziar A. Sharbafi:

*To my wife, Aida, for her devotion and endless support;
her selflessness will always be remembered.*

Contents

3. Conceptual Models of Legged Locomotion

Justin Seipel, Matthew Kvalheim, Shai Revzen, Maziar A. Sharbafi, and André Seyfarth

A Role for Simple Conceptual Models
Justin Seipel

Part II
Control

4.   Control of Motion and Compliance

Katja Mombaur, Heike Vallery, Yue Hu, Jonas Buchli,
Pranav Bhounsule, Thiago Boaventura, Patrick M. Wensing,
Shai Revzen, Aaron D. Ames, Ioannis Poulakakis, and Auke Ijspeert

5.   Torque Control in Legged Locomotion

Juanjuan Zhang, Chien Chern Cheah, and Steven H. Collins

6. Neuromuscular Models for Locomotion

Arthur Prochazka, Simon Gosgnach, Charles Capaday, and Hartmut Geyer

6.1. Introduction: Feedforward vs Feedback in Neural Control: Central Pattern Generators (CPGs) Versus Reflexive Control
Arthur Prochazka and Hartmut Geyer

Part III
Implementation

7.   Legged Robots with Bioinspired Morphology

*Ioannis Poulakakis, Madhusudhan Venkadesan, Shreyas Mandre,
Mahesh M. Bandi, Jonathan E. Clark, Koh Hosoda, Maarten Weckx,
Bram Vanderborght, and Maziar A. Sharbafi*

List of Contributors

Aaron D. Ames
Mechanical and Civil Engineering and Control and Dynamical Systems, California Institute of Technology (Caltech), United States

Maziar A. Sharbafi
School of Electrical and Computer Engineering, College of Engineering, University of Tehran, Iran
Lauflabor Locomotion Laboratory, TU Darmstadt, Germany

Mahesh M. Bandi
Collective Interactions Unit, OIST Graduate University, Tancha, Okinawa, Japan

Pranav Bhounsule
Department of Mechanical Engineering, The University of Texas at San Antonio, San Antonio, TX, United States

Thiago Boaventura
ADRL, ETH Zürich, Zürich, Switzerland

Jonas Buchli
ADRL, ETH Zürich, Zürich, Switzerland

Charles Capaday
Universitätsmedizin Göttingen, Institute for Neurorehabilitation Systems, Georg-August University Göttingen, Göttingen, Germany

Chien Chern Cheah
School of Electric and Electronic Engineering, Nanyang Technological University, Singapore

Jonathan E. Clark

Department of Mechanical Engineering, FAMU/FSU College of Engineering, Tallahassee, FL, United States

Steven H. Collins

Department of Mechanical Engineering, Carnegie Mellon University, United States

Robotics Institute, Carnegie Mellon University, United States

Hartmut Geyer

Robotics Institute, Carnegie Mellon University, Pittsburgh, PA, United States

Simon Gosgnach

Neuroscience and Mental Health Institute, University of Alberta, Edmonton, AB, Canada

Kevin W. Hollander

SpringActive, Inc.

Koh Hosoda

Department of System Innovation, Graduate School of Engineering Science, Osaka University, Toyonaka, Japan

Yue Hu

Optimization, Robotics & Biomechanics, ZITI, IWR, Heidelberg University, Heidelberg, Germany

Auke Ijspeert

Biorobotics Laboratory, EPFL – Ecole Polytechnique Fédérale de Lausanne, Lausanne, Switzerland

Tim Kiemel

School of Public Health, University of Meryland, United States

Matthew Kvalheim

Electrical Engineering and Computer Science, University of Michigan, Ann Arbor, MI, United States

David Lee

School of Life Sciences, University of Nevada, Las Vegas, United States

Dirk Lefeber
Department of Mechanical Engineering, Vrije Universiteit Brussel, Brussels, Belgium

Shreyas Mandre
School of Engineering, Brown University, Providence, RI, United States

Katja Mombaur
Optimization, Robotics & Biomechanics, ZITI, IWR, Heidelberg University, Heidelberg, Germany

Ioannis Poulakakis
Department of Mechanical Engineering, University of Delaware, Newark, DE, United States

Arthur Prochazka
Neuroscience and Mental Health Institute, University of Alberta, Edmonton, AB, Canada

Shai Revzen
Electrical Engineering and Computer Science, University of Michigan, Ann Arbor, MI, United States

Christian Rode
Department of Motion Science, Friedrich-Schiller-Universität Jena, Jena, Germany

Justin Seipel
Purdue University, West Lafayette, IN, United States

André Seyfarth
Lauflabor Locomotion Laboratory, TU Darmstadt, Germany

Tobias Siebert
Institute of Sport and Motion Science, University of Stuttgart, Stuttgart, Germany

Thomas G. Sugar
Fulton Schools of Engineering, The Polytechnic School, Arizona State University, AZ, USA

Heike Vallery
Faculty of Mechanical, Maritime and Materials Engineering, Delft University of Technology, Delft, The Netherlands

Bram Vanderborght
Department of Mechanical Engineering, Vrije Universiteit Brussel, Brussels, Belgium

Madhusudhan Venkadesan
Department of Mechanical Engineering & Materials Science, Yale University, New Haven, CT, United States

Jeffrey Ward
SpringActive, Inc.

Maarten Weckx
Department of Mechanical Engineering, Vrije Universiteit Brussel, Brussels, Belgium

Patrick M. Wensing
Department of Aerospace and Mechanical Engineering, University of Notre Dame, Notre Dame, IN, United States

Juanjuan Zhang
Department of Mechanical Engineering, Carnegie Mellon University, United States

School of Electric and Electronic Engineering, Nanyang Technological University, Singapore

About the Editors

Maziar A. Sharbafi

Maziar A. Sharbafi is an assistant professor in Electrical and Computer Engineering Department of University of Tehran. He is also a guest researcher at the Locomotion Laboratory, TU Darmstadt. He studied control engineering at Sharif University of Technology and University of Tehran (UT) where he has received his Bachelor and Master's degrees, respectively. He started working on bipedal robot control during his PhD studies at University of Tehran which ended in 2007, and worked more on bio-inspired control approaches since he entered Lauflabor in 2011. His current research interests include bio-inspired locomotion control based on conceptual and analytic approaches, postural stability, and the application of dynamical systems and nonlinear control to hybrid systems (e.g., for legged locomotion).

André Seyfarth

André Seyfarth is a full professor for Sports Biomechanics at the Department of Human Sciences of TU Darmstadt and head of the Lauflabor Locomotion Laboratory. After his studies in physics and his PhD in the field of biomechanics, he went as a DFG "Emmy Noether" fellow to the MIT LegLab (Prof. Herr, USA) and the ParaLab at the university hospital Balgrist in Zurich (Prof. Dietz, Switzerland). His research topics include sport science, human and animal biomechanics and legged robots. Prof. Seyfarth was the organizer of the Dynamic Walking 2011 conference ("Principles and Concepts of Legged Locomotion") and the AMAM 2013 conference ("Adaptive Motions in Animals and Machines").

Chapter 1

Introduction

Maziar A. Sharbafi[*,†], André Seyfarth[†]

School of Electrical and Computer Engineering, College of Engineering, University of Tehran, Iran †Lauflabor Locomotion Laboratory, TU Darmstadt, Germany

In human life movements are required to explore and to interact with the world. Movements are necessary for communication and even help anticipate abstract concepts like time. The act or ability of moving from place to place is called locomotion (Merriam-Webster, 2004). Man-made systems can also be designed to move, e.g., wheeled vehicles that can be fast and efficient. However, many of these systems have certain drawbacks such as limited ability to handle gaps or steps, reduced agility and poor coping mechanisms for dealing with uneven terrains. Legged systems are nature's common approach for locomotion on ground. Aristotle was one of the first persons who realized the challenge of designing legged system when he asked "why are man and birds bipeds, but fish footless?" and "why do man and bird, though both bipeds, have an opposite curvature of the legs?" (Aristotle, 2014).

Research on legged locomotion, both in nature and robotics, may enable us to design and construct more agile and efficient moving systems. At the same time, it can also help us understand human movement and control. In turn, this may support developing new approaches for locomotor rehabilitation and assistance. In this respect, findings in biology and robotics can greatly complement each other (Collins et al., 2015). Currently, the principles of animal and human locomotion and their applicability to artificial legged and assistive devices are not fully understood. Given the differences between biological and artificial body design and control, an important question is to what extent should we use biological design and control approaches for building artificial locomotor systems?

Learning from nature does not require mimicking the biological locomotor system in detail. We can already greatly benefit from applying selected design and control principles, such as adding compliant structures to artificial systems or arranging actuators analogous to bi-articular muscles in the human leg.

In recent years, researchers from highly diverse disciplines such as biology, motion science, medicine, and engineering have advanced research on legged

Bioinspired Legged Locomotion. http://dx.doi.org/10.1016/B978-0-12-803766-9.00001-4

locomotion by investigating the underlying principles of body mechanics and related control design (Raibert, 1986; Alexander, 2003; Chevallereau et al., 2013; Winter, 2009; Westervelt et al., 2007; Holmes et al., 2006; Duysens et al., 2002; Duysens and Van de Crommert, 1998; Koditschek et al., 2004). Considering nature as an ingenious teacher, bio-inspired approaches have become increasingly important in the study of legged locomotion (Duysens et al., 2002; Ijspeert, 2008; Seyfarth et al., 2013a). This book aims at providing a comprehensive overview of the biomechanics and control of legged locomotion using perspectives from both biology and engineering. In addition, we introduce a roadmap of the state-of-the-art in studying bio-inspired locomotion.

1.1 WHAT IS BIO-INSPIRED LEGGED LOCOMOTION?

In this book bio-inspiration does not just mean copying structures or controller from nature but instead describes the concepts behind the design and control of legged systems. Since our understanding of legged locomotion in biological systems is not complete, here, bio-inspiration refers to the insight obtained from biology that can be adapted to the needs and capabilities of engineered systems (such as those defined by sensor and actuator properties). The reason behind using biology as an inspiration is that the capabilities of engineered systems may be very different to their biological counterparts. For example, compared to artificial actuators (like electric motors), biological muscles consist of many small actuator units (contractile elements) with distributed properties (such as fast vs. slow twitch muscle fibers) (McMahon, 1984). As a result, a biological motor system is capable of producing versatile movements, spanning tasks of highly different loading and speed conditions. In contrast, state-of-the-art artificial actuators are designed to work optimally during continuous operation at one specific working condition.

Following the presented understanding of bio-inspired legged locomotion, we arranged the book in three general parts:

  i. Locomotion concepts,
  ii. Locomotion control,
  iii. Implementation and applications.

As robotic and biological legged systems have different body structure, biological controllers may not be directly applicable to the engineered system. In particular, because of different properties of biological and artificial actuators (Klute et al., 2002), it is important to extract the logic behind locomotion control in humans and animals rather than simply replicating individual properties and specific control strategies.

In this book we start with bio-inspired locomotion *concepts*. Inspired from nature, we can consider legged locomotion to be composed of three locomotion

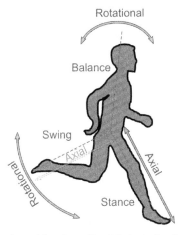

FIGURE 1.1   Main locomotion subfunctions: (i) axial **stance** leg function, (ii) rotational **swing** leg function (an additional axial leg function of the swing leg is used for ground clearance), and (iii) **balance** for maintaining posture.

subfunctions (Seyfarth et al., 2013b): **stance**, **swing**, and **balance**, as shown in Fig. 1.1. Stance function describes the repulsive function of the stance leg (in contact with the ground) to counteract gravity (Seyfarth et al., 2013b). Leg swinging is mainly a rotational movement combined with a complementing axial leg movement to avoid foot scuffing on the ground. Because a major part of the body mass is located at the upper body, the human body is an inherently unstable system unless a controller is continuously keeping balance (Winter, 1995). Therefore, balancing (Pollock et al., 2000) or body posture control (Massion, 1994) is considered to be the third locomotion subfunction required to accomplish stable gaits, especially in bipeds. Template models (Full and Koditschek, 1999) which have a high level of abstraction provide a very useful tool to understand how these subfunctions are controlled and coordinated, both in nature (Blickhan, 1989) and legged robots (Raibert, 1986).

For stable legged locomotion, an appropriate *control* architecture is required to employ the locomotion concepts. Hence, we need to know the corresponding control principles and how to learn from biology to simplify control. In addition, for interaction with humans, lower level force/torque control is beneficial in comparison to position control which might be harmful (Haddadin et al., 2008). On the higher level, legged locomotion requires motor control based on sensory feedback. This control organizes the interaction between the current state of the body and the actuator commands (provided through muscles stimulation in humans) (Duysens et al., 2002), such as task-specific reflex pathways that shape the neuromuscular system dynamics. Since neural and mechanical systems are dynamically coupled (Full and Koditschek, 1999), identifying the interplay of different levels of the neuro-mechanical system is a key challenge in bio-inspired locomotion control.

In order to benefit from bio-inspired locomotion concepts and control that can be used for *implementation* on robots or assistive devices, key characteristics of legged mechanisms need to be identified. Raibert and Hodgins (1993) stated: "We believe that the mechanical system has a mind of its own, governed by the physical structure and laws of physics." Although there is no unique winning body design that is optimal for all types of gaits, identifying general useful mechanisms is crucial. For example, one important lesson learned from nature is that, during legged locomotion, compliance plays a significant role in simplifying control, and enhancing energy efficiency and robustness against perturbations (McMahon, 1985; Full et al., 2000). In addition, body morphology and actuator properties are key aspects in the mechanical design of legged systems and can simplify control and increase energetic efficiency.

1.2 ORGANIZATION OF THE BOOK

This book presents a general overview of legged locomotion research (shown in Fig. 1.2) that consists of a large range of academic disciplines comprising physics, biology, mathematics, robotics, control engineering, computer science, movement science, and biomechanics. The reader is invited to learn more about the related background by studying books concerning different disciplines provided in Table 1.1.

The book is divided into nine chapters, which have the following structure: outline, abstract, introduction, and main contents. The main topics of each chapter are presented in the outline. The abstract shortly describes the relationship between these topics presented in the outline. In the introduction, the reader will find background information on relevant research. The main content is divided into sections. In some chapters they stand alone as separate articles, whereas in others the sections are in closer relation to each other. In the following, we describe the content of the chapters, following the Introduction (**Chapter 1**):

In **Chapter 2**, legged locomotion is described as a composition of three subfunctions: (1) "Stance" leg axial function during contact with ground, (2) "Swinging" the leg during the flight/swing phase rotationally, and (3) "Balancing" the upper body as an interaction between the leg and the upper body. In **Chapter 3**, conceptual models are presented as tools to explain the main features of locomotion.

Chapter 4 focuses on locomotion control and concepts from control theory, such as stability, efficiency, and robustness. In this chapter, we first present several engineering-based approaches and then comment on potential ways in which applying knowledge from biomechanics can be beneficial. Among nu-

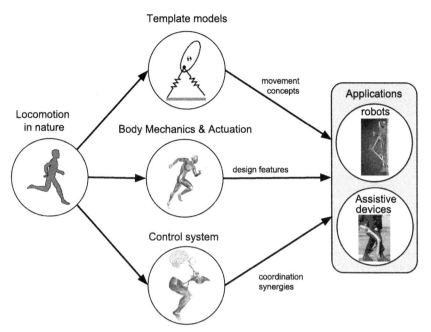

FIGURE 1.2   Bio-inspired legged locomotion. In this book we consider biological locomotion from three perspectives: minimalistic template models, control of locomotion, and implementation of locomotion control on robots and assistive devices through body and actuator properties.

merous engineering control approaches, we discuss the methods that can be employed to mimic human- and animal-like control strategies, such as impedance control (Hogan, 1986), template-based control (Wensing and Orin, 2013), hybrid zero dynamics (HZD) (Westervelt et al., 2007), and central pattern generators (CPG) (Grillner and Wallen, 1985). In order to implement control approaches on a locomotor system, the interaction with environment and circuitry of control are needed to be identified. In **Chapter 5** we focus on torque control which is required to interact with environment. It is more significant when the artificial locomotor system needs to interact with living creatures (e.g., humans). In this chapter we compare the torque-tracking performance of nine control strategies in combination with four high-level controllers that determine the desired torque (implemented on an ankle exoskeleton as a test-bed). A detailed description of muscular system, neural control, their interaction in neuro-muscular systems and also the brain role in locomotion control is presented in **Chapter 6**.

In line with learning from nature, the body morphology, actuators and neuromuscular structure are significant components that need to be employed appropriately to simplify control. **Chapter 7** presents examples that show how

TABLE 1.1 List of books on legged locomotion

Author	Book title	Short description	Field
David A. Winter	Biomechanics and Motor Control of Human Movement	Techniques to measure, analyze and model human movements on mechanical, muscular and neuromuscular level	Human biomechanics
Roger M. Enoka	Neuromechanics of Human Movement	Neural control of human movement in selected tasks, sensory motor system and control of human muscle mechanism	Neuromechanics
Marc H. Raibert	Legged Robots That Balance	Legged locomotion, building useful legged robots, balance and dynamic control	Robotics
Thomas A. McMahon	Muscles, Reflexes and Locomotion	Mathematical model of muscle function, neural control and mechanics of human locomotion	Muscle modeling
R. McNeill Alexander	Principles of Animal Locomotion	Locomotion biomechanics, muscle function, energetics, measurement techniques, and motion types	Animal biomechanics
Christine Chevallereau et al.	Bipedal Robots, Modeling, Design and Building Walking Robots	Gait modeling, gait patterns synthesis and control of bipedal robots, walking, control based on robot modeling, neural network	Bipedal robots

man-made legged systems can benefit from biological body (morphology) design. In **Chapter 8**, we review basic muscle properties and attempts to build actuators with muscle-like behavior. As examples of employing bio-inspired locomotion studies, **Chapter 9** introduces applications of bio-inspired legged locomotion design and control methods to daily life. In addition to a qualitative comparison between engineered and biological locomotor systems, this chapter highlights state-of-the-art research which may help better anticipate potential future research directions in the field.

REFERENCES

Alexander, R.M., 2003. Principles of Animal Locomotion. Princeton University Press.

Aristotle, 2014. ΠΕΡ Ι ΠΟΡ ΕΙΑΣ ΖΩΙΩΝ, eBooks@Adelaide.

Blickhan, R., 1989. The spring-mass model for running and hopping. J. Biomech. 22 (11/12), 1217–1227.

Chevallereau, C., Bessonnet, G., Abba, G., Aoustin, Y., 2013. Bipedal Robots: Modeling, Design and Walking Synthesis. John Wiley & Sons.

Collins, S.H., Wiggin, M.B., Sawicki, G.S., 2015. Reducing the energy cost of human walking using an unpowered exoskeleton. Nature 522 (7555), 212–215.

Duysens, J., Van de Crommert, H.W.A.A., 1998. Neural control of locomotion, part 1: the central pattern generator from cats to humans. Gait Posture 7 (2), 131–141.

Duysens, J., Van de Crommert, H.W., Smits-Engelsman, B.C., Van de Helm, F.C., 2002. A walking robot called human: lessons to be learned from neural control of locomotion. J. Biomech. 35 (4), 447–453.

Full, R.J., Koditschek, D.E., 1999. Templates and anchors: neuromechanical hypotheses of legged locomotion on land. J. Exp. Biol. 202 (23), 3325–3332.

Full, R.J., Farley, C.T., Winters, J.M., 2000. Musculoskeletal dynamics in rhythmic systems: a comparative approach to legged locomotion. In: Biomechanics and Neural Control of Posture and Movement. Springer, New York, pp. 192–205.

Grillner, S., Wallen, P., 1985. Central pattern generators for locomotion, with special reference to vertebrates. Annu. Rev. Neurosci. 8 (1), 233–261.

Haddadin, S., Albu-Schaffer, A., De Luca, A., Hirzinger, G., 2008. Collision detection and reaction: a contribution to safe physical human–robot interaction. In: IEEE/RSJ International Conference on Intelligent Robots and Systems (IROS).

Hogan, N., 1986. Impedance control: an approach to manipulation, part I–III. J. Dyn. Syst. Meas. Control 107, 1–24.

Holmes, P., Full, R.J., Koditschek, D., Guckenheimer, J., 2006. The dynamics of legged locomotion: models, analyses, and challenges. SIAM Rev. 48 (2), 207–304.

Ijspeert, A.J., 2008. Central pattern generators for locomotion control in animals and robots: a review. Neural Netw. 21 (4), 642–653.

Klute, G.K., Czerniecki, J.M., Hannford, B., 2002. Artificial muscles: actuators for biorobotic systems. Int. J. Robot. Res. 21 (4), 295–309.

Koditschek, D.E., Full, R.J., Buehler, M., 2004. Mechanical aspects of legged locomotion control. Arthropod Struct. Develop. 33 (3), 251–272.

Massion, J., 1994. Postural control system. Curr. Opin. Neurobiol. 4 (6), 877–887.

McMahon, T.A., 1984. Muscles, Reflexes and Locomotion. Princeton University Press, Princeton.

McMahon, T.A., 1985. The role of compliance in mammalian running gaits. J. Exp. Biol. 115 (1), 263–282.

Merriam-Webster, I., 2004. Merriam-Webster's Collegiate Dictionary. Merriam-Webster.

Pollock, A.S., Durward, B.R., Rowe, P.J., Paul, J.P., 2000. What is balance? Clin. Rehabil. 14 (4), 402–406.

Raibert, M.H., 1986. Legged Robots that Balance. MIT Press.

Raibert, M.H., Hodgins, J.A., 1993. Legged robots. In: Beer, R., Ritzmann, R., McKenna, T. (Eds.), Biological Neural Networks in Invertebrate Neuroethology and Robotics. Academic Press, Boston, MA, pp. 319–354.

Seyfarth, A., Geyer, H., Lipfert, S., Rummel, J., Minekawa, Y., Iida, F., 2013a. Running and Walking with Compliant Legs. Routledge.

Seyfarth, A., Grimmer, S., Haeufle, D., Maus, H.-M., Peuker, F., Kalveram, K.-T., 2013b. Biomechanical and neuromechanical concepts for legged locomotion. In: Routledge Handbook of Motor Control and Motor Learning. Routledge, pp. 90–112.

Wensing, P.M., Orin, D.E., 2013. High-speed humanoid running through control with a 3D-SLIP model. In: 2013 IEEE/RSJ International Conference on Intelligent Robots and Systems (IROS). IEEE, pp. 5134–5140.

Westervelt, E.R., Grizzle, J.W., Chevallereau, C., Choi, J.-H., Morris, B., 2007. Feedback Control of Dynamic Bipedal Robot Locomotion. Taylor & Francis, CRC Press.

Winter, D., 1995. Human balance and posture control during standing and walking. Gait Posture 3 (4), 193–214.

Winter, D.A., 2009. Biomechanics and Motor Control of Human Movement. John Wiley & Sons.

Part I

Concepts

Chapter 2

Fundamental Subfunctions of Locomotion

Maziar A. Sharbafi, David Lee, Tim Kiemel, and André Seyfarth

Legged locomotion is a complex hybrid, nonlinear and highly dynamic problem. Animals have solved this complex problem as they are able to generate energy efficient and robust locomotion resulted from million years of evolution. However, different aspects of locomotion in biological legged systems such as mechanical design, actuation and control are still not fully understood. Splitting such a complicated problem to simpler subproblems may facilitate understanding and control of legged locomotion. Inspired from template models explaining biological locomotory systems and legged robots, we define three basic locomotor subfunctions: stance leg function, leg swinging and balancing. Combinations of these three subfunctions can generate different gaits with diverse properties. Basic analysis on human locomotion using conceptual models can result in developing new methods in design and control of legged systems like humanoid robots and assistive devices.

PREAMBLE—THINGS TO CONSIDER "BEFORE WALKING"

Animals are integrated collections of "parts" that are good enough to be passed on to the next generation. Because we often project a forward design philosophy on animals, it is all too common to misinterpret animals as optimal designs that should be copied. To the contrary, animals are junkbots that inherit "parts"—really a combination of genetic and epigenetic factors—from their ancestors. Of course, animals' structures can be modified from generation to generation but the evolutionary process is more akin to gathering parts from old VCRs and printers than to the forward-design process used in engineering. Animals build bodies that are good enough for survival in their environment, for competition with other individuals of the same species, and for reproduction to pass their genes on to the next generation—this is the key tenet of evolution by natural selection. Additionally, there are limitations to the body types, shapes, and sizes that a given animal lineage can build. The form and function of an animal's

Bioinspired Legged Locomotion. http://dx.doi.org/10.1016/B978-0-12-803766-9.00003-8

body is subject to developmental, material (i.e., biochemical and tissue-level), and constructional constraints that are inherited from their ancestors.

When considering bio-inspired designs for robots, it is important to look at animals for what they are and how they have been built from the biomaterials available to them. Despite this caveat, nature has produced excellent designs and a level of performance in legged locomotion that is unrivaled by robotic systems. Hence, a guiding principle for robot builders might be to begin with models that capture the fundamental physics and control strategies rather than attempting to mimic the fleshy details of animal locomotion.

In the history of life on Earth, leg-like appendages appear to have evolved independently within many phyla, including Arthropoda, Annelida, Mollusca, Echinodermata, and Chordata (Panganiban et al., 1997). In every case of this messy history, legs originated in an aquatic environment and may have been used for swimming, underwater walking or bounding, grasping, feeding, or holding position in strong currents. The subsequent use of legs for locomotion on land introduced new challenges. Amongst functional changes in ventilation, respiration, desiccation resistance, temperature regulation, and metabolism, these terrestrial pioneers began to move without hydrodynamic forces—instead, using their legs to apply horizontal forces while supporting their full body weight without buoyancy. The transition to terrestrial legged locomotion has been accomplished by insects, arachnids, crustaceans, and vertebrates. A hypothesized role of the first vertebrate limbs were as hold-fasts used by our nearest known lobe-finned fish ancestor, *Tiktaalik*, in fast moving rivers and streams some 375 million years ago (Daeschler et al., 2006). The pelvic bones, femur, tibia, and fibula of *Tiktaalik* are homologous to our own, yet Tiktaalik's appendages had the appearance of strut-like fins (Shubin et al., 2014).

Our ancestral hind limbs have a longer history as aquatic holdfasts than they do as bipedal walking or running legs. The early tetrapods that inherited the appendages of *Tiktaalik* used them to move about quadrupedally on land, as did the early amniotes and mammal-like reptiles. The first mammals used these limbs for quadrupedal terrestrial and arboreal locomotion some 200 million years ago, and the marmoset-like early primates committed to quadrupedal arboreal locomotion about 50 million years ago. Apes descended from quadrupedal primates and began a transition from quadrupedal arborealism to suspensory locomotion, wherein the forelimbs grasp and pull-up on branches while the hind limbs tend to grasp and press against branches, tree trunks, or the ground. The subfamily Ponginae, containing extant orangutans, gorillas, chimpanzees, and humans originated some 12 million years ago from an orangutan-like great ape. The first Hominin with features of habitual bipedalism, *Australopithecus afarensis* appeared in the fossil record about 4 million years ago and *Homo erectus*, with skeletal anatomy nearly the same as our own, appeared just 2 million years ago.

This necessarily stochastic evolutionary history finds us as the only obligate biped amongst primates—and the only striding biped amongst mammals. The only other obligate, striding bipeds living today are birds. Including their thero-pod ancestors, these avian dinosaurs have a 250 million year history as striding bipeds, compared with less than 5 million years in our own lineage.

The example of our own evolutionary history highlights the meandering path we have taken to bipedalism and the short evolutionary distance from our arbo-real ancestors. Does their longer history of bipedal locomotion mean that birds are better "designed" or have a better handle on the physics of bipedalism? No—an organism at any given time is simply good enough to survive, compete, and reproduce in its ecological niche. Nonetheless, it is useful to consider the in-herited design constraints and opportunities, as well as the different ecological niches of birds and humans. For example, natural selection has resulted in fast, economical running of ostriches and economical walking plus long-distance running capabilities of humans—yet these different specializations were shaped in part by the developmental, anatomical, physiological, and behavioral biology inherited from the ostrich and human ancestor. A primary goal of comparative biomechanics is to understand the fundamental physics of locomotion within an evolutionary context. As this understanding progresses, bio-inspired legged robots and robotic prosthetics will transition from copying nature to borrow-ing, in whole or part, its strategies for interacting with the physical world—thus matching or even exceeding the locomotor performance of biological systems.

Bearing these evolutionary caveats in mind, the remainder of this chapter seeks to reveal and interpret the fundamental physics of legged systems. Our approach divides legged locomotion into three subfunctions, which are intrinsi-cally interrelated, yet represent distinct tasks:

- Stance (Chapter 2.1) is the subfunction that redirects the center of mass by exerting forces on the ground.

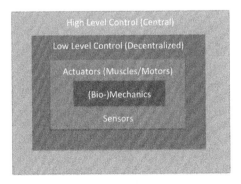

FIGURE 2.0.1   Functional levels of the locomotor system.

- Swing (Chapter 2.2) is the subfunction that cycles the legs between ground contacts.
- Balance (Chapter 2.3) is the subfunction that maintains angular velocity and body attitude within acceptable limits.

Legged locomotion is a complex task with integrated functional levels influencing all three locomotor subfunctions. These levels are mechanical, actuation, sensing, and both low- and high-level control of the animal or machine (Fig. 2.0.1). Our separate treatment of locomotor subfunctions allows interrogation of key functional features at each of these levels.

Chapter 2.1

Stance

David Lee

School of Life Sciences, University of Nevada, Las Vegas, United States

The stance subfunction of legged locomotion considers the ground reaction force exerted on the legs to redirect the body's center of mass from stride to stride. Stance is usually the first subfunction that comes to mind, perhaps because we, large mammals, spend quite a bit of time standing and walking and recognize that our legs keep the rest of our body elevated above the ground during these behaviors. On the other hand, small mammals (of body mass less than about 1 kg body mass), as well as lizards, crocodilians, and amphibians rest their bodies on the ground and tend to walk or run intermittently. From the perspective of these animals, the cycling of the legs from one stance to the next, the *swing* subfunction discussed in the following section might seem equally pervasive. The stance and swing subfunctions of the legs influence each other's dynamics and they also influence the third locomotor subfunction of this chapter, *balance*.

The primary function of the stance leg or legs is to interact with the ground and redirect the body during each stride of locomotion by imposing fluctuations in both the magnitude and direction of the force exerted on the ground. We often think of the stance leg as supporting the body against gravity, even to the extent that knee and ankle extensor muscle groups are sometimes called "antigravity" muscles. The summed vertical force of the stance legs in fact always oscillates about one body weight during steady-speed terrestrial locomotion, such that body weight can be thought of as an "offset."

The stance legs use variations in force to redirect the center of mass, oscillating between braking and propulsive interactions with the ground. The net force in propulsive interactions is in the direction of travel, and that of braking interactions is against the direction of travel. Because of the geometry of the leg–substrate interaction, a forward (protracted) leg tends to cause braking and a backward (retracted) leg tends to cause propulsion.

In any analysis of stance, it is important to consider potential influences of gait, speed, animal size, leg number, and the dynamic interactions between the swing leg(s) and other body segments. The function of a given stance leg is linked to that of other simultaneous stance legs, as well as rotations and translations of leg and body segments. These latter considerations are important for animals and machines that have moving segments with real inertial properties, yet they are neglected in simplified point-mass models that only consider the total ground reaction force acting on the center of mass. Conversely, a stance leg can influence swing and balance by exerting forces and yaw torques, plus roll and pitch torques if the foot is able to grasp or adhere to the substrate. With simultaneous contact of more than one limb, such as during double support of bipedal walking or during quadrupedal or multilegged gait, differential leg forces can produce force couples that also contribute to balance—for example, as the rear wheels of a car resist upward pitch during forward acceleration (Gray, 1968; Murphy and Raibert, 1985; Lee et al., 1999).

2.1.1 EFFECTS OF GAIT

In our treatment of *stance* as one of three locomotor subfunctions, the main objective of the stance legs is to redirect the center of mass, which can be achieved using any of several gaits. Gait is defined by a stereotyped spatiotemporal pattern of leg contacts and oscillations of the center of mass. Redirection of the center of mass may be achieved by using one leg at a time, as during bipedal running with aerial phases, or by using more than one leg at a time, as during all other gaits of bipeds, quadrupeds, and multilegged animals. The collective action of the stance leg or legs exerts oscillating vertical and shear forces to redirect the center of mass during each stride of locomotion.

Vertical center of mass oscillations are achieved using one or more leg at a time, but a given stance leg contributes to only one cycle of vertical oscillation per stride. In symmetrical gaits, which are defined by bilateral (left–right) limb pairs that are one-half stride cycle out of phase (Hildebrand, 1965), two cycles of vertical oscillation are achieved alternately by left and right legs of a pair during each stride. In asymmetrical gaits such as bounding, galloping, and bipedal hopping the collective action of the stance legs achieves a single cycle of vertical oscillation per stride. This achieves a single "gathered suspension" (a flight period with the legs folded under the body) in each stride. Exceptions to this rule for asymmetrical gaits are the fast gallop of cheetahs and greyhounds, as well as the half-bound of rabbits, which include both a "gathered" and "extended suspension"—representing two vertical oscillations per stride (Bertram and Gutmann, 2009). Hence, simultaneous leg forces produce two vertical oscillations per stride during symmetrical gaits and typically only one oscillation per stride during asymmetrical gaits. It is the norm for animals, including ourselves during walking, to exert locomotor forces simultaneously with more than one stance leg. Bipedal running, wherein only one stance leg exerts force at any given time, is the only exception to this rule. However, it may be argued that bipedal hopping of macropods and rodents also falls into this category, considering the two hind legs acting as one.

To affect vertical oscillations of the center of mass, the vertical force exerted by the stance leg or legs alternately rises above and then below body weight during the stride. This is true of walking as well as running. In our bipedal running, vertical force rises above body weight during single-leg support and falls to zero during the aerial phase. Bipedal walking shows an opposite and somewhat counterintuitive pattern where vertical force rises above body weight during the double-leg support at the step-to-step transition and falls below body weight in the middle of single-leg support (Fig. 2.1.1A). Differences between bipedal walking and running can be illustrated by windowing the vertical acceleration in the middle of single-leg stance, showing a trough during walking and a peak during running (Fig. 2.1.1B). For example, comparing humans (black line) to guinea fowl (blue line) shows that guinea fowl use a running gait indicated by shallow peaks in vertical acceleration, whereas humans use a walking gait indicated by shallow troths in vertical acceleration, at two of the intermediate speeds. Because center of mass position is given by the double integral of acceleration with respect to time, the center of mass reaches its lowest vertical position near maximum vertical acceleration—occurring in mid-stance of running and in double-leg support of walking. This difference between running and walking is the basis of a longstanding dichotomy emphasizing that the center of mass reaches its lowest position during mid-stance of running and its highest position during mid-stance of walking (Cavagna et al., 1976, 1977).

Stride dynamics can also be considered in terms of the kinetic and potential energy of the center of mass. Due to the braking impulse during the first half of leg contact, kinetic energy always reaches a minimum near the middle of single-leg stance during both running and walking of bipeds. Pairing this minimum in kinetic energy with the aforementioned potential energy minimum at mid-stance of running and the potential energy maximum at mid-stance of walking provided the impetus to advance two models to characterize running and walking: the spring loaded inverted pendulum (SLIP), a bouncing model with in-phase kinetic and potential energy for running; and a rigid inverted pendulum model with out-of-phase kinetic and potential energy for walking (Cavagna et al., 1977). These two mechanisms have long shaped our understanding of running and walking gaits in bipedal, quadrupedal, and multilegged animals (reviewed by Dickinson et al., 2000).

Bipedal running and hopping as well as quadrupedal or multilegged trotting are well described as "bouncing" gaits, defined by the spring-loaded inverted pendulum (SLIP) model (Blickhan, 1989; McMahon and Cheng, 1990). These gaits show maximum vertical force at mid-stance when the center of mass is at its lowest point, and may or may not include aerial periods between leg contacts. Because SLIP-like gaits may be achieved with leg springs, this provides a mechanism to reduce total energy cost by storing some of the energy elastically in the absorptive phase of early stance and returning it in the generative phase of late stance.

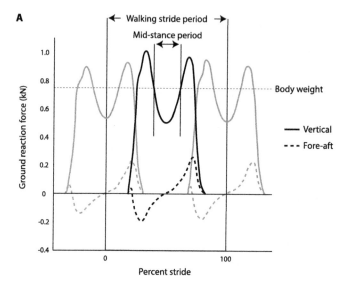

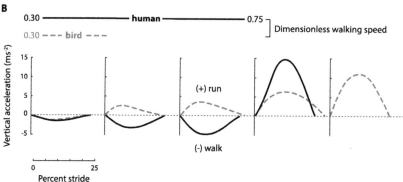

FIGURE 2.1.1 (A) Three human walking steps used to define a stride from mid-stance to mid-stance of the same limb (gray traces). Vertical ground reaction force is below body weight during the mid-stance period of walking. (B) Patterns of mid-stance vertical acceleration across a range of dimensionless walking and running speeds, including humans (black) and guinea fowl (blue). Negative mid-stance acceleration indicates walking and positive, running. Humans maintain negative accelerations up to dimensionless walking speeds as high as 0.75 but guinea fowl switch to running at dimensionless speeds corresponding to moderate human walking. (For interpretation of the references to color in this figure legend, the reader is referred to the web version of this chapter.)

Despite agreement of the bipedal running and quadrupedal or multilegged trotting gaits with the SLIP model, studies showed that bipedal and quadrupedal walking dynamics (e.g., Lee and Farley, 1998; Griffin et al., 2004, Genin et al., 2010) deviate substantially from a rigid inverted pendulum model. This is not unexpected given that the vertical force of a rigid inverted pendulum model reaches a local maximum rather than the necessary minimum at mid-stance (Geyer et al., 2006). Likewise, vertical force will not reach the necessary maximum during the step-to-step transition unless double-leg support is modeled. Hence, it is difficult to reconcile the mechanics of a rigid inverted pendulum with the measured dynamics of walking. In contrast to the rigid inverted pendulum model, Geyer et al. (2006) also showed that simulations of walking on compliant legs can match force patterns observed during human walking, exhibiting a minimum vertical force at mid-stance and a maximum vertical force during double support of the step-to-step transition.

Gait dynamics can be better understood by considering the fundamental physics of the animal's interaction with the substrate. The center of mass reaches its lowest point during the transition between single-leg stances in walking, i.e., during double support at the step-to-step transition. Conversely, the center of mass reaches its lowest point during the middle of single-leg stance in running. Minimum center of mass height coincides with maximum vertical force in both gaits because this is where the center of mass is redirected from falling to rising. Hence, it might be argued that walking and running show similar vertical oscillations—simply achieved by two legs during walking and a single leg during running. However, walking and running in fact show fundamentally different dynamics. To understand what is driving these distinct dynamics, we need to consider the pattern of braking and propulsion during the downward to upward redirection of the center of mass. In SLIP-like bouncing gaits, the force on the center of mass is braking and then propulsive during the downward to upward redirection. Walking is the opposite: propulsive force precedes braking force during the downward to upward redirection. Thus, redirection of the center of mass pairs with opposite patterns of braking and propulsion in walking versus running. This observation provides the impetus for applying collision-based dynamics to legged locomotion.

The guiding principle of collision-based dynamics is that the stance leg or legs seeks to redirect the center of mass with the least mechanical work possible. Mechanical work can be viewed as an extension of D'Alembert's principle of orthogonal constraint, which holds that a force may redirect a mass with zero work as long as force and velocity vectors are kept perpendicular such that their dot-product, mechanical power, is zero (D'Alembert, 1743). Minimizing mechanical power at a given speed minimizes the animal's mechanical cost of transport. Collision-based costs are incurred whenever an animal's force and velocity vectors are not perpendicular. Thus, collision-free dynamics may be achieved only if the legged system is able to maintain a perpendicular relationship between the force and velocity vectors of its center of mass in every instance of stance (Fig. 2.1.2(right)) (Ruina et al., 2005; Lee et al., 2011). Because propulsion precedes braking during the step-to-step

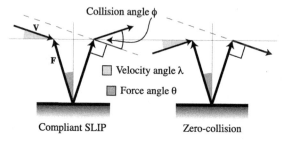

FIGURE 2.1.2   Collision-based dynamics for a SLIP model versus the zero-collision case (Lee et al., 2011).

transition of walking, the force vector can be kept more nearly perpendicular to the velocity vector while the center of mass is redirected from downward to upward, thereby reducing the mechanical work done by the stance legs on the center of mass (Fig. 2.1.2(right)). In contrast, SLIP-like bouncing dynamics cannot minimize the mechanical work required to redirect the center of mass because braking precedes propulsion during downward to upward redirection of the center of mass (Fig. 2.1.2(left)). This violates the principle of minimizing mechanical cost through orthogonal constraint, as seen in the zero-collision case. However, there is evidence in some species that part of the mechanical energy of the SLIP is stored in spring-like tendons that release elastic strain energy later in stance (Biewener, 2005).

Mechanical work is quantified by the mechanical cost of transport, CoT_{mech}, which is the work required to move a unit body weight a unit distance. CoT_{mech} can be determined from center of mass mechanical power—the dot-product of the force vector on the velocity vector. During level, steady-speed locomotion positive and negative work are equal in magnitude. Physiologist and modelers often count only positive mechanical power (e.g., Cavagna et al., 1977; Kuo, 2002), but here we take the absolute value of power to account for both positive (generative) and negative (absorptive) work:

$$CoT_{mech} = \frac{\sum |F \cdot V|}{n \left(mg\bar{v}_y\right)}, \qquad (2.1.1)$$

where $\bar{v}_y$ is the mean forward velocity, g is gravitational acceleration, m is body mass, and n is the number of time-intervals in the summation. From the perspective of a point-mass model (i.e., a model concentrating all of the system's distributed masses at the center of mass and considering only translations), SLIP-like bouncing gaits incur a much greater mechanical cost of transport than gaits, such as walking, that minimize collision-based costs. In humans, for example, the mechanical cost of transport during SLIP-like running is three-times

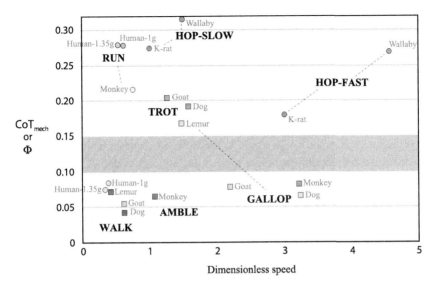

FIGURE 2.1.3 Mechanical cost of transport as a function of dimensionless speed in bipedal and quadrupedal mammals (Lee et al., 2013).

that of walking (Fig. 2.1.3; Lee et al., 2013). The lower mechanical cost of walking is achieved by the combined action of trailing and leading leg forces, which exert propulsive and then braking forces during the downward to upward redirection of the center mass.

As predicted from the theoretical observation that mechanical cost of transport is inversely proportional to the number of collisions (Ruina et al., 2005), quadrupeds more nearly approach zero-collision locomotion than bipeds. Collision-based analysis shows that walking dogs achieve a mechanical cost of transport approximately half that of humans (Fig. 2.1.3). Faster quadrupedal gaits such as the gallop and amble also use sequenced leg contacts and favorable timing of braking and propulsive forces to reduce the work of redirecting the center of mass—but not to the same extent as in quadrupedal walking (Fig. 2.1.3). One might ask if terrestrial animals with six, eight, or ten legs can achieve even lower mechanical costs of transport than quadrupeds, however, most studies of multilegged locomotion in cockroaches and crabs report SLIP-like bouncing gaits akin to quadrupedal trotting (reviewed by Holmes et al., 2006). If multilegged animals combined sequenced leg contacts with propulsive then braking forces during the downward to upward transition, it is plausible that animals with many legs could more smoothly redirect the center of mass, thereby achieving a lower mechanical cost of transport than bipeds or quadrupeds.

Brachiation of gibbons (Usherwood and Bertram, 2003) and siamangs (Michilsens et al., 2012) provide another example of collision-based dynamics that reduce mechanical cost. Using a single arm to redirect the center of mass from downward to upward, these apes achieve mechanical costs of transport lower than those of walking quadrupeds. Collision-based mechanical cost is mitigated more readily during suspensory locomotion because the arm can pull on the overhead support to exert propulsive then braking force during downward to upward redirection of the center of mass. As shown in bipedal, quadrupedal, and brachiation examples, this pattern of force is the hallmark of collision-based dynamics, which seek to redirect the center of mass using the least mechanical work—in accordance with D'Alembert's principle of orthogonal constraint.

2.1.2 EFFECTS OF SIZE

Size is a fundamental determinant of structure and function in animal and machines alike. For the scaling of legged locomotion, the principle of dynamic similarity is a key concept that was introduced four decades ago as a model to predict gait characteristics of bipedal dinosaurs based upon locomotor data from extant birds and humans (Alexander, 1976). The same construct was later applied to quadrupedal mammals (Alexander and Jayes, 1983). The dynamic similarity hypothesis holds that animals of different sizes moving at the same dimensionless speed tend to use the same dimensionless stride lengths, stride frequencies, duty factors and maximum forces. Forward speed is normalized as the Froude number, or preferably the square root of Froude number (McMahon and Cheng, 1990), known as dimensionless speed U:

$$U = \bar{v}_y/\sqrt{gh}, \tag{2.1.2}$$

where $\bar{v}_y$ is the mean forward velocity, g is gravitational acceleration, and h is hip height or leg length. The Froude number represents the ratio of inertial to gravitational acceleration, indicating that a rigid inverted pendulum would escape its circular trajectory at a Froude number or dimensionless speed greater than one. Walking bipeds and quadrupeds, however, abandon walking gaits at Froude numbers greater than about 0.5, corresponding to dimensionless speeds greater than about 0.7 (Alexander, 1984). Dimensionless speeds are determined in the same way for running, hopping, trotting, and galloping gaits and, because these gaits often exceed a dimensionless speed of one, using Froude number instead of its square root would show substantial nonlinearity with increasing speed.

A refinement of Alexander's rigid inverted pendulum calculation predicts a different boundary for escape from a circular arc for a given combination of speed and step length, here defined as the distance between right and left foot-falls. For example, a maximum dimensionless step length (i.e., step length relative to leg length) of 1.15 can be achieved by a rigid-legged walker at a dimensionless speed of 0.7, approximating the fastest walking of humans (Usherwood, 2005). Adding an impulsive step-to-step transitions and a minimum mechanical cost criterion, a subsequent computer optimization study showed that rigid-legged walking is optimal only at dimensionless step lengths less than 0.76 (Srinivasan and Ruina, 2006). Despite observations that humans use intermediate dimensionless step lengths of about 1.0 at a dimensionless speed of 0.7, there is no evidence that human walking follows the circular arc of a rigid inverted pendulum. In fact, experimental data show that vertical oscillations of the center of mass during fast human walking are just 17–28% of those predicted by rigid-legged walking models (Lee and Farley, 1998). In light of these observations, a different mechanism might be found to explain the relationship of dimensionless step length to maximum walking speed of humans.

Dynamic similarity is used to determine equivalent speeds in animals of different leg length such that other effects on locomotion, such as leg number, gait, and phylogeny can be better understood. The principle of dynamic similarity normalizes stride length to hip height h, stride frequency to $\sqrt{g/h}$, and force to body weight mg, predicting equal values of these dimensionless parameters at a given dimensionless speed. Duty factors, calculated as the ratio of foot contact period to stride period, are also predicted to be equal at the same dimensionless speed. When bipedal and quadrupedal gaits are normalized according to dynamic similarity, these dimensionless parameters tend to follow similar, yet sometimes offset trend-lines across a range of dimensionless speeds from 0.2 to 4.0 (Alexander, 1976, 1984, 2004; Alexander and Jayes, 1983). For example, quadrupedal primates tend to use longer stride lengths than other quadrupeds at a given dimensionless speed. Likewise, at the same dimensionless speed, humans walking and running bipedally use much shorter dimensionless stride lengths than do chimpanzees and bonobos during quadrupedal gaits (Aerts et al., 2000). A comparative study of bipedal locomotion in birds spanning three orders of magnitude in size applied dynamic similarity to show that small birds tend to use relatively longer stride lengths and lower frequencies than large birds at a given dimensionless speed (Gatesy and Biewener, 1991). Dynamic similarity is useful for determining equivalent speeds amongst subjects of different size, such as in a human gait study, and it is also well suited to the analysis of species that span a substantial size range. It should be used as an initial hypothesis for understanding size effects in legged locomotion rather than a precise predictive model, for example, across different species or in hopping versus striding locomotion.

Dynamic similarity has also been applied to reduced- and hyper-gravity studies of legged locomotion, wherein gravity, as well as hip height, are free variables influencing dimensionless speed and stride frequency. A study of simulated reduced gravity during human walking, found that dimensionless stride length at a given dimensionless speed decreased as gravity was reduced from 1.0 g to 0.25 g, violating the dynamic similarity prediction that stride length should remain the same at equal dimensionless speeds (Donelan and Kram, 1997). However, in agreement with dynamic similarity, the same study showed that duty factor at a given dimensionless speed was unchanged across gravity conditions. A study of simulated hypergravity during human walking showed qualitative agreement with dynamic similarity, where duty factor increased and dimensionless stride length decreased as dimensionless speed was decreased by a 1.35 g hypergravity condition (Lee et al., 2013). Reduced- and hypergravity studies of human running also show trends consistent with dynamic similarity (Donelan and Kram, 1997; Minetti, 2001). Hence, the principle of dynamic similarity seems robust to changes in gravity with the exception that stride lengths are unexpectedly shortened by reduced gravity conditions during walking. Dynamic similarity therefore remains the primary model to determine comparable speeds whenever animal size or gravity conditions are variable.

Another method of accounting for speed effects that has been used in studies of quadrupedal mammals is to target the trot–gallop transition as a "physiologically equivalent" speed for animals of different size. This approach was introduced in a study showing that stride frequency scales as body mass to the -0.14 power and stride length, to the 0.38 power in mammals from mice to horses (Heglund et al., 1974). A comparable approach measured these parameters at the fastest experimental speeds of running birds and humans to show that stride frequency scales as body mass to the -0.18 power and stride length to the 0.38 power (Gatesy and Biewener, 1991). Trot–gallop transition speeds have also been used to compare oxygen consumption rate and mechanical power (Heglund et al., 1982; Taylor et al., 1982), as well as in vivo bone strain and effective mechanical advantage of muscles about joints (Biewener, 1989, 1990) across quadrupeds of vastly different size. As predicted by dynamic similarity, mechanical cost of transport at the trot–gallop transition speed is invariant across quadrupedal mammals from mice to horses, and also at corresponding speeds of bipedal and multilegged runners (Full, 1989; Full and Tu, 1991). The mechanical cost of transport determined by this allometric analysis is 0.1 based upon positive work alone—doubling this value to account for negative work equals CoT_{mech} as defined in Eq. (2.1.1). Running, hopping, and trotting usually show a CoT_{mech} near 0.2. Yet, as already discussed, mechanical cost of transport is several-fold lower for gaits that use collision-based mitigation of work, such as walking, ambling, and galloping (Fig. 2.1.3).

Animal size also influences the stiffness of the modeled stance leg (or legs) during bipedal running, hopping, and trotting gaits. The spring-loaded inverted pendulum (SLIP) is the simplest two-dimensional model of leg compliance, as it imagines such a virtual spring-loaded leg acting between the center of mass and the ground (McMahon and Cheng, 1990). Because this method depends only on measurement of the total force vector during leg contact and a kinematic estimate of initial virtual leg length, it can be applied to quadrupedal and hexapedal trotting as well as to bipedal running and hopping (Farley et al., 1993; Blickhan and Full, 1993). Considering eight mammals spanning three orders of magnitude in body mass and including hoppers, trotters, and a human runner, allometric analysis showed that virtual leg stiffness scales as body mass to the two-thirds power. This relationship matches the expected leg stiffness based on the ratio of force, which scales in direct proportion to body mass, to length, which scales as body mass to the one-third power when geometric similarity is assumed.

A more explicit experimental approach measures leg stiffness by tracking position of the proximal joint (the hip or shoulder) and modeling a radial leg that extends through the distal-most joint to the ground. This method measures radial leg stiffness by placing an actuator in series with a modeled leg spring and choosing the spring constant that minimizes actuator work. Considering five mammalian species, radial leg stiffness scales approximately as body mass to the two-thirds but is more than 30% stiffer than the virtual leg spring at a given body mass (Lee et al., 2014). The scaling of leg spring constants has also been analyzed using a minimum work criterion in an actuated, damped SLIP model (Birn-Jeffery et al., 2014). This model successfully reproduced the running dynamics of five striding bird species ranging in size from quail to ostriches and found a dimensionless leg stiffness invariant with body mass, as did the analysis of Blickhan and Full (1993). All modeling approaches used so far to investigate the scaling of leg spring stiffness show that stiffness is a function of body mass to the two-thirds power, or equivalently, that dimensionless stiffness is invariant with size.

2.1.3 SUMMARY

This section has examined stance as a locomotor subfunction where a leg or legs redirect the center of mass through simultaneous and/or sequenced contacts with the substrate. The cost of this redirection is determined by collision-based dynamics. Whether there is an aerial phase or not, the summed vertical ground reaction force rises above body weight during part of the stride to redirect the center of mass from downward to upward. This is achieved by a single contact leg in bipedal running and by simultaneous contact of more than one leg in all other gaits. During running, hopping, and trotting, downward to upward redirection of the center of mass is SLIP-like, with braking followed by propulsion. This pattern is reversed in walking, ambling, and galloping gaits, with propulsion followed by braking achieved by sequenced contacts of more than one leg. Mechanical work can be measured using a collision-based approach, which considers the relationship between the center of mass velocity vector and the overall force vector. Whenever these vectors are perpendicular, no work is done on the center of mass. Mechanical work is minimized in this way by walking, ambling, and galloping but not in SLIP-like gaits. Hence, the mechanical cost of transport is about three-fold greater during SLIP-like running, hopping, and trotting compared with gaits that mitigate work. Theory and some experimental evidence suggest that work is increasingly mitigated as the number of sequenced leg contacts increases. The principle of dynamic similarity estimates equivalent speeds for animals of vastly different size by determining a dimensionless speed according to the square root of leg length. This model considers dimensionless stride length, frequency, force, and mechanical cost of transport—predicting equal values of these parameters at a given dimensionless speed. When size spans orders of magnitude, dimensional parameters can also be expressed as power-functions of body mass and this approach has yielded scaling relationships for stride length, frequency, force magnitude, spring stiffness, and mechanical cost of transport. Stance also influences and is influenced by the swing and balance locomotor subfunctions discussed in the remaining sections of this chapter.

REFERENCES

Aerts, P., Van Damme, R., Van Elsacker, L., Duchêne, V., 2000. Spatio-temporal gait characteristics of the hind-limb cycles during voluntary bipedal and quadrupedal walking in bonobos (Pan paniscus). Am. J. Phys. Anthropol. 111 (4), 503.

Alexander, R., 1976. Estimates of speeds of dinosaurs. Nature 261, 129–130.

Alexander, R.M., 1984. The gaits of bipedal and quadrupedal animals. Int. J. Robot. Res. 3 (2), 49–59.

Alexander, R., 2004. Bipedal animals, and their differences from humans. J. Anat. 204 (5), 321–330.

Alexander, R., Jayes, A.S., 1983. A dynamic similarity hypothesis for the gaits of quadrupedal mammals. J. Zool. 201 (1), 135–152.

Bertram, J.E.A., Gutmann, A., 2009. Motions of the running horse and cheetah revisited: fundamental mechanics of the transverse and rotary gallop. J. R. Soc. Interface 6 (35), 549–559.

Biewener, A.A., 1989. Scaling body support in mammals: limb posture and muscle mechanics. Science 245 (4913), 45–48.

Biewener, A.A., 1990. Biomechanics of mammalian terrestrial locomotion. Science 250 (4984), 1097.

Biewener, A.A., 2005. Biomechanical consequences of scaling. J. Exp. Biol. 208 (9), 1665–1676.

Birn-Jeffery, A.V., Hubicki, C.M., Blum, Y., Renjewski, D., Hurst, J.W., Daley, M.A., 2014. Don't break a leg: running birds from quail to ostrich prioritise leg safety and economy on uneven terrain. J. Exp. Biol. 217 (21), 3786–3796.

Blickhan, R., 1989. The spring–mass model for running and hopping. J. Biomech. 22 (11–12), 1217–1227.

Blickhan, R., Full, R.J., 1993. Similarity in multilegged locomotion: bouncing like a monopode. J. Comp. Physiol. A 173 (5), 509–517.

Cavagna, G.A., Thys, H., Zamboni, A., 1976. The sources of external work in level walking and running. J. Physiol. 262 (3), 639.

Cavagna, G.A., Heglund, N.C., Taylor, C.R., 1977. Mechanical work in terrestrial locomotion: two basic mechanisms for minimizing energy expenditure. Am. J. Physiol., Regul. Integr. Comp. Physiol. 233 (5), R243–R261.

Daeschler, E.B., Shubin, N.H., Jenkins, F.A., 2006. A Devonian tetrapod-like fish and the evolution of the tetrapod body plan. Nature 440 (7085), 757–763.

D'Alembert, J.L.R., 1743. Traité de dynamique.

Dickinson, M.H., Farley, C.T., Full, R.J., Koehl, M.A.R., Kram, R., Lehman, S., 2000. How animals move: an integrative view. Science 288 (5463), 100–106.

Donelan, J.M., Kram, R., 1997. The effect of reduced gravity on the kinematics of human walking: a test of the dynamic similarity hypothesis for locomotion. J. Exp. Biol. 200, 3193–3201.

Farley, C.T., Glasheen, J., McMahon, T.A., 1993. Running springs: speed and animal size. J. Exp. Biol. 185 (1), 71–86.

Full, R.J., 1989. Mechanics and energetics of terrestrial locomotion: bipeds to polypeds. In: Energy Transformations in Cells and Animals, pp. 175–182.

Full, R.J., Tu, M.S., 1991. Mechanics of a rapid running insect: two-, four- and six-legged locomotion. J. Exp. Biol. 156 (1), 215–231.

Gatesy, S.M., Biewener, A.A., 1991. Bipedal locomotion: effects of speed, size and limb posture in birds and humans. J. Zool. 224 (1), 127–147.

Genin, J.J., Willems, P.A., Cavagna, G.A., Lair, R., Heglund, N.C., 2010. Biomechanics of locomotion in Asian elephants. J. Exp. Biol. 213 (5), 694–706.

Geyer, H., Seyfarth, A., Blickhan, R., 2006. Compliant leg behaviour explains basic dynamics of walking and running. Proc. R. Soc. Lond. B, Biol. Sci. 273 (1603), 2861–2867.

Gray, J., 1968. Animal Locomotion. Weidenfeld & Nicolson.

Griffin, T.M., Main, R.P., Farley, C.T., 2004. Biomechanics of quadrupedal walking: how do four-legged animals achieve inverted pendulum-like movements? J. Exp. Biol. 207 (20), 3545–3558.

Heglund, N.C., Taylor, C.R., McMahon, T.A., 1974. Scaling stride frequency and gait to animal size: mice to horses. Science 186 (4169), 1112–1113.

Heglund, N.C., Fedak, M.A., Taylor, C.R., Cavagna, G.A., 1982. Energetics and mechanics of terrestrial locomotion. IV. Total mechanical energy changes as a function of speed and body size in birds and mammals. J. Exp. Biol. 97 (1), 57–66.

Hildebrand, M., 1965. Symmetrical gaits of horses. Science 150 (3697), 701–708.

Holmes, P., Full, R.J., Koditschek, D., Guckenheimer, J., 2006. The dynamics of legged locomotion: models, analyses, and challenges. SIAM Rev. 48 (2), 207–304.

Kuo, A.D., 2002. Energetics of actively powered locomotion using the simplest walking model. J. Biomech. Eng. 124 (1), 113–120.

Lee, C.R., Farley, C.T., 1998. Determinants of the center of mass trajectory in human walking and running. J. Exp. Biol. 201 (21), 2935–2944.

Lee, D.V., Bertram, J.E., Todhunter, R.J., 1999. Acceleration and balance in trotting dogs. J. Exp. Biol. 202 (24), 3565–3573.

Lee, D.V., Bertram, J.E., Anttonen, J.T., Ros, I.G., Harris, S.L., Biewener, A.A., 2011. A collisional perspective on quadrupedal gait dynamics. J. R. Soc. Interface, rsif20110019.

Lee, D.V., Comanescu, T.N., Butcher, M.T., Bertram, J.E., 2013. A comparative collision-based analysis of human gait. Proc. R. Soc. Lond. B, Biol. Sci. 280 (1771), 20131779.

Lee, D.V., Isaacs, M.R., Higgins, T.E., Biewener, A.A., McGowan, C.P., 2014. Scaling of the spring in the leg during bouncing gaits of mammals. Integr. Comp. Biol. 54 (6), 1099–1108.

McMahon, T.A., Cheng, G.C., 1990. The mechanics of running: how does stiffness couple with speed? J. Biomech. 23, 65–78.

Michilsens, F., D'Août, K., Vereecke, E.E., Aerts, P., 2012. One step beyond: different step-to-step transitions exist during continuous contact brachiation in siamangs. Biol. Open 1 (5), 411–421.

Minetti, A.E., 2001. Invariant aspects of human locomotion in different gravitational environments. Acta Astronaut. 49 (3), 191–198.

Murphy, K.N., Raibert, M.H., 1985. Trotting and bounding in a planar two-legged model. In: Theory and Practice of Robots and Manipulators. Springer US, pp. 411–420.

Panganiban, G., Irvine, S.M., Lowe, C., Roehl, H., Corley, L.S., Sherbon, B., Wray, G.A., et al., 1997. The origin and evolution of animal appendages. Proc. Natl. Acad. Sci. 94 (10), 5162–5166.

Ruina, A., Bertram, J.E., Srinivasan, M., 2005. A collisional model of the energetic cost of support work qualitatively explains leg sequencing in walking and galloping, pseudo-elastic leg behavior in running and the walk-to-run transition. J. Theor. Biol. 237 (2), 170–192.

Shubin, N.H., Daeschler, E.B., Jenkins, F.A., 2014. Pelvic girdle and fin of Tiktaalik roseae. Proc. Natl. Acad. Sci. 111 (3), 893–899.

Srinivasan, M., Ruina, A., 2006. Computer optimization of a minimal biped model discovers walking and running. Nature 439 (7072), 72–75.

Taylor, C.R., Heglund, N.C., Maloiy, G.M., 1982. Energetics and mechanics of terrestrial locomotion. I. Metabolic energy consumption as a function of speed and body size in birds and mammals. J. Exp. Biol. 97 (1), 1–21.

Usherwood, J.R., 2005. Why not walk faster? Biol. Lett. 1 (3), 338–341.

Usherwood, J.R., Bertram, J.E., 2003. Understanding brachiation: insight from a collisional perspective. J. Exp. Biol. 206 (10), 1631–1642.

Chapter 2.2

Leg Swinging

Maziar A. Sharbafi[*,†] and André Seyfarth[†]

[*]School of Electrical and Computer Engineering, College of Engineering, University of Tehran, Iran [†]Lauflabor Locomotion Laboratory, TU Darmstadt, Germany

For successful locomotion, the swing leg needs to be prepared for the next landing event. For a specific or changing gait condition (gait type, speed, environment), swing leg movement could require achieving a certain foot placement (e.g., to avoid hitting a pothole, to recover from a large perturbation). However, in most cases the actual location of the foot during contact is not a specific target of control. Then, the leg swinging as one locomotion subfunction can rely on the system dynamics to result in a steady gait pattern.

Based on these two mechanisms (foot placement and exploration of natural swing leg dynamics) leg swinging can fulfill different scenarios. In the following we will describe how leg swinging contributes to locomotion, namely how it interacts with other locomotor subfunctions as well as its role in perturbation recovery and switching between the gaits.

At the level of locomotion control, these two mechanisms can be compared to position control vs. passive dynamic walking (Kuo, 2007a). In legged robots, ZMP (zero moment point; Vukobratovic and Borovac, 2004) is one of the most common approaches to achieve stable locomotion in systems employing positional control of leg joints (e.g., in Asimo; Sakagami et al., 2002). ZMP refers to the point inside the base of support about which the ground contact forces exert no moment (see Subchapter 2.3). The idea behind *dynamic walking* relies on passive dynamics of the legs to produce walking, avoids position control, and focuses on producing a cyclic gait (Miura and Shimoyama, 1984; McGeer, 1990a, 1990b; Collins et al., 2005). In passive dynamic walking robots, the stance leg and swing leg behave as an inverted and regular pendulum, respectively (for more explanation about passive dynamic walking see Subchapter 4.6). The required energy to compensate the losses is generated either by gravity when the robot walks on gently sloped terrain (McGeer, 1990a) or minimal actuation (Collins et al., 2005). In Kuo (2007a) the term dynamic walking is defined to refer "specifically to machines designed to harness leg dynamics, using control more to shape and tune these dynamics than to impose prescribed kinematic motions." Such an actuation can be provided by hip torque (Collins et al., 2005) or push off with the trailing leg's ankle (Kuo, 2002). Therefore, the relation between leg swinging with stance control and/or posture control will come to account.

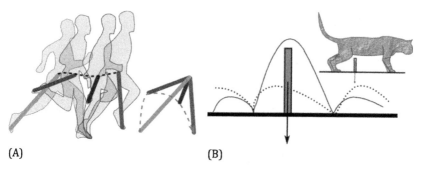

(A)   **(B)**

FIGURE 2.2.1 (A) Swing leg movement in human running with changing leg length and leg angle. (B) Leg swinging in cat walking (McVea and Pearson, 2006). Cats stepped over an obstacle with the forelegs. After a 20 s delay, hind foot toe trajectory showed that the animal remembered the obstacle (solid line), which was lowered while the animal stood still. The dotted line shows a step without the obstacle.

2.2.1 CHARACTERIZING FEATURES OF LEG SWINGING

Leg swinging can be defined as rotational swing leg motion with complementary axial movement (Fig. 2.2.1A), e.g., for ground clearance in walking or for reducing swing leg moment of inertia in running. Indeed, this axial leg shortening is very important in special situations like hurdle running. In that sense, leg swinging can be considered as control of the end effector (here the foot) of a manipulator. In cats, this foot trajectory to overcome obstacles was found to be memorized. The pattern could be restored and realized with the hind limbs after locomotion was interrupted for substantial time as shown in Fig. 2.2.1B (McVea and Pearson, 2006).

However, the main role of leg swinging in (unperturbed) locomotion is its rotational movement, which results in a reorientation of the leg in preparation for the next contact phase. This primitive function of the (load-released) swing leg can already be found in newborns; it is known as the stepping reflex (Siegler et al., 2006).

Swing leg movements and its related control can be characterized by the leg's states at touchdown (landing condition) and the foot trajectory during swing phase (ground clearance of the foot). The leg orientation and leg length at touch-down also contributes to the dynamics of the next contact phase and thus to other locomotion subfunctions (stance and balance). Not only the leg configuration at touchdown but also its changes with time (angular velocity and axial speed of leg shortening/extension) is an important feature to describe gait dynamics. Thus, leg swinging at touchdown is characterized by leg length, leg angle with respect to ground (angle of attack) (Seyfarth et al., 2001), leg angular speed (Seyfarth et al., 2003) and leg shortening speed (Blum et al., 2010). In both walking and running, the swing-leg moves backward towards

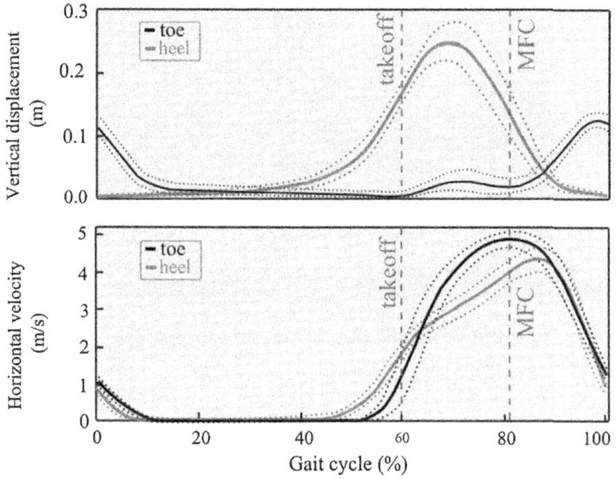

FIGURE 2.2.2 Vertical displacement and horizontal velocity of heel and toe in walking. Minimum foot clearance (MFC) coincides with the moment that the foot travels with maximum horizontal velocity. The figure is adopted from Winter (1992).

the ground before touchdown (late swing-phase) (Muybridge, 1955), called swing leg retraction (SLR). This backward movement is represented by positive angular velocities of the leg (Poggensee et al., 2014) and has a large contribution to gait stability (Seyfarth et al., 2003). SLR supports ground speed matching, helps reduce impact losses during landing (De Wit et al., 2000; Blum et al., 2010) and maintain forward locomotion speed.

The foot trajectory during swing phase needs to provide sufficient ground clearance. In human walking, the foot of the swing leg is aligned horizontally with only small distance (1–2 cm) to the ground which reduces with age increasing (Winter, 1992). In contrast to Winter claim, Mills et al. showed that increasing the variability in minimum toe clearance (MTC) results in high risk of tripping in the elderly while the MTC medians in two young and elderly men were similar (Mills et al., 2008). This helps generate stable and energy efficient gait patterns (Wu and Kuo, 2015). Minimum foot clearance (MFC) is a critical event because the foot gets its maximum horizontal velocity and lowest height simultaneously (shown in Fig. 2.2.2) which increases the danger of tripping in case of hitting an obstacle (Begg et al., 2007).

Judging from human leg muscle activities in the swing leg movement, biarticular hip muscles rectus femoris (RF) and hamstrings (HA) seem to be the main contributors in the swing phase of walking (Nilsson et al., 1985). By modeling these two muscles with biarticular springs, we aim at a better mechanical understanding of their activities in producing stable gait. In addition, such a passive mechanism may also replicate strong correlation observed between RF and

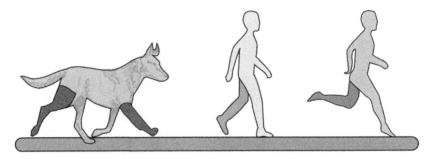

FIGURE 2.2.3 Leg swinging in different gaits. Dark colors show the swing legs.

HA in human swing leg movement (Prilutsky et al., 1998), as a consequence of body mechanics. In Sect. 2.2.3 we introduce a model for swing leg control based on this observation.

2.2.2 LEG SWINGING EFFECTS IN LOCOMOTION

Leg swinging is required for robust locomotion in order to initiate the next step (Fig. 2.2.3). This is a consequence of the limited number of legs which need to hit the ground in a sequential manner. Leg swinging contributes to locomotion dynamics in many ways:

(i) Determining stance phase dynamics as a result of the landing condition,

(ii) Shaping the system states to achieve versatile gaits with selected gait type, footfall pattern, step length, step frequency, robustness and efficiency,

(iii) Distribution of energies in forward, lateral, and vertical directions, e.g., acceleration or changes in locomotion direction,

(iv) Overcoming unwanted ground contacts and perturbation recovery, e.g., obstacle avoidance.

2.2.2.1 Contributing to Stance Phase Dynamics

When determining the initial condition of the next stance leg, swing leg movement significantly affects the stance phase dynamics. It is related not only to the leg configuration, but also to the momentum and angular speed which initiate the new states after impact. Human and animal locomotion experiments show high sensitivity of the initial leg loading during stance to its landing conditions (Moritz and Farley, 2004; Birn-Jeffery and Daley, 2012; Daley and Biewener, 2011). Daley and Biewener (2006) showed that the variation in leg contact angles explains 80% of the variation in stance impulse after an unexpected pothole in guinea fowl running.

2.2.2.2 Trade-off Between Versatility, Robustness, and Efficiency

Redirection of the center of mass speed at touchdown can be used by a suitable swing leg control (foot placement) to stabilize the gait (Townsend, 1985). Furthermore, Winter et al. claim that foot placement is a precise and multifactorial motor control task which is required (beside stance leg control) for stable gaits (Winter, 1992). Donelan et al. showed that both average external mechanical work and metabolic rates increase with the fourth power of step length in human walking (Donelan et al., 2002), which confirms importance of foot placement in energy optimization. Other investigations in humans and quadrupeds show that energy consumption is optimized in many legged animals' locomotion (Alexander, 1984). In the steady state condition, biological locomotors consume the least energy for leg swinging and benefit from passive dynamics. For example, the positive work of human muscles is relatively small (compared to stance) (Neptune et al., 2008), and in hip muscles, they partly function as actively-tunable springs (Doke et al., 2005). However, if perturbation happens or when humans walk on rough terrains, the energy consumption increases considerably to achieve stability (Voloshina et al., 2013). Therefore, one significant contribution of the swing leg adjustment is balancing a trade-off between versatility, robustness, and energy consumption.

As mentioned before, leg retraction is an important feature of leg swinging which has a significant effect on movement stability in quadrupeds (Herr and McMahon, 2001), birds (Daley et al., 2007), bipedal biological locomotors (Muybridge, 1955; Gray, 1968; Blum et al., 2010), and robots (Wisse et al., 2005). This technique for swing leg control reduces foot-velocity with respect to the ground and, as a result decreases landing impact (De Wit et al., 2000). Increasing the leg length in late swing and also angular accelerations (Vejdani et al., 2013) can improve stability and robustness (Blum et al., 2010). In human walking and running, there is a linear relationship between motion speed and swing leg retraction speed and acceleration (Fig. 2.2.4, Poggensee et al., 2014).

2.2.2.3 Distribution of Energies in Forward, Lateral, and Vertical Directions

Adjusting the leg parameters during flight/swing phase is more energy efficient than stance phase because end of the leg (the foot) is free to move. Since the swing leg movement initiates the states of the stance leg in the next step, tuning the system states to select the limit cycle is performed easier during leg swinging. The resulting redirection of the energy in different directions can be used in changing (a) the forward speed, (b) the gait, (c) foot placement on a specific target/position on the ground (e.g., walking on large stones), (d) motion direction, (e) lateral balance, and (f) locomotion on slopes and stairs.

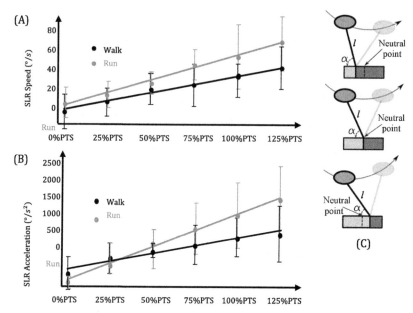

FIGURE 2.2.4 Swing leg retraction (A) speeds and (B) accelerations for human walking and running. Data points are the global means at each speed with errors bars of one standard deviation and associated trend lines for 21 subjects. Figures are adopted from Poggensee et al. (2014). (C) Swing leg adjustment effect on gait speed using neutral point concept. Foot placing at neutral point results in symmetric gait and keeping motion speed. Magnitude and direction of displacement of foot and the neutral point determine the magnitude and direction of forward acceleration. Dark and light boxes show acceleration and deceleration phase, respectively.

(a) Changing the speed (velocity control)

Studies on human walking and running show that locomotion speed significantly influences leg angle of attack and leg retraction speed as two features of swing leg adjustment (Sharbafi et al., 2013, 2016, 2017). A potential role of leg adjustment for control of forward speed is indicated by modeling and robotic studies (Dunn and Howe, 1996). This equally holds for different types of conceptual gait models (explained in detail in Chap. 3 and 4), e.g., inverted pendulum, spring loaded inverted pendulum, or rimless wheels, where angle of attack is a key control property in speed adjustment.

The concept of *neutral point* introduced by Raibert (1986) shows such a relation between swing leg adjustment and motion speed in a simple model of running (describing a monopod or biped hopping robot). Neutral point is the foot position (relative to the hip) at touchdown which results in symmetric leg movement around mid-stance. Then the forward speed is the same at liftoff as it is at touchdown and the average horizontal and angular acceleration over each stride becomes zero. With deviation from the *neutral point*, the body accelerates, with the magnitude and direction of acceleration proportional to the magnitude and direction of the displacement between foot position and neutral point, as shown in Fig. 2.2.4C. Based on that, he suggested the following controller to adjust forward running speed in monopod hopper and biped runner:

$$\alpha = \arccos\left(\frac{\dot{x}T_s + 2k(\dot{x} - \dot{x}_d)}{2l}\right), \tag{2.2.1}$$

in which α, T_s, l, k, $\dot{x}$, and $\dot{x}_d$ are angle of attack, stance time, leg length, constant gain, current and desired speed, respectively. Although this controller has limitations at high speed running, it can perfectly demonstrate the ability of swing leg control in gait speed adjustment in practice.

(b) Changing the gait

Raibert showed that with a simple leg adjustment protocol (beside axial leg force control) a single-legged (monopod) robot can stably hop, run, and even summersault (Raibert, 1986). For switching from one type of locomotion to another (e.g., walking to running) or from standing to dynamic locomotion (and vice versa) leg swinging plays an important role. Two different cases of gait switching are between:

1. Standing and Walking/Running

Gait initiation and termination are two types of switching between standing and locomotion. In gait initiation, the task is changing from keeping balance (placing the body CoM within base of support) to moving body over the ground. During this transition, the upcoming swing leg prepares its push-off by actively loading and then unloading the limb. Consider the inverted pendulum model to represent the human body to explain this phenomenon. Assume the right and left legs are the upcoming swing and stance legs in the first step, respectively. In order to release the right leg from ground contact, an angular momentum is required to shift the center of mass to above the other leg (Winter, 1995). As a result, an increase in GRF is observed at the right leg to generate the required angular momentum and the center of pressure is moved first from middle of the two feet to the right and then to the left foot (Winter, 1995). Compared to gait initiation, COM trajectories are mirrored during gait termination (Jian et al., 1993). This also holds for COP trajectories. *Capture point* and *capture region* concepts (Pratt and Tedrake, 2006) describe how models help calculate a region to place the swing leg to stop in one (or more) step(s) (see Section 2.2.3 for details).

2. Walking and Running

For a given speed (e.g., when moving on a treadmill), either walking or running is preferred (Thorstensson and Robertson, 1987; Hreljac, 1993). If the average speed (i.e., the traveling time) over a specific distance matters, humans sometimes prefer a mixture of walking and running (Long and Srinivasan, 2013). Long and Srinivasan showed that for moving between two fixed points in a specific time, walking and running are preferred for long and short travel time, respectively, but a mixture of walking and running and even standing is chosen for intermediate time. Thus, sometimes steady locomotion may not be energy optimal. This finding was also shown analytically with computational optimization (Long and Srinivasan, 2013). However, humans prefer to switch between walking and running at a specific speed called PTS (preferred transition speed) which barely changes at different positive or negative accelerations (Segers, 2006). Walking is more efficient at speeds below PTS, while running is the more optimal gait for moving faster than PTS. In running at speeds above PTS, muscle activations in the swing leg are lower than in walking at the same speed (Prilutsky and Gregor, 2001). In contrast, in walking at speeds below PTS the muscle activations in stance leg are less than in running at the same speed (Prilutsky and Gregor, 2001). Therefore, costs of swing leg and stance leg movements might be the critical term in determining the energy consumption at high and low speeds, respectively. The transition between walking and running can be related to muscle functions as at maximum waking speed both hip and ankle muscles reach their limits in force production (Neptune and Sasaki, 2005; Prilutsky and Gregor, 2001). These limits are resolved by switching from walking to running at the same speed. Hence, changing gaits support efficient muscle function at different speeds similar to the function of the gear in a bicycle.

(c) Targeting

In steady state gaits with no targets for foot placement (nontargeted gait) swing leg control usually determines the angle of attack (Seyfarth et al., 2001) and leg retraction (Seyfarth et al., 2003; Herr et al., 2002) in a periodic manner. To reach certain targets, swing leg control can be used for foot placement and adjusting the foot orientation. This results in different kinds of foot contacts (e.g., heel strike or fore foot) at touchdown and may help overcome unwanted ground contacts (e.g., with obstacles). Passing over a river by placing the feet on stones or hurdle running are extreme cases of targeting, needing precise swing leg control which do not regularly happen in daily activities. More frequent applications of swing leg adjustment for targeting can be found in locomotion on rough terrains, stepping over obstacles or turning the motion direction. Using fore-foot touchdown for impact avoidance or walking on rough terrains are samples of swing leg strategies for targeting (Lieberman et al., 2010; Voloshina et al., 2013; Pratt and Tedrake, 2006).

(d) Steering

Steering the motion occurs in 3D space. Swing leg control in lateral plane plays the main role in steering the gait besides upper body yaw movement using stance leg. Maus and Seyfarth (2014) showed how lateral leg adjustment can compensate leg asymmetry and develop walking in circles or on a straight line.

(e) Lateral balance with foot placement (walking)

Lateral balancing is much more challenging than balancing in the sagittal plane (Bauby, 2000). McGeer demonstrated that passive walking dynamics allow descending a gentle slope without external power (McGeer, 1990a). So in sagittal plane, the passive leg adjustment and interaction between dynamics of two legs establish a periodic gait down a slight incline, with no external input except gravity. However, this does not hold in frontal plane. Here, a lateral control scheme is needed to withstand perturbations (Kuo, 1999). Baubuy and Kuo (2000) claim that unlike for fore-aft stability high-level neural feedback control is necessary for maintaining lateral stability. The support of body weight requires stabilizing motion dynamics in sagittal plane (MacKinnon and Winter, 1993). However, to achieve stability in frontal plane, proper sensing of lateral motion (like visual and vestibular input) is required for perturbation recovery (Warren et al., 1996; Winter, 1995).

(f) Handling gravity effect (uphill–downhill, stair climbing)

Locomotion on inclined ground or on stairs requires different swing leg adjustment strategies (different trajectories) compared to level ground walking. Foot placement in sloped terrains is more complex than on flat ground as both step length and step height need to be controlled. For example, trajectories generated using a simple pendulum like model of swing leg are not feasible in (stair) ascending because in this case vertical lifting is also required besides horizontal forward movement. In that respect, additional parameters describing the environment (ground) are required. This increases complexity of swing leg control which contributes to both ground clearance and energy management. The control strategies to cope with ground level changes can be identified in locomotion experiments on variable ground height. Grimmer and colleagues showed that leg stiffness and angle of attack are two control parameters to cope with ground level changes (Grimmer et al., 2008). A suitable control strategy to cope with ground level changes is to adapt the leg stiffness to an altered angle of attack. This adaptation is within the J-shaped area in the leg stiffness-attack angle space predicted for stable running in the SLIP model (Seyfarth et al., 2002).

Gravity as an external force influences locomotion control on level and inclined terrains. However, gravitational effects are more critical on sloped grounds or stairs. For example, the largest percentage of falls occurs during stair walking in public places. Here, 80% of these falls on stairs relate to stair descent (Shumway-Cook and Woollacott, 2007). The stance phase in stair ascending is divided into three subphases: (i) weight acceptance, (ii) pull-up, and (iii) forward continuance. The swing phase is divided into (i) foot clearance and (ii) foot placement (Shumway-Cook and Woollacott, 2007), which is similar to level ground gait. The main contributors in foot clearance are tibialis anterior (for foot dorsiflexion) and hamstrings (for knee flexion) while rectus femoris contributes to the second half of the swing phase (similar to level walking Prilutsky and Gregor, 2001). The swing leg—guided by the movement of the pelvis—is lifted and moved forward by hip flexion followed by hip extension and ankle dorsiflexion in preparation of foot placement on the higher step (McFadyen and Winter, 1988). Reducing sensory information (e.g., visual feedback in blind-folded stair gaits) influences swing leg control strategy in walking on stairs much more than in level walking. For example, with limited visual feedback, anticipatory gastrocnemius activation is reduced and the leg is more compliant during landing in stair descending (Craik, 1982). With reduced visual sensory information (e.g., blurred-vision), foot clearance and foot placement become critical control strategies for stair descending. In such conditions, the foot is placed further backwards on the step to increase the safety margin (Simoneau et al., 1991).

2.2.2.4 Recovery from Perturbations

Strategies for perturbation recovery may be divided to three categories: (i) small perturbations that can be recovered by intrinsic muscle behavior (by damping property of the system dynamic) without requiring more activation, (ii) moderate perturbations requiring stance leg muscle activation for achieving posture balance, (iii) large perturbations which need stepping for compensating perturbation and returning to a stable solution (fixed point or limit cycle). In the latter group of perturbations, swing leg adjustment plays the most significant role. Velocity based leg adjustment (see Sect. 2.2.3) are the most common methods used to find the point (region) to place the foot for perturbation recovery of models and machines (robots). Another group is benefiting from system (passive) dynamics such as pendulum-like movement (McGeer, 1990a). After perturbation occurrence, pendulum-like passive dynamics yields shorter steps (than normal steps) which results in lower impacts, but still tolerable in for-aft direction (Kuo and Donelan, 2010).

Eng et al. have investigated the movement strategies and neuromuscular responses to recover from a tripping perturbation in humans (Eng et al., 1994). They have studied perturbations in early and late swing phase of walking. For early swing perturbations, the most common control strategy was an elevating the swing limb while the swing limb lowers in response to the late swing perturbation. In the elevating strategy both swing limb flexion and stance limb extension are involved. Two goals are achieved by this control strategy. Firstly, it removes the limb from the obstacle prior to accelerating it over the obstacle. Secondly, the extensor response of the stance limb generates an early heel-off which increases the CoM height. This provides extra time to extend the swing limb in preparation for the landing. In contrast, swing leg flexion may be dangerous in late swing perturbations, because the swing limb is approaching the ground and the body mass is in front of the stance foot. Instead, the swing leg lowers rapidly (with a flat foot or forefoot landing) which shortens the step length. Hence, the similar recovery strategy by different patterns of muscle activation is generated in early and late perturbations (Eng et al., 1994).

2.2.3 SWING LEG MODELING AND CONTROL

Template models such as the inverted pendulum model (Cavagna et al., 1963; Cavagna and Margaria, 1966) and the spring–mass model (SLIP, spring-loaded inverted pendulum) (Blickhan, 1989; Full and Koditschek, 1999) can help understand principles inherent in human locomotion and to demonstrate them in robotic counterparts. These models concentrate on the description of ground reaction forces (GRF) and center of mass (CoM) trajectories and neglect the effects of swing leg dynamics. In the swing phase of walking, beside ground clearance, the main function of the swing leg is providing an appropriate foot placement, i.e., achieving a suitable leg configuration, a desired angle of attack, and leg retraction in preparation of the next contact phase. Although the swing leg mass also affects whole body motion, in most studies this effect is ignored. In these models, the focus is on COM dynamics and the representation of swing leg movement is reduced to describe an appropriate angle of attack (Kuo, 2007b; Knuesel et al., 2005). In the following, we present an overview of different swing leg adjustments based on such simplified models and we introduce a new model for leg placement based on leg dynamics (Mohammadi et al., 2014).

2.2.3.1 Massless Swing Leg

Using a fixed angle of attack with respect to the ground can stabilize running (Seyfarth et al., 2002) and walking (Geyer et al., 2006). However, the region of possible leg adjustments (regarding leg stiffness, leg angle) for the stable gait is quite limited. The next levels of swing leg adjustment approaches are (1) swing leg retraction with a given leg rotation speed (Herr et al., 2002; Seyfarth et al., 2003) (see Sect. 2.2.2) and (2) adapting the leg angle during leg swinging using state feedback (Pratt and Tedrake, 2006). Leg adjustment strategies may rely on sensory information about the CoM velocity (Raibert, 1986) in which the foot landing position is adjusted based on the horizontal velocity (e.g., Pratt et al., 2006; Poulakakis and Grizzle, 2009; Sato and Beuhler, 2004). Peuker et al. concluded that leg placement with respect to both the CoM velocity and the gravity vectors yielded the most robust and stable hopping and running motions with the SLIP model (Peuker et al., 2012). As a modification of Peuker's approach, Sharbafi et al. (2013) developed a novel VBLA (velocity based leg adjustment) controller. In this controller, the velocity vector (with horizontal and vertical components, $\dot{x}$ and $\dot{y}$) is used to adjust the angle of attack:

$$\alpha = \text{arctg}\left(\frac{\dot{y} - c\sqrt{lg}}{\dot{x}}\right) = \text{arctg}\left(\frac{\dot{y} - kg}{\dot{x}}\right), \tag{2.2.2}$$

in which α, l, and g are the angle of attack (with respect to ground), leg length (hip-to-foot point), and gravitational acceleration, respectively. Here, c is a dimensionless tuning parameter (gain) which can be lumped into the parameter $k = c\sqrt{l/g}$. A comparison of the three methods (VBLA, Raibert and Peuker approaches) showed that VBLA better mimics human-like leg adjustment in perturbed hopping in place, achieves the largest range of running velocities by a fixed set of control parameters, and predicts robust walking in a bipedal SLIP model with extended rigid trunk (Sharbafi and Seyfarth, 2016). This method was successfully implemented on a simulation model of a bioinspired robot (called BioBiped) to generate stable forward hopping with adjustable speeds (Sharbafi et al., 2014). The idea of zeroing horizontal speed (like in recovery from perturbed hopping) with an appropriate swing leg adjustment in walking was presented within the capture point concept (Pratt and Tedrake, 2006). This approach determines a position for foot placement to stop forward motion. For a bipedal system, *capture state* is defined as the state with zero kinetic energy level. By placing the foot (CoP) on a *capture point P*, the controlled motion dynamics moves the states to reach the capture state. The set of all capture points is called capture region. These concepts can be extended to n-step capture point and n-step capture region using leg swinging and recursive definition (Pratt and Tedrake, 2006). If n approaches ∞, the n-step walking can be achieved because the capture region converges to the area on the ground that the foot can be placed at without falling. For implementation on robots, usually simple models like inverted pendulum (Pratt and Tedrake, 2006) or linear inverted pendulum (Pratt et al., 2006) are utilized to find the capture point analytically.

2.2.3.2 Mass in the Swing Leg

Mochon and McMahon presented a model comprising a stiff stance leg and a segmented swing leg (Mochon and McMahon, 1980). Compared with the inverted pendulum model, this model provides a better match of human walking dynamics. Introducing the spring loaded inverted pendulum (SLIP) model, the gait dynamics (GRF and COM movement) of human locomotion can be better represented (Geyer et al., 2006) compared to the inverted pendulum model. However, swing leg movement is still a missing part in SLIP based models. In Sharbafi et al. (2017) a segmented swing leg is added to the SLIP model to represent the swing phase of human-like walking. Such template models help to better understand key features of human walking (e.g. muscle activation patterns and segment motions) which could previously be observed in more complex gait models (Geyer and Herr, 2010).

Judging from human leg muscle activities in the swing leg movement, biarticular hip muscles rectus femoris (RF) and hamstrings (HA) seem to be main contributors for swing leg control in the swing phase of walking (Nilsson et al., 1985). By modeling these two muscles with biarticular springs, better mechanical understanding of their activities in producing stable gait is obtained. In addition, such a passive mechanism may also replicate strong correlation observed between RF and HA in human swing leg movement (Prilutsky et al., 1998), as a consequence of body mechanics. The role of elastic biarticular thigh muscles (represented as springs) on swing leg dynamics can be further investigated, and the appropriate spring parameters and morphology can mimic human swing leg motion in walking. The muscle lever arm ratio, muscle stiffness, and muscle rest lengths influence the COM motion and swing leg behavior. With passive elastic biarticular muscles, walking motion characteristics like swing leg retraction and symmetric stance leg behavior around mid-stance are predicted (Mohammadi et al., 2014). Such a simple bio-inspired control approach can be implemented in robots (Sharbafi et al., 2016). During the swing phase, biarticular muscles can support swing leg rotational movement control while monoarticular muscles (e.g., knee or ankle joints) can provide (axial) leg shortening and lengthening (e.g., leg shortening is required for ground clearance). With such a muscle-specific task allocation, the target of control could be simply setting spring rest lengths to a specific value for each gait condition.

This simple control strategy is able to produce human-like forces and kinematic behavior in walking. It was successfully approved in a simulation model of BioBiped robot for describing forward hopping (Sharbafi et al., 2016). Changing the motion speed can be achieved by adjusting the rest angle of biarticular springs. This provides a simple and efficient swing leg control approach without needing sensory information of the leg configuration. In order to achieve high efficiency during different phases of the gait cycle (e.g., swing phase), non-backdrivable actuators are advantageous. They enable setting the springs' rest lengths to desired values and switching off the motors and to operate with no (or little) resistance when no actuation is needed.

For stable locomotion, swing-leg adjustment needs to complement the other locomotor subfunctions (stance and balance). In future, a better understanding of the interplay of these subfunctions needs to be developed. These insights will help to further improve the design and modular control of locomotor systems both in simulation models and in hardware.

REFERENCES

Alexander, R.McN., 1984. The gaits of bipedal and quadrupedal animals. Int. J. Robot. Res. 3 (2), 49–59.

Bauby, C.E., Kuo, A.D., 2000. Active control of lateral balance in human walking. J. Biomech. 33 (11), 1433–1440.

Birn-Jeffery, A., Daley, M.A., 2012. Birds achieve high robustness in uneven terrain through active control of landing conditions. J. Exp. Biol. 215, 2117–2127.

Begg, Rezaul, Best, Russell, Dell'Oro, Lisa, Taylor, Simon, 2007. Minimum foot clearance during walking: strategies for the minimisation of trip-related falls. Gait Posture 25 (2), 191–198.

Blickhan, R., 1989. The spring–mass model for running and hopping. J. Biomech. 22 (11), 1217–1227.

Blum, Y., Lipfert, S.W., Rummel, J., Seyfarth, A., 2010. Swing leg control in human running. Bioinspir. Biomim. 5, 026006.

Cavagna, G., Saibene, F., Margaria, R., 1963. External work in walking. J. Appl. Physiol. 18 (1), 1–9.

Cavagna, G., Margaria, R., 1966. Mechanics of walking. J. Appl. Physiol. 21 (1), 271–278.

Collins, S., Ruina, A., Tedrake, R., Wisse, M., 2005. Efficient bipedal robots based on passive-dynamic walkers. Science 307 (5712), 1082–1085.

Craik, R.L., 1982. Clinical correlates of neural plasticity. Phys. Ther. 62 (10), 1452–1462.

Daley, M.A., Biewener, A.A., 2006. Running over rough terrain reveals limb control for intrinsic stability. Proc. Natl. Acad. Sci. USA 103, 15681–15686.

Daley, M.A., Felix, G., Biewener, A.A., 2007. Running stability is enhanced by a proximo-distal gradient in joint neuromechanical control. J. Exp. Biol. 210 (3), 383–394.

Daley, M.A., Biewener, A.A., 2011. Leg muscles that mediate stability: mechanics and control of two distal extensor muscles during obstacle negotiation in the Guinea fowl. Philos. Trans. R. Soc. B 366, 1580–1591.

De Wit, B., De Clercq, D., Aerts, P., 2000. Biomechanical analysis of the stance phase during barefoot and shod running. J. Biomech. 33, 269–278.

Doke, J., Donelan, J.M., Kuo, A.D., 2005. Mechanics and energetics of swinging the human leg. J. Exp. Biol. 208 (3), 439–445.

Donelan, J.M., Kram, R., Kuo, A.D., 2002. Mechanical work for step-to-step transitions is a major determinant of the metabolic cost of human walking. J. Exp. Biol. 205, 3717–3727.

Dunn, E.R., Howe, R.D., 1996. Foot placement and velocity control in smooth bipedal walking. In: Proceedings, 1996 IEEE International Conference on Robotics and Automation, vol. 1. IEEE, pp. 578–583.

Eng, Janice J., Winter, David A., Patla, Aftab E., 1994. Strategies for recovery from a trip in early and late swing during human walking. Exp. Brain Res. 102 (2), 339–349.

Full, R.J., Koditschek, D., 1999. Templates and anchors: neuromechanical hypotheses of legged locomotion on land. J. Exp. Biol. 22, 3325–3332.

Geyer, H., Seyfarth, A., Blickhan, R., 2006. Compliant leg behaviour explains basic dynamics of walking and running. Proc. R. Soc. B 273 (1603), 2861–2867.

Geyer, H., Herr, H., 2010. A muscle-reflex model that encodes principles of legged mechanics produces human walking dynamics and muscle activities. IEEE Trans. Neural Syst. Rehabil. Eng. 18 (3), 263–273.

Gray, J., 1968. Animal Locomotion. Weidenfeld & Nicolson.

Grimmer, S., Ernst, M., Guenther, M., Blickhan, R., 2008. Running on uneven ground: leg adjustment to vertical steps and self-stability. J. Exp. Biol. 211 (18), 2989–3000.

Herr, H.M., McMahon, T.A., 2001. A galloping horse model. Int. J. Robot. Res. 20, 26–37.

Herr, H.M., Huang, G., McMahon, T.A., 2002. A model of scale effects in mammalian quadrupedal running. J. Exp. Biol. 205, 959–967.

Hreljac, A., 1993. Preferred and energetically optimal gait transition speeds in human locomotion. Med. Sci. Sports Exerc. 25, 1158–1162.

Jian, Y., Winter, D.A., Ishac, M.G., Gilchrist, L., 1993. Trajectory of the body COG and COP during initiation and termination of gait. Gait Posture 1, 9–22.

Knuesel, H., Geyer, H., Seyfarth, A., 2005. Influence of swing leg movement on running stability. Hum. Mov. Sci. 24, 532–543.

Kuo, A.D., 1999. Stabilization of lateral motion in passive dynamic walking. Int. J. Robot. Res. 18 (9), 917–930.

Kuo, A.D., 2002. Energetics of actively powered locomotion using the simplest walking model. J. Biomech. Eng. 124 (1), 113–120.

Kuo, A.D., 2007a. The six determinants of gait and the inverted pendulum analogy: a dynamic walking perspective. Hum. Mov. Sci. 26 (4), 617–656.

Kuo, A.D., 2007b. Choosing your steps carefully. IEEE Robot. Autom. Mag. 14 (2), 18–29.

Kuo, Arthur D., Donelan, J. Maxwell, 2010. Dynamic principles of gait and their clinical implications. Phys. Ther. 90 (2), 157–174.

Lieberman, Daniel E., Venkadesan, Madhusudhan, Werbel, William A., Daoud, Adam I., D'Andrea, Susan, Davis, Irene S., Mang'Eni, Robert Ojiambo, Pitsiladis, Yannis, 2010. Foot strike patterns and collision forces in habitually barefoot versus shod runners. Nature 463 (7280), 531–535.

Long, L., Srinivasan, M., 2013. Walking, running and resting under time, distance, and speed constraints: optimality of walk-run-rest mixtures. J. R. Soc. Interface 10.

MacKinnon, C.D., Winter, D.A., 1993. Control of whole body balance in the frontal plane during human walking. J. Biomech. 26, 633–644.

Maus, H.M., Seyfarth, A., 2014. Walking in circles: a modelling approach. J. R. Soc. Interface 11 (99), 20140594.

McFadyen, B.J., Winter, D.A., 1988. An integrated biomechanical analysis of normal stair ascent and descent. J. Biomech. 21 (9), 733–744.

McGeer, T., 1990a. Passive dynamic walking. Int. J. Robot. Res. 9 (3), 62–82.

McGeer, T., 1990b. Passive walking with knees. In: Proc. IEEE Int. Robotics Automation Conf. Los Alamitos, CA, pp. 1640–1645.

McVea, D.A., Pearson, K.G., 2006. Long-lasting memories of obstacles guide leg movements in the walking cat. J. Neurosci. 26 (4), 1175–1178.

Mills, Peter M., Barrett, Rod S., Morrison, Steven, 2008. Toe clearance variability during walking in young and elderly men. Gait Posture 28 (1), 101–107.

Miura, H., Shimoyama, I., 1984. Dynamic walking of a biped. Int. J. Robot. Res. 3 (2), 60–74.

Mochon, S., McMahon, T.A., 1980. Ballistic walking. J. Biomech. 13, 49–75.

Mohammadi, A., Sharbafi, M.A., Seyfarth, A., 2014. SLIP with swing leg augmentation as a model for running. In: 2014 IEEE/RSJ International Conference on Intelligent Robots and Systems. IEEE, pp. 2543–2549.

Moritz, C.T., Farley, C.T., 2004. Passive dynamics change leg mechanics for an unexpected surface during human hopping. J. Appl. Physiol. 97, 1313–1322.

Muybridge, E., 1955. The Human Figure in Motion.

Neptune, R.R., Sasaki, K., 2005. Ankle plantar flexor force production is an important determinant of the preferred walk-to-run transition speed. J. Exp. Biol. 208 (5), 799–808.

Neptune, R.R., Sasaki, K., Kautz, S.A., 2008. The effect of walking speed on muscle function and mechanical energetics. Gait Posture 28 (1), 135–143.

Nilsson, J., Thorstensson, A., Halbertsma, J., 1985. Changes in leg movements and muscle activity with speed of locomotion and mode of progression in humans. Acta Physiol. Scand. 123 (4), 457–475.

Peuker, F., Maufroy, C., Seyfarth, A., 2012. Leg adjustment strategies for stable running in three dimensions. Bioinspir. Biomim. 7, 036002.

Poggensee, K.L., Sharbafi, M.A., Seyfarth, A., 2014. Characterizing swing leg retraction in human locomotion. In: International Conference on Climbing and Walking Robots (CLAWAR 2014).

Poulakakis, I., Grizzle,,J.W., 2009. The spring loaded inverted pendulum as the hybrid zero dynamics of an asymmetric hopper. IEEE Trans. Autom. Control 54, 1779–1793.

Pratt, J.E., Tedrake, R., 2006. Velocity-based stability margins for fast bipedal walking. In: Fast Motions in Biomechanics and Robotics. Springer, Berlin, Heidelberg, pp. 299–324.

Pratt, J., Carff, J., Drakunov, S., Goswami, A., 2006. Capture point: a step toward humanoid push recovery. In: 6th IEEE–RAS International Conference on Humanoid Robots, pp. 200–207.

Prilutsky, B.I., Gregor, R.J., Ryan, M.M., 1998. Coordination of two-joint rectus femoris and hamstrings during the swing phase of human walking and running. Exp. Brain Res. 120 (4), 479–486.

Prilutsky, B.I., Gregor, R.J., 2001. Swing-and support-related muscle actions differentially trigger human walk–run and run–walk transitions. J. Exp. Biol. 204 (13), 2277–2287.

Raibert, M., 1986. Legged Robots That Balance. MIT Press, Cambridge, MA.

Sakagami, Y., Watanabe, R., Aoyama, C., Matsunaga, S., Higaki, N., Fujimura, K., 2002. The intelligent ASIMO: system overview and integration. In: Proc. IEEE/RSJ Int. Conf. Intelligent Robots Systems, pp. 2478–2483.

Sato, A., Beuhler, M., 2004. A planar hopping robot with one actuator: design, simulation, and experimental results IROS. In: IEEE/RSJ Int. Conf. on Intelligent Robots and Systems.

Segers, V., Aerts, P., Lenoir, M., De Clercq, D., 2006. Spatiotemporal characteristics of the walk-to-run and run-to-walk transition when gradually changing speed. Gait Posture 24 (2), 247–254.

Seyfarth, A., Günther, M., Blickhan, R., 2001. Stable operation of an elastic three-segment leg. Biol. Cybern. 84 (5), 365–382.

Seyfarth, A., Geyer, H., Günther, M., Blickhan, R., 2002. A movement criterion for running. J. Biomech. 35 (5), 649–655.

Seyfarth, A., Geyer, H., Herr, H., 2003. Swing-leg retraction: a simple control model for stable running. J. Exp. Biol. 206, 2547–2555.

Sharbafi, M.A., Maufroy, C., Ahmadabadi, M.N., Yazdanpanah, M.J., Seyfarth, A., 2013. Robust hopping based on virtual pendulum posture control. Bioinspir. Biomim. 8 (3), 036002.

Sharbafi, M.A., Radkhah, K., von Stryk, O., Seyfarth, A., 2014. Hopping control for the musculoskeletal bipedal robot: BioBiped. In: 2014 IEEE/RSJ International Conference on Intelligent Robots and Systems, pp. 4868–4875.

Sharbafi, M.A., Rode, C., Kurowski, S., Scholz, D., Möckel, R., Radkhah, K., Zhao, G., Mohammadi, A., von Stryk, O., Seyfarth, A., 2016. A new biarticular actuator design facilitates control of leg function in BioBiped3. Bioinspir. Biomim. 11 (4), 046003.

Sharbafi, M.A., Mohammadinejad, A., Rode, C., Seyfarth, A., 2017. Reconstruction of human swing leg motion with passive biarticular muscle models. Human Movement Science 52, 96–107.

Sharbafi, M.A., Seyfarth, A., 2016. VBLA, a swing leg control approach for humans and robots. In: IEEE-RAS International Conference on Humanoid Robots (Humanoids 2016).

Sharbafi, M.A., Seyfarth, A., 2017. How locomotion sub-functions can control walking at different speeds? J. Biomech. 53, 163–170.

Shumway-Cook, A., Woollacott, M.H., 2007. Motor Control: Translating Research Into Clinical Practice. Lippincott Williams & Wilkins.

Siegler, R., Deloache, J., Eisenberg, N., 2006. How Children Develop. Worth Publishers, New York. ISBN 978-0-7167-9527-8, p. 188.

Simoneau, G.G., Cavanagh, P.R., Ulbrecht, J.S., Leibowitz, H.W., Tyrrell, R.A., 1991. The influence of visual factors on fall-related kinematic variables during stair descent by older women. J. Gerontol. 46 (6), M188–M195.

Thorstensson, A., Robertson, H., 1987. Adaptations to changing speed in human locomotion: speed of transition between walking and running. Acta Physiol. Scand. 131, 211–214. http://dx.doi.org/10.1111/j.1748-1716.1987.tb08228.x.

Townsend, M.A., 1985. Biped gait stabilization via foot placement. J. Biomech. 18, 21–38.

Vejdani, H.R., Blum, Y., Daley, M.A., Hurst, J.W., 2013. Bio-inspired swing leg control for spring–mass robots running on ground with unexpected height disturbance. Bioinspir. Biomim. 8 (4), 046006.

Voloshina, A.S., Kuo, A.D., Daley, M.A., Ferris, D.P., 2013. Biomechanics and energetics of walking on uneven terrain. J. Exp. Biol. 216, 3963–3970.

Vukobratovic, M., Borovac, B., 2004. Zero-moment point-thirty-five years of its life. Int. J. Humanoid Robot. 1 (1), 157–173.

Warren, W.H., Kay, B.A., Yilmaz, E.H., 1996. Visual control of posture during walking: functional specificity. J. Exp. Psychol. Hum. Percept. Perform. 22, 818–838.

Winter, David A., 1992. Foot trajectory in human gait: a precise and multifactorial motor control task. Phys. Ther. 72 (1), 45–53.

Winter, D., 1995. Human balance and posture control during standing and walking. Gait Posture 3, 193–214.

Wisse, M., Schwab, A.L., van der Linde, R.Q., van der Helm, F.C.T., 2005. How to keep from falling forward; elementary swing leg action for passive dynamic walkers. IEEE Trans. Robot. 21 (3), 393–401.

Wu, A.R., Kuo, A.D., 2015. Energetic tradeoffs of foot-to-ground clearance during swing phase of walking. In: Dynamic Walking. Columbus Ohio, USA.

Chapter 2.3

Balancing

Tim Kiemel

School of Public Health, University of Meryland, United States

Balance is the control of the body's orientation relative to vertical. To describe the mechanics of balance, we start with some definitions (Vukobratović and Borovac, 2004; Vukobratović et al., 2006; Herr and Popovic, 2008; Maus et al., 2010; Xiang et al., 2010; Goswami, 1999). Balance depends on the *support base*, the smallest convex set on the support surface containing all contact points between the feet and support surface. As an approximation, consider a finite number of contact points q_i with forces f_i acting on the feet (Fig. 2.3.1A). Let $f_{ni}\hat{z}$ be the component of q_i normal to the support surface, where $\hat{z}$ is the vertical basis vector. Define the *center of pressure* (COP) as

$$\text{COP} = \frac{\sum f_{ni}q_i}{\sum f_{ni}} \qquad (2.3.1)$$

and the *ground reaction force* (GRF) as $\sum f_i$. Then the net effect of all contact forces f_i acting at q_i equals the GRF acting at the COP and a moment M about the COP. The horizontal components of M are zero (Goswami, 1999), so the COP is often referred to in the robotics literature as the *zero-moment point* (ZMP). Some authors consider the COP and ZMP to be equivalent (e.g., Goswami, 1999). However, others consider the ZMP to be defined only when the COP is in the interior of the support based (e.g., Vukobratović and Borovac, 2004; Vukobratović et al., 2006) (Fig. 2.3.1A), since the foot may rotate about the COP when the COP lies on the boundary of the support base (Fig. 2.3.1B). To describe such foot rotation, one can define an extension of the ZMP, known as the *foot rotation index* (FRI) (Goswami, 1999) or *fictitious zero-moment point* (FZMP) (Vukobratović and Borovac, 2004), as the point on the support surface where the GRF would need to act to prevent foot rotation. When the FRI/ZMP lies in the interior of the support base, then it coincides with the ZMP = COP and there is no foot rotation (Fig. 2.3.1A).

With these definitions, we can understand the mechanics of balance in terms of how the GRF acting at the COP changes the body's angular momentum $L(t)$ about its center of mass (COM). Herr and Popovic (2008) describe this relationship using the *centroidal moment pivot* (CMP), as illustrated in Fig. 2.3.2 for sagittal-plane motion. The CMP is the intersection of the support surface and the line parallel to the GRF that passes through the COM. For sagittal-plane motion, change in angular momentum $dL_x(t)/dt$ equals the normal component of

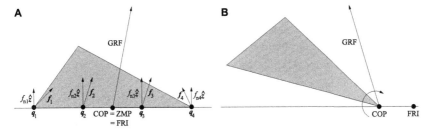

FIGURE 2.3.1 (A) The center of pressure (COP) lies in the interior of the base of support and, thus, coincides with the zero moment point (ZMP) and foot rotation index (FRI). (B) The COP lies on the boundary of the base of support and FRI $\neq$ COP indicates that the foot will rotate about the COP.

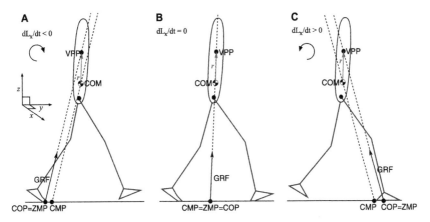

FIGURE 2.3.2 Whole-body mechanics of balance in the sagittal plane.

the GRF times the position of the COP relative to the CMP. Angular momentum changes in the forward (clockwise) or backward (counterclockwise direction) depending on whether the COP is behind the CMP (Fig. 2.3.2A) or ahead of the CMP (Fig. 2.3.2C), respectively.

For periodic locomotion, both $L_x(t)$ and $dL_x(t)/dt$ are periodic with mean 0, allowing angular momentum $L_x(t)$ to be uniquely computed by integrating $dL_x(t)/dt$ (Herr and Popovic, 2008; Elftman, 1939). For human walking, $L_x(t)$ is backwards during mid-stance and increases in the forward direction near the end of single-support when the COP is behind the CMP (Fig. 2.3.2A). $L_x(t)$ obtains its greatest forwards value near heel strike when $dL_x(t)/dt = 0$ (Fig. 2.3.1B). $L_x(t)$ changes in the backward direction when the COP is ahead of the CMP after heel strike (Fig. 2.3.2C), and is again backwards by mid-stance of the opposite leg. This pattern leads to a cancellation of positive and negative values of $L_x(t)$ so that it has mean 0 over the gait cycle. Such temporal cancellation must occur for any periodic gait. In additional, for human walking, there is cancellation of the angular momenta of different body segments at each point in time so that $L_x(t)$ is small throughout the gait cycle (Herr and Popovic, 2008). In other words, the COP and CMT remain close throughout gait cycle, leading to "zero-moment control".

Another method to describe the relationship between the GRF and the COM involves the *virtual pivot point* (VPP) (Maus et al., 2010). In the strictest definition, as illustrated in Fig. 2.3.2, the VPP is a point on the body with a fixed position relative to the COM (i.e., the vector *r* is constant) such that at each point in time, the line through the COP in the direction of the GRF passes through this point. For human walking, there is a VPP above the COM that approximately meets this definition (Maus et al., 2010) (Fig. 2.3.2). Maus et al. (2010) propose a conceptual model in which the VPP above the COM acts like the pivot point of a virtual pendulum, leading to stable balance during locomotion.

Comparing the approaches of Herr and Popovic (2008) and Maus et al. (2010), both examine the relationship between the GRF and the COM. Herr and Popovic (2008) emphasize that the line of action of the GRF passes close to the COM throughout the gait cycle, that is, there is approximate zero-moment control. If zero-moment control were perfect, then the VPP would be the COM. Instead, the small deviations from zero-moment control are such that the VPP is above the COM, leading Maus et al. (2010) to propose virtual pendulum control to explain stable balance during gait.

2.3.1 THE NEURAL CONTROL OF BALANCE: STANDING VS. WALKING

Balance has been extensively studied for standing (Horak and Macpherson, 2010), where the base of support is fixed. The nervous system uses information from various sensory systems (somatosensation, vision, and the vestibular system) to detect deviations away from the desired (nearly) vertical of various body segments, such as the trunk, thighs, and shanks. The nervous system corrects these deviations by modulating the stimulation of muscles acting at various joints, such as the ankles, knees, and hips. Balance during locomotion shares many of the same basic features, although there are important differences due to the changing base of support and rhythmicity inherent in locomotion. Here we consider these similarities and differences, focusing on the example of the control of the trunk relative to vertical.

Fig. 2.3.3 is a schematic representation of the neural control of balance in the general framework of control theory. Here the plant, the entity being controlled, is an input–output process that describes how stimulation of muscles by the nervous system leads to movement. The plant is defined by the biomechanical properties of the body and its mechanical interaction with the environment.

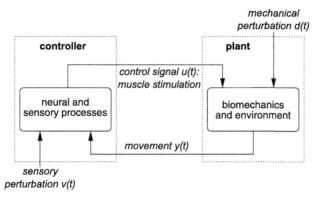

FIGURE 2.3.3 Schematic diagram of the neural control of movement, including standing balance and balance during locomotion.

The controller describes how the nervous system uses sensory information to modulate the stimulation of muscles. Balance is the result of the closed-loop interaction between the plant and the controller.

There are two basic questions about the neural control of balance. First, what is the desired attractor corresponding to stable balance? Second, how does neural feedback stabilize this attractor? For standing, the attractor is a stable fixed point and, thus, is easy to describe: the body, on average, leans slightly forward and there is, on average, a low level of stimulation of appropriate muscles (such as calf muscles) to counteract the small torques due to gravity resulting from the slight forward lean. For walking the attractor is a limit cycle. One must describe, as a function of the phase of gait cycle, the periodic stimulation of each muscle and the periodic movement of each mechanical degree of freedom of the body. For example, Fig. 2.3.4A shows the periodic stimulation, as measured by surface electromyography (EMG), of the erector spinae muscle of the lower trunk, which acts to rotates the trunk backwards and is one of many muscles involved in maintaining trunk orientation (Logan et al., 2017). Similarly, Fig. 2.3.4E shows the periodic motion of the trunk in the sagittal plane. The closed-loop nature of neural control implies that there is a consistency between EMG and kinematic waveforms. The periodic EMG waveforms of all muscles and the properties of the plant predict the periodic waveforms of each mechanical degree of freedom. Conversely, the periodic waveforms of all mechanical degrees of freedom and the properties of the neural controller predict the periodic waveform of each muscle.

One early approach to understanding the relationship among mean EMG and kinematic waveforms was the concept of balancing and unbalancing moments of Winter (1995). The vertical and especially horizontal acceleration of the hip

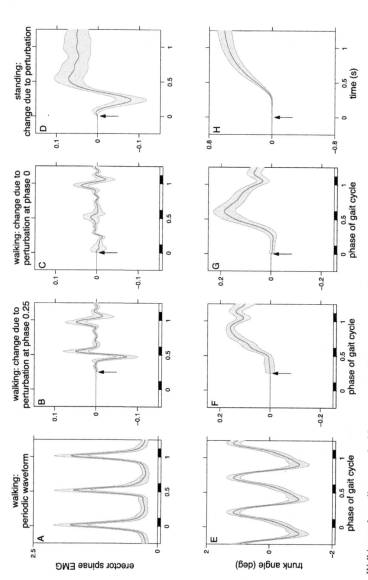

FIGURE 2.3.4 Walking and standing perturbed by movement of the visual scene. Responses to continuous movement of the visual scene were used to infer the effect of a small step in visual-scene position at the times indicated by arrows. Black bars along horizontal axes indicate double-support phases of the gait cycle. Based on data from Kiemel et al. (2011) for standing and Logan and Logan et al. (2017) for walking.

during walking produce an "unbalancing movement" of the trunk about the hip. Muscles must provide a counteracting "balancing" moment to keep the amplitude of trunk-angle oscillation within the observed range of about a degree or two (Fig. 2.3.4E). For example, after heel strike there is a backwards acceleration of the hip, which if not counteracted would produce a large forward rotation of the hip. The stimulation of erector spinae and other trunk extensor muscles near heel strike (Fig. 2.3.4A) produce a counteracting backwards moment on the trunk. Note that stimulation of the erector spinae muscle starts before heel strike, suggesting that the nervous system anticipates the unbalancing moment rather than merely reacting to its effects.

Another approach to understanding the relationship among mean EMG and kinematic waveforms is optimal control modeling (Anderson and Pandy, 2001; Miller, 2014). Given a model of the plant and a cost function, periodic EMG and kinematic waveforms are found that minimize the cost function for walking at a specified speed. The cost function typically penalizes metabolic cost and/or the amount of muscle stimulation. The amount of trunk accelerations predicted by these models roughly approximates observed behavior, although their cost functions do not penalizes acceleration. This calls into question the common belief that balance during locomotion is designed to reduce head accelerations in order to improve visual and vestibular sensing (Winter, 1995).

Sensory feedback control

The trunk during locomotion, as during standing, acts like an inverted pendulum (Winter, 1995) and, thus, is unstable without sensory feedback. Given the wealth of knowledge about how sensory information is used to stabilize standing balance, it is advantageous to consider whether similar principles apply to locomotion, which is much less studied. In particular, system identification methods based on sensory and mechanical perturbations have been used to identify the key features of the plant and the neural controller (Fig. 2.3.2) (van der Kooij et al., 2005; Kiemel et al., 2011) for standing balance. By comparing responses to such perturbations during standing and locomotion, we can gain insight into the similarities and differences between the two types of balance. For example, if the visual scene moves forward, a standing subject interprets this environmental motion as self-motion in the backwards direction. As a result, the nervous system changes levels of muscle stimulation, such as reducing the stimulation of the erector spinae muscles (Fig. 2.3.4D), in order to rotate the trunk forward (Fig. 2.3.4H). The reduction in stimulation is followed by an increase, which acts to limit the forward trunk rotation and eventually, on a longer time scale than shown in Fig. 2.3.4H, bring the trunk back to its original orientation.

If the same visual-scene perturbation is applied to a walking subject early during the single-support phase of the gait cycle, the initial response is similar:

a decrease in erector spinae stimulation (Fig. 2.3.4B) followed by a forward trunk rotation (Fig. 2.3.4F). However, the response is highly dependent on the phase of the gait cycle at which the perturbation occurs. For example, if the perturbation occurs at the beginning of double support, then the initial erector spinal response is greatly reduced (Fig. 2.3.4C). Instead, the stimulations of other muscles respond to the perturbation (not shown), so that the trunk still rotates forward (Fig. 2.3.4G).

This phase dependence occurs because the change in a muscle's stimulation due to a small perturbation occurs during those phases of the gait cycle when the muscle is normally stimulated. For example, the effects of perturbations on erector spinae stimulation occur around the beginning of double support (Fig. 2.3.4B, C) when the periodic stimulation is highest (Fig. 2.3.4A). Thus, the set of muscles that the nervous system can use to provide the earliest response to a perturbation depends on the phase at which the perturbation occurs. If a perturbation occurs during early double support, the erector spinae is in that early-response set. If the perturbation occurs at the beginning of double support, it is not and other muscles are used instead.

In summary, what emerges from the comparison of standing balance and locomotor balance is that both use similar feedback control mechanisms based on sensory information to respond to perturbations, but that locomotor balance has the additional property of phase dependence. This view of feedback control of locomotor balance is generally consistent with how feedback balance control has been implemented in walking models (Aoi et al., 2010; Geyer and Herr, 2010; Song and Geyer, 2015; Günther and Ruder, 2003). In models with a single trunk segment, the position and angular velocity of the trunk is used modulate the stimulation of muscles that act at the hip, but only when the given leg is in stance, giving rise to phase dependency.

REFERENCES

Anderson, F.C., Pandy, M.G., 2001. Dynamic optimization of human walking. J. Biomech. Eng. 123, 381–390. http://dx.doi.org/10.1115/1.1392310.

Aoi, S., Ogihara, N., Funato, T., Sugimoto, Y., Tsuchiya, K., 2010. Evaluating functional roles of phase resetting in generation of adaptive human bipedal walking with a physiologically based model of the spinal pattern generator. Biol. Cybern. 102, 373–387. http://dx.doi.org/10.1007/s00422-010-0373-y.

Elftman, H., 1939. The function of the arms in walking. Hum. Biol. 11, 529–535.

Geyer, H., Herr, H., 2010. A muscle-reflex model that encodes principles of legged mechanics produces human walking dynamics and muscle activities. IEEE Trans. Neural Syst. Rehabil. Eng. 18, 263–273. http://dx.doi.org/10.1109/TNSRE.2010.2047592.

Goswami, A., 1999. Postural stability of biped robots and the Foot-Rotation Indicator (FRI) Point. Int. J. Robot. Res. 18, 523–533. http://dx.doi.org/10.1177/02783649922066376.

Günther, M., Ruder, H., 2003. Synthesis of two-dimensional human walking: a test of the λ-model. Biol. Cybern. 89, 89–106.

Herr, H., Popovic, M., 2008. Angular momentum in human walking. J. Exp. Biol. 211, 467–481. http://dx.doi.org/10.1242/jeb.008573.

Horak, F.B., Macpherson, J.M., 2010. Postural orientation and equilibrium. In: Comprehensive Physiology. John Wiley & Sons, Inc. [cited 2015 Nov. 29].

Kiemel, T., Zhang, Y., Jeka, J.J., 2011. Identification of neural feedback for upright stance in humans: stabilization rather than sway minimization. J. Neurosci. 31, 15144–15153. http://dx.doi.org/10.1523/JNEUROSCI.1013-11.2011.

Logan, D., Kiemel, T., Jeka, J.J., 2017. Using a system identification approach to investigate subtask control during human locomotion. Front. Comput. Neurosci. 10. http://dx.doi.org/10.3389/fncom.2016.00146.

Maus, H.-M., Lipfert, S.W., Gross, M., Rummel, J., Seyfarth, A., 2010. Upright human gait did not provide a major mechanical challenge for our ancestors. Nat. Commun. 1, 70. http://dx.doi.org/10.1038/ncomms1073.

Miller, R.H., 2014. A comparison of muscle energy models for simulating human walking in three dimensions. J. Biomech. 47, 1373–1381. http://dx.doi.org/10.1016/j.jbiomech.2014.01.049.

Song, S., Geyer, H., 2015. A neural circuitry that emphasizes spinal feedback generates diverse behaviours of human locomotion. J. Physiol. 593, 3493–3511. http://dx.doi.org/10.1113/JP270228.

van der Kooij, H., van Asseldonk, E., van der Helm, F.C.T., 2005. Comparison of different methods to identify and quantify balance control. J. Neurosci. Methods 145, 175–203. http://dx.doi.org/10.1016/j.jneumeth.2005.01.003.

Vukobratović, M., Borovac, B., 2004. Zero-moment point—thirty five years of its life. Int. J. Humanoid Robot. 01, 157–173. http://dx.doi.org/10.1142/S0219843604000083.

Vukobratović, M., Borovac, B., Potkonjak, V., 2006. ZMP: a review of some basic misunderstandings. Int. J. Humanoid Robot. 03, 153–175. http://dx.doi.org/10.1142/S0219843606000710.

Winter, D., 1995. Human balance and posture control during standing and walking. Gait Posture 3, 193–214. http://dx.doi.org/10.1016/0966-6362(96)82849-9.

Xiang, Y., Arora, J.S., Abdel-Malek, K., 2010. Physics-based modeling and simulation of human walking: a review of optimization-based and other approaches. Struct. Multidiscip. Optim. 42, 1–23. http://dx.doi.org/10.1007/s00158-010-0496-8.

Chapter 3

Conceptual Models of Legged Locomotion

Justin Seipel, Matthew Kvalheim, Shai Revzen, Maziar A. Sharbafi, and André Seyfarth

This chapter provides an overview of simple conceptual models of locomotion at the scale of whole body movements. First, conceptual models of locomotion are introduced along with a few key empirical observations that support the construction of simple conceptual models. Next, a theoretical perspective is offered based on "templates and anchors" theory, where templates are related to simple conceptual models. Commonly used models of legged locomotion are then presented: The Spring-Loaded Inverted Pendulum (SLIP) model of running and the Inverted Pendulum (IP) model of walking. Legged locomotion is next presented in terms of oscillatory behavior and oscillatory-based analysis. Finally, readers are taken on a tour of a "model zoo" featuring many extensions of the SLIP and IP models to more complex and realistic models.

A Role for Simple Conceptual Models

Justin Seipel

Purdue University, West Lafayette, IN, United States

Legged locomotion of humans and other animals relies on a currently incomprehensible complex of underlying physiological systems. Though we have learned a lot about what is happening inside the body when humans or other animals move, we remain far from a coherent and complete understanding of how all the underlying processes integrate and contribute to whole-body motion.

Despite the overwhelming complexity of the internal processes of legged locomotion, the overall behavior on the level of whole body motion has remarkable coherence and regularity that can be understood using measures and models at the whole-body scale. As a complimentary approach to the direct study of the

Bioinspired Legged Locomotion. http://dx.doi.org/10.1016/B978-0-12-803766-9.00004-X

full complexity of locomotion, it can be helpful to develop relatively simple conceptual models that capture the overall, whole-body characteristics of locomotion. These models may also be more likely to be tractable mentally and mathematically. Further, many simple conceptual models can also be related to physical experiments and corresponding mathematical governing equations that provide powerful capabilities of prediction and scientific analysis. Such models tend to be simple in the sense that they often have a small number of elements and degrees of freedom. They nonetheless often exhibit nonlinear dynamic behavior that requires sophisticated investigation and analysis. Also, the way these models relate to and are applied to biological and robotic systems often requires sophistication.

In this chapter we present conceptual models of whole-body locomotion. Further, these models are shown to be related to physical observations and experiments, as well as mathematical governing equations based on physical laws of motion. Subchapter 3.1 provides an introduction to conceptual models of locomotion and key empirical observations that support simple conceptual model building on a scientific basis. Subchapter 3.2 provides a perspective on "templates and anchors" theory and how it relates to simple conceptual models. Subchapters 3.3 and 3.4 present the Spring-Loaded Inverted Pendulum (SLIP) model of running and the Inverted Pendulum (IP) model of walking. Subchapter 3.5 introduces locomotion in terms of oscillatory behavior and related approaches to analysis. Finally, Subchapter 3.6 presents a "model zoo" featuring many extensions of the SLIP and IP models.

Chapter 3.1

Conceptual Models Based on Empirical Observations

Justin Seipel
Purdue University, West Lafayette, IN, United States

3.1.1   OBSERVING, IMAGINING, AND GAINING INSIGHTS INTO LOCOMOTION

Legged locomotion is in many ways familiar to us. Consider human running, as shown in Fig. 3.1.1. We can recognize the scene of a human in motion, even if it is only a snapshot in time. We can likely recognize the basic anatomical segments of the trunk and limbs as well as the basic patterns of movement that are exhibited. We are likely able to form a kind of mental model of locomotion. Perhaps we can even conjure mental images or a movie related to the overall motions we observe when others run, or the experiences we have when we run.

Despite the familiarity of locomotion, more aspects may remain fuzzy or even foreign to us and become apparent only when explored further with trained observations and/or special tools and techniques. Questions may also help guide observations further. We might ask ourselves: Do we know what is happening when we move? Can we provide an explanation for it? Can we build a system that moves like we do? Do we know how major parts and processes of the body integrate into a coherent movement pattern?

Here we seek to develop a conceptual understanding of locomotion that includes and goes beyond our everyday observations. We also seek to provide a modeling framework that has predictive and design-aiding capabilities. Towards achieving these goals, it can be helpful to develop models of locomotion that provide both simple conceptual understanding as well as clear relationships to physical systems and physical laws.

3.1.2   LOCOMOTION AS A COMPLEX SYSTEM BEHAVIOR

There are significant challenges to developing scientific theories and models of legged locomotion. Movement in humans and other animals relies on a complex integration of skeletal, muscular, neurological, and other physiological systems. The skeletal system is organized with many complex joints, and multiple muscles are organized into groups and can span different numbers of joints and wrap in complex geometries. Also, the connectivity between neurons is beyond anything we understand. As complex as this anatomical perspective already is,

FIGURE 3.1.1 Human locomotion. Modified photograph by William Warby, Heat 1 of the Womens 100 m Semi-Final. Cropped and converted to grayscale. CC BY 2.0.

there remains additional complexity as seen from other perspectives, such as when considering information processes or feedback dynamics. Overall, it is a major challenge to derive simple models directly from the composition of a large number of biophysical parts and integrated processes.

An alternative and complimentary approach to simple model development is to use direct empirical study of overall, whole-body motion, as well as inspiration from mental models and intuition we may have. Other alternatives are possible too, such as attempting to model at an intermediate scale somewhere between the smallest underlying physiological processes and the whole body. All of these approaches can ultimately contribute to the development of a more comprehensive and integrative understanding of biological movement. For now, the focus of this chapter is on simple models that primarily capture overall whole-body movements of legged locomotion and that are related to physical experiments and physical laws.

3.1.3 SOME CHARACTERISTICS OF WHOLE-BODY LOCOMOTION

Models of legged locomotion can be conceived based on direct observations. Here, we focus on empirical observations of both the movement of the main body (trunk) and the corresponding movement of the legs, to reveal overall kinematic patterns that are characteristic of locomotion. Other measurements such as ground reaction forces and energetic consumption can be correlated with body movements to provide insights into kinetic processes influencing motion.

Overall movements of the body can be tracked from one or more points on the trunk, such as a marker on or near the hip (which is in the vicinity of the mass center in humans). The overall motion of the legs may be tracked relative to the

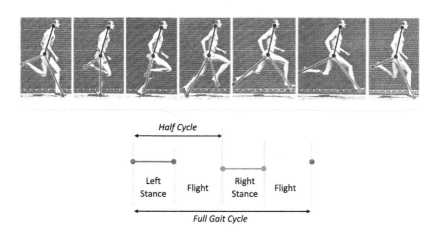

FIGURE 3.1.2 Running sequence. Photographs by E. Muybridge. Markers added by eye.

main body, such as by tracking points on the feet that help indicate whether a leg is in stance or swing, and where it is relative to the body.

Many observations of whole-body movement can be made with the unaided eye, but motion capture and tracking techniques have clarified what is otherwise fuzzy or too fast to see and has enabled significantly greater accuracy and quantification of movement (e.g., photographic techniques developed by E.J. Marey and E. Muybridge enabled and inspired new scientific and artistic works, Marey, 1894; Muybridge, 1979; Silverman, 1996).

An example of tracking and characterizing overall locomotion is provided in Fig. 3.1.2. This illustration and analysis of human running is based on a sequence of Muybridge photographs. The original photographs have been modified with markers (dots) added manually, at the feet, hip, and top of the spine. These markers, connected by lines, indicate major segments of the body: the two legs and trunk. Legs are either functioning in stance (with foot on the ground) or in swing (with foot off the ground), where flight phases of motion occur when both legs are off the ground. The three segments or parts identified here—the trunk, and two legs in stance or swing—relate to differentiated "subfunctions" that integrate together into whole movement. This concept was introduced in Chapter 2 from a motion control perspective, and is discussed again here from the complimentary perspective of three anatomical parts: trunk, stance leg, and swing leg.

3.1.3.1 The Trunk: Bouncing Along

During running, the trunk of the body appears to bounce along (here marked by the hip and top of spine): The trunk bounces or oscillates vertically between the

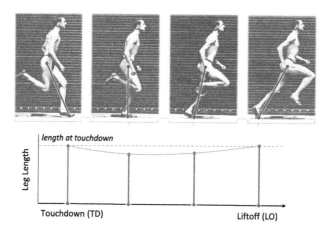

FIGURE 3.1.3 Stance leg length shortening and lengthening.

action of gravity and stance leg forces, all the while making forward progress. Further, the trunk tends to be angled forward of vertical (averaging about 15 degrees in Fig. 3.1.2), and appears to be regulated such that it oscillates slightly about this average. In many cases, locomotion model development is focused on the translational movement of the body and in those cases rotations of the trunk are not included as one may focus on developing a "point-mass" model of the body. An example of a point-mass model of the body is provided in Subchapter 3.3.

3.1.3.2 The Stance Leg: Acting Like a Spring

During the stance phase of running, as shown in Fig. 3.1.3, the stance leg length shortens (compresses) and then lengthens (decompresses) while it pivots about the foot. Observations of ground reaction forces show that the direction of force is significantly aligned with the leg (though not entirely) and that the force magnitude F changes with leg length, emulating Hooke's Law (Blickhan, 1989; Blickhan and Full, 1993):

$$F = k(l_0 - l).$$

Here, k is an effective leg spring stiffness, l_0 is its resting length, and l is the leg length. This emulation of an effective leg spring has been observed across many species, including poly-pedal locomotion where multiple legs can act together as a single virtual leg-spring (Blickhan and Full, 1993). Further, effective leg stiffness behavior has been demonstrated during walking (Geyer et al., 2006). Though these observations provide us with a helpful model of the overall functional behavior of stance legs, and elastic tissues do exist in legs, actual animal legs do not store and return energy the same way as an idealized spring. Animal legs generally require significant energy to operate.

3.1.3.3 The Swing Leg: Recirculating for Touchdown

During locomotion, the swing leg recirculates in order to be placed down at the next touchdown. For this, the prevailing movement of the swing leg is a swinging motion to a position forward of the body. This motion resembles the swinging of a pendulum. In addition to the forward swinging movement, the swing leg also goes through a significant retraction, both at the beginning and at the end of the swing phase: See Fig. 3.1.2. While swing legs can move like a passive pendulum under their own weight, swing legs are also likely to be controlled. Simple models of swing legs could include some combination of passive pendulum-like dynamics and active leg placement control. One highly simplified swing leg model that has been used often results from assuming the swing leg mass is negligible compared with the main body mass, and that the swing leg angle is controlled to follow a prescribed trajectory (as simple as a constant angle) until touchdown. An example of this is provided in Subchapter 3.3.

3.1.4 WHOLE-BODY CONCEPTUAL MODELS AS AN INTEGRATION OF PARTS OR SUBFUNCTIONS

An overall simple conceptual model of locomotion can be arrived at through integrating simple models/functions of the trunk (or main body), stance leg, and swing leg into a whole system. Further, by deriving governing equations of the whole system based on physical laws, we can predict its locomotion behaviors. Later, we present two well-established models of locomotion: the Spring-Loaded Inverted Pendulum (SLIP) model of running in Subchapter 3.3, and the closely associated Inverted Pendulum (IP) model of walking in Subchapter 3.4.

REFERENCES

Blickhan, R., 1989. The spring–mass model for running and hopping. J. Biomech. 22.

Blickhan, R., Full, R.J., 1993. Similarity in multilegged locomotion: bouncing like a monopode. J. Comp. Physiol., A Sens. Neural Behav. Physiol. 173 (5).

Geyer, H., Seyfarth, A., Blickhan, R., 2006. Compliant leg behaviour explains basic dynamics of walking and running. Proc. R. Soc. B 273.

Marey, E.J., 1894. Le movement. G. Masson, Dover.

Muybridge, E., 1979. Muybridge's Complete Human and Animal Locomotion: All 781 Plates from the 1887 Animal Locomotion. Dover Publications, New York.

Silverman, M.E., 1996. Etienne-Jules Marey: 19th Century cardiovascular physiologist and inventor of cinematography. Cardiol. Clin. 19, 339–341.

Chapter 3.2

Templates and Anchors

Matthew Kvalheim and Shai Revzen

Electrical Engineering and Computer Science, University of Michigan, Ann Arbor, MI, United States

3.2.1 A MATHEMATICAL FRAMEWORK FOR LEGGED LOCOMOTION

In this section we present a mathematical framework for analysis and modeling of legged locomotion. This framework is, for most applications, far too general. However, it will serve to provide a precise mathematical foundation, inside which other more practical models and approaches appear as special cases.

The study of legged locomotion is the study of how bodies move through space by deforming appendages we refer to as "legs" and using them to produce reaction forces from the environment that propel the body. Thus, the configuration of the system we seek to study comprises two parts—the location of the body in space, and the "shape" of that body with respect to a frame of reference that travels with the body. In mathematical terms, this means the overall configuration space $\mathcal{Q}$ is:

$$\mathcal{Q} = \text{SE}(3) \times \mathcal{B}, \quad \mathcal{B} \subseteq \mathbb{R}^m \tag{3.2.1}$$

where $\text{SE}(3)$ is the "special Euclidean group of dimension 3", also known as "the space of rigid motions", and $\mathcal{B}$ is taken to be some bounded, continuous, closed, piecewise smooth surface in the space $\mathbb{R}^m$. Let us temporarily use $q = (g, b) \in \mathcal{Q}$ to denote the instantaneous configuration.

In this book we are primarily concerned with legged locomotion that is generated by repeating patterns of motion called "gaits." When an animal or robot executes a gait, it traces out a cycle with b in the shape space $\mathcal{B}$, while at the same time translating and/or rotating the body frame $g \in \text{SE}(3)$ through the world. This form of a mathematical structure, in which a space is given by the Cartesian product of a "base space" (in our case, the shape space of the body) and a group,[1] here $\text{SE}(3)$, is called a (trivial) "principal fiber bundle",[2] or simply a

---

1. More technically, the group is required to be a "Lie group", and each fiber is a "principal homogeneous space" for this Lie group. Readers interested in these technicalities may consult, for example, Steenrod (1951), Husemoller (1994).

2. This bundle is called "trivial" because it is *equal* to a product $\text{SE}(3) \times \mathcal{B}$, whereas in general fiber bundles are spaces which are *locally trivial*, i.e., locally a product in some sense. See, for example, Bloch et al. (2003), Husemoller (1994), Steenrod (1951).

"principal bundle." Subsets of Q of the form $SE(3) \times \{b\}$ for a fixed $b \in B$ are called "fibers." A very readable introduction to the theory of fiber bundles may be found in Chapter 2 of Bloch et al. (2003).

In physics, principal bundles have been used to describe diverse phenomena in which cycles in the base space can be associated with a shift along a fiber. Names for some phenomena in the literature associated with this concept include "Berry phase", "geometric phase", "dynamical phase", "Pancharatnam phase", and "holonomy."

In the study of locomotion, these ideas have been used to describe the maneuvers cats (Marsden et al., 1991) and geckos (Jusufi et al., 2008) use to land on their feet, and the choice of undulatory motions made by snakes and eels (Ostrowski and Burdick, 1998; Hatton and Choset, 2011). When the relationship between shape change and body frame velocity is linear, it is given by a "connection":

$$g^{-1}\dot{g} = A(g, b)\dot{b} \tag{3.2.2}$$

While technical issues and high dimensions of the models create significant difficulties in applying "geometric mechanics" approaches in practice, this theoretical framework can in principal describe legged systems.

3.2.2 TEMPLATES AND ANCHORS: HIERARCHIES OF MODELS

One of the most influential insights allowing legged locomotion systems to be analyzed in practice was articulated in Full and Koditschek (1999), which proposed the use of "templates" for generating refutable, testable hypotheses for legged locomotion. While a "template" is defined as *"the simplest model (least number of variables and parameters) that exhibits a targeted behavior"*, the discussion and more recent treatments of the templates-and-anchors approach follow more closely the concept outlined in Full and Koditschek (1999) on page 3329: *"We will say that a more complex dynamic system is an 'anchor' for a simpler dynamic system if (1) motions in its high-dimensional space 'collapse' down to a copy of the lower-dimensional space of motions exhibited by the simpler system and (2) the behavior of the complex system mimics or duplicates that of the simpler system when operating in the relevant (reduced-dimensional copy of) motion space."* In other words, animals have many degrees of freedom, but move "as if" they have only a few, and limit pose to a behaviorally relevant family of postures. One way to encapsulate this insight mathematically is to presume that animals occupy only a low-dimensional "behaviorally relevant" submanifold of B, the space of possible "poses." As an illustrative example of what this means in practice, consider a photo of a galloping horse. We know that the horse is galloping, because the pose ("shape") of the body that we see in that still image is one which is only used for galloping. In fact, there is a cycle of poses that is associated with that horse galloping, and if environmental circumstances contrive to perturb the horse's body away from appropriate shapes for galloping, it quickly returns to some appropriate galloping pose.

However, the insight extends further: trotting quadrupeds such as horses and dogs, running bipeds such as humans and ostriches, insects like cockroaches employing alternating tripod gaits, and even running decapods like ghost crabs all employ similar center of mass dynamics—the "Spring Loaded Inverted Pendulum (SLIP)" (Dickinson et al., 2000; Blickhan, 1989). All these organisms exhibit similar center of mass dynamics: in each step, they bounce like a pogo stick. The center of mass slows down while descending closer to the ground, reaching its minimum speed at its lowest altitude, while ground reaction force in the normal direction is maximal. The center of mass continues, speeding up as it rises until the body entirely detaches from the ground into an aerial phase of ballistic motion leading to the next step.

In this sense, the SLIP template represents a common governing feature appearing in many organisms when they run quickly. The template is not only a description of a typical subset of poses, but also a low dimensional dynamical model that captures features of the aggregate behavior of the body.

It would be tempting to assume that for every behavior or animal examined, there exists a specific "simplest" template model that governs that behavior. However, Full and Koditschek (1999) had already pointed out that the notion of "simplest" model is problematic, and that both the Lateral Leg Spring (LLS) and the Spring Loaded Inverted Pendulum (SLIP) are templates for running (H_3, H_4 in Full and Koditschek, 1999, Table 1). The specific formal definition of a "template" was left vague.[3]

As an illustrative example, both the "Clock-Torque (CT-)SLIP" (Seipel and Holmes, 2007) and "SLIP with knee" (Seyfarth et al., 2000; Rummel and Seyfarth, 2008) models may be considered to be anchors for the classical sagittal plane spring loaded inverted pendulum (SLIP) model; important features of this model can be further distilled (following Blickhan, 1989) into a vertical hopping model, or alternatively into a compass walker (Usherwood et al., 2008). The three-dimensional pogo stick-like SLIP template (Seipel and Holmes, 2005) may also be viewed as an anchor for the sagittal plane SLIP, but one may be equally justified in reducing this three-dimensional pogo stick to a horizontal plane Lateral Leg Spring (LLS) model (Schmitt and Holmes, 2000a, 2000b)

---

3. This was intentional, based on personal communication with each of the authors.

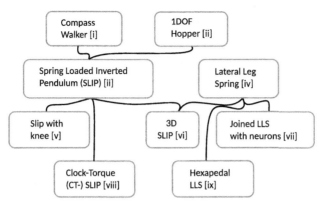

FIGURE 3.2.1 A collection of locomotion models with their template–anchor relationships indicated, showing a partial order structure. [i] Usherwood et al. (2008), [ii] & [iii] Blickhan (1989), [iv] Schmitt and Holmes (2000a, 2000b), [v] Rummel and Seyfarth (2008), Seyfarth et al. (2000), [vi] Seipel and Holmes (2005), [vii] Seipel et al. (2004), [viii] Seipel and Holmes (2007), [ix] Kukillaya and Holmes (2007).

which captures aspects of the horizontal motion such as steering, but ignores the importance of vertical bouncing. Additionally, both the "hexapedal lateral leg spring" (Kukillaya and Holmes, 2007) and "jointed lateral leg spring with neurons" (Seipel et al., 2004) models are extensions of the classical LLS which may be viewed as a template for these models. This hierarchy is depicted in Fig. 3.2.1.

We are led to the conclusion that rather than a template being a unique, ultimate object, "template and anchor" is a relationship between models. A given model Y can be a template for a more anchored model X, while Y itself may be an anchor for a further template Z. We will use the term "template" to imply that this model is "simpler" than its "anchor." Usually, one aspect of this simplicity is a reduction of dimension, and quantities in a template often represent aggregates of quantities from the underlying anchor. For example, both SLIP and LLS reduce the mass distribution of the body to a concentrated mass with or without rotational inertia; both discard modeling the kinetic energy and momentum associated with the legs themselves. An insightful discussion on the design and control of legged robots using template–anchor notions is given in Blickhan et al. (2007).

In the remainder of this chapter we will discuss several of the ways in which a template-and-anchor hierarchy can be constructed to facilitate the understanding of legged locomotion.

As a cautionary note, it should be pointed out that the term "template" has sometimes been used to mean "a spring mass model of center of mass dynamics." In this book, we will use it in the much broader meaning described above.

3.2.3 TEMPLATES IN DYNAMICS, CONTROL, AND MODELING

There are several ways to approach template–anchor relationships which have been used successfully. Mathematicians and physicists studying dynamical systems theory have constructed a variety of notions of dimensionality reduction. From this perspective, the primary object of study is an elaborate mathematical "anchor" model comprising a set of equations, the solutions of which are shown to be approximately or exactly modeled by a simpler "template" model comprising fewer equations with fewer parameters. Several examples of this approach can be found in Holmes et al. (2006). In particular, Kukillaya and Holmes (2007) have shown an example of a hexapedal cockroach model with jointed legs and neuronal control, which can be formally reduced and shown to behave similarly to the far simpler LLS model.

Engineers building robots have looked to templates as "targets of control", i.e., as descriptions of desirable behaviors to be emulated (Westervelt et al., 2003; Revzen et al., 2012; Ames, 2014), or as simplifications to be used for quickly estimating an appropriate control policy (Raibert et al., 1984). Here, the primary object of study is not the template itself, as much as it is the means by which template dynamics are elicited from an anchor.

A more explicit focus on templates is found in work by engineers employing "template-based" strategies for the bio-inspired design and control of robots. Here, the goal is to embed well-known templates in more complex, anchored locomotion systems. Controllers have been designed to embed the dynamics encoded in SLIP and its three-dimensional analog in bipedal robots (Wensing and Orin, 2014; Dadashzadeh et al., 2014). In Poulakakis and Grizzle (2009), the SLIP model is explicitly mathematically embedded as the "hybrid zero dynamics" of an asymmetric version of the SLIP model. Ankarali and Saranli (2011) have used an extended SLIP model involving torque actuation at the hip for designing a controller to achieve underactuated planar pronking in the robot RHex (Saranli et al., 2001). Other researchers have considered the combination of several different templates in the same robot in order to render it capable of achieving multiple goals, such as running/climbing (Miller and Clark, 2015) and running/reorientation/vertical hopping (De and Koditschek, 2015). Lee et al. (2008) have even worked to embed a cockroach-inspired antenna-based wall-following template in a robot with a bio-inspired antenna.

Biomechanists have looked to templates from a data-driven, experiment-centric perspective. Here, the primary objects of study are the locomotion data themselves. The goal is to find low-dimensional models which represent observational data, accounting for both trends and variability with a few meaningful parameters. Data-driven templates have been used successfully to predict how animals recover from perturbations (Revzen et al., 2013) and how humans control and stabilize their running gait (Maus et al., 2014). These ideas are elaborated upon in Section 3.2.4.3.

3.2.4 SOURCES OF TEMPLATES; NOTIONS OF TEMPLATES

Given the three approaches to templates described in the previous section, it is hardly surprising that there are many mathematical notions of being a template-and-anchor pair. In this section we point to some of the literature in the field. The subtle differences and technical caveats associated with applying these notions are outside the scope of our exposition.

3.2.4.1 Dimensionality Reduction in Dynamical Systems

As a simple example, templates exist in the dynamics of linear systems (see, e.g., the textbook of Hirsch and Smale, 1974, for an introduction to linear systems). When a stable Linear Time Invariant (LTI) system of differential equations $\dot{x} = Ax$ has a large "spectral gap"—some modes (projections of solutions onto the generalized eigenspaces of A) collapse much faster than others—the slower modes can justifiably be viewed as a template for the complete higher-dimensional system. This expresses itself as a large difference in the real part of the eigenvalues of the matrix A, with the eigenvalues corresponding to slow template modes having a real part close to zero.

Dynamicists have extended this idea to nonlinear systems in multiple ways using the notion of "invariant manifolds", of which the generalized eigenspaces in the previous example are a special case. A positively (negatively) invariant manifold is a smooth submanifold of the state space of a dynamical system for which any initial condition belonging to this submanifold remains in the submanifold as it evolves forward (backward) in time. An invariant manifold is a smooth submanifold of the state space of a dynamical system which is both negatively and positively invariant; in other words, an invariant manifold is a union of trajectories. (Positively) invariant manifolds are often useful notions of templates—here, the template appears in a form which guarantees that the anchor dynamics restricted to template states are invariant, meaning that if the anchor begins in a state belonging to the template it can no longer escape back to exhibiting more complex behaviors. An excellent survey of the many ways invariant manifold methods have been useful in science and engineering is given in Chapter 1 of Wiggins (1994).

One well-known class of invariant manifolds which can be used to form useful templates are the asymptotically stable normally hyperbolic invariant manifolds (NHIMs) (Hirsch et al., 1970; Eldering, 2013; Wiggins, 1994); by "asymptotically stable", we mean that they attract all nearby trajectories asymptotically. Special cases of NHIMs include hyperbolic fixed points and hyperbolic periodic orbits (Hirsch and Smale, 1974). A particularly nice property of NHIMs is that they persist under small smooth perturbations of the equations defining the dynamical system (Hirsch et al., 1970), and the compact invariant manifolds which persist under smooth perturbations are normally hyperbolic (Mané, 1978). This makes NHIMs useful from a modeling perspective. Since physical measurements cannot determine parameters of a mathematical model with perfect accuracy, any physically meaningful feature of a mathematical model must persist under small perturbations.

Viewed as infinite-dimensional dynamical systems, even certain partial differential equations admit a template-like structure both through the theory of normal hyperbolicity (Bates et al., 1998, 2000) and the related theory of "inertial manifolds", the second class of (positively) invariant manifolds we mention here (Constantin et al., 2012; Foias et al., 1988b). Inertial manifolds, when they exist, are finite-dimensional positively invariant manifolds containing the global attractor of a (possibly infinite-dimensional) dynamical system and attracting all solutions at an exponential rate (Foias et al., 1988b). If an inertial manifold exists for a given partial differential equation, it governs the long-term dynamics. Examples of systems having an inertial manifold include dissipative systems such as those that appear in elasticity and fluid dynamics (Constantin et al., 2012). Techniques for computationally producing approximate inertial manifolds have been studied (Foias et al., 1988a).

"Center manifolds" are the third class of invariant manifolds we mention here. We briefly describe the most basic notion of center manifold at the level of generality relevant for our discussion; see, for example, the discussion in Section 3.2 of Guckenheimer (1983) for more details. Given a system of differential equations $\dot{x} = f(x)$ and a stable equilibrium point x_0 with $f(x_0) = 0$, the eigenvalues of the linearization $Df(x_0)$ split into collections of eigenvalues having negative and zero real part. These collections of eigenvalues respectively determine stable and center subspaces. The center manifold theorem states that there exist "stable" and "center" invariant manifolds respectively tangent to these subspaces. Trajectories in the stable manifold approach x_0 exponentially in positive time. While the stable manifold is always unique, in general the center manifold need not be. Center manifolds may also be defined for periodic orbits (see Theorem 4 of Section 3.5 in Perko, 2001) and more general attractors (Chow et al., 2000). Center manifolds and NHIMs have somewhat similar spectral properties, but they differ in that NHIMs have an instrinsic global definition whereas center manifolds are only defined locally. This local definition manifests itself in the fact that center manifolds are in general nonunique (see Section 1.1.2 of Eldering, 2013 for more discussion).

All of the classes of (positively) invariant manifolds we have mentioned have the property that they attract all nearby states. The template is stable in the sense that anchor states which are near template states will asymptotically approach the template. However, one important reason these notions of templates are so useful is more subtle than this; not only do nearby anchor states approach these invariant manifold templates, they approach *specific trajectories* in the template. This provides justification for the approximation of anchor dynamics by template dynamics. For inertial manifolds, this property is known in the literature as "asymptotic completeness" (Robinson, 1996), and the fact that center manifolds have this property is shown, for example, in Carr (1982). For NHIMs, this property is often noted by referring to the existence of an "invariant foliation" or "invariant fibration" of the basin of attraction of the invariant manifold (Hirsch et al., 1970), and is also sometimes referred to as "asymptotic phase" in the literature (Bronstein and Kopanskii, 1994) (we also refer to this as "dynamical phase" in Subchapter 3.5). Guckenheimer (1975) contains a simpler discussion of the properties of asymptotic phase for the special case of exponentially stable limit cycles.

As an illustrative example of the utility of invariant manifold notions of templates in the analysis of legged locomotion, consider an oscillator. As explained in Subchapter 3.5, an oscillator, by definition, consists of the dynamics in the basin of attraction of an exponentially stable periodic orbit (also known as a limit cycle).[4] The image of the periodic orbit, or set of points traced out by the limit cycle, is itself a normally hyperbolic invariant manifold. Defining the anchor to be the dynamics on the entire basin of attraction, a template may be taken to consist of the dynamics restricted to the image of the periodic orbit. Explicitly, the existence of asymptotic phase on the basin of attraction implies that each anchor state will asymptotically coalesce with a (in this case, unique) template state which may be represented by assigning to each anchor state a number $\theta \in [0, 2\pi)$. This is the "phase oscillator" template explained in Subchapter 3.5. However, for many practical applications, this particular template approximation of the anchor dynamics may be too coarse. As explained in Subchapter 3.5, the theory of normal forms (Bronstein and Kopanskii, 1994) shows that large "spectral gaps" in the "Floquet multipliers" of an oscillator yield additional invariant "slow manifolds" corresponding to slow "Floquet modes." Anchor states will again asymptotically approach particular template trajectories in such a way that the dynamics restricted to such an invariant manifold constitutes a good template approximation of the anchor dynamics. Physically, the limit cycle may

---

4. As in Subchapter 3.5, an "oscillator" is a *deterministic* system as defined here, while in Subchapter 3.5 we use the term "rhythmic system" to refer to a *nondeterministic* system resulting from perturbations of a (deterministic) oscillator by ("relatively small") noise.

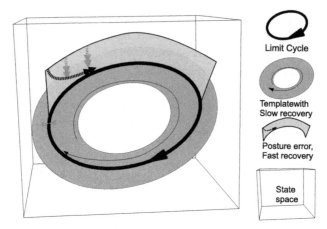

Limit Cycle

Templatewith
Slow recovery

Posture error,
Fast recovery

State
space

FIGURE 3.2.2 An example of an invariant manifold template–anchor relationship in the context of modeling legged locomotion shape-space dynamics by an oscillator. The collection of states corresponding to "Floquet multipliers" with relatively large magnitude form an invariant "slow manifold." The states belonging to this invariant manifold may be thought of as the states which return slowly to an unperturbed gait, modeled by the limit cycle. Taking the anchor to be the dynamics on the entire state space, the dynamics restricted to this invariant slow manifold may serve as a template. Alternatively, the dynamics restricted to the states traced out by the limit cycle itself may serve as a "phase oscillator" template which is a coarser approximation of the anchor dynamics.

be viewed as representing a perfectly periodic gait subject to no environmental or neuromuscular perturbations. The invariant slow manifold template may then be viewed as the collection of anchor states having "slow recovery" when perturbed from this steady gait. Any anchor states not belonging to this template will quickly return to the template and may be viewed as "posture errors." This is illustrated in Fig. 3.2.2.

Yet another source of templates comes from mechanical models possessing symmetries. Roughly speaking, a differential equation is said to possess a "symmetry" if it is invariant under the action of a "Lie group" (Lee, 2012) on its state space. For the case of mechanical systems, reduction tools such as Noether's Theorem from geometric mechanics (Abraham and Marsden, 1978; Bloch et al., 1996) yield conserved quantities (e.g., energy, momentum, angular momentum) which constrain trajectories of the dynamical system to lower-dimensional submanifolds. Dynamics restricted to these lower-dimensional submanifolds form templates for the original anchored mechanical system, and one can understand the behavior of the template in terms of the anchor and vice-versa. We note that other reduction methods in the spirit of Galois' work on algebraic equations (Dummit and Foote, 2004) also exist for the analysis of ordinary differential equations possessing symmetries but not necessarily arising from mechanical systems (Olver, 2000).

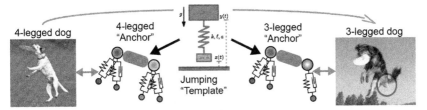

FIGURE 3.2.3 An uninjured dog and a 3-legged dog can both jump to catch a Frisbee. The ability to do so well can be expressed (red arrow) by taking the dynamics anchored in the 4-legged morphology, abstracting them as a jumping template and embodying this template in a 3-legged anchor reflecting the new morphology. The quality of this abstraction and embodiment can be quantified in a formal way within the framework of approximate bisimulation.

An even greater focus on templates as approximations can be found in the theory of "bisimulation" appearing in its original form in the study of discrete state transition systems in computer science (Park, 1981). Intuitively, two systems are bisimilar if they cannot be distinguished by an "external observer." Bisimulation has been generalized to apply to continuous-time and hybrid dynamical systems. In fact, the template notions previously mentioned in this section are bisimulations of their anchor dynamics, which follows from Proposition 11 of Haghverdi et al. (2005). Bisimulation provides a formalism for discussing templates and anchors for situations more general than the case in which the template is an invariant submanifold of the anchor.

Despite the level of generality afforded by the framework of bisimulation, requiring bisimilarity between models as a criterion for template–anchor relationships can sometimes arguably be too restrictive for modeling physical systems. Bisimilarity relations are not necessarily robust to noise, measurement error, or other perturbations to physical models. Recent work has extended the notion of bisimilarity by providing a definition of "approximate bisimulation" (Girard and Pappas, 2007). The utility of approximate bisimulation lies in its ability to quantify the quality of approximation by one mathematical model of another. In particular, the language of approximate bisimulation can be used to quantify the degree to which some mathematical model is a template for another anchored model. As a simple example, a double pendulum with one small mass m and one large mass M can be approximated by a single pendulum of mass $M + m$. For a more interesting example, consider the following. Animals, such as dogs, are able to instantiate the same template despite seemingly catastrophic injury such as limb loss. Fig. 3.2.3 illustrates the approximate template–anchor relationships between relevant models for this case. These examples are hardly surprising from the perspective of mechanical intuition, but the theory of approximate bisimulation renders these observations formal, testable, and quantifiable in a computational framework.

3.2.4.2   Templates Based on Mechanical Intuition

By far, the most prolific source of models intended as templates has been the insight of researchers. As described in Section 3.2.2, the insight of Blickhan led to the introduction of the SLIP template in his seminal work (Blickhan, 1989). Despite being an energetically conservative model without control inputs, the SLIP has enjoyed enormous success in making tractable the tasks of animal locomotion analysis (Section 3.2.2) and robot design and control (Section 3.2.3). The success of the sagittal plane spring-loaded inverted pendulum as a mathematical model inspired various extensions of SLIP, such as CT-SLIP (Seipel and Holmes, 2007), as well as three-dimensional (Seipel and Holmes, 2005), bipedal (Geyer et al., 2006), and segmented versions of SLIP (Seyfarth et al., 2000; Rummel and Seyfarth, 2008). Other templates such as LLS (Schmitt and Holmes, 2000a, 2000b) and its extensions (described in Section 3.2.2) such as a model with additional joints and neuronal interactions (Seipel et al., 2004) and a hexapedal version of LLS (Kukillaya and Holmes, 2007) were developed to specifically capture the horizontal component of locomotion. Thoughtful consideration of modeling has produced a plethora of additional templates of varying levels of complexity appropriate for other situations. Inspired by the climbing aptitude of insects and geckos, Goldman et al. (2006) proposed a template for describing rapid vertical climbing. Observations of cockroaches using their antennae to follow walls motivated the introduction of an antenna-based wall following template (Cowan et al., 2006). Human walking inspired a template based on the notion of "virtual pivot points" (Maus et al., 2010). Examples of other templates proposed for specific classes of models include a quadrupedal running template for robotic systems with articulated torsos (Cao and Poulakakis, 2013) and a kinematic template proposed for an eight-legged miniature octopedal robot assumed to be in quasi-static motion (Karydis et al., 2015).

Typically, templates have been proposed without explicitly formulating the anchor model to which they relate, although there are exceptions. In other work, there is an emphasis placed on exploring relationships between various templates and their anchors. To name but a few examples, see Seipel and Holmes (2005, 2006), as well as Chapter 5 of Holmes et al. (2006) and the references therein.

3.2.4.3   Data-Driven Model Reduction

Data-driven dimensionality and model reduction has emerged as an industrious and interdisciplinary field of research, having broad applications to the science and engineering fields and drawing upon techniques from optimization, statistics, dynamical systems theory, and machine learning. Classical approaches to dimensionality reduction include linear subspace projection methods such as "principal component analysis" and "factor analysis" (Jolliffe, 2002); one active area of current research concerns nonlinear dimensionality reduction approaches such as "manifold learning" (Lee and Verleysen, 2007), which generalize linear projection methods by replacing linear subspaces with submanifolds. Projection methods such as these identify a small (relative to the dimensionality of the raw data) collection of parameters which may accurately represent the raw data, and this collection of parameters is optimal in some sense depending on the projection method used. Such a small parameter set may accurately capture the spatial information present in time series data and motivate the construction of reduced-order spatio-temporal mathematical models. Givon et al. (2004) contains a review of several other algorithmic approaches to dimensionality reduction, focusing on methods specifically aimed at model reduction of general dynamical systems.

In the context of legged locomotion, there has been work on the construction of templates directly motivated from data. Operating under the assumption that the underlying mathematical model is an oscillator (see Subchapter 3.5), several researchers have performed nonparametric system identification of biomechanical systems (Ankarali, 2015; Wang, 2013; Revzen, 2009; Hurmuzlu and Basdogan, 1994; Hurmuzlu et al., 1996). Researchers have additionally attempted to find nonlinear coordinate systems directly from data in which oscillator dynamics are linear (see Revzen and Kvalheim, 2015, and references therein for more mathematical detail), and have coined the term "Data-Driven Floquet Analysis" (DDFA) to collectively refer to the computation of this linearizing coordinate system and to other oscillator system identification methods (Revzen, 2009). The linearizing coordinate change of DDFA can be viewed as a special case of finding linearizing "observables", which are themselves eigenfunctions of the "Koopman operator" (Rowley et al., 2009; Koopman, 1931), and may in some cases be computed using "Dynamic Mode Decomposition" (Schmid, 2010) and its extensions. Using the techniques of DDFA, Revzen and Guckenheimer (2011) present a method for identifying appropriate dimensions of reduced-order models of legged locomotion and other rhythmic systems directly from noisy data and without explicit knowledge of governing equations. By exploiting the structure of the stability basin of an oscillator, they determine a candidate dimension for the slow manifold by examining the magnitudes of eigenvalues of Poincaré return maps. This candidate dimension serves as an upper bound for the dimension of a statistically significant template.

In the specific context of human walking, Wang used analysis of Poincaré maps to show a relationship between upper body/trunk motion and foot placements, providing a rigorous data-driven derivation of human walking features previously conjectured (Wang, 2013). Maus et al. (2014) performed DDFA on human running data and showed that while the SLIP template predicts within-step kinematics of the center of mass, it fails to predict stability and behavior beyond one step. Furthermore, insights derived from DDFA enable Maus et al. (2014) to identify that swing-leg ankle states are important predictors of human locomotion beyond those present in the SLIP template. Augmenting the SLIP model with these predictors, the authors construct a model shown to have predictive power superior to SLIP for the available subject population.

3.2.5 CONCLUSION

The answer to the question of "which notion of template–anchor relationship should be used?" depends on one's goals and on practical limitations of the application in mind.

As an example of one end of the spectrum, mathematicians wanting to explore mathematical relationships need to write down or use existing equations of motion, which may make many assumptions about the underlying physics and/or biology of a locomoting system. In this case, various templates may be amenable to discovery by theoretical consideration. For example, invariant manifolds may be found "by hand" or numerical methods. Alternatively, reduction tools from geometric mechanics and the theory of Lie groups may be used to produce templates if symmetries are present in the equations of motion. Notions such as bisimulation and approximate bisimulation from computer science are used to formalize template–anchor notions in some areas of the literature.

On the opposite end of the spectrum, experimental biologists deal with actual data and do not have access to explicit mathematical models *a priori*. For this reason, researchers have worked on data-driven methods of system identification and model reduction. As outlined in Section 3.2.4.3, many algorithms have been used in attempts to tackle this problem for real-world systems in general, and several researchers have worked on methods aimed specifically toward legged locomotion. In particular, there has been some success in using Data Driven Floquet Analysis both using data to directly explain previously conjectured features of human locomotion and in motivating new templates of human running which may outperform SLIP as predictive models.

In between these two extremes, engineers and control theorists need methods to obtain practical models amenable to computation for which they can produce their own template "targets of control" to achieve desirable behaviors in robotic systems. Some engineers have used "template-based" methods in the bio-inspired design and control of robots, attempting directly to embed the low-dimensional dynamics of classical templates such as SLIP in high-dimensional anchored robots in order to achieve useful behaviors.

There are a myriad of notions and examples of "templates and anchors" outlined in this chapter, and many of these notions appear, at least at first glance, to be quite distinct. Many engineers, scientists, and mathematicians may benefit from exposure to these ideas.

REFERENCES

Abraham, R., Marsden, J.E., 1978. Foundations of Mechanics. Benjamin/Cummings Publishing Company, Reading, MA.

Ames, Aaron D., 2014. Human-inspired control of bipedal walking robots. IEEE Trans. Autom. Control 59 (5), 1115–1130. http://dx.doi.org/10.1109/TAC.2014.2299342.

Ankarali, M.M., 2015. Variability, Symmetry, and Dynamics in Human Rhythmic Motor Control. PhD thesis. Johns Hopkins University, Baltimore, MD, USA.

Ankarali, M.M., Saranli, U., 2011. Control of underactuated planar pronking through an embedded spring-mass Hopper template. Auton. Robots 30 (2), 217–231. http://dx.doi.org/10.1007/s10514-010-9216-x.

Bates, P., Lu, K., Zeng, C., 1998. Existence and Persistence of Invariant Manifolds for Semiflows in Banach Space, vol. 645. American Mathematical Soc.

Bates, P., Lu, K., Zeng, C., 2000. Invariant foliations near normally hyperbolic invariant manifolds for semiflows. Trans. Am. Math. Soc. 352 (10), 4641–4676. http://dx.doi.org/10.1090/S0002-9947-00-02503-4.

Blickhan, R., 1989. The spring–mass model for running and hopping. J. Biomech. 22 (11–12), 1217–1227. http://dx.doi.org/10.1016/0021-9290(89)90224-8.

Blickhan, R., Seyfarth, A., Geyer, H., Grimmer, S., Wagner, H., Gunther, M., 2007. Intelligence by mechanics. Philos. Trans. R. Soc., Math. Phys. Eng. Sci. 365 (1850), 199–220.

Bloch, A.M., Krishnaprasad, P.S., Marsden, J.E., Murray, R.M., 1996. Nonholonomic mechanical systems with symmetry. Arch. Ration. Mech. Anal. 136 (1), 21–99.

Bloch, A.M., Baillieul, J., Crouch, P., Marsden, J.E., Krishnaprasad, P.S., Murray, R.M., Zenkov, D., 2003. Nonholonomic Mechanics and Control, vol. 24. Springer.

Bronstein, A.U., Kopanskii, A.Y., 1994. Smooth Invariant Manifolds and Normal Forms, 1st edition. World Scientific Publishing, Salem, MA. ISBN 981021572X.

Cao, Q., Poulakakis, I., 2013. Quadrupedal bounding with a segmented flexible torso: passive stability and feedback control. Bioinspir. Biomim. 8 (4), 046007. http://dx.doi.org/10.1088/1748-3182/8/4/046007.

Carr, J., 1982. Applications of Centre Manifold Theory, vol. 35. Springer Science & Business Media.

Chow, S.-N., Liu, W., Yi, Y., 2000. Center manifolds for smooth invariant manifolds. Trans. Am. Math. Soc. 352 (11), 5179–5211. http://dx.doi.org/10.1090/S0002-9947-00-02443-0.

Constantin, P., Foias, C., Nicolaenko, B., Temam, R., 2012. Integral Manifolds and Inertial Manifolds for Dissipative Partial Differential Equations, vol. 70. Springer Science & Business Media.

Cowan, N.J., Lee, J., Full, R.J., 2006. Task-level control of rapid wall following in the American cockroach. J. Exp. Biol. 209 (9), 1617–1629. http://dx.doi.org/10.1242/jeb.02166.

Dadashzadeh, B., Vejdani, H.R., Hurst, J., 2014. From template to anchor: a novel control strategy for spring–mass running of bipedal robots. In: 2014 IEEE/RSJ International Conference on Intelligent Robots and Systems (IROS 2014). IEEE, pp. 2566–2571.

De, A., Koditschek, D.E., 2015. Parallel composition of templates for tail-energized planar hopping. In: 2015 IEEE International Conference on Robotics and Automation (ICRA). IEEE, pp. 4562–4569.

Dickinson, M.H., Farley, C.T., Full, R.J., Koehl, M.A.R., Kram, R., Lehman, S., 2000. How animals move: an integrative view. Science 288 (5463), 100–106.

Dummit, D.S., Foote, R.M., 2004. Abstract Algebra, vol. 1984. Wiley, Hoboken.

Eldering, J., 2013. Normally Hyperbolic Invariant Manifolds. Atlantis Studies in Dynamical Systems, vol. 2. http://dx.doi.org/10.2991/978-94-6239-003-4.

Foias, C., Jolly, M.S., Kevrekidis, I.G., Sell, G.R., Titi, E.S., 1988a. On the computation of inertial manifolds. Phys. Lett. A 131 (7), 433–436. http://dx.doi.org/10.1016/0375-9601(88)90295-2.

Foias, C., Sell, G.R., Temam, R., 1988b. Inertial manifolds for nonlinear evolutionary equations. J. Differ. Equ. 73 (2), 309–353. http://dx.doi.org/10.1016/0022-0396(88)90110-6.

Full, R.J., Koditschek, D.E., 1999. Templates and anchors: neuromechanical hypotheses of legged locomotion on land. J. Exp. Biol. 202 (23), 3325–3332.

Geyer, H., Seyfarth, A., Blickhan, R., 2006. Compliant leg behaviour explains basic dynamics of walking and running. Proc. R. Soc. Lond. B, Biol. Sci. 273 (1603), 2861–2867. http://dx.doi.org/10.1098/rspb.2006.3637.

Girard, A., Pappas, G.J., 2007. Approximation metrics for discrete and continuous systems. IEEE Trans. Autom. Control 52 (5), 782–798. http://dx.doi.org/10.1109/TAC.2007.895849.

Givon, D., Kupferman, R., Stuart, A., 2004. Extracting macroscopic dynamics: model problems and algorithms. Nonlinearity 17 (6), R55.

Goldman, D.I., Chen, T.S., Dudek, D.M., Full, R.J., 2006. Dynamics of rapid vertical climbing in cockroaches reveals a template. J. Exp. Biol. 209 (15), 2990–3000. http://dx.doi.org/10.1242/jeb.02322.

Guckenheimer, J.M., 1975. Isochrons and phaseless sets. J. Math. Biol. 1, 259–273. http://dx.doi.org/10.1007/BF01273747.

Guckenheimer, J.M., 1983. Nonlinear Oscillations, Dynamical Systems, and Bifurcations of Vector Fields, 1st edition. Springer-Verlag, New York, NY. ISBN 0387908196.

Haghverdi, E., Tabuada, P., Pappas, G.J., 2005. Bisimulation relations for dynamical, control, and hybrid systems. Theor. Comput. Sci. 342 (2), 229–261. http://dx.doi.org/10.1016/j.tcs.2005.03.045.

Hatton, R.L., Choset, H., 2011. Geometric motion planning: the local connection, stokes theorem, and the importance of coordinate choice. Int. J. Robot. Res. 30 (8), 988–1014.

Hirsch, M.W., Smale, S., 1974. Differential Equations, Dynamical Systems, and Linear Algebra, 1st edition. Academic Press, New York, NY. ISBN 0123495504.

Hirsch, M.W., Pugh, C.C., Shub, M., 1970. Invariant manifolds. Bull. Am. Math. Soc. 76 (5), 1015–1019.

Holmes, P., Full, R.J., Koditschek, D.E., Gukenheimer, J.M., 2006. The dynamics of legged locomotion: models, analyses, and challenges. SIAM Rev. 48 (2), 206–304. http://dx.doi.org/10.1137/S0036144504445133.

Hurmuzlu, Y., Basdogan, C., 1994. On the measurement of dynamic stability of human locomotion. J. Biomech. Eng. 116 (1), 30–36. http://dx.doi.org/10.1115/1.2895701.

Hurmuzlu, Y., Basdogan, C., Stoianovici, D., 1996. Kinematics and dynamic stability of the locomotion of post-polio patients. J. Biomech. Eng. 118 (3), 405–411. http://dx.doi.org/10.1115/1.2796024.

Husemoller, D., 1994. Fibre Bundles. Graduate Texts in Mathematics, vol. 20.

Jolliffe, I., 2002. Principal Component Analysis. Wiley Online Library. http://dx.doi.org/10.1002/9781118445112.stat06472.

Jusufi, A., Goldman, D.I., Revzen, S., Full, R.J., 2008. Active tails enhance arboreal acrobatics in geckos. Proc. Natl. Acad. Sci. 105 (11), 4215–4219.

Karydis, K., Liu, Y., Poulakakis, I., Tanner, H.G., 2015. A template candidate for miniature legged robots in quasi-static motion. Auton. Robots 38 (2), 193–209. http://dx.doi.org/10.1007/s10514-014-9401-4.

Koopman, B.O., 1931. Hamiltonian systems and transformation in Hilbert space. Proc. Natl. Acad. Sci. USA 17 (5), 315.

Kukillaya, R.P., Holmes, P.J., 2007. A hexapedal jointed-leg model for insect locomotion in the horizontal plane. Biol. Cybern. 97 (5–6), 379–395. http://dx.doi.org/10.1007/s00422-007-0180-2.

Lee, J., Sponberg, S.N., Loh, O.Y., Lamperski, A.G., Full, R.J., Cowan, N.J., 2008. Templates and anchors for antenna-based wall following in cockroaches and robots. IEEE Trans. Robot. 24 (1), 130–143. http://dx.doi.org/10.1109/TRO.2007.913981.

Lee, J.A., Verleysen, M., 2007. Nonlinear Dimensionality Reduction. Springer Science & Business Media.

Lee, J.M., 2012. Introduction to Smooth Manifolds, vol. 218. Springer Science & Business Media.

Mané, R., 1978. Persistent manifolds are normally hyperbolic. Trans. Am. Math. Soc. 246, 261–283. http://dx.doi.org/10.1090/S0002-9947-1978-0515539-0.

Marsden, J.E., O'Reilly, O.M., Wicklin, F.J., Zombros, B.W., 1991. Symmetry, stability, geometric phases, and mechanical integrators. Nonlinear Sci. Today 1 (1), 4–11.

Maus, H.-M., Lipfert, S.W., Gross, M., Rummel, J., Seyfarth, A., 2010. Upright human gait did not provide a major mechanical challenge for our ancestors. Nat. Commun. 1, 70. http://dx.doi.org/10.1038/ncomms1073.

Maus, H.-M., Revzen, S., Guckenheimer, J.M., Ludwig, C., Reger, J., Seyfarth, A., 2014. Constructing predictive models of human running. J. R. Soc. Interface 12. http://dx.doi.org/10.1098/rsif.2014.0899.

Miller, B.D., Clark, J.E., 2015. Towards highly-tuned mobility in multiple domains with a dynamical legged platform. Bioinspir. Biomim. 10 (4), 046001. http://dx.doi.org/10.1088/1748-3190/10/4/046001.

Olver, P.J., 2000. Applications of Lie Groups to Differential Equations, vol. 107. Springer Science & Business Media.

Ostrowski, J., Burdick, J., 1998. The geometric mechanics of undulatory robotic locomotion. Int. J. Robot. Res. 17 (7), 683–701.

Park, D., 1981. Concurrency and Automata on Infinite Sequences. Springer. http://dx.doi.org/10.1007/BFb0017309.

Perko, L., 2001. Differential Equations and Dynamical Systems, 3rd edition. Springer-Verlag, New York, NY. ISBN 0387951164.

Poulakakis, I., Grizzle, J.W., 2009. The spring loaded inverted pendulum as the hybrid zero dynamics of an asymmetric hopper. IEEE Trans. Autom. Control 54 (8), 1779–1793.

Raibert, M.H., Brown, H.B., Chepponis, M., 1984. Experiments in balance with a 3d one-legged hopping machine. Int. J. Robot. Res. 3 (2), 75–92. http://dx.doi.org/10.1177/027836498400300207.

Revzen, S., 2009. Neuromechanical Control Architectures of Arthropod Locomotion. PhD thesis. University of California, Berkeley.

Revzen, S., Guckenheimer, J.M., 2011. Finding the dimension of slow dynamics in a rhythmic system. J. R. Soc. Interface. http://dx.doi.org/10.1098/rsif.2011.043.

Revzen, S., Kvalheim, M., 2015. Data driven models of legged locomotion. SPIE Defense+ Security. International Society for Optics and Photonics, 94671V. http://dx.doi.org/10.1117/12.2178007.

Revzen, S., Ilhan, B.D., Koditschek, D.E., 2012. Dynamical trajectory replanning for uncertain environments. In: 2012 IEEE 51st Annual Conference on Decision and Control (CDC). IEEE, pp. 3476–3483. http://dx.doi.org/10.1109/CDC.2012.6425897.

Revzen, S., Burden, S.A., Moore, T.Y., Mongeau, J.-M., Full, R.J., 2013. Instantaneous kinematic phase reflects neuromechanical response to lateral perturbations of running cockroaches. Biol. Cybern. 107 (2), 179–200. http://dx.doi.org/10.1007/s00422-012-0545-z.

Robinson, J.C., 1996. The asymptotic completeness of inertial manifolds. Nonlinearity 9 (5), 1325.

Rowley, C.W., Mezić, I., Bagheri, S., Schlatter, P., Henningson, D.S., 2009. Spectral analysis of nonlinear flows. J. Fluid Mech. 641, 115–127. http://dx.doi.org/10.1017/S0022112009992059.

Rummel, J., Seyfarth, A., 2008. Stable running with segmented legs. Int. J. Robot. Res. 27 (8), 919–934. http://dx.doi.org/10.1177/0278364908095136.

Saranli, U., Buehler, M., Koditschek, D.E., 2001. RHex: a simple and highly mobile hexapod robot. Int. J. Robot. Res. 20 (7), 616–631. http://dx.doi.org/10.1177/02783640122067570.

Schmid, P.J., 2010. Dynamic mode decomposition of numerical and experimental data. J. Fluid Mech. 656, 5–28. http://dx.doi.org/10.1017/S0022112010001217.

Schmitt, J., Holmes, P., 2000a. Mechanical models for insect locomotion: dynamics and stability in the horizontal plane–I. Theory. Biol. Cybern. 83 (6), 501–515. http://dx.doi.org/10.1007/s004220000181.

Schmitt, J., Holmes, P., 2000b. Mechanical models for insect locomotion: dynamics and stability in the horizontal plane–II. Application. Biol. Cybern. 83 (6), 517–527. http://dx.doi.org/10.1007/s004220000180.

Seipel, J.E., Holmes, P., 2005. Running in three dimensions: analysis of a point-mass sprung-leg model. Int. J. Robot. Res. 24 (8), 657–674. http://dx.doi.org/10.1177/0278364905056194.

Seipel, J., Holmes, P., 2006. Three-dimensional translational dynamics and stability of multi-legged runners. Int. J. Robot. Res. 25 (9), 889–902. http://dx.doi.org/10.1177/0278364906069045.

Seipel, J., Holmes, P., 2007. A simple model for clock-actuated legged locomotion. Regul. Chaotic Dyn. 12 (5), 502–520. http://dx.doi.org/10.1134/S1560354707050048.

Seipel, J.E., Holmes, P.J., Full, R.J., 2004. Dynamics and stability of insect locomotion: a hexapedal model for horizontal plane motions. Biol. Cybern. 91 (2), 76–90. http://dx.doi.org/10.1007/s00422-004-0498-y.

Seyfarth, A., Blickhan, R., Van Leeuwen, J.L., 2000. Optimum take-off techniques and muscle design for long jump. J. Exp. Biol. 203 (4), 741–750.

Steenrod, N.E., 1951. The Topology of Fibre Bundles, vol. 14. Princeton University Press.

Usherwood, J.R., Szymanek, K.L., Daley, M.A., 2008. Compass gait mechanics account for top walking speeds in ducks and humans. J. Exp. Biol. 211 (23), 3744–3749. http://dx.doi.org/10.1242/jeb.023416.

Wang, Y., 2013. System Identification Around Periodic Orbits With Application to Steady State Human Walking. PhD thesis. The Ohio State University.

Wensing, P.M., Orin, D., 2014. 3d-slip steering for high-speed humanoid turns. In: 2014 IEEE/RSJ International Conference on Intelligent Robots and Systems (IROS 2014). IEEE, pp. 4008–4013. http://dx.doi.org/10.1109/IROS.2014.6943126.

Westervelt, E.R., Grizzle, J.W., Koditschek, D.E., 2003. Hybrid zero dynamics of planar biped walkers. IEEE Trans. Autom. Control 48 (1), 42–56. http://dx.doi.org/10.1109/TAC.2002.806653.

Wiggins, S., 1994. Normally Hyperbolic Invariant Manifolds in Dynamical Systems, 1st edition. Springer, New York, NY. ISBN 9781461243120.

Chapter 3.3

A Simple Model of Running

Justin Seipel

Purdue University, West Lafayette, IN, United States

3.3.1 RUNNING LIKE A SPRING-LOADED INVERTED PENDULUM (SLIP)

Humans and other animals run in a way that loosely resembles a pogo-stick bouncing along; see Fig. 3.3.1 for an illustration of human running. This behavior is approximately captured in the spring–mass, or Spring-Loaded Inverted Pendulum (SLIP) model of running; see Fig. 3.3.2. Further, passive dynamic running mechanisms can also embody SLIP-like running; see Fig. 3.3.3.

There are several features of running behavior that are in common for animal, robot, and SLIP model running: During the *stance phase* of running, the body first moves downwards, reaching a minimum height at or near mid-stance, then moves upwards, all the while pivoting about the foot of the stance leg. After the stance leg lifts off, the trunk continues to rise during a *flight phase* of motion, until reaching a maximum height apex, then falls until the swing leg touches down to start the next stance. During stance, the length from trunk to the foot of the stance leg compresses (shortens) and then decompresses (lengthens), roughly in proportion to the ground reaction force acting on the leg, effectively like a spring. During flight, when all legs are off the ground, the leading swing leg (with foot off the ground) is moved into position for the next foot touchdown.

The overall behavior of running, as summarized here, can be captured in simple conceptual models of locomotion such as the SLIP model (e.g., spring–mass "SLIP" models by Blickhan, 1989 and McMahon and Cheng, 1990 and other SLIP models introduced in Subchapters 3.2 and 3.6). SLIP models are often low-dimensional models, commonly using a single point-mass representing the body and a single massless leg that can represent key stance and swing leg functions during both stance and flight phases, respectively. Please see Subchapters 3.2 and 3.6 for an overview of SLIP-based models with varying degrees of complexity and realism. More realistic SLIP models might explicitly include more aspects of locomotion such as the movements of both legs during all phases of movement, as well as rotations and translations of the body/trunk. However, with increased realism there is often a trade-off in model complexity. Here a simple point-mass implementation of the SLIP model of running is presented.

FIGURE 3.3.1 Illustration of running (modified chronophotograph by Étienne-Jules Marey).

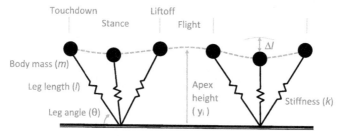

FIGURE 3.3.2 An illustration of the Spring-Loaded Inverted Pendulum (SLIP) running. Here a running sequence is shown with descriptive labels of key events and phases of motion: touchdown, stance phase, liftoff, and flight phase prior to the next touchdown. The position of the body mass (m) at any instant during stance is indicated by the leg length (l) and leg angle (θ). During flight, the leg angle is controlled to be equal to the value β upon the next touchdown, where the leg length at touchdown is l_o. The leg stiffness (k) and the leg compression (Δl) are also indicated. The maximum height at apex is indicated by y_i (note that the subscript i indicates the ith apex event, to be followed by the ($i + 1$)th event). The timing of touchdown and liftoff events are close to those in Fig. 3.3.1 but do not exactly correspond.

3.3.1.1 Physical Mechanisms and Robots Related to the SLIP Model

The concept of pogo-stick locomotion or SLIP locomotion has also been influenced by the work of mechanicians and roboticists who were inspired by human and other animal motion to produce running machines and robots. For example, dynamic legged robots as described in Raibert (1986), and the Robotic Hexapod RHex as recorded in Saranli et al. (2001). More recently, passive running mechanisms have been demonstrated, with elements that directly relate to the spring-loaded inverted pendulum model. For example, the passive locomotion mechanism of Owaki et al. (2010) exhibits running-like behavior; see Fig. 3.3.3. Such running mechanisms and robots are essentially like pogo sticks bouncing along, and are similar to the Spring-Loaded Inverted Pendulum model of running.

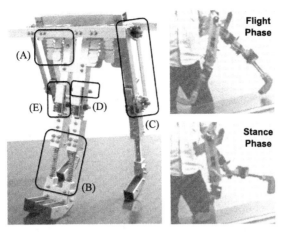

FIGURE 3.3.3 A passive dynamic legged mechanism. Images here are reproduced from Owaki et al. (2010) and displayed in a new arrangement. (Left panel) Image of a passive dynamic running mechanism, shown in a static position, with key system elements labeled: (A) hip springs to facilitate leg rotational oscillations, (B) leg springs to facilitate leg compression oscillations, (C) parallel link mechanism to synchronize the two outer legs, (D) shock absorber to dampen impact, and (E) a knee hyperextension mechanism to enable a form of mechanical support at the knee during stance, hypothesized to be needed in the absence of muscles or other actuators acting to transfer load across the knee joint (Owaki et al., 2010). (Right-top panel) A photograph captured during the flight phase, demonstrating one of the characteristic features of dynamic legged locomotion with a level of energy surpassing normal walking behavior. (Right-bottom panel) A photograph captured during the first half of stance, where the stance leg spring is clearly compressed. Note that in order for this system to maintain a steady stable gait, it runs on an inclined plane (here, running on an inclined treadmill). This mechanism exhibits flight phases and some characteristics of running, though currently does not produce maximum leg compression near mid-stance as is characteristic of SLIP running. Despite these differences with the classical assumptions and behaviors of the SLIP model, here we can see basic SLIP principles embodied in a physical system.

3.3.2 MATHEMATICAL AND PHYSICS-BASED SLIP MODEL

The SLIP model of running can be described in a more precise mathematical form based on physical laws of motion. The SLIP model, as presented in this chapter (Fig. 3.3.2), is composed of a point mass m representing the body, an effective leg stiffness k representing a massless stance leg, and a massless leg during flight representing a swing leg. Here, we derive the mathematical equations governing the motion of the SLIP model, based on a similar but more detailed presentation in Shen and Seipel (2016).

During the stance phase of motion the body mass m moves forward pivoting about the foot of the stance leg, and can be described by the leg length l and angle θ. The Lagrangian L of the system, a description of the system's kinetic energy T and potential energy V, is

$$L = T - V = \frac{1}{2}m\left(\dot{l}^2 + \left(l\dot{\theta}\right)^2\right) - \frac{1}{2}k\left(l - l_0\right)^2 - mgl\sin\theta.$$

Application of the Euler–Lagrange Equation to L yields the following equations governing stance:

$$m\ddot{l} = ml\dot{\theta}^2 - k\left(l - l_0\right) - mg\sin\theta,$$
$$ml^2\ddot{\theta} = -mgl\cos\theta - 2ml\dot{l}\dot{\theta}.$$

The stance phase of motion ends when the stance leg reaches its uncompressed length $l = l_0$. This event is called liftoff. After liftoff, the flight phase of motion follows.

During the flight phase of motion, the mass center is only affected by gravity, and so the motion is most simply described in terms of the height: $\ddot{y} = -g$. The horizontal component of velocity is constant during flight. During flight, the angle for the next touchdown leg is set to a specified value β and held there in preparation for the touchdown event, when the foot reaches the ground ($y = l_0 \sin\beta$) and the flight phase ends. After touchdown, a new stance phase follows and the gait pattern repeats.

The governing equations of stance and flight, together with the event equations defining liftoff and touchdown, can be solved to determine the overall locomotion solutions of the SLIP model. In general, a numerical approach to solving the governing equations is used, though analytical solutions are possible for approximations of the SLIP model (e.g., Ghigliazza et al., 2005; Saranlı et al., 2010; Schwind and Koditschek, 2000; Geyer et al., 2005; Robilliard and Wilson, 2005; Altendorfer et al., 2004; Shen and Seipel, 2016).

3.3.2.1 Ground Reaction Forces During Stance

In addition to computing the solutions of the governing equations to yield position and velocity, other quantities such as ground reaction forces can be computed and predicted. Ground reaction forces are often measured in locomotion experiments and can provide insights into the kinetics of locomotion. A comparison of experimentally measured ground reaction forces of human running and predictions made by the SLIP model are presented in Fig. 3.3.4 (a modified reproduction of plots from Geyer et al., 2006). This demonstrates that multiple key features of ground reaction force in both the fore-aft (horizontal) and the vertical directions can be accurately predicted by the SLIP model.

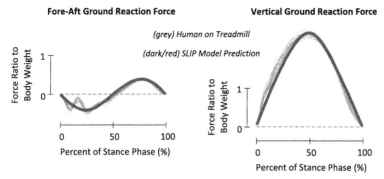

FIGURE 3.3.4 The ground reaction forces of human running and SLIP model predictions. Human experimental traces and SLIP model traces are reproduced from Geyer et al. (2006).

3.3.2.2   Stride Maps: Behavior Investigated Step-by-Step

The dynamic nature of locomotion is often studied using a stride map: a function that governs how the system states, like position and velocity, change from one step to the next. In general, this is constructed using a Poincaré Return Map. In less precise terms, this is like taking a snapshot of the system at either a set interval of time, or alternatively, every time a well-defined event occurs (e.g., every time a foot touches down, or every time the trunk mass reaches a maximum height apex). The mapping that results is often referred to as a *stride map*.

3.3.2.3   Stability of Locomotion

The stability of running solutions can be determined using the stride map, which is a common approach for SLIP models. For a more general discussion of stability and analysis methods, please see Full et al. (2002), Strogatz (1994), or Guckenheimer and Holmes (1983). A common technique is to find periodic solutions and then determine whether small deviations to the periodic motion will lead to the system diverging away from the periodic motion or returning to it. This can be approximated by linearizing the stride map and evaluating it with respect to the periodic solution being investigated. The eigenvalues of the resulting linear system will indicate the kind of local stability that occurs in the neighborhood of the periodic locomotion (where if the magnitude of all eigenvalues is less than one there exists asymptotic stability; if greater than one, unstable; if equal to one, further analysis is needed). For example, for the SLIP model presented above, asymptotically stable periodic running exists for a wide range of system parameters, as described in Geyer et al. (2005) and reproduced here in Fig. 3.3.5. In this figure, reproduced from Geyer et al. (2005), an apex-to-apex stride map is used. Here, two fixed points are shown, each representing differ-

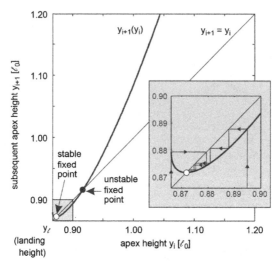

FIGURE 3.3.5 Here, a mapping from one apex to the next is displayed, reproduced from Geyer et al. (2005). There are two fixed points where the same apex height repeats each step, indicating a periodic locomotion solution. However, the stability of these two solutions differs. The stable fixed point is demonstrated by the inset figure, where an example sequence of steps is shown converging upon the stable fixed point value. Note that the analysis in Geyer et al. (2005) makes use of dimensionless parameters and some naming conventions that are different than those used here.

ent periodic locomotion solutions. One represents a stable limit cycle, or stable periodic locomotion. The other fixed point is an unstable periodic locomotion solution of the SLIP model. The SLIP model also exhibits other behaviors, such as higher period locomotion (Ghigliazza et al., 2005).

3.3.3 SOME INSIGHTS INTO RUNNING AIDED BY SLIP-BASED MODELS

3.3.3.1 Adaptive, Resilient Locomotion Based on Open-Loop Stability

An aspect of locomotion theory influenced by SLIP or pogo-stick models is our understanding of how locomotion is regulated or controlled in animals, and how it could be regulated in robots or assistive devices. In particular, SLIP models have demonstrated that largely uncontrolled dynamics of running can be self-stabilizing, requiring minimal control sensing or actuation. Understanding how open-loop stability properties of running integrate with more active feedback and actuation layers of locomotion is still far from being understood (perhaps partly due to the complexity of neuromechanical systems). Nonetheless, many simple SLIP model analyses have demonstrated both basic stability properties (e.g., Ghigliazza et al., 2005; Geyer et al., 2005) but also improved stability properties by including features we know represent realistic biological strategies, such as swing leg placement control (e.g., Knuesel et al., 2005 and other studies introduced in Subchapter 3.6), and inclusion of forcing and damping (e.g., Shen and Seipel, 2012). More examples of controlled and actuated SLIP models are presented in Subchapters 3.2 and 3.6.

3.3.3.2   Reducing Energetic Costs through Compliant Interaction

SLIP models of running have demonstrated clearly the theoretical possibility of locomotion with relatively small energetic cost (due to efficient energy storage in compliant legs and low-mass, low-impact legs that are idealized in many SLIP models). Robots, and prosthetic devices in particular, can be designed to efficiently store and return energy using elegant elastic structures inspired by SLIP models. Though the SLIP model is a highly idealized conception of running, and we know that animal and robot running generally involves many forms of energetic loss and actuation, the SLIP model can nonetheless provide insights into theoretical limiting cases that can influence and challenge our thinking about locomotion.

3.3.3.3   Momentum Trading to Benefit Stability

Another perspective on the mechanics of running is based upon momentum of the body (its mass times velocity) and angular momentum about the stance foot. During locomotion, there are transitions between flight phases where forward linear momentum is conserved, and stance phases where angular momentum is nearly conserved (or approximately conserved in the case of negligible gravity). At events like liftoff and touchdown, we can think of the system transitioning between these two modes. Whatever momentum was being conserved in one phase now gets "traded" or otherwise exchanged such that part of it contributes to a new conserved form of momentum. This has been referred to as "momentum trading" (e.g., Holmes et al., 2006). Without this aspect of the switching (or hybrid) dynamics of locomotion, the stability properties of an energy conserving SLIP system would not be possible (Holmes et al., 2006). The regulation of locomotion might be thought about in part as the regulation of traded momentum, from one step to the next.

3.3.3.4   Useful Inefficiency: Inefficiency can Benefit Robustness

An inefficient use of energy might sometimes be beneficial for creating more robust stability of legged locomotion. A common aspect of physical running, though less commonly represented in the simplest of SLIP models, is a significant energetic cost. While it is physically possible to demonstrate entirely passive SLIP-based running mechanisms, even in these cases there are energy losses that are overcome by using an inclined plane (e.g., Owaki et al., 2010). In other words, some non-negligible amount of positive work on the system and negative work on the system appears to be a common feature of periodic running locomotion. Further, recent studies have suggested that this could play a substantial role in the stability of locomotion, helping to generate significantly greater robustness (e.g., it can contribute to significantly larger basins of attraction, Shen and Seipel, 2012). In addition to regulating momentum, locomotion might also be thought about as regulating the flow of energy from step to step.

REFERENCES

Altendorfer, R., Koditschek, D.E., Holmes, P., 2004. Stability analysis of legged locomotion models by symmetry-factored return maps. Int. J. Robot. Res. 23.

Blickhan, R., 1989. The spring–mass model for running and hopping. J. Biomech. 22.

Full, R.J., Kubow, T., Schmitt, J., Holmes, P., Koditschek, D., 2002. Quantifying dynamic stability and maneuverability in legged locomotion. Integr. Comp. Biol. 42 (1).

Geyer, H., Seyfarth, A., Blickhan, R., 2005. Spring–mass running: simple approximate solution and application to gait stability. J. Theor. Biol. 232.

Geyer, H., Seyfarth, A., Blickhan, R., 2006. Compliant leg behaviour explains basic dynamics of walking and running. Proc. R. Soc. B 273.

Ghigliazza, R.M., Altendorfer, R., Holmes, P., Koditschek, D., 2005. A simply stabilized running model. SIAM Rev. 47 (3).

Guckenheimer, J., Holmes, P., 1983. Nonlinear Oscillations, Dynamical Systems, and Bifurcations of Vector Fields. Springer-Verlag.

Holmes, P., Full, R.J., Koditschek, D., Guckenheimer, J., 2006. Dynamics of legged locomotion: models, analyses and challenges. SIAM Rev. 48 (2).

Knuesel, H., Geyer, H., Seyfarth, A., 2005. Influence of swing leg movement on running stability. Hum. Mov. Sci. 24.

McMahon, T.A., Cheng, G.C., 1990. The mechanics of running: how does stiffness couple with speed? J. Biomech. 23.

Owaki, D., Koyama, M., Yamaguchi, S., Ishiguro, A., 2010. A two-dimensional passive dynamic running biped with knees. In: Proceedings of IEEE ICRA.

Raibert, M., 1986. Legged Robots That Balance. MIT Press.

Robilliard, J.J., Wilson, A.M., 2005. Prediction of kinetics and kinematics of running animals using an analytical approximation to the planar spring–mass system. J. Exp. Biol. 208.

Saranli, U., Buehler, M., Koditschek, D.E., 2001. RHex: a simple and highly mobile hexapod robot. Int. J. Robot. Res. 20 (7).

Saranlı, U., Arslan, Ö., Ankaralı, M.M., Morgül, Ö., 2010. Approximate analytic solutions to non-symmetric stance trajectories of the passive spring-loaded inverted pendulum with damping. Nonlinear Dyn. 62 (4).

Schwind, W.J., Koditschek, D.E., 2000. Approximating the stance map of a 2-DOF monoped runner. J. Nonlinear Sci. 10 (5).

Shen, Z.H., Seipel, J.E., 2012. A fundamental mechanism of legged locomotion with hip torque and leg damping. Bioinspir. Biomim. 7 (4).

Shen, Z., Seipel, J., 2016. A piecewise-linear approximation of the canonical spring-loaded inverted pendulum model of legged locomotion. J. Comput. Nonlinear Dyn. 11 (1).

Strogatz, S.H., 1994. Nonlinear Dynamics and Chaos: With Applications to Physics, Biology, Chemistry, and Engineering. Westview Press.

Chapter 3.4

Simple Models of Walking

Justin Seipel

Purdue University, West Lafayette, IN, United States

3.4.1 WALKING LIKE AN INVERTED PENDULUM

The movements of human walking, and animal walking more generally, have been likened to the motion of an inverted pendulum (e.g., Alexander, 1976; Mochon and McMahon, 1980); see Figs. 3.4.1 and 3.4.2 for illustrations of human and inverted pendulum walking, respectively. Further, some walking mechanisms and robots have exhibited similar walking behavior and are sometimes constructed in ways that are mechanically analogous to an inverted pendulum; see Fig. 3.4.2 for one example by McGeer (1990).

The overall motions of the trunk, stance leg, and swing leg of walking humans, walking mechanisms and robots, and the inverted pendulum model share many similarities: During the stance phase of human walking, when a single leg is on the ground, the body tends to rise and then fall as it pivots about the foot. This is similar to the way an inverted pendulum moves about its pivot. Walking is also described as a pattern or *gait* with alternating left and right legs (or sets

FIGURE 3.4.1 Illustration of human walking based on a modified chrono-photograph taken by Étienne-Jules Marey. A leg length has been superimposed on the original image, approximately from the hip to an approximate center of pressure for heel-to-toe walking. Overall, this approximates a vaulting or pendular motion of the body about the foot.

 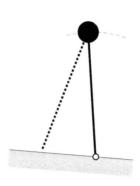

FIGURE 3.4.2 (left) A modified photograph of a passive walking mechanism, and (right) an illustration of an inverted pendulum model that represents a physics-based mathematical model. The original walking mechanism photograph is from McGeer (1990).

of legs). This characteristic walking pattern can also be exhibited by *bipedal* inverted pendulum models.

In biological and robot walking, when the *stance leg* is on the ground, the other leg, a *swing leg*, swings forward into position for *touch down*. Swing leg touchdown typically occurs before the current stance leg will *lift off* (also called take off), which leads to a double stance phase. This is followed by another single leg stance phase, and the overall pattern repeats. This process can be approximated in bipedal inverted pendulum models. For mathematical simplicity, in some inverted pendulum models the double stance phase is assumed to occur in an instant.

Overall, the basic walking movements of humans and other animals can be approximated by bipedal inverted pendulum locomotion, and can also be embodied physically in walking mechanisms. Though the concept of a bipedal inverted pendulum is dramatically simple when compared to walking humans or other animals, it can nonetheless help us understand and predict many aspects of walking.

3.4.2 PASSIVE WALKING MECHANISMS: PHYSICAL MODELS AND PHYSICS-BASED MATH MODELS

The notion of walking like an inverted pendulum can be investigated via experimental study of simple physical "inverted pendulum" walking mechanisms, such as walking toys and other *passive* dynamic walking mechanisms (e.g., McGeer, 1990; Coleman and Ruina, 1998; Collins et al., 2001). Many of these mechanisms are passive in the sense that they do not have active power elements such as motors to drive locomotion. Passive walkers generally maintain a steady gait by walking down an inclined plane, though some do use small actuators to maintain nearly passive walking (e.g., Collins et al., 2005; Bhounsule et al., 2014). In many of these walking mechanisms, one can directly observe a physical bipedal inverted pendulum mechanism in action (Fig. 3.4.2). In this way, a walking mechanism can be considered as a kind of *physical model* of human and animal locomotion that strongly demonstrates a role that passive dynamics can play in locomotion. Mathematical models of bipedal inverted pendulum walking can be closely associated with physical walking mechanisms, via application of the laws of mechanics, or they can be developed in more direct relationship to empirical studies of biological locomotion, via motion capture, ground reaction forces and other techniques.

3.4.3 MATHEMATICAL EQUATIONS GOVERNING A BIPEDAL INVERTED PENDULUM (IP) MODEL

Now that we have discussed many of the foundational concepts of walking that are embodied in bipedal inverted pendulum models, here we explicitly present a mathematical model for a bipedal inverted pendulum. In particular, we present the mathematical equations that describe the mechanics, hybrid dynamics, and control of a bipedal inverted pendulum. We present one particular inverted pendulum model of walking in order to provide a simple example of explicit mathematical governing equations. Please see Subchapter 3.6 for an overview of multiple established Inverted Pendulum models, including models that are more complex than the one presented here.

Here we present a highly simplified version of the inverted pendulum model based closely on a previous study by Wisse et al. (2006). Specifically, we assume that the swing leg has negligible mass, that the swing leg is controlled to touch down with a prescribed angle and leg retraction speed, that the time spent in double stance phase is negligible compared to the total stride, and that stance leg liftoff and swing leg touchdown are instantaneous. Other inverted pendulum models have relaxed some of these assumptions. For example, several models have included the effects of swing leg mass (e.g., Coleman and Ruina, 1998 and others reviewed in Subchapter 3.6), and more recent extensions of the inverted pendulum model have included leg compliance, enabling a substantial double stance phase as well as more accurate prediction of ground reaction forces (e.g., Geyer et al., 2006). The particular and highly simplified Inverted Pendulum model presented here, along with its corresponding equations and figures, are reproductions of the particular model and results presented by Wisse et al. (2006).

3.4.3.1 Behavior Within a Single Stance Phase

The mathematical equations governing stance for a simple inverted pendulum model can be derived by applying laws of physics. Common approaches include

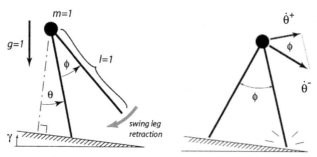

FIGURE 3.4.3   A simple Inverted Pendulum model of walking. The figure is reproduced from Wisse et al. (2006). The model is simplified via fixing mass, gravity, and leg length in the governing equations to be equal to one. Reductions in the total number of system parameters can also be gained through formal nondimensionalization techniques.

applying Newton's Second Law of Mechanics to a Free Body Diagram of forces acting on the system, or applying the Euler–Lagrange Equation to a description of the system's energy. Here, we use the Euler–Lagrange approach: Looking at the simple bipedal IP system modeled in Fig. 3.4.3, where the system rotates with an angle θ, and where the mass of the swing leg is assumed to be negligible, we can describe the kinetic energy T, potential energy V, and resulting Lagrangian L as follows:

$$L = T - V = \frac{1}{2}ml^2\dot{\theta}^2 - mgl\cos(\theta - \gamma).$$

Here m is the mass, l is the leg length, g is gravity, and γ is the angle of the inclined plane. We then apply the Euler–Lagrange Equation of mechanics to yield the following differential equation governing motion of the system, in terms of the angle θ:

$$\ddot{\theta} = gl^{-1}\sin(\theta - \gamma).$$

This equation, along with the initial conditions at the beginning of stance, determine the motion of the inverted pendulum during stance. For the model shown in Fig. 3.4.3, the model analysis was simplified by taking the leg length, gravity, and mass to be equal to one. More explicit nondimensionalization could yield a similar simplification. Note also that the angle θ used here has a different reference than that used for the Spring-Loaded Inverted Pendulum model presented in Subchapter 3.3.

3.4.3.2   Stance Leg Liftoff and Swing Leg Touchdown

The termination of the stance phase of a given leg is often defined as when the foot loses contact with the ground (which can also be related to the ground reaction force). Realistically, for normal walking this would occur sometime after the current swing leg touches down and so a substantial double stance phase would occur. However, for simple inverted pendulum models the double stance phase is assumed to be infinitesimal in duration. Therefore, here we allow the termination of stance to occur approximately at the same instant as the swing leg touches down. By determining when the swing leg touches down in such a simplified model, we automatically determine the time when the stance leg terminates.

The touchdown of the swing leg can be defined as when the foot of the swing leg reaches the ground, or when the distance between the swing leg foot and ground reaches zero. Further, to avoid counting glancing contacts as a swing leg touchdown, one can also require that the velocity of the foot is pointing into the ground when this distance reaches zero. The distance between the swing foot and ground will generally depend on complicated dynamics and control of the swing leg. For the simplified model discussed here, it is assumed that there is a swing leg controller that maintains a prescribed trajectory of the swing leg, with a prescribed retraction angular velocity near the time of touchdown. In this scenario, the swing leg actually first swings past what will become the touchdown angle and then *retracts* towards the desired angle for touchdown. If touchdown were delayed or occurred early, it would result in a different touchdown angle. In this simple model, the stability of walking can be influenced by this effect.

3.4.3.3 The Mechanics of Switching from One Stance Leg to the Next

During the infinitesimal double stance phase of motion of this simple walking model, the current stance leg lifts off and the swing leg touches down at the same time. During this period, even if it occurs over an infinitesimal period as assumed in simple walking models, there is a change in momentum of the system such that the velocity which was heading downwards at the end of one stance will change and head upwards in order to vault over the next stance leg. In order to model this process, an impulse–momentum equation can be used. We assume the simplified model presented here is entirely passive, so the leg lifting off is not able to apply an impulse during this sequence (other inverted pendulum models include active toe-off impulses which have been shown to help reduce overall energetic cost, Kuo, 2002). We are left to assume that for the system to have a velocity direction consistent with the circular arc of the new stance leg; there must be a net impulse that makes it so. It is reasonable to assume that much of this impulse occurs along the length of the touchdown leg, and so we assume a touchdown impulse entirely aligned with the touchdown leg that cancels all momentum in that direction. The remaining momentum is perpendicular to the new stance leg. From the application of impulse–momentum equations it is worked out that the angular velocity just after the leg switching is less than that just before the leg switching, depending on the angle ϕ between the two legs:

$$\dot{\theta}^+ = \cos(\phi)\dot{\theta}^-,$$

where $\dot{\theta}^+$ is the angular speed of the inverted pendulum the instant just after the leg switching process, and $\dot{\theta}^-$ is just before.

3.4.3.4   Stride Maps: Behavior Investigated Step-by-Step

One of the key methods currently used to investigate walking behavior is to see how the states of the system, such as positions and velocities, change at discrete intervals from step-to-step. This creates a mapping of the system states from one stride to the next, and can be constructed using a Poincaré Return Map. In less precise terms, this is like taking a snapshot of the system at either a set interval of time, or alternatively, every time a well-defined event occurs (e.g., every time a foot touches down, or every time the trunk mass reaches a maximum height apex). The mapping that results is often referred to as a *stride map*. For systems that are integrable, a return map (or stride map) can be written in closed-form mathematical expressions. However, it is also common to numerically integrate governing equations to produce a stride map, especially for more complex models of locomotion.

3.4.3.5   Stability of Locomotion

The dynamic stability of locomotion is often of interest when studying Inverted Pendulum models of walking. Surprisingly, such systems can exhibit stable locomotion even when no active control is present. Many physical parameters of the walking system could potentially affect stability in important ways, and understanding the underlying passive dynamics can also benefit the design of controllers that can be added to the system. Here we present only one simple example: We investigate how stability of walking depends on the swing leg retraction speed, as an example to highlight how stability is studied for such a simple walking model. For more general discussion of stability and analysis methods, please see Full et al. (2002), Strogatz (1994), or Guckenheimer and Holmes (1983).

A common approach to measure stability is to use the stride map to find periodic walking motions and then determine whether small deviations to the periodic motion will lead to the system diverging away from the periodic motion or returning to it. This is done systematically by analytically or numerically

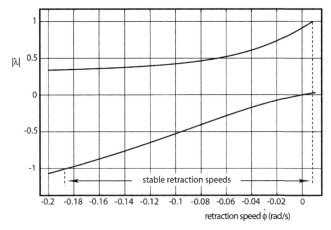

FIGURE 3.4.4 Example result of a stability analysis of inverted pendulum models of walking. The result shown here is reproduced from Wisse et al. (2006) for this example. It shows how the stability of the system can be influenced by the speed of swing leg retraction. In particular, this study found that there is a range of swing leg retraction speeds for which stable walking motions were found (where the magnitude of the two eigenvalues of the system are both less than one).

calculating a linearization of the stride map, evaluated with respect to the periodic solution being investigated. This yields a linear discrete dynamic system that approximates the stride map. The eigenvalues of this linear system will indicate the kind of local stability that occurs in the neighborhood of the periodic locomotion being investigated (where if the magnitude of all eigenvalues is less than one there exists asymptotic stability; if greater than one, unstable; if equal to one, further analysis is needed). For example, for the model presented above it was found that asymptotically stable periodic walking exists for a range of the leg retraction speeds: This result is reproduced here in Fig. 3.4.4 from Wisse et al. (2006).

3.4.4 SOME INSIGHTS INTO WALKING AIDED BY INVERTED PENDULUM MODELS

Walking mechanisms, and the associated study of the inverted pendulum model of walking, have been influential and have provided insights regarding the mechanics and control of walking. They have led to theory about the flow of energy during the walking cycle, both what is physically possible and insights on what may be happening in biological systems. Empirical study of these walking mechanisms and related theoretical study of inverted pendulum models have also suggested that the regulation of walking could rely in part on its passive dynamics. One such insight was that passive dynamic walking is a process that is statically unstable about a given resting point, but yet has dynamic stability over a walking cycle (see, for example, the work by Coleman and Ruina, 1998). Following are a few concepts about walking that have been influenced by simple walking models. These concepts are largely presented using a mechanics perspective of walking, though they have implications for control. These concepts are complementary to those presented for the Spring-Loaded Inverted Pendulum model in Subchapter 3.3.

3.4.4.1   Walking Includes a Pendular Flow of Energy

One basic discovery, inspired by mechanical analysis of inverted pendulum motion, is that kinetic energy and potential energy are exchanged during walking. For a passive inverted pendulum system, energy is conserved during stance, and so we have the basic theoretical mechanics result: Kinetic Energy + Potential Energy = Constant. Since the total system energy is constant, any change of kinetic energy corresponds with an equal and opposite change in potential energy. This means that as the mass center rises the speed slows, corresponding to a decrease in kinetic energy that is equal to the increase in potential energy. This occurs until apex, at the highest point where the mass center is directly above the pivot. Assuming that the initial kinetic energy (and corresponding mass center speed) is enough for the inverted pendulum to reach and pass through apex, then after apex the mass center falls. As the mass center falls, the speed increases, corresponding to an increase in kinetic energy that is equal to the decrease in potential energy. Not all levels of energy in the system will permit a "gait" or motion of the inverted pendulum. For a passive inverted pendulum walker, the initial value of kinetic energy needs to be larger than the increase in potential energy needed to reach apex, in order for the system to pass through apex with a nonzero speed.

3.4.4.2   Walking Includes the Catching of Repeated Falls

Walking can also be likened to controlled falling, where the body falls and pivots about the stance foot, only to be caught by the next stance leg. This view is consistent with the energetic concepts of inverted pendulum motion during stance and adds to it the importance of the *placement of the swing leg* to switch from one leg to the next. This requires a bipedal inverted pendulum walking system that is able to transition from one leg to another. While one leg is in stance, the other is in a swing phase. The swing leg is often assumed to follow a passive dynamic trajectory under the influence of its own weight, though in reality it is also likely regulated (e.g., Coleman and Ruina, 1998). A simplified modeling approach is to assume the swing leg is controlled to follow a trajectory based on time, phase, or system states relative to the main body and/or ground (e.g., Wisse et al., 2006).

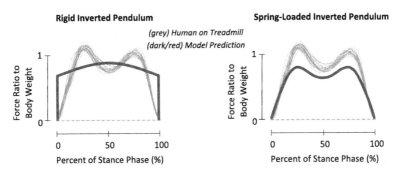

FIGURE 3.4.5 Walking vertical ground reaction force predicted by the IP and SLIP models. Human experiment and model traces reproduced from Geyer et al. (2006).

3.4.4.3 Momentum is Exchanged During Double Stance

Walking generally includes a significant "double stance" phase where the body is supported by two legs. For inverted pendulum models, it is often assumed that double support happens over a negligible period of time. In inverted pendulum models the legs are often assumed to be rigid, and this requires that double support phases vanish since movement is otherwise kinematically restricted. Despite this simplification, physical insights can still be gained regarding how the two legs act during the transition from one stance leg to the next. For example, it makes an energetic difference what the order is of the different possible impulses from the two legs (e.g., a toe-off impulse occurring before the touchdown leg impulse can reduce energetic costs, as presented in Kuo, 2002). However, a more accurate analysis could be gained by relaxing the rigid leg assumption and enabling a substantial-duration double support phase. For example, Geyer et al. (2006) showed that a bipedal Spring-Loaded Inverted Pendulum model can more accurately predict the ground reaction forces of human walking, including the portion during which double-support phases occur; see Fig. 3.4.5. This approach has an added benefit of utilizing the SLIP modeling framework already associated with running.

3.4.5 INTEGRATION OF WALKING AND RUNNING MODELS

Though we have so far studied walking as separate from running, walking and running can be viewed as expressions of the same legged locomotion system, whether of humans, other animals, or robotic systems. It is also possible to predict both walking and running in a single mathematical model without adding much additional complexity compared with the IP and SLIP models. This can be achieved by blending the bipedal nature of the IP model and the effective leg spring of SLIP, into a bipedal SLIP model, as demonstrated in Geyer et al.

(2006). There it was also demonstrated that many predictions of human walking are improved, such as predictions of ground reaction forces: For example, as reproduced in Fig. 3.4.5, it is apparent that the bipedal SLIP model captures well the characteristic shape of human walking ground reactions. Other extensions to the SLIP model have also demonstrated how two gaits might arise from one simple SLIP-based mechanism. For example, a clock-torqued SLIP model was inspired by the robot RHex, as well as cockroaches, and demonstrates that a range of walking and running behaviors result from simple clock parameter adjustments (Seipel and Holmes, 2007). Though the simplest walking models and the simplest running models may continue to have many uses and may be preferred for the sake of simplicity, a more integrated modeling framework of walking and running also has many potential advantages and uses. Regardless of the model used, it is likely helpful to remember that walking and running are behaviors that can arise from similar underlying processes, and that the study of one can often inform the other.

REFERENCES

Alexander, R., 1976. Mechanics of bipedal locomotion. In: Davies, P.S. (Ed.), Perspectives in Experimental Biology. Pergamon Press, Oxford, UK.

Bhounsule, P.A., Cortell, J., Grewal, A., Hendriksen, B., Karssen, J.G.D., Paul, C., Ruina, A., 2014. Low-bandwidth reflex-based control for lower power walking: 65 km on a single battery charge. Int. J. Robot. Res. 33 (10).

Coleman, M.J., Ruina, A., 1998. An uncontrolled toy that can walk but cannot stand still. Phys. Rev. Lett. 80 (16).

Collins, S.H., Wisse, M., Ruina, A., 2001. A three-dimensional passive-dynamic walking robot with two legs and knees. Int. J. Robot. Res. 20.

Collins, S., Ruina, A., Tedrake, R., Wisse, M., 2005. Efficient bipedal robots based on passive dynamic walkers. Sci. Mag. 307.

Full, R.J., Kubow, T., Schmitt, J., Holmes, P., Koditschek, D., 2002. Quantifying dynamic stability and maneuverability in legged locomotion. Integr. Comp. Biol. 42 (1).

Geyer, H., Seyfarth, A., Blickhan, R., 2006. Compliant leg behaviour explains basic dynamics of walking and running. Proc. R. Soc. B 273.

Guckenheimer, J., Holmes, P., 1983. Nonlinear Oscillations, Dynamical Systems, and Bifurcations of Vector Fields. Springer-Verlag, New York.

Kuo, K.A.D., 2002. Energetics of actively powered locomotion using the simplest walking model. J. Biomech. Eng. 124.

McGeer, T., 1990. Passive dynamic walking. Int. J. Robot. Res. 9 (2).

Mochon, S., McMahon, T., 1980. Ballistic walking. J. Biomech. 13.

Seipel, J., Holmes, P., 2007. A simple model for clock-actuated legged locomotion. Regul. Chaotic Dyn. 12 (5), 502–520.

Strogatz, S.H., 1994. Nonlinear Dynamics and Chaos: With Applications to Physics, Biology, Chemistry, and Engineering. Westview Press.

Wisse, M., Atkeson, C.G., Kloimwieder, D.K., Diehl, M., Mombaur, K., 2006. Dynamic stability of a simple biped walking system with swing leg retraction. In: Fast Motions in Biomechanics and Robotics: Optimization and Feedback Control. Springer, Berlin, Heidelberg.

Chapter 3.5

Locomotion as an Oscillator

Shai Revzen and Matthew Kvalheim

Electrical Engineering and Computer Science, University of Michigan, Ann Arbor, MI, United States

3.5.1 LOCOMOTION AS AN OSCILLATOR

Virtually all animals, and even more so bipeds such as humans, move in a rhythmic way at moderate to high speeds of their available range of speeds. We refer to these as "rhythmic", rather than the more mathematically strict term "periodic", which is reserved for systems that have a precisely defined period within which motions repeat exactly. We define a rhythmic system as a stochastic system whose underlying deterministic part (the "drift" in the language of Stochastic Differential Equations) has an exponentially stable periodic solution. The cycles of legged locomotion, known as "strides", typically vary from each other in duration and geometry of motion. As animals move slightly faster or slower, their limbs follow similar trajectories at slightly higher or lower rates. Even at a given stride frequency animal motions exhibit variability. At least to casual observation, it seems this variability (normalized for body size) is greater in smaller animals, in animals using more legs for propulsion, and in animals moving more slowly.

Taking the "templates and anchors" perspective of Subchapter 3.2, we can rephrase this observation as a statement that the so-called "phase oscillator" is the simplest template of most moderate-speed legged locomotion. In other words, the simplest model of legged locomotion is the timely progression through a repeating sequence of body postures, which happens also to include interaction with the ground that produces propulsion. For this chapter, we will refer to this cycle as the "gait cycle."

The phase oscillator template of locomotion can be modeled as a curve in the configuration space of the animal's body, and a velocity associated with every point on that curve. Alternatively, it can be modeled as a periodic function of time, e.g., using a Fourier series model of the body configuration as a function of "phase."

Under sufficiently small perturbations of the environment or body posture, animal motions recover to the gait cycle after few steps. This suggests that the slightly richer structure of an "asymptotically stable oscillator" ("oscillator" for short) applies just as universally. From a mathematical perspective, an oscillator is the differential equation that governs motion within the stability basin of a

FIGURE 3.5.1 Obtaining a prediction of future motion using a phase estimate. (Left to right) Starting with cockroach foot motions in the body frame of reference (first plot), we focused on the fore-aft motions as a function of time (second plot), and computed the velocity as a function of position (middle six plots; nondimensionalized by z-scoring). We combined these linearly to a single "foot oscillation state" with +1 coefficients for one tripod and −1 for the other. This gave rise to a state which collectively describes the phase of the gait as a whole (large oval plot). The polar angles of the plots of each leg and of the combined state follow a linear trend over multiple strides (rightmost plot). All plots shown are from the same time segment of a single experiment tracking a *Blaberus* cockroach running at moderate speed.

cycle, i.e., the gait cycle and all bodily states that allow for the gait cycle to be recovered.

There is a rich mathematical literature on the structure of oscillators. If we restrict our attention in that literature to those oscillators that are "structurally stable" and "generic", i.e., oscillators which are physically observable and whose dynamics would change only a little if properties of the body and environment change slightly, all smooth oscillators share several properties. One of the most important of these is that oscillators have a phase coordinate for the entire stability basin. This phase specializes to, and is therefore consistent with, the phase oscillator phase on the gait cycle itself. Any perturbations of the animal away from the gait cycle will typically result in a phase shift that will persist after the animal returns to the gait cycle. Since all observables of a rhythmic system must themselves be rhythmic, we may aim to estimate the phase of an animal's phase oscillator template ("dynamical phase" hereon, called "asymptotic phase" in Subchapter 3.2 – not to be confused with "dynamical phase" in geometric mechanics) by observing neuromechanical quantities such as body configurations, speeds of various body parts, forces and torques, and EMG or other neuronal measurements.

Once a method of phase estimation is available, predicting phase as a function of time should produce a linear trend if the phase oscillator template is compatible with the observations (see Fig. 3.5.1). By subtracting this linear trend from the instantaneous phase estimate we can obtain the "residual phase" which can be used to identify how oscillations change under the influence of external perturbations (Revzen et al., 2009).

Dynamical phase is the only dynamical variable of the phase oscillator template. The study of how that phase responds to the body and environment allows us to eliminate possible neuromechanical control architectures, e.g., by separating out responses that could be achieved only with changes to descending neural

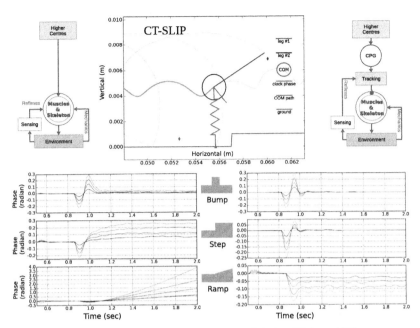

FIGURE 3.5.2 Clocked, Torqued SLIP model (CT-SLIP, Seipel and Holmes, 2007) set to parameters of a *Blaberus* cockroach running gait (upper center) with two different control architectures (upper left & right diagrams), was subjected to an assay of three perturbations (lower center). In each case, the magnitude of the perturbation is varied, producing qualitatively different residual phase response curves (plots lower left & right) for the two architectures. Results show that phase alone can be used to differentiate the neuromechanical control architecture (Revzen et al., 2009).

signals, and responses that could occur for solely mechanical reasons (Revzen et al., 2009). An example of such an analysis is shown in Fig. 3.5.2.

3.5.2 STRIDE REGISTRATION AS PHASE ESTIMATION

Estimating dynamical phase can also be seen as a way of representing methods of "stride registration." Whenever we observe a rhythmically moving animal, we encounter the problem of stride registration: which samples of stride n represent "the same" state in the gait cycle as which samples of stride $n + 1$? Whenever investigators construct a notion of a gait cycle, they implicitly define such a registration method. In each such class of states which are "the same" in this sense, there is one distinguished representative which lies on the gait cycle itself. Because it lies on the cycle, it is a state of the phase oscillator template of that animal motion. Thus we see that any stride registration method corresponds to a choice of assigning phase to data samples.

Typical stride registration methods in the literature include linearly interpolating once-per-stride events in time, e.g., heel strike (Jindrich and Full, 2002;

Ting et al., 1994) or anterior extreme position of a limb (Cruse and Schwarze, 1988). Some work in robot control has attempted to parameterize a target gait using a hip-to-heel angle, or other combination of internal angles (Chevallereau et al., 2003; Sreenath et al., 2011). By construction, these driving variables are a form of step registration as well.

The advantage of phase-based stride registration becomes clear if we assume a state independent measurement noise, and that observed motions are perturbations around a core phase oscillator template. Estimating the phase oscillator's phase and using it for binning and averaging the measurements ensures that all equal sized bins have (asymptotically) the same number of samples. Thus the bin average estimates provided are homoscedastic and standard statistical hypothesis testing tools can be used to test for treatment effects. If any other stride registration method is used the bin averages will be heteroscedastic, and require much more refined statistical techniques.[5]

Let us compare the process of naïve stride registration and a dynamical phase-based one. For the former, we define an event detector function which has positive zero crossings when the desired event occurs, e.g., for heel-strike based stride registration we take the time and force pairs (t_i, f_i) from a force plate under the running human and renormalize to $(t_i, 1 - f_i/(mg))$. We then detect the positive crossing times $\{c_k\}$ and form the piecewise linear function of time $p(\cdot)$ such that $p(c_k) = k$ are its knot points. We now select a number of bins N_b and put the (multidimensional) data sample (t_i, d_i) in the bin $b_i := \lfloor N_b(p(t_i) - \lfloor p(t_i) \rfloor) \rfloor$. We estimate the period of the gait cycle τ by taking a central statistic such as the median of $\{c_{k+1} - c_k\}$. Taking a representative such as sample average of the data in each bin in an appropriate way for the data itself, we obtain the model that at time t the gait cycle places the animal at body configuration given by the representative of the bin $\lfloor N_b(t \bmod \tau)/\tau \rfloor$.

A dynamical phase-based stride registration would consist of first training or deriving a phase estimator that gives a phase p_i for every data sample (t_i, d_i). Using that phase estimate instead of $p(t_i)$, i.e., by taking $b_i := \lfloor N_b(p_i - \lfloor p_i \rfloor) \rfloor$, we proceed with the same approach to obtain bin representatives.

It should be noted that in many cases, producing the gait cycle model at a given phase does not require binning, and can instead be done by building a Fourier series model of animal properties $d(t)$ as a function of phase using a Fourier series of some order N_f:

$$x(\varphi) = \sum_{k=-N_f}^{N_f} a_k e^{i2\pi k\varphi}, \qquad (3.5.1)$$

5. Phase defines a measure on the cycle which is flow invariant, and thus averaging a function of state with respect to the phase measure along trajectories does not introduce additional variance due to the dynamics—only the preexisting measurement noise.

where

$$a_k := \int_{\text{all } t} e^{-2\pi i k t/\tau} d(t) \frac{dp}{dt}(t)\, dt. \tag{3.5.2}$$

3.5.3   RECOVERY FROM PERTURBATIONS

The structurally stable, generic oscillators that we use as models of locomotion share an additional property: they can be "linearized exactly." The core insight dates to the late 19th century, when Gaston Floquet showed that linear time periodic (LTP) differential equations can be solved by writing their solutions as a periodic part multiplying the solutions for a linear time invariant part (Floquet, 1883). This insight extends from LTP systems to oscillators because one may view the dynamics of the oscillator as a perturbation of the dynamics of its phase oscillator template, which is time periodic. The theory of "Normal Forms" (Bronstein and Kopanskii, 1994; Lan and Mezić, 2013) shows that Floquet's result does in fact extend to the entire stability basin of the oscillator.

In other words, the oscillators that appear in locomotion problems can be rewritten with respect to appropriately chosen coordinates such that they are linear time invariant (LTI) systems in the new coordinates. In these linearizing coordinates, the tools of linear systems theory and control theory can be brought to bear, telling us that the long-term dynamics are governed by a single "system matrix" A which describes the LTI equation of motion. For a gait cycle with period τ, the matrix norm of $e^{\tau A}$ provides a bound on how quickly perturbations decay back to the unperturbed gait, with the magnitude decreasing by at least a factor of $|e^{\tau A}|$ every stride. It is important to note that in the linearizing coordinates, the results apply to both large and small perturbations; if return map Jacobians are used without a full coordinate change, the result only applies to small perturbations.

The Floquet Normal Form provides even more detailed insight. Every perturbation to the state of the animal can be rewritten in terms of a linear combination $\{\xi_k\}$ of the eigenvectors $\{v_k\}$ of A, $x(0) = \sum_k \xi_k v_k$, and will thus evolve as

$$x(t) = \sum_k e^{\lambda_k t} \xi_k v_k. \tag{3.5.3}$$

The "Floquet Multipliers" $e^{\lambda_k \tau}$ are invariant to the choice of coordinates,[6] and can therefore be computed in the original coordinates we use to obtain our measurements. Computing Floquet Multipliers is thus the method of choice for determining the stability of smooth oscillators (see Fig. 3.5.3).

---

6. This follows because the matrices involved in different coordinate representations are similar (conjugate) to each other and thus have the same eigenvalues.

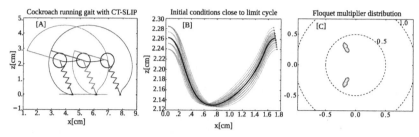

FIGURE 3.5.3   A CT-SLIP (Seipel and Holmes, 2007) model of a running *Blaberus* cockroach alternates between right and left foot touchdown events [A]. The center of mass bounces vertically every step, exhibiting a limit cycle ([B], heavy black line). At "apex", with vertical velocity zero and going negative, it is convenient to define a Poincaré section. This section is 2D, consisting of height (z) and horizontal velocity (v_x) of the center of mass. An ensemble of initial conditions at apex, varying in both z and v_x ([B], colored lines) can be integrated to the next apex ([B], colored dots). Using linear regression, the affine map taking apex states to the next apex can be estimated, and its eigenvalues—the Floquet Multipliers—computed. Bootstrap analysis can further be used to get a distribution of eigenvalues and produce confidence bounds for the estimate ([C], 1000 bootstrap computations from the ensemble in [B]). For this gait the eigenvalues are a complex conjugate pair, of magnitude less than 0.5. This tells us that the oscillator is very robustly stable, and (small) perturbations decay in magnitude by better than a factor of 2 every step. (For interpretation of the references to color in this figure, the reader is referred to the web version of this chapter.)

3.5.4   SUBSYSTEMS AS COUPLED OSCILLATORS

The entire argument presented above for treating a phase oscillator as a template for animal locomotion applies equally well to parts of an animal's body. The partitioning of the animal into subsystems can be physiological, e.g., viewing the nervous system as one or more oscillators as well as viewing the musculoskeletal system as one or more oscillators. It can also follow morphology, e.g., treating each limb as an oscillator. In all such cases, one ends up with a notion of "subsystem phases" (Revzen et al., 2009), and of an animal locomotion template consisting of coupled phase oscillators.

Which gait an animal is employing at any given time can be ascertained from the relative phases of the legs (see Fig. 3.5.4).

3.5.5   LEGGED LOCOMOTION OSCILLATORS ARE HYBRID DYNAMICAL SYSTEMS

The theory of oscillators, as described hereto, was developed for "smooth" dynamical systems—ones for which the equations of motion are at least continuously differentiable. Unfortunately, the models used for legged locomotion rarely satisfy this requirement. Typically, the equations of motion of a legged system depend strongly on which legs are in ground contact. Indeed, the very dimension of the system or the number of mechanical degrees of

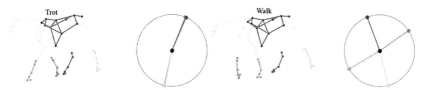

FIGURE 3.5.4 Coupled oscillator description of horse gaits. A body reference frame was fixed to the horse (black dots) and the motion of six markers on each leg was described as a phase oscillator and thereby reduced to a single phase per leg. The phases of all four legs are shown (radial lines in circles; color corresponds to animal leg) for both trotting (left) and walking (right) [figure from Yu et al., 2016].

freedom may change as contact varies. For legged systems, we must extend our scope to the study of "Hybrid Dynamical Systems." Several subtly different definitions of Hybrid Systems exist in the literature (Alur et al., 1993; Burden et al., 2015, 2016; Goebel et al., 2009; Holmes et al., 2006; Nerode and Kohn, 1993), but all share several features: (1) the solutions of the Hybrid System are referred to as "executions", rather than "trajectories"; (2) dynamics are defined over several "domains" and are smooth within each domain; (3) "reset maps" link domains to each other, and an execution may go through a reset map by taking its value in one domain, applying the map and using the image as the initial condition in the new domain; (4) the sets of points in each domain over which reset maps may be applied are called "guards." As a concrete example which is also of interest to legged locomotion, assume we have two masses linked by a vertical spring and constrained to bounce in the vertical direction in earth gravity above level ground. While both masses are in the air, the dynamics are the smooth ballistic motion of the two masses, with the additional internal force of the connecting spring. The flight domain is four-dimensional, with two mechanical degrees of freedom (DOF). Assume further that when a mass hits the ground, it loses all kinetic energy in a plastic collision. Thus, with the lower mass on the ground, we may use a two-dimensional, one DOF model. Adding the assumption that at length 0 the spring exerts enough force to lift the top mass from the ground, we have a hybrid system with 2 domains and 4 reset maps (see Fig. 3.5.5).

One may readily envision that with the addition of a periodic actuation force applied by the spring, the system may enter a range of persistent hopping at some constant amplitude which balances the energy lost by m colliding with the ground with the energy injected by the actuator.

While the core results of oscillator theory and Floquet theory do not apply to this system as stated, since it is not a smooth oscillator, recent results (Burden et al., 2015) show that after two cycles this system becomes restricted to a 2D surface in the 4D ballistic domain, such that the motions in the stance domain

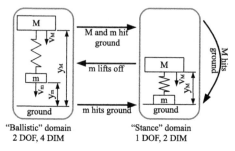

FIGURE 3.5.5 An example of a hybrid dynamical system model. A vertically bouncing pair of masses (M, m) connected by a spring can be modeled with two domains (rounded frames) and four reset maps (labeled arrows). In the ballistic flight domain, the system is 4-dimensional since the state contains position and velocity for each mass. In the stance domain, the lower mass (m) is stationary on the ground, and the state is two dimensional, consisting only of position and velocity of one mass. Reset maps take states in which masses collide with ground to the associated stance state, and take states in which the lower mass would detach from the ground, from stance into ballistic motion.

and in the ballistic domain can be "stitched together" using a function that is smooth everywhere except the guards, and leading to dynamics that are smooth in the new coordinates. In fact, this equivalence[7] to smooth systems extends to multilegged locomotion gaits in which many legs hit the ground at once—a class of models which was only analyzed recently (Burden et al., 2016). Thus, we find that once some technical complications are addressed, the long-term behavior of hybrid oscillator models that arise in legged locomotion is the same as that of the more familiar smooth oscillators.

3.5.6 ADVANCED APPLICATION: DATA DRIVEN FLOQUET MODELS

One of the strengths of the oscillator perspective on locomotion is the ability to identify properties of feasible locomotion models from observational data (Revzen, 2009; Revzen and Kvalheim, 2015; Wang, 2013). This approach has been called "Data Driven Floquet Analysis (DDFA)" and consists of a collection of numerical methods that attempt to reconstruct the oscillator dynamics of the putative legged locomotion oscillator directly from observational data.

One application of DDFA is the identification of plausible dimensions for template models. As described in Subchapter 3.2, multiple models with varying levels of detail may exist for a given legged locomotion behavior. Viewed as an oscillator, the same behavior has a set of Floquet multipliers, the magnitudes of which define a set of decay rates. Each Floquet multiplier is associated with a "Floquet mode"—a specific phase-dependent way of the motions being offset

---

7. Formally, a piecewise smooth and everywhere continuous conjugacy.

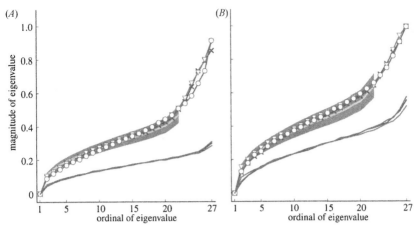

FIGURE 3.5.6 (Reproduced from Fig. 8 of Revzen and Guckenheimer, 2008) Comparison of Floquet multiplier magnitude distributions obtained from running cockroaches. Since this analysis is done at a specific phase in the cycle, magnitudes are plotted for three different phases (0.79, 1.57, and 3.14 radians in red, green, and blue, respectively). Experimental motion data is marked with markers; unmarked lines come from surrogates—randomly paired crossings of the surface on which the Floquet multipliers are computed—and offer a null hypothesis which demonstrates that meaningful cycle-to-cycle dynamics exist. A 21-dimensional random effects model selected by the algorithm of Revzen and Guckenheimer (2008) (gray confidence band with green center-line) shows the portion of the Floquet multiplier magnitudes that can be explained by random effects. In this 27-dimensional dataset, the template dynamics are therefore at most 6-dimensional. (For interpretation of the references to color in this figure legend, the reader is referred to the web version of this chapter.)

from the limit cycle. For example, a Floquet multiplier of magnitude 0.5 would be associated with a mode that decays by a factor of two every cycle. Floquet modes evolve independently of each other, and thus any subset of modes is, in principle at least, a reduced-dimension model of the dynamics.

By the very requirement that they describe the long-term dynamics of locomotion, templates will thus comprise modes that correspond to the larger Floquet multipliers. This observation allows Floquet multipliers computed from experimental data to be sorted by magnitude and compared with the Floquet multipliers of a null (random effect) model (Revzen and Guckenheimer, 2011). The multipliers that cannot be accounted for by random effects may be counted, and provide an upper bound on the dimension of a template model that can reasonably be supported with those data (see Fig. 3.5.6).

The Floquet models obtained from DDFA may be used to extend existing models of locomotion by identifying additional states that improve prediction. In the case of human running, while the SLIP model has an excellent fit to observations (Ludwig et al., 2012), it fails to predict stability properties, and is in fact

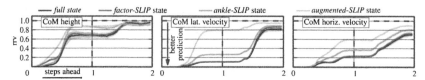

FIGURE 3.5.7 (Reproduced from Fig. 5, Maus et al., 2014) ability of various models to explain observed quantities in human running data, plotted as "relative remaining variance (rrv)": the ratio of residual variance to data variance. An rrv of 1 means no predictive ability; rrv of 0 is perfect prediction. The "full state" DDFA model, and the "factor-SLIP" model derived from it are better predictors than the "augmented SLIP" model which is itself slightly more powerful than classical Spring-Loaded Inverted Pendulum models. The structure of the data driven factor-SLIP suggested adding ankle states to the system, leading to the physically meaningful ankle-SLIP model and capturing most of the potential prediction gains of the DDFA full state model.

unstable at some of the range of running gait parameters humans use. In attempting to predict step-to-step running dynamics, Maus et al. (2014) showed linear feedback using an augmented SLIP model whose state consists of SLIP state variables and all SLIP parameters was less effective at predicting future states than a DDFA-derived linear model. By subjecting the DDFA model to factor analysis, five governing linear factors were obtained for a state with nearly 200 dimensions. Examination of the weights in these factors suggested that adding an ankle state could extend SLIP and give large improvements in prediction (see Fig. 3.5.7). This showed that DDFA modeling may be used to incrementally extend existing analytical models for specific goals, e.g., maximizing predictive ability.

3.5.7 SUMMARY

At intermediate speeds, limit cycle oscillators are a useful reduced model of legged locomotion. The rich theory and tools available for analysis of oscillator dynamics provide a uniform language for expressing and understanding gaits.

Future work includes the substantial space for improvement in the numerical algorithms used for DDFA and development of algorithms that require shorter time series. Better algorithms for identifying parameters of coupled oscillator models of locomotion are needed, as most of the coupled oscillator methods from the physics literature (Pikovsky et al., 2003) assume far weaker coupling and far lower phase noise. New directions from Koopman Theory (Budišić et al., 2012) suggest a reframing of DDFA in terms of decomposition of oscillator dynamics into Koopman modes, although numerical algorithms for accomplishing this goal are in their infancy. Finally, little to no work exists on the identification and numerical analysis of hybrid oscillators, as the theory of such oscillators is a recent addition to the field.

This work was funded in part by ARO grant #W911NF-14-1-0573 "Morphologically Modulated Dynamics" to S. Revzen.

REFERENCES

Alur, R., Courcoubetis, C., Henzinger, T.A., Ho, P., 1993. Hybrid Automata: An Algorithmic Approach to the Specification and Verification of Hybrid Systems. Springer. http://dx.doi.org/10.1007/3-540-57318-6_30.

Bronstein, A.U., Kopanskii, A.Y., 1994. Smooth Invariant Manifolds and Normal Forms, 1st edition. World Scientific Publishing, Salem, MA. ISBN 981021572X.

Budišić, M., Mohr, R., Mezić, I., 2012. Applied koopmanisma. Chaos 22 (4), 047510.

Burden, S.A., Revzen, S., Sastry, S.S., 2015. Model reduction near periodic orbits of hybrid dynamical systems. IEEE Trans. Autom. Control 60 (10), 2626–2639. http://dx.doi.org/10.1109/TAC.2015.2411971.

Burden, S.A., Revzen, S., Sastry, S.S., Koditschek, D.E., 2016. Event-selected vector field discontinuities yield piecewise-differentiable flows. SIAM J. Appl. Dyn. Syst. 15 (2), 1227–1267.

Chevallereau, C., Abba, G., Aoustin, Y., Plestan, F., Westervelt, E.R., Canudas-de Wit, C., Grizzle, J.W., 2003. Rabbit: a testbed for advanced control theory. IEEE Control Syst. 23 (5), 57–79. http://dx.doi.org/10.1109/MCS.2003.1234651.

Cruse, H., Schwarze, W., 1988. Mechanisms of coupling between the ipsilateral legs of a walking insect (carausius morosus). J. Exp. Biol. 138, 455–469.

Floquet, G., 1883. Sur les équations différentielles linéaires à coefficients périodiques. Ann. Sci. Ec. Norm. Super. 12, 47–88.

Goebel, R., Sanfelice, R.G.s., Teel, A.R., 2009. Hybrid dynamical systems. IEEE Control Syst. 29 (2), 28–93. http://dx.doi.org/10.1109/MCS.2008.931718.

Holmes, P., Full, R.J., Koditschek, D.E., Gukenheimer, J.M., 2006. The dynamics of legged locomotion: models, analyses, and challenges. SIAM Rev. 48 (2), 206–304. http://dx.doi.org/10.1137/S0036144504445133.

Jindrich, D.L., Full, R.J., 2002. Dynamic stabilization of rapid hexapedal locomotion. J. Exp. Biol. 205 (18), 2803–2823.

Lan, Y., Mezić, I., 2013. Linearization in the large of nonlinear systems and Koopman operator spectrum. Physica D 242 (1), 42–53.

Ludwig, C., Grimmer, S., Seyfarth, A., Maus, H.-M., 2012. Multiple-step model-experiment matching allows precise definition of dynamical leg parameters in human running. J. Biomech. 45 (14), 2472–2475. http://dx.doi.org/10.1016/j.jbiomech.2012.06.030.

Maus, H.-M., Revzen, S., Guckenheimer, J.M., Ludwig, C., Reger, J., Seyfarth, A., 2014. Constructing predictive models of human running. J. R. Soc. Interface 12. http://dx.doi.org/10.1098/rsif.2014.0899.

Nerode, A., Kohn, W., 1993. Models for Hybrid Systems: Automata, Topologies, Controllability, Observability. Springer. http://dx.doi.org/10.1007/3-540-57318-6_35.

Pikovsky, A., Rosenblum, M., Kurths, J., 2003. Synchronization: A Universal Concept in Nonlinear Sciences, vol. 12. Cambridge University Press.

Revzen, S., 2009. Neuromechanical Control Architectures of Arthropod Locomotion. PhD thesis. University of California, Berkeley.

Revzen, S., Guckenheimer, J.M., 2008. Estimating the phase of synchronized oscillators. Phys. Rev. E. http://dx.doi.org/10.1103/PhysRevE.78.051907.

Revzen, S., Guckenheimer, J.M., 2011. Finding the dimension of slow dynamics in a rhythmic system. J. R. Soc. Interface. http://dx.doi.org/10.1098/rsif.2011.043.

Revzen, S., Kvalheim, M., 2015. Data driven models of legged locomotion. SPIE Defense+ Security. International Society for Optics and Photonics, 94671V. http://dx.doi.org/10.1117/12.2178007.

Revzen, S., Koditschek, D.E., Full, R.J., 2009. Towards testable neuromechanical control architectures for running. In: Progress in Motor Control. Springer, pp. 25–55. http://dx.doi.org/10.1007/978-0-387-77064-2_3.

Seipel, J., Holmes, P., 2007. A simple model for clock-actuated legged locomotion. Regul. Chaotic Dyn. 12 (5), 502–520. http://dx.doi.org/10.1134/S1560354707050048.

Sreenath, K., Park, H.-W., Poulakakis, I., Grizzle, J.W., 2011. A compliant hybrid zero dynamics controller for stable, efficient and fast bipedal walking on Mabel. Int. J. Robot. Res. 30 (9), 1170–1193. http://dx.doi.org/10.1177/0278364910379882.

Ting, L.H., Blickhan, R., Full, R.J., 1994. Dynamic and static stability in hexapedal runners. J. Exp. Biol. 197 (1), 251–269.

Wang, Y., 2013. System Identification Around Periodic Orbits With Application to Steady State Human Walking. PhD thesis. The Ohio State University.

Yu, M.Y., Liedtk, A., Revzen, S., 2016. Trotting horses synchronize their legs during the second half of stance. Integr. Comp. Biol. 56, E247.

Chapter 3.6

Model Zoo: Extended Conceptual Models

Maziar A. Sharbafi* and André Seyfarth[†]
**School of Electrical and Computer Engineering, College of Engineering, University of Tehran, Iran* [†]*Lauflabor Locomotion Laboratory, TU Darmstadt, Germany*

In the previous chapters we explained how abstractions and simplification can help understand locomotion principles. For this, several locomotion models with reduced representation of the human body were introduced. In general, the description of legged systems can be based on

- **Highly simplified models** (e.g., template models) which focus on the principal dynamics of the movement using only few parameters, or
- **More detailed simulation models** (e.g., muscle–skeletal models) like OpenSim (http://opensim.stanford.edu) and AnyBody (http://www.anybodytech.com/) with a high number of degrees of freedom (DOF) and with many model parameters.

In this chapter we will describe how simplified models can be subsequently extended in order to increase the level of more detail of the simulation models.

Whereas complex simulation models are often directly related to the structure of the human body (body segments corresponding to bones, muscles, tendons and other soft tissues), the design of conceptual simplified models highly depends on mechanical intuition like in the inverted pendulum (IP) model (Cavagna et al., 1963), the lateral leg spring (LLS) model (Schmitt and Holmes, 2000), or the spring-loaded inverted pendulum (SLIP) model (Blickhan, 1989; McMahon and Cheng, 1990). These models are focusing on describing the *axial leg function* as a simple telescopic leg spring, with either a constant leg length during stance (IP model) or a leg force proportional to the amount of leg compression (LLS or SLIP model). The assumption of spring-like leg function can be found (in approximation) experimentally both in animals (Blickhan and Full, 1993) and humans (Lipfert, 2010) during steady state locomotion. However, there are also clear deviations in the locomotion dynamics that are not well described by these simple models.

The key limitations of both the IP model and the SLIP model as the most common template models for legged locomotion are summarized in Table 3.6.1. Corresponding model extensions that are suitable to overcome these limitations are also presented. It is important to note that we only select elementary extensions of the model, however, also combinations of the model extensions

TABLE 3.6.1 Extensions of the Template Models to Resolve the Limitations in Explaining Locomotion Features

Limitation of model	Extension	Extended model	Description
Focus on a single leg	Second leg	Bipedal SLIP (B-SLIP)	Required to study different gaits (e.g., walking and running)
		rimless Wheel, IP with swing leg dynamics	
	More legs	Quadrupedal SLIP (Q-SLIP)	For animal or infant locomotion
Focus on axial leg function	Rigid trunk	SLIP with trunk (T-SLIP)	Enables control of body posture
		IP with trunk (e.g., bisecting)	
	Foot segment	SLIP with rigid flat / curved foot (F-SLIP)	Enables roll-over function of foot with shift in center of pressure (COP) during contact
		IP with rigid flat /curved foot	
	Leg segments	2-segment leg with thigh and shank	Leg geometry influences transfer between joint torque and leg force
		3-segment leg with thigh, shank and foot	
	Hip spring model	Hip spring between both legs	Tuning of leg swing with stance leg
Prescribed axial leg function	Varying leg parameters in stance phase	E-SLIP	Permits energy stability with change in leg spring parameters in midstance
		VLS-SLIP	Permits energy stability with variable leg spring parameters during stance
		LIP	Permits changes in leg length and leg force
Mass-less leg	Add leg masses	M-SLIP	Considers leg masses in stance leg
		Passive dynamic walker or Acrobot model	

TABLE 3.6.1 (*continued*)

Limitation of model	Extension	Extended model	Description
Focus on sagittal plane	Lateral movements of COM	3D SLIP	Permits 3D running and walking with lateral leg placements
		3D LIP	Permits 3D running and walking with lateral leg placements
Purely mechanical description	Add muscle dynamics	Leg with muscle model	Actuation of the leg through muscle forces with optional reflex pathways

are possible to consider, like XT-SLIP (Sharbafi et al., 2013a) which is an extended SLIP model with trunk (T-SLIP), and added leg mass (M-SLIP) or the ballistic walking model presented of Mochon and McMahon (1980). Model extensions can address either mechanics or control of the system. Another class of model extensions comprises muscles (e.g., single-joint and two-joint muscles with muscle fiber-tendon dynamics) and neural circuits (e.g., sensory feedback pathways) describing muscle stimulation and integration of sensory signals. A sophisticated extension of the SLIP model including muscles, reflex pathways, and segmented legs is the gait model of Geyer and Herr (2010), which originates on the neuro-muscular model introduced by Geyer et al. (2003).

The extensions of IP and SLIP models described in this subchapter (Table 3.6.1) are shown in Fig. 3.6.1 and Fig. 3.6.2, respectively. The reasoning of the different extensions in both templates is often similar. In the following, we will describe selected model extensions in more detail. We will start with model extensions regarding the leg structure, followed by models describing the dynamics of the trunk and finally models including lateral leg placements and locomotion in 3D.

3.6.1 MORE DETAILED REPRESENTATIONS OF THE LEG

In the aforementioned template models, a point mass sits on top of a rigid or compliant massless leg. The focus of these models is on CoM movement, which considers the stance leg movement as the first locomotion subfunction and partially leg swinging (the second locomotion subfunction). The following extensions in the leg structure addressing more features in each of these two subfunctions will be presented:

- **Stance leg:** (a) adding leg mass, inertia and damping, (b) adaptation of leg parameters during motion and (c) increasing number of segments

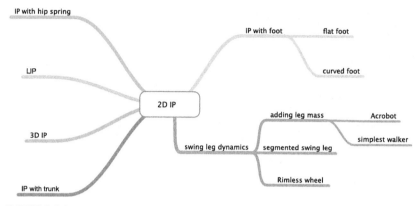

FIGURE 3.6.1 Extensions of the sagittal inverted pendulum (2D IP) model with selected added model features: hip spring between both legs, foot (flat or curved) attached to the lower end of the IP, swing leg dynamics by adding leg masses, segmented swing leg or rimless wheel model, linear inverted pendulum (LIP) with leg force law, including lateral movements (3D IP), and adding a trunk. Different control policies can be applied to each of these model extensions, e.g., the capture point concept for LIP model (Pratt et al., 2006).

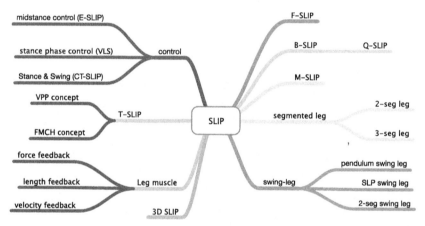

FIGURE 3.6.2 Extensions of the sagittal SLIP model with selected added model features: foot segment (F-SLIP); the number of legs (bipedal B-SLIP, quadrupedal Q-SLIP, etc.); leg masses (M-SLIP); segmented legs, 2 and 3 leg segments; swing leg dynamics as a pendulum, spring loaded pendulum SLP or two-segmented swing leg; control for varying leg spring properties during stance (VLS-SLIP), at mid-stance (E-SLIP) or continuous control during step (CT-SLIP); added trunk (T-SLIP); muscle-like leg function (leg muscle) and different reflex pathways (force, length, and velocity feedback) and lateral movements (3D SLIP). Each of these model extensions can be considered as a separate or in combination with others, e.g., BT-SLIP. Gray color indicates control features of SLIP based models.

- **Swing leg:** (a) addition of one or more legs (b) increasing number of segments in swing leg, (c) adding leg mass

The simplest way to extend the locomotion template models regarding the leg function is adding additional massless legs to the body. With this more advanced swing leg adjustment approaches (e.g., leg retraction; Wisse et al., 2006) can be developed. Second groups of extensions can be from the control point of view. For example adaptation of the stance leg parameters during gait (E-SLIP, Ludwig et al., 2012) or a continuous unified controller for swing and stance phase (CT-SLIP, Seipel and Holmes, 2007) which enhances the model abilities in reproducing more biological and also more stable gaits while keeping the model complexity. Third, the legs can be represented in a more physical way by considering leg mass, inertia or damping. Finally, the number of segments can approach the numbers in human/animal legs. In the following, we explain some of these extensions in related models.

3.6.1.1 Extending the Number of Limbs (B-SLIP, Q-SLIP)

Inspired by the work on the SLIP model, Herr et al. (2002) developed a quadrupedal SLIP model to describe trotting and galloping in several animals (chipmunk, dog, goat, and horse). The model was extended with a compliant trunk (described by neck and back stiffness). Hip and shoulder were actively powered resulting in a running pattern that was similar in kinematics (e.g., limb angles) and kinetics (e.g., peak force, limb stiffness) to experimental data. Interestingly, swing leg retraction was identified as a key feature required obtaining stable running in the model (Seyfarth et al., 2003). A very similar quadrupedal SLIP model with rigid trunk was created following the design of the Scout II robot (Poulakakis et al., 2005). The predicted stable bounding gait is in close agreement with the behavior observed in the robot. Later on, the model and robot dynamics were extended to galloping (Smith and Poulakakis, 2004).

Geyer et al. (2006) extended the sagittal SLIP model to a bipedal version (B-SLIP), which was the first model capable of predicting walking and running gait within the same model setup. In this model, the only parameter change required to achieve different gaits was the system energy. For moderate speeds (around 1 m/s) walking patterns with double humped profiles of the ground reaction force are found. In contrast, with higher locomotion speed, a running gait with single-humped patterns of the ground reaction force is observed. Surprisingly, the model predicts further walking gaits with more than two humps for lower energies. Such gaits (e.g., with three-humped force patterns) can indeed be found in human locomotion like in amputees' gaits or in the development of gait during early childhood (Gollhofer et al., 2013).

3.6.1.2 Rimless Wheel

The inverted pendulum model is developed to explain walking. Because of the rigidity of the stance leg, no flight phase exists to represent running or hopping. Therefore, the minimum number of legs in this model is 2. The addition of the number of legs for this model is not common because it is usually applied to understand human gaits or developing passive dynamic robots except for increasing stability in 3D, e.g., McGeer passive walker (McGeer, 1990). However, rimless wheel model can be considered as an extension of inverted pendulum model with adding more legs, which are coupled with a fixed angle between limbs (McGeer, 1990). More explanations about the passive dynamic walking model and rimless wheel can be found in Section 4.4.

3.6.1.3 Stance Leg Adaptation (VLS and E-SLIP)

In the SLIP model, stance leg parameters like leg stiffness and angle of attack are often set to a specific value. This usually represents the steady-state (average) gait pattern during locomotion. However, leg function varies from step to step (e.g., in response to ground level changes, Daley et al., 2007; Müller et al., 2010, and also during the stance phase Riese et al., 2013). Such variations in leg parameters can be represented in extensions of the SLIP model in order to better match experimental data. With this also deviations from the conservative spring-like leg function can be described which may lead to also energetically stable gait patterns.

During human locomotion, there is a tendency towards higher leg stiffness during leg loading (leg shortening) compared to unloading (leg lengthening). Additionally, the leg length is often larger at takeoff compared to touchdown (Lipfert et al., 2012). There are several possible explanations for that, e.g., eccentric force enhancement during leg compression or the role of leg segmentation (Maykranz et al., 2009; see F-SLIP model below).

Based on the SLIP model, two simple approaches were introduced to address changes in leg parameters during stance phase:

1) In the variable leg spring (VLS) model a continuous change of leg parameters over time is assumed (Riese and Seyfarth, 2011). For stable hopping, a decrease in leg stiffness and a continuous increase in rest length of the leg spring (Fig. 3.6.3A) were required in the model unless sufficient leg damping is provided. This is in line with experimental findings on changes in stance leg parameters during human locomotion (Lipfert et al., 2012).
2) In the E-SLIP model, a sudden change in leg parameters at midstance is considered (Fig. 3.6.3B) without a sudden drop or increase in leg force. This model permits to consider step-to-step changes in system energy as found in human running (Ludwig et al., 2012).

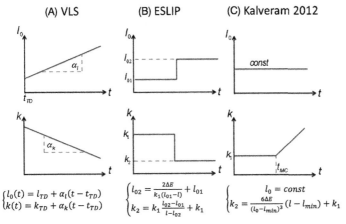

FIGURE 3.6.3 Extensions of the SLIP model with adjustable leg springs parameters (stiffness k and rest length l_0). In the variable leg spring VLS approach, leg stiffness and rest length change linearly with time (Riese and Seyfarth, 2011). In the two other approaches (ESLIP and Kalveram et al., 2012), a fixed amount of energy ΔE is added during stance phase after maximum leg compression (t_{MC}) when the leg length reaches its minimum value (l_{min}). The equations show the parameters used in the figure.

3) Following the approach of Kalveram et al. (2012), leg stiffness can be changed during leg extension (leg unloading) such that a defined amount of energy is injected to the leg. This approach inspired the control of Marco hopper robot as well as the Marco-2 hopper robot with segmented leg (Oehlke et al., 2016).

Changes in leg parameters in steady-state movements were observed experimentally at a global (leg) level (Lipfert et al., 2012; Lipfert, 2010 for walking and running; Riese et al., 2013 for hopping; Seyfarth et al., 1999 for take-off phase in long jump) as well as at local (joint or muscle) level (Peter et al., 2009 AMS for running). So far, it is still unclear whether and how limb stiffness is adjusted at a global (leg) level or a local (joint, muscle) level. It remains for future research to investigate in more detail how changes in state variables (angles, angular velocities) and environmental changes (e.g., changed ground properties) effect these adjustments of leg parameters during stance phase.

3.6.1.4 Clock-Torque SLIP (CT-SLIP)

In order to keep the simplicity of the SLIP model and increasing the ability to predict more features of legged animal and robot locomotion dynamics, the CT-SLIP model was developed by Seipel and Holmes (2007). In this model, a reference clock drives the leg movement (using a PD controller for stance leg) while damping is added to the stance leg. In this model, the same mechanism

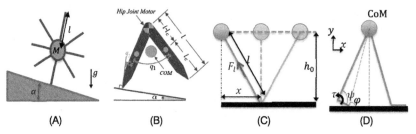

FIGURE 3.6.4 Extensions of IP model (A) rimless wheel, (B) Acrobot, (C) LIP, and (D) actuated ankle.

(continuous leg rotation) is utilized to control the leg in both stance and swing phases. This model, which is inspired by RHex robot (Saranli et al., 2001), can address hip actuation, more realistic take off and touch down (with leg retraction) and more importantly, a more robust gait compared to the SLIP model, without increasing dimension of the model (Seipel and Holmes, 2007).

3.6.1.5 Linear Inverted Pendulum Mode (LIPM)

The original IP model does not consider displacements in axial leg direction during stance and thus forces the CoM to move on a circular arch. Introducing a prismatic joint in stance leg converts the IP into a SLIP model if the generated force is proportional to the leg length. In that respect SLIP can be considered as an extension of IP with additional leg spring.

For a long time, many studies in walking were related to the CoM movement described by the inverted pendulum paradigm and the six determinants of gait (Kuo, 2007). Based on minimization of CoM displacement, the six determinants of gait theory (Saunders et al., 1953) result in no vertical CoM excursion in walking. In 1991, the linear inverted pendulum model was introduced by Kajita and Tani (1991), in which the leg force is determined to compensate gravity, resulting in zero vertical acceleration. The ground reaction force can only act along the leg axis (CoP–CoM line) and the vertical element of the leg force should be equal to the body weight (Mg). According to parameters shown in Fig. 3.6.4C, the required leg force (F_l) to achieve the CoM height (h_0) when the horizontal distance between CoM and CoP equals x is computed as follows:

$$F_l = \frac{Mg}{h_0}l = \frac{Mg\sqrt{x^2 + h_0^2}}{h_0}. \tag{3.6.1}$$

Following this approach, leg force is predicted to increase with leg lengthening, which is opposite to experimental findings (Lipfert et al., 2012) and the concept of spring-like leg function.

The LIPM model was used to develop capture point concept (Pratt et al., 2006) as a method for leg adjustment to reach zero forward speed at vertical leg configuration within one step (see Section 2.2 for details). Some versions of this model consider a rotation around the CoM by using an upper body (e.g., Kajita and Tani, 1991 with a constant angular velocity) or a flywheel with torque control (Pratt et al., 2006).

3.6.1.6 Addition of Leg Mass to IP (Acrobot, Simplest Walking Model)

The "pure" IP model with massless legs (Alexander, 1976; Hemami and Golliday, 1977; Wisse et al., 2006) is rarely utilized in walking analysis. Addition of a point mass to each leg can simplify control (e.g., based on passive swing leg movement) and also makes the model more realistic. The resulting model was called "the simplest walking model" (Garcia et al., 1998) or the "compass gait model" (Goswami et al., 1996). The compass gait concept was already pointed out by Borelli (1680) in his famous book "De Motu Animalium." This popular model is able to represent walking without the need for active control of the swing leg. The stability of the predicted gait was well analyzed (Goswami et al., 1996). Investigation of the limit cycle stability for walking on slope with this model versus parameter variations (slope, normalized leg mass, and leg length) demonstrate that a wide range of solutions gradually evolves through a regime of bifurcations from stable symmetric gaits to asymmetric gaits and eventually arriving at an apparently chaotic gait where no two steps are identical (Goswami et al., 1998).

Different leg mass locations are considered, like the leg's CoM position (at about the center of the leg) like in the passive dynamic walking model (McGeer, 1990) or small masses at tip toes (Garcia et al., 1998). A very similar model compared to the simplest walking model is the Acrobot model (Westervelt et al., 2007). In this model the mass is distributed along the leg and not concentrated at the hip. In general, addition of leg mass (i) can simplify control, (ii) enables passive walking down a shallow slope, (iii) permits describing leg swinging (another locomotion subfunction), but at the same time it (iv) requires control, e.g., of hip torques when walking on flat terrain to stabilize the gait and to compensate for energy losses (energy management). Please find more explanations on passive dynamic walking models in Subchapter 4.6.

3.6.1.7 Addition of Mass to SLIP Leg (M-SLIP)

In the M-SLIP (Peuker et al., 2012) model, the leg is represented by a rigid leg segment and a prismatic spring attached at the distal part of this segment.

In stance phase, the spring is aligned in leg axis by a viscoelastic rotational coupling between the rigid segment and the leg spring. During swing phase, the prismatic leg spring is bent such that the leg segment can freely swing forward. The leg angle is adjusted by setting the rest angle of the rotational hip spring and this leg angle switches between two values depending on the state (one stiffness for swing phase and another one for stance phase).

With leg masses, the gait dynamics is more realistic but also more complex (e.g., landing impacts). Compared to the SLIP model, the predicted solutions for stable running of a one-legged system with leg masses (M-SLIP) are shifted towards flatter angles of attack (Peuker et al., 2012). In an alternating, bipedal M-SLIP model, however, the inertial effects of both legs are compensating each other such that the region of stable running is similar to the one observed in the SLIP model. This indicates that also the model with leg masses can inherit solutions of the SLIP model. At the same time, leg inertia of the leg with mass permits creating swing-leg trajectories (e.g., by introducing a hip torque) that were not represented by the original SLIP model. In this model, a PD (proportional, derivative) controller is used for hip torque control in walking. It is similar to the hip spring model developed by Dai Owaki for running (Owaki et al., 2008).

3.6.1.8 Extending SLIP with Leg Segments (F-SLIP, 2-SEG, 3-SEG)

Biological limbs are designed as a serial arrangement of leg segments with muscles spanning the leg joints. In the following, benefitting from muscle properties, we describe a number of extended SLIP models using muscle mechanics, which lead to a more detailed representation of leg segments and describe their effects on the gait dynamics.

In the **F-SLIP model** (Maykranz et al., 2009), the prismatic leg spring is extended distally by a rigid foot segment, which is attached by a rotational foot spring (similar to the ankle joint). This model permits to describe a shift of the center of rotation along the foot segment as found in heel to toe running or in human walking. Similar to the VLS model it can describe an increase in leg length from touchdown to take-off due to the asymmetric arrangement of the foot, pointing forward (Maykranz and Seyfarth, 2014). The resulting force–length curve of the leg indicates a drop in leg stiffness from early to late stance phase. This is a consequence of the mechanical action of the compliantly attached foot segment. In late stance the foot joint (ankle) is leaving the ground resulting in a more realistic representation of the push-off in human locomotion. Surprisingly, the F-SLIP model is able to predict running well, but has limited capability to generate walking patterns.

Another segmented model extending the SLIP model is the 2-segment model by Rummel and Seyfarth (2008). Here, the leg spring is replaced by a rotational (knee) spring attached between the two massless leg segments (upper and lower leg). The analysis of the model shows that running solutions can be observed for different rest angles of the knee spring with clear deformations in the predicted regions for stable locomotion. With extended rest angles (knee joint angles between 150–170 degrees), a larger region of angles of attack compared to the SLIP model result in stable running. In contrast to the SLIP model, knee stiffness needs to be increased for faster running. This increase of knee stiffness with speed was also found experimentally (Rummel and Seyfarth, 2008; Lipfert, 2010). There has been a number of similar leg models with two segments presented with muscle-like joint function, e.g., for describing jumping (Alexander, 1990a, 1992; Seyfarth et al., 2000) and hopping tasks (Geyer et al., 2003).

Finally, in a three-segment model including foot, shank and thigh, the adjustment of joint stiffness for spring-like leg function was investigated by Seyfarth et al. (2001). This simulation study shows that a shared loading of knee and ankle requires not only a proper distribution of knee and ankle stiffness but also additional mean to avoid joint buckling or overextension. The following means for achieving stable leg function could be identified (Seyfarth et al., 2006):

1) Elastic two-joint connection between ankle and knee (e.g., gastrocnemius muscle)
2) Asymmetric segment lengths with shorter foot and asymmetric joint configurations (extended knee, bent ankle)
3) Joint constraints (e.g., heel contact by calcaneous) prevents too large ankle bending and avoids knee overextension
4) Nonlinear progressive joint stiffness (with larger nonlinearity in knee compared to ankle)
5) Transition from a zig-zag mode to a bow configuration of the leg (like in spiders)

The transfer of this mechanical three-segment leg model to a muscle–skeletal model was presented by Geyer and Herr (2010).

3.6.1.9 Ankle Actuated IP

The position dependence of passive ankle joint mechanics was shown in Weiss et al. (1986). Considering a flat foot and elastic element to model ankle torque, Ahn developed an ankle actuated IP model (Ahn, 2006) (see Fig. 3.6.4D). In this model, the constraint of having instantaneous double support in IP is resolved. A rotational spring with rest angle equal to π starts working for the trailing leg

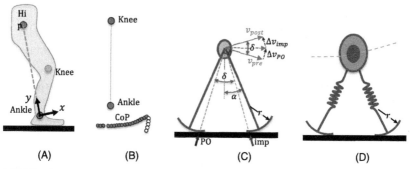

FIGURE 3.6.5 Curved foot model (A) knee–ankle–foot (KAF) coordinate system (adopted from Hansen et al., 2004), (B) CoP motion in KAF coordinate system (adopted from Hansen et al., 2004), (C) IP with curved feet (adopted from Adamczyk et al., 2006). Push-off (PO) produces positive work which should be equal to the negative work consumed at impact (imp) for periodic gait. (D) SLIP with curved feet (Whittington and Thelen, 2009).

after touchdown of the other leg until take-off or reaching straight ankle angle:

$$\tau = \begin{cases} k(\pi - \psi) & \psi < \pi, \\ 0 & \psi \geq \pi. \end{cases} \tag{3.6.2}$$

This preloaded ankle spring (Fig. 3.6.4B) injects energy and supports push-off.

3.6.1.10 Curved Feet Model

In bipedal locomotion the center of pressure (CoP) is not fixed on the ground like assumed in gait template models (with point contact). Extending the model with flat feet is a possible solution for introducing a moving CoP during ground contact. This model extension needs additional ankle torque control (Ahn, 2006). A simpler solution, which can generate human-like CoP movement without requiring ankle torque control, is using curved feet (McGeer, 1990). In human walking, a circular path with a radius of curvature 30% of leg length represents the CoP trajectory in knee–ankle–foot (KAF) coordinate system shown in Fig. 3.6.5A, B (Hansen et al., 2004). Similar ratio between the foot curvature radius and the leg length was found in the case of different prosthetic legs (Curtze et al., 2009) and also agrees closely with McGeer's robot (McGeer, 1990).

In line with these findings, extended IP and SLIP models with curved feet were used to investigate foot function in walking. Inspired by McGeer's robot (McGeer, 1990), Kuo presented an IP model with arc-shaped feet, called "Anthropomorphic Model" (Kuo, 2001). This model was used to predict the preferred speed–step length relationship (Kuo, 2001) and later to predict the effects of changing the radius of curvature on cost of transport (Adamczyk et al., 2006). In the latter study, the mechanical work and metabolic activities of human body are measured in walking experiments with shoes having different curvatures showing that using 33% of the leg length as the curved foot radius results in the minimum mechanical work and metabolic rate activities (Adamczyk et al., 2006). Considering impulsive push-off as the energy source to compensate losses at impact (Fig. 3.6.5C), it is shown that having this curvature will appear energetically advantageous for plantigrade and human walking, partially due to decreased work for step-to-step transitions (losses is reduced with any nonzero radius (Fig. 3.6.5C, as $\delta < 2\alpha$). In Whittington and Thelen (2009), a new SLIP model extended with curved feet (Fig. 3.6.5D) illustrates stable gait for an interval of feet curvature and shows that increase of foot radius up to one-third of the leg length decreases the maximum amount of the ground reaction force. All these studies show that extending template models with arc-shaped feet is useful for analyzing gait dynamics and energetics.

3.6.2   UPPER BODY MODELING

For posture control, the third locomotion subfunction, we need to extend the template models by adding an upper body, e.g., by a rigid trunk. With this additional degree of freedom, developing a controller for balancing is required. Therefore, several models were developed to address posture balance based on template models. In extended IP with a rigid torso, often traditional control engineering methods are used for keeping the torso upright (McGeer, 1988; Grizzle et al., 2001; Gregg and Spong, 2009). In Wisse et al. (2004) a passive model is presented, in which the upper body is aligned mechanically creating a bisecting angle between the two legs. In the following we explain bioinspired SLIP-based models for posture control based on by human/animal locomotion: VPP, compliant hip, and FMCH.

3.6.2.1   Virtual Pivot Point (VPP)

In the SLIP model the body dynamics is described by a point-mass. This model can only describe leg force pointing to this point-mass which differs from GRF patterns in human (or other bipedal) gaita. During locomotion, the forces acting on the body are not necessarily directed to the center of mass (COM). For instance, in human walking the stance leg forces point to a slightly above the COM. In order to describe such deviations of the leg force from the leg axis (from contact point at ground to COM), the point-mass needs to be replaced by an extended body, e.g., a rigid trunk (Fig. 3.6.6A). To study the control of a hopping robot, Poulakakis and Grizzle (2009) extended the SLIP model with a rigid upper body. They used the hybrid zero dynamics (HZD) approach to successfully control the system. Maus et al. (2010) applied the same extension

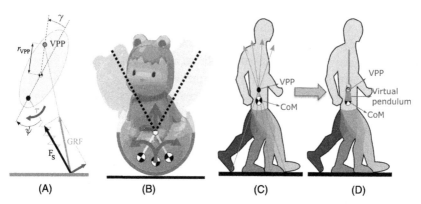

FIGURE 3.6.6   (A) T-SLIP model and VPP control concept introduced by Maus et al. (2010). (B) Virtual pendulum point (VPP) concept shown illustrated in a Roly Poly toy, (C) GRF vectors and VPP concept in human walking. (D) Virtual pendulum concept in human locomotion.

(rigid upright trunk) to a bipedal SLIP model to implement the virtual pivot point (VPP) concept. This approach assumes leg forces to intersect at a fixed location above the body COM to keep postural balance like a Roly Poly toy (Fig. 3.6.6B, C). Both stable walking and running could be predicted by this model (Fig. 3.6.6A) and the predicted hip torques are similar to those observed in human walking. As a result, the inverted pendulum model of locomotion can be transferred to a periodic movement, modeled by a regular virtual pendulum (VP) as shown in Fig. 3.6.6C, D.

3.6.2.2   Force Modulated Compliant Hip (FMCH)

The key idea behind the FMCH approach is to substitute the VPP concept with a structural model that has physical representation. With that we want to keep the basic idea of a virtual pendulum (VP) thorugh adaptable hip compliance. Therefore, first, the hip joint (between trunk and leg) of the T-SLIP model was equipped with a passive spring (Fig. 3.6.7A) simulating the effects of extensor and flexor muscles resulting in stable walking (Rummel and Seyfarth, 2010). Stable running and hopping could be predicted by implementing the virtual pendulum (VP) concept using passive hip springs (Sharbafi et al., 2013b). In this study, the quality of posture control based on of the passive compliant hip was compared to the virtual pendulum posture controller (VPPC) and also the hybrid zero dynamics (HZD, Westervelt et al., 2007, see Section 4.7). The robustness and control quality (e.g., settling time) with passive hip springs are worse than VPPC, but sufficient considering passivity of the control. Then, by applying the leg force to modulate hip compliance within FMCH model, a large improvement in balance control was achieved. It results in human-like posture balance and provides a mechanical explanation for the VPP concept

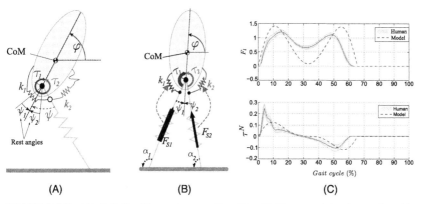

(A)   (B)   (C)

FIGURE 3.6.7 (A) T-SLIP with passive compliant hip. (B) Force modulated compliant hip (FMCH) model for walking. (C) Comparison of the leg force and the hip torque in human walking and FMCH model at normal waling speed.

(Sharbafi and Seyfarth, 2014, 2015). In this force modulated compliant hip (FMCH) model (shown in Fig. 3.6.7B) the hip torque (τ) is a product of a constant (c), the leg force (F_s) and the difference between the hip to the leg angle (ψ), and its rest angle (ψ_0) as follows:

$$\tau = cF_s(\psi - \psi_0). \tag{3.6.3}$$

It is mathematically shown that the required torque in VPPC is precisely approximated by FMCH in a range of hip and leg movements which are representative for human gaits (Sharbafi and Seyfarth, 2015). Fig. 3.6.7C shows the leg force and hip torque developed by the FMCH model for stable walking at normal walking speed (1.4 m/s) compared to the human experimental results. In addition to explain human gaits, this concept can be utilized in bipedal robot control and assistive devices (e.g., exoskeleton).

This model can be considered as a candidate for neuro-mechanical template for posture control. The model suggests a sensory pathway originating at a force sensor of the leg extensor muscle (e.g., in the knee) and a gain factor (constant c). In contrast to the neural system, no processing delays are considered in the FMCH model. Also, the muscle function is reduced to an activation-dependent tunable spring. These are clear simplifications compared to neuro-muscular processing of sensory data.

3.6.3 EXTENSION TO 3D

In order to extend the models to 3D, in addition to increasing the system degrees of freedom and enlargement of the state space the lateral leg placement is the main challenge. In the following the 3D SLIP and IP models are presented.

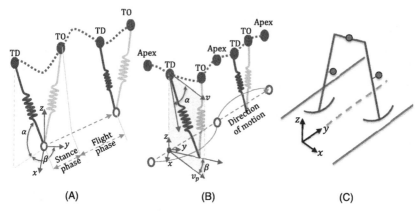

FIGURE 3.6.8 (A) 3D SLIP model with leg adjustment w.r.t. world coordinate (Seipel and Holmes, 2005). The desired leg orientation is defined by α and β. (B) 3D SLIP model with leg adjustment in body coordinate (Peuker et al., 2012). In this approach the velocity vector (v) defines the body coordinate frame. The vector v_p is the projection of v in the horizontal plane (x, y). The body lateral plane is defined by v and v_p. The desired leg orientation l is defined by α and β the angle between v and projection of l in the frontal plane and the angle between v_p and the projection of l in the horizontal plane, respectively. (C) 3D passive dynamic walking model (McGeer, 1990).

3.6.3.1 3D SLIP

Like in 2D SLIP, stable gaits require a proper leg adjustment. Interestingly, stable gaits are not predicted with a given step length (or step frequency) but by adjusting the leg angle for landing (angle of attack) with respect to gravity. This indicates that the locomotion pattern is rather an outcome than a target of control. For instance, if a subject runs on a treadmill, the variability of the gait pattern increases when the very same preferred step length or step frequency is provided (markings on the belt, metronome) as targets for locomotion (Ludwig et al., 2010).

In 2005 Seipel and Holmes published a paper investigating running stability predicted by a new SLIP model extended to 3D by including a lateral leg placement at touchdown (Seipel and Holmes, 2005). The lateral leg angle was selected with alternating direction (left or right) with respect to a desired running direction (Fig. 3.6.8A). Surprisingly, no stable running patterns were predicted by this novel 3D SLIP model. Later, Peuker et al. (2012) introduced a velocity-based leg adjustment. Here, the leg angle was laterally adjusted within the plane spanned by the COM velocity vector and gravity vector. With this change in the coordinate frame for swing leg adjustment, stable running solutions were predicted for a huge range of angle of attack and lateral leg angles before landing.

In 2014, Maus and Seyfarth developed an extended bipedal SLIP model for 3D walking. The simulation results of this 3D walking model reveals that changes in leg adjustments between the two legs can result in walking in curves (Maus and Seyfarth, 2014). However, there are combinations of leg parameter adjustments between the two legs which still results in straight walking (with a fixed direction of progression) for even asymmetric leg configurations regarding leg stiffness and angle of attack. The predicted asymmetric walking patterns can be neutrally stable. This means that the direction of walking will change if a sudden lateral push is applied to the body, however after the perturbation the walking direction remains constant. This outcome is similar to the predictions of the lateral leg spring (LLS) model of Schmitt and Holmes (2000) that operates in the horizontal plane only.

3.6.3.2 3D IP

The focus of conceptual model based gait analyses is on 2D motion in sagittal plane. Moving from side to side to modulate lateral foot placement and rotating about the vertical (yaw) axis at the ankles are the two observations in biological legged locomotion (humans). One of the first attempts for such extensions was 3D model of passive dynamic walking (Fig. 3.6.6C) incorporating both roll and yaw rotation (McGeer, 1993). However, they found that the model couldn't stably walk without control. Representing the theoretical stability of a walking machine that rocks side to side without yaw motion, Kuo could stabilize the passively unstable system by a simple control scheme inheriting much of the passive behavior (Kuo, 1999).

In Zijlstra and Hof (1997) a 3D inverted pendulum model was utilized to explain human walking in 3D space with a sinusoidal left-right movement of CoM. Using such a 3D compass gait model, Gregg and Spong (2009) extended the planar walking into directional 3D dynamic walking (e.g., moving on a circle) by controlled reduction approach. Other extensions like 3D LIPM (Kajita et al., 2001), 3D IP+torso (Gregg and Spong, 2009), the generalized 3D IP (Sakka et al., 2010), and 3-segmented IP based model with small actuation at ankle (Wisse et al., 2001) are instances of studies to build an anthropomorphic 3D model for stable walking based on inverted pendulum model. Recently, 3LP, a 3D linear IP-based model including torso and swing dynamics, was presented by Faraji and Ijspeert (2016) to represent all three subfunctions of legged locomotion with a IP based model. In addition, they could predict nonlinear speed frequency relationship as one optimality trends in human walking.

3.6.4 EXTENSION WITH MUSCLE MODELS

In the previous sections we focused on mechanical representations of legged locomotion. Compared to human locomotion, the complex interactions within the biological (e.g., human) body were described based on highly simplified models with only a few lumped parameters. For instance, leg stiffness is such a common parameter. It summarizes the complex interaction of segmented body mechanics with active and passive compliant structures (e.g., muscles, ligaments, tendons, connective tissues, etc.) and the environment (e.g., compliant ground contact). In this section we present a number of gait model extensions, which additionally take muscle–tendon dynamics into account.

For spring-like leg operation, a concerted interplay between many components in the biological body is required including:

- Active muscle forces based on muscle properties (force–length and force–velocity relations) and muscle activation dynamics (see Subchapter 8.1),
- Connective tissues
- Titin filaments (see Subchapter 8.1)
- Muscle lever arm geometry at joints
- Tendon compliance
- Geometry of segmented legs
- Muscle arrangement in relation to joints (e.g., two-joint muscles)
- Interface mechanics to the environment (e.g., foot–ground interaction)

In a first approach, the geometry of the leg was represented by two leg segments (see above, e.g., Rummel and Seyfarth, 2008). To generate leg force, joint torque can be introduced by mechanical components (e.g., rotational spring; Alexander, 1990b) or by an extensor muscle spanning the joint (Alexander, 1990a; Seyfarth et al., 2000). Due to the eccentric force enhancement in muscles, leg force becomes larger during leg compression (muscle lengthening) compared to leg extension (muscle shortening) if constant muscle activation is assumed. To generate continuous movements like in hopping or running, a modulation of muscle stimulation with time is required with lower muscle activation during eccentric phase, compared to concentric phase. This even holds if the leg geometry is ignored and the muscle is directly replacing the leg spring. Here, repulsive leg function (like in hopping and running) can be achieved by an appropriate muscle stimulation pattern (feedforward control; Häufle et al., 2010), by using sensory feedback pathways, or by a combination of both (Häufle et al., 2012). The combination of feedforward and feedback provides superior stability and perturbation rejection compared to feedforward and feedback schemes in isolation.

In the two-segment leg model with a knee extensor muscle and the neural control of Geyer et al. (2003) it was shown that similar leg function as described in the spring–mass model was predicted by a positive force feedback applied to the leg muscle. This spring-like leg function emerges after 1–2 hopping cycles and recovers quickly after perturbations (e.g., ground level changes).

With more than two leg segments different arrangements of muscles can be considered including one- and two-joint muscles (Seyfarth et al., 2001) spanning ankle, knee, and hip joints. There has been a long scientific debate about the specific role of two-joint (biarticular) muscles, including their ability to coordinate the action of adjacent joints ((Doorenbosch and van Ingen Schenau, 1995)), transfer of energy (Sharbafi et al., 2016), reduced energy and peak power requirements of joint actuation (Grimmer et al., 2012). Another suggested function of two-joint muscles is their ability to direct (orient) leg force (Doorenbosch and van Ingen Schenau, 1995; Sharbafi et al., 2016).

Muscle function is largely supported by the action of compliant structures arranged in series (e.g., tendon) and in parallel to the muscle fibers (e.g., titin). Serial elastic elements can reduce the muscle fiber displacement and speed during stretch-shortening-cycles (e.g., in jumping, running or walking; Seyfarth et al., 2000). In contrast, parallel elastic elements (Rode et al., 2009) help reduce the muscle fiber force, but keep the elongation and the speed of the fibers unchanged. Both elastic elements can largely reduce the energy and the peak power requirements of the muscle during movement.

The potential role of sensory feedback for achieving stable locomotion was demonstrated in a 7-segment neuromuscular human walking model presented by Geyer and Herr (2010). In this model, seven muscles were represented in each leg. The muscles were controlled by tuning the corresponding reflex parameters (sensor source, gain, delay). The model was continuously extended during the last years and can predict human-like walking and running at different speeds and in different environmental conditions (e.g., stairs, slopes, curves). This model is described in more detail in Subchapter 6.5.

REFERENCES

Adamczyk, P.G., Collins, S.H., Kuo, A.D., 2006. The advantages of a rolling foot in human walking. J. Exp. Biol. 209 (20), 3953–3963.

Ahn, J., 2006. Analysis of Walking and Balancing Models Actuated and Controlled by Ankles. Doctoral dissertation. Massachusetts Institute of Technology.

Alexander, R.McN., 1976. Mechanics of bipedal locomotion. Persp. Exp. Biol. 1, 493–504.

Alexander, R.M., 1990a. Optimum take-off techniques for high and long jumps. Philos. Trans. R. Soc. Lond. B, Biol. Sci. 329 (1252), 3–10.

Alexander, R., 1990b. Three uses for springs in legged locomotion. Int. J. Robot. Res. 9 (2), 53–61.

Alexander, R.M., 1992. Simple models of walking and jumping. Hum. Mov. Sci. 11 (1), 3–9.

Blickhan, R., 1989. The spring-mass model for running and hopping. J. Biomech. 22 (11–12), 1217–1227.

Blickhan, R., Full, R.J., 1993. Similarity in multilegged locomotion: bouncing like a monopode. J. Comp. Physiol., A Sens. Neural Behav. Physiol. 173 (5), 509–517.

Borelli, G.A., 1680. De Motu Animalium. English translation by P. Maquet, Springer-Verlag, Berlin. 1989.

Cavagna, G., Saibene, F., Margaria, R., 1963. External work in walking. J. Appl. Physiol. 18, 1–9.

Curtze, C., Hof, A.L., van Keeken, H.G., Halbertsma, J.P., Postema, K., Otten, B., 2009. Comparative roll-over analysis of prosthetic feet. J. Biomech. 42 (11), 1746–1753.

Daley, M.A., Felix, G., Biewener, A.A., 2007. Running stability is enhanced by a proximo-distal gradient in joint neuromechanical control. J. Exp. Biol. 210 (3), 383–394.

Doorenbosch, C.A., van Ingen Schenau, G.J., 1995. The role of mono- and bi-articular muscles during contact control leg tasks in man. Hum. Mov. Sci. 14 (3), 279–300.

Faraji, S., Ijspeert, A.J., 2016. 3LP: a linear 3D-walking model including torso and swing dynamics. arXiv preprint arXiv:1605.03036.

Garcia, M., Chatterjee, A., Ruina, A., Coleman, M., 1998. The simplest walking model: stability, complexity, and scaling. ASME J. Biomech. Eng. 120 (2), 281–288.

Geyer, H., Seyfarth, A., Blickhan, R., 2003. Positive force feedback in bouncing gaits? Proc. R. Soc. Lond. B, Biol. Sci. 270 (1529), 2173–2183.

Geyer, H., Seyfarth, A., Blickhan, R., 2006. Compliant leg behaviour explains basic dynamics of walking and running. Proc. R. Soc. Lond. B, Biol. Sci. 273 (1603), 2861–2867.

Geyer, H., Herr, H., 2010. A muscle-reflex model that encodes principles of legged mechanics produces human walking dynamics and muscle activities. IEEE Trans. Neural Syst. Rehabil. Eng. 18 (3), 263–273.

Gollhofer, A., Taube, W., Nielsen, J.B., 2013. Biomechanical and neuromechanical concepts for locomotion. In: Routledge Handbook of Motor Control and Motor Learning. Routledge. Chapter 5.

Goswami, A., Thuilot, B., Espiau, B., 1996. Compass-Like Biped Robot, Part I: Stability and Bifurcation of Passive Gaits. Doctoral dissertation. INRIA.

Goswami, A., Thuilot, B., Espiau, B., 1998. A study of the passive gait of a compass-like biped robot symmetry and chaos. Int. J. Robot. Res. 17 (12), 1282–1301.

Gregg, R.D., Spong, M.W., 2009. Bringing the compass-gait bipedal walker to three dimensions. In: IEEE/RSJ International Conference on Intelligent Robots and Systems, pp. 4469–4474.

Grimmer, M., Eslamy, M., Gliech, S., Seyfarth, A., 2012. A comparison of parallel- and series elastic elements in an actuator for mimicking human ankle joint in walking and running. In: 2012 IEEE International Conference on Robotics and Automation (ICRA). IEEE, pp. 2463–2470.

Grizzle, J.W., Abba, G., Plestan, F., 2001. Asymptotically stable walking for biped robots: analysis via systems with impulse effects. IEEE Trans. Autom. Control 46 (1), 51–64.

Häufle, D.F.B., Grimmer, S., Seyfarth, A., 2010. The role of intrinsic muscle properties for stable hopping—stability is achieved by the force–velocity relation. Bioinspir. Biomim. 5 (1), 016004.

Häufle, D.F.B., Grimmer, S., Kalveram, K.T., Seyfarth, A., 2012. Integration of intrinsic muscle properties, feed-forward and feedback signals for generating and stabilizing hopping. J. R. Soc. Interface 9 (72), 1458–1469.

Hansen, A.H., Childress, D.S., Knox, E.H., 2004. Roll-over shapes of human locomotor systems: effects of walking speed. Clin. Biomech. 19 (4), 407–414.

Hemami, H., Golliday, C.L., 1977. The inverted pendulum and biped stability. Math. Biosci. 34 (1), 95–110.

Herr, H.M., Huang, G.T., McMahon, T.A., 2002. A model of scale effects in mammalian quadrupedal running. J. Exp. Biol. 205 (7), 959–967.

Kajita, S., Tani, K., 1991. Study of dynamic biped locomotion on rugged terrain-derivation and application of the linear inverted pendulum mode. In: Proceedings of IEEE International Conference on Robotics and Automation, pp. 1405–1411.

Kajita, S., Matsumoto, O., Saigo, M., 2001. Real-time 3D walking pattern generation for a biped robot with telescopic legs. In: IEEE International Conference on Robotics and Automation, 2001. Proceedings 2001 ICRA, vol. 3. IEEE, pp. 2299–2306.

Kalveram, K.T., Häufle, D.F., Seyfarth, A., Grimmer, S., 2012. Energy management that generates terrain following versus apex-preserving hopping in man and machine. Biol. Cybern. 106 (1), 1–13.

Kuo, A.D., 1999. Stabilization of lateral motion in passive dynamic walking. Int. J. Robot. Res. 18 (9), 917–930.

Kuo, A.D., 2001. A simple model of bipedal walking predicts the preferred speed–step length relationship. J. Biomech. Eng. 123 (3), 264–269.

Kuo, A.D., 2007. The six determinants of gait and the inverted pendulum analogy: a dynamic walking perspective. Hum. Mov. Sci. 26 (4), 617–656.

Lipfert, S.W., 2010. Kinematic and Dynamic Similarities Between Walking and Running. Kovač.

Lipfert, S.W., Günther, M., Renjewski, D., Grimmer, S., Seyfarth, A., 2012. A model-experiment comparison of system dynamics for human walking and running. J. Theor. Biol. 292, 11–17.

Ludwig, C., Lipfert, S., Seyfarth, A., 2010. Variability in human running is not reduced by metronome signals. In: Proceedings International Society of Biomechanics (ISB) Conference.

Ludwig, C., Grimmer, S., Seyfarth, A., Maus, H.M., 2012. Multiple-step model-experiment matching allows precise definition of dynamical leg parameters in human running. J. Biomech. 45 (14), 2472–2475.

Maus, H.M., Seyfarth, A., 2014. Walking in circles: a modelling approach. J. R. Soc. Interface 11 (99), 20140594.

Maus, H.M., Lipfert, S.W., Gross, M., Rummel, J., Seyfarth, A., 2010. Upright human gait did not provide a major mechanical challenge for our ancestors. Nat. Commun. 1, 70.

Maykranz, D., Seyfarth, A., 2014. Compliant ankle function results in landing-take off asymmetry in legged locomotion. J. Theor. Biol. 349, 44–49.

Maykranz, D., Grimmer, S., Lipfert, S., Seyfarth, A., 2009. Foot function in spring mass running. In: Autonome Mobile Systeme 2009. Springer, Berlin, Heidelberg, pp. 81–88.

McGeer, T., 1988. Stability and Control of Two-Dimensional Biped Walking. Technical Report, 1, Center for Systems Science, Simon Fraser University, Burnaby, BC, Canada.

McGeer, T., 1990. Passive dynamic walking. Int. J. Robot. Res. 9 (2), 62–82.

McGeer, T., 1993. Passive dynamic biped catalogue, 1991. In: Experimental Robotics II. Springer, Berlin, Heidelberg, pp. 463–490.

McMahon, T.A., Cheng, G.C., 1990. The mechanics of running: how does stiffness couple with speed? J. Biomech. 23, 65–78.

Mochon, S., McMahon, T.A., 1980. Ballistic walking. J. Biomech. 13 (1), 49–57.

Müller, R., Grimmer, S., Blickhan, R., 2010. Running on uneven ground: leg adjustments by muscle pre-activation control. Hum. Mov. Sci. 29 (2), 299–310.

Oehlke, J., Sharbafi, M.A., Beckerle, P., Seyfarth, A., 2016. Template-based hopping control of a bio-inspired segmented robotic leg. In: 2016 6th IEEE International Conference on Biomedical Robotics and Biomechatronics (BioRob), pp. 35–40.

Owaki, D., Osuka, K., Ishiguro, A., 2008. On the embodiment that enables passive dynamic bipedal running. In: IEEE International Conference on Robotics and Automation, 2008. ICRA 2008. IEEE, pp. 341–346.

Peter, S., Grimmer, S., Lipfert, S.W., Seyfarth, A., 2009. Variable joint elasticities in running. In: Autonome Mobile Systeme 2009. Springer, Berlin, Heidelberg, pp. 129–136.

Peuker, F., Seyfarth, A., Grimmer, S., 2012. Inheritance of SLIP running stability to a single-legged and bipedal model with leg mass and damping. In: IEEE RAS & EMBS International Conference on Biomedical Robotics and Biomechatronics (BioRob), pp. 395–400.

Poulakakis, I., Smith, J.A., Buehler, M., 2005. Modeling and experiments of untethered quadrupedal running with a bounding gait: the Scout II robot. Int. J. Robot. Res. 24 (4), 239–256.

Poulakakis, I., Grizzle, J.W., 2009. The spring loaded inverted pendulum as the hybrid zero dynamics of an asymmetric hopper. IEEE Trans. Autom. Control 54 (8), 1779–1793.

Pratt, J., Carff, J., Drakunov, S., Goswami, A., 2006. Capture point: a step toward humanoid push recovery. In: 2006 6th IEEE–RAS International Conference on Humanoid Robots. IEEE, pp. 200–207.

Riese, S., Seyfarth, A., 2011. Stance leg control: variation of leg parameters supports stable hopping. Bioinspir. Biomim. 7 (1), 016006.

Riese, S., Seyfarth, A., Grimmer, S., 2013. Linear center-of-mass dynamics emerge from non-linear leg-spring properties in human hopping. J. Biomech. 46 (13), 2207–2212.

Rode, C., Siebert, T., Herzog, W., Blickhan, R., 2009. The effects of parallel and series elastic components on the active cat soleus force-length relationship. J. Mech. Med. Biol. 9 (01), 105–122.

Rummel, J., Seyfarth, A., 2008. Stable running with segmented legs. Int. J. Robot. Res. 27 (8), 919–934.

Rummel, J., Seyfarth, A., 2010. Passive stabilization of the trunk in walking. In: International Conference on Simulation, Modeling and Programming for Autonomous Robots (SIMPAR).

Sakka, S., Hayot, C., Lacouture, P., 2010. A generalized 3D inverted pendulum model to represent human normal walking. In: 2010 10th IEEE–RAS International Conference on Humanoid Robots. IEEE, pp. 486–491.

Saranli, U., Buehler, M., Koditschek, D.E., 2001. RHex: a simple and highly mobile hexapod robot. Int. J. Robot. Res. 20 (7), 616–631.

Saunders, J.B. de C.M., Inman, V.T., Eberhart, H.D., 1953. The major determinants in normal and pathological gait. J. Bone Jt. Surg. 35, 543–558.

Schmitt, J., Holmes, P., 2000. Mechanical models for insect locomotion: dynamics and stability in the horizontal plane I. Theory. Biol. Cybern. 83 (6), 501–515.

Seipel, J., Holmes, P., 2005. Running in three dimensions: analysis of a point-mass sprung-leg model. Int. J. Robot. Res. 24 (8), 657–674.

Seipel, J., Holmes, P., 2007. A simple model for clock-actuated legged locomotion. Regul. Chaotic Dyn. 12 (5), 502–520.

Seyfarth, A., Friedrichs, A., Wank, V., Blickhan, R., 1999. Dynamics of the long jump. J. Biomech. 32 (12), 1259–1267.

Seyfarth, A., Blickhan, R., Van Leeuwen, J.L., 2000. Optimum take-off techniques and muscle design for long jump. J. Exp. Biol. 203 (4), 741–750.

Seyfarth, A., Günther, M., Blickhan, R., 2001. Stable operation of an elastic three-segment leg. Biol. Cybern. 84 (5), 365–382.

Seyfarth, A., Geyer, H., Herr, H., 2003. Swing-leg retraction: a simple control model for stable running. J. Exp. Biol. 206 (15), 2547–2555.

Seyfarth, A., Geyer, H., Blickhan, R., Lipfert, S., Rummel, J., Minekawa, Y., Iida, F., 2006. Running and walking with compliant legs. In: Fast Motions in Biomechanics and Robotics. Springer, Berlin, Heidelberg, pp. 383–401.

Smith, J.A., Poulakakis, I., 2004. Rotary gallop in the untethered quadrupedal robot scout II. In: IEEE/RSJ International Conference on Intelligent Robots and Systems (IROS), pp. 2556–2561.

Sharbafi, M.A., Maufroy, C., Ahmadabadi, M.N., Yazdanpanah, M.J., Seyfarth, A., 2013a. Robust hopping based on virtual pendulum posture control. Bioinspir. Biomim. 8 (3), 036002.

Sharbafi, M.A., Ahmadabadi, M.N., Yazdanpanah, M.J., Mohammadinejad, A., Seyfarth, A., 2013b. Compliant hip function simplifies control for hopping and running. In: IEEE/RSJ International Conference on Intelligent Robots and Systems (IROS).

Sharbafi, M.A., Seyfarth, A., 2014. Stable running by leg force-modulated hip stiffness. In: 5th IEEE RAS/EMBS International Conference on Biomedical Robotics and Biomechatronics, pp. 204–210.

Sharbafi, M.A., Seyfarth, A., 2015. FMCH: a new model for human-like postural control in walking. In: IEEE/RSJ International Conference on Intelligent Robots and Systems (IROS), pp. 5742–5747.

Sharbafi, M.A., Rode, C., Kurowski, S., Scholz, D., Möckel, R., Radkhah, K., Zhao, G., Mohammadi, A., von Stryk, O., Seyfarth, A., 2016. A new biarticular actuator design facilitates control of leg function in BioBiped3. Bioinspir. Biomim. 11 (4), 046003.

Weiss, P.L., Kearney, R.E., Hunter, I.W., 1986. Position dependence of ankle joint dynamics. II. Active mechanics. J. Biomech. 19, 737–751.

Westervelt, E.R., Grizzle, J.W., Chevallereau, C., Choi, J.H., Morris, B., 2007. Modeling, analysis, and control of robots with passive point feet. In: Feedback Control of Dynamic Bipedal Robot Locomotion. CRC Press, pp. 43–79.

Wisse, M., Schwab, A.L., Van der Linde, R.Q., 2001. A 3D passive dynamic biped with yaw and roll compensation. Robotica 19 (03), 275–284.

Wisse, M., Schwab, A.L., van der Helm, F.C., 2004. Passive dynamic walking model with upper body. Robotica 22 (06), 681–688.

Wisse, M., Atkeson, C.G., Kloimwieder, D.K., 2006. Dynamic stability of a simple biped walking system with swing leg retraction. In: Fast Motions in Biomechanics and Robotics. Springer, Berlin, Heidelberg, pp. 427–443.

Whittington, B.R., Thelen, D.G., 2009. A simple mass-spring model with roller feet can induce the ground reactions observed in human walking. J. Biomech. Eng. 131 (1), 011013.

Zijlstra, W., Hof, A.L., 1997. Displacement of the pelvis during human walking: experimental data and model predictions. Gait Posture 6 (3), 249–262.

Part II

Control

Chapter 4

Control of Motion and Compliance

Katja Mombaur, Heike Vallery, Yue Hu, Jonas Buchli,
Pranav Bhounsule, Thiago Boaventura, Patrick M. Wensing,
Shai Revzen, Aaron D. Ames, Ioannis Poulakakis, and Auke Ijspeert

INTRODUCTION

The control of legged locomotion and in particular of biped locomotion is a complex task. This is due to the fact that legged locomotion is often unstable, underactuated, redundant, nonlinear, and with complex hybrid dynamics. Indeed, legged locomotion is often (statically) **unstable** because the projection of the center of mass moves at times outside of the small polygons of support provided by feet on the ground. A locomotion controller therefore needs to ensure *dynamic* stability and prevent that the robot does not fall over while locomoting. Legged locomotion is **underactuated** because, unlike robot manipulators that are attached to a base, the feet of legged robots are not bolted to the ground and the leg actuators have therefore no direct control of the orientation and position of the main body. Legged robots are furthermore often **redundant** systems with more actuators and more actuated degrees of freedom than would in principle be needed to adjust the six degrees of freedom that determine the position and orientation of the main body. In addition, each of these actuators can perform infinitely many different actuation patterns many combinations of which lead to the same overall locomotion output, such as a step of a given length or a gait at a given speed. A locomotion controller has therefore to solve this redundancy problem and choose among many possible control actions for each motor to move the body forward. Furthermore, the relationships between actuator commands and the movement of the main body are highly **nonlinear**. Unlike differential drive wheeled robots in which the forward speed of the robot is linearly related to the rotation velocities of the motors, a locomotion controller for a legged robot needs to implement complex nonlinear, typically periodic, relationships between motor commands and desired movements of the main body. Finally, legged locomotion presents complex **hybrid dynamics**, which refers

Bioinspired Legged Locomotion. http://dx.doi.org/10.1016/B978-0-12-803766-9.00006-3
135

to the fact that the number and points of contact of the body with the ground change over time. Typically, a biped robot can switch between zero (when running), one, two, or more (when other parts of the body touch the environment) points of contacts with the environment, and this leads to drastic changes in the dynamics and the equations of motion. Problems with hybrids dynamics are difficult to model properly and make it hard to plan control actions in advance. Taken together these properties make it very difficult to design robust locomotion controllers in particular for complex unstructured terrains.

In this chapter, we will present several important concepts in locomotion control, as well as several examples of control approaches. In that respect, first we explain the basic concepts of locomotion control such as stability, robustness and efficiency. Insertion of compliance as a bioinspired term in the robot body complicates control while providing advantages in terms of robustness and efficiency. Impedance control can be considered as control technique to deal with complexity added in compliant robots or to provide virtual compliance by control even if the structure is rigid. The second part of this chapter includes control approaches which are bio-inspired or can be translated to implement bioinspired concepts.

The first subchapter (by K. Mombaur and H. Vallery) reviews different concepts to evaluate stability and robustness of locomotion considering nominal walking situations as well as the reaction to larger external perturbations. Stability criteria discussed include those based on characteristic points or on dynamic quantities of the system evaluated separately at each point of a motion as well as those evaluating the entire trajectory or limit cycle of periodic motion. Robustness criteria aim to assess the size of its stability region or basin of attraction of a stable solution. Different experimental and computational approaches to evaluate the ability to recover from large perturbations or pushes are outlined.

The second subchapter (by K. Mombaur) discusses how optimal control can serve as guiding principle of legged locomotion in humans and robots. It starts by giving different examples for dynamic multibody system models of the locomotor systems which represent an important basis for optimization studies. Different possible objective function and optimization problem formulation for humans and robots and their numerical solution are discussed. In addition, the chapter focuses on the inverse problem, i.e., the identification of the optimality criterion underlying a recorded human motion, which is called the inverse optimal control problem, also presenting several example solution approaches. It concludes by summarizing current research questions related to the optimization of locomotion.

The third subchapter (by K. Mombaur, Y. Hu, and J. Buchli) discusses different concepts of compliance in human and robot locomotion including constant and variable compliance. Compliance provides many advantages for dynamic motions, such as increased efficiency, stability, adaptability and softness, but also introduces many challenges for the design and control of locomotion. The chapter outlines how optimal control can be used to tune compliance parameters and control compliant actuators, based on different models.

The fourth subchapter (by J. Buchli and T. Buenaventura) focuses on the topic of impedance control in locomotion. It reviews concepts of whole-body motion modeling and task space as an important basis for this research and discusses the interaction of the human or robots with the environment. Passive impedance (by hardware) as well as active impedance (by control) is described. The paper introduces different methods and concepts to emulate appropriate impedances for a legged robot either for modelling or for function. This technique can be also used to emulate muscle models in order to mimic biological properties in legged locomotion.

The fifth subchapter (by P. Wensing and S. Revzen) reviews control approaches for locomotion based on so-called template models. Template models are very simple models consisting only of a single (or very few masses) to describe the locomotion system which can already serve to extract basis characteristics of locomotion and to be considered as target for control. As template and anchor concept is discussed in detail in Chapter 3, here the focus is on how these models can be employed to design controllers. Therefore, after selecting an appropriate template model, establishing its relation to the anchor model through implementation in controller (called anchoring the template models) is the focus of this subchapter.

The sixth subchapter (by P. Bhounsule) describes passive dynamic walking (PDW) robots that use a purely mechanical approach to motion control and walk down inclined slopes in a highly efficient and natural way without any actuators and sensors. Passive dynamic walking concepts have also inspired minimally powered passive walking. The most important feature of PDW based controllers is that PDW robots benefit from natural dynamics of the system and mimic human walking with minimum efforts. After reviewing elementary stability concepts and summarizing passive dynamic walking results for the simplest 2D walking model, the chapter discusses how minimal control can be used to create almost passive walking robots on level ground. It also summarizes the current state-of-the-art of powered passive dynamic robots with respect to energy-efficiency, stability, robustness, versatility, mechanical design, estimation, and robot complexity, and formulates the grand challenges for the future.

The seventh subchapter (by A.D. Ames and I. Poulakakis) presents hybrid zero dynamics (HZD) as a control synthesis framework for feedback laws that realize dynamic walking and running in a class of robotic bipeds. The goal is to develop reliable and efficient legged locomotion controllers that combine provable stability and performance guarantees with fast and elegant motions and fully exploit the capabilities of the system. The idea of hybrid zero dynamics is to reduce the complexity of whole-body control by task encoding through the enforcement of lower-dimensional target dynamics. This subchapter introduces the core concepts of this control approach and its application walking and running robots, discussing different models, theory and implementation. Using human locomotion data, this technique can be extended to implement a bioinspired controller through the notion of partial hybrid zero dynamics (PHZD). With that, biological locomotor behaviors are formulated by few virtual constraints that can be implemented on robots or assistive devices (e.g., prostheses).

The last subchapter (by A. Ijspeert) presents a control approach based on the concept of central pattern generators, which are neural circuits found in animals that can produce coordinated rhythmic patterns necessary for locomotion. CPGs implemented as coupled nonlinear oscillators can similarly serve as building blocks of locomotion controllers for legged robots. Interestingly, the use of robots could potentially also help answering some open questions in neuroscience about the respective roles of CPGs versus reflexes in human locomotion.

Chapter 4.1

Stability and Robustness of Bipedal Walking

Katja Mombaur* and Heike Vallery[†]

*Optimization, Robotics & Biomechanics, ZITI, IWR, Heidelberg University, Heidelberg, Germany
[†]Faculty of Mechanical, Maritime and Materials Engineering, Delft University of Technology, Delft, The Netherlands

Abstract. *In this sub-chapter, we review different criteria to evaluate stability and robustness of locomotion in nominal situations as well as during large perturbations. We discuss different stability measures and control concepts that are based on characteristic points or on dynamic quantities of the system evaluated separately at each point of the motion. In addition, we discuss stability criteria for periodic motions that are based on the entire trajectory, as e.g. the study of the limit cycle properties of the motion. Another important topic is to asses the robustness of a solution which defines the stability margins of a stable solution, i.e. the size of its stability region, for which we present different approaches. As a last topic, we consider the robustness with respect to large perturbations or pushes while walking and standing, outlining different experimental and computational results.*

4.1.1 INTRODUCTION

Stability is one of the most important properties of bipedal locomotion. Intuitively, "stable" gait can be defined as any gait that does not lead to a fall. This implies that the definition of stability should relate to the set of all states that a bipedal walker can experience and still avoid falling, the so-called "viability kernel" (Wieber, 2002). However, this definition is difficult to handle in practice, due to the size of this set and the absence of systematic tools to synthesize controllers from the definition. Another intuitive, but a bit more focused definition of "stable" gait is that of persisting also in the presence of perturbations, i.e., the ability of a person, animal or robot to continue the motion that was planned before the perturbation. In the first definition case, also recovery actions would be acceptable that may completely modify the motion or even the type of motion, e.g., recovering from a perturbation while walking by coming to a full stop.

Both types of stability are hard to achieve. Even though humans usually learn to walk without any problems during the first years of their life, the underlying mechanisms, and especially the natural human stability control, are not yet fully understood. The difficulty of achieving stable walking can easily be seen in humanoid robotics where even the most expensive and advanced robots in the world, participating in the DARPA robotics challenge in 2015 suffered from many falls during walking motions that would not be very challenging for an able-bodied human. A better understanding of stability would not only be useful for robotics, but also for many medical applications. Stability concepts are required for a better control of prostheses, orthoses, and exoskeletons, or for Functional Electrical Stimulation. For individuals with neurological or orthopedic impairments and for members of the elderly generation exhibiting mobility problems, it is important to asses the general as well as immediate risk of falling.

In this chapter, we will not be able to address all these issues since they describe a whole roadmap for interdisciplinary research of many years. The focus of this chapter is on computational criteria that may help assess and control the stability and reaction to small and large perturbations of different types of robots, humans, and humans using wearable technical devices. We also describe some stability-related experiments in biomechanics and robotics, and we touch upon control approaches based on these criteria.

With respect to the different views on stability, we will adopt the following terminology from here on:

- **Stability** refers to the property of a motion to persist even under small perturbations;
- **Robustness** of a solution addresses the question of the size of the corresponding stability regions;
- **Push recovery or large perturbation recovery** describes situations where an active reaction with a switch of motions and motion strategy is required in order to avoid a fall.

This chapter is organized as follows: In Sections 4.1.2 and 4.1.3 we discuss different stability criteria. While Section 4.1.2 focuses on pointwise criteria (in the temporal as well as sometimes in the spatial sense), Section 4.1.3 discusses the limit cycle approach, which studies the entire motion or at least a full cycle at once. Section 4.1.4 outlines different criteria to analyze the robustness of stable motions. In Section 4.1.5, we discuss research on push recovery including computational approaches and experiments. Section 4.1.6 contains some final remarks.

4.1.2 STABILITY CRITERIA RELATED TO INSTANTANEOUS PROPERTIES OF THE WALKING SYSTEM

In this section, we will review several classical criteria that are used to define stability of walking motions. Here we focus on those criteria that can be evaluated independently at every instant of a walking motion, just looking at the

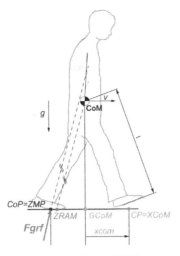

FIGURE 4.1.1 Walking model with center of mass (CoM), projected center of mass (GCoM), point of zero rate of angular momentum (ZRAM), center of pressure (CoP)/zero moment point (ZMP), virtual pivot point (VPP), and capture point (CP)/extrapolated center of mass (XCoM).

current state of the system or some simplified model of it. In contrast to that, in the next section we discuss criteria which are based on the study of an entire motion or limit cycle. Due to their instantaneous nature, the measures of walking stability discussed here can be directly used to derive control laws for robot walking. Many of these stability measures result in characteristic points which are summarized in Fig. 4.1.1.

4.1.2.1 Projected Center of Mass

The projected center of mass (GCoM) is the projection of the center of mass (CoM) of the walking system to the ground (Fig. 4.1.1). It can be used to define the static stability of a system where it is put in relationship to the polygon of support (PoS), i.e., the convex hull of all contact points of the system with the ground. If the GCoM lies within the polygon of support and if the velocity and acceleration of the CoM is negligible, the system is statically stable. This measure has been heavily used in the context of slower multilegged robots such as six- or eight-legged crawling robots (e.g., Berns et al., 1994), e.g., resulting in the tripod gait for six-legged robots. It is not relevant for bio-inspired dynamic bipedal walking or even quadrupedal walking using bio-inspired gaits like pace, trot, or gallop. While statically stable bipedal walking is in principle possible, it requires very large feet and very slow walking speeds, slowly shifting the GCoM from one foot to the other, which is not interesting from the perspective of this book.

4.1.2.2   Zero Moment Point

The most commonly used stability concept in the field of humanoid robots is the Zero Moment Point (ZMP) (Vukobratović and Borovac, 2004). The ZMP is the point on the ground where the resulting torques of inertia and gravity forces of the robot about the horizontal axes become zero. For nonsliding motions on level ground, and in the absence of external forces acting on the system, besides ground reaction forces, it is equivalent to the Center of Pressure (CoP) which is the point where the resulting ground reaction force applies to the foot (or feet in the case of multi-foot contact). This is shown in Fig. 4.1.1. While the CoP is computed based on external forces, the ZMP is defined using inertia properties and internal accelerations of the segments of the robot:

$$p_{ZMP, Q} = \frac{n \times M_Q^{gi}}{R_{gi} \cdot n}, \tag{4.1.1}$$

where $p_{ZMP, Q}$ denotes the position vector of the ZMP with respect to a general fixed point Q on the contact surface, n the vector normal to the contact surface, R_{gi} the sum of the gravity and inertia force at the center of mass, and M_Q^{gi} the moment at point Q caused by acceleration, gravity and change in angular momentum of the segments (Sardain and Bessonnet, 2004). The ZMP can be interpreted as an augmented GCoM that additionally takes the effects of translational and rotational acceleration into account.

In humanoid robotics, the ZMP is generally not computed using the full model of the robot with all segments, but with simplified robots in order to speed up computation in the context of control algorithms. The most popular model is the table–cart model of Kajita et al. (2003) that approximates the human as a point-mass in the pelvis center and only considers horizontal motion of this point. The position of the table cart ZMP on the floor ($p_{ZMP, x}$, $p_{ZMP, y}$) in forward and sidewards directions x and y then can be directly computed from the center of mass position x, y, z (with constant height $z = z_c$) and accelerations:

$$p_{ZMP, x} = x - \frac{z_c}{g}\ddot{x}, \tag{4.1.2}$$

$$p_{ZMP, y} = y - \frac{z_c}{g}\ddot{y}. \tag{4.1.3}$$

In order to stabilize gaits of humanoid robots, ZMP-based control concepts aim at keeping the simplified ZMP well within the PoS. While the precise ZMP can not travel outside the polygon of support, stability properties of the system get unpredictable when the ZMP is at the edge of the PoS. Controllers typically apply a large safety margin to the real boundary such that the ZMP is only allowed to lie in a much smaller subset of the PoS. E.g. in the case of the humanoid robot HRP-2, the standard stabilizing controller aims to maintain the table cart ZMP within a few centimeters around the projection of the ankle joint. Other examples for ZMP-based humanoid robot control are Asimo (Sakagami et al., 2002), and also some of the control approaches used in the DRC (Kuindersma et al., 2014; Wang et al., 2014).

While this approach is very easy to apply, it leads to quite conservative and not very human-like gaits with strong knee flexion. Computations have shown that relaxing strict ZMP constraints results in more dynamic gaits with more stretched knees, i.e., a higher pelvis position (Koch et al., 2012).

Dynamic human walking is characterized by ZMPs and COPs approaching the edge of the PoS frequently e.g. for level ground walking traveling from the very back to the very front of the foot. Also during very dynamic balancing motions, including recovery motions of the arms and the upper body, the ZMP / CoP reach the boundary of the PoS at many different places and for extended periods of time. One major problem of the ZMP criterion, even if precisely computed based on the whole-body model, is that it is not possible to predict if the system will fail in the next second or if it can be stabilized. Part of this is based on the fact that it only takes momentary snapshots.

4.1.2.3 Capture Point or Extrapolated Center of Mass

The capture point (CP) (Pratt et al., 2006) or extrapolated center of mass (xCOM) (Hof et al., 2005) is another point on the ground that can be used to quantify stability, and to directly guide foot placement. It is based on a highly simplified model of gait, the linear inverted pendulum. By solving the differential equations of this model, the point on the ground can be found where the CoP needs to be placed in order to make the system come to a full halt above it. For a walker with leg length l walking at a speed v under the influence of gravity g, the point is located at a distance of v/ω_0, with eigenfrequency $\omega_0 = \sqrt{g/l}$, in front of the CoM (Fig. 4.1.1). To achieve continuous walking, the leg can be placed slightly behind of this point (in walking direction, i.e. closer), for example, by constant offset control (Hof et al., 2005). Also for this strategy, experimental evidence was found in humans (McAndrew Young et al., 2012). However, healthy young subjects, healthy elderly subjects and elderly fallers seem to exhibit different stability margins with respect to this point (Lugade et al., 2011). The capture point and the concept of capturability also play an important role in the context of push recovery, i.e., the recovery of large perturbations (see below). It has also been used to generate stabilizing controllers for devices that observe (Paiman et al., 2016) or assist human walking (Vallery et al., 2012; Monaco et al., 2017).

4.1.2.4 Virtual Pivot Point

The Virtual Pivot Point (VPP) (Maus et al., 2010) represents a hypothesis on human balance control. It is based on the observation that many animals and also approximately the human, direct the ground reaction force vector towards a point above the body's center of mass throughout the gait cycle (Fig. 4.1.1). This way, the body mimics a physical pendulum, with the VPP as hinge joint. As opposed to an inverted pendulum, a regular hanging pendulum is not inherently unstable, and does not require active state feedback control throughout the entire gait cycle. Experimental evidence for this hypothesis was found (Maus et al., 2010), and models based on the VPP showed high coefficients of determination for predicted ground reaction force direction and whole-body angular momentum (Maus et al., 2010). The point can be used to command hip control torques to direct the ground reaction force vector. Later, the same group found that the emergence of a virtual pivot point can also be explained by feedback control of hip muscles (Sharbafi and Seyfarth, 2015). It has also been used to observe human walking (Paiman et al., 2016).

4.1.2.5 Angular Momentum

Another quantity that is assumed to play an important role for the stability of a walking motion is the total angular momentum about the center of mass, also called the centroidal momentum. Since the orientation of humans and robots during locomotion – as well as in any other upright form of movement – remains more or less vertical and there is no continued rotation about any horizontal axis as in somersaults, the average angular momentum about the frontal and sagittal axis must be zero. The same is true for the longitudinal axis in straight walking and running, but changes for curved walking and spinning motions and jumps. In stability research, the variations of angular momentum during locomotion are studied. As described by basic laws of physics, angular momentum changes over the locomotion cycle under the action of external forces, typically ground reaction forces. During aerial phases of running and jumping, the total angular momentum remains constant.

The total angular momentum about the CoM is computed taking the contributions of all segments of the body into account:

$$H_C = \sum_{i=1}^{n}(r_i \times m_i \dot{r}_i) + \sum_{i=1}^{n}(\Theta_i \omega_i), \qquad (4.1.4)$$

where Θ_i is the inertia matrix of segment i with respect to its center of mass, and m_i its mass, r_i the distance vector from the total CoM to the segment CoM, $\dot{r}_i$ the corresponding relative velocity, and ω_i the angular velocity vector of segment i.

The angular momentum during walking as well as the contributions of the different segments have been investigated in Popovic et al. (2004), Herr and Popovic (2008). The values of the total angular momentum remain quite small throughout the cycle, but they are clearly not zero everywhere, which is sometimes assumed. In Heidelberg, we have obtained similar results for the angular momentum of neutral and emotionally modified locomotion (Felis, 2015). For example, the angular momentum about the frontal axis (i.e., for rotations in the sagittal plane) oscillates up to 4 Nm/s which amounts to about 10% of the angular momentum of a highly dynamic somersault motion in platform diving (Koschorreck and Mombaur, 2011), so it is not negligible.

The fact that the angular momentum is not zero everywhere and that it actually changes is also in accordance with the observations on the virtual pivot point above. The angular momentum change in the sagittal plane is generated by the moment of the ground reaction force about the center of mass, which is nonzero since the VPP lies above the CoM, resulting in a nonzero lever arm. The direction of the moment changes since the ground contact occurs in front of or behind the GCoM (see Fig. 4.1.1).

The absolute angular momentum and the change of angular momentum about the different axes can also be used as objective functions in the generation of optimal locomotion for human or robot models; compare with Subchapter 4.2. The results we have obtained for human models in Heidelberg show that it seems to be an important criterion in a multiobjective function to generate human-like walking (Felis and Mombaur, 2016) and that it is certainly not the unique criterion of human walking (otherwise it would also be zero everywhere).

4.1.2.6   Zero Rate of Angular Momentum Point

Linked to the properties discussed in the previous paragraph, also a characteristic point has been defined: The Zero Rate of Angular Momentum (ZRAM) point is an indirect measure of the rate of change of angular momentum that is currently being generated by the moment of the ground reaction force vector (Goswami and Kallem, 2004). It is located at the intersection of the ground and a line that passes through the body's center of mass and is parallel to the line of action of the ground reaction force vector. Given that the moment of all external forces about the center of mass of a mechanical system is equal to the rate of change of angular momentum of this system about its center of mass (the centroidal momentum), this point is an intuitive measure of how much the body's rotational movement is currently changing: The further away this point is located from the actual center of pressure, the higher the moment that is generated, and the more the body's centroidal momentum will change. In order to maintain constant centroidal momentum, the walker needs to direct the ground reaction force vector exactly towards its center of mass. Oscillations of the ZRAM are expected, as they correspond to the changes in angular momentum discussed above and relate to the self-stabilizing characteristics of the Virtual Pivot Point, dictating that the ground reaction force vector does not pass exactly through the center of mass.

4.1.3 STABILITY CRITERIA FOR LIMIT CYCLES

Another fundamentally different way to study stability is to look at the entire motion or at least an entire walking step or double step at once and evaluate how small perturbations would affect this particular motion. This approach follows the mathematical stability theory of Lyapunov and has been frequently used in robotics in the field of (passive-)dynamic walking. The stability measures and limit cycle analyses discussed in this section do not provide direct guidelines for control, but rather give hints for robot design, and they offer practically relevant, global information on whether a walker is going to fall or not.

Passive-dynamic robots (see, e.g., McGeer, 1990, 1991, 1992; Coleman and Ruina, 1998 and Subchapter 4.6 of this book) are purely mechanical walking devices, typically with stiff or kneed legs, that have neither motors nor sensors and control systems and walk down inclined slopes. Stability of their walking motions is based on their intelligent mechanical design which is capable of executing limit cycles, i.e., self-stable periodic motions. There are also more recent versions of dynamic walking robots with little actuation and little feedback (Collins et al., 2005) which are still based on the same dynamic principles. Very often they have curved feet which are only in point or line contact with the ground, so, e.g., the ZMP criteria discussed above would be of no help. The same is true for dynamic human walking or running where very only small and rapidly changing contact areas exist.

4.1.3.1 Definition of Stability and Orbital Stability in the Sense of Lyapunov

According to Lyapunov, a solution of a nonautonomous (i.e., explicitly time-dependent) nonlinear system[1] is **stable** if small perturbations of the trajectory result in perturbed trajectories that always stay in a finite neighborhood of the original one, and it is **asymptotically stable** if additionally the perturbed solutions converge to the unperturbed one for $t \to \infty$ (e.g., Cronin, 1994), see Fig. 4.1.2 left. In the case of autonomous systems (i.e., systems with no explicit

---

1. Note that this mathematical concept of autonomy/nonautonomy is not equivalent of the use of these terms in robotics and may seem counterintuitive from a robotics perspective.

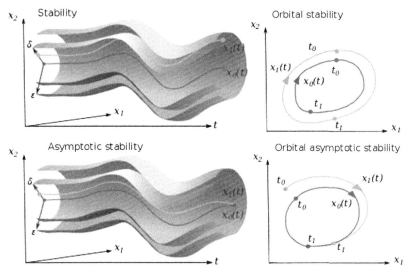

FIGURE 4.1.2 Definition of stability and asymptotic stability in the sense of Lyapunov (left) and of orbital stability and orbital asymptotic stability (right).

time dependency) the notions of **orbital stability** and **orbital asymptotic stability** (see Fig. 4.1.2 right) become important which are similar to the previous ones, with the exception that time shifts of the solution by the perturbation (also called orbital shifts) may occur and persist or even grow, but are not considered, i.e., only the distance between the orbits as a whole is important and not the distance of corresponding time points. The strictly passive-dynamic robots without any actuation fall in this category.

4.1.3.2 Stability Analysis of Walking Using Lyapunov's First Method

The concept of Lyapunov stability can especially well be applied to periodic motions, also called limit cycles, and is therefore very interesting for the study of walking and running. In this case, the propagation of perturbations over one walking cycle, typically one step, is considered. The study uses the Poincaré map that maps the states at the beginning of this cycle to the states at the end of this cycle. A periodic solution of the dynamic equations corresponds to a fixed point of the Poincaré map (Fig. 4.1.3). For the investigation of stability of a motion, the Jacobian of the Poincaré map, which in mathematics is called monodromy matrix, transfer matrix or sensitivity matrix, is computed as

$$X(T) = \frac{dx(T)}{dx(0)}, \tag{4.1.5}$$

Time-dependent Poincaré sections   *State-dependent Poincaré sections*

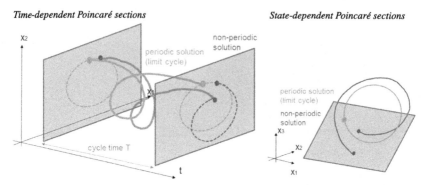

FIGURE 4.1.3 Poincaré maps corresponding to nonautonomous systems with Poincaré sections in regular time intervals (left) and to autonomous systems with state-dependent Poincaré sections (right).

and contains the sensitivities – or first order derivatives – of the states at the end of the cycle with respect to their initial values. Lyapunov's first method (e.g., Cronin, 1994; Hsu and Meyer, 1968) states that a periodic solution with cycle time T of a periodic nonautonomous system

$$\dot{x}(t) = f(t, x(t)) \text{ with } f(t, \cdot) = f(t + T, \cdot) \qquad (4.1.6)$$

is asymptotically stable if all eigenvalues of the monodromy matrix satisfy

$$|\lambda_i(X(T))| < 1. \qquad (4.1.7)$$

These eigenvalues are sometimes also called the Floquet multipliers in the field of passive dynamic robots even though, strictly speaking, Floquet theory in which Floquet multipliers are defined, addresses the problem of linear differential equations, while robots are described by nonlinear differential equations. Lyapunov's first method provides the extension of Floquet theory to nonlinear systems by studying the variational system about a particular solution. If not all state variables x are periodic, which, e.g., is the case if one variable describes the forward direction of walking, then the nonperiodic directions have to be eliminated from the matrix by projection prior to applying this stability criterion. In the original matrix this direction is associated with an eigenvalue of 1, meaning simply that the same motion could be performed from an arbitrary starting point and that this shift would be conserved. Also in the case of autonomous systems, such as passive dynamic walking systems, there is always an invariant eigenvalue of 1, describing that perturbations along the orbit are conserved, The system is orbitally asymptotically stable if eq. (4.1.7) holds for all other eigenvalues. As shown in Mombaur et al. (2005a), the Lyapunov stability criterion can be used not only for simple systems of type (4.1.6), but also for hybrid multiphase systems which are required to describe general walking and running motions of humans and robots (see Subchapter 4.2). If there are phases with fewer degrees of freedom, as it is often the case in walking models due to additional foot contacts, there are two zero eigenvalues for the overall cycle corresponding to the constrained position and velocity of one degree of freedom. This describes that the perturbations associated with this degree of freedom are naturally damped out.

This criterion has been thoroughly investigated for passive dynamic robots by several authors (Garcia et al., 1998; McGeer, 1990; Goswami et al., 1996; Coleman, 1998; Hurmuzlu, 1993) to analyze the stability of a given motion and the effect of parameter variations. The same criterion can also be used in the optimal control context to generate the most stable systems by minimizing the maximum eigenvalue (also called the spectral radius) of the monodromy matrix for passive as well as actuated open-loop controlled systems. With this approach it was possible to generate open-loop stable systems of different complexity (Mombaur et al., 2005a, 2005b; Mombaur, 2009), also see Subchapter 4.2 of this book.

Related to this criterion are formulations based on other norms of the monodromy matrix. Examples for this are the spectral radius and the 1- and ∞-norms which are both upper bounds on the spectral radius (see the theorem of Hirsch (e.g., Stoer and Bulirsch, 1990)) and therefore represent stricter measures of stability. Mombaur (2001) shows the application of these criteria to walking and hopping motions.

4.1.3.3 Applicability of Limit Cycle Stability Concepts to Feedback-Controlled Robots and Humans

The examples discussed so far in this section, to which the analytical concept of Lyapunov stability has been applied, include passive dynamic walking robots and other simple, essentially open-loop controlled robot configurations. However, it is possible to generalize this concept to a system with feedback control. Lyapunov's first method can be applied to any periodic solution of a nonlinear system with a periodic right-hand side. This requires also including the equations of all feedback controllers into the system to be studied, in addition to the mechanical equations previously treated. Thus it could, in principle, also be used to generate closed-loop stable limit cycles for humanoid robots, e.g., by means of optimization if a full description of the system with mechanics and controllers in terms of differential equations is available. We are, however, not aware that this has already been tried. It should also be noted that the computational effort involved in analyzing and in particular in optimizing stability in the sense of Lyapunov is very high for systems with many degrees of freedom (see Mombaur, 2009).

Lyapunov stability can, in principle, also be pursued as a path to explain stability of human walking and running. The gait of some passive-dynamic robots which are Lyapunov stable looks much more human-like than the one of most complex humanoid robots. It is assumed that open-loop stability plays an important role in human walking and running robots. Obviously humans are not entirely open-loop controlled as they also use important reflexes and higher-level controllers for the control of walking and running. Therefore, in order to be able to thoroughly apply Lyapunov analysis to explain stability in human gaits, full models would be required, addressing the essential components of the human neuro-musculo-skeletal system. In Mombaur (2009), open-loop stable running motions have been generated for a multibody system model with human kinematic and dynamic properties driven by torques in the joints as a proof of concept. The resulting motion was in fact not human-like since important components for explaining human stability were missing. The first important components to add in his context are muscle models, since it has been discovered that for biological system the force generating mechanisms inside the muscles, in particular the force–length relationship of the muscle contributes to self-stability of the system, in addition to favorable dynamics (Blickhan et al., 2007). In addition, all relevant feedback loops of the nervous system would have to be added. Such a model is obviously not available, yet, and it will take a significant amount of research to establish something close to this goal, requiring combined efforts from very different research fields. Research on stability of models with simplified control loops might also prove to be very interesting.

There have also been attempts to experimentally measure stability in the sense of Lyapunov for human walking, at first by Hurmuzlu and Basdogan (1994). Poincaré sections have been identified by characteristic events such as heel strike, heel off, or toe off. The data collected has been analyzed using a simplified model just taking trunk angle and overall leg angles into account. No active sensitivity computation or generation of perturbed trajectories is implemented in these experimental conditions, but instead the natural variability of the motion is used. The computed eigenvalues are also very much influenced by measurement noise. Also there has been a lot of discussion about the relevance of this analysis to evaluate the global or momentary risk of a person falling, but no clear conclusion has been reached. Many experimental studies, however, focus on more active perturbation studies; see below.

4.1.4 ROBUSTNESS MEASURES OF WALKING

The concepts of Lyapunov stability and orbital stability discussed in the previous section define stability of a motion with respect to small perturbations of the trajectory. But how small is the small perturbation that a particular walking motion can sustain? The values of the computational criteria, e.g., the absolute size of the maximum eigenvalue, only describe how quickly the perturbation will decay from one cycle to the next (in the particular norm defined by the eigenvalues). It does not say anything about the absolute size of sustainable perturbations. This important property is termed robustness of a solution, which defines the stability margins of a solution. Quantifying robustness for walking motions of humans and robots is a very difficult issue for several reasons. Walking systems have multiple degrees of freedom, and perturbations may occur in any of these position and velocity variables individually or in all possible combinations of them, which results in a very high number of possibilities. Perturbations may also be induced by the environment, e.g., the floor conditions, or there may be perturbations in the actuation of the walking systems. Since these systems are highly nonlinear and coupled, the full nonlinear dynamics have to be considered (either in the model or the real world) to investigate how perturbations affect all variables of the system over time. It is not a priori clear how long perturbed motions should be observed to judge if the perturbed solution goes back to the original one or not. A thorough test of robustness also checks the closeness to constraints, e.g., the distance of the swing foot to the ground, which is not taken care of in any of the stability criteria but which may lead to premature contact of the foot and failure of the motion when a perturbation is applied.

Note that we are adopting here, in general, the engineering definition of the term robustness by looking at absolute sizes of possible perturbations. In mathematics, the term robustness denotes first-order sensitivity information of the quantity investigated, in the case of Lyapunov stability it would be the derivative of the maximum eigenvalue with respect to changes in the variables. This mathematical type of robustness is, however, less relevant for practical implementations.

Due to the above mentioned difficulties, there is no straightforward way to compute robustness of a walking motion numerically and in an objective way. In this section, we present several approaches to quantify robustness applied in the past.

4.1.4.1 Robustness Analysis via the Basin of Attraction

The Basin of Attraction offers a global perspective to analyze robustness of limit-cycle walking (Schwab and Wisse, 2001): It represents the set of all initial conditions on the Poincaré section from which the system will converge to the fixed point(s). This results in a region surrounding the fixed point(s), and any point outside this Basin of Attraction will eventually result in a fall. For most practically relevant systems, this method requires numerical simulation of walking movements starting with many initial conditions, to establish the basin of attraction. Mostly, this is done in a grid-based fashion, for example, using cell mapping of the Poincaré section. Evaluating the stride function of the walker for initial conditions of a particular cell center yields a final condition that is either contained within the same cell, or within another one. Cells pointing to themselves are called sink cells. By repeating this process, each cell can be associated with a sequence of subsequent cells, ending either in a sink cell or in a repetitive cycle. The cells that either end in a repetitive cycle or in a fixed point define the basin of attraction. Sink cells can be unfeasible or represent a fixed point of a limit cycle. Analyzing initial conditions within a small neighborhood of the fixed point of a limit can be used to confirm if this limit cycle is stable. Robustness can be analyzed in terms of the size of the basin of attraction and the distance of the fixed point from its boundaries.

A drawback of this method is that it requires substantial amounts of data, usually obtained via model-based simulations. Therefore, it is not suitable to analyze robustness of a physical system via experiments in the real world. However, it has extensively been used to study robustness of the simplest walking model. For example, the Basin of Attraction of the simplest walker confirmed the practical observation that the set of perturbed states from which it can safely recover is very small (Schwab and Wisse, 2001). The same type of analysis demonstrated that sufficiently fast swing leg control can always prevent the walker from falling forward (Wisse et al., 2005).

4.1.4.2   Robustness Analysis via the Gait Sensitivity Norm

The Gait Sensitivity Norm has been developed (Hobbelen and Wisse, 2007) as a more practical measure of robustness. This method quantifies the effect of a set of disturbances e on a set of gait indicators g using the H_2-norm, which indicates how much energy amplification occurs in the system:

$$\mathbf{N}_{GSN} = \left\| \frac{\partial g}{\partial e} \right\|_2 . \qquad (4.1.8)$$

Disturbances e can be chosen freely to include all those that are relevant to the designer, ranging from impulsive disturbances to continuous ones. Also, the gait parameters g can be tailored such that they quantify the quality of gait or are directly related to failure modes. For example, floor irregularities can be chosen as disturbance and step duration as gait parameter (Hobbelen and Wisse, 2007).

Like for the basin of attraction, also this robustness measure requires the dynamic system response, obtained either from simulations or from practical experiments. However, only a smaller, more targeted, and practically relevant set of data is needed.

4.1.4.3   Robustness Analysis Based on Lyapunov's Second Method

More famous than Lyapunov's first method, which is the basis for the limit cycle analysis in Section 4.1.3.2 above, is the second method of Lyapunov (see, e.g., Cronin, 1994). This method does not rely on any linearization or first order sensitivity information, but instead uses a so-called Lyapunov function $V(t, x)$ to determine if the solution $x \equiv 0$ of the nonlinear differential equation $\dot{x} = f(t, x(t))$, i.e., $0 = f(t, 0)$ is stable.

The Lyapunov function $V(t, x)$ which can be computed at all points along the solution represents a generalization of the potential energy function. The potential energy of a mechanical system (e.g., a pendulum) is minimal at a stable equilibrium (the lowest point) and maximal at an unstable equilibrium (the highest point). A Lyapunov function is defined by the following properties: it is zero at $x = 0$, is positive definite for $x \neq 0$, and has a negative semidefinite derivative with respect to time, $\dot{V}$. Of course, the Lyapunov function can be more complex than the potential energy and it is a priori unclear what this function looks like and if it exists at all. Lyapunov's second method only states that IF such a function exists, then the trivial solution $0 = f(t, 0)$ is stable. In detail, it distinguishes the cases of just negative semidefiniteness of $\dot{V}$ (solution is stable), negative definiteness (solution is asymptotically stable), and $\dot{V}(x) \leq -\alpha V(x)$ and $V(x) \geq b|x|^\beta$, $(\alpha, \beta, b > 0)$ (solution is exponentially stable). It can also be applied to nontrivial solutions by subtracting the solution and looking at the deviation from this solution.

So Lyapunov's second method is a classical stability criterion and would therefore fit in the stability sections of this chapter. However, it is very difficult to apply this method for stability analysis or improvement in a practical system, since it requires the construction of a suitable Lyapunov function for each of these systems. So far, such functions have only been found for certain classes of systems, e.g., the total energy is a Lyapunov function for Hamiltonian systems. This need for construction, however, makes it difficult to use Lyapunov's second method for an automated analysis of stability of motions or even for optimization of stability.

Nevertheless, very promising approaches have been made using Lyapunov's second method for controller design and robustness analysis of the closed loop system which is why we are mentioning the method here. Papachristodoulou and Prajna (2002) has proposed an approach to construct Lyapunov functions based on sums of squares. It was also extended to systems with contacts and discontinuities (Posa et al., 2013). So far, this method has been applied successfully to develop controllers for given equilibrium solution (Majumdar et al., 2013). Wieber (2002) has proposed a stability margin in the sense of Lyapunov that computes the distance of all points of a given solution to the next nonviable solution (i.e., a solution that leads to a fall) for a given controller, using the Lyapunov function.

An extension of these approaches constructing Lyapunov functions using sums of squares also to the generation of new stable solutions and the simultaneous development of suitable controllers might be very promising, but is also a challenging task.

4.1.4.4 Pseudospectra for Robustness Analysis of the Matrix Spectrum

One way to investigate part of the robustness of the stability analysis based on Lyapunov's first method uses so-called pseudospectra of the monodromy matrix. This robustness analysis studies how the spectral radius changes with changes in the matrix entries which in turn are caused by perturbations in the independent variables. A pseudospectrum of a matrix is a tool that helps asses the first part of this dependency, namely the absolute changes of the spectrum produced by changes in the matrix: Pseudospectra of a matrix can be defined in terms of the spectra of all nearby matrices which result from given maximum perturbations:

$$\Lambda_\varepsilon(C) = \{z \in \mathbb{C} : z \in \Lambda(C + E) \text{ for some E with } \|E\| < \varepsilon\} \qquad (4.1.9)$$

where Λ denotes the spectrum of a matrix. An overview of the theory of pseudospectra, including equivalent definitions and useful tools, is given in Trefethen and Embree (2005). In Fig. 4.1.4, we give an example of a pseudospectrum of an open-loop stable solution for the flic flac robot, taken from Mombaur (2006) and also presented in Subchapter 4.2. In the example, the monodromy matrix is of dimension 10 and hence has 10 eigenvalues, two of which are zero due to a loss of degrees of freedom in positions and velocity in the contact phase, and one turns out to be quite small for this particular solution. The 8 nonzero eigenvalues are real. All eigenvalues of the original matrix are shown as black dots in the figure, and they clearly lie within the unit circle. The colored lines show, for different values of ε, the boundaries of the regions in which the corresponding eigenvalues of the perturbed matrices will be situated.

Since the monodromy matrix is a nonsymmetric matrix which has no orthogonal basis of eigenvectors and may exhibit bad transients or high sensitivity of

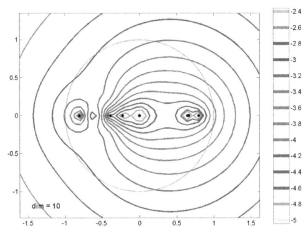

FIGURE 4.1.4 An example pseudospectrum analysis of the monodromy matrix related to an open-loop stable periodic solution for a bipedal gait, computed with Eigtool (Trefethen and Embree, 2005). The stable region is inside the unit circle (in gray).

eigenvalues with respect to the matrix entries, the pseudospectra provide a first and very useful hint on the relevance of the computed spectral radius for the stability of a practical solution. However, it does not, obviously, contain any information about the sensitivity of the matrix entries with respect to perturbations of the free variables, so it cannot be considered as a full robustness criterion.

4.1.5 RECOVERY FROM LARGE PERTURBATIONS AND PUSHES

The criteria and concepts treated in the previous sections concern the stability and robustness of locomotion, when investigating under which conditions a planned motion might still be executed in a stable way. Another important question in the context of walking is how a human or a robot reacts to large perturbations which require a significant change in strategy in order to avoid falling. In this case, the original motion plan is replaced by some recovery strategy which may be, e.g., a stopping motion to come to a full rest or a significantly modified walking motion. So in contrast to the previous question, a fundamental change of the motion or even the type of motion is perfectly acceptable as long as the human or robot is not falling.

Once a bipedal system is pushed with a significant impulse, the gait patterns are considerably disturbed. Humans and also many controllers for humanoids change their behavior and exhibit dedicated recovery strategies in this situation, instead of relying only on intrinsic self-stabilization of their nominal gait.

During standing, different types of strategies have been found in humans, depending on the strength of the perturbation. The first mechanism consists in mechanical viscoelastic properties of tendons and joints (Rietdyk et al., 1999). These properties depend on the muscles' activation levels, so stiffness can be increased via anticipatory co-contraction. However, these torques are not the result of active contraction in response to the perturbation (Rietdyk et al., 1999). Only later in time, due to neural delays, active contraction can move the CoP by means of ankle torques (Sardain and Bessonnet, 2004), also called the "ankle strategy". Slightly larger perturbations are countered by the "hip strategy", which moves the upper body in the opposite direction with respect to the lower body, changing the body's angular momentum. Ankle and hip strategies are dominant during stance, and only for larger perturbations humans step out (Kuo and Zajac, 1993; Horak, 1987). A full model of physiological balance control during stance has also been successfully implemented on a robotic platform (Mergner, 2007). In many studies, arm movements are ignored; however, for dynamic motions they are very important. Kuindersma et al. (2014) showed how rapid arm movements can be used for push recovery while standing.

During walking, foot placement dominates recovery behavior in humans (Hof, 2008; Townsend, 1985). Recovery actions by the arms also seem to play a role to counter perturbations during gait: Even though arm swing may not aid local stability, it was found that dedicated recovery actions of the arms increase robustness to perturbations (Bruijn et al., 2010). This also indicates that there might be a switch in strategy between normal gait and push recovery. Ferber et al. (2002) and Tang et al. (1998) studied which muscles are most relevant to sustain large perturbations. Pijnappels et al. (2005) investigated how important early reactions of the stance leg are for a successful recovery after tripping.

To investigate recovery mechanisms in humans, diverse experimental setups have been devised, particularly treadmills that can apply perturbations (Engelhart et al., 2012). Also cable systems with winches (Pidcoe and Rogers, 1998; Mergner et al., 2003; Fritschi et al., 2014) or weights (Mouchnino et al., 2012) have been used to apply perturbations to the upper body during standing and gait. In the future, also wearable perturbation devices would be possible to manipulate angular momentum of the human body (Li and Vallery, 2012; Lemus et al., 2017 Chiu and Goswami, 2014). In the KoroiBot project (Mombaur, 2013–2016), we have also performed various human push recovery experiments while standing and while walking (see, e.g., Kaul et al., 2015; Schemschat et al., 2016a) which are all included in the KoroiBot Motion Capture Database (2016). Different pushing devices have been developed and used to apply variable external pushes, including pushing sticks equipped with sensors to measure the size and direction of the pushing forces applied. The goal of this study was to understand how humans recover from large perturbations and then transfer this knowledge to humanoid robots.

To analyze robustness to pushes, most of the above measures cannot be applied. For example, all limit cycle related analyses specifically consider the return to this particular motion and cannot handle the concept of alternative recovery motions. The concept of the gait sensitivity norm can be applied to push recovery as an approach to determine critical push directions. Dedicated analysis tools that consider the theoretical ability of a biped to avoid falling from any arbitrary initial condition (in particular, after large pushes) are still rare. This is explicable because this would in fact be a question of formal verification, which can involve huge computational efforts for higher-dimensional, nonlinear, and hybrid systems (Althoff and Krogh, 2014).

For a linear inverted pendulum model, a particular type of reachability analysis has been suggested, namely the above introduced capture point concept has been extended to the principle of "capturability" (Koolen et al., 2012). The so-called "capture region" is the reachable set of possible next footholds such that a walker can come to a full stop in a given number n or fewer future steps. The principle was defined in Koolen et al. (2012) as "the ability of a system to come to a stop without falling by taking n or fewer steps, given its dynamics and actuation limits." According to this definition, the 0-step capture region equals the walker's current support polygon, whereby the instantaneous capture point (ICP) has to be contained within it, and it has to be possible for the Center of Pressure to be placed upon it instantaneously. Next, 1-step capturability requires the existence of a step position within the reachable area (considering a constraint of maximum step length l_{max}), such that the system becomes 0-step capturable then. Similarly, for n-step capturability, a foothold must exist such that the system becomes $(n-1)$-step capturable.

In contrast to the simple model underlying the capture point concept, we have investigated the use of whole-body models and optimal control to generate reactions to push recovery while walking. On the one hand, dynamic models of humans are fitted to experimental human data in the least squares sense to further analyze the dynamics underlying push recovery during walking motions (Schemschat et al., 2016a). On the other hand, optimization is used for movement synthesis of push recovery, in this case based on a human model and minimizing the deviation from the original walking cycle as well as the efficiency of the motion (Schemschat et al., 2016b). The optimization criteria could be adjusted to the particular situation or demand, putting more emphasis on the continuation of the previous motion or on efficiency, or on any other criterion added. This optimization approach can also be applied to humanoid robots based on the respective whole-body models. This would require special techniques to solve such big optimal control problems for humanoid models online, e.g., as suggested in Clever et al. (2017).

Well-known robotics approaches to recover from pushes include the reactive stepping method by Morisawa et al. (2009) to generate the position of the CoM and the ZMP as well as the placement of the foot after a perturbation. Wieber (2006) showed how model predictive control (MPC) can be used to generate a recovery step online to counteract perturbations. Stephens and Atkeson (2010) used MPC for step recovery control with force-controlled joints. A hierarchical push recovery strategy using reinforcement learning was proposed (Yi et al., 2011). A momentum-based reactive stepping controller was developed in Yun and Goswami (2011). A new optimization-based approach for push recovery in case of multiple noncoplanar contacts was introduced in Mansour et al. (2011).

The most impressive experiments on push recovery for humanoid robots while walking have been achieved by Boston Dynamics with their robots Petman (Nelson et al., 2012) and Atlas. Both robots are powered by hydraulics, are extremely strong and can counteract very large perturbations during walking applied by external pushes or strong perturbations of the environment. This has been demonstrated in various YouTube videos as well as during the DARPA challenge. Unfortunately, for confidentiality reasons, the rare publications on these robots only give sparse technical information about the control strategy. Some work by other authors on making the Atlas walk on rough terrain has been published e.g. in Feng et al. (2013).

4.1.6 DISCUSSION & OUTLOOK

In this chapter, we have discussed different possibilities to study stability, robustness, and the ability to recover from large perturbations that can be used to study gaits in robotics and biomechanics.

Even though this is a crucial topic for robot locomotion as well as for physiological and pathological human locomotion, no uniquely accepted and generally applicable criteria for stability and robustness exist. As we have shown, many approaches have been developed and many criteria have been formulated, but none of them can so far fully explain the stability of truly dynamic human locomotion. As a consequence, there is also no control approach yet that can make a humanoid robot walk like a human, or control a prosthesis or orthosis in the way a human would control the respective limb. Some of the criteria and the corresponding control concepts are too conservative to result in truly humanlike movement and other criteria require model information that does not exist, yet. The development of good and reliable stability and robustness measures for fast dynamic locomotion will be an important research topic for the next years.

Once such general criteria for stability and robustness are established, they also have to be included in the motion generation and motion optimization process, typically combining them with the other motion performance related criteria.

For better stability and robustness, it will also be important to better link the design and control process of robots and of technical devices. Inspired by the work in passive dynamic walking robots, the mechanics and inherent stability of typical motions to be executed should already be taken into account in the design phase. Self-stabilizing mechanical elements might also be used on humanoid robots. With the shift to more compliance in robots, also the self-stabilizing properties of springs could be exploited. Such efforts could be supported by simple parameter studies, but also by extensive model-based simulations and optimization to evaluate all choices.

Capture point approaches have been used with success to solve different push recovery tasks in robotics. It has to be investigated in the future how powerful and generalizable the capturability concept is and in which situations the discussed whole-body approaches might be useful for push recovery.

One of the topics which has not been discussed in this chapter is the role of variability in the context of walking stability and robustness. In human movement, there always is some variability from step to step, and the assumption of a perfect limit cycle as it was used for some of the criteria does, of course, not hold precisely. However, for healthy adult gait, steady-state walking is typically very close to a limit cycle. There have been many investigations trying to relate the variability of a walking motion but so far no uniform picture appeared. While in elderly people there is a high variability and also a higher risk of falling, there are many children who also walk in a variable way, yet are very stable at the same time. It also should be noted that in general one tries to link variability to the general walking performance and the global risk of falling, and not to the imminent risk of falling. So it seems that variability is not useful as a basis for controller decisions.

REFERENCES

Althoff, M., Krogh, B.H., 2014. Reachability analysis of nonlinear differential-algebraic systems. IEEE Trans. Autom. Control 59 (2), 371–383.

Berns, K., Piekenbrock, St., Dillmann, R., 1994. Learning control of a six-legged walking machine. In: ASME Press Series.

Blickhan, R., Seyfarth, A., Geyer, H., Grimmer, S., Wagner, H., Günther, M., 2007. Intelligence by mechanics. Philos. Transact. A Math. Phys. Eng. Sci. 365 (1850).

Bruijn, S.M., Meijer, O.G., Beek, P.J., van Dieën, J.H., 2010. The effects of arm swing on human gait stability. J. Exp. Biol. 213 (23), 3945–3952.

Chiu, J., Goswami, A., 2014. Design of a wearable scissored-pair control moment gyroscope (SP-CMG) for human balance assist. In: Proceedings of the 38th ASME Mechanisms and Robotics Conference.

Clever, D., Harant, M., Mombaur, K., Naveau, M., Stasse, O., Endres, D., 2017. COCoMoPL: a novel approach for humanoid walking generation combining optimal control, movement primitives and learning and its transfer to the real robot HRP-2. IEEE Robot. Autom. Lett. 2 (2), 977–984.

Coleman, M.J., 1998. A Stability Study of a Three-Dimensional Passive-Dynamic Model of Human Gait. PhD thesis. Cornell University.

Coleman, M.J., Ruina, A., 1998. An uncontrolled walking toy that cannot stand still. Phys. Rev. Lett. 80 (16), 3658–3661.

Collins, S.H., Ruina, A.L., Tedrake, R., Wisse, M., 2005. Efficient bipedal robots based on passive-dynamic walkers. Science 307, 1082–1085.

Cronin, J., 1994. Differential Equations – Introduction and Qualitative Theory. Marcel Dekker.

Engelhart, Denise, van Asseldonk, Edwin, Grin, Lianne, van der Kooij, Herman, 2012. Lateral balance control during walking: foot placement and stability in response to external perturbations. In: Proceedings of ISPGR. Trondheim, Norway.

Felis, M.L., 2015. Modeling Emotional Aspects of Human Locomotion. PhD thesis. University of Heidelberg.

Felis, M.L., Mombaur, K., 2016. Synthesis of full-body 3-D human gait using optimal control methods. In: Proceedings of IEEE-RAS International Conference on Robotics and Automation.

Feng, S., Xinjilefu, X., Huang, W., Atkeson, C.G., 2013. 3D walking based on online optimization. In: 2013 13th IEEE-RAS International Conference on Humanoid Robots (Humanoids). IEEE, pp. 21–27.

Ferber, Reed, Osternig, Louis R., Woollacott, Marjorie H., Wasielewski, Noah J., Lee, Ji-hang, 2002. Reactive balance adjustments to unexpected perturbations during human walking. Gait Posture 16, 238–248.

Fritschi, M., Jelinek, H.F., McGloughlin, T., Khalaf, K., Khandoker, A.H., Vallery, H., 2014. Human balance responses to perturbations in the horizontal plane. In: Engineering in Medicine and Biology Society (EMBC), 2014 36th Annual International Conference of the IEEE, pp. 4058–4061.

Garcia, M., Chatterjee, A., Ruina, A., 1998. Speed, efficiency, and stability of small-slope 2-D passive dynamic walking. In: Proceedings of IEEE International Conference on Robotics and Automation. Leuven, Belgium.

Goswami, A., Espiau, B., Keramane, A., 1996. Limit cycles and their stability and passive bipedal gaits. In: Proceedings of IEEE International Conference on Robotics and Automation, pp. 246–251.

Goswami, A., Kallem, V., 2004. Rate of change of angular momentum and balance maintenance of biped robots. In: International Conference on Robotics and Automation, vol. 4, pp. 3785–3790.

Herr, H., Popovic, M., 2008. Angular momentum in human walking. J. Exp. Biol. 211, 467–481.

Hobbelen, D., Wisse, M., 2007. A disturbance rejection measure for limit cycle walkers: the gait sensitivity norm. Trans. Robot. 23 (6), 1213–1224.

Hof, At L., 2008. The 'extrapolated center of mass' concept suggests a simple control of balance in walking. Hum. Mov. Sci. 27, 112–125.

Hof, A.L., Gazendam, M.G.J., Sinke, W.E., 2005. The condition for dynamic stability. J. Biomech. 38, 1–8.

Horak, Fay B., 1987. Clinical measurement of postural control in adults. Phys. Ther. 67 (12), 1881–1885.

Hsu, J.C., Meyer, A.U., 1968. Modern Control Principles and Applications. McGraw-Hill.

Hurmuzlu, Y., 1993. Dynamics of bipedal gait: part II – stability analysis of a planar five-link biped. J. Appl. Mech. 60, 337–343.

Hurmuzlu, Y., Basdogan, C., 1994. On the measurement of dynamic stability of human locomotion. Trans. ASME J. Biomech. Eng. 116, 30.

Kajita, S., Kanehiro, F., Kaneko, K., Fujiwara, K., Harada, K., Yokoi, K., Hirukawa, H., 2003. Biped walking pattern generation by using preview control of zero-moment point. In: IEEE International Conference on Robotics and Automation, Proceedings, ICRA'03, vol. 2. IEEE, pp. 1620–1626.

Kaul, L., Mandery, C., Busch, D., Asfour, T., 2015. Experimental analysis of human push recovery by stepping. In: Proceedings of IEEE-RAS International Conference on Humanoid Robots.

Koch, K.H., Mombaur, K., Souères, P., 2012. Optimization-based walking generation for humanoids. In: IFAC-SYROCO 2012.

Koolen, T., de Boer, T., Rebula, J., Goswami, A., Pratt, J., 2012. Capturability-based analysis and control of legged locomotion, part 1: theory and application to three simple gait models. Int. J. Robot. Res. 31 (9), 1094–1113.

KoroiBot Motion Capture Database, 2016. https://koroibot-motion-database.humanoids.kit.edu/. Last visited January 2017.

Koschorreck, J., Mombaur, K., 2011. Modeling and optimal control of human platform diving with somersaults and twists. Optim. Eng. 12 (4), 29–56.

Kuindersma, S., Permenter, F., Tedrake, R., 2014. An efficiently solvable quadratic program for stabilizing dynamic locomotion. In: Robotics and Automation (ICRA), IEEE International Conference on.

Kuo, A.D., Zajac, F.E., 1993. Human standing posture: multi-joint movement strategies based on biomechanical constraints. Prog. Brain Res. 97, 349–358.

Lemus, D., van Frankenhuyzen, J., Vallery, H., 2017. Design and evaluation of a balance assistance control moment gyroscope. ASME J. Mech. Robot. [Online before print].

Li, D., Vallery, H., 2012. Gyroscopic assistance for human balance. In: Proceedings of the 12th International Workshop on Advanced Motion Control (AMC). Sarajevo, Bosnia and Herzegowina.

Lugade, V., Lin, V., Chou, L-S., 2011. Center of mass and base of support interaction during gait. Gait Posture 33, 406–411.

Majumdar, Anirudha, Ahmadi, Amir Ali, Tedrake, Russ, 2013. Control design along trajectories with sums of squares programming. In: International Conference on Robotics and Automation (ICRA), 2013.

Mansour, D., Micaelli, A., Escande, A., Lemerle, P., 2011. A new optimization based approach for push recovery in case of multiple noncoplanar contacts. In: 2011 11th IEEE-RAS International Conference on Humanoid Robots, pp. 331–338.

Maus, H.M., Lipfert, S.W., Gross, M., Rummel, J., Seyfarth, A., 2010. Upright human gait did not provide a major mechanical challenge for our ancestors. Nat. Commun. 1 (6).

McAndrew Young, P.M., Wilken, J.M., Dingwell, J.B., 2012. Dynamic margins of stability during human walking in destabilizing environments. J. Biomech. 46, 1053–1059.

McGeer, T., 1991. Passive dynamic biped catalogue. In: Chatila, R., Hirzinger, G. (Eds.), Proceedings of the 2nd International Symposium of Experimental Robotics, Toulouse. Springer-Verlag, New York.

McGeer, T., 1992. Principles of walking and running. In: Advances in Comparative and Environmental Physiology. Springer-Verlag, Berlin.

McGeer, T., 1990. Passive dynamic walking. Int. J. Robot. Res. 9 (2), 62–82.

Mergner, T., 2007. Modelling sensorimotor control of human upright stance. Prog. Brain Res. 165, 283–297.

Mergner, T., Maurer, C., Peterka, R.J., 2003. A multisensory posture control model of human upright stance. Prog. Brain Res. 142, 189–201.

Mombaur, K.D., 2001. Stability Optimization of Open-loop Controlled Walking Robots. PhD thesis. University of Heidelberg. www.ub.uni-heidelberg.de/archiv/1796. VDI-Fortschrittbericht, Reihe 8, No. 922, ISBN 3-18-392208-8.

Mombaur, K.D., 2006. Performing open-loop stable flip-flops – an example for stability optimization and robustness analysis of fast periodic motions. In: Fast Motions in Robotics and Biomechanics – Optimization and Feedback Control. In: Lecture Notes in Control and Information Science. Springer.

Mombaur, K.D., 2009. Using optimization to create self-stable human-like running. Robotica 27, 321–330. Published online June 2008.

Mombaur, K., 2013–2016. The KoroiBot project. http://www.koroibot.eu.

Mombaur, K.D., Bock, H.G., Schlöder, J.P., Longman, R.W., 2005a. Open-loop stable solution of periodic optimal control problems in robotics. ZAMM - J. Appl. Math. Mech./Z. Angew. Math. Mech. 85 (7), 499–515.

Mombaur, K.D., Longman, R.W., Bock, H.G., Schlöder, J.P., 2005b. Open-loop stable running. Robotica 23 (01), 21–33.

Monaco, V., Tropea, P., Aprigliano, F., Martelli, D., Parri, A., Cortese, M., Molino-Lova, R., Vitiello, N., Micera, S., 2017. An ecologically-controlled exoskeleton can improve balance recovery after slippage. Sci. Rep. 7, 46721.

Morisawa, Mitsuharu, Harada, Kensuke, Kajita, Shuuji, Kaneko, Kenji, Solà, Joan, Yoshida, Eiichi, Mansard, Nicolas, Yokoi, Kazuhito, Laumond, Jean-Paul, 2009. Reactive stepping to prevent falling for humanoids. In: 9th IEEE-RAS International Conference on Humanoid Robots, Humanoids 2009. Paris, France, December 7–10, 2009, pp. 528–534.

Mouchnino, L., Robert, G., Ruget, H., Blouin, J., Simoneau, M., 2012. Online control of anticipated postural adjustments in step initiation: evidence from behavioral and computational approaches. Gait Posture 35 (4), 616–620.

Nelson, Gabe, Saunders, Aaron, Neville, Neil, Swilling, Ben, Bondaryk, Joe, Billings, Devin, Lee, Chris, Playter, Robert, Raibert, Marc, 2012. PETMAN: a humanoid robot for testing chemical protective clothing. J. Robotics Soc. Jpn. 30 (4), 372–377.

Paiman, C., Lemus, D., Short, D., Vallery, H., 2016. Observing the state of balance with a single upper-body sensor. Front. Robot. AI 3.

Papachristodoulou, A., Prajna, S., 2002. On the construction of Lyapunov functions using the sum of squares decomposition. In: Proceedings of IEEE Conference on Decision and Control.

Pidcoe, P.E., Rogers, M.W., 1998. A closed-loop stepper motor waist-pull system for inducing protective stepping in humans. J. Biomech. 31 (4), 377–381.

Pijnappels, Mirjam, Bobbert, Maarten F., van Dieën, Jaap H., 2005. How early reactions in the support limb contribute to balance recovery after tripping. J. Biomech. 38 (3), 627–634.

Popovic, M., Englehart, A., Herr, H., 2004. Angular momentum primitives for human walking: biomechanics and control. In: Proceedings of the IEEE/RSJ International Conference on Intelligent Robots and Systems.

Posa, Michael, Tobenkin, Mark, Tedrake, Russ, 2013. Lyapunov analysis of rigid body systems with impacts and friction via sums-of-squares. In: Proceedings of the 16th International Conference on Hybrid Systems: Computation and Control. ACM, pp. 63–72.

Pratt, J., Carff, J., Drakunov, S., Goswami, A., 2006. Capture point: a step toward humanoid push recovery. In: Proceedings of IEEE-RAS International Conference on Humanoid Robots.

Rietdyk, S., Patla, A.E., Winter, D.A., Ishac, M.G., Little, C.E., 1999. Balance recovery from mediolateral perturbations of the upper body during standing. J. Biomech. 32, 1149–1158.

Sakagami, Y., Watanabe, R., Aoyama, C., Matsunaga, S., Higaki, N., Fujimura, K., 2002. The intelligent ASIMO: system overview and integration. In: Intelligent Robots and System, 2002. IEEE/RSJ International Conference on, pp. 2478–2483.

Sardain, Philippe, Bessonnet, Guy, 2004. Forces acting on a biped robot. Center of pressure-zero moment point. IEEE Trans. Syst. Man Cybern. 34, 630–637.

Schemschat, R.M., Clever, D., Felis, M.L., Chiovetto, E., Giese, M., Mombaur, K., 2016a. Joint torque analysis of push recovery motions during human walking. In: Proceedings of 6th IEEE

RAS/EMBS International Conference on Biomedical Robotics and Biomechatronics (BioRob). UTown, Singapore.

Schemschat, R.M., Clever, D., Mombaur, K., 2016b. Optimal push recovery for periodic walking motions. In: 6th IFAC Workshop on Periodic Control Systems PSYCO, vol. 49. Elsevier, Einhoven, Netherlands, pp. 93–98.

Schwab, A., Wisse, M., 2001. Basin of attraction of the simplest walking model. In: Design Engineering Technical Conferences.

Sharbafi, Maziar A., Seyfarth, Andre, 2015. FMCH: a new model to explain postural control in human walking. In: Proceedings of the 2015 IEEE/RSJ International Conference on Intelligent Robots and Systems (IROS).

Stephens, B., Atkeson, C., 2010. Push recovery by stepping for humanoid robots with force controlled joints. In: Proceedings of IEEE-RAS International Conference on Humanoid Robotics.

Stoer, J., Bulirsch, R., 1990. Numerische Mathematik 2. Springer.

Tang, P.-F., Woollacott, Majorie H., Chong, Raymond K.Y., 1998. Control of reactive balance adjustments in perturbed human walking: roles of proximal and distal postural muscle activity. Exp. Brain Res. 119 (2), 141–152.

Townsend, M.A., 1985. Biped gait stabilization via foot placement. Biomechanics 18 (1), 21–38.

Trefethen, L.N., Embree, M., 2005. Spectra and Pseudospectra. Princeton University Press.

Vallery, H., Bögel, A., O'Brien, C., Riener, R., 2012. Cooperative control design for robot-assisted balance during gait. Automatisierungstechnik 60, 715–720.

Vukobratović, Miomir, Borovac, Branislav, 2004. Zero-moment point — thirty five years of its life. Int. J. Humanoid Robot. 1 (01), 157–173.

Wang, Hongfei, Zheng, Yuan F., Jun, Youngbum, Oh, Paul Y., 2014. DRC-hubo walking on rough terrains. In: 2014 IEEE International Conference on Technologies for Practical Robot Applications, TePRA 2014. Woburn, MA, USA, April 14–15, 2014, pp. 1–6.

Wieber, P.B., 2002. On the stability of walking systems. In: International Workshop on Humanoid and Human Friendly Robotics.

Wieber, Pierre-Brice, 2006. Trajectory free linear model predictive control for stable walking in the presence of strong perturbations. In: 2006 6th IEEE-RAS International Conference on Humanoid Robots. Genova, Italy, December 4–6, 2006, pp. 137–142.

Wisse, M., Schwab, A.L., van der Linde, R.Q., van der Helm, F.C.T., 2005. How to keep from falling forward: elementary swing leg action for passive dynamic walkers. Robotics 21 (3), 393–401.

Yi, Seung-Joon, Zhang, Byoung-Tak, Hong, D., Lee, D.D., 2011. Learning full body push recovery control for small humanoid robots. In: 2011 IEEE International Conference on Robotics and Automation, pp. 2047–2052.

Yun, S.k., Goswami, A., 2011. Momentum-based reactive stepping controller on level and non-level ground for humanoid robot push recovery. In: 2011 IEEE/RSJ International Conference on Intelligent Robots and Systems, pp. 3943–3950.

Chapter 4.2

Optimization as Guiding Principle of Locomotion

Katja Mombaur

Optimization, Robotics & Biomechanics, ZITI, IWR, Heidelberg University, Heidelberg, Germany

Abstract. *It is a common assumption that optimization is a guiding principle of human and animal movements, also including locomotion in many situations. Reasons for this can be found in the evolution of the human body and its gait in general, as well as in the lifelong learning and training process of every individual. In this subchapter, we discuss how this principle can be exploited to use optimal control approaches for the analysis and generation of walking and running motions of humans and robots. We give different examples for dynamic multibody system models of the locomotor systems which represent an important basis for these studies. We present the formulation and numerical solution of optimal control problems for locomotion generation for selected objective functions. Results of optimal control problems for human and robot locomotion are presented. In addition, we are interested in the inverse problem, i.e., the identification of the optimality criterion underlying a recorded motion, which is called the inverse optimal control problem. The formulation and numerical solution of these problems for locomotion examples are discussed, and example results of inverse optimal control for human locomotion based on whole-body models are shown. We also discuss some current research questions related to the optimization of locomotion.*

4.2.1 INTRODUCTION

Nature optimizes in many ways. A fundamental reason behind this can be found in the principles of evolution based on selection and survival of the fittest (Darwin, 1859; Rosen, 1967). Popular examples for optimality in nature are optimal or close to optimal structures such as honeycombs (minimizing material for mesh generation in given volume), optimal materials or material surfaces such as shark skin (minimizing drag (Ng and Luo, 2016)) which are studied in the field of bionics to develop technical devices and materials copying biology. Also many behaviors of humans and animals are optimal or nearly optimal. This also includes many types of motions which are assumed to be optimal due to epigenesis and phylogenesis with the optimization criterion being adapted to the particular situation. Optimization effects can be found in the mechanical properties of the locomotion systems (Alexander, 1984, 1996), but also in the closed-loop sensory motor system (Todorov, 2004). In this article we want to discuss how optimization serves as a guiding principle of biological locomotion with a special focus on bipedal locomotion of humans and humanoids or other bipedal robots. We also discuss how – as a logical consequence – mathematical optimization and optimal control techniques can be used in the study of bipedal locomotion.

Humans and other anthropomorphic systems have many degrees of freedom (DOF) and are highly redundant, i.e., a given motion task, e.g., performing a step of a given length, can be executed in an infinite number of ways, involving different combinations of motions of the individual joints. Redundancy is an advantage if one way of performing a movement is not possible due to some external constraint and another one can be chosen. But it is complex to explain which of the many redundant ways is preferred by the human body or should be selected for a robot. How is the natural way of performing a motion distinguished from unnatural ones? What is the most efficient, the fastest, the smoothest way a movement can be executed? These are only some of the questions to which optimization can give an answer.

In addition, anthropomorphic and other biological or bio-inspired motions are underactuated in the general case. This means that not all DOF are directly actuated – usually only the internal DOF (or most of them), but not the overall position and orientation of the body in space. This one only results indirectly from the combination of the internal joint's actions and the interaction of the body with the environment (in the case of locomotion essentially the contact of the feet with the floor). The question arising in this context is how the actuated DOF must be controlled in order to make the entire system move in a desired way. This is also a problem that can be answered by optimal control. An optimal control problem in mathematics is an optimization problem for a dynamic process, i.e., a process described by a differential equation. The term control here does not necessarily refer to feedback control; the process dynamics can describe either an open-loop or a closed-loop system.

Optimal control approaches have been used by researchers from different fields to study locomotion. Human movement studies to generate optimal walking and running have been performed in Ackermann and van den Bogert (2010), Schultz and Mombaur (2010), Felis and Mombaur (2016), Geyer and Herr (2010) with a focus on biomechanics. In computer animation, optimization based approaches serve the goal to generate realistically looking motions for human shapes and fantasy anthropomorphic characters (Geijtenbeek et al., 2013; Wang et al., 2012; Sok et al., 2007; Tassa et al., 2012) where not all physical constraints have to be satisfied. More precision is required in robotics, where optimal control is also used to generate walking motions for specific humanoid robots (Buschmann et al., 2007; Atkeson and Liu, 2013; Lengagne et al., 2011; Miossec et al., 2006; Suleiman et al., 2007; Koch et al., 2012a, 2012b, 2014). In all these cases, the dynamic model description as well as the objective function formulation have been established a priori and the problems have been solved by means of forward optimal control.

However, the inverse problem is also very interesting in biomechanics and all fields that are concerned with understanding human or other biological behavior: Given a specific human movement for which motion capture data has been recorded and a defined model used for its description, which is the underlying objective function that gives rise to this movement? This is called an inverse optimal control problem. For some types of motions, e.g., in sports, this question is easy to answer since the objective function corresponds to the voluntary goal of the human subjects, like maximizing running speed or jump height. However, for many other types of motion, this optimization takes place unconsciously and is guided to some extent by the mechanics of the body, and here the question is far from trivial. Inverse optimal control problems are challenging since they require the solution of an identification problem inside an optimal control problem. They are related to some of the problems solved in reinforcement learning, since the inverse optimal control essentially learns the objective function.

This chapter is organized as follows. Since whole-body models describing the forward and inverse dynamics of locomotion form an important basis for this research on optimal control, these models are described in Section 4.2.2, highlighting the hybrid nature of locomotion with multiple phases and discontinuous impacts. In Section 4.2.3, we describe the formulation of bipedal gait generation as an optimal control problem and its numerical solution. Section 4.2.4 summarizes some results for optimal control of walking and running for robot and human models. In Section 4.2.5, we present the inverse optimal control problem looking at mathematical formulation and solution approaches. Section 4.2.6 discusses special considerations one has to take when formulating a inverse optimal control problems for specific motion problems, as well some applications to human locomotion. In Section 4.2.7, we give a final discussion and outlook for future research.

4.2.2 FORWARD AND INVERSE DYNAMICS MODELS OF LOCOMOTION

In this section, we present different dynamic whole-body models of locomotion. As all models, the whole-body models discussed in this sections are obviously simplifications of the real world. However, the goal is to establish models that are as precise as possible while still allowing for acceptable computation times.

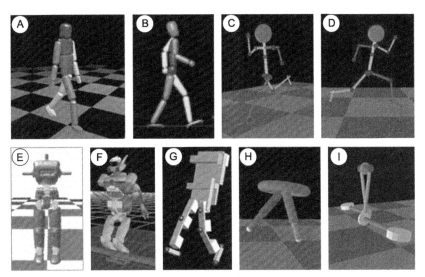

FIGURE 4.2.1 Whole-body dynamic models of human and robot locomotion which include realistic description of the dynamic behavior of the systems: models of human walking in (A) 3D and (B) 2D (Felis et al., 2015; Felis, 2015) and of human running in (C) 3D and (D) 2D (Schultz and Mombaur, 2010), as well as models of walking humanoid robots, such as HeiCub, a variant of iCub (E), (F) HRP-2 (Koch, 2015), and (G) Leo (Schuitema et al., 2010), and models of (H) a two-legged open-loop stable simulated robot (Mombaur et al., 2005c) and (I) of the Tinkertoy robot (Mombaur, 2001; Coleman et al., 2001).

In contrast to template models, discussed in Subchapter 4.5 of this book, whole-body dynamic models aim to capture the dynamics of the essential segments and DOF of the system, i.e., the human or the robot, that are relevant for the motion considered. In the case of whole-body models of humans, e.g., this does not mean that all bony segments and all of the more than 200 DOF have to be considered, or that each bone has to be modeled separately. If walking motions are studied then it is usually not required to model the DOF of the fingers or all motion possibilities of the spine. For human locomotion, whole body models of around 12–14 segments with 35–40 DOF in 3D or 9–15 DOF in the sagittal plane seem suitable.

Fig. 4.2.1 shows different examples of whole-body dynamic models for systems of different origin and complexity. In the first row, whole-body models for running and walking in 3D and 2D are shown. The second row contains pictures of dynamic models of current humanoid robots, as well as models of other bipedal robots.

One important feature of bipedal walking and running motions is that they consist of several phases of motions, each of which is characterized by different types of contacts of the human/robot with the environment. Generally speak-

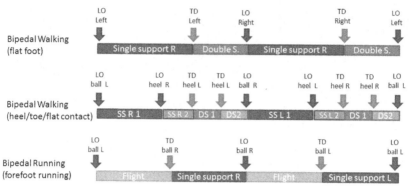

FIGURE 4.2.2 Phase orders for biological and robotics walking and running motions (TD = touchdown, LO = liftoff).

ing, running motions consist of alternating single foot contact phases and flight phases, and walking consists of alternating single and double support phases. Depending on in how much detail foot contact is described, we may also identify subphases of a foot contact considering if the heel, toes, or entire sole, etc., are in contact with the ground. Fig. 4.2.2 shows the phase orders for human and robotic walking and human running. For four-legged systems, the number of possible combinations of foot contacts, and therefore the number of different phases, is even higher than for bipeds. For humans and humanoids, there may be additional contacts between the hands and the environment, e.g., when climbing stairs using a handrail. In any case, the different contacts change the dynamics of the system, such that each phase is described by its own set of differential equations. In addition, there may be discontinuities between phases when a contact occurs instantaneously by fully inelastic impact, and velocities are subject to quite instantaneous changes. We have discovered in previous research that modeling an impact as fully inelastic gives a quite good approximation of reality if humans and robots walk on normal ground and with standard feet and soles. Only for very soft terrains like mattresses or trampolines or when walking on very flexible soles a compliant modeling of the ground gets more realistic. The dynamic models of these multibody systems have to correctly describe these hybrid dynamics properties. In the following paragraph, we will describe the general form of such models.

For the description of walking systems, different choices of coordinates are possible. A typical choice of coordinates for a biological walking system is to use the position and orientation of the base body (often the pelvis), and additionally all internal DOF at the joints. To describe the different phases, the same or different sets of coordinates can be used. We usually stick to the same set of coordinates for all phases.

If minimal coordinates q are chosen for a phase, i.e., the number of coordinates is equal to the number of DOF of the system in this phase (which would apply to the flight phase in running for the coordinate choice described above), the motion is described by a set of ordinary differential equations of the following form:

$$M(q, p)\ddot{q} + N(q, \dot{q}, p)\dot{q} = F(q, \dot{q}, p, \mathcal{M}), \qquad (4.2.1)$$

Here M is the mass or inertia matrix, N the vector of nonlinear effects, and F the vector of all external forces (including gravity, joint torques $\mathcal{M}$, drag, etc.). F may also include the action of passive elements, e.g., of spring–damper elements which may be linear (with diagonal stiffness and damping matrices K and D and rest length vector q_0),

$$F_{kd} = K(q - q_0) - D\dot{q}, \qquad (4.2.2)$$

or generally nonlinear.

For a redundant choice of coordinates q, i.e., when the dimension of q is larger that the number of DOF, the coupling can be described by a constraint of the form $g(q) = 0$ and a corresponding constraint force in the differential equation. For the coordinate choice above, this applies to all phases with some foot–ground contact where different types of contacts apply to different sets of constraints. This results in a system of differential algebraic equations (DAE) of index 3 for the equations of motion and can be transformed into a DAE of index 1 by index reduction:

$$\dot{q} = v, \qquad (4.2.3)$$

$$\dot{v} = a, \qquad (4.2.4)$$

$$\begin{pmatrix} M(q, p) & G(q, p)^T \\ G(q, p) & 0 \end{pmatrix} \begin{pmatrix} a \\ \lambda \end{pmatrix} = \begin{pmatrix} -N(q, v) + F(q, v, p, \mathcal{M}) \\ \gamma(q, v, p) \end{pmatrix}, \qquad (4.2.5)$$

with acceleration $a = \ddot{q}$ and Lagrange multipliers λ. The matrix G is the Jacobian of the position constraints $G = (\partial g / \partial q)$, and γ the corresponding Hessian matrix $\gamma = -((\partial G / \partial q) \dot{q}) \dot{q}$. The Lagrange multipliers are equivalent to the negative of the contact forces resulting from the corresponding constraints. The fact that ground contact during walking and running is unilateral, i.e., the ground can only push against the foot, and not pull, can be taken into account by formulating an appropriate constraint on λ. In addition, it must be guaranteed that the original position and velocity constraints of the system are still satisfied. This is achieved by respecting the two invariant equations:

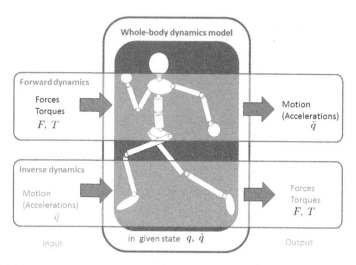

FIGURE 4.2.3   Forward and inverse dynamics use of the same whole-body model.

$$g_{pos} = g(q(t), p) = 0, \qquad (4.2.6)$$

$$g_{vel} = G(q(t), p) \cdot \dot{q}(t) = 0. \qquad (4.2.7)$$

It should be noted that these same equations can be used to solve the forward dynamics or the inverse dynamics problem as shown in Fig. 4.2.3. The task of forward dynamics is to compute the resulting acceleration of the system, given a specific vector of external forces, and at a state (i.e., position and velocity) of the system. The inverse dynamics task does the opposite: it computes, for a specific acceleration (again at a given position and velocity), the external forces that are required to produce these accelerations. The equations are the same in both cases, just the set of known and unknown variables changes. In Section 4.2.3, we will discuss how optimal control can be used if none of the two sets of variables is known a priori. For this approach, both forward and inverse dynamics formulations can be applied.

For walking and running, changes between the different motion phases described above usually do not happen at predefined time points, but depend on the states of the system. For example, touchdown with a particular foot takes place when the lowest point of this foot gets as low as the ground. Liftoff, on the other hand, occurs when the vertical contact force (and thus also the corresponding Lagrange multiplier in Eq. (4.2.5)) becomes zero. This can be described by the zero of the following so-called switching function:

$$s(q(\tau_s), v(\tau_s), p) = 0. \qquad (4.2.8)$$

The velocity discontinuities caused by inelastic impacts can be computed as

$$
\begin{pmatrix} M(q,p) & G(q,p)^T \\ G(q,p) & 0 \end{pmatrix} \begin{pmatrix} v_+ \\ \Lambda \end{pmatrix} = \begin{pmatrix} M(q)v_- \\ 0 \end{pmatrix}, \qquad (4.2.9)
$$

where v_+ are the unknown velocities after impact and v_- are the corresponding velocities immediately before impact. Matrices M and G are the same as in Eq. (4.2.5). Note that for the choice of compliant impacts there would be no discontinuities in the velocities but only in the right-hand side of the system since the floor contact would be modeled by an external spring–damper force. There would also be no loss in DOF due to the contact, as it occurs in the fully inelastic contact case.

For modeling a specific walking system, the above equations of motions, impact equations and switching functions have to be established. For very simple systems, such as (H) and (I) in Fig. 4.2.1 as well as for template models, it is still possible to set up these equations by hand, for systems like (G) this gets already very challenging, and for systems (A)–(F) it is practically impossible. Instead, dynamic modeling tools can be used to set up the model of the multibody system. It is not necessary to provide symbolic equations of motion, but functions that either establish the different parts of the equations of motions, e.g., mass matrix M, constraint Jacobian G, etc., or functions that directly give an answer to the forward dynamics or the inverse dynamics problem for a given input.

The automatic model generator of HuMAnS by Wieber et al. (2006) uses Maple to generate explicit code to compute the entries of matrices M and G, vector N and γ. In our group, we have developed two different modeling tools which both are, in principle, suited to model general multibody systems, but in our case mainly serve the primary purpose of generating whole-body human and humanoid models. The Rigid Body Dynamics Library (RBDL) (Felis et al., 2015, 2016) is based on an order n recursive algorithm by Featherstone (2007). The other tool, Dynamod, Koch (2015) is based on explicit code generation in a similar way as HuMAnS. The examples shown in Fig. 4.2.1 (A), (B) and (E) were generated using RBDL; (C) and (D) using RBDL; and (F) using DynaMod. On top of the RBDL code, we also established the human metamodel HeiMan 3D (Felis et al., 2015; Felis, 2015) which has been used, e.g., to set up the 3D human walking models (A) and (B) in Fig. 4.2.1.

Once the specific equations of motion for a given system have been established, another important step is the determination of the correct model parameters for geometry and inertia. For human models, one possibility is the use of tabular anthropometric data, e.g., from de Leva (1996) or Winter (2004), which only uses the subject's total height and mass and then uses regression formula to compute the specific segment data. In Ho Hoang and Mombaur (2015), we have proposed an adjustment of this data to take into account specific characteristics of the elderly generation. If a precise match between dynamic model and real human data or good prediction quality is to be achieved, the parameters have to be adjusted to the individual properties of the subject in a better way. In Felis et al. (2015) we have described an approach to establish subject specific dynamic human models based on kinematic measurements. A more precise and sophisticated approach to generate subject-specific geometric and inertial data based on MRI images has also been developed in our group by Sreenivasa et al. (2016). For robot models, the geometric and inertial parameters of the segments are usually known from the robot design process. In addition, there may by dynamic parameters linked to friction and damping in the robot as well as compliance in the joints (see above) or in external contacts.

The equations above, which describe the dynamics of the multibody system, i.e., the relationship between motion and forces/torques, can be extended by additional models describing the generation of forces/torques or the different feedback loops of the walking system. With respect to the first, muscle models could be integrated in the case of human walking models describing the link between muscle activation or excitation and muscle force. In this case, each torque would have to be replaced by the action of at least two muscles – agonist and antagonist – in many cases more, so the number of controls of the system would be at least doubled. The complexity of the evaluation of the right-hand side of the multibody system equations would be significantly increased. If also activation dynamics are included for each muscle, the number of state variables would grow by the number of controls. In the case of robot walking, models of the motors or other actuators could be considered, which would also add another differential equation per motor. However, this is not the subject of this chapter and is instead discussed in other chapters of this book.

Legged locomotion is often a periodic or quasiperiodic form of motion. It is therefore often desirable to formulate periodicity constraints to the model on all velocity variables v and a reduced set of position variables q_{red}, only eliminating the coordinate describing the person's or the robot's direction of motion. Usually, these gaits are also symmetric with identical left and right steps, such that the periodicity constraints are not applied over the full cycle of two steps (i.e., a full stride), but over one step, including the formulation of a shift of sides in the model. In the examples for locomotion presented in Section 4.2.4, we will also look at more unusual kind of locomotion including somersaulting and flip-flopping which includes rotations about the medio-lateral axis (which will be taken into account in the formulation of the periodicity constraints by applying an appropriate delta to the respective angle).

4.2.3 FORMULATING LEGGED LOCOMOTION AS OPTIMAL CONTROL PROBLEM

The generation of legged locomotion can be formulated as an optimization problem based on the dynamic models of the motions described in the previous section. There are two reasons for this, the first one caused by biological inspiration, the second one by considerations of practical solvability of the problem.

1. As described in the introduction, optimality is a guiding principle of nature, and movements which are performed frequently or trained in the context of sports tend to be optimal with respect to some optimality criterion. It is therefore logical to use mathematical optimization in the study of movement in order to mimic this biological optimization process. Criteria to be optimized in the motion generation process can be biologically inspired in the case of humans, and biologically but also technically motivated in the case of robots.

2. From a practical perspective, optimization helps to solve the feasibility issue and the redundancy issue of walking at the same. time. As outlined above, if locomotion (or any other type of motion) is to be generated by means of simulation, either all torque/force histories would have to be fully known for the forward dynamics problem or all acceleration histories would have to be known for the inverse dynamics problem. However, in practice, often none of them is known in advance, and due to the complexity of the problem pure trial and error usually leads to infeasible motions. Optimization can help solve this feasibility problem by computing motions and joint torques, etc., that satisfy all the different equality and inequality constraints imposed on the system and the motion task, i.e., result in a feasible motion. In addition, optimization solves the redundancy problem, i.e., it determines out of the many – sometimes infinite number of – ways to perform a given motion task, the one that is optimal with respect to the chosen optimization criterion.

In mathematical terms, the task of solving an optimization problem for a dynamic process model of the form described in Section 4.2.2 is called an optimal control problem. This is not a control problem in the sense of feedback control, but the task is to simultaneously optimize trajectories and inputs (open-loop control variables) which are all unknown functions in time. To be more precise, due to the hybrid dynamics nature of the locomotion models discussed above, we are facing a multiphase optimal control problem with discontinuities. We assume here that we know the order of phases in the case locomotion (which is usually true for a given gait), but that the durations of all these phases are unknown. This can be formulated as follows:

$$\min_{x(\cdot),u(\cdot),p,\tau} \sum_{j=1}^{n_{ph}} \left(\int_{\tau_{j-1}}^{\tau_j} \phi_j(x(t),u(t),p)\,dt + \Phi_j(\tau_j,x(\tau_j),p) \right) \quad (4.2.10)$$

s. t. $\dot{x}(t) = f_j(t,x(t),u(t),p)$ for $t \in [\tau_{j-1},\tau_j]$,

$$j = 1,\ldots,n_{ph},\ \tau_0 = 0,\ \tau_{n_{ph}} = T, \quad (4.2.11)$$

$$x(\tau_j^+) = x(\tau_j^-) + J(\tau_j^-,x(\tau_j^-),p) \quad \text{for} \quad j = 1,\ldots,n_{ph},$$

$$(4.2.12)$$

$$g_j(t,x(t),u(t),p) \geq 0 \qquad \text{for} \quad t \in [\tau_{j-1},\tau_j], \quad (4.2.13)$$

$$r_{eq}(x(0),\ldots,x(T),p) = 0, \quad (4.2.14)$$

$$r_{ineq}(x(0),\ldots,x(T),p) \geq 0. \quad (4.2.15)$$

In these equations, $x(t)$ is the vector of state variables (including positions q and velocities v of Section 4.2.2), and $u(t)$ is the vector of control variables of the system (which in the cases discussed here are the joint torques). Note that we are using the dynamic equations in the forward sense with torques as input, but as outlined above this does not mean that torques have to be known, since controls and states are determined simultaneously by the optimal control problem. It would also be possible to formulate the optimal control problem based on inverse dynamics, but in this case another choice of control variables would be made (e.g., equivalent to accelerations $u = \ddot{q}$). p is the vector of free model parameters. τ is the vector of phase switching times, with the overall time of the motion being $T = \tau_{n_{ph}}$.

Eqs. (4.2.11) and (4.2.12) are place-holders for the hybrid system dynamics with continuous and discrete motion phases of Section 4.2.2. Here, we use ODEs for simplicity of presentation; but as shown above, we usually face DAE models for most or part of the phases. With respect to constraints, special care must be taken to formulate all relevant constraints, but not more than necessary since it is favorable to use all available freedom in the system to truly optimize the motion. Eq. (4.2.13) describes all continuous inequality constraints, such as simple lower and upper bounds on all variables as well as more complex relations between several variables. In addition, there are coupled and decoupled pointwise equality (4.2.14) and inequality constraints (4.2.15), e.g., start and end point constraints on the states, phase switching conditions, or periodicity constraints.

For the formulation of the objective function (4.2.10), we can distinguish two different types of terms: Lagrange-type objective functions (first term) take the form of an integral, while Mayer-type objective functions Φ_j (second term) only depend on values at the end of the respective phase. Typical Mayer-type objective functions are minimum phase times, minimum total maneuver time, or maximum distance traveled. Typical examples of Lagrange-type objective functions include different types of energy or effort minimization, e.g., minimization of mechanical energy or minimization of weighted torques squared, minimization of muscle excitations or activations (to different powers), efficiency maximization, etc. Also some of the stability criteria (compare Subchapter 4.1 such as angular momentum minimization or the zero moment point (ZMP) and capture point common in robotics can be described by Lagrange-type functions.

Other criteria related to stability may get much more complex. For example, if stability is formulated using Lyapunov's first method for nonlinear periodic systems, one has to compute the spectral radius ρ of the monodromy $X(T)$ for which $\rho := |\lambda_i(X(T))|_{max} < 1$ must hold. The monodromy matrix is the Jacobian of the Poincaré map which maps the states from one cycle to the next. No matter if this criterion is used in the objective function (to make the eigenvalue as small as possible) or in a constraint (to keep it strictly below one), this requires that sensitivity information of the trajectory is computed in the objective function or the constraint, respectively, i.e., in a place where usually only trajectory information is required. This means that an augmented form of the original system has to be treated in the constraints of the optimal control problem, consisting of the original dynamics and impact equations along with the corresponding variational differential equation and the update formulas for the sensitivity matrices at impacts. For a state variable vector of dimension n_x, this results in n_x^2 additional dynamic equations and variables. In addition, the spectral radius criterion may have a problem of ill-conditioning at points of multiple maximum eigenvalues, since the monodromy matrix is a nonsymmetric matrix. The spectral radius becomes nondifferentiable, and sometimes even non-Lipschitz at these points, which may occur no matter if the criterion is used as an objective function or as a constraint. Nonetheless, it has been possible to find significant descent and even an optimum using this criterion and the methods made for smooth criteria which are described below. A detailed discussion of stability optimization goes beyond the scope of the paper, but can be found in Mombaur (2009). For an earlier version of the stability optimization code using a bi-level formulation, see, e.g., Mombaur et al. (2005a), in which stability is optimized in the upper level changing model parameters and another criterion – e.g., related to energy – is minimized in the lower level adapting states and controls.

Optimal control problems are harder to solve than standard nonlinear optimization problems since they are infinite-dimensional problems, meaning that the unknown variables are not just n-dimensional vectors, but rather functions of time. This concerns the state variables of the dynamic model (typically including at least all position and velocity variables) as well as the control or input variables (depending on the model considered the torques, the muscle activations or excitations or the input variables of the controllers). In addition, there may be additional free variables of the optimal control problems which are not functions of time, e.g., free model parameters.

There are, in principle, three different ways to handle the infinite-dimensionality of the control functions: dynamic programming, direct and indirect methods. The basic idea of dynamic programming is to transform the continuous time problem into a sequence of problems to be solved at every time step in time and thus to compute the optimal control – and then the state – for the next interval. It uses the principle of optimality of subarcs and leads to the Hamilton–Jacobi–Bellman equation (Bertsekas, 2005).

For the direct approach – also called first-discretize-then-optimize approach – the free control functions are discretized by replacing them by a finite number of free parameters, e.g., in terms of piecewise constant or piecewise linear function approximations. After this discretization, we obtain a boundary value problem for the state variables. In contrast, indirect methods, also called first-optimize-then-discretize methods, formulate the necessary optimality conditions of the infinite-dimensional optimal control problem. This also results in a boundary value problem. This type of methods includes Pontryagin's Maximum Principle (Pontryagin et al., 1962). While the indirect approach is more precise since it keeps the full space of control variables, the direct method is much better suited to solve practical problems. For both the direct and indirect method, the solution of the boundary value problem, which corresponds to the treatment of the infinite-dimensionality of the state variables, can be solved by shooting methods or by collocation methods (Ascher et al., 1998).

Collocation uses a discretization of states on the so-called collocation points. The original continuous dynamic equation only has to be satisfied at the collocation points, which results in the formulation of a finite number of constraints instead of the differential equations.

Shooting methods use a different approach in which states are parameterized, not discretized. The optimization process only manipulates the states at the so-called shooting points, but the full dynamics of the system are simultaneously evaluated using integrators, i.e., they "shoot" trajectories from these starting points. While single shooting only uses the parameterized states at the initial time point, multiple shooting uses multiple points, from which independent integrations are started. Continuity between these integrated trajectory pieces and the next multiple shooting point, i.e., the start of the next trajectory piece, is guaranteed by additional constraints. After these two discretization/parameterization steps, both direct collocation and direct shooting methods result in nonlinear programming problems (NLP), which can be solved by general or special structure exploiting NLP techniques, e.g., special sequential quadratic programming (SQP) methods. Efficient direct multiple shooting methods have been developed by Bock and Plitt (1984), Leineweber et al. (2003), and

Houska et al. (2011), while von Stryk developed a direct collocation method (von Stryk, 1994). Both approaches are used to generate locomotion by different researchers, but no systematic comparison of the two methods exists so far.

In our research and for the solutions presented later in this article, we built upon the direct multiple shooting methods by Bock mentioned above, and as implemented in the code MUSCOD (Leineweber et al., 2003). This approach has been extended to mechanical DAEs with discontinuities of the above form.

4.2.4   APPLICATION OF OPTIMAL CONTROL TO GENERATE LOCOMOTION IN HUMANS AND ROBOTS

In this section, we present several examples for the application of optimal control in the generation of walking and running motions and some more exotic forms of locomotion. All motions are generated using the multiphase hybrid dynamics formulation and the numerical solution method based on direct multiple shooting discussed in the previous section.

As a first example, we show optimal motions for the simple dynamic bipedal model (H) of Fig. 4.2.1. Its configuration reminds of the early MIT leg lab robots by Raibert (1986). The main purpose of discussing this model is to demonstrate how optimal control can generate different types of motions for the same model by just replacing some equality and inequality constraints. As shown in the figure, the model consists of a central body and two prismatic legs. We only consider 2D motions in the sagittal plane, i.e., there are three global DOF described by the 2D position and the orientation of the central body. Physically, the model has four internal DOF (relative angles between central body and legs as well as the prismatic joints of the two legs), powered by torques and linear series elastic actuators, respectively. Since the lower part of the leg is assumed to be massless, the prismatic DOF is not there when the respective leg is in the air, so we have a total of 5 DOF for the flight phase and 4 DOF for single leg contact. The motions that we discuss here are (also compare Fig. 4.2.4):

- A symmetric and periodic running motion with alternating single leg contact phases and flight phases (Mombaur et al., 2005c);
- A symmetric and periodic somersaulting motion. It also has alternating single leg contact and flight phases, but the system performs a full 360° rotation about the medio-lateral axis in every step (Mombaur et al., 2005b);
- A periodic flic-flac motion, also with alternating flight and single leg contact phases, this time only performing 180° rotations between steps (Mombaur, 2006). Note that while the pictures show two legs and two "arms" to resemble the human flic-flac motions, the dynamic model actually is identical to the previous ones with just two legs, one being turned upwards.

(A) Periodic running motion

(B) Periodic somersaulting motion

(C) Periodic flic flac motion

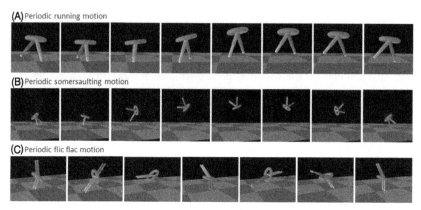

FIGURE 4.2.4   Three different types of motion for the same dynamic model imposed by different boundary conditions and different constraints: running (Mombaur et al., 2005c), somersaulting (Mombaur et al., 2005b), and flic-flacs (Mombaur, 2006).

The phase types and phase order in all three cases have been the same. The different motions are produced simply by changing the periodicity constraints to incorporate the imposed rotations. In addition, inequality constraints for collision avoidance between central body and legs have to be slightly changed for the leg turned upward in the flic-flac case, since it now must stay on the other side of the torus. It would, of course, also be possible to generate walking motions for the same model; however, in this case also the type and order of phases would have to be changed, with additional two-leg contact phases and no flight phases. The main optimization criterion applied in the solutions shown above was a maximization of stability with the result that all these solutions are open-loop stable. This has been achieved using the bilevel approach mentioned previously where stability is optimized in the upper level with the additional criterion in the lower level to minimize actuator inputs, i.e., some kind of energy. More details can be found in the cited papers. Other optimization criteria could also be applied, such as cost of transport, etc.

The second example that we are presenting is an optimality study of the full humanoid robot HRP-2 of Fig. 4.2.1F performed by Koch in his PhD thesis (Koch, 2015). HRP-2 is a full-size humanoid robot from Japan developed by Kawada and AIST (Kaneko et al., 2004). This robot has 6 global and 30 internal DOF which are essentially position controlled, but with an elasticity in the ankle joints. The goals of generating optimal motions for the robot are:

- To have an approach that selects all properties of the motion at the same time (i.e., all joint trajectories, step height, length width, actuator inputs and outputs, contacts with the environment, etc.);

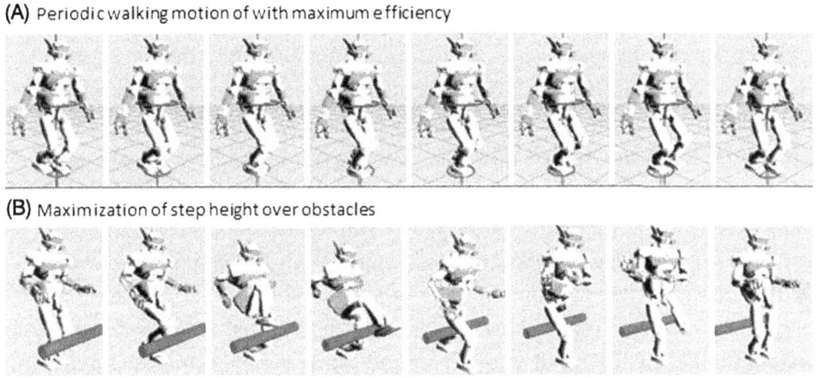

(A) Periodic walking motion of with maximum efficiency

(B) Maximization of step height over obstacles

FIGURE 4.2.5 Examples for simulated cyclic and non-cycle optimized motions for HRP-2: (A) a periodic walking motion maximizing efficiency (Koch et al., 2012b) and (B) a step over an obstacle with maximum height, starting and stopping at bipedal resting position (Koch, 2015).

- To be able to generate different styles of robot motions for some given task;
- To bring the robot towards its extremes in terms of kinematics, dynamics, performance, etc.

The dynamic model captures all kinematic and dynamic features of the mechanics of HRP-2. One problem, however, is that HRP-2 can, in fact, not execute all motions that its mechanics alone would permit. The robot uses a high level control system that is actually very conservative in its judgment and imposes strict constraints, e.g., on the trajectory of the ZMP. For formulating optimal control problems, it is important to consider also all these constraints as far as possible, i.e., as far as they are known. As most humanoid robots, the robot walks with full flat foot contacts, so only single support and double support have to be described. The summary of the study of periodic walking motions for HRP-2 is given in Koch et al. (2012a, 2012b), see Fig. 4.2.5A. We have investigated different objective functions in this context, some of which are biologically inspired, i.e., try to mimic what humans are doing and others are more technically or performance oriented. We have studied the minimization of joint torques squared, maximization of forward velocity, maximization of postural stability, maximization of efficiency, and minimization of angular velocities. The criteria have a quite important impact on the style of the walking motion. Periodicity and symmetry constraints are imposed on the walking motions. In addition, we have looked at the effect of different constraints such as restrictions on the ZMP to stay very close to the ankle joint. It could be shown that relaxing the ZMP constraints generates a significantly more upright gait. The second type of problems we have considered is the generation of the largest possible steps over obstacles in the attempt to bring the robot to its extremes

(Koch et al., 2014), see Fig. 4.2.5B. It could be shown that for such problems with very challenging dynamics it is necessary to model all tiny details of dynamic behavior and the influence of the controllers. As we discovered, ignoring, e.g., the ankle elasticity in this context leads to infeasible motions since the elasticity in fact slows the motion down. It was also not possible to simply use the primary optimization criterion which was to maximize the step height. Instead, several other criteria had to be included via reverse engineering to take into account all preferences of the control system. With this approach it was possible to generate a new motion with record step height of 20 cm which has been executed on the real robot. As a comparison, the optimal solution for the full dynamic model of the robot with all torque limits but ignoring the ankle elasticity and some control systems characteristics had previously led to a step height of 47 cm which then, however, turned out to be infeasible (Koch, 2015).

The third motion generation example we are discussing concerns wholebody motions for human models. We have studied walking and running motions. All systems discussed here are equipped with torque actuators at each of the internal DOF replacing the action of the human muscles. Running models (compare Fig. 4.2.1C and D), as they have been developed in Schultz and Mombaur (2010) describe 2D and 3D sprinting as a sequence of alternating flight and single leg contact phases. The 3D model consists of 12 rigid bodies, and has 25 DOF in flight – 6 global DOF associated with the position and orientation of the pelvis and 19 internal DOF related to internal joint angles. The model is powered by torques and parallel spring–damper elements in all joints. We assume point contact with the ball of the foot which is a realistic assumption for fast running. We have imposed the average running speed of 10 m/s which is not too far from world class sprinting speed and minimized a combination of active joint torques squared and torque derivatives squared because it selects – out of the infinitely many ways to run at this speed – the one that does it with the smallest effort and the smallest change in effort over time per joint. Especially the first part of the criterion seems to lead to a good approximation of human-like dynamic behavior. Step and phase times as well as spring–damper parameters have been left free for optimization. The resulting running motions look very realistic; see Fig. 4.2.6A for the 3D model. We have also investigated stability optimization to generate open-loop stable running using the one-level approach discussed in Section 4.2.3 (Mombaur, 2009), which, however, did not lead to a very human-like motion. The obvious (and not very surprising) conclusion of this is that human running is not open-loop stable, but depends on some crucial feedback signals. These loops could, in principle, be included in the model to perform the same type of stability study for the closed loop system; however, we are not aware of any model that captures all relevant feedback loops of walking as well as the entire biomechanical part of the system in a satisfactory way.

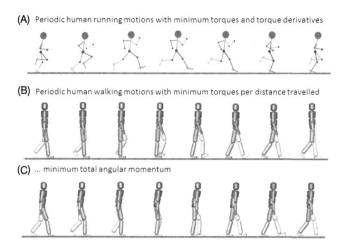

(A) Periodic human running motions with minimum torques and torque derivatives

(B) Periodic human walking motions with minimum torques per distance travelled

(C) ... minimum total angular momentum

FIGURE 4.2.6 Optimized running and walking motions for whole-body human models.

Human multi body system walking models in 2D and 3D have been developed by Felis in his thesis (Felis, 2015); see Fig. 4.2.3A and B. The formulation of the optimal control problem is similar as for the humanoid robot HRP-2. What gets more challenging here is the description of foot contact since the human capability to roll over the foot has been considered. This means that each foot contact is divided into heel only, flat foot, and ball only contact. Of course, a further refinement of this foot contact model is possible to include more flexible foot curvatures, which is subject of current research. We have investigated the effect of different optimization criteria on the motions, such as the minimization of joint torques squared, the maximization of walking efficiency, the minimization of total angular momentum, and the maximization of step frequency, all leaving the step and phase times as well as average velocity free (Felis and Mombaur, 2016), see Fig. 4.2.6B and C. We also heuristically investigated different possible combinations of these criteria, some of which get quite close to generating a human-like walking motions. Also the effect of fixing walking speed to different values has been studied. In order to get even closer to experimental recordings of true human walking, the approach presented in the next two sections is required.

4.2.5 WHAT IS THE COST FUNCTION OF HUMAN LOCOMOTION? THE INVERSE OPTIMAL CONTROL PROBLEM

Following the fundamental hypothesis of optimality of human movement, a central question is what precisely is optimized for a given task, i.e., what is the

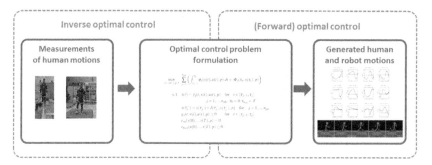

FIGURE 4.2.7   Inverse optimal control problems vs. optimal control problems.

exact objective function for the problem of interest. In contrast to the previous sections, where we have started from a mathematical problem formulation with a mechanical model and a given objective function and looked for a solution, we now start from an observed real movement.

As described in other chapters in this book, human movement can be experimentally observed and quite precisely measured by different techniques, such as motion capture systems (optical systems or inertial measurement units), force plates for ground reaction forces, EMG measurements for muscle activity, etc. From these measurements, the state variables' trajectories are fully or partly known, and sometimes also control histories can be (partly) measured. The inverse optimal control problem then consists in determining, from this information, the optimization criterion that has produced this observed solution. This can be a single criterion, but usually it is a combination of different weighted criteria. This inverse optimal control problem is a hard problem since it consists in solving a parameter estimation problem within an optimal control problem. This naturally results in a bilevel formulation with the parameter identification problem in the upper level and the optimal control problem solution in the lower level. We will further discuss this below.

Fig. 4.2.7 shows the connection between optimal control and inverse optimal control problems where the previous starts from the optimal control formulation and the latter one leads to it. It also illustrates how they can build on top of each other: as soon as the optimization criterion is identified for a class of movements, this criterion can be used to predict other movements of the same class under conditions that have not been studied in experiments.

The term *inverse optimal control* for the identification of an objective function in an optimal control problem from measurements has been coined by Kalman (1964) for linear problems. Later other types of problems have been discussed, e.g., inverse optimization in combinatorial problems (Heuberger, 2004). The class of nonlinear inverse optimal control problems receives a lot of attention in the mathematical community, especially in the context of reformulating the problem as a so-called MPEC (mathematical program with equilibrium constraints) where essentially the optimal control problem is replaced by the corresponding first order optimality conditions (Luo et al., 1996). Work is performed on the theoretical side (e.g., Ye, 2005; Dempe and Gadhi, 2007), but also some successful algorithms have been implemented for direct methods based on multiple shooting (Hatz et al., 2012, 2014) and collocation (Albrecht et al., 2010).

Important related work has been performed by Liu et al. (2005) who study realistic movement generation for character animation by physics-based models: here the problem was not to find the objective function, but to identify unknown model parameters from motion capture data using a nonlinear inverse optimization technique. Other related work comes from the field of learning control and reinforcement learning (Atkeson and Schaal, 2010), which is very a popular approach in robotics and which in the same way as optimization assumes that a function is approved over the course of learning. Although the terminology is different, learning system and inverse optimal control both assume that there is an objective function optimized by the process and use an estimation of performance index that describes the progress of learning or identification (Aghasadeghi and Bretl, 2011). A fundamental difference is that in optimization usually all steps are performed on a computer using simulation models, while learning also takes place over real world experiences. Sometimes learning methods are combined with optimization methods to solve problems of the inverse optimal control type (e.g., Levine and Koltun, 2013 and Dörr et al., 2015).

We have proposed an approach that keeps the original bilevel form (Mombaur et al., 2010) and works for general nonlinear inverse optimal control problems, even with multiple phases and discontinuities of the type that we are facing for studying walking motions. It uses a combination of an efficient direct optimal control technique for the lower level and a gradient-free optimization technique for the upper level. A new implementation of this method has been performed recently.

As an important prerequisite for the solution of the inverse optimal control problem, we assume that we can establish a set of reasonable independent base functions $\Psi_i(t)$ for the objective function. For this, in principle, also mathematical base functions, such as Fourier or polynomial bases, could be used. However, the objective of inverse optimal control is to determine the objective function in terms of some physically meaningful expressions, not in terms of some purely mathematical functions. We therefore chose physically motivated base functions, e.g., related to energy, efficiency, mechanical work, speed, smoothness of motions, etc. This will be discussed in more detail below. The choice of good base functions for a particular problem is crucial. On the one hand, a fit can only be as good as the available objective functions permit; on the other hand, the base functions must not be redundant with respect to their effect on the motion, otherwise no solution of the inverse optimal control problem is possible since there would be an infinite number of possibilities of combining the two objective functions with the same effect. The contributions of all these base functions $\Psi_i(t)$ to the overall objective function can be expressed by a priori unknown weight factors α_i. With this we have expressed the inverse optimal control problem as a problem to determine the weight factors α that result in the best possible fit to experimental data. Mathematically, this is written as

$$\min_{\alpha} \quad \sum_{j=1}^{m} \|x'^*(t_j; \alpha) - x'_M(t_j)\|^2, \qquad (4.2.16)$$

where $x'^*(t; \alpha)$ results from the solution of

$$\min_{x,u,p,T} \quad \sum_{i=1}^{n} \alpha_i \int_0^T \Psi_i(x(t), u(t), p)dt \qquad (4.2.17)$$

$$\text{s. t.} \quad \dot{x}(t) = f(t, x(t), u(t), p), \qquad (4.2.18)$$

$$g(t, x(t), u(t), p) \geq 0, \qquad (4.2.19)$$

$$r_{eq}(x(0), \ldots, x(T), p) = 0, \qquad (4.2.20)$$

$$r_{ineq}(x(0), \ldots, x(T), p) \geq 0. \qquad (4.2.21)$$

The two levels mentioned above are clearly visible: in the upper level, we aim to minimize the distance between the measured motion and the computed one by optimizing over the vector of weight parameters α. In the lower level, we solve a (forward) optimal control problem for the current iterate of α in order to compute the solution x^* to evaluate the objective function of the higher level problem. The problem formulation shown here only uses single phase problems for simplicity of presentation; however, it is no problem to extend this formulation to the type of multiphase optimal control problems discussed in the previous section by replacing the constraints of the lower part by Eqs. (4.2.11)–(4.2.15). Weight factors corresponding to base functions that turn out to be not relevant for the problem under investigation should go to zero during the solution of the problem. Obviously, there still is one redundancy in the problem formulation: analytically, the lower level problem would be exactly the same if all weight factors were multiplied by the same scalar (numerically, differences might occur due to scaling). This can be addressed by either fixing one of the weights (of which one is certain that it will not vanish) to a predefined value, e.g., to 1.0. The other option is to add to (4.2.20) a condition that the sum of all α_i is equal to a constant. We usually use the previous one, but have also used the latter one with equal success.

With x' we denote the subset of the state vector for which measurement information can be directly or indirectly gained (also see the discussion in the next session). $x'_M(t_j)$ denotes these measurements, typically taken at discrete time points t_j, and $x'^*(t_j)$ the corresponding parts of the optimal state trajectory x^*, evaluated at these same points. The task of the upper level is to determine the weight factors α that result in the best fit between the computational model and the measurements in the least squares sense. For each function evaluation here the lower level optimal control problem has to be solved, i.e., the upper level function cannot be expected to satisfy any conditions on smoothness or differentiability that are usually assumed to be valid for many gradient based optimization algorithms. Since derivative information is hard to obtain, we prefer to apply a derivative-free optimization technique, i.e., it only requires function evaluations and no explicit gradient information. The derivative-free optimization code BOBYQA (Bound Optimization BY Quadratic Approximation) by Michael Powell (2009) which can also handle bounds on the free parameters performs particularly well in this context. In the new implementation, also the code COBYLA is used. The task of the lower level is to efficiently solve the forward optimal control problem which occurs in each iteration of the upper level. For this, it is important to use an efficient method since this problem needs to be solved frequently; and we can use the direct boundary value problem approach MUSCOD that has been described in Section 4.2.3.

This method has already been used to study optimality criteria of different problems, among them the whole-body locomotion examples addressed in the next section, but also other applications such as yoyo playing (Mombaur and Sreenivasa, 2010), interaction and locomotion paths (Mombaur et al., 2010), etc.

4.2.6 APPLICATION OF INVERSE OPTIMAL CONTROL TO ANALYZE OPTIMALITY IN HUMAN LOCOMOTION

The goal of this section is to illustrate how the methods described above can be used to identify optimality criteria of human movement. We present some of the inverse optimal control work on whole-body human locomotion performed in the ORB (Optimization, Robotics & Biomechanics) research group. The models used in this context are the models in the first row of Fig. 4.2.3; here we specifically mention work performed with models (B) and (C). For a more detailed overview of inverse optimal control for human locomotion based on different modeling levels (i.e., whole-body models, template models, and overall locomotion path generation), see the recent article (Mombaur and Clever, 2017).

To actually formulate and solve the inverse optimal control problem (4.2.16)–(4.2.21) for a specific movement and data set, several decisions have to be taken first from a biomechanical as well as a technical perspective. These include:

- How many motions and how many subjects should be considered at once?
  In principle, it is possible to formulate inverse optimal control problems either for (i) individual motions of one subject, (ii) several motions of the same subject, or (iii) several motions of several subjects. The choice depends on the particular biomechanical question asked, i.e., if individual behavior is to be analyzed or compared, or if an average behavior is to be modeled. For (i), the problem formulation is given above. For (ii), either average data of the different motions can be used, again resulting in the problem formulation above or each motion can be analyzed separately, resulting in the simultaneous solution of multiple optimal control problems in the lower level and an additional sum over all motions for computing the overall fit in the upper level. For (iii), also the model parameters have to be adjusted to each of the subjects, such that here definitely multiple lower level optimal control problems have to be solved simultaneously. If for each subject data for multiple motions is available, then this can be treated with the options described under (ii). Examples for some of the cases will be given below.
- How should the fit between model and data be computed?
  As shown in Eq. (4.2.16), the upper level fit is based on the least squares distance between modeled and measured state variable trajectories. However, in practice generally the state variables of the system are not directly measured. As described in Section 4.2.2, a typical choice of state variables is to use position and orientation of the pelvis as well as all internal joint coordinates, as well as the corresponding velocities. However, no biomechanical measurement device will directly measure these quantities. The most standard case is to use marker-based systems which measure Cartesian coordinates of markers at specified positions of the human body (or to be more precise, on some approximately defined positions of the skin that exhibits relative motion with respect to the skeleton, which makes the problem harder). Other options include, e.g., IMUs which measure acceleration and rotational velocity information at other specific points. This means that transformations have to be performed either of the recorded data or of the model variables. For marker-based motions, the fit can be achieved either on marker position level (requiring that corresponding marker positions are computed in the model) or on the joint level (requiring that pseudomeasured joint angles are derived from the measured marker positions). On the one hand, the latter is not so desirable since this transformation involves the use of another model for the transformation and induces errors that are not visible anymore afterwards. On the other hand, it is much better in terms of analyzing deviations in the fit and drawing conclusions for potential adjustment of the

model. So no clear decision can be taken, and both approaches are possible. In the examples below, both approaches have been pursued. If the fit is performed on the state variables, appropriate scaling of variables may be necessary, depending on the particular choice and relative sizes.

- Which base functions should be considered in the objective function for human locomotion?

As discussed above, choosing appropriate base functions is one of the fundamental choices that have to be made for each inverse optimal control problem. Popular candidates for the study of human locomotion in different contexts (healthy or pathological walking) are:

1. Minimization of integral joint torques squared (related to motion effort) (see results presented in Section 4.2.4);
2. Minimization of absolute mechanical work at all joints (integral of absolute mechanical power);
3. Minimization of different types of energy (per time);
4. Minimization/maximization of time (for a walking distance);
5. Maximization of step length;
6. Minimization/maximization of step frequency;
7. Minimization of interval of jerk squared (jerk of a particular point or angular jerks at joints, as suggested in Flash and Hogan (1984) for arm movements);
8. Minimization of rotational motion of the head (so-called head stabilization) (Pozzo et al., 1990);
9. Minimization of head acceleration;
10. Minimization of length of walking path;
11. Minimization or maximization of walking speed;
12. Minimization of angular momentum about center of mass;
13. Maximization of postural stability;
14. Maximization of other stability criteria, e.g., related to capture point.

For each particular walking problem, the interesting set of such physical base functions has to be determined. As mentioned above, special care needs to be taken to choose nonredundant base functions. To give a simple example of redundant cost functions, for a system moving straight at a fixed speed, minimization of walking time and minimization of length of walking path would be equivalent. On the other hand, the minimization of time to perform a given task and the minimization of energy spent per time are usually completely independent. Unfortunately, the situation is not always as clear, and the cases of local as well as global redundancy have to be considered.

- How can the above choice of physical base functions be used in the inverse optimal control formulation?

An important point which is often not carefully considered is how to transfer the choice of physical base objective functions taken above into terms of the parameterized objective function in the mathematical problem formulation. Many of the functions listed above refer to quantities at the individual joints. In these cases the contributions of all joints may be different, also taking into account, e.g., the different strength of the actuators at these joints. Therefore in the extreme case, each joint should get its own term and own weight factor α_i in the objective function. This means that, e.g., just the choice of the objective function related to joint torques would result, for a model with, e.g., 15 internal DOF, in 15 terms and 15 unknowns in the upper level for the inverse optimal control problem. However, in some cases synergies can be exploited, e.g., by assuming equivalent weights for left and right body half, or by assuming similar weights for different parts of the body (e.g., lower body, upper body).

As a first example, we discuss the 3D running model (Fig. 4.2.1C) with 25 DOF and 19 internal DOF for which already forward optimal control solutions have been shown in Fig. 4.2.6A. In contrast to the forward problem which addressed high speed sprinting, this inverse optimal control problem studies slower running (jogging) at controlled speed of 10 km/h on a treadmill. We therefore impose the running speed as a constraint of type (4.2.20) to the problem. We still keep the assumption of forefoot running, i.e., there is no flat foot ground contact, only point-like contact with the ball of foot. Running data has been collected by our collaboration partners at the University of Rennes using a Vicon system at 100 Hz and 43 markers within the Locanthrope project. For this example, the fit between modeled and recorded motion is achieved on the level of marker positions. Overall, the parameterized objective function studied has 13 terms: 10 terms related to the minimization of torques (function 1 above) at the 19 internal joints, but exploiting symmetry, 1 related to the maximization of step length (function 4) and 2 related to head velocity (function 7). Results are reported in Mombaur et al. (2013), and summarized in Mombaur and Clever (2017). None of the parameters goes to zero, i.e., all components seem to play a role. However, the fit for the best possible objective function is not as good as desired. Reasons for this are the quite coarse model in the upper body which does not allow performing all motions that the human body can do, as well as – potentially – the fact that additional objective function terms should be added. Both items are addressed in current research.

The second example addresses human walking motions and uses the dynamic walking model in the sagittal plane Fig. 4.2.1B. It consists of 14 segments with a total of 16 DOF of which 13 are internal DOF. Detailed results are presented in Clever et al. (2016b), and a summary can be found in Mombaur and

Clever (2017). The focus here is on level ground unconstrained walking; however, inverse optimal control studies of more complex walking situations with constrained are underway. The study has been performed based on 6 subjects which have very different height. Vicon data at 100 Hz has been recorded within the KoroiBot project by our collaboration partners at the University of Tuebingen and archived in the KoroiBot database (KoroiBot Motion Capture Database, 2016; Mandery et al., 2015). The parameterized objective function used here consists of 7 terms: 4 of them concern again the torque criterion (function 1), but this time the same weight is used for entire parts of the body (hip, lower legs, arms, torso + head). In addition, there is 1 term for head stabilization (function 4), 1 for step length maximization (function 4), and 1 for step frequency maximization (function 5). In this study we compute the least squares fit based on position state variables, i.e., the coordinates of the model, which on the experimental side are derived from marker data using the MMM framework (Terlemez et al., 2014). In total seven motions have been considered, two for one of the subjects, and one each for all other subjects, where the motion consisted in all cases of one (right) step. Each of these motions has been considered as a separate inverse optimal control problem, corresponding to case (i) since the walking styles of the persons looked too different to combine them into one single description. This allowed us to compare results. For each of the motions, a set of weights could be determined, but as expected they are not identical but show a significant correlation. We can therefore hypothesize that the optimization criterion of a subject for a class of motions can be seen as a combination of a general objective function term and a term describing the personal movement style.

4.2.7 DISCUSSION & OUTLOOK

In this chapter, we have analyzed the role of optimality in bipedal locomotion. For this, two different types of problems have been considered, namely (forward) optimal control and inverse optimal control, both of which are based on complex dynamic locomotion models of humans and robots. For inverse optimal control, also experimental data of human locomotion has been taken into account.

Using the optimal control approaches discussed in Section 4.2.3, we could successfully generate various types of locomotion for different models and different orders of phases and constraint sets as shown in Section 4.2.4. In all these cases, motion generation by pure simulation and trial and error would have been practically impossible or at least not feasible in a reasonable amount of time. Different objective functions and constraints allow generating motions of various styles and types and satisfying different goals. Appropriate formulations can be used to bring (humanoid) robots towards their technical limits. Some mixed criteria have been heuristically found for human walking and running models that make the resulting motions look quite human-like.

The last statement leads us to inverse optimal control problems, the goal of which is to identify in a more precise way the underlying objective function of experimental recordings of human locomotion and which have been discussed in Sections 4.2.5 and 4.2.6. Inverse optimal control problems can be formulated for one or several data sets, motion types or subjects, depending on the underlying hypothesis of how large the area of validity of the investigated optimization law is. Inverse optimal control problems have been solved for different types of walking and running motions (among other types of motions) and different sets of data revealing important components of the objective functions. First studies comparing different subjects have shown that there is a correlation between objective function components of different subjects, but that also every subject is characterized by its own personal walking styles and individual motion variations.

An important issue which has not been directly addressed in this chapter is that optimal control problems for motion generation in robotics should ultimately be solved online, i.e., in real time. This would allow (humanoid) robots to generate and control walking motions on the spot when perceiving a new environment. The optimal control problems based on hybrid whole-body models, e.g., for the robot HRP-2, as presented in Section 4.2.4, take several hours to solve on a standard computer; hence alternative approaches are required.

Online optimal control, also known as (nonlinear) model-predictive control, would be the method of choice; however, there is no approach around that can solve optimal control problems for the mentioned hybrid multibody system models in real time while respecting all constraints.

Impressive results in this direction using online optimization have been achieved in computer graphics (e.g., Wang et al., 2012; Tassa et al., 2012), which also includes online reactions to events. In this case also some shortcuts are taken by simplifying constraints since the essential goal is to make the animations look good, but in robotics more realism of constraints is required to make solutions feasible for the real robot. Koenemann et al. (2015) work with a model of the robot HRP-2, but also seem to ignore several dynamic constraints.

Model-predictive control currently is used for simple models such as (linear) inverted pendulum models (Kajita et al., 2003) which also leads to quite agile behavior; see, e.g., Faraji et al. (2014), or 6D centroidal body models of walking (Kudruss et al., 2015). In the KoroiBot project and follow-ups, we are pursuing research on NMPC for walking motions to move to more complex robot models. This not only involves efficient solution of optimal control problems but also the treatment of receding horizons with changing phases of a priori unknown order. We also investigate how to best combine model-based optimal control and model-free robot learning at the example of the bipedal robot Leo (Schuitema et al., 2010).

Another approach pursued in KoroiBot is the combination of optimal control and movement primitives. The basic idea, as presented in Koch et al. (2015), is to use several optimal control problem solutions based on the full model and all constraints to learn movement primitives using Gaussian process models and Bayesian binning. These primitives are then used to assemble new walking steps on the spot, as required by the environment, which is feasible in just a few seconds. It could be shown that already with a small number of movement primitives it is possible to generate new walking motions which are very close to dynamic feasibility and to optimality of the criterion used for the training data. It has been applied to generate periodic walking motions at new step lengths but also adaptive walking patterns with changing step length (Clever et al., 2016a). Overall this seems to be a very promising approach to implement optimality principles on complex humanoid robots.

ACKNOWLEDGEMENTS

The research leading to these results has received funding from the EU seventh Framework Program (FP7/2007-2013) under grant agreement no 611909 (KoroiBot) and the German Excellence Initiative.

REFERENCES

Ackermann, M., van den Bogert, A.J., 2010. Optimality principles for model-based prediction of human gait. J. Biomech. 43 (6), 1055–1060.

Aghasadeghi, N., Bretl, T., 2011. Maximum entropy inverse reinforcement learning in continuous state spaces with path integrals. In: Proceedings of IEEE/RSJ IROS.

Albrecht, S., Passenberg, C., Sobotka, M., Peer, A., Buss, M., Ulbrich, M., 2010. Optimization criteria for human trajectory formation in dynamic virtual environments. In: Haptics: Generating and Perceiving Tangible Sensations. In: LNCS, Springer.

Alexander, R.M., 1984. The gaits of bipedal and quadrupedal animals. Int. J. Robot. Res. 3 (2), 49–59.

Alexander, R.M., 1996. Optima for Animals. Princeton University Press.

Ascher, U.M., Mattheij, R.M.M., Russel, R.D., 1998. Numerical Solution of Boundary Value Problems for Ordinary Differential Equations. Prentice Hall.

Atkeson, C.G., Liu, C., 2013. Trajectory-based dynamic programming. In: Modeling, Simulation and Optimization of Bipedal Walking. In: Cognitive Systems Monographs, vol. 18. Springer, Berlin, Heidelberg, pp. 1–15.

Atkeson, C.G., Schaal, S., 2010. Learning control in robotics. IEEE Robot. Autom. Mag. 17, 20–29.

Bertsekas, Dimitri P., 2005. Dynamic Programming and Optimal Control, vol. I, 3rd edition. Athena Scientific, Belmont, MA, USA.

Bock, H.G., Plitt, K.-J., 1984. A multiple shooting algorithm for direct solution of optimal control problems. In: IFAC World Congress, pp. 242–247.

Buschmann, T., Lohmeier, S., Bachmayer, M., Ulbrich, H., Pfeiffer, F., 2007. A collocation method for real-time walking pattern generator. In: Proceedings of the IEEE-RAS International Conference on Humanoid Robots.

Clever, D., Harant, M., Koch, K.H., Mombaur, K., Endres, D.M., 2016a. A novel approach for the generation of complex humanoid walking sequences based on a combination of optimal control and learning of movement primitives. In: Robotics and Autonomous Systems.

Clever, D., Schemschat, R.M., Felis, M.L., Mombaur, K., 2016b. Inverse optimal control based identification of optimality criteria in whole-body human walking on level ground. In: Proceedings of International Conference on Biomedical Robotics and Biomechatronics (BioRob2016).

Coleman, M.J., Garcia, M., Mombaur, K.D., Ruina, A., 2001. Prediction of stable walking for a toy that cannot stand. Phys. Rev. E 2.

Darwin, C., 1859. On the Origin of Species.

de Leva, P., 1996. Adjustments to Zatsiorsky–Seluyanov's segment inertia parameters. J. Biomech. 29, 1223–1230.

Dempe, S., Gadhi, N., 2007. Necessary optimality conditions for bilevel set optimization problems. Global Optim. 39 (4), 529–542.

Dörr, A., Ratliff, N., Bohg, J., Toussaint, M., Schaal, S., 2015. Direct loss minimization inverse optimal control. In: Proceedings of Robotics Science and Systems (RSS).

Faraji, S., Pouya, S., Ijspeert, A., 2014. Robust and agile 3D biped walking with steering capability using a footstep predictive approach. In: Robotics Science and Systems (RSS).

Featherstone, R., 2007. Rigid Body Dynamics Algorithms. Springer, New York, NY.

Felis, M., 2015. Modeling Emotional Aspects of Human Locomotion. PhD thesis. University of Heidelberg.

Felis, M.L., 2016. RBDL: an efficient rigid-body dynamics library using recursive algorithms. Auton. Robots, 1–17.

Felis, M.L., Mombaur, K., 2016. Synthesis of full-body 3-D human gait using optimal control methods. In: IEEE International Conference on Robotics and Automation (ICRA 2016).

Felis, M.L., Mombaur, K., Berthoz, A., 2015. An optimal control approach to reconstruct human gait dynamics from kinematic data. In: IEEE/RAS International Conference on Humanoid Robots (Humanoids 2015), pp. 1044–1051.

Flash, T., Hogan, N., 1984. The coordination of arm movements: an experimentally confirmed mathematical model. J. Neurosci. 5, 1688–1703.

Geijtenbeek, T., van de Panne, M., van der Stappen, A.F., 2013. Flexible muscle-based locomotion for bipedal creatures. ACM Trans. Graph. 32 (6).

Geyer, H., Herr, H., 2010. A muscle-reflex model that encodes principles of legged mechanics produces human walking dynamics and muscle activities. IEEE Trans. Neural Syst. Rehabil. Eng. 18 (3), 263–273.

Hatz, K., 2014. Efficient Numerical Methods for Hierarchical Dynamic Optimization with Application to Cerebral Palsy Gait Modeling. PhD thesis. University of Heidelberg.

Hatz, K., Schlöder, J.P., Bock, H.G., 2012. Estimating parameters in optimal control problems. SIAM J. Sci. Comput. 34 (3).

Heuberger, C., 2004. Inverse combinatorial optimization: a survey on problems, methods, and results. J. Comb. Optim. 8 (3), 329–361.

Ho Hoang, K.-L., Mombaur, K., 2015. Adjustments to de Leva-anthropometric regression data for the changes in body proportions in elderly humans. J. Biomech. 48 (13), 3732–3736.

Houska, B., Ferreau, H.J., Diehl, M., 2011. ACADO toolkit – an open source framework for automatic control and dynamic optimization. Optim. Control Appl. Methods 32 (3), 298–312.

Kajita, S., Kanehiro, F., Kaneko, K., Fujiwara, K., Harada, K., Yokoi, K., Hirukawa, H., 2003. Biped walking pattern generation by using preview control of zero-moment point. In: IEEE Int. Conf. on Robotics & Automation.

Kalman, R., 1964. When is a linear control system optimal? Trans. ASME J. Basic Eng., Ser. D, 51–60.

Kaneko, K., Kanehiro, F., Kajita, S., Hirukawaa, H., Kawasaki, T., Hirata, M., Akachi, K., Isozumi, T., 2004. Humanoid robot HRP-2. In: Proceedings of the IEEE International Conference on Robotics & Automation.

Koch, K.H., 2015. Using Model-based Optimal Control for Conceptional Motion Generation for the Humannoid Robot HRP-2 14 and Design Investigations for Exo-Skeleton. PhD thesis. University Heidelberg.

Koch, K.H., Clever, D., Mombaur, K., Endres, D.M., 2015. Learning movement primitives from optimal and dynamically feasible trajectories for humanoid walking. In: IEEE/RAS International Conference on Humanoid Robots (Humanoids 2015), pp. 866–873.

Koch, K.H., Mombaur, K., Souères, P., 2012a. Optimization-based walking generation for humanoids. In: IFAC-SYROCO 2012.

Koch, K.H., Mombaur, K., Souères, P., 2012b. Studying the effect of different optimization criteria on humanoid walking motions. In: Noda, I., Ando, N., Brugali, D., Kuffner, J.J. (Eds.), Simulation, Modeling, and Programming for Autonomous Robots. In: Lecture Notes in Computer Science, vol. 7628. Springer, Berlin, Heidelberg, pp. 221–236.

Koch, K.H., Mombaur, K., Souères, P., Stasse, O., 2014. Optimization based exploitation of the ankle elasticity of HRP-2 for overstepping large obstacles. In: IEEE/RAS International Conference on Humanoid Robots (Humanoids 2014).

Koenemann, A., Del Prete, A., Tassa, Y., Todorov, E., Stasse, O., Bennewitz, M., Mansard, N., 2015. Experiments with mujoco on HRP-2. In: IEEE/RSJ International Conference on Intelligent Robots and Systems (IROS 2015).

KoroiBot Motion Capture Database, 2016. https://koroibot-motion-database.humanoids.kit.edu/. Last visited May 2016.

Kudruss, M., Naveau, M., Stasse, O., Mansard, N., Kirches, C., Souères, P., Mombaur, K., 2015. Optimal control for whole-body motion generation using center-of-mass dynamics for predefined multi-contact configurations. In: IEEE/RAS International Conference on Humanoid Robots (Humanoids 2015), pp. 684–689.

Leineweber, D.B., Bauer, I., Bock, H.G., Schlöder, J.P., 2003. An efficient multiple shooting based reduced SQP strategy for large-scale dynamic process optimization – part I: theoretical aspects – part II: software aspects and applications. Comput. Chem. Eng. 27, 157–174.

Lengagne, S., Kheddar, A., Yoshida, E., 2011. Generation of optimal dynamic multi-contact motions: application to humanoid robots. IEEE Trans. Robot.

Levine, S., Koltun, V., 2013. Guided policy search. In: ICML.

Liu, C.K., Hertzmann, A., Popovic, Z., 2005. Learning physics-based motion style with inverse optimization. ACM Transactions on Graphics (SIGGRAPH 2005) 24.

Luo, Z.-Q., Pang, J.-S., Ralph, D., 1996. Mathematical Programs with Equilibrium Constraints. Cambridge University Press.

Mandery, C., Terlemez, Ö., Do, M., Vahrenkamp, N., Asfour, T., 2015. The KIT whole-body human motion database. In: IEEE International Conference on Advanced Robotics (ICAR 2015), pp. 329–336.

Miossec, S., Yokoi, K., Kheddar, A., 2006. Development of a software for motion optimization of robots – application to the kick motion of the HRP-2 robot. In: IEEE Int. Conf. on Robotics & Automation.

Mombaur, K.D., 2009. Using optimization to create self-stable human-like running. Robotica 27, 321–330. Published online June 2008.

Mombaur, K.D., Clever, D., 2017. Inverse optimal control as a tool to understand human movement. In: Geometric and Numerical Foundations of Movements. In: STAR Series. Springer, Berlin, Heidelberg. vol. 117.

Mombaur, K.D., Olivier, A.H., Crétual, A., 2013. Forward and inverse optimal control of bipedal running. In: Modeling, Simulation and Optimization of Bipedal Walking. In: Cognitive Systems Monographs, vol. 18. Springer, Berlin, Heidelberg, pp. 165–179.

Mombaur, K.D., Sreenivasa, M., 2010. Inverse optimal control as a tool to understand human yoyo playing. In: ICNAAM 2010.

Mombaur, K.D., Truong, A., Laumond, J.-P., 2010. From human to humanoid locomotion – an inverse optimal control approach. Auton. Robots 28 (3).

Mombaur, K.D., 2001. Stability Optimization of Open-loop Controlled Walking Robots. PhD thesis. University of Heidelberg. www.ub.uni-heidelberg.de/archiv/1796. VDI-Fortschrittbericht, Reihe 8, No. 922, ISBN 3-18-392208-8.

Mombaur, K.D., 2006. Performing open-loop stable flip-flops – an example for stability optimization and robustness analysis of fast periodic motions. In: Fast Motions in Robotics and Biomechanics – Optimization and Feedback Control. In: Lecture Notes in Control and Information Science. Springer.

Mombaur, K.D., Bock, H.G., Schlöder, J.P., Longman, R.W., 2005a. Open-loop stable solution of periodic optimal control problems in robotics. ZAMM - J. Appl. Math. Mech./Z. Angew. Math. Mech. 85 (7), 499–515.

Mombaur, K.D., Bock, H.G., Schlöder, J.P., Longman, R.W., 2005b. Self-stabilizing somersaults. IEEE Trans. Robot. 21 (6).

Mombaur, K.D., Longman, R.W., Bock, H.G., Schlöder, J.P., 2005c. Open-loop stable running. Robotica 23 (01), 21–33.

Ng, E.Y.K., Luo, Y. (Eds.), 2016. Bio-Inspired Surfaces and Applications. World Scientific.

Pontryagin, L.S., Boltayanskii, V.G., Gamkrelidze, R.V., Mishchenko, E.F., 1962. The Mathematical Theory of Optimal Processes. Wiley.

Powell, M.J.D., 2009. The BOBYQA algorithm for bound constrained optimization without derivatives. Report No. DAMTP 2009/NA06. Centre for Mathematical Sciences, University of Cambridge, UK.

Pozzo, T., Berthoz, A., Lefort, L., 1990. Head stabilization during various locomotor tasks in humans. Exp. Brain Res. 82 (1), 97–106.

Raibert, M.H., 1986. Legged Robots That Balance. MIT Press, Cambridge.

Rosen, R., 1967. Optimality Principles in Biology. Springer Science+Business Media, LLC.

Schuitema, E., Wisse, M., Ramakers, T., Jonker, P., 2010. The design of LEO: a 2D bipedal walking robot for online autonomous reinforcement learning.

Schultz, G., Mombaur, K., 2010. Modeling and optimal control of human-like running. Trans. Mechatron. 15 (5).

Sok, K.W., Kim, M.M., Lee, J.H., 2007. Simulating biped behaviors from human motion data. ACM Trans. Graph. 26 (3), 107.

Sreenivasa, M., Chamorro, C.J.G., Alvarado, D.G., Rettig, O., Wolf, S.I., 2016. Patient-specific bone geometry and segment inertia from MRI images for model-based analysis of pathological gait. J. Biomech. 49 (9), 1918–1925.

Suleiman, W., Yoshida, E., Laumond, J.-P., Monin, A., 2007. On humanoid motion optimization. In: IEEE-RAS Int. Conf. on Humanoid Robots, pp. 180–187.

Tassa, Y., Erez, T., Todorov, E., 2012. Synthesis and stabilization of complex behaviors through online trajectory optimization. In: IEEE/RSJ Int. Conf. on Intelligent Robots and Systems.

Terlemez, Ö., Ulbrich, S., Mandery, C., Do, M., Vahrenkamp, N., Asfour, T., 2014. Master motor map (MMM) – framework and toolkit for capturing, representing, and reproducing human motion on humanoid robots. In: IEEE/RAS International Conference on Humanoid Robots (Humanoids 2014), pp. 894–901.

Todorov, E., 2004. Optimality principles in sensorimotor control. Nat. Neurosci. 7 (9), 907–915.

von Stryk, O., 1994. Numerische Lösung optimaler Steuerungsprobleme: Diskretisierung, Parameteroptimierung und Berechnung der adjungierten Variablen. Number 441 in Fortschritt-Berichte VDI, Reihe 8: Mess-, Steuer- und Regelungstechnik, VDI Verlag, Düsseldorf.

Wang, J.M., Hamner, S.R., Delp, S.L., Koltun, V., 2012. Optimizing locomotion controllers using biologically-based actuators and objectives. ACM Trans. Robot. 31 (4). Article 25.

Wieber, P.-B., Billet, F., Boissieux, L., Pissard-Gibollet, R., 2006. The HuMAnS toolbox, a homogenous framework for motion capture, analysis and simulation. In: Internal Symposium on the 3D Analysis of Human Movement.

Winter, D.A., 2004. Biomechanics and Motor Control of Human Movement. John Wiley & Sons.

Ye, J.J., 2005. Necessary and sufficient optimality conditions for mathematical programs with equilibrium constraints. J. Math. Anal. Appl. 307 (350–369).

Chapter 4.3

Efficiency and Compliance in Bipedal Walking

Katja Mombaur[‡], Yue Hu[‡], and Jonas Buchli[§],

[‡]*Optimization, Robotics & Biomechanics, ZITI, IWR, Heidelberg University, Heidelberg, Germany*
[§]*ADRL, ETH Zürich, Zürich, Switzerland*

Abstract. *In this chapter we discuss different concepts of compliance in locomotion including constant and variable compliance. Compliance provides many advantages for dynamic motions, such as increased efficiency, stability, adaptability and softness, but also introduces many challenges for the design and control. Optimal control provides a promising approach to optimally tune compliance parameters and control compliant actuators. Different models for control are discussed. We give several examples for understanding the use of compliance in human motion and for optimizing compliance in different types of robots.*

4.3.1 INTRODUCTION

Compliance plays an important role in locomotion of humans and animals. There are many sources of compliance in biological systems including muscles, tendons, soft tissue, etc. Compliance of joints can even be modulated by different levels of cocontraction of agonist and antagonist muscles. There are many advantages of compliance: it allows a better adaptation to a given task; it permits storing energy and releasing it at another point in the locomotion cycle and therefore can help in increasing the efficiency of locomotion (Alexander, 1988). In addition, compliance can improve the stability of a system since it generates natural backdriving forces on a mechanical level, if compliance is properly controlled for a motion. Compliant external surfaces result in softer contacts and more safety in interactions.

In robotics, there has also been a strong shift in the past decades from rigid position controlled joints to torque controlled and compliant actuators and the use of passive compliant elements. The aim is to also benefit from the advantages listed above by introducing compliance in the robot design. However, properly selecting design parameters of compliant elements and control inputs for compliant actuators is no easy task due to the high number of degrees of freedom (DOF) of the locomotion systems and the high number of parameters to tune. For the solution of this problem in robotics applications, optimal control provides an efficient tool.

The goal of this chapter is to give an overview of the benefits of optimal control for the proper tuning of compliance in robot motions as well as an approach to understand compliance in humans. We summarize several studies on compliance in locomotion and related dynamic motions that have been performed independently by the different authors, addressing the optimization of variable and constant compliance. This chapter is thus complementary to Subchapters 4.4 and 4.2 on impedance control and optimal control, respectively. The focus here is on the compliance in joints, not in external contact surfaces. The studies presented here therefore require locomotion models with all relevant joints or the real system, but no simplified template models.

In recent years there has been an increasing interest in compliance aspects in human and animal locomotion. While it was initially considered relevant mainly for running, it has been analyzed also in walking motions and was demonstrated to be essential by means of simple spring–mass models where entire legs are replaced by linear springs on level ground (Geyer et al., 2006) and rough terrains (Liu et al., 2015), or variable stiffness springs (Visser et al., 2013). Other studies were focused on the role of biarticular muscles (Iida et al., 2008; Mombaur, 2014) and tendons (Endo et al., 2006) by using spring–damper models.

Compliance at the joint level also plays a central role in locomotion (Latash and Zatsiorsky, 1993), where many researchers address the analysis of joint stiffness by studying the torque angle relationship of the leg joints (Weiss et al., 1986a, 1986b), namely hip, knee, and ankle. The aim of many of these studies is to look for a possible way to recreate the same walking motion with simple mechanisms such as linear springs, which has been demonstrated to be possible in certain phases of walking (Shamaei et al., 2013a, 2013b, 2013c) and running (Günther and Blickhan, 2002). Joint torques are computed with inverse dynamics and a statistical analysis on a high number of subjects of both gender was carried out, but without identifying differences between them, which was then done by Gabriel et al. (2008). However, studies have shown that compliance in humans is variable and modulates due to the cocontraction of agonist and antagonistic muscles acting on the joints during the execution of movements (Ferris et al., 1998; Hogan, 1984).

Humans are able to adjust the impedance (in addition to their kinematic plans), both in terms of direction and magnitude to the requirements of a task. This can be demonstrated by subjecting a person to a random force fields in reaching movements and observing how the measured impedance at the hand changes (Burdet et al., 2001; Selen et al., 2009; Franklin et al., 2007). These experiments demonstrate, as expected from control theoretical principles, that impedance adaptation can serve at least two purposes: (1) It can be used to stabilize a kinematic task against random perturbation by *increasing* the stiffness. (2) By *lowering* the stiffness the body or parts, it can be decoupled by from external disturbances. While such experiments for practical reasons have most commonly been done with the upper extremities based on general control theoretical principles, we can conjecture that such principles are general and apply to the lower extremities as well. Task requirements might be reaching a certain goal (i.e., kinematic/geometric constraints), not falling over (i.e., stability, a mixed dynamic/geometric constraint), speed (translating into time or velocities as constraints). More often than not at any given time several such goals might be important and are complicating finding the right task achievement. Recent results prove more and more convincingly, however, that optimal control can find the motions and associated controllers to satisfy such difficult planning and control problems. The above mentioned task constraints are usually either handled as soft constraints via the cost function or as "real" constraints via transcription.

Humans in daily life walk in many different environments, the most common ones are level ground, up and down slopes of different inclinations, stairs of different sizes, and different types of rough terrain. So in order to better understand locomotion, it is necessary to analyze walking in all these different scenarios. But, despite the large amount of literature on stiffness at the joint level, most of them are focused on level ground walking, with only a few works on other walking scenarios. In biomechanics there is some work on the analysis of kinematics and kinetics of slope walking (Franz et al., 2012; Silder et al., 2012) and stair climbing (Andriacchi et al., 1980; Amirudin et al., 2014), but there is a lack of studies focused on joint stiffness in this context.

In robotics, the main objective related to compliance is to gather fundamental information to develop compliant control and actuation principles, some of which have elastic elements with fixed stiffness and others with variable stiffness.

In Mombaur et al. (2009) three different hypotheses about the use of compliance have been formulated in bipedal locomotion and demonstrated in the examples of multibody system models of different complexity using optimization techniques. It was postulated that an optimally tuned compliance can significantly reduce the cost of transport, can produce naturally looking motions, and can also improve the stability of locomotion.

Compliance control, and in particular variable compliance, can be implemented in several ways in robots. One of the most flexible approaches is emulating compliance with an outer position loop around an inner force/torque control loop (Semini et al., 2015) (see also Subchapter 4.4. Here the gains of the position controller then correspond to impedances with the position gain being the stiffness and the derivative gain being the damping. The challenge now is to find suitable impedances for any given time, thus amounting to finding a gain schedule. Variable impedance can also be realized with semipassive systems.

Here two major categories exist: (1) active change of the spring constant (Wolf et al., 2016) and (2) dynamic adjustment of the impedance via the series motor in a series elastic actuator (Pratt and Williamson, 1995a). The first approach has the typical limitation that the spring constant cannot be instantaneously changed and that the devices are usually still too bulky to be well integrated into complex robotic systems, e.g., such as small robots, prostheses, or exoskeletons. The second approach has the limitation that strict control theoretical stability bounds significantly limit the achievable compliance levels.

As robots started to come out from the factories and enter human populated environments, humanoid robots using compliant actuators started to appear (Pratt and Williamson, 1995a), such as the Lucy robot using pneumatic artificial muscles (Verrelst et al., 2005), the Roboray using tendon driven actuators (Kim et al., 2012), and M2V2 (Pratt and Krupp, 2008), the COMAN (Colasanto et al., 2012), WALK-MAN (Tsagarakis et al., 2016) and iCub (Metta et al., 2010; Parmiggiani et al., 2012) using Series Elastic Actuators (SEA) (Pratt and Williamson, 1995b). In particular, the COMAN humanoid robot showed to be able to perform stable walking with SEA (Li et al., 2012; Moro et al., 2014; Dallali et al., 2012). The introduction of compliant actuators has the aim of absorbing impacts, and facilitating the generation of human-like movements and energy efficiency. The often quoted motivation of increased safety by adding compliant elements does not hold as a general principle (Semini et al., 2015).

The rest of this chapter is organized as follows. In Section 4.3.2, we present different models of compliance in locomotion system, including different types of constant and variable compliance. Section 4.3.3 describes the possibilities and benefits of using optimal control for compliance studies. In Section 4.3.4 we present some examples of optimal control based studies of compliance in human locomotion. Section 4.3.5 focuses on examples for compliance optimization in robot motions. Section 4.3.6 finally formulates a conclusion and some research perspectives.

4.3.2 DIFFERENT MODELS OF COMPLIANCE

In this section, we discuss different types of compliance from the modeling perspective. It is assumed that the whole walking system is described as a rigid multibody system (MBS) powered by joint torques. The equations of motions of a typical walking system take the following form:

$$\begin{pmatrix} M & G^T \\ G & 0 \end{pmatrix} \begin{pmatrix} \ddot{q} \\ \lambda \end{pmatrix} = \begin{pmatrix} -C + \tau \\ \gamma \end{pmatrix}, \qquad (4.3.1)$$

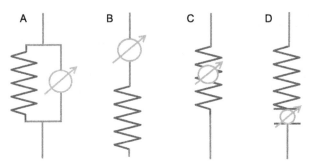

FIGURE 4.3.1 Different types of compliance model discussed in this chapter: (A) constant parallel compliance, (B) constant serial compliance, (C) variable compliance with stiffness modulation, (D) variable compliance with rest length modulation.

with mass matrix M, vector of nonlinear effects also including gravity C, vector of joint torques (or forces) τ, including active and passive components, accelerations $\ddot{q}$, Lagrange multipliers λ, constraint Jacobian G, and constraint Hessian γ. Here we use a slightly less general formulation than in Subchapter 4.2, which assumes that there is no drag, and no external pushing forces, etc., and that the actuation is applied directly at the joints.

Specific model properties of some walking models will be discussed later in the example sections. We are now considering different ways how compliance can influence these total joint torques τ, distinguishing between models of constant and variable compliance.

4.3.2.1   Constant Compliance

We start by discussing models of constant compliance which can enter either in parallel or in series with the active torque/forces. These two concepts are illustrated in Fig. 4.3.1A and B, respectively.

In the case of **constant parallel compliance**, the torques or forces generated by the springs are added to the active torques or forces:

$$\tau_{P,i}(t) = \tau_{active,i}(t) + \tau_{spring,i}(t)$$
$$= \tau_{active,i} + k_{p,i}(\phi_{0,i} - \phi_i(t)), \qquad (4.3.2)$$

as shown here for the torque at joint i with joint angle ϕ_i, assuming a linear spring with spring coefficient $k_{p,i}$. $\phi_{0,i}$ denotes the joint angle at which the spring is not deformed, which for simplicity we call "rest length", even though it strictly is an angle in the rotational case. This combined $\tau = \tau_P$ enters the right-hand side of Eq. (4.3.1). In the same way, also a linear damper with free damping coefficient $b_{p,i}$ could be added in parallel to the torque and the spring:

$$\tau_{P,i}(t) = \tau_{active,i}(t) + \tau_{spring,i}(t) + \tau_{damp,i}(t)$$
$$= \tau_{active,i} + k_{p,i}(\phi_{0,i} - \phi_i(t)) - b_{p,i}\dot{\phi}_i(t). \qquad (4.3.3)$$

We only mention damping here once, but it can, of course, be added to the other compliance models discussed below in the same way. Also nonlinear springs and dampers might be chosen. The free spring (and potentially damper) parameters can be tuned by optimization, e.g., to achieve a higher efficiency or stability. Examples of this type of can be found in robots where simple mechanical springs are put in parallel to the motor actions, and damping or velocity friction occurs naturally in each joint. In humans, passive tissues of the muscles as well as in the joints may act as parallel compliance and damping.

In the case of **constant serial compliance**, the active and passive terms are not added, but the torque/force generated by the active element is transmitted to the link via the spring, which obviously is deformed by the load (see Subchapter 4.4 for a discussion of the input/output relationships of compliant elements). In robotics, this kind of compliance appears when a rigid actuator is assembled in series with a transmission system with some compliance, or mechanical springs put in series with the actuators. An example for the latter are the series elastic actuators (SEA) introduced in (Pratt and Williamson, 1995a) and used in the iCub robot (Metta et al., 2010), also discussed in the example sections. In humans, the tendons are a prominent example of elastic elements that work in series with the actuators, the muscles. There are several muscles in the human body in which the influence of the series elastic element should not be neglected, e.g., in the Gastrocnemius muscle/Achilles tendon complex. Serial constant compliance can be modeled by augmenting the equations of motions (4.3.1) to

$$\begin{pmatrix} M & 0 & G^T \\ 0 & R & 0 \\ G & 0 & 0 \end{pmatrix} \begin{pmatrix} \ddot{q} \\ \ddot{\theta} \\ \lambda \end{pmatrix} = \begin{pmatrix} -C + \tau_C \\ \tau_S - \tau_C \\ \gamma \end{pmatrix}. \qquad (4.3.4)$$

As before, $\ddot{q}$ describes the accelerations of the mechanical parts, corresponding to the chosen coordinates, and $\ddot{\theta}$ describes the accelerations of the points between actuators and springs. τ_S denotes the active torques requires in this serial context and τ_C the coupling torques transmitted by the spring. The matrix R is a diagonal matrix describing the inertia of the rotors in robots and of muscles, etc., in humans. It is important to note that the serial compliance significantly increases the complexity of the problem by doubling the number of variables used for the description of all actuated joints with additional compliance.

4.3.2.2 Variable Compliance

In addition to the model of constant compliance discussed so far, also models of variable compliance are of big interest for studying locomotion in robotics as well as in biomechanics. In these cases, the actuators are not built in series or in parallel to springs, but they are acting inside the spring, directly changing the spring's property. Note that this kind of spring is not conservative any more as an ideal passive spring, but the actuation allows putting energy in as well as taking it out of the system.

There are two principally different ways to generate variable compliance: (i) by modulation of the stiffness coefficient, or (ii) by modulation of the rest length of the spring, as shown in Fig. 4.3.1C and D, respectively. Here we explain both approaches starting from the model of a linear torsional spring. The same approaches could be applied to nonlinear spring models.

Compliance modulation by stiffness coefficient modulation can be modeled as follows:

$$\tau_{SM,i}(t) = k_{v,i}(t)(\phi_{0,i} - \phi_i(t)). \tag{4.3.5}$$

Here compliance $k_{v,i}(t)$ of each joint can change in time, but the rest length $\phi_{0,i}$ remains constant and the current stiffness coefficient is multiplied by the deviation of the current joint angle from the rest angle of the spring. The MACCEPA actuators (Van Ham et al., 2007), for example, follow the principle of changing the stiffness of the spring by two electric motors. Variable stiffness also plays an important role in humans who have the capability of flexibly adjusting joint stiffness via different levels of co-contraction. Higher stiffness is, e.g., used to counteract large forces and to help maintain or stay close to a position, while a small stiffness is used in more efficient, dynamic, and flexible motions that do not have to resist external forces.

The other case, **modulation of compliance by rest length modulation** can be modeled for each joint as follows:

$$\tau_{RM,i}(t) = k_{r,i}(\phi_{0,i} + u_i(t) - \phi_i(t)). \tag{4.3.6}$$

Here we assume that the stiffness is constant $k_{r,i}$, but that the actuator can modulate the rest length of the spring $\phi_{0,i} + u_i(t)$ from the original value $\phi_{0,i}$ by changing $u_i(t)$. Such a concept can also be implemented using SEA or extensions which implement the concept of equilibrium point control. Instead of controlling the force of the actuator, the rest length of the spring is actively controlled (Van Ham et al., 2009; Robinson, 1997). Examples of this type of compliance and how in can be optimized are discussed in Section 4.3.5.2. There also is a discussion about the role of rest length modulation in human motions.

4.3.2.3 Extension of Compliance Models to Coupled Joints

So far, we have essentially discussed compliance at the level of separate joints. However, an important property of biological locomotion systems is the coupling of joints via compliant bi-articular muscle tendon structures, such as the rectus femoris at the front of the thighs or the hamstrings at the back of the thighs, which all couple hip and knee joints. Concepts of bi-articular tendons have also been shown to work well in simple robot systems (Iida et al., 2008).

In principle, all models discussed above can be extended to situations with bi-articular coupling. We will not detail this for all cases, but describe as an example how the model based on stiffness modulation can be extended to bi-articular joints (Mombaur, 2014; Hu and Mombaur, 2016):

$$\tau_i(t) = k_{B,i,j}(t)(\phi_{0,i,j} - \phi_i - \phi_j) + k_{v,i}(t)(\phi_{0,i} - \phi_i(t)). \qquad (4.3.7)$$

The first term describes the coupling torque which depends on the angles of both joints, an individual rest length parameter $\phi_{0,i,j}$, and the individual variable stiffness coefficient $k_{B,i,j}(t)$. In this model, the same term would be added for the bi-articular effects on the other joint of the pair. This is a quite simple approach to formulate the coupling just based on torques, but more sophisticated approaches exist, especially in the context of full muscle modeling, e.g., based on the forces that are transmitted via the bi-articular muscle tendon structure and the real lever arms which might be different in the joints and even changing with the state of the system (see, e.g., Sherman et al., 2013; Scholz et al., 2014), as well as creating normal forces in the joints. The model presented above will be used in the studies of human compliance shown later.

4.3.3 USING OPTIMAL CONTROL FOR COMPLIANCE STUDIES

In this section, we discuss how optimal control methods can be used to tune compliance in walking systems for design and control optimization for both constant and variable compliance models. Fig. 4.3.2 shows the components of an optimal control problem. We do not go into the detailed description and solution methods of an optimal control problem and its solution since this has already been treated in Subchapter 4.2 on optimization. However, it should be emphasized that optimal control, in contrast to any other method, can simultaneously optimize all compliant components of the locomotion system for an optimal support of the dynamic motion considered. For systems with many DOF and compliance models with many possible variations, no manual tuning is feasible and such an automated optimization is absolutely crucial.

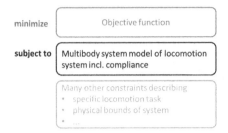

FIGURE 4.3.2 Components of an optimal control problem for optimization of design and control of compliant models.

In optimal control problems, we distinguish the following types of variables: t, denoting time, $x(t)$ denoting the vector of system states, which are free functions in time, $u(t)$ denoting the vector of control or input variables of the system which are also free functions in time, and p the vector of free model parameters. In the case of constant compliance, the compliance parameters to be tuned would be part of the free model parameters p to be optimized. In the case of variable compliance, it is useful to choose the variable to be modulated (i.e., the rest length or the stiffness coefficient) or suitable derivatives of them as free control variables $u(t)$ to be optimized.

The constraints of the optimal control problem (second and third block in Fig. 4.3.2 make sure that the physical properties of the system (the human, the robot, the animal) are correctly taken into account, including the properties of the compliance model chosen, and that the desired locomotion task is fully described. Here it is also important to correctly capture the limits of compliance modulating actuators or for the choice of passive compliant elements in the model.

Examples for objective functions (firs block in the figure) in the context of compliance optimization are:

- For models with constant parallel compliance like in Eqs. (4.3.2) or (4.3.3), different measures related to energy consumption or efficiency could be optimized, such as the integral over the weighted sum of active torques squared $\int_0^T \tau_{active}^T W \tau_{active} dt$ or the cost of transport. This results in an optimal choice of the parameters of the parallel springs (and dampers) within the chosen bounds to support this criterion.
- Performance related criteria, such as a maximization of walking speed or jumping height or width, could be considered for all compliance models, in all cases resulting in compliance that is tuned for this particular criterion.
- For investigating compliance in human motions, based on motion recordings, the fit of the dynamic model with respect to the data can be optimized. Depending on the spring model, there still might be some redundancy in this

formulation (e.g., in the parallel constant compliance case, the same total torque could be produced by an infinite number of combinations of active and passive torques). In this case, any other optimization criterion, e.g., minimizing the active part above, could be added as regularization term (i.e., with a small weight) to distinguish otherwise redundant solutions.

- For variable compliance, it might be interesting to investigate the role of compliance modulation or rest length modulation (for the first example, see Section 4.3.4.2). For this a minimization of the derivatives of the respective terms, which can be chosen as controls, would be suitable. This can also be combined with other criteria.
- Also different types of stability criteria can be used in the objective functions, such as eigenvalues defining Lyapunov stability, ZMP or capture point criteria (also compare Subchapter 4.1 on stability).
- With compliant systems, position control is not straightforward, and also not a suitable thing to do. If, however, it is important to exactly reach or pass through some points, they can be considered as point constraints of the problem; if points should be approximately reached or given paths should be approximately followed, they may be considered in an objective function, e.g., in form of a least squares formulation. This also works for deviations from a desired step/hopping length, height, or width.

4.3.4 OPTIMIZATION-BASED COMPLIANCE STUDIES IN HUMANS

In this section, we present some examples of optimization-based studies of compliance in human motion distinguishing between models of constant compliance and variable compliance.

4.3.4.1 Constant Parallel Compliance Models for Running and Walking

Here we discuss an example for the optimal exploitation of constant parallel compliance by optimization. We consider human running motions for which we set up models in the sagittal plane as well as in 3D, also taking sidewards motions as well as yaw and roll directions int account (Schultz and Mombaur, 2010). The results discussed here concern the sagittal plane running model as shown in Fig. 4.3.3 which has nine segments (a combined pelvis–torso segment, and two thighs, shanks, feet, and arms). It has mono-articular springs in all joints in parallel to the actuators, but no dampers. For more information, see Mombaur (2009).

The optimization studies performed with this model do not involve any human recordings, but aim at generating running motions by optimization only.

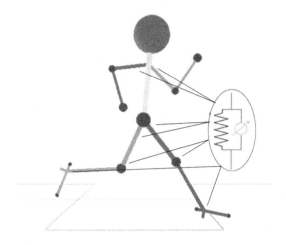

FIGURE 4.3.3 Model of human running in the sagittal plane with constant parallel compliance at each joint.

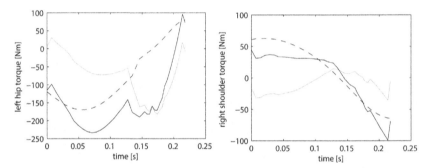

FIGURE 4.3.4 Comparison of active (green line / light gray in printed version), spring torques (black dashed line) and total torques (black solid line) for two joints of the running model.

We have imposed the average running speed of 10 m/s and then searched for the solution with minimum active torques squared. Fig. 4.3.4 shows the results already presented in Mombaur et al. (2009). As shown at the examples of two joints, the passive torques produced by the springs contribute significantly to the total torque, allowing for a reduced active torque. The average savings for these two joints in the particular running motion studied amount to approximately 50%, computed in terms of absolute values of torques.

For the same model, we have also investigated the existence of open-loop stable running motions, i.e., motions that are stable without any feedback (see also the chapter on stability). In this case the resulting spring values are obviously very different since they are principally tuned for generating backdriving forces to the trajectory, in the same sense as a P-controller with a constant set point would do. See Mombaur (2009) for an account on how different objective functions influence the parameter values of the model.

More recently, these studies have been extended to 3D models of human walking with 34 DOF (Felis and Mombaur, 2016), also showing that a significant amount of the effort can be taken over by passive elements. We also have investigated how much passive parallel compliance can contribute to recovery actions from pushes while walking (Kopitzsch and Mombaur, 2017).

4.3.4.2 Compliance Modulation in Human Walking in Different Situations

For the study of compliance in human walking, we have looked at the compliance modulation at the joint level by considering mono- and bi-articular springs with the possibility to modulate stiffness (see models (4.3.6) and (4.3.7)). As many state-of-the-art walking mechanisms use elastic elements with constant stiffness, we are interested in the actual influence of stiffness modulation on the walking gait, addressing the following questions:

1. Is stiffness modulated during human walking?
2. If it modulates, how would a reduction of the modulation influence the walking gait?

This section summarizes the results published previously in Hu and Mombaur (2016, 2017). For this study, a simple model such as the spring-loaded inverted pendulum cannot be used, but we need a model with human topology including the most essential human joints, as in Fig. 4.3.5. Some simplifying assumptions are made, as we restrict the present study to the sagittal plane, and therefore only look at the pitch joints. The model was introduced in Hu and Mombaur (2016) and consists of a 2-dimensional (2D) rigid multibody model with 14 segments and 16 DOF, of which 13 are internal DOF and 3 are for the free floating base. The foot model is a flat foot described by two contact points, one on the toe and one on the heel. Torsional springs are introduced in each of the leg joints, i.e., hip, knee, and ankle. A bi-articular spring is inserted between the hip and knee joints in each leg corresponding to the bi-articular muscles of humans. A damper is inserted in each ankle to avoid possible oscillations that might occur at liftoff. The stiffness of all the mono- and bi-articular springs is variable and represents the actuation of the leg joints, where torques are generated by varying the stiffness of each spring in time. The modulation of the stiffness, the rest position of each spring, and the damping factor of the ankle are determined by the optimization process.

To answer our first question, we have identified stiffness histories for human walking motions from measurements. Here, the stiffnesses of the springs rep-

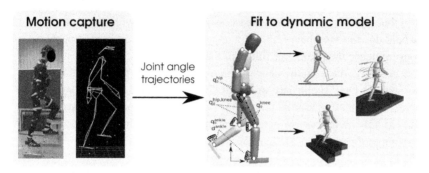

Motion capture   **Fit to dynamic model**

Joint angle trajectories

FIGURE 4.3.5 Joint angle trajectories are extracted from motion capture data and mapped onto a 2D human model with springs in the leg joints. The same model is used for walking on level ground, up a slope, and stairs climbing.

resent the control inputs of the system for the leg joints in the optimal control problem, while in the upper body joint torques are used as controls. We proceed by mapping joint angle trajectories obtained from motion capture data to our model by taking into account dynamic constraints and contact sequences corresponding to the different walking phases. The data sets considered each consist of a single step for four different male subjects walking in three different environments: level ground, walking up a slope, and climbing up stairs. For each subject and environment, one recording is considered from which a periodic step is extracted, in which the right leg is the stance leg and the left one the swing leg. For the mapping procedure, the model is scaled according to the height and weight of the subjects in order to obtain reasonable fittings, and then a least squares optimization problem is solved to minimize the fitting error.

The optimal control problem solutions provide us with stiffness profiles for each of the subjects and environment. The profiles show that stiffness is *not* constant over the whole walking gait, and that it could only be approximated as constant in certain phases (e.g., double support phase). In particular, modulations in a range between 0 and 1500 [Nm/rad] could be observed in many joints in different environments. The results show common features, but overall present very different properties among the walking scenarios and can hardly be generalized for walking in all environments. In the case of level ground, the stiffness is higher in the stance leg knee joint during single support as the whole weight of the body is on the stance leg, while in the swing leg it presents a high spike (up to $\sim$1200 [Nm/rad]) in the ankle joint after the impact of the foot toe with the ground due to the collision. In walking up the slope the average behavior is similar to the level ground case, where the stiffness has high values in the knee joint of the stance leg during single support, while the stiffness of the swing leg is much lower than in the level ground case, the spike on the ankle joint after impact is also much smaller (up to ~300 [Nm/rad]). When climbing stairs the stiffness values are overall smaller than in the previous two scenarios, however, the phases are also different, as climbing stairs does not involve separate toe-tip contacts but it is assumed that the swing leg strikes on the step with a flat foot. However, also in this case, spikes are observed after impacts. The stiffness values of the bi-articular coupling spring are much lower than the mono-articular springs: the maximum values are around 150 [Nm/rad], with highest values during stair climbing. In Hu et al. (2014) a model with fewer DOF was used, and bi-articular couplings were not included. In this case, in some phases the joint stiffness reached higher values than those obtained with the current model with bi-articular couplings. This means that the introduction of bi-articular coupling could reduce the stiffness since forces are summed up, but the modulations are still high and can hardly be defined as constant through all the walking phases in all the three scenarios.

From these observations, we can state that stiffness has high modulations over the walking cycle, where it has higher values in the stance leg during single support and present jumps after impacts. We are therefore interested in answering our second question concerning the effect of a reduced modulation on the walking gait, and if a gait were still reproducible if stiffness were assumed to be constant. To answer this question, we introduced the derivative of stiffness as controls of the optimal control problem, treating the stiffness as additional state instead. In the upper body we keep the joint torques as control inputs. Additionally to the minimization of fitting errors, we added two more objective functions consisting of the minimization of stiffness derivatives to reduce modulation and the minimization of stiffness jumps between phases which were introduced as slack variables in the optimal control problem.

The results are obtained for the four subjects and three environments with combinations with different weights of the objective functions. In particular we imposed increasing weights on the reduction of modulation and on the stiffness jumps to analyze their effects. In Fig. 4.3.6 we show a subset of results obtained for the three environments for one of the subjects. We can observe that with high weight on minimization of stiffness derivatives the stiffness profiles are damped until becoming almost constant (with modulation in the range of 1 [Nm/rad]) in different phases, but can still assume very high values and high jumps between phases. When jumps are minimized, the stiffness over the whole walking cycle is almost constant but can still have big values. To analyze the influence of these profiles on the gait, we looked at the average fitting errors of the leg joint angle trajectories. In the case of level ground and walking up a slope, the gait can be still vaguely approximated, but with bigger deviations from the original joint trajectories than in the case with variable stiffness. In the case of stair climbing, the modulation has higher influence on the gait, as the deviations are much bigger, where the joint with biggest error is the hip joint of both stance and swing legs. In every environment, however, deviations are on average smaller if at least stiffness jumps across phases are allowed, even though it can be assumed that human stiffness does not exhibit such jumps.

From the observations made on fitting errors and the obtained stiffness profiles, we can say that in generating walking motions, stiffness modulation should be taken into account where continuous modulation should be a desired feature. However, this is hard to achieve with real life systems such as state-of-the-art variable stiffness actuators. We showed that on level ground and walking up a slope it is possible to recreate walking motions with moderate deviations from the original motion also with constant stiffness over the whole walking cycle, but with different stiffness values in different joints. The gait could be improved if stiffness can be changed for different walking phases, in particular when complex walking environments are involved. A trade-off between constant and highly modulated stiffness consists in the possibility of switching the stiffness value when transitioning into a different walking phase. The range of this switch is, however, high as we can observe from Fig. 4.3.6.

To answer the two questions we can conclude that:

1. Stiffness modulates during human walking and can assume high values.
2. When reducing the modulation to have almost constant stiffness, a gait that is different but not too far from the original gait can be generated for walking on level ground and slopes, but that large differences occur for the stair climbing case.

These conclusions are based on previous assumptions and simplifications of the model, which we cannot completely neglect in analyzing the outcomes, such as the model in 2D with motions on the sagittal plane only, the flat foot with only two contact points and the structure of the springs introduced in the joints and the bi-articular coupling which already result in a nonzero error in the fit between model and experiments for the best possible fit. Despite the simplifications, the results we obtained still give an important insight into how stiffness modulates in human walking and how the modulation influences the walking gait considering all effects in the sagittal plane, which is the dominant plane for walking.

4.3.5 OPTIMIZATION-BASED COMPLIANCE STUDIES IN ROBOTS

In this section, we discuss examples of the use of different types of compliance in robots and specifically focus on the use of optimal control to select the best possible compliance parameters and control inputs with respect to the chosen criterion.

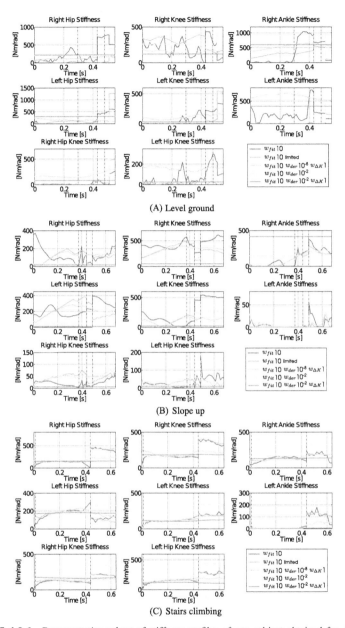

FIGURE 4.3.6 Representative subset of stiffness profiles of one subject obtained for different combinations of weights on the objective functions: w_{fit} for fitting error, w_{der} for stiffness derivatives, $w_{\Delta K}$ for stiffness jumps. Limited means that the stiffness derivatives are constrained within the boundaries ± 1000 [Nm/rad·s], while in the first case the boundaries are 100 times higher. Please note that all the plots have different scaling. (For interpretation of the colors in this figure, the reader is referred to the web version of this chapter.)

4.3.5.1 Constant Serial Compliance in Robots

Beside the advantages introduced by the compliant actuators in humanoid robots, the complexity of motion generation and control of the robots also increases significantly. In this section, we discuss robots with constant serial compliance, as described in Eq. (4.3.4).

One important example is the new design of the iCub (Metta et al., 2010) robot legs (Parmiggiani et al., 2012), which was derived from the COMAN humanoid robot (Colasanto et al., 2012), with the aim to perform walking and balancing motions. A novelty with respect to previous iCub versions was the introduction of serial elasticity in the knee and ankle pitch joints, with spring stiffness of 350 [Nm/rad]. This is similar to the series elastic actuators (Pratt and Williamson, 1995b), but no measurements of the force via the spring deformation are taken. A peculiarity of these actuators is that the spring can be unmounted, making the actuator rigid and giving the possibility of comparing the performances with and without compliance in the actuators.

We use the HeiCub robot (Hu et al., 2016b), a reduced version of the full iCub robot located in Heidelberg, with 3 DOF torso and 6 DOF legs. Before studying walking motions, we want to address the simpler but still challenging task of squatting with serial elasticity, by using model based optimal control. The objective is to study the effect of serial elasticity from the motion generation perspective in order to compare different objective functions and identify properties and parameters for the more complex task of walking.

The model is reduced to a 2 or 3 DOF model in the sagittal plane as in Fig. 4.3.7, as the squatting motion does not have significant influence on the frontal and transversal planes. The full dynamic properties of the robot and the elasticity in the knee and ankle pitch joints are taken into account in the problem formulation, i.e., all segments are included but the joints are treated as fixed except for the hip pitch, knee and ankle pitch joints. In this way the generated motions can be transferred to the real robot. The hip joint, which is actually rigid, is modeled here with very high stiffness. The motions are generated with respect to a set of objective functions including maximization of squat range, and minimization of joint torques, joint and motor accelerations, and mechanical work. Stability is modeled based on the Zero Moment Point (ZMP), which has to lie inside the polygon of support.

During the squatting motion, the robot first goes down from a fully stretched position and then goes back up, hence the squatting problem is formulated as an optimal control problem with two continuous phases, where the states are the generalized joint and motor positions and velocities, and the control inputs are the joint torques. The motion is defined as a periodic motion which starts and ends in the same conditions, which means that it could ideally be repeated an

FIGURE 4.3.7   The model of HeiCub reduced to 2 or 3 DOF. The motion is performed in two phases. Numerical results are implemented on the real robot.

infinite number of times. The robot does not follow any predefined trajectory, but the whole-body trajectories are left free to be determined by the optimal control problem solution.

We computed the maximum achievable squat range by formulating it as objective function. In the case of the 2 DOF model, the hip pitch joint was kept fixed at 30 degrees, and the maximum obtained squat range with this setup is of 3.3 [cm], which is extremely small considering that the robot legs are 51 [cm] long. This results from the poor balancing capabilities introduced by the fixed hip joint. In the case of the 3 DOF model, the results look much better: the achievable range is of 18.1 [cm], which corresponds to more than 35% of the leg length of HeiCub. All following computational results presented are for the model with 3 DOF and the determined squat range.

Motions are generated for both the models with and without SEA with a fixed time of 1.5 [s] per phase to compare the different objectives. In the case of rigid actuators, when other objective functions than torque minimization are considered, the torque has high variations in all three joints. The acceleration is subject to a sudden change when the phase changes from squat down to lift up, due to the higher average torque and mechanical work required to perform the motion at lower speed. This could be compensated by adjusting the weights of the objective functions: giving a higher weight to the acceleration term the fast change would be smoothened, but this would, of course, have a negative influence on the torque and work terms. In the case of series elasticity, the behavior is completely different than in the rigid actuator case. Due to the springs, the acceleration variations are attenuated on the joint side. This happens also in the hip joint, mainly in the transition between the two phases, where accelerations of the motors are much higher. Even if a very high stiffness was assigned to the hip pitch to emulate a rigid actuator, the effect of flexibility is still present. The elastic joints, i.e., knee and ankle pitch, are moved first and the rigid joint, i.e., hip pitch, is the last to move. Due to this delay in moving the hip joint, a high velocity is required for the hip pitch joint. This is due to the higher average effort required to move the hip pitch joint at lower speed and to the presence of the springs in the other joints.

The results obtained for both cases with the combination of all objectives was implemented on the real robot. When using the rigid actuators, the positions are tracked accurately in all the joints, but a higher error can be observed on the hip pitch joint, where a small delay is present. This is because the hip pitch and roll joints are coupled and therefore they cannot achieve the same velocity as the other joints. In the case of serial springs, we used a control mode where the tracking feedback loop is closed on the motor side and the spring is left to its natural oscillations. In this case both joint and motor positions could not be accurately tracked, where a bigger error is present on the joint positions, which resulted to be closer to the motor positions. This means that in the real robot the stiffness of the joint could be actually higher than in the model, where the effect could also be due to other elements, such as the stiffness of the transmission system, friction, and damping. As in optimal control models good parameter values are very important, these unknown parameters could only be identified on the actual robot via repeated experiments to further adjust the model.

The goal of this study on squatting was to identify suitable models, parameters and constraints for more complex problems with serial compliance, such as walking. Numerical results have shown that rigid joints should be treated as such rather than as elastic with high stiffness values in optimal control, as this might lead to stiff problems as expected. The experiments on the real robot have shown that the actual stiffness of the joints with serial elasticity is most probably different from the one set in the model, as in the real system many other factors also play important roles in addition to the mechanical spring. Walking motions are commonly generated with reduced models such as the inverted pendulum (Kajita et al., 2003), which has been done also for the HeiCub robot without the use of series elasticity (Hu et al., 2016a). Despite the successful walking motions, reduced models do not respect and also do not allow exploiting the whole-body dynamics of the robots.

As we did for the squat motion, here we aim to use optimal control with the whole-body model of the robot including also serial elasticity, where the problem is formulated in a similar way. The main differences lie in the model, where all the DOF are considered (i.e., 15 internal DOF and 6 DOF for the floating base), and the dynamics of the system is hybrid, i.e., there is a combination of continuous and discrete phases where discontinuities at impacts occur. A higher number of phases are also needed, since three types of steps occur with different phases: the initialization step (or maybe more than one) with which the robot starts the walking motion, the periodic steps in which the robot is in a cyclic motion, and the ending step(s) in which the robot comes to a stop. Constraints include all limitations of the robot acquired during the previous experiments as well as stability constraints. Several objective functions can be used in order to obtain different walking motions: minimization of the joint torques, joint accelerations, motor accelerations (model with SEA only), mechanical work, maximization of step length, walking velocity. The use of serial elasticity during walking is expected to contribute to energy savings and shock absorption – if the passive spring is properly chosen for the walking task. Therefore a thorough analysis of the performance needs to be carried out.

4.3.5.2   Variable Rest Length Results in Robots

In this section, we discuss a simulation model of a two-legged running robot that consists of a toroidal trunk and two telescopic legs and performs motions in the sagittal plane. The robot has four actuators in total, two torque actuators in the hips between trunk and leg with parallel constant spring damper elements, and two series elastic actuators in the telescopic legs, which modulate the rest lengths of the springs. The robot is inspired by the early MIT hopping and running robots. This precise robot, as shown in the simulation models, has never been built (but a simplified version has been (Lang et al., 2009), which, however, used different types of actuators and was not powerful enough to perform the computed motions). The robot model has 7 DOF in the plane when in the air, which can be reduced to 5 DOF if the lower parts of the legs are assumed to be massless. During contact, this reduces to 4 DOF (see Fig. 4.3.8). For more information about the model, see Mombaur et al. (2005).

As shown in Subchapter 4.2 on optimization, different types of periodic motions have been generated for this model, including running, somersaults and flic-flacs, just by changing a few locomotion task constraints. The focus in this context was the optimization of open-loop stability. Here, we focus on the behavior of the series elastic actuator and the modulation of the rest length for these results. In Fig. 4.3.9, we compare the SEA actuation for two different motions of the model taken from Mombaur et al. (2005b): the most stable somersaulting motion which also turns out to be very efficient, and the most efficient running motion which, however, is open-loop unstable (but was chosen for comparison since the stable running motion Mombaur et al., 2005 was quite inefficient).

In both cases, the maximum change added to the original rest length was chosen as 0.2, and both curves reach this maximum. In the running motion, the contact phase takes more than half of the step time, while in somersaulting it is only around 25%. The total step time is increased by 16% for the somersault. It should be noted that an increase of rest length in a compressed spring adds

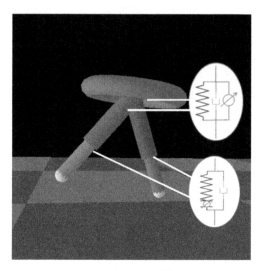

FIGURE 4.3.8 Model of bipedal running robot with SEA.

energy to the system, and a decrease removes energy; however, if the decrease takes place in the flight phase it passes without any effect, due to the assumption of massless lower legs. The SEA actuation of the somersaults is characterized by a very quick peak at the beginning and a somewhat longer second peak until after takeoff. In the running case, there is only one slowly ascending peak, which starts only in the second part of the contact phase, and also lasts until after takeoff, but results in a smaller takeoff speed in vertical direction The right part of the figure shows that the hopping height is much larger for the somersault, which results in the longer flight times mentioned above. The edges in both rest length actuation profiles come from the piecewise constant discretization of the controls chosen for the optimal control problem solution in this case. In the case of a real robot, the representation could be adjusted to the physical properties of the robot's actuators.

4.3.5.3 Variable Compliance in Robots

As discussed in the introduction, variable compliance control can be implemented in several ways in robots. The gains of the position controller then correspond to impedances with the position gain being the stiffness and the derivative gain being the damping. The challenge now is to find suitable impedances for any given time, thus amounting to finding a gain schedule.

Notwithstanding the differences in the different approaches to implement variable impedance, optimal control methods are very well suited to exploit the capabilities of adjusting the impedance in all these systems in a principled way.

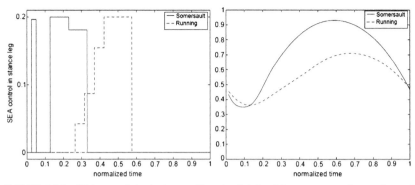

FIGURE 4.3.9 SEA controls in the legs and hopping height of the torus center for running and somersaulting motions.

Hwangbo et al. (2015) show how the impedance of a legged robot with series elastic actuation (Fig. 4.3.10A) can be optimized such that it becomes more robust to nonperceived obstacles. The method employed is a model free, black-box optimization algorithm. Since this method does not rely on a model of the system, the optimization can be executed directly based on evaluating the candidate solutions on hardware. The result is a highly time varying impedance schedule linked to takeoff and touchdown events of the leg (Fig. 4.3.10B). The impedance shows a negative change of stiffness at touchdown. The same is observed in human gait models (Blum et al., 2007). Model based optimal control can as well be used to find impedance schedule. Optimal control is especially powerful if it is used to unify path planning and control gain design as demonstrated in de Crousaz et al. (2015). A fully optimized gait pattern with variable impedances for a quadruped robot is demonstrated in Farshidian et al. (2017). Here Sequential Linear Optimal (SLQ) control is used to find the optimal control.

A model-free method called reinforcement learning (Sutton and Barto, 1998) can be used to find impedance schedules in case model knowledge is insufficient for model based optimal control. In Buchli et al. (2011) it is shown how such an approach finds strongly time varying gain schedules for different robots and tasks by trials of the system and given high level task-unspecific feedback on the quality of task achievement via a cost function (i.e., reinforcement learning). Thus a robot can learn suitable impedances, which is particularly interesting in hard to model situations, typical for biologically inspired robots interacting with complex environments.

When subjected to random force fields humans have a specific way of shaping both the directional impedance as well as the chosen reference path for the hand suitable for the characteristics of the external force experienced. The re-

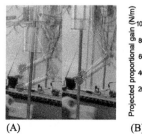

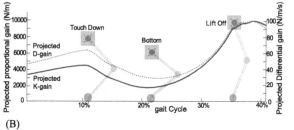

(A)   (B)

FIGURE 4.3.10   (A) A blind hoping legged robot asked to stabilize hopping height and thus to reject random obstacles placed under the foot. (B) Gain schedule at the foot over one hoping cycle. Reprinted from Hwangbo et al. (2015).

sults in Stulp et al. (2012) show that such variable impedance reinforcement learning approaches in principle replicate the impedance and motion learning observed in humans.

The introduced model-based and model-free methods can be combined to develop methods that can benefit both from domain knowledge through model based optimal control and the advantages of model-free learning to improve over the initial model based controller and adapt to the real system. Derived from the first principle of optimality, optimal control and reinforcement learning both share the same mathematical foundation. Farshidian et al. (2014) demonstrates such a pipeline on an instable torque controlled robot. First, a model-based controller and plan is derived with SLQ. This plan is then refined on the real robot using Path Integral Reinforcement Learning.

4.3.6   DISCUSSION & OUTLOOK

In this chapter we have presented several studies on constant and variable compliance in human and robotic locomotion that were based on optimal control as a very helpful tool.

In the study of human walking, we have used variable compliance models as well as models with fixed parallel compliance in the joints in addition to torque actuators. For the constant parallel compliance we could show for various motions that the constant parameter can be tuned in order to best support a desired motion, e.g., in the sense of minimizing and energy-related cost function. In this case, the compliance optimally stores and releases energy in order to minimize the overall energy expenditure. However, it should be noted that the determined parameters are optimal just for this particular motion and don't necessarily work so well for another one, so they would have to be changed. In humans, this can be interpreted by the fact that only part of the passive torque comes from truly passive tissue and that the rest of the compliance part comes from adjustable compliance of the muscles. The studies of variable compliance in human walking based on motion capture recordings have shown that stiffness variation is a characteristic of human walking in different situations. A thorough analysis in other environments with 3D models would be very interesting, as other anatomical planes and rotations would also be considered. This would also allow studying situations as step stone walking, or (lateral) stability studies in beam walking. In this case a revision of the model with more bi-articular couplings might be needed. Stiffness variation might be a desirable feature in wearable robotics such as exoskeletons and prosthesis, as their goal is to support the patients to walk like healthy subjects. Therefore efficient methods have to be developed that do not only allow solving optimal control problems for the variable stiffness problems as offline data fitting problems, but also as online motion generation problems.

Transferring the concept of constant parallel stiffness to robots, the issue of the constant parameters that are only tuned for one particular motion would have to be addressed: either one could implement switchable springs that are only meant to support a set of extremal motions for which they are specially tuned, or one could look for a whole set of typical situations for which support should be provided, e.g., extreme push recovery situations, and then the set of constant springs that is optimally tuned on average for all of them could be implemented.

The actual advantage of variable stiffness, and also constant stiffness for humanoid robots, still needs further investigation. Even though there are many conceptual advantages of compliance, the challenges with it still have to be overcome before compliance can be truly exploited by robots. So far there are only a few bipeds with compliance, and those with variable stiffness that have successfully performed walking are even rarer. A systematic comparison in walking tasks should be performed to actually conclude on the best possible choices of objective functions, and the optimized compliance for this difficult task, and to draw general conclusions about the advantages and disadvantages of compliance in bipedal locomotion. A lot of research will be performed in this area in the coming years.

REFERENCES

Alexander, R. McNeill, 1988. Elastic Mechanisms in Animal Movement. Cambridge University Press.

Amirudin, A.N., Parasuraman, S., Kadirvel, A., Ahmed Khan, M.K.A., Elamvazuthi, I., 2014. Biomehcanics of hip, knee and ankle joint loading during ascent and descent walking. Proc. Comput. Sci. 42, 336–344.

Andriacchi, T.P., Andersson, G.B., Fermier, R.W., Stern, D., Galante, J.O., 1980. A study of lower-limb mechanics during stair-climbing. J. Bone & Joint Surgery 62 (5), 749–757.

Blum, Y., Rummel, J., Seyfarth, A., 2007. Advanced Swing Leg Control for Stable Locomotion. In: Autonome Mobile Systeme. Springer.

Buchli, J., Stulp, S., Theodorou, E., Schaal, S., 2011. Learning variable impedance control. Int. J. Robot. Res. 30 (7), 820–833.

Burdet, E., Osu, R., Franklin, D.W., Milner, T.E., Kawato, M., 2001. The central nervous system stabilizes unstable dynamics by learning optimal impedance. Nature 414 (6862), 446–449.

Colasanto, L., Tsagarakis, N.G., Caldwell, D.G., 2012. A compact model for the compliant humanoid robot COMAN. In: IEEE International Conference on Biomedical Robotics and Biomechatronics, pp. 688–694.

Dallali, H., Kormushev, P., Li, Z., Caldwell, D.G., 2012. On global optimization of walking gaits for the compliant humanoid robot, COMAN using reinforcement learning. Cybern. Inf. Technol. 12 (3), 39–52.

de Crousaz, C., Farshidian, F., Neunert, M., Buchli, J., 2015. Unified motion control for dynamic quadrotor maneuvers demonstrated on slung load and rotor failure tasks. In: IEEE International Conference on Robotics and Automation, pp. 2223–2229.

Endo, K., Paluska, D., Herr, H., 2006. A quasi-passive model of human leg function in level-ground walking. In: IEEE/RSJ International Conference on Intelligent Robots and Systems, pp. 4935–4939.

Farshidian, F., Neunert, M., Buchli, J., 2014. Learning of closed-loop motion control. In: IEEE/RSJ International Conference on Intelligent Robots and Systems (IROS), pp. 1441–1446.

Farshidian, F., Neunert, M., Winkler, A., Rey, G., Buchli, J., 2017. An efficient optimal planning and control framework for quadrupedal locomotion. In: IEEE International Conference on Robotics and Automation (ICRA). Preprint on Arxiv.

Felis, M.L., Mombaur, K., 2016. Synthesis of full-body 3-D human gait using optimal control methods. In: IEEE International Conference on Robotics and Automation (ICRA 2016).

Ferris, D.P., Louie, M., Farley, C.T., 1998. Running in the real world: adjusting leg stiffness for different surfaces. Proc. R. Soc. Lond. B, Biol. Sci. 265 (1400), 989–994.

Franklin, D.W., Liaw, G., Milner, T.E., Osu, R., Burdet, E., Kawato, M., 2007. The end-point stiffness of the arm is directionally tuned to instability in the environment. J. Neurosci.

Franz, J.R., Lyddon, N.E., Kram, R., 2012. Mechanical work performed by the individual legs during uphill and downhill walking. J. Biomech. 45, 257–262.

Gabriel, R.C., Abrantes, J., Granata, K., Bulas-Cruz, J., Melo-Pinto, P., Filipe, V., 2008. Dynamic joint stiffness of the ankle during walking: gender-related differences. Phys. Ther. Sport 9, 16–24.

Geyer, H., Seyfarth, A., Blickhan, R., 2006. Compliant leg behaviour explains basic dynamics of walking and running. Proc. R. Soc. B, 2861–2867.

Günther, M., Blickhan, R., 2002. Joint stiffness of the ankle and the knee in running. J. Biomech. 35, 1459–1474.

Van Ham, R., Sugar, T., Vanderborght, B., Hollander, K.W., Lefeber, D., 2009. Compliant actuator designs. IEEE Robot. Autom. Mag. 16 (3), 81–94.

Hogan, N., 1984. Adaptive control of mechanical impedance by coactivation of antagonist muscles. IEEE Trans. Autom. Control 29 (8), 681–690.

Hu, Y., Eljaik, J., Stein, K., Nori, F., Mombaur, K., 2016a. Walking of the iCub humanoid robot: implementation and performance analysis. In: IEEE-RAS International Conference on Humanoid Robots (Humanoids).

Hu, Y., Felis, M., Mombaur, K., 2014. Compliance analysis of human leg joints in level ground walking with an optimal control approach. In: IEEE-RAS International Conference on Humanoid Robots (Humanoids 2014).

Hu, Y., Mombaur, K., 2016. Analysis of human leg joints compliance in different walking scenarios with an optimal control approach. In: 6th IFAC International Workshop on Periodic Control Systems (PSYCO 2016).

Hu, Y., Mombaur, K., 2017. Influence of compliance modulation on human locomotion. In: IEEE International Conference on Robotics and Automation (ICRA).

Hu, Y., Nori, F., Mombaur, K., 2016b. Squat motion generation for the humanoid robot iCub with Series Elastic Actuators. In: 2016 6th IEEE International Conference on Biomedical Robotics and Biomechatronics (BioRob). IEEE, pp. 207–212.

Hwangbo, J., Gehring, C., Sommer, H., Siegwart, R., Buchli, J., 2015. Policy learning with an efficient black-box optimization algorithm. Int. J. Humanoid Robot. 12 (03), 1550029.

Iida, F., Rummel, J., Seyfarth, A., 2008. Bipedal walking and running with spring-like biarticular muscles. J. Biomech. 41 (3), 656–667.

Kajita, S., Kanehiro, F., Kaneko, K., Fujiwara, K., Harada, K., Yokoi, K., Hirukawa, H., 2003. Biped walking pattern generation by using preview control of zero-moment point. In: IEEE International Conference on Robotics and Automation, Proceedings. ICRA'03 2003, vol. 2. IEEE, pp. 1620–1626.

Kim, J., Lee, Y., Kwon, S., Seo, K., Kwak, H., Lee, H., Roh, K., 2012. Development of the lower limbs for a humanoid robot. In: 2012 IEEE/RSJ International Conference on Intelligent Robots and Systems (IROS). IEEE, pp. 4000–4005.

Kopitzsch, R.M., Mombaur, K., 2017. Using spring–damper elements to support human-like push recovery motions. In: RAAD 2017, Conf. on Robotics in Alpe–Adria-Region.

Lang, T., Mombaur, K., Friedmann, S., Schlöder, J., 2009. Modeling and optimal control of a pogo-leg running robot. In: Proceedings of CLAWAR. Istanbul, Turkey.

Latash, M.L., Zatsiorsky, V.M., 1993. Joint stiffness: myth or reality? Hum. Mov. Sci. 12, 653–692.

Li, Z., Tsagarakis, N.G., Caldwell, D.G., 2012. Walking trajectory generation for humanoid robots with compliant joints: Experimentation with COMAN humanoid. In: 2012 IEEE International Conference on Robotics and Automation (ICRA). IEEE, pp. 836–841.

Liu, Y., Wensing, P.M., Orin, D.E., Zheng, Y.F., 2015. Trajectory generation for dynamic walking in a humanoid over uneven terrain using a 3D-actuated dual-SLIP model. In: IEEE/RSJ IROS, pp. 374–380.

Metta, G., Natale, L., Nori, F., Sandini, G., Vernon, D., Fadiga, L., Von Hofsten, C., Rosander, K., Lopes, M., Santos-Victor, J., et al., 2010. The iCub humanoid robot: an open-systems platform for research in cognitive development. Neural Netw. 23 (8), 1125–1134.

Mombaur, K., 2009. Using optimization to create self-stable human-like running. Robotica 27, 321–330. Published online June 2008.

Mombaur, K., Scheint, M., Sobotka, M., 2009. Optimal control and design of legged robots with compliance. Automatisierungstechnik 57 (7).

Mombaur, K.D., Longman, R.W., Bock, H.G., Schlöder, J.P., 2005. Open-loop stable running. Robotica 23 (01), 21–33.

Mombaur, K., 2014. A study on optimal compliance in running. In: Dynamic Walking Conference. Zürich.

Moro, F.L., Tsagarakis, N.G., Caldwell, D.G., 2014. Walking in the resonance with the COMAN robot with trajectories based on human kinematic motion primitives (kmps). Auton. Robots 36 (4), 331–347.

Parmiggiani, A., Metta, G., Tsagarakis, N., 2012. The mechatronic design of the new legs of the iCub robot. In: 2012 12th IEEE-RAS International Conference on Humanoid Robots (Humanoids). IEEE, pp. 481–486.

Pratt, G.A., Williamson, M.M., 1995a. Series elastic actuators. In: IEEE International Conference on Intelligent Robots and Systems (IROS).

Pratt, G.A., Williamson, M.M., 1995b. Series elastic actuators. In: 1995 IEEE/RSJ International Conference on Intelligent Robots and Systems, Human Robot Interaction and Cooperative Robots, pp. 399–406.

Pratt, J., Krupp, B., 2008. Design of a bipedal walking robot. In: SPIE Defense and Security Symposium. International Society for Optics and Photonics, p. 69621F.

Robinson, D., 1997. Walking through the Leg Lab: series elasticity and virtual model control. In: Workshop II: New Approaches in Dynamic Walking and Climbing Machines, ICAR 97, 8th International Conference on Advanced Robotics, pp. 77–87.

Scholz, A., Stavness, I., Sherman, M., Delp, S., Kecskeméthy, A., 2014. Improved muscle wrapping algorithms using explicit path-error Jacobians. In: Computational Kinematics. Springer, pp. 395–403.

Schultz, G., Mombaur, K., 2010. Modeling and optimal control of human-like running. IEEE/ASME Trans. Mechatron. 15 (5), 783–792.

Selen, L.P., Franklin, D.W., Wolpert, D.M., 2009. Impedance control reduces instability that arises from motor noise. J. Neurosci. 29 (40), 12606–12616.

Semini, C., Barasuol, V., Boaventura, T., Frigerio, M., Focchi, M., Caldwell, D.G., Buchli, J., 2015. Towards versatile legged robots through active impedance control. Int. J. Robot. Res. 34 (7), 1003–1020.

Shamaei, K., Sawicki, G.S., Dollar, A.M., 2013a. Estimation of quasi-stiffness and propulsion work of the human ankle in the stance phase of walking. PLOS ONE 8.

Shamaei, K., Sawicki, G.S., Dollar, A.M., 2013b. Estimation of quasi-stiffness of the human hip in the stance phase of walking. PLOS ONE 8.

Shamaei, K., Sawicki, G.S., Dollar, A.M., 2013c. Estimation of quasi-stiffness of the human knee in the stance phase of walking. PLOS ONE 8.

Sherman, M.A., Seth, A., Delp, S.L., 2013. What is a moment arm? Calculating muscle effectiveness in biomechanical models using generalized coordinates. In: ASME 2013 International Design Engineering Technical Conferences and Computers and Information in Engineering Conference. American Society of Mechanical Engineers. V07BT10A052.

Silder, A., Besier, T., Delp, S.L., 2012. Predicting the metabolic cost of incline walking from muscle activity and walking mechanics. J. Biomech. 45, 1842–1849.

Stulp, F., Buchli, J., Ellmer, A., Mistry, M., Theodorou, E.A., Schaal, S., 2012. Model-free reinforcement learning of impedance control in stochastic environments. IEEE Trans. Auton. Ment. Dev. 4 (4), 330–341.

Sutton, R.S., Barto, A.G., 1998. Introduction to Reinforcement Learning, 1st edition. MIT Press, Cambridge, MA, USA.

Tsagarakis, N.G., Caldwell, D.G., Bicchi, A., Negrello, F., Garabini, M., Choi, W., Baccelliere, L., Loc, V., Noorden, J., Catalano, M., et al., 2016. WALK-MAN: a high performance humanoid platform for realistic environments. J. Field Robotics (JFR).

Van Ham, R., Vanderborght, B., Van Damme, M., Verrelst, B., Lefeber, D., 2007. Maccepa, the mechanically adjustable compliance and controllable equilibrium position actuator: design and implementation in a biped robot. Robot. Auton. Syst. 55 (10), 761–768.

Verrelst, B., Van Ham, R., Vanderborght, B., Daerden, F., Lefeber, D., Vermeulen, J., 2005. The pneumatic biped "Lucy" actuated with pleated pneumatic artificial muscles. Auton. Robots 18 (2), 201–213.

Visser, L.C., Stramigoli, S., Carloni, R., 2013. Control strategy for energy-efficient bipedal walking with variable leg stiffness. In: IEEE International Conference on Robotics and Automation, pp. 5624–5629.

Weiss, P.L., Kearney, R.E., Hunter, L.W., 1986a. Position dependence of ankle joint dynamics-I. Passive mechanics. Biomechanics 19, 727–735.

Weiss, P.L., Kearney, R.E., Hunter, L.W., 1986b. Position dependence of ankle joint dynamics-II. Active mechanics. Biomechanics 19, 737–751.

Wolf, S., Grioli, G., Eiberger, O., Friedl, W., Grebenstein, M., Höppner, H., Burdet, E., Caldwell, D.G., Carloni, R., Catalano, M.G., Lefeber, D., Stramigioli, S., Tsagarakis, N., Van Damme, M., Van Ham, R., Vanderborght, B., Visser, L.C., Bicchi, A., Albu-Schäffer, A., 2016. Variable stiffness actuators: review on design and components. IEEE/ASME Trans. Mechatron. 21 (5), 2418–2430.

Chapter 4.4

Impedance Control for Bio-inspired Robots

Jonas Buchli and Thiago Boaventura
ADRL, ETH Zürich, Zürich, Switzerland

Abstract. *In this subchapter we introduce the basic concepts required to reason about the physics of the interaction of the robot with its environment and means how to control this interaction.*

It is often useful to think about a robot as a collection of rigid bodies connected by articulated links, leading to a model in the form of a rigid body dynamics system. Being a model, it is not fully reflecting reality (e.g., it neglects finite stiffness of the links), but allows one to reason about some of the important governing dynamics of a robot. Such modeling has been tremendously productive and insightful both for robotics (Featherstone and Orin, 2000) and modeling of human and animal biomechanics (Khatib et al., 2009; Burdet et al., 2013). We will use the rigid body dynamics model in this chapter to discuss the effect of dynamic interaction of a robot with its environment. We will show that with the methods and concepts introduced here one can emulate appropriate impedances for a legged robot either for modeling or for function, e.g., such as locomotion and manipulation.

4.4.1 RIGID BODY DYNAMICS

A single rigid body possesses 6 degrees of freedom (DoF) and can be described by the basic equation of motion of a rigid body:

$$\mathbf{M}(\mathbf{q})\ddot{\mathbf{q}} + \mathbf{c}(\mathbf{q}, \dot{\mathbf{q}}) + \mathbf{f} = \mathbf{0}. \tag{4.4.1}$$

Here $\mathbf{M}(\mathbf{q}) \in \mathcal{R}^{6 \times 6}$ is the inertia matrix of the body, $\mathbf{q} \in \mathcal{R}^6$ represents the 6 states of a rigid body (3 Cartesian positions and 3 rotations), $\mathbf{c}(\mathbf{q}, \dot{\mathbf{q}}) \in \mathcal{R}^6$ is the Coriolis force vector, and $\mathbf{f} \in \mathcal{R}^6$ is a generalized force/torque vector, accounting for all external forces and torques acting on the system, including gravity.[2]

---

2. We use the following notation: upper case bold for matrices, lower case bold for vectors, and lower case for scalars. Also, we elected $\mathcal{R}^n$ to denote vectors composed of n "floating point numbers" representing physical quantities that are not necessarily in $\mathbb{R}$. Thus they do not generally constitute vector spaces and thus in general $\mathcal{R}^n \notin \mathbb{R}^n$. The same argument is extended to matrices. See, e.g., also Featherstone (2007) for an in-depth discussion.

A system of uncoupled n_b bodies would possess $n_b \times 6$ DoF. Every joint that we introduce to connect two such bodies together introduces at least one constraint on the relative motion of the bodies, imposing a loss of degrees of freedom of the overall system. We can thus describe the *state* of the system by $n = (n_b \times 6) - c_j$ variables, where c_j is the number of constraints introduced by the joints. A set of coordinates that capture these DoF are commonly also known as minimum coordinates. In robotics, links are typically connected with revolute or prismatic joints, which allow only a single DoF by introducing $c_j = 5$ constraints each: 3 relative Cartesian movement directions and 2 relative rotations for a rotational joint; and 2 Cartesian directions, 3 rotations for a prismatic joint. Thus, for a set of $n_j = (n_b - 1)$ revolute or prismatic joints, we have $n = \big((n_j + 1) \times 6\big) - \big(5 \times n_j\big) = n_j + 6$. This means that an articulated rigid body system with n_j prismatic and/or revolute joints can be fully described using $n_j + 6$ variables.

Example. Consider an articulated "robot" consisting of two links. The two bodies are connected with a revolute joint. Thus the total number of DoF of the mechanism is $n = 7$. A possible choice of a minimal coordinate system in this case is $[x_1, y_1, z_1, \alpha_1, \beta_1, \gamma_1, \theta_{12}]$, i.e., the position and orientation of one of the bodies with respect to an inertial frame, and the angle of the revolute joint connecting the two bodies.

Using a choice of minimal coordinates, we can write the equation of motion of such a system constituted by rigid bodies connected together as

$$\mathbf{M(q)\ddot{q}} + \mathbf{c(q, \dot{q})} + \mathbf{g(q)} + \mathbf{f(q, \dot{q}}, t) = \mathbf{0} \qquad (4.4.2)$$

with $\mathbf{M(q)} \in \mathcal{R}^{n \times n}$, $\mathbf{c(q, \dot{q})} \in \mathcal{R}^n$, $\mathbf{f(q, \dot{q}}, t) \in \mathcal{R}^n$. Note that we have split up $\mathbf{f}$ and made the gravitational forces explicit in the new term $\mathbf{g(q)} \in \mathcal{R}^n$ for convenience. The physics of the system still looks very much like a rigid body dynamics; however, the properties of the system are now state dependent, and thus in general also a function of time. It is beyond the scope of this chapter to show how the quantities $\mathbf{M(q)}$ and $\mathbf{c(q, \dot{q})}$ are derived. There are many great text books that describe different methods to do so. In particular, we would like to point the interested reader to Featherstone (2007), where highly efficient algorithms to calculate these quantities are described. The key message is that these quantities can efficiently and automatically be derived by computer code. See Frigerio et al. (2012, 2016) for an efficient, yet easy to use implementation of these concepts.

For the modeling of rigid body systems, we typically use a few coordinate systems that have a special importance. First, the already mentioned *inertial coordinate system* that represents the "resting" world, which does not move.

It is also called *world coordinate system*. Second, there is the *joint coordinate system* that describes the position of the links by using only the joint position (e.g., angles) with respect to a certain reference, which is usually the position of the previous body in the kinematic chain. This is usually a more intuitive and "natural" representation of the robot.

There are other coordinate systems that may be relevant for modeling and control of rigid body systems. For instance, there is the coordinate system fixed to an arbitrarily chosen reference body, named *base*, which is often the heaviest or biggest body of the robot. This coordinate system is called *base coordinate system*, which moves as the base moves. There is also the *motor coordinate system*, which is usually the same as the joint coordinates, but not necessarily. Also, especially in bio-inspired robots one often finds the case that a motor system (e.g., an artificial muscle or tendon based actuation system) spans two DoF, e.g., to mimic biarticular muscles, or that some joints deliberately remain unactuated. Such cases can readily be modeled with the notions of task or operational spaces, as introduced in the next sections. In case of a muscle/tendon actuated system, however, it can be useful to reintroduce a dedicated *muscle coordinate system*, see Demircan et al. (2012) for an example.

Example. Consider the simple 2D floating base legged robot shown in Fig. 4.4.1. Given the constraints inserted by the revolute joints, the position of each link of the robot can be fully defined by the (x_i, z_i) position of a certain point (e.g., the center of mass) fixed to it. Since the base is floating and can also rotate, it needs an additional angular position θ_b in order to be fully defined. Thus, to fully define the configuration of the entire robot using only the world frame, we would need 7 variables: $x_b, z_b, \theta_b, x_1, z_1, x_2, z_2$. By using the joint coordinate system for the links we have the minimal representation with 5 variables: $x_b, z_b, \theta_b, \theta_1, \theta_2$.

A particularly useful set of coordinates for mobile robots with arms and legs are the joint coordinates $\mathbf{q_j}$ and the position and orientation $\mathbf{x_b}$ of the base frame:

$$\mathbf{q} = [\mathbf{x_b} \quad \mathbf{q_j}]^T \tag{4.4.3}$$

where

$$\mathbf{x_b} = [x, y, z, \alpha, \beta, \gamma]^T,$$
$$\mathbf{q_j} = [\theta_1, \dots, \theta_{n_j}]^T.$$

This leads to the formulation of a *floating base rigid body system* (Sentis and Khatib, 2006; Dubowsky et al., 1999; Mistry et al., 2010), that is, a system of

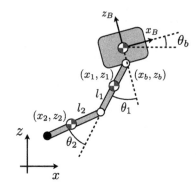

FIGURE 4.4.1 Schematics for a planar legged robot with 2 revolute joints. Based on the inertial Cartesian frame $[x, z]$ the positions of base and links (x_i, z_i), with $i = \{b, 1, 2\}$, can be defined. The orientation of the base is given by θ_b. Other coordinate systems are also shown, such as the joint polar coordinate system $[\theta_1, \theta_2]$ and the base Cartesian coordinate system $[x_B, z_B]$. These coordinate systems can be used separately or in combination to model and control a robot.

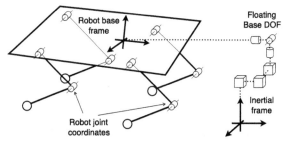

FIGURE 4.4.2 3D representation of three typical coordinate systems used in modeling biological inspired robots. The robot *joint coordinates* $\mathbf{q_j}$ describe the (internal) positions of the links with respect to the *base frame* $\mathbf{x_b}$, which in turn is measured relatively to the *inertial frame*. Together these coordinate systems build a complete minimal coordinate system for a mobile robot.

coupled rigid bodies, i.e., a robot, that is not fixed to any point in the environment, see Fig. 4.4.2:

$$\mathbf{M(q)\ddot{q}} + \mathbf{c(q, \dot{q})} + \mathbf{g(q)} + \mathbf{J_c(q)}^T \mathbf{f_c} + \mathbf{f_m} + \mathbf{f(q, \dot{q}, t)} = 0. \qquad (4.4.4)$$

In the description above, we made the effect of the contact forces of the robot end-effectors with the environment explicit rather than adding them up in the general external force vector $\mathbf{f}$ by writing them as $\mathbf{J_c(q)}^T \mathbf{f_c}$, where $\mathbf{J_c(q)} \in \mathcal{R}^{n_c \times n}$ is the configuration-dependent contact Jacobian, and $\mathbf{f_c} \in \mathcal{R}^{n_c}$ the contact forces vector of the n_c contacts. In particular, this form makes it clear that it is not necessary to derive different rigid body dynamics models for each contact state the robot may face (e.g., when the foot is in stance or flight phase in a legged robot). Instead, the different contact states are reflected in here by different choices of the contact Jacobian $\mathbf{J_c}$. This is a highly flexible and useful approach both for simulation (Coros et al., 2011) and control (Barasuol et al., 2013) since the contact Jacobian can be selected algorithmically at run-time by a program.

Furthermore, given a robot usually has actuators to drive some or all of its joints, a term of motor forces $\mathbf{f_m}$ can also be made explicit. We elect to write this term as

$$\mathbf{f_m} = \mathbf{S}^T \boldsymbol{\tau} \tag{4.4.5}$$

where $\boldsymbol{\tau} \in \mathcal{R}^{n_j}$ is the vector of joint torques, and $\mathbf{S} \in \mathcal{R}^{n_j \times n}$ is a binary selection matrix reflecting the fact that only some of the DoF of a robot can be directly actuated.

Example. The selection matrix $\mathbf{S}$ for a 3D robot with fully actuated joint space, would be written using $\mathbf{S} = \begin{bmatrix} \mathbf{0}_{n_j \times 6} & \mathbf{I}_{n_j \times n_j} \end{bmatrix}^T$. This is because the 6 DoF of the *base* of the robot can never be directly actuated.

Finally, all further external forces acting on the robot links can be summed up and accounted for in $\mathbf{f}$. This includes, e.g., joint friction, and eventually forces due to passive mechanical components fixed to the robot structured, such as springs and dampers.

Example. Consider adding a rotational spring to the first revolute joint of the 2D robot of Fig. 4.4.1, and that it is the only "external" force acting on it. The external force $\mathbf{f}$ would be modeled as $\mathbf{f} = [\mathbf{0}_{1 \times 3} \quad f_{spr} \quad 0]$ where $f_{spr} = k_{spr}(\theta_1 - \theta_{1_0})$, θ_{1_0} is the spring resting length, and k_{spr} the spring stiffness.

A detailed discussion of the mathematics of rigid body dynamics systems is beyond the scope of this chapter. A great in depth discussion of rigid body dynamics and constraints can be found in Featherstone (2007).

4.4.2   TASK/OPERATIONAL SPACES

Ultimately, instead of the dynamics in joint space, one would rather like to know what the "physics" of the robot looks like at other points of interest, e.g., the hand, the foot, the Center of Mass (CoM), etc. Especially, we would like to understand the behavior of the interaction with the environment at such chosen points. It turns out that the dynamics at these points can be straightforwardly derived from the above introduced general dynamics.

As we will see shortly, there are two key quantities of interest to understand the physics governing interaction in any rigid body dynamics system: the forces and the velocities. We thus need to understand how to transform these quantities from one coordinate system into another one. The key is choosing a coordinate frame (the "task space") that describes the point of interest and the directions of motion at this point. Then, properties of the coordinate transform can be used to map the equations describing the physics between the two systems. In the interest of space and explanatory power, in the following we will use an informal, illustrative, yet not entirely mathematically precise way of deriving the two relationships.

For the time being, we will assume that there exists a coordinate transform between the two chosen systems: $\mathbf{x} = T(\mathbf{q})$. Firstly, let us now derive the velocity $\dot{\mathbf{x}}$ in the new coordinate system:

$$\dot{\mathbf{x}} = \frac{d\mathbf{x}}{dt} = \frac{dT(\mathbf{q})}{dt}, \tag{4.4.6}$$

which using the chain rule can be written as

$$\dot{\mathbf{x}} = \frac{\partial T(\mathbf{q})}{\partial \mathbf{q}} \dot{\mathbf{q}} = \mathbf{J}_x^q \dot{\mathbf{q}}. \tag{4.4.7}$$

The partial derivative $\mathbf{J}_x^q = \frac{\partial T(\mathbf{q})}{\partial \mathbf{q}} = \frac{\partial \mathbf{x}}{\partial \mathbf{q}}$ is called the *Jacobian* matrix. To stress which of the two involved coordinate systems we are referring to, colloquially we often say that it is the *Jacobian from coordinate system q to x*, and denote this fact in the notation. As we will see, the Jacobian has a central role in understanding almost any aspect of rigid body dynamics, including impedance behavior.

Secondly, let us derive the relationship of the (generalized) forces in the two coordinate systems for a nonredundant rigid body system (i.e., a system with the same number of degrees of freedom in task and joint spaces). Here we use a simple variational principle. The work being done in both coordinate-systems must be the same. Thus any small change in the state of the system described in one coordinate system must amount to the same work in the other system:

$$\mathbf{f}^T \delta\mathbf{x} = \boldsymbol{\tau}^T \delta\mathbf{q}. \tag{4.4.8}$$

Applying the definition of the Jacobian,

$$\mathbf{J} = \frac{\delta\mathbf{x}}{\delta\mathbf{q}}, \tag{4.4.9}$$

we can rewrite this as

$$\mathbf{f}^T \mathbf{J}\delta\mathbf{q} = \boldsymbol{\tau}^T \delta\mathbf{q}. \tag{4.4.10}$$

This statement has to be valid for all nonzero "virtual displacements", and thus we get

$$\boldsymbol{\tau} = \mathbf{J}^T \mathbf{f}. \tag{4.4.11}$$

For redundant systems, where the "robot" has more degrees of freedom at the joint than at the task space, the relationship of these generalized forces is not unique anymore. In this case, there is a nullspace where joint motions do not produce any motion or work at the task space. For more details on redundant systems and on how to handle this nullspace, please refer to, e.g., Mistry and Righetti (2011), Sentis (2007).

4.4.3 IMPEDANCE & ADMITTANCE

With the above introduced foundational concepts on the physics of a rigid body system with its environment, let us now model them as impedances and admittances.

We are seeking to understand the dynamic interaction (i.e., a functional description of the time evolution in the form of a differential or integral equation) of a certain point on the robot with its environment. This is captured by the dynamic relationship of the flow and effort variables, which describe the power flow at this point. In order to be able to fully describe the time evolution of power, one needs to understand the states of the system, which in this case are energy storages. In any given mechanical system there are two types of energy storage elements: one type for kinetic energy and one type for potential energy.

To understand the basic physics of these elements and the fundamental difference between the two, let us take a look at the simplest linear storage elements: an *ideal* mass m (kinetic energy) and an *ideal* spring (potential energy) following Hooke's law with spring constant k.

We can see that there is a fundamental difference between these two elements. For the *mass*, the velocity is causally dependent on the force imposed on the it, while for the *spring* the force is causally dependent on the velocity imposed on the spring. Also, for a mass and a spring, the energy stored is directly determined by the *state* "velocity" and the physical parameter "mass" ($e_k = (\frac{m}{2})\mathbf{v}^2$), and by the *state* "force" and the physical parameter "spring constant" ($e_p = (\frac{1}{2k})\mathbf{f}^2$), respectively. We thus realize that one of the most foundational principles of physics, causality, dictates two distinct classes of elements, which we elect to define as *admittances* and *impedances*, respectively. We thus come to a first alternative definition of admittances as stores of kinetic energy, and impedances as stores of potential energy.

Example. From the above definition it becomes clear that gravity acts as a configuration-dependent *impedance* on the robot bodies. Consider a planar 1-DOF fixed-base robot with a single revolute joint. The Jacobian matrix $\mathbf{J}$ of

TABLE 4.4.1 Pure mass and spring dynamics

	Mass	Spring
Governing differential equation	$\dot{\mathbf{v}} = \frac{1}{m}\mathbf{f}$	$\dot{\mathbf{f}} = k\mathbf{v}$
Governing integral equation	$\mathbf{v} = \frac{1}{m}\int \mathbf{f}dt$	$\mathbf{f} = k\int \mathbf{v}dt$

TABLE 4.4.2 Admittance and impedance definitions

	Admittance	Impedance
Integral definition	$\mathbf{v} = \int f_A(\mathbf{f})dt$	$\mathbf{f} = \int f_I(\mathbf{v})dt$
Differential definition	$\dot{\mathbf{v}} = f_A(\mathbf{f})$	$\dot{\mathbf{f}} = f_I(\mathbf{v})$

the link center of mass $\mathbf{p}_G = [x_G\ z_G]^T$ would be given by

$$\mathbf{J} = \begin{bmatrix} \frac{dx_G}{d\theta} \\ \frac{dz_G}{d\theta} \end{bmatrix} = \begin{bmatrix} -l\sin\theta \\ l\cos\theta \end{bmatrix}$$

where l is the magnitude of the vector $\mathbf{p}_G$, and θ the angle between $\mathbf{p}_G$ and a horizontal line. The gravitational force acting on the center of mass is given by $\mathbf{f} = [f_x\ f_z]^T = [0\ -mg]^T$, where m is the total mass of the link and g is the gravity acceleration magnitude. The respective joint torque is then $\tau = \mathbf{J}^T\mathbf{f} = -lmg\cos\theta$. This derivation clearly illustrates that the reflected torque caused by gravity can be interpreted as a nonlinear joint spring. An equivalent derivation of the above torque equation can be done considering the cross product $\tau = \mathbf{p}_G \times m\mathbf{g}$. For robots with more degrees of freedom, the reflected torque at the a certain joint will depend not only on the position of the link attached to it, but also on the position of all the preceding and subsequent links. For floating base robots, the reflected gravity torques will depend also on position and orientation of the base.

To write the definitions in general form we must express power as $p(t) = \mathbf{v}(t) \cdot \mathbf{f}(\mathbf{v}(t))$ in case of an impedance or $p(t) = \mathbf{f}(t) \cdot \mathbf{v}(\mathbf{f}(t))$ in case of an admittance, which will thus lead to an integral equation, as shown in Table 4.4.2. We list an equivalent differential definition, which is the common definition of impedance and admittance in the literature. As can be seen from these definitions, there is no requirement for the admittance function f_A and the impedance function f_I being linear, or even continuous.

Also from the introductory discussion above and the definitions in Table 4.4.1 and Table 4.4.2, it is easy to intuitively grasp that the two elements have a natural definition of input and outputs. This is a reflection of the causality underlying the physical nature of such mechanical elements: *inputs* can be non-differentiable, but the *outputs* cannot. The reasoning behind this is that in order to have an instant (e.g., step-like) response at the output, the input to the dynamic system would need to have an infinite magnitude. For instance, for an impedance (i.e., a spring) to have a step-like force response, an infinite velocity would be required. While of course we can mathematically express such infinite quantities, they are a clear sign that important aspects of the physical reality are being neglected. This is particularly important if these relationships are used in closed loop control, where causality must be observed and is thus very important for impedance control.

This simple causality analysis leads to a second alternative and equivalent definition of admittance and impedance based on their input/output relation. An impedance, defined at a *particular location* of a mechanical system, is a function describing the force produced when a motion is imposed on the system at this location. In analogy, an admittance is defined as the function describing the motion that a mechanical system exhibits if a force is applied to a given location.

From this discussion, we realize that without further assumptions we cannot transform an admittance into an impedance and vice versa, as this in general leads to a causality inversion. If, however, we make some assumptions, common ones are continuity and linearity, we can define algebraic relationships between the two that are useful for mathematical treatment (such as standard and well known treatment of linear systems for control (Franklin, 1993)). In this case, the admittances and impedances can be defined as transfer functions from force to velocity and velocity to force, respectively. These definitions can also be further generalized and applied to other physical quantities via the definition of flow and effort, and bond graph modeling (Hogan, 1985a), as shown in Table 4.4.3.

Example. Applying Newton's second law to a damped spring–mass system, we get the dynamics $\mathbf{f} = m\ddot{\mathbf{x}} + b\dot{\mathbf{x}} + k(\mathbf{x} - \mathbf{x_0})$, where $\mathbf{x_0}$ is the spring resting length vector. We can then apply the Laplace transform to this ordinary differential equation to get the impedance transfer function $z(s) = \frac{f(s)}{x(s)} = ms + b + \frac{k}{s}$. Since for linear systems the admittance is the inverse of the impedance, we can also write the admittance transfer function $a(s) = \frac{1}{z(s)} = \frac{1}{ms+b+\frac{k}{s}}$.

We now want to turn back and apply the concept of impedance to a robot modeled as a general articulated rigid body dynamics system as introduced in the beginning of this chapter.

4.4.4 IMPEDANCE OF A ROBOT

The definition of an impedance is always linked to a chosen location of interest in a mechanical system and thus a given system can exhibit many different

TABLE 4.4.3 Linear impedances & admittances

	Impedance	Admittance
Transfer function definition	$z(s) = \frac{f(s)}{v(s)}$	$a(s) = \frac{1}{z(s)} = \frac{v(s)}{f(s)}$
Spring	$\frac{k}{s}$	$\frac{s}{k}$
Mass	ms	$\frac{1}{ms}$
Damper	b	$\frac{1}{b}$
Spring–mass–damper	$ms + b + \frac{k}{s}$	$\frac{1}{ms+b+\frac{k}{s}}$

impedances (or admittances); of course, some of them are functionally linked through the rigid body dynamics.

In other words, using the mathematics of task spaces we can transform the impedance given in one coordinate system into another. For example, we might be interested in the reflected motor inertia/admittance at the output of a gear train, or the impedance seen at the foot of a robot controlled via motors in its joints.

Example. Consider passive springs placed at the revolute joints of a robotic arm or leg such that $\boldsymbol{\tau} = \mathbf{K}_\theta(\mathbf{q} - \mathbf{q_0}) \Rightarrow \dot{\boldsymbol{\tau}} = \mathbf{K}_\theta \dot{\mathbf{q}}$. Considering the task space relationship $\dot{\mathbf{q}} = \mathbf{J}^{-1}\dot{\mathbf{x}}$, we can write $\dot{\boldsymbol{\tau}} = \mathbf{K}_\theta \mathbf{J}^{-1}\dot{\mathbf{x}}$. Now, given the task space generalized forces relationship, we have $\boldsymbol{\tau} = \mathbf{J}^T \mathbf{f} \Rightarrow \dot{\boldsymbol{\tau}} = \dot{\mathbf{J}}^T \mathbf{f} + \mathbf{J}^T \dot{\mathbf{f}}$. For a fixed posture (i.e., $\dot{\mathbf{J}} = \mathbf{0}$) we have $\dot{\boldsymbol{\tau}} = \mathbf{J}^T \dot{\mathbf{f}}$. Via simple manipulations and substitutions, we can write $\dot{\mathbf{f}} = \mathbf{J}^{-1^T} \mathbf{K}_\theta \mathbf{J}^{-1}\dot{\mathbf{x}} = \mathbf{K_x}\dot{\mathbf{x}}$, where the term $\mathbf{K_x} = \mathbf{J}^{-1^T} \mathbf{K}_\theta \mathbf{J}^{-1}$ can be seen as the reflected task space stiffness. In the same way, one can derive, e.g., the reflected motor inertia at a certain task-space.

In general, we can write the behavior of the original rigid body dynamics system (4.4.2) in the chosen new coordinate system of the task space as

$$\boldsymbol{\Lambda}(\mathbf{x})\ddot{\mathbf{x}} + \boldsymbol{\mu}(\mathbf{x}, \dot{\mathbf{x}}) + \mathbf{p}(\mathbf{x}) = \mathbf{f}. \qquad (4.4.12)$$

The inertia $\boldsymbol{\Lambda}(x)$ in task space is configuration dependent and can be shown to be (Khatib, 1987)

$$\boldsymbol{\Lambda}(x) = \mathbf{J}^{-T}(\mathbf{q})\mathbf{M}(\mathbf{q})\mathbf{J}^{-1}(\mathbf{q}). \qquad (4.4.13)$$

Important to realize is that the impedance seen at a certain point of a robot is in general not only dependent on its joint configuration and joint controller, but as well the contact situation with the environment. A simple thought experiment illustrates this fact: Consider the linear admittance presented by a point mass floating in free space which is $A(s) = \frac{1}{Ms}$. Now consider the same body resting against the earth, the admittance seen is now the combined admittance of the particle and the large planet (which is for practical matters infinity). The methods we introduced let us readily reason through such "reflected" inertia in a general case of an articulated robot in contact with the environment.

4.4.5 IMPEDANCE CONTROL

The impedance at a specific contact point on a robot can be achieved in two general ways: *passively* or *actively* (Mason, 1981; Whitney, 1985). *Passive impedance* is obtained through hardware and can be attributed to mechanical elements, such as: the limited stiffness of the robot's links; the compliance of the actuator transmission (e.g., springs, gearboxes, harmonic drives, hydraulic oil, air, etc.); and the softness of the robot "skin" (e.g., a layer of rubber). On the other hand, *active impedance* is achieved via the control of the robot states and the manipulation of joints. These two approaches (and the following) can be combined.

Let us now turn the discussion to the question how one, through the appropriate choice of a controller, can actively influence the impedance of a robot. From the sections above it becomes clear that there are several ways of adjusting the robot impedance: (1) through applying motor forces in reaction to changes in the system state, (2) through configuring the robot into a certain pose, and finally, (3) by placing the robot into a given contact situation. In general, we can combine all three of these methods to achieve a certain behavior of a robot. Let us look in turn into these three different options.

4.4.5.1 Impedance Control Through Joint Control

A well-known and common way of regulating the dynamic behavior of a robot is by controlling its joints. There are several different choices and flavors of impedance controllers. For instance, they can be designed in a task space (e.g., impedance control Hogan, 1985b, using operational space control Khatib, 1987, and virtual model control Pratt et al., 2001), or in joint space (Boaventura et al., 2015). Although the space of interest may change among these approaches, they have an important common characteristic: they rely on a state feedback controller which usually produces *desired joint torques* as output. Thus, a particularly useful way to implement such impedance controllers on real robots is by using SISO low-level joint torque controllers to track the generated torque commands (Boaventura, 2013; Focchi, 2013). As the design of these joint torque controllers goes beyond the scope of this chapter, we will focus instead on presenting few of the above mentioned impedance control approaches. For now, we assume the robot has high-fidelity joint torque sources.

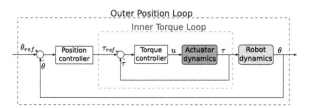

FIGURE 4.4.3 Joint cascade impedance control architecture: an outer loop feeds back the position and creates the torque reference τ_{ref} for the inner torque loop, which calculates the input command u to the actuator. Both outer and inner controllers can feedback additional states in case a model-based controller is employed.

An often used implementation of an impedance controller consists of a cascaded control architecture composed by an inner torque loop and outer position feedback loop that sets the desired impedance (Whitney, 1987), as depicted in Fig. 4.4.3. Here, both control loops are SISO, that is, each joint is controlled individually and does not explicitly take the states of the other joints into account (Boaventura et al., 2015; Luca et al., 2006). It is also common to use a MIMO approach for the outer impedance loop, using e.g., a LQR (linear quadratic regulator) controller (Stengel, 1994; Mason et al., 2014) or a nonlinear feedback linearization controller (Slotine and Li, 1991), where the intrinsic joint coupling can more easily be taken into account. The output of the MIMO controller is usually a vector of joint torques, which are sent as desired values to, usually SISO, joint torque controllers.

To illustrate the design of the outer impedance loop, consider a simple ideal mass m that is driven by a motor force $\mathbf{f_m}$ in a plane. We would like to set a desired dynamic behavior for this mass such that, when interaction with an external entity, this contact force $\mathbf{f_c}$ would be given by

$$\mathbf{f_c} = b_d \Delta\dot{\mathbf{x}} + k_d \Delta\mathbf{x} - m_d\ddot{\mathbf{x}} \tag{4.4.14}$$

where m_d, b_d, and k_d the desired mass, damping, and stiffness, respectively. The operator Δ represents the difference between the desired and the actual values, e.g., $\Delta\mathbf{x} = \mathbf{x_d} - \mathbf{x}$. In this case, the expected acceleration of the mass is

$$\ddot{\mathbf{x}} = m_d^{-1}(b_d \Delta\dot{\mathbf{x}} + k_d \Delta\mathbf{x} - \mathbf{f_c}). \tag{4.4.15}$$

The dynamics of the real mass is determined by Newton's second law $m\ddot{\mathbf{x}} + \mathbf{f_m} + \mathbf{f_c} = 0$. Solving for the motor force and substituting the desired acceleration above, we get the impedance control command,

$$\mathbf{f_m} = -m\, m_d^{-1}(b_d \Delta\dot{\mathbf{x}} + k_d \Delta\mathbf{x} - \mathbf{f_c}) - \mathbf{f_c}. \tag{4.4.16}$$

Example. In case the control designer does not wish to change the inertia of the system, i.e., $m = m_d$, the impedance control law reduces to $\mathbf{f_m} = -(b_d \Delta \dot{\mathbf{x}} + k_d \Delta \mathbf{x})$, which is equivalent to a position PD feedback controller.

We can also easily map the impedance control law derived above to an articulated multi DOF rigid body system. Using the definition of the Jacobian matrix, Eq. (4.4.7), and the desired linear acceleration, Eq. (4.4.15), we can write

$$\ddot{\mathbf{q}} = \mathbf{J}^{-1} \left(\ddot{\mathbf{x}} - \dot{\mathbf{J}}_c \dot{\mathbf{x}} \right) = \mathbf{J}_c^{-1} \left(\mathbf{M}_{\mathbf{x_d}}^{-1} \left(\mathbf{B}_{\mathbf{x_d}} \Delta \dot{\mathbf{x}} + \mathbf{K}_{\mathbf{x_d}} \Delta \mathbf{x} - \mathbf{f_c} \right) - \dot{\mathbf{J}}_c \dot{\mathbf{x}} \right) \quad (4.4.17)$$

where $\mathbf{M}_{\mathbf{x_d}}$, $\mathbf{B}_{\mathbf{x_d}}$, and $\mathbf{K}_{\mathbf{x_d}}$ are the desired *task space* inertia, damping, and stiffness matrices, respectively. Using the rigid body dynamics, Eq. (4.4.4), this leads to the traditional articulated impedance control law for a fully actuated robot, as presented by Hogan (1985b)

$$\mathbf{f_m} = -\mathbf{M}\mathbf{J}_c^{-1} \left(\mathbf{M}_{\mathbf{x_d}}^{-1} \left(\mathbf{B}_{\mathbf{x_d}} \Delta \dot{\mathbf{x}} + \mathbf{K}_{\mathbf{x_d}} \Delta \mathbf{x} - \mathbf{f_c} \right) - \dot{\mathbf{J}}_c \dot{\mathbf{x}} \right) - \mathbf{g} - \mathbf{c} - \mathbf{J}_c^{T} \mathbf{f_c} - \mathbf{f}.$$
$$(4.4.18)$$

Note that this impedance control law requires the measurement of the interaction force $\mathbf{f_c}$ at the contact points.

It is also possible to set a desired impedance in *joint space* similarly to the task space case above by setting a desired joint acceleration as

$$\ddot{\mathbf{q}} = \mathbf{M}_{\theta_d}^{-1} \left(\mathbf{B}_{\theta_d} \Delta \dot{\mathbf{q}} + \mathbf{K}_{\theta_d} \Delta \mathbf{q} - \mathbf{J}_c^{T} \mathbf{f_c} \right),$$

which leads to the following control law:

$$\mathbf{f_m} = -\mathbf{M}\mathbf{M}_{\theta_d}^{-1} \left(\mathbf{B}_{\theta_d} \Delta \dot{\mathbf{q}} + \mathbf{K}_{\theta_d} \Delta \mathbf{q} - \mathbf{J}_c^{T} \mathbf{f_c} \right) - \mathbf{g} - \mathbf{c} - \mathbf{J}_c^{T} \mathbf{f_c} - \mathbf{f} \quad (4.4.19)$$

where $\mathbf{M}_{\theta_d}$, $\mathbf{B}_{\theta_d}$, and $\mathbf{K}_{\theta_d}$ are the joint space desired inertia, damping, and stiffness, respectively.

4.4.5.2 Impedance Control Through Kinematic Configuration Control

Here the kinematic configuration of the robot is adjusted such that a certain impedance is realized. For example, a humanoid robot can chose a stretched out position of the arm to apply more force to an object through directly coupling the inertia of the main body with the output point at the arm. In other words, the object will see the main body as a large part of the reflected inertia at the point of contact with the robot.

Example. In this example we show how one can use the posture of the robot to control its impedance/admittance. Consider the 2-DOF planar legged robot depicted in Fig. 4.4.1. The fixed-base Jacobian matrix $\mathbf{J}_\theta$ for the foot can be written as

$$\mathbf{J}_\theta = \begin{bmatrix} l_1 \cos(\theta_1) + l_2 \cos(\theta_1 + \theta_2) & l_2 \cos(\theta_1 + \theta_2) \\ -l_1 \sin(\theta_1) - l_2 \sin(\theta_1 + \theta_2) & -l_2 \sin(\theta_1 + \theta_2) \end{bmatrix}.$$

As shown in Eq. (4.4.13), the inertia $\mathbf{\Lambda}(\mathbf{x})$ at the task space depends on the inverse of $\mathbf{J}_\theta$. Since near singularities (i.e., $\theta_2 \to 0$ in this case) we have $\det(\mathbf{J}_\theta) \to 0$, the apparent inertia $\mathbf{\Lambda}$ at these regions tends to infinity, and so does the impedance.

If we consider the base is now floating, with extra translational 2 DOFs in x and z directions (rotation not considered, i.e., $\theta_b = 0$), the floating base Jacobian matrix $\mathbf{J}_\mathbf{B}$ would be

$$\mathbf{J}_\mathbf{B} = \left(\begin{array}{cc|c} 1 & 0 & \\ 0 & 1 & \mathbf{J}_\theta \end{array} \right).$$

Due to the matrix inversion in Eq. (4.4.13), it is not straightforward to gain insight into the inertia matrix in an analytic form. However, if we consider the arm in the singular position ($\theta_2 = 0$) and vertical ($\theta_1 = 0$), the inertia matrix $\mathbf{\Lambda}$ at the foot would be given by

$$\mathbf{\Lambda} = \begin{bmatrix} f(p) & 0 \\ 0 & M_b + m_1 + m_2 \end{bmatrix}$$

where $f(p)$ is a general function of a set of parameters p that include the length, center of mass position, mass, and moment of inertia of the links, as well as the base mass M_b. As we can see, the inertia seen at the foot in z direction in this singular position is equal the sum of all the masses, which is a very intuitive result that demonstrates how one can use kinematics to regulate the impedance of a robot.

4.4.5.3 Impedance Control Through Contact Control

Finally, the robot can chose a certain contact configuration to increase the impedance. For instance, a humanoid robot might deliberately chose to rest its forearms against a solid object such as a table in order to have a higher output impedance at the hands and a decoupling of the trunk and the hand to achieve higher manipulation precision. Humans tend to use such a strategy in fine manipulation tasks requiring high precision.

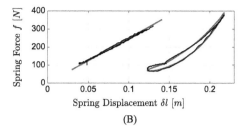

(A) (B)

FIGURE 4.4.4 (A) Schematic of the virtual elements implemented in the HyQ robot leg, which uses a stiff hydraulic actuation system. (B) Linear spring and exponential-spring-damper emulation. The black line shows the ideal behavior, the red the emulated behavior. The work done by these components can readily be evaluated from the available joint data. (For interpretation of the colors in this figure, the reader is referred to the web version of this chapter.)

4.4.6   EMULATION OF MUSCLE MODELS

The presented methods of impedance control can be used to emulate a wide range of "virtual models", it including models from biology. For example, the joint actuator models can be inspired from biology and include the dynamics and properties of muscles, tendon systems, etc. In Fig. 4.4.4B we show the emulation of a linear spring $\mathbf{f} = 2500\delta\mathbf{l}$ and an exponential spring–damper $\mathbf{f} = 17e^{10\delta\mathbf{l}} + 50\dot{\mathbf{l}}$ with the above introduced impedance control on a leg of the HyQ robot, Fig. 4.4.4A (Boaventura et al., 2012; Semini et al., 2011). Thus, such approaches allow one to study the behavior of muscle models in real world behaviors. The energy consumption of the emulated springs, dampers, muscles and other virtual models and flows between them can readily be analyzed from the available joint data, thus amounting in a study of *virtual metabolics*.

An often used muscle model is the Hill model (Hill, 1938), which can be written as (Shadmehr and Wise, 2004)

$$\dot{\mathbf{f}}_{\mathbf{m}} = f(\mathbf{v}_{\mathbf{m}}, \mathbf{l}_{\mathbf{m}}, \mathbf{f}_{\mathbf{m}}) \qquad (4.4.20)$$

where $\mathbf{f}_{\mathbf{m}}$ is the muscle force vector, and $\mathbf{l}_{\mathbf{m}}$ and $\mathbf{v}_{\mathbf{m}}$ are muscle length and contraction vector rate, respectively. This relation not only fits the definition of an impedance, but also describes a state dependent torque that can directly be applied to a torque controlled robot by calculating the muscle length and contraction rate using a Jacobian from joint space to a "virtual" muscle space. The Jacobian emulates the function of the muscle attachment points and the resulting lever arms. See Burdet et al. (2013) for an in-depth introduction and treatment of impedance control of humans using the concepts presented in this chapter among others.

REFERENCES

Barasuol, V., Buchli, J., Semini, C., Frigerio, M., De Pieri, E., Caldwell, D., 2013. A reactive controller framework for quadrupedal locomotion on challenging terrain. In: IEEE International Conference in Robotics and Automation (ICRA), pp. 2554–2561.

Boaventura, T., 2013. Hydraulic Compliance Control of the Quadruped Robot Hyq. PhD dissertation. Istituto Italiano di Tecnologia and University of Genoa, Italy.

Boaventura, T., Semini, C., Buchli, J., Frigerio, M., Focchi, M., Caldwell, D.G., 2012. Dynamic torque control of a hydraulic quadruped robot. In: IEEE International Conference in Robotics and Automation (ICRA), pp. 1889–1894.

Boaventura, T., Buchli, J., Semini, C., Caldwell, D.G., 2015. Model-based hydraulic impedance control for dynamic robots. IEEE Trans. Robot. Autom. 31 (6), 1324–1336.

Burdet, E., Franklin, D., Milner, T., 2013. Human Robotics – Neuromechanics and Motor Control. MIT Press.

Coros, S., Karpathy, A., Jones, B., Lionel, R., Van De Panne, M., 2011. Locomotion skills for simulated quadrupeds. ACM Trans. Graph. 30 (4), 59.

Demircan, E., Besier, T., Khatib, O., 2012. Muscle force transmission to operational space accelerations during elite golf swings. In: 2012 IEEE International Conference on Robotics and Automation (ICRA), pp. 1464–1469.

Dubowsky, S., Sunada, C., Mavroidis, C., 1999. Coordinated motion and force control of multi-limbed robotic systems. Auton. Robots 6 (1), 7–20. http://dx.doi.org/10.1023/A:1008816424504 [Online].

Featherstone, R., 2007. Rigid Body Dynamics Algorithms. Springer, New York.

Featherstone, R., Orin, D., 2000. Robot dynamics: equations and algorithms. In: IEEE International Conference on Robotics and Automation (ICRA), vol. 1, pp. 826–834.

Focchi, M., 2013. Strategies to Improve the Impedance Control Performance of a Quadruped Robot. PhD dissertation. Istituto Italiano di Tecnologia and University of Genoa, Italy.

Franklin, G., 1993. In: Franklin, G., Powell, J., Emami-Naeini, A. (Eds.), Feedback Control of Dynamic Systems, 3rd ed. Addison-Wesley Longman Publishing, Boston.

Frigerio, M., Buchli, J., Caldwell, D.G., 2012. Code generation of algebraic quantities for robot controllers. In: 2012 IEEE/RSJ International Conference on Intelligent Robots and Systems (IROS). IEEE, pp. 2346–2351.

Frigerio, M., Buchli, J., Caldwell, D., Semini, C., 2016. Robcogen: a code generator for efficient kinematics and dynamics of articulated robots, based on domain specific languages. J. Softw. Eng. Robot. (JOSER) 7 (1), 36–54.

Hill, A., 1938. The heat of shortening and the dynamic constant of muscle. Proc. R. Soc. B.

Hogan, N., 1985a. Impedance control – an approach to manipulation. I – theory. II – implementation. III – applications. ASME Trans. J. Dyn. Syst. Meas. Control 107, 1–24.

Hogan, N., 1985b. Impedance control: an approach to manipulation: part II – implementation. ASME Trans. J. Dyn. Syst. Meas. Control 107, 8–16.

Khatib, O., 1987. A unified approach for motion and force control of robot manipulators: the operational space formulation. IEEE J. Robot. Autom. 3 (1), 43–53.

Khatib, O., Demircan, E., De Sapio, V., Sentis, L., Besier, T., Delp, S., 2009. Robotics-based synthesis of human motion. J. Physiol. 103 (3), 211–219.

Luca, A.D., Albu-Schaffer, A., Haddadin, S., Hirzinger, G., 2006. Collision detection and safe reaction with the DLR-III lightweight manipulator arm. In: IEEE/RSJ International Conference on Intelligent Robots and Systems, pp. 1623–1630.

Mason, M.T., 1981. Compliance and force control for computer controlled manipulators. IEEE Trans. Syst. Man Cybern. 11 (6), 418–432.

Mason, S., Schaal, S., Righetti, L., 2014. Full dynamics LQR control for bipedal walking. In: IEEE-RAS International Conference on Humanoid Robots, pp. 374–379.

Mistry, M., Righetti, L., 2011. Operational space control of constrained and underactuated systems. In: Proceedings of Robotics: Science and Systems. Los Angeles, CA, USA.

Mistry, M., Buchli, J., Schaal, S., 2010. Inverse dynamics control of floating base systems using orthogonal decomposition. In: IEEE Int. Conference on Robotics and Automation (ICRA), pp. 3406–3412.

Pratt, J., Chew, C., Torres, A., Dilworth, P., Pratt, G., 2001. Virtual model control: an intuitive approach for bipedal locomotion. Int. J. Robot. Res. 20 (2), 129–143.

Semini, C., Tsagarakis, N.G., Guglielmino, E., Focchi, M., Cannella, F., Caldwell, D.G., 2011. Design of HyQ – a hydraulically and electrically actuated quadruped robot, IMechE Part I. J. Syst. Control Eng. 225 (6), 831–849.

Sentis, L., 2007. Synthesis and Control of Whole-Body Behaviors in Humanoid Systems. PhD dissertation. Stanford University.

Sentis, L., Khatib, O., 2006. A whole-body control framework for humanoids operating in human environments. In: IEEE International Conference on Robotics and Automation (ICRA), pp. 2641–2648.

Shadmehr, R., Wise, S., 2004. Supplementary material – a simple muscle model. In: Computational Neurobiology of Reaching and Pointing. MIT Press.

Slotine, J., Li, W., 1991. Applied Nonlinear Control. Prentice Hall, New Jersey.

Stengel, R. (Ed.), 1994. Optimal Control and Estimation. Courier Dover Publications, New York.

Whitney, D., 1987. Historical perspective and state of the art in robot force control. Int. J. Robot. Res. 6 (1), 3–14.

Whitney, D.E., 1985. Historical perspective and state of the art in robot force control. In: IEEE International Conference on Robotics and Automation (ICRA), vol. 2, pp. 262–268.

Chapter 4.5

Template Models for Control

Patrick M. Wensing[¶] **and Shai Revzen**[∥]

[¶]*Department of Aerospace and Mechanical Engineering, University of Notre Dame, Notre Dame, IN, United States* [∥]*Electrical Engineering and Computer Science and Ecology and Evolutionary Biology, University of Michigan, Ann Arbor, MI, United States*

4.5.1 INTRODUCTION

Within integrative biology, template models have enabled us to gain insight and understanding into a variety of complex motor control tasks. Indeed, we see that despite the relative complexity of our musculoskeletal systems, locomotion is often well modeled conceptually with relatively few degrees of freedom. Species from crabs to kangaroos bounce in a dynamically similar fashion, well described by the Spring-Loaded Inverted Pendulum (SLIP) template model. While the SLIP only captures the role of spring-leg operation in governing the center of mass (CoM) dynamics, in principle, increasingly complex template models may be sought and developed to *describe* increasingly rich aspects of motion. However, the importance of conceptual models is hypothesized to extend well beyond merely these *descriptive* capabilities.

In addition, template models may be used as *prescriptive* models of locomotion, providing low dimensional dynamic targets for closed-loop control in biological and robotic systems alike. It has been hypothesized that the cognitive processes regulating locomotion may include reasoning centered around reduced dimensional subsets of our full dynamics – template dynamics that capture the most salient aspects of a locomotory behavior (Full and Koditschek, 1999). This conjecture is a central tenant of the templates and anchors hypothesis from Chapter 3. These principled reductions capture universal characteristics of locomotion and may play an important role in enabling animals to generalize dynamic performance across such a wide range of scenarios in nature. This section aims to illuminate how conceptually similar reductions may be applied within control of our robots, enabling versatile locomotion with template models for control.

Before we begin, why should one consider the use of these reductive models for control when performance guarantees in the full state space may be provided by other control techniques in this chapter? At present, many methods with performance guarantees are not yet computationally viable for real-time application in the high-dimensional state space of our robots. Some methods, for instance those in Subchapter 4.7, provide certificates of stability and robustness through offline analysis. However, these guarantees are often only valid in a narrow region of the state space. The hybrid and nonlinear dynamics in our high-degree-of-freedom (DoF) robots challenge the development of guarantees that hold more broadly. These challenges collectively motivate the use of template models for control: by addressing an important subset of the dynamics, template-based control provides computational and analytical advantages that enable real-time computation and simplify control system analysis.

As we start to view template models for control, it is important to understand the fundamental differences from the use of template models in biology. Within biology, it is a role of the integrative biologist to *discover* reduced dimensional template dynamics embedded in human or animal motion. These template dynamics may take the form

$$\dot{\mathbf{x}}_T = \mathbf{f}(\mathbf{x}_T)$$

where $\mathbf{f}(\cdot)$ captures the effects of closed-loop sensorimotor control, and $\mathbf{x}_T$ represents the state of the template. Rather than discover existing dynamics, within robotics, it is instead the goal to *synthesize* closed-loop dynamics through real-time control. For an uncontrolled robot, it is possible that no template dynamics exist in advance. As a result, the specific aim of template-based control is to *realize* the behavior of a template through feedback in the full model. To guide the design of this feedback, we view templates themselves as controlled dynamic systems

$$\dot{\mathbf{x}}_T = \mathbf{f}(\mathbf{x}_T, \mathbf{u}_T) \tag{4.5.1}$$

with template control inputs $\mathbf{u}_T$ used to shape the closed-loop response. By strategically crafting closed-loop template controls that achieve a high-level goal (a desired running speed, or recovery from a push disturbance, for instance), anchoring a template imparts satisfaction of these performance objectives in the full model.

4.5.1.1 A Design Process for Template-Based Control

Despite the wide range of applications for template models in control, the development of template-based controllers generally follows a common workflow. This design process can be broken down into three rough steps as depicted in Fig. 4.5.1: template selection, template control, and establishing the template/anchor relationship. Template selection entails the choice or design of a reduced dimensional control dynamic system that captures the challenges of a motion control task while respecting the limitations of any target hardware. Following this selection, reduced dimensional control strategies may be designed for the template. Properties of the closed-loop dynamics for this template may

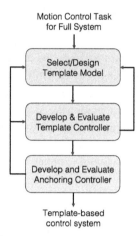

Motion Control Task
for Full System

Select/Design
Template Model

Develop & Evaluate
Template Controller

Develop and Evaluate
Anchoring Controller

Template-based
control system

FIGURE 4.5.1   Three step design process to employ template models for control. The steps proceeding top to bottom represent a linear progression of designing a template-based control system. At each stage in the process, design decisions from earlier steps should be refined, as represented by the upward flows.

then be used to guide control in a high-DoF robot towards establishing a template/anchor relationship.

We further detail each of these steps in the sections that follow. Examples of developing walking and running controllers for a humanoid are presented to clarify the design steps. Despite the rather linear progression in the presentation of the examples, we note that the design process in general should be iterative. In practice, insights gained at each step should be used to continually inform refinements to decisions made earlier in the process. For example, following the development of a SLIP-based template controller in step (2), attempted application to a robot with high-impedance transmissions in step (3) may require redesign of the template model selected from step (1). At the conclusion of the process, the final output is a real-time control system, which may be used in a physical or simulated robot.

4.5.2   TEMPLATE MODEL SELECTION

In the first step of the process, an appropriate template for the motion control task must be selected or designed. This selection may come from study into motor control for biological systems, from previously applied models in robotics, or from personal insight into the fundamental physics of the motor control task. There is no explicitly right or wrong template model for a given task, and within robotics one need not be confined to those models that have appeared in biology. Indeed, fundamental differences between mechanical actuators and materials

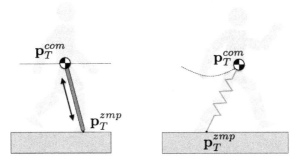

FIGURE 4.5.2 The linear inverted pendulum (LIP) and spring loaded inverted pendulum (SLIP) models are commonly employed templates for walking and running control in legged robots.

compared to biological muscles and tissues necessitates strategically principled bio-inspiration. As a result, the validity of a template model in robotics may only be judged based on the results of the final template-based control system. This may be viewed as a downside. However, we see a similar story for the use of template models in biology. Across both realms, a template model can never explicitly be proven as *the* correct template, but instead must be judged based on its usefulness to understand or control new physical behaviors.

Within the control of locomotion to date, popular template models have generally been simple *physical* dynamic systems. That is to say, systems whose dynamics follow the laws of Newtonian physics. Such models include the linear inverted pendulum model (LIP) as commonly used for walking control (Kajita et al., 2001, 2003; Herdt et al., 2010), the SLIP model used in running control (Blickhan, 1989; Seipel and Holmes, 2005; Garofalo et al., 2012; Wensing and Orin, 2013b), and many others. Examples below detail these common center of mass (CoM) template models further, highlighting their benefits to control dynamic walking (Kuindersma et al., 2015) and running (Wensing and Orin, 2013b). We will follow these examples though the design process of template models for control applied to a high-DoF robot in the sections to come. Through this development $\mathbf{x}_T \in \mathbb{R}^{n_T}$ and $\mathbf{u}_T \in \mathbb{R}^{m_T}$ will denote the state and controls for the template.

4.5.2.1 Linear CoM Models for Walking

Example 1 (LIP Model for Walking). The Linear Inverted Pendulum (LIP) Model, shown in Fig. 4.5.2, captures the connection between the Center of Mass (CoM) and ZMP dynamics under an assumption of constant walking height. It is a commonly selected template to develop walking controllers. The model consists of a point mass m located at a position $\mathbf{p}_{com} = [x, y, z]^T \in \mathbb{R}^3$. The position of the mass is intended to represent the CoM of a system with more degrees of freedom. The model evolves according to forces $\mathbf{f}_{zmp} = [f_x, f_y, f_z]^T \in \mathbb{R}^3$ that act at a ZMP $\mathbf{p}_{zmp} = [p_x, p_y, p_z]^T \in \mathbb{R}^3$. For walking on level ground, p_z remains constant. As is standard with point-mass template models, the LIP assumes that ground reaction forces (GRFs) create no moment about the CoM. This condition defines the line of action for the GRF and requires $f_x/f_z = (x - p_x)/(z - p_z)$ with similar conditions for f_y. As a result, the template dynamics take the form:

$$m\ddot{x} = \frac{f_z}{z - p_z}(x - p_x), \tag{4.5.2}$$

$$m\ddot{y} = \frac{f_z}{z - p_z}(y - p_y), \tag{4.5.3}$$

$$m\ddot{z} = f_z - mg, \tag{4.5.4}$$

where g is the gravitational constant. Under the assumption of a constant height, $\dot{z} = \ddot{z} \equiv 0$. Thus, letting $\omega = \sqrt{g/(z - p_z)}$, we obtain

$$\ddot{x} = \omega^2(x - p_x), \tag{4.5.5}$$

$$\ddot{y} = \omega^2(y - p_y). \tag{4.5.6}$$

These dynamics are unstable, with poles on the real axis at $s = \pm\omega$. The ZMP positions in the plane can be viewed as the control input $\mathbf{u}_T = [p_x, p_y]^T \in \mathbb{R}^2$ for this system, with state $\mathbf{x}_T = [x, y, \dot{x}, \dot{y}]^T$. For this template, a constraint that the ZMP must remain in a support polygon, discussed in Section 4.1.2.2, can be expressed directly through constraints on $\mathbf{u}_T(t)$.

For use in the next section, we note that the LIP dynamics can be expressed in state space:

$$\dot{\mathbf{x}}_T(t) = \mathbf{A}\,\mathbf{x}_T(t) + \mathbf{B}\,\mathbf{u}_T(t) \tag{4.5.7}$$

$$= \begin{bmatrix} \mathbf{0} & \mathbf{I} \\ \omega^2\mathbf{I} & \mathbf{0} \end{bmatrix} \mathbf{x}_T(t) + \begin{bmatrix} \mathbf{0} \\ -\omega^2\mathbf{I} \end{bmatrix} \mathbf{u}_T(t), \tag{4.5.8}$$

with the CoM accelerations considered as an output $\mathbf{y}_T = [\ddot{x}, \ddot{y}]^T \in \mathbb{R}^2$, where

$$\mathbf{y}_T(t) = \mathbf{C}\mathbf{x}_T(t) + \mathbf{D}\mathbf{u}_T(t) \tag{4.5.9}$$

$$= \begin{bmatrix} \omega^2\mathbf{I} & \mathbf{0} \end{bmatrix} \mathbf{x}_T(t) + \begin{bmatrix} -\omega^2\mathbf{I} \end{bmatrix} \mathbf{u}_T(t). \tag{4.5.10}$$

Example 2 (ZMP/CoM Dynamics for Walking with Nonconstant Height). In some cases, such as walking up stairs, the assumption of a fixed CoM height may be prohibitive within a template model for walking control. In the case that a nonconstant desired CoM height $z(t)$ is known in advance, the remaining dynamic equations for $x(t)$ and $y(t)$ are no longer time invariant. We again assume that the forces create no moment about the CoM. Considering that $f_z(t) = m g + m \ddot{z}(t)$, redefinition of $\omega := \omega(t)$ to be time varying according to

$$\omega(t) = \sqrt{\frac{g + \ddot{z}(t)}{z(t) - p_z}}$$

results in an linear time varying (LTV) system for the lateral (x, y) dynamics:

$$\ddot{x} = \omega(t)^2 (x - p_x), \tag{4.5.11}$$

$$\ddot{y} = \omega(t)^2 (y - p_y). \tag{4.5.12}$$

Despite the additional complexity, constraints on the ZMP are still readily addressed through constraints directly on $\mathbf{u}_T(t) = [p_x, p_y]^T$.

4.5.2.2 SLIP Models for Running

Example 3 (A Passive SLIP Model for Humanoid Running). The Spring-Loaded Inverted Pendulum (SLIP) model, described in Subchapter 3.3, is a commonly selected template for control of running and hopping robots. In the SLIP, a point mass m alternates between periods of flight and stance. In flight, the point mass experiences ballistic physics $\ddot{\mathbf{p}}_T^{com} = \mathbf{g}$, where $\mathbf{g} \in \mathbb{R}^3$ is the gravity vector.

During flight, the leg may be repositioned through touchdown angles (θ, ϕ) in the forward and lateral directions. For application to humanoid control, these touchdown angles may be given with respect to an estimated hip position, $\mathbf{p}_T^{hip}$, at a fixed offset from the CoM as shown in Fig. 4.5.3. The hip offset may change from step to step to account for the leg in stance. In comparison to defining touchdown angles in spherical coordinates relative to the CoM, this alternate touchdown angle definition provides closer correspondence with the virtual leg angles in a robot that includes hip separation. As an alternative, the Cartesian position of the foot may be used directly as a control input in flight. This requires more careful consideration of virtual leg length constraints, but provides a close correspondence between the LIP and SLIP in terms of their states and controls.

SLIP stance begins at touchdown (TD) of its virtual leg, wherein a Hookean spring of stiffness k_s and initial length ℓ_0 imparts conservative forces on the mass. The stance dynamics follow

$$m \ddot{\mathbf{p}}_T^{com} = k_s (\ell_0 - \|\boldsymbol{\ell}\|) \hat{\boldsymbol{\ell}} + m \mathbf{g} \tag{4.5.13}$$

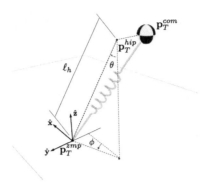

FIGURE 4.5.3 Leg angle definitions for 3D-SLIP applied to humanoid control. Leg angles are given with respect to a virtual hip position that exists at a fixed offset from the CoM during flight.

where $\boldsymbol{\ell} := \mathbf{p}_T^{com} - \mathbf{p}_T^{zmp} \in \mathbb{R}^3$ represents the virtual leg, $\hat{\boldsymbol{\ell}} \in \mathbb{R}^3$ the unit vector along the leg, and ℓ_0 the rest length computed at touchdown. Often, the ZMP position $\mathbf{p}_T^{zmp}$ is fixed following touchdown for simplicity even when the SLIP acts as a template for a robot with flat feet. Stance ends at liftoff (LO), wherein the model transitions back to flight. This energetically passive SLIP model can be controlled through touchdown angle inputs $\mathbf{u}_T[n] = [\theta, \phi]^T$ at each step n.

A unique event for each step (e.g. apex during flight) may be used to define a Poincaré section for study of the step-to-step dynamics. Letting $\mathbf{x}_T[n] = [z, \dot{x}, \dot{y}]^T$ at the nth apex, we define the step-to-step dynamics through

$$\mathbf{x}_T[n+1] = \mathbf{f}(n, \mathbf{x}_T[n], \mathbf{u}_T[n]) \qquad (4.5.14)$$

where $\mathbf{f}(n, \mathbf{x}_T, \mathbf{u}_T)$ is a Poincaré return map.

Example 4 (An Active SLIP Model to Enable Energetic Transitions). As one of a variety of potential extensions for this model, the examples in this chapter will consider modulating the stiffness of the SLIP leg once per step. Other valid actuation schemes for the SLIP may consider changes to its rest leg length, which are equivalent to the addition of a linear actuator in series with the spring. For simplicity, we consider a fixed stiffness k_{s_1} before maximum compression of the spring in stance, and a stiffness k_{s_2} following maximum compression. This consideration modulates the total energy E by $\Delta E = \frac{1}{2}(k_{s_2} - k_{s_1})(\ell_0 - \|\boldsymbol{\ell}\|)^2$, enabling changes in speed and height from step to step. Considering these stiffnesses as control variables along with the leg touchdown angles, the active SLIP may be controlled by selecting

$$\mathbf{u}_T[n] = [\theta, \phi, k_{s_1}, k_{s_2}]^T \qquad (4.5.15)$$

at each step n. As in the LIP, a key aspect of the template is that the forces emanate from a well defined point $\mathbf{p}_T^{zmp}$ that must reside under the supporting contacts.

4.5.2.3 Perspectives of Template Model Selection

Each of these template models have a notable characteristic in that they follow the laws of Newtonian physics. Templates that follow the laws of Newtonian physics have benefits when targeted to legged robots, which themselves must follow the laws of physics. Physical template models may often easily be restricted to operate in a regime that is dynamically feasible for the target system. For instance, control inputs across each of these pendular models directly influence the center of pressure. Constraints for this point to remain within a support polygon are easy to formulate, which can be used to simplify template planning and control. To contrast, for a target system such as a humanoid, constraints on *its* control inputs (joint torques) to satisfy center of pressure constraints are much more complex, and in general are nonlinearly dependent on state.

Despite these benefits afforded in physical template models, such physicality is not a requirement. With such freedom, template models may be judiciously crafted to possess linear (Kajita et al., 2003; Kuindersma et al., 2014), integrable (Mordatch et al., 2010), polynomial (Park et al., 2015), or other simplified dynamics. These simplifications may be sought to further facilitate analysis and control. We again stress that, within robotics, the applicability of template models need not be grounded in biological or exact physical plausibility. Instead templates should ultimately be assessed based on the additional performance that they bring to motion control in physical robots.

Selections of control inputs $\mathbf{u}_T$ also have important reachability implications for the template. For instance, the selection of controls for a passive SLIP precludes the possibility to reach any states at higher or lower total energy levels. As template controls are designed to improve reachability properties, they need not match those controls available the full system. Design choices should be made, however, such that the resultant template dynamics are able to be replicated in the full model given its associated control authority.

4.5.3 TEMPLATE MODEL CONTROL

After selecting a candidate template for a motor control task, the second step in the process is to develop a control system for the template itself. The key idea in using template models for control is that the solution of this template control problem can provide guiding principles to solve the control problem for the full system. General methods to design control systems for template models are the same as those that might otherwise applied for the full model. As a benefit, approaches that might not scale (in time or space complexity) to the full model may become practical when considered in the scope of a template. Trajectory optimization for optimal control (Diehl et al., 2006) may offer applicability for real-time model-predictive control, feedback motion planning libraries (Tedrake et al., 2010) may scale to the dimensionality of template models, and feedforward knowledge-bases may be developed through offline computation (Wu and Geyer, 2013). Template model control has enabled many contemporary systems (Kuindersma et al., 2015; Feng et al., 2013; Pratt et al., 2012; Takenaka et al., 2009; Rezazadeh et al., 2015) to skirt Bellman's curse of dimensionality (Bellman, 1957) while maintaining real-time computation.

4.5.3.1 Control of Linear CoM Models for Walking

Many center of mass (CoM) templates have led to template controllers realized in modern walking humanoids. CoM templates driven by the zero moment point (ZMP) in discrete time (Kajita et al., 2003; Dimitrov et al., 2011) and continuous time (Tedrake et al., 2015) enable real-time optimal control solutions, highlighted in the example below. Whether in the LIP model, or its extension with a fixed CoM height trajectory, linear template dynamics simplify the development of optimal controllers. Other methods have used differential dynamic programming (DDP) to solve the optimal control problem locally when vertical CoM dynamics are allowed to vary but are not fixed a priori (Feng et al., 2013).

Example 5 (ZMP Preview Control Through LQR). A common method of walking control for humanoid robots attempts to control the ZMP position underneath the feet. The linear inverted pendulum template model from Example 1 can enable efficient computation of ZMP controllers for use in a high-DoF robot. Let us assume that a desired ZMP trajectory $\mathbf{u}_T^d(t)$ has been generated in advance and that an assumption of constant CoM height is reasonable. Due to the unstable pole in the LIP dynamics, blindly using this ZMP trajectory as input to the LIP template will result in an unstable CoM motion, which must be handled through feedback in a ZMP controller.

ZMP Preview control attempts to predict (and minimize) future CoM motions and ZMP errors through model predictive control (MPC). This main idea has been pursued by many authors (Kajita et al., 2003; Dimitrov et al., 2011; Tedrake et al., 2015) with different solution methods and formulations. Following the techniques in Kuindersma et al. (2015) and Tedrake et al. (2015), this problem can be formulated as one of continuous time linear optimal control

$$J^*(t_0, \mathbf{x}_0) = \min_{\mathbf{u}(t)} \int_{t_0}^{\infty} \|\mathbf{y}_T(t)\|_{\mathbf{R}}^2 + \|\mathbf{u}_T(t) - \mathbf{u}_T^d(t)\|_{\mathbf{Q}}^2 \, dt \qquad (4.5.16)$$

$$\text{s.t. } \dot{\mathbf{x}}_T(t) = \mathbf{A}\,\mathbf{x}_T(t) + \mathbf{B}\,\mathbf{u}_T(t), \qquad (4.5.17)$$

$$\mathbf{y}_T(t) = \mathbf{C}\mathbf{x}_T(t) + \mathbf{D}\mathbf{u}_T(t), \tag{4.5.18}$$

$$\mathbf{x}_T(t_0) = \mathbf{x}_0, \tag{4.5.19}$$

where $J^*(t_0, \mathbf{x}_0)$ represents the optimal cost-to-go for an optimal ZMP preview controller. The matrices $\mathbf{R} = \mathbf{R}^T > 0$ and $\mathbf{Q} = \mathbf{Q}^T > 0$ are positive definite matrices that encode the relative importance of minimizing CoM accelerations and ZMP tracking errors through the weighted ℓ_2-norms $\|\mathbf{y}\|_{\mathbf{R}} = \sqrt{\mathbf{y}^T \mathbf{R} \mathbf{y}}$ and $\|\mathbf{u}\|_{\mathbf{Q}} = \sqrt{\mathbf{u}^T \mathbf{Q} \mathbf{u}}$, respectively.

We note that due to the invertibility of $\mathbf{D}$, $\mathbf{u}_T$, or $\mathbf{y}_T$ could either be viewed as the control for optimization. Regardless, this is a standard LQR problem (Tedrake et al., 2015) with an optimal cost-to-go of the form:

$$J^*(t, \mathbf{x}_T) = \mathbf{x}_T^T \mathbf{S}_1(t)\mathbf{x}_T + \mathbf{x}_T^T \mathbf{s}_2(t) + s_3(t). \tag{4.5.20}$$

Details to analytically form $\mathbf{S}_1(t)$, $\mathbf{s}_2(t)$, and $s_3(t)$ are provided in (Tedrake et al., 2015) for the interested reader and may be derived as an exercise. The optimal control law is given by the Hamilton–Jacobi–Bellman (HJB) equation (Bertsekas, 2005) for $\mathbf{u}_T$ or $\mathbf{y}_T$ as

$$\mathbf{u}_T^*(t, \mathbf{x}_T(t))$$

$$= \underset{\mathbf{u}_T}{\mathrm{argmin}} \left(\|\mathbf{C}\mathbf{x}_T(t) + \mathbf{D}\mathbf{u}_T\|_{\mathbf{R}}^T + \|\mathbf{u}_T - \mathbf{u}_T^d(t)\|_{\mathbf{Q}}^2 + \frac{\mathrm{d}}{\mathrm{d}t} J^*(t, \mathbf{x}_T(t)) \right), \text{ or} \tag{4.5.21}$$

$$\mathbf{y}_T^*(t, \mathbf{x}_T(t)) = \underset{\mathbf{y}_T}{\mathrm{argmin}}\, L(t, \mathbf{x}_T(t), \mathbf{y}_T) \tag{4.5.22}$$

$$= \underset{\mathbf{y}_T}{\mathrm{argmin}} \left(\|\mathbf{y}_T(t)\|_{\mathbf{R}}^T + \|\mathbf{D}^{-1}(\mathbf{y}_T - \mathbf{C}\mathbf{x}_T(t)) - \mathbf{u}_T^d(t)\|_{\mathbf{Q}}^2 + \frac{\mathrm{d}}{\mathrm{d}t} J^*(t, \mathbf{x}_T(t)) \right), \tag{4.5.23}$$

which balances instantaneous costs with long-term costs encoded in the optimal-cost-to-go (Kuindersma et al., 2015; Tedrake et al., 2015). The next subsection will explore how this optimal control solution can be lifted into a more complex robot model.

Capture point methods provide a different perspective to controlling the CoM. The capture point (CP) was originally introduced as a point on the ground where a robot would have to step to bring its CoM to a complete stop (Pratt et al., 2006, 2012; Koolen et al., 2012). This concept is illustrated in Fig. 4.5.4. For the linear inverted pendulum model, the CP $\boldsymbol{\xi} = [\xi_x, \xi_y, 0]^T$ is a composite variable in a sense that it incorporates both position and velocity information:

$$\xi_x = x + \frac{1}{\omega}\dot{x} \tag{4.5.24}$$

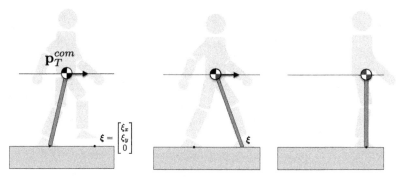

FIGURE 4.5.4 A capture point is a place where the foot may be placed such that the model may be brought to a complete stop. For the LIP model, the capture point is a linear combination of the CoM (x, y) position and the $(\dot{x}, \dot{y})$ velocity.

with similar definition in y (Koolen et al., 2012). Beyond seeing the capture point as a place to step, it should be noted that the capture point may be viewed as a bandwidth-tuned CoM look-ahead. Indeed, the term $1/\omega$ in (4.5.24) represents a time constant for the LIP dynamics. Performing a linear change of variables $(x, \dot{x}) \rightarrow (x, \xi_x)$ results in the system

$$\dot{x} = -\omega (x - \xi_x), \qquad (4.5.25)$$

$$\dot{\xi}_x = \omega (\xi_x - p_x). \qquad (4.5.26)$$

In the case that the ZMP is placed at the capture point, (4.5.26) provides $\dot{\xi} = 0$ while (4.5.25) implies that the (x, y) CoM exponentially converges to the capture point. Thus, (4.5.25) describes the system dynamics due to the *stable* pole of the LIP at $s = -\omega$, again providing an attractor for the CoM to the capture point. In contrast, (4.5.26) describes the dynamics from the *unstable* pole of the LIP at $s = +\omega$. These unstable capture point dynamics, however, represent a controllable subsystem, motivating the development of explicit capture point controllers.

As a result of the cascaded structure in the CoM/CP dynamics, control of the second-order LIP dynamics (4.5.5) may be instead pursued through control of the first-order capture point dynamics (4.5.26). Under proper tracking control of the capture point itself, stable attraction of the CoM to the capture point provides CoM tracking as a byproduct. As a result of this first-order structure, the instantaneous capture point and its 3D extension, also called the divergent component of motion (Takenaka et al., 2009; Englsberger et al., 2015), have provided an analytically clean framework to consider CoM control. Capture point-based methods are able to plan CoM trajectories directly for push recovery and walking over uneven terrains through closed-form analysis that is judiciously enabled by this reductive change of variables. Insightful geometric interpretations from the capture point and its implications for push recovery are elegantly covered in Koolen et al. (2012) with extensions to 3D in Englsberger et al. (2015).

4.5.3.2 Control for SLIP-Based Models

Within the domain of running, SLIP-based template models have provided many principles for adjusting leg behaviors to control dynamic balance. Due to the hybrid structure of the SLIP dynamics, control of the SLIP may be managed both through the selection of discrete control variables (i.e., touchdown angles) as well as continuous actions (i.e., changes in leg stiffness, nominal rest length, etc.). As in Example 4, continuous controls may often be parameterized by a discrete set of variables to simplify Poincaré analysis. Wu and Geyer (2013) developed deadbeat controllers for the 3D-SLIP using offline optimization to determine leg touchdown angles for ground height disturbance rejection. For online computation, the lack of an analytical expression for the Poincaré return map presents computational challenges. Carver (2003) and Wensing and Orin (2013b) developed locally deadbeat controllers using a linearized analysis of the Poincaré return map to reactively handle disturbances. Other authors have developed approximations to the return map to accelerate online control computations (Arslan et al., 2009; Geyer et al., 2005; Piovan and Byl, 2016). The following example highlights the ability to develop SLIP-based footstep controllers through linearized analysis.

Example 6 (Approximate Deadbeat Control of the 3D-SLIP). In this example, we will develop a stabilizing controller for the actuated 3D-SLIP model running on level terrain. Following Example 4, let us assume that we have precomputed nominal controls $\mathbf{u}_T^d[n]$ and associated apex states $\mathbf{x}_T^d[n]$ that follow the discrete dynamics:

$$\mathbf{x}_T^d[n+1] = \mathbf{f}(n, \mathbf{x}_T^d[n], \mathbf{u}_T^d[n]). \tag{4.5.27}$$

In this example, we seek to find a control law $\mathbf{u}_T[n] = \boldsymbol{\pi}(n, \mathbf{x}_T[n])$ that provides local asymptotic tracking $\mathbf{x}_T[n] \to \mathbf{x}_T^d[n]$ to the nominal state trajectory as $n \to \infty$. As one of many possible approaches, constructing a deadbeat controller would remove all tracking error within a single step. This would require a policy such that

$$\mathbf{x}_T^d[n+1] = \mathbf{f}(n, \mathbf{x}_T[n], \boldsymbol{\pi}(n, \mathbf{x}_T[n])). \tag{4.5.28}$$

Finding a policy satisfying this equation exactly requires detailed computation that may not be viable online. We demonstrate an approximate solution here.

Taking a Taylor expansion of (4.5.14) around the nominal trajectory provides

$$\tilde{\mathbf{x}}_T[n+1] = \mathbf{A}[n]\tilde{\mathbf{x}}_T[n] + \mathbf{B}[n]\tilde{\mathbf{u}}_T[n] + o\left(\left\|\left(\tilde{\mathbf{x}}_T[n], \tilde{\mathbf{u}}_T[n]\right)\right\|\right) \quad (4.5.29)$$

where $\tilde{\mathbf{x}}_T[n] = \mathbf{x}_T[n] - \mathbf{x}_T^d[n]$, $\tilde{\mathbf{u}}_T[n] = \mathbf{u}_T[n] - \mathbf{u}_T^d[n]$,

$$\mathbf{A}[n] = \left.\frac{\partial \mathbf{f}}{\partial \mathbf{x}_T}\right|_{(n, \mathbf{x}_T^d[n], \mathbf{u}_T^d[n])}, \text{ and} \quad (4.5.30)$$

$$\mathbf{B}[n] = \left.\frac{\partial \mathbf{f}}{\partial \mathbf{u}_T}\right|_{(n, \mathbf{x}_T^d[n], \mathbf{u}_T^d[n])}. \quad (4.5.31)$$

When the matrix $\mathbf{B}[n]$ has full row rank, the deadbeat condition (4.5.28) can be satisfied locally through the selection of any feedback law $\tilde{\mathbf{u}}_T[n] = \mathbf{K}[n]\tilde{\mathbf{x}}_T[n]$ satisfying

$$0 = \mathbf{A}[n] + \mathbf{B}[n]\mathbf{K}[n]. \quad (4.5.32)$$

The control policy in original coordinates

$$\pi\left(\mathbf{x}_T[n], n\right) = \mathbf{u}_T^d[n] + \mathbf{K}[n]\left(\mathbf{x}_T[n] - \mathbf{x}_T^d[n]\right) \quad (4.5.33)$$

then admits local asymptotic tracking to the desired trajectory.

This simple approach can automatically capture many of the powerful heuristics that enabled dynamic gaits in Raibert's machines (Raibert, 1986). For instance, considering the actuated 3D-SLIP from Examples 1 and 2 with parameters $m = 72.5$ kg, $\ell_h = 0.97$ m, and $\mathbf{p}_T^{hip} = \mathbf{p}_T^{com} + [0, 12 \text{ cm} \cdot (-1)^n, 0]^T$, the controls

$$\theta^d[n] = 0.4 \text{ rad}, \qquad \phi^d[n] = 0 \text{ rad}, \quad (4.5.34)$$

$$k_{s_1}^d[n] = 12.7 \text{ kN/m}, \quad \cdot \quad k_{s_2}^d[n] = 12.7 \text{ kN/m} \quad (4.5.35)$$

can be used to generate a left-right symmetric 2-step periodic running gait for a nominal speed of $\dot{x} = 3.5$ m/s (Wensing and Orin, 2013b). For such a gait, a feedback policy $\mathbf{K}[n]$ that satisfies (4.5.32) can also be developed to be 2-step periodic. For a left-foot step (i.e., $\mathbf{p}_T^{hip} = \mathbf{p}_T^{com} + [0, 12 \text{ cm}, 0]^T$ with coordinates given in Fig. 4.5.3), one such feedback gain is

$$\begin{bmatrix} \tilde{\theta} \\ \tilde{\phi} \\ \tilde{k}_{s_1} \\ \tilde{k}_{s_2} \end{bmatrix} = \underbrace{\begin{bmatrix} -0.51 & 0.13 & -0.01 \\ -1.95 & -0.08 & 0.90 \\ 36.9 & 13.2 & 0.86 \\ -36.9 & -13.2 & -0.86 \end{bmatrix}}_{\mathbf{K}[n]} \begin{bmatrix} \tilde{z} \\ \dot{\tilde{x}} \\ \dot{\tilde{y}} \end{bmatrix} \quad (4.5.36)$$

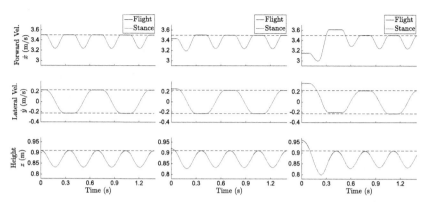

FIGURE 4.5.5 (Left) Nominal 2-step periodic 3D-SLIP gait for running at 3.5 m/s. (Center) Local deadbeat performance of the SLIP controller with small disturbances. (Right) Asymptotic tracking performance for larger disturbances with the local deadbeat SLIP controller.

where an additional constraint $\tilde{k}_{s_1} = -\tilde{k}_{s_2}$ has been employed to resolve the fact that four control variables exist to satisfy three deadbeat constraints. Looking at the bottom two rows shows the control response to a state having excess potential energy in z, or excess kinetic energy in $\dot{x}$ or $\dot{y}$. In all cases, the feedback law is one that employs a stiffer spring at touchdown, and softens at maximum compression in order to remove the excess energy. Other heuristics exist in each column. The first column shows that a running state that begins too high should be countered by placing the foot further under the CoM at touchdown. Similar leg placement heuristics are shown in the second and third columns. We note that these automatically tuned heuristics can provide guiding principles, not just for template control, but for humanoid control as we will pursue in the next subsection.

To demonstrate the performance of the control law (4.5.33), Fig. 4.5.5 shows the nominal state trajectory, the state trajectory subject to a small disturbance wherein local deadbeat behavior is approximately observed, and response to a larger perturbation wherein asymptotic tracking is recovered.

It should be noted that certain operating regimes of the 2D-SLIP model have been shown to possess so-called self-stable behavior. That is to say, for a fixed touchdown angle θ and leg stiffness k_s, open-loop stable gaits have been shown to exist (Seyfarth et al., 2002; Ghigliazza et al., 2003). While these results are an interesting finding, they do not extend to 3D (Seipel and Holmes, 2005) and represent a lower bound on the domain of attraction and robustness achievable by the addition of feedback.

More recently, the bipedal SLIP model has been proposed as an alternative walking template (Geyer et al., 2006) for the CoM. Recent work has developed controllers for this template based on analysis of its Poincaré return map

(Vejdani et al., 2015; Liu et al., 2015) similar to in the 3D-SLIP for running. This template includes rich nonlinear hybrid dynamics, due to two virtual spring legs making and breaking contact. As a result, control strategies for this model have often required offline computation with detailed knowledge bases used for real-time control.

4.5.3.3 Beyond Tracking Control for Pendular Models

Beyond these traditional CoM models, the computer graphics community has employed many other nontraditional template models for physics-based simulation of virtual characters. Mordatch et al. (2010) used a translational LIP with decoupled vertical SLIP to provide a relaxation of the SLIP model for locomotion planning with evolutionary search. da Silva et al. (2008) used iLQR with a three-link model as a template for the body and leg CoM motions. Ye and Liu (2010) applied differential dynamic programming to a CoM and angular momentum model that included an integrated centroidal angular momentum (Orin et al., 2013) as a nonphysical surrogate angular state.

Many of these cited works have focused template control on asymptotic tracking guarantees. Moving forward, template control warrants investigation to achieve other important control specifications such as robustness, viability, and yet others. Terrain robustness specifically has been a focus of much work within SLIP frameworks (Ernst et al., 2009; Wu and Geyer, 2013; Liu et al., 2016). Other work has advocated for viability (Wieber, 2008) as a more appropriate goal in legged systems. For a state to be viable intuitively means that it can be controlled to avoid any undesirable regions of the state space (such as those in which the robot has fallen to the ground). Indeed, in robotics when we say a system is "stable" it rarely is meant rigorously in a Lyapunov sense, but rather is more loosely meant in a sense of not falling down. Viability theory offers potential to bring rigor to this loose specification. Its rigorous technical definitions lead, in principle, to methods for identifying viable regions of the state space. Such verifications are beyond the reach of current computational methods for high-DoF systems. Yet, a notion of viability drives the definition of capturability (Koolen et al., 2012), and motivates the use of low-DoF template models to pursue viability more broadly (Sherikov et al., 2015).

4.5.4 ESTABLISHING A TEMPLATE/ANCHOR RELATIONSHIP

In the last step of the process, a high-DoF control system must be developed to establish the template in the anchor dynamics. Control of the template itself does not consider the issue of how the various actuators in the full system should be recruited. Thus, a high-DoF control problem exists to realize any template/anchor (T/A) relationship. Fig. 4.5.6 provides a rough diagram of how a solution to this realization fits into a full solution for real-time template-based control in a high-DoF anchor.

An initial part of this design stage entails the selection of which template/anchor notion to pursue. As alluded to in Subchapter 3.2, there are a wide range of specifications for what a proper T/A relationship may entail. At one extreme, the template states may be represented by a normally hyperbolic invariant manifold (NHIM) embedded in the state space of the anchor. To satisfy this notion of the T/A relationship, a diffeomorphic copy of the controlled template dynamics must be rendered on the NHIM in closed-loop. Towards realizing this condition, the template may be sought as the hybrid zero dynamics (HZD) (Westervelt et al., 2003; Poulakakis and Grizzle, 2009) of a judiciously crafted output regulation problem. HZD methods will be described further in Subchapter 4.7, and offer promise to bring the full scope of the templates and anchors hypothesis to bear in experimental machines.

Simpler methods may strive to replicate only select aspects of the template within the anchor dynamics. For instance, rather than the SLIP state encoding a target whole body state for the anchor (as in the NHIM notion of the T/A relationship), it may instead simply encode a target state for the anchor CoM. In this light, we will denote "anchor features" as any features of motion in the anchor whose control is informed by the template. Existing methods for controlling anchor features to match a template differ considerably in the degree of replication quality that they provide. In methods that only loosely achieve the desired template dynamics, this inaccuracy itself may represent a disturbance to any template-level controller. In methods that more precisely achieve the target template dynamics, a greater burden may exist on the template to operate in a manner that is replicable in the full system.

4.5.4.1   Realizing Template Dynamics Through Task-Space Control

Task-space or operational-space control provides a formal framework to pursue an exact realization of a template's continuous dynamics in a more complex anchor system. The state of the anchor system can be given as $[\mathbf{q}, \dot{\mathbf{q}}]^T \in \mathbb{R}^{n_A}$ with configuration $\mathbf{q}$. In legged robots, the configuration $\mathbf{q}$ generally possesses the structure $\mathbf{q} = [\mathbf{q}_b, \mathbf{q}_j]$ where $\mathbf{q}_b \in SE(3)$ is the configuration of a floating-base and $\mathbf{q}_j \in \mathbb{R}^{n_j}$ is the configuration of the internal joints. We denote $\mathbf{x}_A(\mathbf{q}, \dot{\mathbf{q}}) \in \mathbb{R}^{n_T}$ (same dimensions as $\mathbf{x}_T$) as the anchor feature state (i.e., the projection of the anchor state onto those features that are captured in the template). For instance, for CoM templates where $\mathbf{x}_T = [\mathbf{p}_T^{com}, \dot{\mathbf{p}}_T^{com}]$, the projection would extract the CoM position and velocity $\mathbf{x}_A := [\mathbf{p}_A^{com}, \dot{\mathbf{p}}_A^{com}]$ of the anchor

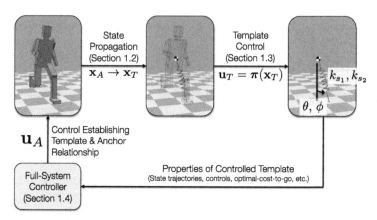

FIGURE 4.5.6 Real-time template-based control diagram. The relevant features of the anchor system $\mathbf{x}_A = \mathbf{g}(\mathbf{q}, \dot{\mathbf{q}})$ are propagated to the template $\mathbf{x}_T$. A control system applied to the template $\mathbf{u}_T = \boldsymbol{\pi}(\mathbf{x}_T)$ gives rise to target properties of the closed-loop template dynamics. These properties are lifted into the full system by a high-dimensional controller that coordinates many actuators $\mathbf{u}_A$ to realize a template/anchor relationship.

system. For physical template models more broadly, this relationship is often clear, however, for more abstract templates, this assignment itself may present another degree of design freedom.

As shown in Fig. 4.5.6, this projected state $\mathbf{x}_A$ is propagated to the template $\mathbf{x}_A \rightarrow \mathbf{x}_T$ to begin the application of template-based control. As one of many possibilities, let us assume that such propagation happens discretely, with trajectories of the controlled template $\mathbf{x}_T(t)$ provided between updates. For instance, in the SLIP model, such propagation might only occur at the Poincaré section, with $\mathbf{x}_T(t)$ representing a controlled SLIP trajectory for the next step. Controlling the full system trajectory $\mathbf{x}_A(t)$ to asymptotically anchor the template $\mathbf{x}_A(t) \rightarrow \mathbf{x}_T(t)$ is then a well studied problem within the context of the operational-space and whole-body control literature.

Replicating the Dynamics of CoM Templates

A common whole-body control solution applied to the replication of template dynamics is to minimize the replication errors at a dynamic level through real-time optimization in closed-loop. To build towards this solution, we reintroduce the standard dynamic equations of motion

$$\mathbf{H}(\mathbf{q})\ddot{\mathbf{q}} + \mathbf{C}(\mathbf{q}, \dot{\mathbf{q}})\dot{\mathbf{q}} + \mathbf{G}(\mathbf{q}) = \mathbf{S}_a^T \boldsymbol{\tau} + \mathbf{J}_s(\mathbf{q})^T \mathbf{F}_s \qquad (4.5.37)$$

where $\mathbf{H}$, $\mathbf{C}\dot{\mathbf{q}}$, and $\mathbf{G}$ are the familiar mass matrix, velocity product terms, and gravitational terms, respectively. Here $\mathbf{F}_s$ collects ground reaction forces (GRFs) for appendages in support and $\mathbf{J}_s$ is a combined support Jacobian. The matrix $\mathbf{S}_a = [\mathbf{0}_{n_j \times 6} \ \mathbf{1}_{n_j \times n_j}]$ is a selection matrix for the actuated joints.

For CoM templates with anchor features $\mathbf{x}_A = [\mathbf{p}_A^{com}, \dot{\mathbf{p}}_A^{com}]$, an optimization problem can first be formulated to partial-feedback linearize the CoM dynamics of the anchor under whole-body constraints. Given a desired acceleration for the CoM $\ddot{\mathbf{p}}_A^{com,d}$, the goal of the whole-body controller is to select joint torques $\mathbf{u}_A := \boldsymbol{\tau}$ that most closely realize this commanded acceleration. This can be formulated as an optimization problem

$$\min_{\ddot{\mathbf{q}}, \boldsymbol{\tau}, \mathbf{F}_s} \frac{1}{2} \left\| \mathbf{J}_{com} \ddot{\mathbf{q}} + \dot{\mathbf{J}}_{com} \dot{\mathbf{q}} - \ddot{\mathbf{p}}_A^{com,d} \right\|^2 \tag{4.5.38}$$

$$\text{subject to } \mathbf{H}\ddot{\mathbf{q}} + \mathbf{C}\dot{\mathbf{q}} + \mathbf{G} = \mathbf{S}_a^T \boldsymbol{\tau} + \mathbf{J}_s^T \mathbf{F}_s, \tag{4.5.39}$$

$$\mathbf{J}_s \ddot{\mathbf{q}} + \dot{\mathbf{J}}_s \dot{\mathbf{q}} = \mathbf{0}, \tag{4.5.40}$$

$$\mathbf{F}_s \in \mathcal{C}, \tag{4.5.41}$$

where $\mathbf{J}_{com} \in \mathbb{R}^{3 \times (n_j + 6)}$ is the CoM Jacobian, $\mathbf{F}_s \in \mathbb{R}^{6n_f}$ are ground reaction forces for n_f feet in planar contact, and $\mathcal{C}$ is the convex cone of forces that can be created through the available contacts (Wensing and Orin, 2013a). The constraint (4.5.40) enforces that any supporting contacts must not move. While ground forces $\mathbf{F}_s$ may often be viewed as the Lagrange multipliers associated with this constraint, their solution is not unique when multiple feet are in planar contact. As a result, the optimization enforces that at least one set of these Lagrange multipliers must satisfy frictional constraints.

If the optimization problem can be solved to an optimal objective function value of 0, then the current contacts provide the necessary control authority to exactly realize the commanded dynamics. When this is the case, solution of the optimization problem in closed-loop will provide a feedback linearization from the commanded dynamics $\ddot{\mathbf{p}}_A^{com,d}$ to the anchor feature $\mathbf{p}_A^{com}$. As a result, a common approach to achieve template tracking is to select the commanded feature dynamics according to the law

$$\ddot{\mathbf{p}}_A^{com,d} = \ddot{\mathbf{p}}_T^{com} + k_D (\dot{\mathbf{p}}_T^{com} - \dot{\mathbf{p}}_A^{com}) + k_P (\mathbf{p}_T^{com} - \mathbf{p}_A^{com}) \tag{4.5.42}$$

where k_P and k_D are positive definite gain matrices. This selection provides an asymptotically stable second-order dynamic for the anchoring error $\mathbf{e}(t) = \mathbf{p}_T^{com}(t) - \mathbf{p}_A^{com}(t)$. If the template includes only kinematic features and their associated derivatives, then the above development is general through the use of an appropriate task Jacobian. If the template includes purely velocity-dependent features, such as angular momentum, then minimal modifications to the above development can be employed (Wensing and Orin, 2013a).

Exploiting Redundancy

It should be noted that there is significant redundancy remaining after only fulfilling the commanded anchor feature dynamics. Often the use of this flexibility is needed to track aspects of the full system that are absent in the template. For instance, in running with the SLIP template, the lack of a swing leg in the template requires coordinated swing leg torques in the full model to move the swing leg into place for the following step.

These torques may be found algorithmically through solving a subsequent optimization problem that respects the optimal anchor feature dynamics from (4.5.38). Given desired accelerations $\ddot{\mathbf{p}}_1^d \in \mathbb{R}^{n_1}$ for another feature of motion, such as a swing foot position, modified torques can be derived from the solution to

$$\min_{\ddot{\mathbf{q}}, \boldsymbol{\tau}, \mathbf{F}_s} \frac{1}{2} \left\| \mathbf{J}_1 \ddot{\mathbf{q}} + \dot{\mathbf{J}}_1 \dot{\mathbf{q}} - \ddot{\mathbf{p}}_1^d \right\|^2 \tag{4.5.43}$$

$$\text{subject to } \mathbf{H} \ddot{\mathbf{q}} + \mathbf{C} \dot{\mathbf{q}} + \mathbf{G} = \mathbf{S}_a^T \boldsymbol{\tau} + \mathbf{J}_s^T \mathbf{F}_s, \tag{4.5.44}$$

$$\mathbf{J}_s \ddot{\mathbf{q}} + \dot{\mathbf{J}}_s \dot{\mathbf{q}} = \mathbf{0}, \tag{4.5.45}$$

$$\mathbf{J}_{com} \ddot{\mathbf{q}} + \dot{\mathbf{J}}_{com} \dot{\mathbf{q}} = \ddot{\mathbf{p}}_A^*, \tag{4.5.46}$$

$$\mathbf{F}_s \in \mathcal{C}, \tag{4.5.47}$$

where $\ddot{\mathbf{p}}_A^*$ is the optimal anchor feature acceleration resulting from first solving (4.5.38). By solving (4.5.38) and then (4.5.43) in a cascaded fashion, a strict prioritization is being given to tracking the anchor features above all else. When a strictly ordered hierarchy exists between yet further tasks, this approach may be extended through additional cascaded solves or through dedicated hierarchical solvers (Escande et al., 2014).

As an alternative, soft priorities can be implemented through solving a single optimization problem with an objective that is a weighted combination of the task command errors:

$$\min_{\ddot{\mathbf{q}}, \boldsymbol{\tau}, \mathbf{F}_s} w_A \left\| \mathbf{J}_{com} \ddot{\mathbf{q}} + \dot{\mathbf{J}}_{com} \dot{\mathbf{q}} - \ddot{\mathbf{p}}_A^{com,d} \right\|^2 + \sum_i w_i \left\| \mathbf{J}_i \ddot{\mathbf{q}} + \dot{\mathbf{J}}_i \dot{\mathbf{q}} - \ddot{\mathbf{p}}_i^d \right\|^2 \tag{4.5.48}$$

$$\text{subject to } \mathbf{H} \ddot{\mathbf{q}} + \mathbf{C} \dot{\mathbf{q}} + \mathbf{G} = \mathbf{S}_a^T \boldsymbol{\tau} + \mathbf{J}_s^T \mathbf{F}_s, \tag{4.5.49}$$

$$\mathbf{J}_s \ddot{\mathbf{q}} + \dot{\mathbf{J}}_s \dot{\mathbf{q}} = \mathbf{0}, \tag{4.5.50}$$

$$\mathbf{F}_s \in \mathcal{C}. \tag{4.5.51}$$

In comparison to using a strict prioritization, soft prioritization requires fewer invocations of an optimization solver, but can suffer from numeric conditioning issues if a large disparity in weights is desired. Overall, these optimization-based whole-body control methods perform well in practice, and have been a

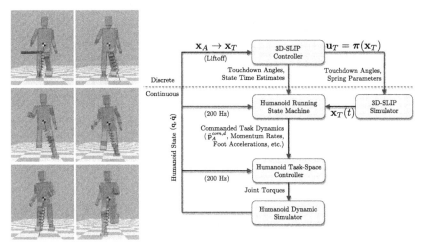

FIGURE 4.5.7 Template-based whole-body response to a push disturbance and associated whole-body control system. Following a push (red arrow in the upper-left subfigure), the disturbed state of the humanoid is projected and propagated to the SLIP template $\mathbf{x}_A \rightarrow \mathbf{x}_T$. SLIP control $\mathbf{u}_T = \boldsymbol{\pi}(\mathbf{x}_T)$ suggests touchdown angles for the upcoming steps and a dynamically feasible CoM trajectory $\mathbf{x}_T(t)$ for a template-based recovery in the following stance. (For interpretation of the colors in this figure, the reader is referred to the web version of this chapter.)

workhorse for modern humanoids in recent years. However, they are fundamentally single-step model predictive control schemes where the actions that are instantaneously greedy are designed to play out favorably in the long term. The template controllers from Section 4.5.3 readily admit formal guarantees on their performance. However, when coupled to optimization-based whole body controllers such as (4.5.38), currently little can be proven or guaranteed about the long-term behavior of these systems.

Example 7 (Closed-Loop Control of High-Speed Humanoid Running).
Fig. 4.5.7 shows the results of a full template-based control system applied to humanoid push recovery while running at 3.5 m/s with the SLIP controller developed in Example 6. At each liftoff, anchor features are propagated to the SLIP template. Controlled SLIP trajectories $\mathbf{x}_T(t)$ then provide physics-based recovery motions. At each step, (4.5.33) provides touchdown angles and matched spring stiffnesses k_{s_1} and k_{s_2} to provide approximate deadbeat tracking back to the nominal gait within one step. Due to system features, such as ground/foot impacts that are not captured in the SLIP, this full template-based controller does not experience the strong deadbeat behavior shown in Example 6. Fig. 4.5.8 shows the CoM response in the simulation experiment demonstrating that the template recovery motions are realized in the anchor.

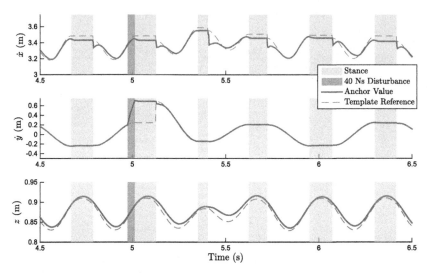

FIGURE 4.5.8 Response of a full template-based control system for a humanoid following a lateral push disturbance of 40 Ns. At each step a CoM reference trajectory is generated based on the results of the local deadbeat template controller from Example 6. At each transition from flight to stance, an impact impulse in the full system, which is due to leg mass not captured in the SLIP, provides a persistent disturbance to the full template-based control system. The system nominally recovers to its steady-state gait within 4 steps.

Since the SLIP assumes massless legs that can be instantly repositioned in flight, a state machine is used in the humanoid to coordinate leg motions. This state machine does so through providing foot acceleration commands to an optimization-based controller as in (4.5.43). Foot acceleration commands similarly take a PD form as in (4.5.42). However, rather than the template providing reference trajectories, cubic spline references are generated online to achieve the desired virtual leg configuration at touchdown. To resolve remaining redundancy, a centroidal angular momentum (Orin et al., 2013) rate of change command and pose acceleration command (Wensing and Orin, 2013b) are also provided to (4.5.43). The optimization is solved at rate of 200 Hz, and resulting joint torques τ are applied in dynamic simulation. Further details can be found in Wensing and Orin (2013b).

4.5.4.2 Lifting Other Properties of Template Control

The previous example included feedback to the template on a discrete basis. Other methods continuously resolve the behavior of the full system with an optimal reaction from the template by lifting properties of an optimal template controller. For instance, optimal controllers for template models as in Example 5 can be used to guide CoM control in a humanoid. Following the construction of the optimal-cost-to-go from Example 5, the whole-body QP (4.5.48) from the previous subsection could be modified as (Kuindersma et al., 2015)

$$\min_{\ddot{q},\tau,\mathbf{F}_s} \ w_A L(t, \mathbf{x}_A, \ddot{\mathbf{p}}_A^{com}) + \sum_i w_i \|\mathbf{J}_i \ddot{q} + \dot{\mathbf{J}}_i \dot{q} - \ddot{\mathbf{x}}_i^c\|^2 \qquad (4.5.52)$$

$$\text{subject to } \mathbf{H}\ddot{q} + \mathbf{C}\dot{q} + \mathbf{G} = \mathbf{S}_a^T \tau + \mathbf{J}_s^T \mathbf{F}_s, \qquad (4.5.53)$$

$$\mathbf{J}_s \ddot{q} + \dot{\mathbf{J}}_s \dot{q} = \mathbf{0}, \qquad (4.5.54)$$

$$\mathbf{F}_s \in \mathcal{C}, \qquad (4.5.55)$$

where again $L(t, \mathbf{x}_A, \ddot{\mathbf{p}}_A^{com})$ from (4.5.22) (Kuindersma et al., 2015) balances instantaneous template control costs with long-term costs encoded in its optimal-cost-to-go. The benefit of this formulation over (4.5.48) comes from the fact that when the optimal template dynamics are not instantaneously feasible, deviations from the optimal template dynamics are not created equal. Some deviations, while equal in magnitude, may be more costly than others in the long term. This subtlety is addressed by the long-term costs encoded in the optimal cost-to-go, while the error norm penalty in (4.5.48) manages no such long-term trade-off. A summary of the whole-body control results for the MIT DRC team, which used this approach in hardware with ATLAS, can be found in Kuindersma et al. (2015, 2014). In particular, the formulation of minimizing $L(t, \mathbf{x}_A, \ddot{\mathbf{p}}_A^{com})$ was cited to provide practical robustness over other methods that simply enforce decreasing the template-based optimal-cost-to-go over time.

Overall, these optimization-based control methods guarantee template tracking error will asymptotically approach zero when sufficient control authority exists. If there are unmodeled aspects of the anchor system that affect $\mathbf{x}_A$ but are not captured in the template, these unmodeled aspects represent a persistent disturbance to the template-based control system. Otherwise, asymptotic tracking across these examples is accomplished in (4.5.48) and (4.5.52), through carefully selecting joint torques that compensate for nonlinear joint-space dynamics and replace them with those commanded for the locomotion features. Tracking performance, as a result, is predicated on the correctness of the dynamic model itself and thus may exhibit sensitivity to modeling errors. Other less model-intensive methods are able to overcome these drawbacks but provide a looser match between template and anchor features.

4.5.4.3 Anchoring the Template Through Less Model-Intensive Methods

Virtual model control is a less model-intensive method to track locomotion features over time. Virtual model control (VMC) uses virtual components to formulate fictitious forces that govern interaction between real and virtual systems (Pratt et al., 2001). These fictitious forces are then realized through static joint torque mappings, often using the Jacobian transpose. Virtual systems may represent springs and dampers, or other general mechanisms that are designed to produce virtual interactions and regulate the state of the real system. Thus, in a sense, virtual model control is a descendant of simple impedance control, yet more general interaction dynamics may be authored through design of virtual models. VMC has been applied widely in quadruped robots to stabilize dynamic locomotion plans (Park and Kim, 2015; Semini et al., 2015; Coros et al., 2011; Winkler et al., 2014) and in dynamic balance for humanoids (Stephens and Atkeson, 2010). When physical template models are used in a template controller, the template itself can often be viewed as a virtual model. When this is the case, VMC provides a direct method to approximately realize the template dynamics. By only relying on kinematic data from the real system to resolve virtual template forces, VMC is essentially model-free at a dynamic level. Thus, without compensation for dynamic effects, VMC relies on heuristic gain tuning to find controller parameters that provide suitable performance at each operating point. Due to the changes in neglected dynamic forces across operating regimes, however, performance claims for these controllers can often only be verified empirically.

4.5.4.4   Template-Inspired Mechanical Design

Throughout this section, rigid-body dynamics (4.5.37) have been assumed with torque sources modeled at the joints. Yet, many modern machines do include other forms of impedance (compliance and damping) at the joints and more generally across the body structures. The natural dynamics of these structures may point toward the selection of a particular template that can significantly reduce the burden of imparting a template/anchor relationship through closed-loop control. For instance, the bipedal-SLIP model was used to develop high-level template-based controllers for ATRIAS (Rezazadeh et al., 2015). Lower-level control mechanisms did not explicitly attempt to impart bipedal-SLIP dynamics to the machine. However, since ATRIAS was designed to exhibit dynamics that embody the bipedal-SLIP (Ramezani et al., 2014), the robot has been shown to exhibit qualitative similarity to its template in terms of its ground reaction forces. A similar success story may be found in Raibert's early hoppers, whose control systems where inspired by SLIP-based laws and implemented on physical SLIP-type robots with air springs (Raibert, 1986). Further intersection of template-based control and template-inspired design presents interesting future prospects to simplify closed-loop control of legged machines.

4.5.5 CONCLUSIONS

This section has described the design process for using template models for control. We have discussed the three main sub-problems that must be integrated to realize a template-based control system. First, a template pertinent to the locomotion task must be identified. There is no right or wrong template for a given task and those from biology need only serve as inspiration. Next, control methodologies for the template itself must be identified and developed. In some cases, the template may not provide sufficient control flexibility to achieve desired controller specifications, requiring modifications to the template under consideration. Following the solution of a template control problem, controlled template dynamics must be retargeted to the full system, addressing the coordination of many actuators to achieve low-dimensional specifications. Many methods exist to realize these target template dynamics, which vary in the degree of replication quality that they impart. The application of template models for control does require the intuition of a human designer. However, its focus on the most important characteristics of locomotion has enabled reactive control architectures in many experimental robots, and may yet play a key role in unlocking the full potentials of legged machines.

REFERENCES

Arslan, O., Saranli, U., Morgul, O., 2009. An approximate stance map of the spring mass hopper with gravity correction for nonsymmetric locomotions. In: IEEE International Conference on Robotics and Automation, ICRA'09, pp. 2388–2393.

Bellman, R., 1957. Dynamic Programming. Princeton University Press, Princeton, NJ.

Bertsekas, D.P., 2005. Dynamic Programming and Optimal Control. Athena Scientific.

Blickhan, R., 1989. The spring–mass model for running and hopping. J. Biomech. 22 (11–12), 1217–1227.

Carver, S., 2003. Control of a Spring Mass Hopper. PhD thesis. Cornell University, Ithaca, NY.

Coros, S., Karpathy, A., Jones, B., Reveret, L., van de Panne, M., 2011. Locomotion skills for simulated quadrupeds. In: ACM SIGGRAPH 2011 Papers, SIGGRAPH '11. ACM, New York, NY, USA, pp. 59:1–59:12.

da Silva, M., Abe, Y., Popović, J., 2008. Interactive simulation of stylized human locomotion. In: ACM SIGGRAPH 2008 Papers. Los Angeles, California, pp. 82:1–10.

Diehl, M., Bock, H., Diedam, H., Wieber, P.-B., 2006. Fast direct multiple shooting algorithms for optimal robot control. In: Diehl, M., Mombaur, K. (Eds.), Fast Motions in Biomechanics and Robotics. In: Lecture Notes in Control and Information Sciences, vol. 340. Springer, Berlin, Heidelberg, pp. 65–93.

Dimitrov, D., Sherikov, A., Wieber, P.-B., 2011. A sparse model predictive control formulation for walking motion generation. In: 2011 IEEE/RSJ International Conference on Intelligent Robots and Systems (IROS), pp. 2292–2299.

Englsberger, J., Ott, C., Albu-Schaffer, A., 2015. Three-dimensional bipedal walking control based on divergent component of motion. IEEE Trans. Robot. 31, 355–368.

Ernst, M., Geyer, H., Blickhan, R., 2009. Spring-legged locomotion on uneven ground: a control approach to keep the running speed constant. In: 12th Int. Conf. on Climbing and Walking Robots (CLAWAR), pp. 639–644.

Escande, A., Mansard, N., Wieber, P.-B., 2014. Hierarchical quadratic programming: fast online humanoid-robot motion generation. Int. J. Robot. Res. 33 (7), 1006–1028.

Feng, S., Xinjilefu, X., Huang, W., Atkeson, C.G., 2013. 3D walking based on online optimization. In: Proc. of the IEEE-RAS International Conference on Humanoid Robots.

Full, R.J., Koditschek, D.E., 1999. Templates and anchors: neuromechanical hypotheses of legged locomotion on land. J. Exp. Biol. 202 (23), 3325–3332.

Garofalo, G., Ott, C., Albu-Schaffer, A., 2012. Walking control of fully actuated robots based on the bipedal SLIP model. In: IEEE Int. Conf. on Robotics and Automation, pp. 1456–1463.

Geyer, H., Seyfarth, A., Blickhan, R., 2005. Spring–mass running: simple approximate solution and application to gait stability. J. Theor. Biol. 232 (3), 315–328.

Geyer, H., Seyfarth, A., Blickhan, R., 2006. Compliant leg behaviour explains basic dynamics of walking and running. Proc. R. Soc. B, Biol. Sci. 273 (1603), 2861–2867.

Ghigliazza, R.M., Altendorfer, R., Holmes, P., Koditschek, D.E., 2003. Passively stable conservative locomotion. SIAM J. Appl. Dyn. Syst.

Herdt, A., Diedam, H., Wieber, P.-B., Dimitrov, D., Mombaur, K., Diehl, M., 2010. Online walking motion generation with automatic foot step placement. Adv. Robot. 24 (5–6), 719–737.

Kajita, S., Kanehiro, F., Kaneko, K., Yokoi, K., Hirukawa, H., 2001. The 3D linear inverted pendulum mode: a simple modeling for a biped walking pattern generation. In: Proc. of the IEEE/RSJ Int. Conf. on Intelligent Robots and Systems, vol. 1, pp. 239–246.

Kajita, S., Kanehiro, F., Kaneko, K., Fujiwara, K., Harada, K., Yokoi, K., Hirukawa, H., 2003. Biped walking pattern generation by using preview control of Zero-Moment Point. In: Proceedings of the IEEE International Conference on Robotics and Automation (ICRA). Taipei, Taiwan, vol. 2, pp. 1620–1626.

Koolen, T., de Boer, T., Rebula, J., Goswami, A., Pratt, J., 2012. Capturability-based analysis and control of legged locomotion, Part 1: theory and application to three simple gait models. Int. J. Robot. Res. 31 (9), 1094–1113.

Kuindersma, S., Permenter, F., Tedrake, R., 2014. An efficiently solvable quadratic program for stabilizing dynamic locomotion. In: Proc. of the IEEE International Conference on Robotics and Automation, pp. 2589–2594.

Kuindersma, S., Deits, R., Fallon, M., Valenzuela, A., Dai, H., Permenter, F., Koolen, T., Marion, P., Tedrake, R., 2015. Optimization-based locomotion planning, estimation, and control design for the atlas humanoid robot. In: Autonomous Robots, pp. 1–27.

Liu, Y., Wensing, P.M., Orin, D.E., Zheng, Y.F., 2015. Dynamic walking in a humanoid robot based on a 3D actuated dual-slip model. In: 2015 IEEE International Conference on Robotics and Automation (ICRA), pp. 5710–5717.

Liu, Y., Wensing, P.M., Schmiedeler, J.P., Orin, D.E., 2016. Terrain-blind humanoid walking based on a 3-d actuated dual-slip model. IEEE Robot. Autom. Lett. 1, 1073–1080.

Mordatch, I., de Lasa, M., Hertzmann, A., 2010. Robust physics-based locomotion using low-dimensional planning. In: ACM SIGGRAPH 2010 papers. Los Angeles, California, pp. 71:1–8.

Orin, D.E., Goswami, A., Lee, S.-H., 2013. Centroidal dynamics of a humanoid robot. Auton. Robots 35 (2), 161–176.

Park, H.-W., Kim, S., 2015. Quadrupedal galloping control for a wide range of speed via vertical impulse scaling. Bioinspir. Biomim. 10 (2), 025003.

Park, H.-W., Wensing, P., Kim, S., 2015. Online planning for autonomous running jumps over obstacles in high-speed quadrupeds. In: Proceedings of Robotics: Science and Systems. Rome, Italy.

Piovan, G., Byl, K., 2016. Approximation and control of the slip model dynamics via partial feedback linearization and two-element leg actuation strategy. IEEE Trans. Robot. 32, 399–412.

Poulakakis, I., Grizzle, J., 2009. The spring loaded inverted pendulum as the hybrid zero dynamics of an asymmetric hopper. IEEE Trans. Autom. Control 54, 1779–1793.

Pratt, J., Chew, C.-M., Torres, A., Dilworth, P., Pratt, G., 2001. Virtual model control: an intuitive approach for bipedal locomotion. Int. J. Robot. Res. 20 (2), 129–143.

Pratt, J., Koolen, T., de Boer, T., Rebula, J., Cotton, S., Carff, J., Johnson, M., Neuhaus, P., 2012. Capturability-based analysis and control of legged locomotion, part 2: application to m2v2, a lower-body humanoid. Int. J. Robot. Res. 31 (10), 1117–1133.

Pratt, J., Carff, J., Drakunov, S., Goswami, A., 2006. Capture point: a step toward humanoid push recovery. In: IEEE-RAS Int. Conf. on Humanoid Robots. Genova, Italy, pp. 200–207.

Raibert, M.H., 1986. Legged Robots that Balance. MIT Press, Cambridge, MA, USA.

Ramezani, A., Hurst, J.W., Hamed, K.A., Grizzle, J., 2014. Performance analysis and feedback control of ATRIAS, a three-dimensional bipedal robot. J. Dyn. Syst. Meas. Control 136 (2), 021012.

Rezazadeh, S., Hubicki, C., Jones, M., Peekema, A., Van Why, J., Abate, A., Hurst, J., 2015. Spring–mass walking with ATRIAS in 3D: robust gait control spanning zero to 4.3 kph on a heavily underactuated bipedal robot 10. V001T04A003.

Seipel, J.E., Holmes, P., 2005. Running in three dimensions: analysis of a point-mass sprung-leg model. Int. J. Robot. Res. 24 (8), 657–674.

Semini, C., Barasuol, V., Boaventura, T., Frigerio, M., Focchi, M., Caldwell, D.G., Buchli, J., 2015. Towards versatile legged robots through active impedance control. Int. J. Robot. Res. 34 (7), 1003–1020.

Seyfarth, A., Geyer, H., Günther, M., Blickhan, R., 2002. A movement criterion for running. J. Biomech. 35 (5), 649–655.

Sherikov, A., Dimitrov, D., Wieber, P.B., 2015. Balancing a humanoid robot with a prioritized contact force distribution. In: 2015 IEEE-RAS 15th International Conference on Humanoid Robots (Humanoids), pp. 223–228.

Stephens, B.J., Atkeson, C.G., 2010. Dynamic balance force control for compliant humanoid robots. In: IEEE/RSJ Int. Conf. on Intelligent Robots and Systems. Taipei, Taiwan, pp. 1248–1255.

Takenaka, T., Matsumoto, T., Yoshiike, T., 2009. Real time motion generation and control for biped robot-1st report: walking gait pattern generation. In: IEEE/RSJ International Conference on Intelligent Robots and Systems, pp. 1084–1091.

Tedrake, R., Manchester, I.R., Tobenkin, M., Roberts, J.W., 2010. LQR-trees: Feedback motion planning via sums-of-squares verification. Int. J. Robot. Res. 29 (8), 1038–1052.

Tedrake, R., Kuindersma, S., Deits, R., Miura, K., 2015. A closed-form solution for real-time ZMP gait generation and feedback stabilization. In: 2015 IEEE-RAS 15th International Conference on Humanoid Robots (Humanoids), pp. 936–940.

Vejdani, H.R., Wu, A., Geyer, H., Hurst, J.W., 2015. Touch-down angle control for spring-mass walking. In: 2015 IEEE International Conference on Robotics and Automation (ICRA), pp. 5101–5106.

Wensing, P.M., Orin, D.E., 2013a. Generation of dynamic humanoid behaviors through task-space control with conic optimization. In: IEEE Int. Conf. on Rob. and Automation. Karlsruhe, Germany, pp. 3103–3109.

Wensing, P.M., Orin, D.E., 2013b. High-speed humanoid running through control with a 3D-SLIP model. In: Proc. of the IEEE/RSJ Int. Conf. on Intelligent Rob. and Sys. Tokyo, Japan, pp. 5134–5140.

Westervelt, E., Grizzle, J., Koditschek, D., 2003. Hybrid zero dynamics of planar biped walkers. IEEE Trans. Autom. Control 48, 42–56.

Wieber, P.-B., 2008. Viability and predictive control for safe locomotion. In: IEEE/RSJ International Conference on Intelligent Robots and Systems, IROS 2008, pp. 1103–1108.

Winkler, A., Havoutis, I., Bazeille, S., Ortiz, J., Focchi, M., Dillmann, R., Caldwell, D., Semini, C., 2014. Path planning with force-based foothold adaptation and virtual model control for torque controlled quadruped robots. In: 2014 IEEE International Conference on Robotics and Automation (ICRA), pp. 6476–6482.

Wu, A., Geyer, H., 2013. The 3-D spring–mass model reveals a time-based deadbeat control for highly robust running and steering in uncertain environments. IEEE Trans. Robot. 29 (5), 1114–1124.

Ye, Y., Liu, C.K., 2010. Optimal feedback control for character animation using an abstract model. In: ACM SIGGRAPH 2010. New York, NY, USA, pp. 74:1–9.

Chapter 4.6

Control Based on Passive Dynamic Walking

Pranav Bhounsule

Department of Mechanical Engineering, The University of Texas at San Antonio, San Antonio, TX, United States

4.6.1 INTRODUCTION

How much control is needed to create walking gaits for legged robots? The passive dynamic walking paradigm suggests that movement in a legged robot requires no control because walking can emerge purely from the mechanics of the legs. Passive dynamic walking robots are machines that use their natural dynamics, i.e., their mass distribution and geometry, to move downhill with no actuation or control.

The concept of passive dynamic walking is about a century old as evidenced by a number of patents on downhill walking toys (Fallis, 1888; Bechstein, 1912; Mahan and Moran, 1909; Wilson, 1938). The Wilson Walker is shown in Fig. 4.6.1A. It has two legs, each of which connects to a body by a hinge joint. When launched correctly, the toy is able to walk stably down a slight incline. Specifically, the sideways rocking of the body lifts a foot off the ground. The off-ground foot then swings forward to complete a step. The same sequence is repeated with the other foot, thus enabling steady downhill locomotion.

The Wilson walker inspired McGeer (1990) to create the first passive dynamic walking machine. His robot, called the Dynamite, had four legs with knees but arranged in pairs so that the inner two and outer two legs alternate during walking (see Fig. 4.6.1B for a replica made at Cornell University). Like the Wilson walker, Dynamite was able to walk stably downhill when launched with the right set of initial conditions. But the configuration of the legs limits the walking only to the sagittal or the front–back plane. Collins et al. (2001) created a 3D passive dynamic robot with two kneed legs and two swinging arms (see Fig. 4.6.1C and D). Their design had swinging arms coupled to the legs and feet with guide rails to stabilize side-to-side (roll motion) and turning (yaw motion). Owaki et al. (2011) built the first successful passive dynamic running robot. Their design had four legs with knees arranged in pairs (two inner- and outer-legs coupled to each other), an axial spring in each of the legs to cushion collisions, a spring between the legs to aid hip swing, and arc shaped feet. The robot was able to successfully run 36 steps on downhill ramp with slope of 0.22 rad. All these robots have the common feature that they use their natural

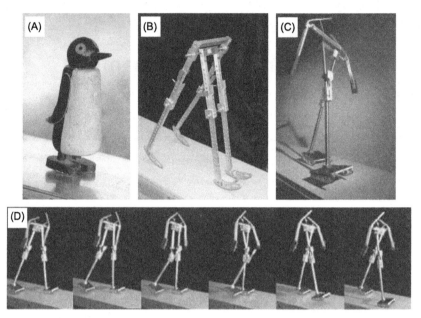

FIGURE 4.6.1 (A) The Wilson walker; (B) A copy of McGeer's passive dynamic walker built at Cornell University; (C) A 3D passive dynamic walker with arms from Cornell University. These figures are from Collins et al. (2005). (D) A sequence of snapshots during walking of the 3D passive dynamic walker shown in (C). The figure is from Collins et al. (2001).

dynamics and gravity to descend downhill. Since these robots use no motors, they are very energy-efficient. However, the most striking aspect is that their motion looks natural and graceful, like that of a human. Indeed, Mochon and McMahon (1980) have shown that the leg swing in human walking is dictated greatly by the natural dynamics with very little control. This suggests that perhaps humans exploit their natural dynamics to walk while expending negligible amounts of energy. We think that these two aspects, the energy-efficiency and the biological relevance, makes it appealing and interesting to study the role of passive dynamics in creating functional legged robots.

The rest of the chapter is written as follows. We describe the simplest passive dynamic walker in Sect. 4.6.2 and provide necessary details for analyzing its motion. This model is a nice starting point for beginners in the field. Next in Sect. 4.6.3, we describe techniques to enable passive-dynamic walking on level ground with or without control. The discussion and challenges in creating passive-dynamics based robots are in Sect. 4.6.4. Finally, the conclusions follow in Sect. 4.6.5.

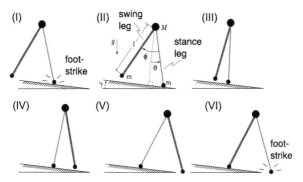

FIGURE 4.6.2 A typical step of the simplest walker. (For interpretation of the colors in this figure, the reader is referred to the web version of this chapter.)

4.6.2 PASSIVE DYNAMIC WALKING ON A SLOPE

The first known simulation of a passive dynamic walking model was done by McGeer (1990). Two other well-known papers are those by Goswami et al. (1998), who called it the compass-gait walker (reminiscent of the compass tool used in drawing), and by Garcia et al. (1998), who created an extremely simplified model and called it the simplest walker. Garcia's model had a point mass at the hip and massless legs. After nondimensionalizing velocity, the model has a single parameter, the ramp slope. The simplicity of this model makes it very attractive for learning about passive dynamic models. We present the analysis used in Garcia et al. (1998) in the next section. The MATLAB code for simulating the simplest walker and for general mass distribution round feet walker is available in the paper by Bhounsule (2014b). Another tutorial paper on passive dynamics is by Wisse and Schwab (2005).

4.6.2.1 Model Description and Equations of Motion

Fig. 4.6.2II shows a model of the simplest walker. The model consists of a mass M at the hip and a point mass m at each of the two feet. Each leg has length ℓ, gravity g points downwards, and the ramp slope is γ. The leg in contact with the ramp is called the stance leg (thin red line) while the other leg is called the swing leg (thick blue line). The angle made by the stance leg with the normal to the ramp is θ (counterclockwise is positive) and the angle made by the swing leg with the stance leg is ϕ (clockwise is positive). Fig. 4.6.2 a single walking step for the walker. The walker starts in (I), the state in which the front leg is the stance leg and the trailing leg is the swing leg. A sequence of snapshots that make up a single step are shown from (II) to (V). Finally in (VI), the swing leg collides with the ground and becomes the new stance leg. At this point, we have a complete gait cycle, i.e., the walker configuration in (VI) is the same as (I). Note that between (III) and (IV), there is foot scuffing because the swing leg passes through the ground. We ignore foot scuffing in the model but an experimental prototype needs to have a mechanism to create foot clearance during swing. Foot clearance can be created by having actuated ankles (Bhounsule et al., 2014a) or by adding knees to the walker (McGeer, 1993).

A single step of the walker consists of the following sequence:

$$\underbrace{\text{Single Stance phase} \longrightarrow \text{Foot–ground contact event} \longrightarrow \underbrace{\text{Foot-strike phase} \longrightarrow \text{Single Stance}}_{\text{one step/period-one limit cycle}}}$$

(4.6.1)

Next, we state the equations of motion for the phases and events described in Eq. (4.6.1) and provide a brief explanation on the derivation. Please see the appendix for more details on the derivation.

Single Stance Phase (Continuous Dynamics)

In this phase of motion, the stance leg pivots and rotates about the stationary foot, while the swing leg pivots and rotates about the hinge connecting the two legs. The assumptions are: the stance leg does not slip, there is no hinge friction, and foot scuffing is ignored. The equations for this phase are:

$$\ddot{\theta} = \sin(\theta - \gamma), \tag{4.6.2}$$

$$\ddot{\phi} = \sin(\theta - \gamma) + \{\dot{\theta}^2 - \cos(\theta - \gamma)\}\sin(\phi). \tag{4.6.3}$$

Eqs. (4.6.2) and (4.6.3) are obtained by doing an angular momentum balance about stance foot contact point and hip hinge respectively, followed by nondimensionalizing the time with $\sqrt{\ell/g}$ and applying the limit, $m/M \to 0$.

Foot–Ground Contact Event

The swing leg contacts the ground when the following condition is met:

$$\phi = 2\theta. \tag{4.6.4}$$

Foot-Strike Phase (Discontinuous Dynamics)

In this phase of motion, the legs exchange their roles. That is, the current swing leg becomes the new stance leg and the current stance leg becomes the new swing leg. The assumptions are: the swing leg has a plastic collision (no slip and no bounce) with the ground, the collision is instantaneous, and there is no double support phase. The equations for this phase are:

$$\theta^+ = -\theta^-, \tag{4.6.5}$$

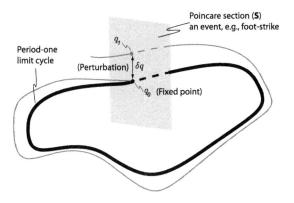

FIGURE 4.6.3 A Poincaré map is used to find walking solutions and to analyze stability.

$$\phi^+ = -\phi^- = -2\theta^-, \tag{4.6.6}$$

$$\dot{\theta}^+ = \cos(2\theta^-)\dot{\theta}^-, \tag{4.6.7}$$

$$\dot{\phi}^+ = \left(1 - \cos(2\theta^-)\right)\cos(2\theta^-)\dot{\theta}^-, \tag{4.6.8}$$

where the superscripts $-$ and $+$ denote the instance just before and just after foot-strike, respectively. The switching of the leg angles is given by Eqs. (4.6.5) and (4.6.6). The angular rates of the legs after foot-strike are obtained by using conservation of angular momentum about the impending foot-strike point and the hinge joint at the hip to obtain Eqs. (4.6.7) and (4.6.8), respectively. Then, time is nondimensionalized using $\sqrt{\ell/g}$ and the limit, $m/M \to 0$, is applied.

4.6.2.2 Analysis Using Poincaré Return Map

A Poincaré return map is used to find steady-state walking motions and to analyze motion stability (Garcia et al., 1998; McGeer, 1990; Strogatz, 2014). In Fig. 4.6.3, the gray region is the Poincaré section and denotes an instance in the walking motion (e.g., before foot-strike, after foot-strike, and mid-stance).

We assume the Poincaré section to be the instance just after foot-strike. Let $\mathbf{q_0} = \{\theta_0^+, \dot{\theta}_0^+, \phi_0^+, \dot{\phi}_0^+\}$ be the state after foot-strike. Then, there is a function $\mathbf{S}$ that takes the initial condition, $\mathbf{q_0}$, and returns the state after one step, $\mathbf{q_1}$. The function $\mathbf{S}$ is called the stride map. Thus, the Poincaré map is $\mathbf{q_1} = \mathbf{S(q_0)}$. There is an initial condition $\mathbf{q_0}$ such that

$$\mathbf{q_0} = \mathbf{S(q_0)}. \tag{4.6.9}$$

The above condition defines a period-one limit cycle. In other words, the initial condition after foot-strike, $\mathbf{q_0}$, defines a walker state that maps onto itself after one step. Similarly, one can find a period-two limit cycle by applying the function S twice, and so on.

In general, it is not possible to find S and the state q_0 analytically, so one needs to resort to numerical techniques. To compute S, we first integrate the equations of motion in the single stance phase (Eqs. (4.6.2) and (4.6.3)) till the foot-strike event (Eq. (4.6.4)), and apply the leg support exchange conditions (Eqs. (4.6.5)–(4.6.8)). Finally, to find four initial conditions in q_0, the zeros of Eq. (4.6.9) ($q_0 - S(q_0) = 0$) are found. The zeros can be found by root finding techniques such as Newton–Raphson's method. In our experience, a good initial guess is of paramount importance for the root finder to give quick results. To find good initial conditions, we recommend simulating and animating a single step to see if it is close to repeating and then use those as a guess for the root finder (also see Wisse and Schwab, 2005).

After obtaining q_0, the stability of the period-one limit cycle is analyzed. To do this, one needs to compute the eigenvalues of Jacobian of the Poincaré map, S. To obtain the Jacobian, we used the central difference with a step size of 10^{-5}. The limit cycle is stable if the magnitude of the biggest eigenvalue is less than 1 and unstable otherwise (Garcia et al., 1998; McGeer, 1990; Strogatz, 2014).

We give benchmark results for a ramp slope, $\gamma = 0.009$, the only free parameter in this model. Using the method described above, there are two period-one limit cycles. Table 4.6.1, first row, gives the two limit cycles. Table 4.6.1, second row, gives the eigenvalues of each of the fixed points, q_0. As seen from the table, the middle column is the stable limit cycle because the biggest eigenvalue is inside the unit circle while the third column from left is the unstable limit cycle because the biggest eigenvalue is outside the unit circle. Thus one limit cycle is stable and the other is unstable. Fig. 4.6.4 shows the angular position of the stance and swing leg as a function of time for the stable limit cycle and phase portrait of the stable limit cycle.

4.6.2.3 Passive Dynamic Walking in 3-Dimensions

McGeer (1993) and Garcia (1999) analyzed a 3D model with four degrees of freedom (roll or side-to-side, pitch or front–back, yaw or turning on the stance leg and interleg pitch angle between stance and swing leg). However, both of them were unable to find a stable walking gait. Kuo (1999) considered a simpler 3D model without the yaw degree of freedom. After doing an exhaustive search, he found that one eigenvalue was always greater than one. This eigenvalue associated with this unstable gait was in the roll direction and was due to a mismatch in the roll velocity at ground contact condition. Further, he demonstrated that several simple strategies such as: applying a torque in the yaw direction, spin-

TABLE 4.6.1 Fixed points (first row and denoted by q_0), eigenvalues using central difference (second row and denoted by λ), for the simplest walker for slope, $\gamma = 0.009$. The fixed points are accurate to 12 decimal places. The eigenvalues computed by central difference and with perturbation size of 10^{-5} and are accurate to 5 decimal places

Variable	Stable solution	Unstable solution
State, q_0	$\begin{bmatrix} 0.200310900544287 \\ -0.199832473004977 \\ 0.400621801088574 \\ -0.015822999948318 \end{bmatrix}$	$\begin{bmatrix} 0.193937369810184 \\ -0.203866927442010 \\ 0.387874739620369 \\ -0.015144260853192 \end{bmatrix}$
Eigenvalues, λ	$\begin{bmatrix} 0 \\ 0.000000001586465 \\ -0.190099639402167 - i0.557599274284362 \\ -0.190099639402167 + i0.557599274284362 \end{bmatrix}$	$\begin{bmatrix} -0.000000000000002 \\ -0.000000005231481 \\ 0.459589047035257 \\ 4.003865226079296 \end{bmatrix}$

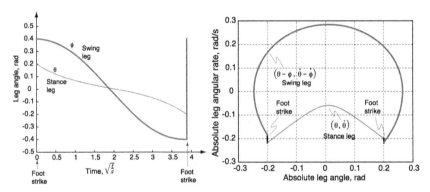

FIGURE 4.6.4   (Left) Stance leg and swing leg angle as a function of time for one step of the simplest walker. (Right) Phase portrait for one step of the simplest walker for slope, $\gamma = 0.009$.

ning a reaction wheel, moving the upper body slightly, and controlling the lateral foot placement, all have the effect of stabilizing the roll motion while preserving the passive dynamics.

Collins et al. (2001) were able to create a stable, 3D passive dynamic machine by adding swinging arms (see Fig. 4.6.1). Coleman and Ruina (1998) created a nonanthropomorphic walker with ellipsoidal feet that was able to walk stably downhill. Though Coleman and Ruina were able to explain the stability of their walker using Poincaré based methods (Coleman et al., 2001), it is not clear what design parameters are critical in achieving stable three-dimensional passive dynamic walking.

(A) Passive walker with upper body

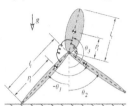

(B) Passive walker with inertial device

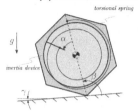

FIGURE 4.6.5 Collisionless walking models: (A) bipedal walking model with upper body coupled to the legs through torsional springs (Gomes and Ruina, 2011), (B) rimless walking model with inertial device with torsionally coupled spring (Gomes and Ahlin, 2015).

4.6.3 POWERED BIPEDAL ROBOTS INSPIRED FROM PASSIVE DYNAMICS

In walking robots, energy is lost each time the foot hits the ground (unless special mechanism is used to prevent collisional losses). In order to sustain steady walking, this energy needs to be supplied through external means. In case of passive dynamic robots walking downhill, this energy is supplied by gravity. These facts suggests two different approaches to enable level ground walking: (1) prevent energy loss during collision by suitable robot design (see Sect. 4.6.3.1), and (2) use an actuator to supply the lost energy (see Sect. 4.6.3.2). The rest of this section will highlight some of the methods to enable almost-passive walking on level ground.

4.6.3.1 Collisionless Walking

One way to enable level ground walking with passive models is to find means of reducing the collisional losses at foot-strike to zero. Gomes and Ruina (2011) created a passive dynamic walking model which had an upper-body that was coupled to each leg through a torsional spring (see Fig. 4.6.5A). They found internal oscillatory modes of the upper body that ensures that the swing leg contacts the ground with zero velocity. Thus, the robot is able to sustain walking on level ground without external energy input. However, note that the motion of the robot is unstable because even the slightest perturbation will create a collisional loss at foot-strike and the robot will be off the limit cycle. Thus there are no stable (asymptotic, uniform, etc.) solutions for collisionless locomotion models. Also, the model requires the swing foot to stick to the ground and later release for swing. Gomes and Ahlin (2015) have created a physical prototype of a rimless wheel, another passive dynamic model (McGeer, 1990), that can demonstrate nearly collisionless walking. Their device consists of the rimless wheel coupled to an inertial wheel through a torsional spring. Between the mid-

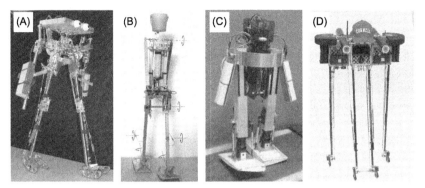

FIGURE 4.6.6 Powered walkers inspired from passive dynamics: (A) Cornell powered biped, (B) Delft powered biped, and (C) MIT learning biped. These figures are from Collins et al. (2005), and (D) Cornell Ranger (Bhounsule et al., 2014a).

dle to the end of a step, the torsional spring transfers the energy of the rimless wheel to the inertial wheel thereby reducing the wheel velocity to almost zero just before the next spoke makes contact with the ground. The torsional spring then transfers the stored energy back to the wheel from start to the middle of the step speeding up the rimless wheel. This energy transfer ensures walking on level ground without collisional losses.

4.6.3.2 Actuating Passive Dynamic Walking Robots

In robots where collisionless walking is not possible, one can add one or more actuators to enable level ground walking. Fig. 4.6.6 shows powered bipedal robots based on passive dynamic walking principles. The Cornell biped (Fig. 4.6.6A) has five internal degrees of freedom (two ankles, two knees, and a hip), the arms are mechanically linked to the opposite leg, and the upper body is kinematically constrained so that its midline bisects the hip angle through a hip bisection mechanism. The robot is electrically powered by an ankle push-off that is triggered when the opposing foot hits the ground. The Delft biped (Fig. 4.6.6B) is similar to Cornell biped, but is powered by pneumatic hip actuation and has a passive ankle. The MIT learning biped (Fig. 4.6.6C) is based on the simpler ramp-walker passive hip, is powered by two servo motors in each ankle, and uses reinforcement learning to automatically acquire the controller (Collins et al., 2005). The Cornell Ranger (Fig. 4.6.6D) has three internal degrees of freedom (one hip and two ankles) and is electrically powered. More details on control of Ranger are discussed later in this section. Next, we review control schemes that preserve the natural dynamics while enabling walking on level ground.

Virtual passive dynamic walking is able to recreate downhill walking by adding a virtual gravity field using ankle and hip actuators. In passive dynamics walking with a downhill slope of γ, gravity makes an angle of γ with the direction perpendicular to the ramp. Thus, the component of gravity normal to the ramp is $g\cos(\gamma)$ and along the ramp is $g\sin(\gamma)$. But since γ is relatively small, one can approximate the normal component as g and horizontal component as $g\gamma$. However, if the slope was zero (level ground walking), then the component normal to the ground would be g and it would be 0 in the horizontal direction. From the above arguments we see that the walker on level ground is missing a horizontal component of $g\gamma$. Thus, the idea behind virtual passive dynamic walking control is to *use actuators* to create a virtual gravitational field such that the horizontal component is $g\gamma$ and leave the vertical component unaffected (Asano et al., 2000). The resulting motion is very similar to passive dynamic walking on slope γ but it is on level ground. However, this requires both, an ankle as well as a hip actuator.

Another way to achieve almost passive dynamic walking is to track a constant mechanical energy. The key idea is that passive dynamic robots are able to maintain a periodic walking motions because their mechanical energy (i.e., kinetic + potential energy) is constant between steps. Thus to recreate passive dynamic walking on level ground, one can use the actuators to track this mechanical energy (Goswami et al., 1997). Further, each slope has a different total mechanical energy. Thus, by tracking the total mechanical energy for a given slope, the walking motion can be made slope independent. A key point here is that the tracking gains need to be kept low to ensure that the natural dynamics of the passive gait is preserved.

Yet another way of preserving passive dynamic walking is to use on–off or bang–bang control to supply the energy lost during collision. Camp (1997) presented a 2D knee-less model with two legs and two powered ankles that used such an actuation scheme. The ankle motor is turned on when the swing leg reaches a prescribed angle and shut-off at the instance of foot-strike. The walker exhibits a variety of stable and unstable limit cycles as the motor stall torque is varied. The stall torque is thus analogous to the ramp of the passive dynamic walker. An extreme case of this type of control is to use an impulse type control to power walking (Formalsky, 1995). An impulse is provided at the beginning of the swing phase and no actuation is provided for the rest of the step. By choosing appropriate impulse at the beginning of swing phase the robot is able to walk stably.

Low gain proportional-derivative (PD) controllers can be used to create passive-dynamic like walking gaits on level ground. Typical implementation involves dividing the walking step into set of states or a state machine, and having different PD controllers and set-points for different states (Braun and Goldfarb,

2009; Dertien, 2006). The gains on the PD controller are weak so that they do not interfere with the natural dynamics of the legs.

Instead of using continuous feedback to track the mechanical energy, one can use feedback at discrete times in the walking step. For instance, when a passive dynamic robot walks on level ground without any control whatsoever, the end-of-step state will be different from the start-of-step state because of the collisional losses. The error can be used to derive feedback control law that nullifies the difference (Miura and Shimoyama, 1984). This type of control is called once per step control because the feedback error and corrections are based on sampling the state once per step. Bhounsule et al. (2014a) took a similar approach to stabilize the robot Ranger (see Fig. 4.6.6D) which walked a distance of 40.5 miles nonstop on a single battery charge. The stabilization is in addition to the energy-optimal trajectory controller that is set up on the robot. The Poincaré map for Ranger is about the mid-stance position. The energy-optimal trajectory is linearized about the Poincaré map. In the linearized equations, the state variables are the stance leg velocity, swing leg position and velocity at mid-stance and the control actions are the foot placement and ankle push-off. The linearized equations are used to set up a discrete linear quadratic regulator to reduce the errors in the state at the Poincaré section (Bhounsule et al., 2014b). We provide more details in the next section.

4.6.3.3 Discrete-Decision Continuous Action Control

Next, we present a controller formulation that does discrete, event-based, intermittent control that is able to preserve much of the passive dynamics of walking robots (also see Bhounsule et al., 2014b). We illustrate the problem with a hypothetical example and then show how it can be used to control a bipedal robot.

Control Problem

Let the state of the full, possibly nonlinear, system be $x(t)$, the control be $u(t)$ and the continuous system dynamics defined by F with $\dot{x} = F(x, u)$. Further, assume the system has a desirable nominal trajectory $\bar{x}(t)$ associated with a nominal baseline control $\bar{u}(t)$,

$$\dot{\bar{x}} = F(\bar{x}, \bar{u}). \tag{4.6.10}$$

The feedforward command $\bar{u}(t)$ in the above equation is open-loop and does not stabilize the system adequately, or perhaps at all. For example, even with perfect initial conditions, modeling errors, actuator imperfections and disturbances will cause the system to too-much, or catastrophically ("failure"), deviate from the nominal trajectory. So we add a feedback control that supplements u with

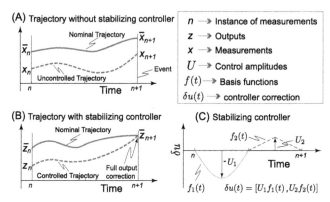

FIGURE 4.6.7 Schematic example: (A) the nominal (solid red) and deviated (dashed blue) trajectory, for some dynamic variable x of interest. We measure the state x at the start of a continuous interval, namely at section n; (B) the new deviated trajectory in target variables z after switching on our feedback controller. In this example, feedback controller nulls (zeros) the output z at the end of the interval, illustrating a "dead-beat" controller. (C) The feedback motor program has two control actions: a sinusoid for first half cycle and a hat function for the second half of the cycle. These shapes are arbitrary and different from each other in form only for illustrative purposes. They could overlap in time. We choose the amplitudes U_1 and U_2 of the two functions at the start of the interval depending on the error $(x - \bar{x})$. By a proper choice of the amplitudes U_1 and U_2 deviations are, in this example, fully corrected in between measurements. The choice of trigger for event n, the choice of sensor measurements x, the choice of output variables z, and the control shape functions $f(t)$ are offline design choices.

a control δu to adequately brings the system back to the nominal trajectory. In this case, we do feedback at discrete times and the control commands are simple feedforward control functions over the interval. This differs from common continuous feedback control because we only sense key quantities and only at occasional times.

Schematic Example

We illustrate the event-based intermittent feedback control idea with a schematic example. Consider the nominal trajectory of a second-order system shown as a solid red color line in Fig. 4.6.7. Let n and $n + 1$ be instances of time at which we are taking measurements from sensors. The time interval between the measurements n and $n + 1$ is typically on the order of the characteristic time scale of interest (and *not* the shortest time our computational speed allows). Let us assume that we take two measurements, $x_n = [x_1 \; x_2]'$ (e.g., a position and velocity) at time n. We want to regulate two outputs: z_1 and z_2 (some attributes of the state x_n) at time $n + 1$.

Assume that, due to external disturbances, the system has deviated from its nominal trajectory. We show the trajectory as a dashed blue color line in

Fig. 4.6.7A. Now, the state of the system is $\bar{x}_n$ ($\neq x_n$) at time n. When feedback corrections are absent, the relevant output $\bar{z}_{n+1}$ ($\neq z_{n+1}$) whose components, in notational shorthand, are $[\bar{z}_1 \ \bar{z}_2]'$.

Our feedback controller measures deviations at time n ($\delta x_n = x_n - \bar{x}_n$) and uses actuation to reduce the deviations in output variables ($\delta z_{n+1} = z_{n+1} - \bar{z}_{n+1}$). For illustration, we choose two control actions, $\delta \mathbf{u}_n = [U_1 f_1(t) \ U_2 f_2(t)]'$, a half-sinusoid and a hat function, each active for half the time between time $n + 1$ and n (Fig. 4.6.7C). The controller adjusts the amplitudes (U_1 and U_2) of the two control functions, based on measured deviations δx_n, to regulate the deviated outputs δz_{n+1}. For example, with a proper choice of the amplitudes, it should be possible to fully correct the deviations in the output variables, as seen in Fig. 4.6.7B.

In the simplest cases, we linearize the map from the measurement section n to the section $n + 1$. The sensitivities of the dynamic state to the previous state and the controls $\mathbf{U}_n = [U_1 \ U_2]'$ are: $\mathbf{A} = \partial x_{n+1}/\partial x_n$, $\mathbf{B} = \partial x_{n+1}/\partial \mathbf{U}_n$, $\mathbf{C} = \partial z_{n+1}/\partial x_n$, and $\mathbf{D} = \partial z_{n+1}/\partial \mathbf{U}_n$. The brute-force way of calculating the sensitivity matrices $\mathbf{A}, \mathbf{B}, \mathbf{C}$ and $\mathbf{D}$ is by numerical finite-difference calculations. We then have, for our linearized discrete system model:

$$\delta x_{n+1} = \mathbf{A}\delta x_n + \mathbf{B}\mathbf{U}_n, \tag{4.6.11}$$

$$\delta z_{n+1} = \mathbf{C}\delta x_n + \mathbf{D}\mathbf{U}_n. \tag{4.6.12}$$

Again, the δx_n are a list of measured deviations, the δz_n are a list of deviations which we wish to control, the $\mathbf{U}$ are the activation amplitudes (2 in our example above). For simplicity, assume full state measurement, the controller architecture is thus

$$\mathbf{U}_n = -\mathbf{K}\delta x_n, \tag{4.6.13}$$

where $\mathbf{K}$ is a constant gain matrix. We choose the gains $\mathbf{K}$ to meet or optimize various goals using a discrete linear quadratic regulator (DLQR).

For most systems, ones that have the needed controllability, it is possible to find shape functions $f_1(t)$ and $f_2(t)$ so that the matrix $\mathbf{B}$ is nonsingular. In the same way that a square matrix is generically nonsingular, n random shape functions for an n order system should (generically) lead to a nonsingular $\mathbf{B}$ and thus the possibility of 1-step dead-beat control. Of course, the matrix $\mathbf{B}$ can be more or less well conditioned depending on how independent the shape functions are from each other.

Discrete Linear Quadratic Regulator (DLQR)

One can use a DLQR to any goal function z of the state. In DLQR (Ogata, 1995), we seek to minimize the cost function J_{dlqr} defined as

$$J_{\text{dlqr}} = \sum_{n=0}^{n=\infty} \left(\delta z_{n+1}{}^T \mathbf{Q}_{zz} \delta z_{n+1} + \mathbf{U}_n{}^T \mathbf{R}_{UU} \mathbf{U}_n \right), \qquad (4.6.14)$$

where $\mathbf{Q}_{zz}$ and $\mathbf{R}_{UU}$ are matrices that weight the different components of δz_{n+1} and $\mathbf{U}_n$ ($\mathbf{R}_{UU}$ must be positive definite and $\mathbf{Q}_{zz}$ positive semidefinite). The weights $\mathbf{Q}_{zz}$ and $\mathbf{R}_{UU}$ are design parameters picked to give reasonably fast return to nominal values but without unduly high gains (which might tend to lead to control command that are beyond safety limits). They are often given as diagonal for simplicity.

Putting Eq. (4.6.12) into Eq. (4.6.14) and rearranging gives

$$J_{\text{dlqr}} = \sum_{n=0}^{n=\infty} \left(\delta x_n{}^T \mathbf{Q} \delta x_n + 2 \delta x_n{}^T \mathbf{N} \mathbf{U}_n + \mathbf{U}_n{}^T \mathbf{R} \mathbf{U}_n \right), \qquad (4.6.15)$$

where $\mathbf{Q} = \mathbf{C}^T \mathbf{Q}_{zz} \mathbf{C}$, $\mathbf{N} = \mathbf{D}^T \mathbf{Q}_{zz} \mathbf{C}$, and $\mathbf{R} = \mathbf{D}^T \mathbf{R}_{zz} \mathbf{D} + \mathbf{R}_{UU}$. J_{dlqr} can be minimized with a linear state feedback, $\mathbf{U}_n = -\mathbf{K} \delta x_n$ with gain $\mathbf{K}$ found by solving the standard Riccati equation (Ogata, 1995).

Other Goals

The same linear control architecture given by Eq. (4.6.13) could have gains $\mathbf{K}$ chosen to optimize or achieve other criteria that do not fit into standard basic linear control formalisms. For example, there could be a weight on the sparseness of $\mathbf{K}$, on nonquadratic costs for error and control over some range of initial conditions, on the basin of attraction for the nonlinear system, etc. To calculate $\mathbf{K}$ one might then require more involved optimization calculations, but the structure of the resultant controller would be preserved. Similarly the choice of shape functions could be subject to optimization on independence, smoothness, maximizing control authority, etc.

Factors to Consider While Designing the Controller

The systems we are interested in controlling are not those in which we do measure control quality by how closely a target is followed; clearly, the type of intermittent control we discuss here is not optimal for that. Rather, we are interested in preventing total system failure. For walking or for an inverted pendulum, falling down is failure. To slightly generalize, by failure we mean that the system state has moved outside a particular target region surrounding the target point. How is this region defined? In practice, it is the region outside of which nonlinear effects lead to divergence of the solution to points much farther from the target (e.g., falling down). Sticking to the linear model, the user has to supply the target region based on intuitions, experience, or nonlinear modeling. Some issues in the controller design include:

1. Selecting a suitable section or instance of time to take measurements — this instant should be when the dynamic-state estimation is reasonably accurate, and when dynamic-state errors which cause failure are evident;
2. Selecting measurement variables (x_n) that are well-predict system failure;
3. Picking output variables (z_n) that can well-correct against system failure; and
4. Picking actuator shape profiles ($f(t)$'s) that have large, and relatively independent, effects on the target variables, and are also sufficiently smooth for implementation with real motors.

We next discuss the above points with in the context of a walking robot.

Example: Controlling a Bipedal Walking Robot

For a 2D bipedal robot walking at steady speed, here is how we can go about designing a discrete controller (Bhounsule, 2014a). A typical walking step of a bipedal robots includes two phases: a smooth continuous phase in which the entire robot vaults over the grounded leg, and a nonsmooth discontinuous phase in which the legs exchange roles.

1. Suitable section or instance of time to take measurements: Any instant not-close to support-exchange is a good time for measurement. This is because the measurements are typically noisy during the nonsmooth support change (heel-strike collision).
2. Suitable measurement variables (x_n) that are representative of system failure: The state of the lower body is most important for walking balance, so good measurement variables are the state (position and velocity) of the stance leg.
3. Suitable output variables (z_n) that also correlate with system failure: Step time, step length are important quantities to regulate during walking, and they serve as good output variables.
4. Suitable actuator shape profiles ($f(t)$'s) that have large and relatively independent effects on the target variables: For leg swing, for example, two torque profiles, one with large amplitude near the start of the interval, and one with large amplitude near the end, yield good control authority over position and velocity of the swing leg at the end of the interval.

Once the above quantities are picked, we can check the system controllability. If the system is not well controllable (correction of reasonable disturbances requires unreasonable actuation amplitudes), the first likely fix is picking better actuation shape functions.

As noted, we used this discrete feedback control idea to stabilize steady walking gait of a bipedal robot leading to energy-efficiency record and long distance 65 km walking record (Bhounsule, 2012; Bhounsule et al., 2014a; Ruina, 2012).

Computing the Linearization

For linear control approaches, the gain selection depends on having the linearized map Eqs. (4.6.11) and (4.6.12) from Eq. (4.6.10). We assume we have a system, or computational model of the system, with which we can perform numerical experiments. To get the matrices $\mathbf{A}$ and $\mathbf{C}$, we can perturb x_n element-wise and use finite difference to compute these matrices. Similarly, to get matrices $\mathbf{B}$ and $\mathbf{D}$, we can put in small amplitudes of the controls $\mathbf{U}_n$ and use finite difference to compute the sensitivities.

4.6.4   DISCUSSION AND CHALLENGES

4.6.4.1   Energy Efficiency and Dynamic Walking

Energy-efficiency for a variety of locomotion/mobility modes is quantified by total cost of transport (TCOT) (Tucker, 1970) and the mechanical cost of transport (MCOT) which are defined as follows:

$$\text{TCOT} = \frac{\text{total energy used per step}}{\text{weight} \times \text{step length}}, \tag{4.6.16}$$

$$\text{MCOT} = \frac{\text{mechanical energy used per step}}{\text{weight} \times \text{step length}}. \tag{4.6.17}$$

The total energy includes the mechanical energy and other energy-terms like dissipation in the resistive elements of electric motors, energy to power the electronics (e.g., sensors, computers). For passive dynamic walkers, the total energy is equal to the mechanical energy and is equal to the tangent of the ramp slope. Thus, $\text{MCOT} = \tan(\gamma) = \text{TCOT}$, where γ is the ramp slope. McGeer's Dynamite had a $\text{TCOT} = \text{MCOT} = 0.025$ (McGeer, 1990). Some of the most energy-efficient powered legged robots are: Collins biped ($\text{TCOT} = 0.2$, $\text{MCOT} = 0.055$) (Collins and Ruina, 2005); Cornell Ranger ($\text{TCOT} = 0.19$, $\text{MCOT} = 0.04$) (Bhounsule et al., 2014a); and Cargo ($\text{TCOT} = 0.1$) (Guenther and Iida, 2017). To put these numbers in perspective, humans have a $\text{TCOT} = 0.3$ (Atzler et al., 1925)[3] and $\text{MCOT} = 0.05$ (Margaria, 1968). Note that both TCOT and

---

3. The TCOT is computed using the total metabolic energy. However, if only the energy to walk is taken into account then human TCOT is 0.2.

MCOT are functions of the step size and step velocity and the above values correspond to the lowest energy values at a specific step size and step velocity (Bertram and Ruina, 2001).

4.6.4.2 Stability and Robustness

Passive dynamic-based walkers have shown poor stability and robustness characteristics. The most well-known method of computing stability of passive dynamic-based robots is using the eigenvalues of the limit cycle (see Sect. 4.6.2.2). The walking motion is stable if the magnitude of the biggest eigenvalue is less than 1 and unstable otherwise. In particular, an eigenvalue equal to 0 implies that all disturbances are nullified in a single step. Thus a values closer to zero implies greater stability. However, passive dynamic robots have rarely demonstrated an eigenvalue less than $0.6 > 0$ (Bhounsule et al., 2014a). One way of stabilizing the passive dynamic-based walkers is to develop a controller that sets the eigenvalue to a desired value, also known as pole placement (Kuo, 1999; Bhounsule et al., 2014a, 2014b). Another option is to minimize the biggest eigenvalue during the controller design phase (Mombaur et al., 2001).

A commonly used metric for robustness of passive dynamics-based walkers is the maximum change in height that the robot can withstand without falling (Wisse et al., 2005). One can nondimensionalize the change in height with the leg length to compare different robots. The maximum step-down (normalized by leg length) for passive dynamics-based robots from TU Delft are: Max 1%, Denise 1%, and Mike 2% (Hobbelen, 2008), indicating poor robustness to terrain variation. Kim and Collins (2017) have found that adding random disturbances rather than a single disturbance is a better indicator of stability. They have also found that to get consistent results, one needs to evaluate stability (ability to not fall) over 100 steps. Kelly and Ruina (2015) provide a technique for creating asymptotically stable and robust using Lyapunov function. But all the approaches so far evaluate the robustness after controller design. A challenge then is to come up with a technique to design a controller for a given robustness.

4.6.4.3 Versatility, Maneuverability, and Agility

Versatility refers to the ability of the bipedal robot to stand, walk, turn, and climb stairs (Kuo, 2007). Maneuverability is the robot's ability to turn its body or change the heading (Full et al., 2002; Jindrich and Full, 1999), and agility is defined as the robot's ability to change its velocity (Bowling, 2011). Passive dynamics-based robots have demonstrated very limited versatility, agility, and maneuverability. There does not seem to be any fundamental limitation in addressing these metrics except that very limited work has been done in this regard.

4.6.4.4 Mechanical Design

Proper tuning of the mass distribution, inertia, and leg geometry is vital to enable unactuated passive dynamic walking down a ramp. We discuss the issues next.

The natural frequency of the swinging leg should be such that it is able to swing forward to break the forward fall about the stance leg. The natural frequency depends on the leg inertia and the location of the center of mass of the leg. The pendulum swing time is directly proportional to the inertia of the leg and inversely proportional to the location of the center of mass of the leg. Thus, by increasing the inertia or moving the center of mass near the torso increases the swing time and which increases the natural frequency of walking. If the natural frequency increases too much then there will be no passive walking solutions. However, moving the center of mass away from the pin joint will increase the energy loss at foot-strike, leading to energy-inefficiency. Thus, there is a trade-off in locating the center of mass on the legs. Another key parameter is the offset of the center of mass with respect to the line joining the hip joint and the foot contact point. Simulations have shown that the existence of walking solutions are extremely sensitive to the mass fore-aft offset.

Adding an upper body increases the energy-efficiency and stability of a 2D model of walking but adds more complexity to the walker (Wisse et al., 2007). One way of reducing the complexity is to kinematically couple the upper body to the legs through a hip bisection mechanism. The hip bisection mechanism ensures that the angle of the upper body is the average of the angle between the two legs. However, it is conjectured that the hip bisection mechanism could potentially reduce the energy efficiency because of the need to actively counteract effects of the torso on the trailing leg following collision (private communication, Steve Collins).

A circular shaped foot is more energy-efficient than a point foot. As the radius of curvature of the foot increases, the collisional losses at foot-strike decreases, thereby increasing energy-efficiency. When the radius of curvature of the foot is equal to the leg length, there is a collision free support transfer between the legs, provided the center of mass is also at the hip joint. Such a walker is called a synthetic wheel (McGeer, 1990) and can walk on level ground without using external energy.

Walking robots also need a mechanism that will enable ground clearance during leg swing. One technique is to use sideways rocking to allow for ground clearance (e.g., see Wilson Walker, Fig. 4.6.1A). To enable rocking, the bottom of the feet are made circular in the longitudinal as well as lateral direction with the center of both arcs approximately at the same place (Kuo, 1999). In addition, the leg mass, center of mass, and inertia needs to be tuned so that the lateral and longitudinal swing leg motion have the correct frequency which is dependent on the slope and dynamics of the rest of the walker. Another technique of creating ground clearance is to use knees but needs proper design (e.g., a latching mechanism) to prevent knee buckling. As both these methods add additional degrees of freedom, it also decreases the range of passive walking solutions.

Finally, friction in joints need to be as little as possible. Simulations with passive dynamic walkers have shown that passive dynamic walking solutions disappear as the friction increases (McGeer, 1990). For a passive inspired powered robot it is vital for the motors to be back-drivable to allow for passive leg swing.

4.6.4.5 Estimation

Good control depends on good estimates of the robot state and perhaps of the external disturbances. For example, to create energy-efficient walking with ankle actuation, the timing of push-off is critical. Push-off before heel-strike is four times cheaper than push-off after heel-strike (Kuo, 2002; Ruina et al., 2005). However, to do push-off just before heel-strike one needs good estimates of the time to heel-strike, which depends on the stance and swing leg angles and the terrain. Since it is next to impossible to have a precise estimate of all these things, it is not possible to determine the exact time to heel-strike. A compromise is to start the rear ankle push-off as the front foot hits the ground so as to achieve an overlap between the two. Sometimes it might be necessary to know the robot state just after heel-strike (e.g., if control is based on instance after heel-strike). However, the robot is vibrating at the instance after heel-strike which makes it challenging to do state estimation. Finally, almost all passive dynamic robots walk blindly. If these robots have to walk in practical scenarios such as in the presence of obstacles or stepping stones, it is crucial to incorporate vision based estimation and modify the control algorithm accordingly.

4.6.4.6 Higher Dimensional Systems

Most successful passive dynamics-based walkers have a few degrees of freedom, typically between 3 to 6. It is not obvious how to extend passive dynamics control approach to high dimensional systems such as humanoids which have 10+ degrees of freedom. Most humanoids are versatile but not quite energy-efficient (TCOT of Honda's ASIMO is around 3.2 and that of Boston Dynamics' PETMAN/ATLAS is around 5 (Bhounsule et al., 2014a)). Creating energy-efficient and versatile humanoids will dramatically increase their practicality.

4.6.5 CONCLUSION

Passive dynamic walking is an attractive concept because of the low energy usage and the naturalness in the motion. However, the major drawbacks of passive-dynamics robots are: limited robustness, limited versatility, and limited agility/maneuverability which restricts their applications to simple systems and simple scenarios. How to create walking machines that meet all the above metrics is clearly an important, but unsolved challenge.

ACKNOWLEDGEMENT

This work was partially supported by NSF grant number IIS-1566463.

APPENDIX 4.6.6

4.6.6.1   Derivation of Equations of Motion for the Simplest Walker

The equations of motion for the simplest walker were given in Section 4.6.2. We provide more details here.

Single Stance Phase

The equations of motion in single stance phase are given below:

$$\mathbf{A}_{ss}\mathbf{X}_{ss} = \mathbf{b}_{ss}, \tag{4.6.18}$$

$$\mathbf{A}_{ss} = \begin{bmatrix} -\ell^2 \left(M + 2m - 2m\cos(\phi)\right) & -\ell^2 m \left(\cos(\phi) - 1\right) \\ l^2 m \left(\cos(\phi) - 1\right) & \ell^2 m \end{bmatrix}, \quad \mathbf{X} = \begin{bmatrix} \ddot{\theta} \\ \ddot{\phi} \end{bmatrix},$$

$$\mathbf{b}_{ss} = \begin{bmatrix} Mg\ell\sin(\gamma - \theta) - \ell^2 m\dot{\phi}^2 \sin(\phi) - g\ell m\sin(\gamma - \theta + \phi) & + g\ell m\sin(\gamma - \theta) + 2\ell^2 m\dot{\theta}\,\dot{\phi}\sin(\phi) \\ \ell^2 m\dot{\theta}^2 \sin(\phi) - g\ell m\sin(\gamma - \theta + \phi) \end{bmatrix}.$$

To reduce them to the simplest walker Eqs. (4.6.2) and (4.6.3), we nondimensionalize time with $\sqrt{\ell/g}$ and take the limit $m/M \to 0$.

Next, we give more details about the derivation of the equation for single stance. Let $\vec{H}_{/X}$ and $\vec{M}_{/X}$ denote the rate of change of angular momentum and external torque about the point X, respectively. The first and second lines in the above equation are obtained by equating the angular momentum to the external torque about the foot in touch with the ground, C_1, and the hip, H, respectively. These points of interest are shown in Fig. 4.6.8A. We obtain the following equations:

$$\dot{\vec{H}}_{/C_1} = \vec{M}_{/C_1}, \tag{4.6.19}$$

$$\dot{\vec{H}}_{/H} = \vec{M}_{/H}. \tag{4.6.20}$$

The above two equations can be written as:

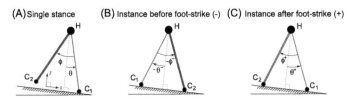

(A) Single stance   **(B)** Instance before foot-strike (-) **(C)** Instance after foot-strike (+)

FIGURE 4.6.8 (A) Simplest walker in single stance phase. This caricature is used to derive equation for single stance mode. (B, C) Simplest walker at an instance just before and after foot-strike, respectively. These two caricatures are used to relate angles and velocities after foot-strike with those before foot-strike.

$$\vec{r}_{H/C_1} \times M\vec{a}_H + \vec{r}_{C_2/C_1} \times m\vec{a}_{C_2} = \vec{r}_{H/C_1} \times M\vec{g} + \vec{r}_{C_2/C_1} \times m\vec{g}, \quad (4.6.21)$$

$$\vec{r}_{C_2/H} \times m\vec{a}_{C_2} = \vec{r}_{C_2/H} \times m\vec{g}, \quad (4.6.22)$$

where

$$\vec{g} = g\,\hat{j}\,\cos(\gamma) - g\,\hat{i}\,\sin(\gamma), \quad (4.6.23)$$

$$\vec{a}_H = -\hat{i}\left(l\ddot{\theta}\cos(\theta) - l\dot{\theta}^2\sin(\theta)\right) - \hat{j}\left(l\cos(\theta)\,\dot{\theta}^2 + l\ddot{\theta}\sin(\theta)\right), \quad (4.6.24)$$

$$\vec{a}_{C_2} = -\hat{i}\big(l\ddot{\theta}\cos(\theta) - l\cos(\theta - \phi)\,(\ddot{\theta} - \ddot{\phi}) - l\dot{\theta}^2\sin(\theta)$$
$$+ l\sin(\theta - \phi)\,(\dot{\theta} - \dot{\phi})^2\big)\dots$$
$$- \hat{j}\big(l\ddot{\theta}\sin(\theta) + l\dot{\theta}^2\cos(\theta) - l\sin(\theta - \phi)\,(\ddot{\theta} - \ddot{\phi})$$
$$- l\cos(\theta - \phi)\,(\dot{\theta} - \dot{\phi})^2\big), \quad (4.6.25)$$

$$\vec{r}_{H/C_1} = \hat{j}\,l\,\cos(\theta) - \hat{i}\,l\,\sin(\theta), \quad (4.6.26)$$

$$\vec{r}_{C_2/C_1} = \hat{j}\,(l\cos(\theta) - l\cos(\theta - \phi)) - \hat{i}\,(l\sin(\theta) - l\sin(\theta - \phi)), \quad (4.6.27)$$

$$\vec{r}_{C_2/H} = \hat{i}\,l\,\sin(\theta - \phi) - \hat{j}\,l\,\cos(\theta - \phi). \quad (4.6.28)$$

To create an actuated model, a hip torque and an ankle torque needs to be added to the first and second line of $\mathbf{b}_{ss}$ in Eq. (4.6.18), respectively.

Foot-Strike Phase

The angles after foot-strike are obtained by comparing the angles in Fig. 4.6.8B with that in Fig. 4.6.8C. These are given by:

$$\theta^+ = -\theta^-, \quad (4.6.29)$$

$$\phi^+ = -\phi^- = -2\theta^-. \quad (4.6.30)$$

The angular velocities after foot-strike are given by:

$$\mathbf{A}_{hs}\mathbf{X}_{hs} = \mathbf{b}_{hs}, \tag{4.6.31}$$

$$\mathbf{A}_{hs} = \begin{bmatrix} \ell^2 \left(M + 2m - 2m\cos(\phi)\right) & \ell^2 m\left(\cos(\phi) - 1\right) \\ -l^2 m\left(\cos(\phi) - 1\right) & -\ell^2 m \end{bmatrix},$$

$$\mathbf{X}_{hs} = \begin{bmatrix} \dot{\theta}^+ \\ \dot{\phi}^+ \end{bmatrix}, \qquad \mathbf{b}_{hs} = \begin{bmatrix} M\ell^2\dot{\theta}^-\cos(\phi^-) \\ 0 \end{bmatrix}. \tag{4.6.32}$$

To reduce the above two equations to the simplest walker Eqs. (4.6.7) and (4.6.8), we nondimensionalize time with $\sqrt{\ell/g}$ and take the limit $m/M \to 0$.

Next, we show how to obtain the above velocities after heel-strike. Let $\vec{H}_{/X}^-$ and $\vec{H}_{/X}^+$ denote the angular momentum about the point X before and after foot-strike, respectively. The first and second lines in the above equation are obtained by equating the angular momentum about the foot that is about to touch the ground, C_1, and the hip, H, respectively, to get the following equations:

$$\vec{H}_{/C_2}^- = \vec{H}_{/C_1}^+, \tag{4.6.33}$$

$$\vec{H}_{/H}^- = \vec{H}_{/H}^+. \tag{4.6.34}$$

Note that for the instance after foot-strike the contact points C_1 and C_2 are swapped. The above equation can be written as:

$$\vec{r}_{H/C_2}^- \times M\vec{v}_H^- + \vec{r}_{C_1/C_2}^- \times m\vec{v}_{C_1}^- = \vec{r}_{H/C_1}^+ \times M\vec{v}_H^+ + \vec{r}_{C_2/C_1}^+ \times m\vec{v}_{C_2}^+, \tag{4.6.35}$$

$$\vec{r}_{C_1/H}^- \times m\vec{v}_{C_1}^- = \vec{r}_{C_2/H}^+ \times m\vec{v}_{C_2}^+, \tag{4.6.36}$$

where

$$\vec{r}_{H/C_2}^- = \hat{j}\,\ell\cos(\theta^- - \phi^-) - \hat{i}\,\ell\sin(\theta^- - \phi^-), \tag{4.6.37}$$

$$\vec{r}_{C_1/C_2}^- = \hat{i}\left(\ell\sin(\theta^-) - \ell\sin(\theta^- - \phi^-)\right) - \hat{j}\left(\ell\cos(\theta^-) - \ell\cos(\theta^- - \phi^-)\right), \tag{4.6.38}$$

$$\vec{r}_{H/C_1}^+ = \hat{j}\,\ell\cos(\theta^+) - \hat{i}\,\ell\sin(\theta^+), \tag{4.6.39}$$

$$\vec{r}_{C_2/C_1}^+ = \hat{j}\left(\ell\cos(\theta^+) - \ell\cos(\theta^+ - \phi^+)\right) - \hat{i}\left(\ell\sin(\theta^+) - \ell\sin(\theta^+ - \phi^+)\right), \tag{4.6.40}$$

$$\vec{r}_{C_1/H}^- = \hat{i}\,\ell\sin(\theta^-) - \hat{j}\,\ell\cos(\theta^-), \tag{4.6.41}$$

$$\vec{r}_{C_2/H}^+ = \hat{i}\,\ell\sin(\theta^+ - \phi^+) - \hat{j}\,\ell\cos(\theta^+ - \phi^+), \tag{4.6.42}$$

$$\vec{v}_H^- = -\hat{i}\,\ell\dot{\theta}^-\cos(\theta^-) - \hat{j}\,l\dot{\theta}^-\sin(\theta^-), \tag{4.6.43}$$

$$\vec{v}_{C_1}^- = 0, \tag{4.6.44}$$

$$\vec{v}_H^+ = -\hat{\imath}\, \ell\, \dot{\theta}^+ \cos\left(\theta^+\right) - \hat{\jmath}\, \ell\, \dot{\theta}^+ \sin\left(\theta^+\right), \tag{4.6.45}$$

$$\vec{v}_{C_2}^+ = \left(-\hat{\imath}\, \left(l\, \dot{\theta}^- \cos\left(\theta^-\right) - \ell \cos\left(\theta^- - \phi^-\right)\right)\right.$$

$$\left. - \hat{\jmath}\, \left(\ell\, \dot{\theta}^- \sin\left(\theta^-\right) - \ell \sin\left(\theta^- - \phi^-\right)\right)\right)\left(\dot{\theta}^- - \dot{\phi}^-\right). \tag{4.6.46}$$

REFERENCES

Asano, F., Yamakita, M., Furuta, K., 2000. Virtual passive dynamic walking and energy-based control laws. In: Proc. of IEEE/RSJ International Conference on Intelligent Robots and Systems. Takamatsu, Japan, vol. 2, pp. 1149–1154.

Atzler, E., Herbst, R., Lehmann, G., Muller, E., 1925. Arbeitsphysiologische Studien. Pflugers Archiv European J. Physiol. 208 (1), 184–244.

Bechstein, B.U., 1912. Improvements in and relating to toys. UK Patent 7453.

Bertram, J.E.A., Ruina, A., 2001. Multiple walking speed–frequency relations are predicted by constrained optimization. J. Theor. Biol. 209 (4), 445–453.

Bhounsule, P.A., 2014a. Control of a compass gait walker based on energy regulation using ankle push-off and foot placement. Robotica FirstView 1–11, 6.

Bhounsule, P.A., 2014b. Numerical accuracy of two benchmark models of walking: the rimless spoked wheel and the simplest walker. In: Dynamics of Continuous, Discrete and Impulsive Systems, Series B: Applications and Algorithms, pp. 137–148.

Bhounsule, P.A., Cortell, J., Grewal, A., Hendriksen, B., Karssen, J.G.D., Paul, C., Ruina, A., 2014a. Low-bandwidth reflex-based control for lower power walking: 65 km on a single battery charge. Int. J. Robot. Res. 33 (10), 1305–1321.

Bhounsule, P.A., Ruina, A., Stiesberg, G., 2014b. Discrete-decision continuous-actuation control: balance of an inverted pendulum and pumping a pendulum swing. J. Dyn. Syst. Meas. Control 137 (5), 9.

Bhounsule, P.A., 2012. A Controller Design Framework for Bipedal Robots: Trajectory Optimization and Event-Based Stabilization. PhD thesis. Cornell University.

Bowling, A., 2011. Impact forces and agility in legged robot locomotion. J. Vib. Control 17 (3), 335–346.

Braun, D.J., Goldfarb, M., 2009. A controller for dynamic walking in bipedal robots. In: Proceedings of 2009 IEEE International Conference on Intelligent Robots. St. Louis, MO, USA.

Camp, J., 1997. Powered "Passive" Dynamic Walking. Master's thesis. Cornell University.

Coleman, M.J., Garcia, M., Mombaur, K., Ruina, A., 2001. Prediction of stable walking for a toy that cannot stand. Phys. Rev. E 64 (2), 022901.

Coleman, M.J., Ruina, A., 1998. An uncontrolled walking toy that cannot stand still. Phys. Rev. Lett. 80 (16), 3658–3661.

Collins, S., Ruina, A., Tedrake, R., Wisse, M., 2005. Efficient bipedal robots based on passive-dynamic walkers. Science 307 (5712), 1082.

Collins, S.H., Ruina, A., 2005. A bipedal walking robot with efficient and human-like gait. In: Proceeding of 2005 International Conference on Robotics and Automation. Barcelona, Spain.

Collins, S.H., Wisse, M., Ruina, A., 2001. A three-dimensional passive-dynamic walking robot with two legs and knees. Int. J. Robot. Res. 20 (7), 607.

Dertien, E., 2006. Dynamic walking with Dribbel. IEEE Robot. Autom. Mag. 13 (3), 118–122.

Fallis, G.T., 1888. Walking toy. U.S. Patent No. 376588.

Formalsky, A.M., 1995. Impulsive Control for Anthropomorphic Biped. Springer.

Full, R.J., Kubow, T., Schmitt, J., Holmes, P., Koditschek, D., 2002. Quantifying dynamic stability and maneuverability in legged locomotion. Integr. Comp. Biol. 42 (1), 149–157.

Garcia, M., Chatterjee, A., Ruina, A., Coleman, M., 1998. The simplest walking model: stability, complexity, and scaling. J. Biomech. Eng. 120 (2), 281–288.

Garcia, M.S., 1999. Stability, Scaling, and Chaos in Passive-Dynamic Gait Models. PhD thesis. Cornell University.

Gomes, M.W., Ruina, A., 2011. Walking model with no energy cost. Phys. Rev. E 83 (3), 032901.

Gomes, M.W., Ahlin, K., 2015. Quiet (nearly collisionless) robotic walking. In: 2015 IEEE International Conference on Robotics and Automation (ICRA). IEEE, pp. 5761–5766.

Goswami, A., Espiau, B., Keramane, A., 1997. Limit cycles in a passive compass gait biped and passivity-mimicking control laws. Auton. Robots 4 (3), 273–286.

Goswami, A., Thuilot, B., Espiau, B., 1998. A study of the passive gait of a compass-like biped robot: symmetry and chaos. Int. J. Robot. Res. 17 (12), 1282–1301.

Guenther, F., Iida, F., 2017. Energy efficient monopod running with large payload based on open loop parallel elastic actuation. Tran. Robot. 33 (1), 102–113.

Hobbelen, D.G.E., 2008. Limit Cycle Walking. TU Delft, Delft University of Technology.

Jindrich, D.L., Full, R.J., 1999. Many-legged maneuverability: dynamics of turning in hexapods. J. Exp. Biol. 202 (12), 1603–1623.

Kelly, M., Ruina, A., 2015. Non-linear robust control for inverted-pendulum 2D walking. In: 2015 IEEE International Conference on Robotics and Automation (ICRA). IEEE, pp. 4353–4358.

Kim, M., Collins, S., 2017. Once-per-step control of ankle push-off work improves balance in a three-dimensional simulation of bipedal walking. IEEE Trans. Robot. 33 (2), 406–418.

Kuo, A.D., 1999. Stabilization of lateral motion in passive dynamic walking. Int. J. Robot. Res. 18 (9), 917–930.

Kuo, A.D., 2002. Energetics of actively powered locomotion using the simplest walking model. Trans. ASME J. Biomech. Eng. 124 (1), 113–120.

Kuo, A.D., 2007. Choosing your steps carefully. IEEE Robot. Autom. Mag. 14 (2), 18–29.

Mahan, J.J., Moran, J.F., 1909. Toy. U.S. Patent No. 1007316.

Margaria, R., 1968. Positive and negative work performances and their efficiencies in human locomotion. Eur. J. Appl. Physiol. Occup. Physiol. 25 (4), 339–351.

McGeer, T., 1990. Passive dynamic walking. Int. J. Robot. Res. 9 (2), 62.

McGeer, T., 1993. Passive dynamic biped catalogue, 1991. In: Experimental Robotics II. Springer, pp. 463–490.

Miura, H., Shimoyama, I., 1984. Dynamic walk of a biped. Int. J. Robot. Res. 3 (2), 60.

Mochon, S., McMahon, T.A., 1980. Ballistic walking. J. Biomech. 13 (1), 49–57.

Mombaur, K.D., Bock, H.G., Schloder, J.P., Longman, R.W., 2001. Human-like actuated walking that is asymptotically stable without feedback. In: International Conference on Robotics and Automation. Seoul, Korea, pp. 4128–4133.

Ogata, K., 1995. Discrete-Time Control Systems, vol. 2. Prentice Hall, Englewood Cliffs, NJ.

Owaki, D., Koyama, M., Yamaguchi, S., Kubo, S., Ishiguro, A., 2011. A 2-D passive-dynamic-running biped with elastic elements. IEEE Trans. Robot. 27 (1), 156–162.

Ruina, A., 2012. Cornell ranger 2011, 4-legged bipedal robot. http://ruina.tam.cornell.edu/research/topics/locomotion_and_robotics/ranger/Ranger2011/, or google search for cornell ranger.

Ruina, A., Bertram, J.E.A., Srinivasan, M., 2005. A collisional model of the energetic cost of support work qualitatively explains leg sequencing in walking and galloping, pseudo-elastic leg behavior in running and the walk-to-run transition. J. Theor. Biol. 237 (2), 170–192.

Strogatz, S.H., 2014. Nonlinear Dynamics and Chaos: With Applications to Physics, Biology, Chemistry, and Engineering. Westview Press.

Tucker, V.A., 1970. Energetic cost of locomotion in animals. Comp. Biochem. Physiol. 34 (4), 841–846.

Wilson, J.E., 1938. Walking toy. US Patent 2140275.

Wisse, M., Hobbelen, D.G.E., Schwab, A.L., 2007. Adding an upper body to passive dynamic walking robots by means of a bisecting hip mechanism. IEEE Trans. Robot. 23 (1), 112–123.

Wisse, M., Schwab, A.L., 2005. First steps in passive dynamic walking. In: Climbing and Walking Robots. Springer, pp. 745–756.

Wisse, M., Schwab, A.L., van der Linde, R.Q., van der Helm, F.C.T., 2005. How to keep from falling forward: elementary swing leg action for passive dynamic walkers. IEEE Trans. Robot. 21 (3), 393–401.

Chapter 4.7

Hybrid Zero Dynamics Control of Legged Robots

Aaron D. Ames** and Ioannis Poulakakis[††]

**Mechanical and Civil Engineering and Control and Dynamical Systems, California Institute of Technology (Caltech), United States [††]Department of Mechanical Engineering, University of Delaware, Newark, DE, United States*

4.7.1 BIPEDAL ROBOTS WITH HZD CONTROLLERS

The method of hybrid zero dynamics (HZD) offers a paradigm for designing feedback control laws that induce reliable dynamically-stable walking and running motions on bipedal robots, all the while providing analytically tractable guarantees of performance. The method has been introduced by Jessy Grizzle, Eric Westervelt, and their collaborators in Grizzle et al. (2001), Westervelt et al. (2003), Westervelt (2003), Morris and Grizzle (2009), where geometric nonlinear control tools are developed to generate provably stable limit-cycle walking motions in a class of bipedal robots by dealing directly with their underactuated and hybrid nature; the book (Westervelt et al., 2007) provides an integrative perspective. At its core, the method relies on restricting the dynamics of the robot on a lower-dimensional attractive and invariant subset of its state space. This is achieved by defining a set of holonomic output functions with the control objective being to drive these outputs to zero. Through this process, a lower-dimensional dynamical system emerges from the closed-loop dynamics of the robot that governs the existence and stability properties of its behavior.

Beyond its theoretical value, the method has been successful in experimentally generating robust walking motions on the planar bipedal robot Rabbit (shown in Fig. 4.7.1); see Chevallereau et al. (2003), Westervelt et al. (2004) for details regarding these experiments. Rabbit's successful walking experiments prompted the extension of the HZD method to stabilize bipedal running (Chevallereau et al., 2005). However, while initial experiments have been successful in exciting running on Rabbit, the resulting motions could not be sustained due to actuator limitations (Morris et al., 2006). Given that elastic energy storage elements – e.g., in the form of tendons in animals (Alexander, 1988) or springs (Raibert, 1986) in robots – play a significant role in the realization of running motions, subsequent research efforts on the HZD method concentrated on its implementation on compliant robots. Based on theoretical tools developed in Morris (2008), Morris and Grizzle (2009), the notion of compliant hybrid zero dynamics has been introduced in Poulakakis (2008),

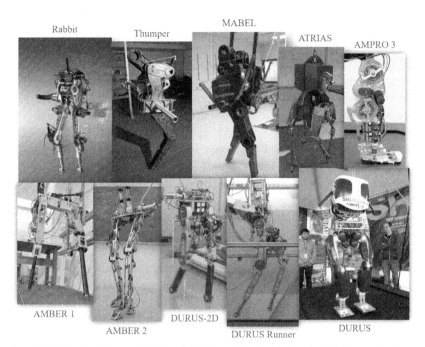

Rabbit   Thumper   MABEL   ATRIAS   AMPRO 3

AMBER 1   DURUS-2D   AMBER 2   DURUS Runner   DURUS

FIGURE 4.7.1 A collection of robots and robotic assistive devices for which HZD-based methods have successfully resulted in stable bipedal locomotion.

Poulakakis and Grizzle (2009b) and further refined in Sreenath (2011) to induce robust walking (Sreenath et al., 2011) and running (Sreenath et al., 2013) motions in experiments with the compliant bipedal robot MABEL[4] shown in Fig. 4.7.1. In addition to these experiments, MABEL provided an excellent platform for validating advanced locomotion controllers for accommodating unexpected large ground-height variations (Park et al., 2013), and for testing alternative Lyapunov-based HZD control schemes as in Ames et al. (2014a) that afford greater flexibility in incorporating constraints such as actuator torque saturation (Galloway et al., 2015). These ideas have been translated to other robots, including the use of Lyapunov-based HZD techniques to realize walking on the planar robot DURUS-2D (Cousineau and Ames, 2015) (shown in Fig. 4.7.1), along with running on the same platform (Ma et al., 2017).

Building upon the successes of HZD, which focused on underactuated walking due to its clear separability from methods that require full actuation (e.g., zero moment point (ZMP) based frameworks (Kajita et al., 2003, 2006;

---

4. MABEL and its monopedal version Thumper depicted in Fig. 4.7.1 have been designed and constructed by Professor J. Hurst in a collaborative effort between The University of Michigan and Carnegie Mellon University; see Grizzle et al. (2009) for an overview and Hurst et al. (2007), Hurst and Rizzi (2008), Hurst (2008) for details relevant to the underlying design philosophy.

Vukobratović et al., 2006; Goswami, 1999a, 1999b)), the HZD methodology has been proven to be extensible to walking behaviors that more clearly resemble humans. When locomoting, humans naturally display different discrete phases that correspond to their changes in contact with the world (Ackermann, 2007). In the context of robots, this multi-contact behavior can be represented by a hybrid dynamical system model of walking consisting of discrete domains, wherein dynamics of the robot change discretely as a function of its contact with the ground; this results in phases of full, under- and over-actuation (Ames et al., 2011; Vasudevan et al., 2013). The ability of HZD to handle underactuation motivated its extension to this multi-domain locomotion scenario. In particular, using human locomotion data as inspiration, the framework of human-inspired control extended HZD to the full-actuated case through the notion of *partial* hybrid zero dynamics (PHZD) (Ames, 2014). Combining this approach with the HZD methods for underactuated walking, e.g., on AMBER 1, resulted in the ability to consider multi-domain HZD, thereby achieving multi-contact walking behaviors on bipedal robots, including ATRIAS (Hereid et al., 2014), AMBER 2 (Zhao et al., 2014b, 2015a), and the prosthesis AMPRO (Zhao et al., 2016b) (see Fig. 4.7.1). These results, both formally and experimentally, indicated that HZD can provide a mathematical framework for realizing human-like walking behaviors on robotic systems.

The last frontier for HZD-based methods was their extension to three-dimensional (3D) walking and realization on humanoid robots; the challenges and approaches are described in detail in Grizzle et al. (2014, 2010). This will enable us to engage dynamically moving bipeds in motion planning tasks, including navigation in environments cluttered by obstacles as in Gregg et al. (2012), Motahar et al. (2016), Veer et al. (2017). While 3D robot walking utilizing HZD had long proved feasible in simulation, and had even proven realizable on small-scale humanoids like the NAO robot (Ames et al., 2012a; Powell et al., 2013), bridging the gap between this theory and the experimental realization on full-scale humanoid robots is a difficult task. This is, at its core, a function of the fact that HZD uses the entire dynamics of the robot to generate walking gaits in the context of a constrained nonlinear programming problem. When this optimization problem can be solved, it results in dynamic and efficient gaits – yet as the complexity of the robot increases, solving the problem becomes more difficult making the translation to hardware evermore challenging. During the design and development of the humanoid robot DURUS (shown in Fig. 4.7.1), structure in nonlinear optimization problem necessary to generate gaits was discovered and exploited to allow for rapid gait generation; bringing the time needed to obtain a stable walking gait from hours to a few minutes (Hereid et al., 2016); importantly, due to the presence of springs in the ankles of DURUS, this was done in the context of multi-domain walking that exploits

PHZD to achieve stability. Not only was stable walking achieved, but it was done so in a sustained fashion with the end result being the public demonstration of DURUS at the DARPA Robotics Challenge, wherein it walked continuously for over 2 1/2 hours covering over 2 km – all on a single 1.1 kWh battery (Reher et al., 2016a). Additionally, these results were extended to the case of multi-contact walking with natural heel–toe behaviors thereby demonstrating human-like humanoid locomotion (Reher et al., 2016b). Moreover, in both cases, the walking realized on DURUS was the most efficient walking realized on a bipedal humanoid robot (Collins and Ruina, 2005). The final lessons from the realization of HZD on a variety of platforms was that it provides a powerful method for realizing dynamic walking behaviors on bipedal and humanoid robots.

4.7.2 MODELING LEGGED ROBOTS AS HYBRID DYNAMICAL SYSTEMS

Walking and running behaviors can be modeled as distinguished periodic orbits of mechanical systems that are strongly nonlinear and hybrid in nature. For example, a simplified walking cycle consists of successive phases of single support (swing phase) and double support (impact phase). On the other hand, running comprises phases where a leg is in contact with the ground (stance phase) and phases where the system is in the air following a ballistic motion under the influence of gravity (flight phase). This combination of continuous dynamics and discrete transitions among them is characteristic of legged locomotion and it gives rise to hybrid system models, which are the focus of this section.

4.7.2.1 Continuous Dynamics

Let Q be the configuration space of a robot with n degrees of freedom, i.e., $n = \dim(Q)$, with coordinates $q \in Q$; examples of coordinate choices for various bipeds are shown in Fig. 4.7.2. For the sake of definiteness, it may be necessary to choose Q to be a subset of the actual configuration space of the robot so that global coordinates can be defined,[5] i.e., such that Q is embeddable in $\mathbb{R}^n$, or more simply $Q \subset \mathbb{R}^n$. Consider the equations of motion for a robot given in the general form by the Euler–Lagrange equations (Murray et al., 1994; Spong et al., 2006):

$$D(q)\ddot{q} + C(q, \dot{q})\dot{q} + G(q) = Bu \qquad (4.7.1)$$

where $D(q)$ is the mass matrix, and $C(q, \dot{q})\dot{q}$, $G(q)$ are vectors containing the centrifugal and Coriolis forces and the gravitational forces, respectively, and

---

5. At various points of this chapter we will assume that certain matrix functions have full rank; it may be necessary to carefully choose Q to satisfy these conditions.

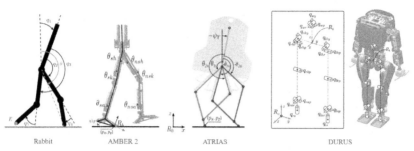

| Rabbit | AMBER 2 | ATRIAS | DURUS |

FIGURE 4.7.2 Examples of the configuration space for a collection of bipedal and humanoid robots; in this case, Rabbit (Westervelt et al., 2003), AMBER 2 (Zhao et al., 2014b), ATRIAS (Hereid et al., 2014) and DURUS (Hereid et al., 2016).

$B \in \mathbb{R}^{n \times m}$ is the actuation matrix which determines the way in which the torque inputs, $u \in U \subset \mathbb{R}^m$, actuate the system (where here U is the set of admissible control inputs). Importantly, the actuation matrix changes based upon the actuation type of the robot: in the case of *full actuation*, this matrix is full rank, while in the case of *underactuation* this matrix has rank $m < n$ indicating that it is not possible to actuate all of the degrees of freedom of the system, and if $m > n$ the system is *overactuated*, i.e., there is more control authority than degrees of freedom in the system.

Selecting the state vector x to include the configuration variables and the corresponding rates, that is $x = [q^T \ \dot{q}^T]^T \in \mathbb{R}^{2n}$, and noticing that

$$\frac{d}{dt}\begin{bmatrix} q \\ \dot{q} \end{bmatrix} = \begin{bmatrix} \dot{q} \\ -D(q)^{-1}(C(q,\dot{q})\dot{q} + G(q)) \end{bmatrix} + \begin{bmatrix} 0 \\ D(q)^{-1}B \end{bmatrix} u \qquad (4.7.2)$$

results in the following state-space form of the continuous dynamics (4.7.1):

$$\dot{x} = f(x) + g(x)u. \qquad (4.7.3)$$

4.7.2.2   Discrete Dynamics

In the basic walking model – more advanced models are discussed in Section 4.7.2.4 below – the continuous dynamics (4.7.3) represents the swing phase, which evolves until the leg hits the ground, thereby resulting in an impact. It is this impact that is the basis for the hybrid dynamical system model that underlies walking and running motions (Westervelt et al., 2007; Grizzle et al., 2014; Ames et al., 2011; Haddad et al., 2006). In particular, we consider the height of the swing foot and the surface defined by this height being zero. If $p_{\text{toe}}^{\text{v}}$ denotes the height of the toe of the swing leg, then the surface

$$S = \{x \in TQ \mid p_{\text{toe}}^{\text{v}}(q) = 0 \text{ and } \dot{p}_{\text{toe}}^{\text{v}}(x) < 0\} \qquad (4.7.4)$$

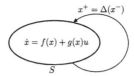

FIGURE 4.7.3 Representation of the hybrid model of the biped as a system with "impulse" effects.

is the *switching surface* (or the *guard* in the terminology of hybrid systems). Upon reaching the switching surface, the system undergoes an impact, which in basic walking models represents an instantaneous double support phase. The end result is an update law

$$x^+ = \Delta(x^-) \qquad (4.7.5)$$

mapping the pre-impact states, x^-, to the post-impact states, x^+. This impact model includes both a "change of coordinates" in the configuration variables corresponding to swapping the swing and stance legs, together with a discrete change in the velocity of the system determined by a plastic impact of the swing foot with the ground (which causes the stance foot to leave the ground and thus become the swing foot). More details on deriving the impact map Δ can be found in the book (Westervelt et al., 2007).

4.7.2.3 Hybrid Control System

The end result of these constructions is a *system with impulsive effects* or a *hybrid control system*:

$$\mathscr{HC} : \begin{cases} \dot{x} = f(x) + g(x)u, & x \in D \backslash S, \\ x^+ = \Delta\left(x^-\right), & x^- \in S, \end{cases} \qquad (4.7.6)$$

where $D = \{x \in TQ \mid p_{\text{toe}}^{\text{v}}(q) > 0\}$ is the *domain* of the system, i.e., we require the swing foot to be above the ground. Note that the domain is often restricted to the admissible domain through the inclusion of friction constraints. The system (4.7.6) is depicted in Fig. 4.7.3. We also note that sometimes systems of this form are written as a tuple

$$\mathscr{HC} := (D, U, S, \Delta, (f, g)), \qquad (4.7.7)$$

as is more common in the hybrid systems literature (Lygeros et al., 2003; Goebel et al., 2009; van der Schaft and Schumacher, 2000; Lamperski and Ames, 2013).

4.7.2.4 Advanced Models of Locomotion

The hybrid system model of a walking robot that we have considered so far includes only a single continuous and discrete domain. For robots with more complex mechanical characteristics, i.e., springs and nontrivial feet, hybrid system models with more complex discrete structure are needed. Interestingly enough, these hybrid models naturally relate to human locomotion models, i.e., humans tend to display a specific discrete domain structure when walking, wherein their contact points with the ground change throughout the gait (Ames et al., 2011; Vasudevan et al., 2011, 2013). Other examples of locomotion models evolving on multiple domains can be found in running motions, due to the alternation between stance and flight phases (Poulakakis and Grizzle, 2009a; Sreenath et al., 2013). These more complex models for locomotion can again be modeled as a hybrid system.

The key element to advanced models of robotic walking and running is an oriented graph, $\Gamma = (V, E)$, that indicates how the contact points change throughout the course of a gait, i.e., the vertices of this graph (V) correspond to different collections of contact points with the ground, and the transitions (described by edges E) occur when these contact points change; see Fig. 4.7.4 for examples in the case of multi-contact locomotion and Fig. 4.7.8 for a running model. The end result is a hybrid control system model of the form

$$\mathcal{HC} := (\Gamma, D, U, S, \Delta, FG) \tag{4.7.8}$$

where, in this case, $D = \{D_v\}_{v \in V}$ is a collection of domains, $U = \{U_v\}_{v \in V}$ is a collection of admissible inputs, $S = \{S_e\}_{e \in E}$ is a set of switching surfaces, $\Delta = \{\Delta_e\}_{e \in E}$ is a set of impact maps with $\Delta_e : S_e \subset D_{\text{source}(e)} \to D_{\text{target}(e)}$ and $FG = \{(f_v, g_v)\}_{v \in V}$ is a collection of control systems of the form (4.7.3). It is important to note that the degree of actuation changes for each domain, i.e., on some domains the system may be underactuated, on some it might be fully actuated, and on others it can be overactuated. The specific methods for constructing hybrid system models as they relate to the changing contact points of the robot can be found in Ames et al. (2011).

To provide a concrete example, consider the multicontact model of the bipedal robot AMBER 2 shown in Fig. 4.7.4. As described in Zhao et al. (2014b, 2015a), this model consists of three domains, D_{v^+}, D_{v^i}, and D_{v^-}, that depend on how the robot's contact points (heel and toe) change throughout the course of a step. The dynamics on each of these domains changes with the change in contact points. Importantly, each of these domains display a different actuation type: D_{v^+} is overactuated, D_{v^i} is fully actuated, and D_{v^-} is underactuated. Other examples of multidomain walking, and the

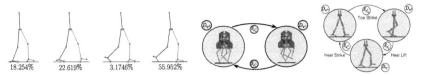

18.254%   22.619%   3.1746%   55.952%

FIGURE 4.7.4   (Left) The discrete domains, and the percentage of the step spent in a domain, for a human walking (as obtained from experimental data), (Middle) the discrete two domain structure associate with the hybrid system model of ATRIAS, and (Right) the discrete three-domain structure for AMBER 2 when walking with articulated feet (Zhao et al., 2014b).

application of hybrid zero dynamics to these systems to achieve robotic running, will be discussed in Section 4.7.6.2. Finally, an interesting class of multidomain hybrid models emerges naturally in the context of planning the motion of dynamically walking bipeds amidst obstacles (Gregg et al., 2012; Motahar et al., 2016). This can be achieved through the sequential composition of primitive limit-cycle walking motions each stabilized through HZD as in Motahar et al. (2016), Veer et al. (2017).

4.7.3   VIRTUAL CONSTRAINTS FOR LOCOMOTION

Central to the HZD approach is the introduction of *virtual* constraints. These constraints represent relations among the robot's degrees of freedom that correspond to preferred postures during the realization of a walking or running gait. They are formulated as functions of the configuration variables of the form $h(q) = 0$, $q \in Q$, and can thus be interpreted as *holonomic* constraints, the enforcement of which effectively restricts the robot's motion on low-dimensional surfaces embedded in its higher-dimensional state space. It should be emphasized however, that the key difference with the classical notion of holonomic constraints from analytical mechanics (Goldstein et al., 2002) is that virtual holonomic constraints are imposed on the system via its actuators, not via workless constraint forces.

4.7.3.1   Virtual Constraints

In our setting, we would like to "force" a set of coordinates – those over which we have control – to follow desired patterns. Doing so both enforces certain patterns with regard to walking motions and reduces the overall dimensionality of the system to a reduced dimensional space, thus giving rise to a lower-dimensional dynamical system, namely the *zero dynamics*. Mathematically, we consider the difference between an actual output, y_a, and a desired output, y_d, expressed via

$$y(q) := y_a(q) - y_d(\tau(q), \alpha) \in \mathbb{R}^m, \qquad (4.7.9)$$

where the desired function begins as a function of time, $y_d(t, \alpha)$, dependent on a parameter set α, and converted to a function of the configuration variables through a *parameterization of time* often chosen to be of the general form,

$$\tau(q) = \frac{\theta(q) - \theta^+}{\theta^- - \theta^+}, \tag{4.7.10}$$

where $\theta : Q \to \mathbb{R}$ is a phase variable, $\theta^+ = \theta(q^+)$ is its value post-impact, $\theta^- = \theta(q^-)$ is its value pre-impact, and therefore $\tau : Q \to [0, 1]$ throughout the course of a step. To provide some intuition, as Fig. 4.7.2 indicates, the phase variable θ can be chosen to correspond to the angle of the line connecting the hip with the toe of the support leg, which is a monotonically increasing quantity that captures "progression" of the support leg into the step; see Westervelt et al. (2007) for details. It is important to emphasize that the outputs (4.7.9) depend only on the configuration variables, hence the term virtual holonomic constraints.

4.7.3.2 Designing Virtual Constraints for Locomotion Tasks

Based upon the framework of virtual constraints, the main idea is to consider a vector of output variables y_a in (4.7.9), with one output for each actuator. These outputs capture quantities that are of interest, e.g., angles in the system or other geometric relationships, such as the position of the center of mass or the height of the swing foot. The goal is to drive these outputs to evolve according to a collection of desired behaviors as represented by y_d, which is a function of the phase variable τ and a set of parameters α that allow "tuning" the constraints according to desired specifications. The art of gait design is to pick y_d so that it displays certain properties so that driving $y \to 0$ in (4.7.9) guarantees stability of the system. A concrete way of selecting y_d in (4.7.9) is through the use of Beziér polynomials of degree M, i.e., for $i = 1, \ldots, m$,

$$y_d(\tau(q), \alpha)_i = \sum_{k=0}^{M} \frac{M!}{k!(M-k)!} \alpha_{k,i} \tau(q)^k (1 - \tau(q))^{M-k}. \tag{4.7.11}$$

The use of Beziér polynomials is only one choice of functions for the design of the desired evolution y_d in (4.7.9), which offers some flexibility in imposing desired boundary conditions on the different phases that compose a cyclic locomotion pattern. More details about certain key properties of these polynomials and on how to use them in the context of gait design can be found in Westervelt et al. (2007).

Virtual constraints, designed via Beziér polynomials, provides a computationally efficient way of constructing virtual constraints. Yet, since the desired behavior is given by polynomials, they do not necessarily capture the virtual constraints present in human walking. That is, we could instead seek *human-inspired* virtual constraints whose design is inspired by human locomotion data. In this light, human data suggests that for certain collections of outputs, y_a, appear to act like the time solution to a mass–spring–damper system, i.e., humans appear to display simple "linear" behavior when the proper collection of virtual constraints are considered (Sinnet et al., 2011a; Ames, 2012; Powell et al., 2012; Huihua et al., 2012; Zhao et al., 2014b). This motivates the following *mass–spring–damper* desired output:

$$y_d(\tau(q), \alpha)_i = y_{\text{MSD}}(\tau(q), \alpha)_i$$
$$:= e^{-\alpha_{4,i} \tau(q)} (\alpha_{k,i} \cos(\alpha_{k,i} \tau(q)) + \alpha_{k,i} \sin(\alpha_{k,i} \tau(q))) + \alpha_{k,i} \tag{4.7.12}$$

for $i = 1, \ldots, m$, which is simply the time solution to a linear mass–spring–damper system, i.e., a second order linear system. Human data has been calculated from a variety of actual output combinations, y_a, and it has been shown that y_{MSD} accurately describes (with high correlation) these outputs; examples include the position of the hip, the position of the center of mass, and the knee angles (Ames, 2014; Sinnet et al., 2014).

Another class of virtual constraints developed for *fully actuated* walking robots considers both velocity modulating and position modulating virtual constraints. More specifically, in this case we can modulate both the position of the robot – through the virtual constraints defined in (4.7.9) – and its traveling speed. Moreover, we would like to do this in a general fashion that will allow for different collections of virtual constraints depending on the bipedal robot being considered and the desired behavior to be achieved. To regulate the velocity of the robot in an explicit fashion, we consider the following virtual constraints (Ames, 2014):

$$y_1(q, \dot{q}) = \frac{\partial \theta(q)}{\partial q} \dot{q} - v, \tag{4.7.13}$$

$$y_2(q) = y_{2,a}(q) - y_{2,d}(\tau(q), \alpha) \tag{4.7.14}$$

where y_2 are the position modulating outputs as defined in (4.7.9), $\theta : Q \to \mathbb{R}$ is the phase variable of the virtual constraint (4.7.9), and v is the desired velocity. For example, we may wish to explicitly control the forward velocity of the center of mass to regulate the robot's speed; in this case θ would be the position of the center of mass, and v would be the desired velocity. In doing so, it is often useful to consider the following modified form for the parameterizations of time:

$$\tau(q) = \frac{\theta(q) - \theta(q^+)}{v}, \tag{4.7.15}$$

where q^+ is the post-impact configuration of the robot. Therefore, τ directly couples the phase of the robot to the forward progression of the velocity modulating output. As in the case of purely position modulating virtual constraints (4.7.9), the goal is to construct a controller that drives $y_1 \to 0$ and $y_2 \to 0$ to force the robot to progress forward in a desired fashion while displaying the coupling dictated by y_2.

4.7.4   USING FEEDBACK CONTROL TO IMPOSE VIRTUAL CONSTRAINTS

As discussed in Section 4.7.3, the goal is to drive $y \to 0$ in order to force the actual outputs, y_a, to the desired outputs, y_d, i.e., in order to achieve $y_a \to y_d$. This objective can be achieved through the use of a core tool in nonlinear control: *feedback linearization* (Sastry, 1999). The end result is a controller that drives the system to the *zero dynamics* surface and renders this surface invariant through the continuous dynamics. Therefore, applying this control law implies that the full dynamics of the robot will ultimately evolve on a low dimensional space *for the continuous dynamics*. The next section will discuss how to achieve this through the full hybrid dynamics of the robot and, thereby, realize periodic walking and running motions in the hybrid models of Section 4.7.2.

4.7.4.1   Feedback Linearization

The goal of feedback linearization is to uncover a relationship between the output and the control input. This is achieved by differentiating the output until this relationship is revealed. To be concrete, let us consider differentiating y in (4.7.9) along solutions of the continuous dynamics (4.7.3). We have

$$\dot{y}(q,\dot{q}) = \frac{\partial y(q)}{\partial q}\dot{q}. \tag{4.7.16}$$

Since none of the inputs appear in this equation, we differentiate a second time to obtain

$$\ddot{y}(q,\dot{q}) = \frac{\partial}{\partial q}\left(\frac{\partial h(q)}{\partial q}\dot{q}\right)\dot{q} + \frac{\partial y(q)}{\partial q}\ddot{q} \tag{4.7.17}$$

and substituting in the dynamics (4.7.1) yields

$$\ddot{y}(q,\dot{q}) = \underbrace{\frac{\partial}{\partial q}\left(\frac{\partial h(q)}{\partial q}\dot{q}\right)\dot{q} + \frac{\partial y(q)}{\partial q}\left[-D^{-1}(q)\left(C(q,\dot{q})\dot{q} + G(q)\right)\right]}_{L_f^2 y(q,\dot{q})}$$

$$+ \frac{\partial y(q)}{\partial q} D^{-1}(q)B\ u \tag{4.7.18}$$
$$\underbrace{\qquad\qquad\qquad\qquad}_{L_g L_f y(q,\dot{q})}$$

where, since we differentiated y twice to obtain the input, the virtual constraint is a *relative degree two output* in the terminology of nonlinear control (Isidori, 1995; Sastry, 1999). In the context of the mixed position and velocity modulating outputs (4.7.13)–(4.7.14), the result will be a mixed relative degree, as will be discussed in more detail in Section 4.7.4.3 below.

To obtain a controller that drives $y \to 0$ we consider (4.7.18) which can be written in terms of x as

$$\ddot{y}(x) = L_f^2 y(x) + L_g L_f y(x)u , \tag{4.7.19}$$

where $L_g L_f y(x) \in \mathbb{R}^{m \times m}$ is the *decoupling matrix* that is assumed to be invertible. Therefore, selecting

$$u(x, \mu) = \left(L_g L_f y(x)\right)^{-1} \left[-L_f^2 y(x) + \mu\right], \tag{4.7.20}$$

where $\mu \in \mathbb{R}^m$ is an auxillary input, results in a linear relationship between the second derivative of y and the new input μ, as in

$$\ddot{y} = \mu. \tag{4.7.21}$$

That is, the end result is a linear control system of the form

$$\begin{bmatrix} \dot{y} \\ \ddot{y} \end{bmatrix} = \underbrace{\begin{bmatrix} 0 & I \\ 0 & 0 \end{bmatrix}}_{F} \begin{bmatrix} y \\ \dot{y} \end{bmatrix} + \underbrace{\begin{bmatrix} 0 \\ I \end{bmatrix}}_{G} \mu \tag{4.7.22}$$

where $I \in \mathbb{R}^{m \times m}$ is the identity matrix. Therefore, the control law

$$\mu_\epsilon(y, \dot{y}) = -\frac{K_P}{\epsilon^2} y - \frac{K_D}{\epsilon} \dot{y} \qquad \Rightarrow \qquad \begin{bmatrix} \dot{y} \\ \ddot{y} \end{bmatrix} = \frac{1}{\epsilon} \underbrace{\begin{bmatrix} 0 & \epsilon I \\ -\frac{1}{\epsilon} K_P & -K_D \end{bmatrix}}_{F_{cl(\epsilon)}} \begin{bmatrix} y \\ \dot{y} \end{bmatrix}$$

$$\tag{4.7.23}$$

where K_P and K_D are chosen so that $F_{cl}(\epsilon)$ is stable (Hurwitz) for all $0 < \epsilon < 1$. Note that here ϵ forces the system to converge at a user defined rate. Therefore, the control law

$$u^*(x) = \left(L_g L_f y(x)\right)^{-1} \left[-L_f^2 y(x) - \frac{K_P}{\epsilon^2} y - \frac{K_D}{\epsilon} \dot{y}\right] \tag{4.7.24}$$

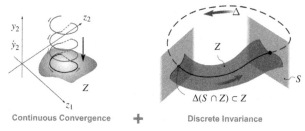

Continuous Convergence ✚ Discrete Invariance

FIGURE 4.7.5 Illustration of the key concepts related to hybrid zero dynamics: continuous convergence to a low dimensional zero dynamics surface Z, coupled with a hybrid invariance condition: $\Delta(S \cap Z) \subset Z$.

drives the output y and its derivative $\dot{y}$ to zero exponentially fast at a rate of $\frac{1}{\epsilon}$.

4.7.4.2 Zero Dynamics

The control law introduced in Section 4.7.4.1 drives $y(x) \to 0$ and $\dot{y}(x) \to 0$. That is, it drives the continuous dynamics to the *zero dynamics surface*, illustrated in Fig. 4.7.5 and defined by

$$Z := \{x \in D \mid y(x) = 0 \text{ and } \dot{y}(x) = 0\}, \tag{4.7.25}$$

where the dimension of the surface is the degree of underactuation of the system $2(n - m)$; it is reminded that n is the number of the degrees of freedom of the robot and m is the number of actuators. One can find (local) coordinates for the zero dynamics surface, Z, given by $z : D \to \mathbb{R}^{2(n-m)}$ such that $(z, y, \dot{y}) : D \to \mathbb{R}^{2n}$ is a (local) diffeomorphism. To provide a concrete example, suppose that $m = n - 1$ and the angle between the robot and the ground is the first coordinate, q_1. Then, necessarily, $\theta(q)$ is a function of q_1 and $y(q)$ is independent of q_1. We can, therefore, pick coordinates for the zero dynamics as follows (Westervelt et al., 2007):

$$z_1(q) = \theta(q),$$
$$z_2(q) = D(q)_{(1,*)}\dot{q}$$

where $D(q)_{(1,*)}$ is the first row of the inertia matrix in (4.7.1).

Utilizing the coordinates for the zero dynamics and letting $\eta = (y, \dot{y})$ be the coordinates for the dynamics transversal to Z, the system can be represented as

$$\dot{\eta} = \widehat{f}(\eta, z) + \widehat{g}(\eta, z)u, \tag{4.7.26}$$
$$\dot{z} = w(\eta, z)$$

where

$$\widehat{f}(\eta(x), z(x)) = \begin{bmatrix} \dot{y}(x) \\ L_f^2 y(x) \end{bmatrix}, \qquad \widehat{g}(\eta(x), z(x)) = \begin{bmatrix} 0 \\ L_g L_f y(x) \end{bmatrix},$$

which are written here in the original x coordinates by using the fact that η is defined in terms of y and $\dot{y}$, which, in turn, are functions of x; note that these expressions can be converted to the (η, z) coordinates through the (local) diffeomorphism relating (η, z) with x. Additionally, the control law (4.7.20) with the auxiliary input μ chosen as in (4.7.23) and expressed in terms of the η coordinates results in a linear, time-invariant system $\dot{\eta} = F_{cl}(\epsilon)\eta$ describing the dynamics transversal to Z. Effectively, the feedback control law (4.7.24) of Section 4.7.4.1 ensures that the zero dynamics surface Z is attractive and invariant under the continuous time dynamics of the system – that is, $\eta \to 0$ and $\eta(0) = 0$ implies that $\eta(t) \equiv 0$ for all future times $t \geq 0$. As a result, the *zero dynamics* – that is, the maximal dynamics compatible with the output being identically equal to zero – can be written as

$$\dot{z} = w(0, z). \tag{4.7.27}$$

It is worth mentioning that, to arrive at (4.7.27), the number of outputs equals the number of inputs. In other words, all the inputs available for control are "slaved" to drive the outputs to zero. Depending on the controller's objectives, however, it is possible to define the vector of outputs y so that its dimension is smaller than the dimension of the input vector u, and keep the remaining control inputs for additional control *within* the zero dynamics, which now becomes *controlled*. This may increase the dimension of the zero dynamics, but it provides greater flexibility for developing control action. Examples of this approach include the stabilization of running motions on compliant robots by "shaping" compliance within the zero dynamics (Poulakakis and Grizzle, 2007a, 2007b, 2009b), or by incorporating active force control (Sreenath et al., 2013); see Section 4.7.6.4 below for more details. On a final note, the presence of exogenous inputs in the zero dynamics may result from externally applied forces, giving rise to *forced* zero dynamics, as in Veer et al. (2015, 2016). For example, this is the case when a bipedal robot physically collaborates with a leading external agent – another robot or a human – to transport an object in their workspace (Motahar et al., 2015b). In this case, the objective of the feedback controller is to adapt the robot's locomotion pattern to the externally applied force (Veer et al., 2015, 2016).

4.7.4.3 Partial Zero Dynamics

We can also consider the case in which there is a velocity modulating output (Ames, 2014), i.e., where there are virtual constraints of the form (4.7.13) and (4.7.14). In this case, differentiating y_1 until the control input appears, as was done in (4.7.18) yields

$$\dot{y}_1(q,\dot{q}) = \underbrace{\frac{\partial}{\partial q}\left(\frac{\partial\theta(q)}{\partial q}\dot{q}\right)\dot{q} + \frac{\partial\theta(q)}{\partial q}\left[-D^{-1}(q)\left(C(q,\dot{q})\dot{q} + G(q)\right)\right]}_{L_f y_1(q,\dot{q})}$$

$$+ \underbrace{\frac{\partial\theta(q)}{\partial q}D^{-1}(q)B}_{L_g y_1(q,\dot{q})}\, u \tag{4.7.28}$$

because y_1 depends on the angular velocity in (4.7.13). Since the control input appears after differentiating once, it implies that y_1 has relative degree one. Therefore, in the case of human-inspired output combinations we mixed relative degree one and relative degree two outputs. That is, we can combine (4.7.28) with (4.7.18) (with y replaced by y_2) to obtain

$$\begin{bmatrix} \dot{y}_1 \\ \ddot{y}_2 \end{bmatrix} = \underbrace{\begin{bmatrix} L_f y_1(q,\dot{q}) \\ L_f^2 y_2(q,\dot{q}) \end{bmatrix}}_{L_f(x)} + \underbrace{\begin{bmatrix} L_g y_1(q,\dot{q}) \\ L_g L_f y_2(q,\dot{q}) \end{bmatrix}}_{A(x)} u \tag{4.7.29}$$

where $A(x)$ is the decoupling matrix that must be full rank. Therefore, analogously to (4.7.20), we have

$$u(x,\mu) = A(x)^{-1}\left(-L_f(x) + \mu\right)$$

$$\Rightarrow \quad \begin{bmatrix} \dot{y}_1 \\ \ddot{y}_2 \end{bmatrix} = \mu \tag{4.7.30}$$

$$\Rightarrow \quad \begin{bmatrix} \dot{y}_1 \\ \dot{y}_2 \\ \ddot{y}_2 \end{bmatrix} = \underbrace{\begin{bmatrix} 0 & 0 & 0 \\ 0 & 0 & I \\ 0 & 0 & 0 \end{bmatrix}}_{F}\begin{bmatrix} y_1 \\ y_2 \\ \dot{y}_2 \end{bmatrix} + \underbrace{\begin{bmatrix} 1 & 0 \\ 0 & 0 \\ 0 & I \end{bmatrix}}_{G}\mu.$$

Here $\mu = \begin{bmatrix} \mu_1 \\ \mu_2 \end{bmatrix}$ where μ_1 is the input to $\dot{y}_1$ and μ_2 is the input to $\ddot{y}_2$. As in (4.7.23), we can pick

$$\mu(y_1, y_2, \dot{y}_2) = \begin{bmatrix} -\frac{1}{\epsilon}y_1 \\ -\frac{K_P}{\epsilon^2}y_2 - \frac{K_D}{\epsilon}\dot{y}_2 \end{bmatrix}$$

$$\Rightarrow \quad \underbrace{\begin{bmatrix} \dot{y}_1 \\ \dot{y}_2 \\ \ddot{y}_2 \end{bmatrix} = \frac{1}{\epsilon} \begin{bmatrix} -1 & 0 & 0 \\ 0 & 0 & \epsilon I \\ 0 & -\frac{1}{\epsilon}K_P & -K_D \end{bmatrix} \begin{bmatrix} y_1 \\ y_2 \\ \dot{y}_2 \end{bmatrix}}_{F_{cl}(\epsilon)} \quad (4.7.31)$$

such that $F_{cl}(\epsilon)$ is Hurwitz. Therefore, we have defined a controller that achieves $y_1 \to 0$, $y_2 \to 0$ and $\dot{y}_2 \to 0$ (at a rate of $\frac{1}{\epsilon}$). Moreover, because we considered the velocity modulating virtual constraint, we have that $\dot{\theta}(q) \to v$. Hence, the velocity of the system – as it is captured by the rate $\dot{\theta}(q)$ of the phase variable – converges to the desired value.

As was discussed in Section 4.7.4.2, we can consider the surface that the system converges to under the feedback control law introduced in (4.7.29). In this case, while our virtual constraints consist of both relative degree one and relative degree two outputs, we will only consider the surface that the relative degree two outputs converge to, and then study the behavior of the relative degree one outputs on this surface. In particular, consider the *partial hybrid zero dynamics* surface given by Ames (2014)

$$PZ := \{x \in D \mid y_2(x) = 0 \text{ and } \dot{y}_2(x) = 0\}, \quad (4.7.32)$$

which is rendered attractive by (4.7.31). Writing $\eta = (y_2, \dot{y}_2)$, then we can again write the system in the form given by (4.7.26). The advantage of the partial zero dynamics can be seen easiest in the case of full actuation, i.e., when $n = m$, as is the case for many humanoid robots. In this fully-actuated case, the z dynamics in (4.7.26) become *controlled* with the ankle torque of the robot available to propel the robot forward. This is evidenced by the fact that, in this case, the coordinates for z can be chosen as $z_1 = \theta(q)$ and $z_2 = \dot{\theta}(q, \dot{q})$ wherein, by (4.7.30) and (4.7.31), the q dynamics become linear:

$$\dot{z}_1 = z_2. \quad (4.7.33)$$

$$\dot{z}_2 = -\frac{1}{\epsilon}(z_2 - v).$$

That is, on the surface PZ the system evolves according to linear dynamics that drive $\dot{\theta} \to v$. It will be seen later that this ensures a stable walking gait in the case of full actuation.

4.7.5 GENERATING PERIODIC MOTIONS

The goal of gait synthesis is to generate periodic walking gaits for a bipedal robot, along with the feedback controller that enforces these periodic motions. This is where *hybrid zero dynamics (HZD)* provides a powerful framework

(Grizzle et al., 2001; Westervelt et al., 2003; Westervelt, 2003; Morris and Grizzle, 2009). In particular, the feedback controller introduced in Section 4.7.4 rendered the zero dynamics surface, Z, both attractive and invariant *for the continuous dynamics*. Yet, when the system reaches the switching surface S it will be "thrown away" from the zero dynamics surface. This has the potential to destabilize the system even if the dynamics in Z are stable. That is, the discrete dynamics in the hybrid system can destroy continuous-time invariance and destabilize the system, even if the continuous dynamics is well behaved. This is the core idea behind hybrid zero dynamics. By ensuring *hybrid invariance* of the zero dynamics (see Fig. 4.7.5),

$$\Delta(S \cap Z) \subset Z, \qquad (4.7.34)$$

it prevents the system from being destabilized through impact – in fact, the main result of hybrid zero dynamics is that the condition (4.7.34) implies stability of the overall dynamics provided that the zero dynamics are stable. This section will establish the fundamental results related to hybrid zero dynamics.

4.7.5.1   Hybrid Zero Dynamics

Under the influence of the controllers discussed in Section 4.7.4, the "open-loop" hybrid control system (4.7.6) takes the form of the "closed-loop" hybrid dynamical system

$$\mathcal{H} : \begin{cases} \dot{x} = f_{cl}(x), & x \in D \backslash S, \\ x^+ = \Delta\left(x^-\right), & x^- \in S, \end{cases} \qquad (4.7.35)$$

where

$$f_{cl}(x) = f(x) + g(x)u(x)$$

and $u(x)$ is the feedback controller given in (4.7.24).

Recall that the feedback controller in (4.7.24) rendered the zero dynamics surface, Z, given in (4.7.25) exponentially stable. Yet, it may be the case that for $x^- \in Z$, the post-impact state of the system $x^+ = \Delta(x^-) \notin Z$. This implies that the pre-impact state is "thrown-away" from the zero dynamics surface. Therefore, if the impacts occur at a rate faster than the controller can stabilize the system, the end result is that the impacts will destabilize the system. Therefore, the core condition that we enforce is *hybrid invariance*, i.e.,

$$\Delta(S \cap Z) \subset Z \quad \text{or, equivalently,} \quad x^- \in S \cap Z \quad \Rightarrow \quad x^+ = \Delta(x^-) \in Z.$$

$$(4.7.36)$$

Enforcing (4.7.36) requires the proper choice of virtual constraints. In particular, recall that the controller in (4.7.24) was synthesized from the virtual constraints given in (4.7.9) which, in turn, depended on the parameter set $\alpha \in \mathbb{R}^k$ with k the total number of parameters of the desired outputs, $y_d(\tau(q), \alpha)$. Therefore, the zero dynamics surface Z depends on the parameters α so that changing the values of these parameters changes the shape of the surface. This allows us to explicitly shape the zero dynamics surface to enforce the hybrid invariance condition (4.7.36), and it can be done systematically in the context of a nonlinear constrained optimization problem of the form

$$\alpha^* = \underset{\alpha \in \mathbb{R}^k}{\text{argmin}} \ \text{Cost}(\alpha) \qquad \text{(HZD Optimization)}$$

$$\text{s.t.} \quad \Delta(S \cap Z) \subset Z \qquad \text{(HZD)}$$

where Cost is a user defined cost function that can be chosen to produce walking gaits with desirable properties. The details of the optimization problem are beyond the scope of this chapter, but can be found in Westervelt et al. (2007), Ames (2014). We only mention here that any physically relevant constraints – such as constraints on torque, angular velocity, ground reaction forces and friction limitations – can be added to the optimization. Adding such constraints ensures the physical realizability of the resulting walking gait (Zhao et al., 2014b; Hereid et al., 2016).

The choice of cost function in (HZD Optimization) can determine the "shape" of the resulting gait, i.e., the overall behavior of the gait. For example, a common choice is a cost function that minimizes the overall torque while maximizing the distance traveled, i.e.,

$$\text{Cost}(\alpha) = \frac{1}{\text{step length}} \int_0^T \|u(\alpha)\|^2 dt \qquad (4.7.37)$$

where $u(\alpha)$ is the feedback controller calculated for a given parameter set. Another common choice of cost function, especially in the context of *human-inspired control* (Ames, 2014), is to use the difference between the outputs as calculated from human data and the desired functions with parameters, α, seeded from human data (Ames, 2012); the end result is typically "human-like" walking gaits. It is interesting to note that, with the proper choice of constraints, one often sees similar gaits independent of specific cost functions as the constraints tend to be a large factor in the resulting look of the gait. Finally, if efficiency is the goal, a cost function that minimizes the cost of transport can be selected (Reher et al., 2016a).

Given constraint parameters that yield well-defined hybrid zero dynamics, the end result is that the system evolves on the zero dynamics surface during the continuous dynamics and this surface is invariant through impact. Therefore, when supplied with initial conditions in Z, the dynamics of the system evolves according to the *restricted hybrid system*:

$$\mathcal{H}|_Z : \begin{cases} \dot{z} = w(0, z), & z \in Z \backslash S \cap Z, \\ z^+ = \Delta_Z(z^-), & z^- \in S \cap Z, \end{cases} \quad (4.7.38)$$

with w the zero dynamics given in (4.7.27), and $\Delta|_z : S \cap Z \to Z$ the restriction of the impact map to Z. This system is low-dimensional, e.g., for one degree of underactuation it is a two-dimensional system, and the behavior of this system dictates the behavior of the full order dynamics, $\mathcal{H}$, regardless of the dimension of the full order dynamics. For example, for a 23 degree of freedom humanoid robot (as is the case for DURUS), $\mathcal{H}$ will be a 46 dimensional hybrid system, but its behavior will still be completely determined by the behavior of the restricted hybrid system, which may have dimension as low as two (Reher et al., 2016b).

To examine how the behavior of the restricted hybrid system $\mathcal{H}|_Z$ affects the behavior of the full-order hybrid model $\mathcal{H}$, we consider periodic orbits corresponding to walking gaits of interest. In particular, for the full-order dynamics (4.7.6), let $\phi_t^{f_{cl}}(x_0)$ be the (unique) solution to the continuous dynamics $\dot{x} = f_{cl}(x)$ at time $t \geq 0$ with initial condition x_0 (where we assume local Lipschitz continuity of $f_{cl}(x)$). For $x^* \in S$ we say that $\phi_t^{f_{cl}}$ is hybrid periodic if there exists a $T > 0$ such that $\phi_T^{f_{cl}}(\Delta(x^*)) = x^*$. Given a hybrid periodic solution, we are interested in considering the stability of the corresponding hybrid periodic orbit,

$$\mathcal{O} = \{\phi_t^{f_{cl}}(\Delta(x^*)) : 0 \leq t \leq T\}.$$

To study the stability of this orbit, we first consider the *time-to-impact function*

$$T_I(x) = \inf\{t > 0 : \phi_t^{f_{cl}}(\Delta(x)) \in S, \text{ with } x \in S\},$$

which is well defined by the implicit function applied to the function $H(t, x) = p_{toe}^v(\phi_t^{f_{cl}}(\Delta(x)))$ (where p_{toe}^v is the vertical position of the toe used to define S in (4.7.4)) since it satisfies $H(T, x^*) = 0$. The end result is the *Poincaré map* which is a $P : S \to S$ which is well defined in a neighborhood of x^*, and is given by

$$P(x) = \phi_{T_I(x)}^{f_{cl}}(x).$$

Importantly, the stability of the periodic orbit $\mathcal{O}$ is equivalent to the stability of the Poincaré map viewed as a discrete time dynamical system $x_{n+1} = P(x_n)$ with fixed point $x^* = P(x^*)$, i.e.,

$\mathcal{O}$ is exponentially stable

$\Leftrightarrow$ x^* is an exponentially stable equilibrium point of P,

see Morris and Grizzle (2009) for a proof, and Sastry (1999), Wendel and Ames (2010, 2012), Burden et al. (2011) for more details on Poincaré maps. These constructions can also be applied to the restricted hybrid system given in (4.7.38). That is, given a hybrid periodic orbit $\mathcal{O}_Z \subset Z$, there is the associated *restricted Poincaré map* $P_Z : Z \cap S \to Z \cap S$ which determines the stability of $\mathcal{O}_Z$. Moreover, given a hybrid periodic orbit $\mathcal{O}_Z$, since Z is an invariant subspace of D it follows that $\mathcal{O} = \iota(\mathcal{O}_Z)$ is a hybrid periodic orbit; here $\iota : Z \hookrightarrow D$ is the canonical embedding.

We now have the technical machinery to state the main result for hybrid zero dynamics. Intuitively, this results states that

$$\mathcal{O}_Z \text{ is exponentially stable} \quad \Rightarrow \quad \mathcal{O} = \iota(\mathcal{O}_Z) \text{ is exponentially stable.}$$

More formally, we have the following fundamental theorem of hybrid zero dynamics (Westervelt et al., 2007, 2003; Grizzle et al., 2001; Morris and Grizzle, 2009):

Theorem 1 (Hybrid Zero Dynamics). *Consider the hybrid control system $\mathscr{H}\mathscr{C}$ given in (4.7.6) with the control law in (4.7.24) applied to obtain the hybrid system $\mathscr{H}$ given in (4.7.35), and assume hybrid zero dynamics (4.7.34), $\Delta(S \cap Z) \subset Z$. If there exists a locally exponentially stable hybrid periodic orbit $\mathcal{O}_Z$ of the restricted hybrid system $\mathscr{H}|_Z$, then there exists an $\bar{\epsilon} > 0$ such that for all $\bar{\epsilon} > \epsilon > 0$ the hybrid periodic orbit $\mathcal{O} = \iota(\mathcal{O}_Z)$ is locally exponentially stable for the full-order hybrid system $\mathscr{H}$.*

The importance of this result is that the zero dynamics provides a *substantially* lower dimensional surface in which to search for stable periodic orbits. In fact, in the case when the robot has one degree of underactuation ($m = n - 1$) closed form expressions can be obtained that guarantee the existence and stability of a hybrid periodic orbit $\mathcal{O}_Z$; this can be added directly to the optimization problem in (HZD Optimization) as a constraint to guarantee that any parameter set produced by the optimization is exponentially stable for the full-order dynamics of the system. In the case of full actuation, even stronger conclusions can be reached.

4.7.5.2   Partial Hybrid Zero Dynamics

In the context of gait generation for humanoid robots, one often has the luxury of dealing with a fully actuated system; in this case, partial hybrid zero dynamics (PHZD) provides a useful tool for gait generation. In particular, for PHZD we consider virtual constraints of the form (4.7.13) and (4.7.14) that allow for virtual configuration constraints on the robot via y_2, along with the ability to regulate the forward progression of the robot via y_1. Applying this controller developed in Section 4.7.4.3 to the hybrid control system $\mathcal{HC}$ given in (4.7.6) results in a closed-loop hybrid dynamical system $\mathcal{H}$ as in (4.7.35) except, in this case, $u(x)$ is given in (4.7.30). Additionally, we established that this controller resulted in the corresponding partial zero dynamics surface PZ being both attractive and invariant. Therefore, if this surface is invariant through impact:

$$\Delta(S \cap PZ) \subset PZ \qquad \text{(PHZD)}$$

the end result is *partial hybrid zero dynamics*. As in the case of hybrid zero dynamics, we can consider the optimization problem (HZD Optimization) with the constraint (HZD) replaced with (PHZD). Given a parameter set that solves this optimization problem, we have the corresponding restricted hybrid system $\mathcal{H}|_P Z$. The advantage, in this case, is that the dynamics $\dot{z} = w(0, z)$ take the simple linear form given in (4.7.33). Since the y_2 dynamics are also linear by choice of controller, the entire hybrid system becomes a *linear* hybrid dynamical system. The structural properties associated with the PHZD motivates the following key partial hybrid zero dynamics result (Ames, 2014).

Theorem 2 (Partial Hybrid Zero Dynamics). *Let $\mathcal{HC}$ given in (4.7.6) be fully actuated, with the control law in (4.7.30) applied to obtain a hybrid system $\mathcal{H}$, and assume partial hybrid zero dynamics (PHZD): $\Delta(S \cap PZ) \subset PZ$. Then, there exists a locally exponentially stable hybrid periodic orbit $\mathcal{O}_{PZ}$ of the restricted hybrid system $\mathcal{H}|_{PZ}$, and an $\bar{\epsilon} > 0$ such that for all $\bar{\epsilon} > \epsilon > 0$ the hybrid periodic orbit $\mathcal{O} = \iota(\mathcal{O}_{PZ})$ is locally exponentially stable for the full-order hybrid system $\mathcal{H}$.*

That is, in the case of fully actuated robots (such as traditional humanoids), we have the following intuitive representation of Theorem 2:

$$\text{Fully Actuated} + \Delta(S \cap PZ) \subset PZ$$
$$\Rightarrow \quad \mathcal{O} = \iota(\mathcal{O}_{PZ}) \text{ is exponentially stable}$$

or, equivalently, the existence of parameters α in (4.7.14) that yield partial hybrid zero dynamics implies a stable walking gait for fully actuated robots. It is important to note that PHZD can also be applied to robots with compliance (or underactuation) – e.g., the humanoid robot DURUS (Hereid et al., 2016; Reher et al., 2016a) – provided that these compliant elements are "normal" to the actuators that allow for forward progression of the robot. In this case, there will exist nontrivial passive dynamics in the partial zero dynamics surface and, therefore, a periodic orbit must be found in this surface to guarantee the existence of a stable periodic orbit in the full-order dynamics.

4.7.5.3 Control Lyapunov Functions

The methods presented thus far involved using feedback to linearize the dynamics of the robotic systems, wherein a linear system was defined to stabilize the virtual constraints. Yet, the control law presented intrinsically ignores the natural dynamics of the system – the main component of the feedback linearization process is to cancel out the dynamics using (4.7.20). Instead of canceling out the nonlinear dynamics, we can leverage them in the context of *control Lyapunov functions*. This has the benefit of yielding an entire class of controllers that stabilize the system. Additionally, these controllers will stabilize periodic orbits in the system in a pointwise optimal fashion.

Let us return to the canonical form of the dynamics obtained before the system was feedback linearized, i.e., the nonlinear system given in (4.7.26). Recall that, for the dynamics in this form, $\eta = (y, \dot{y})$ characterizes the controlled variables of the system while z describes the dynamics encompassed the passive component of the robot. Let us denote by Y the space with η as coordinates so that $D = Y \times Z$. A continuously differentiable function $V_\epsilon : Y \to \mathbb{R}_{\geq 0}$ is a *rapidly exponentially stabilizing control Lyapunov function (RES-CLF)* (Ames et al., 2012b, 2014a) if there exist positive constants $c_1, c_2, c_3 > 0$ such that for all $1 > \epsilon > 0$,

$$c_1 \|\eta\|^2 \leq V_\epsilon(\eta) \leq \frac{c_2}{\epsilon^2} \|\eta\|^2, \tag{4.7.39}$$

$$\inf_{u \in U} \left[L_{\hat{f}} V_\epsilon(\eta, z) + L_{\hat{g}} V_\epsilon(\eta, z)u + \frac{c_3}{\epsilon} V_\epsilon(\eta) \right] \leq 0 \tag{4.7.40}$$

for all $(\eta, z) \in Y \times Z$.

The existence of a RES-CLF yields a family of controllers that rapidly exponentially stabilize the system to the zero dynamics. In particular, we can consider the control values

$$K_\epsilon(\eta, z) = \{u \in U : L_{\hat{f}} V_\epsilon(\eta, z) + L_{\hat{g}} V_\epsilon(\eta, z)u + \frac{c_3}{\epsilon} V_\epsilon(\eta) \leq 0\}, \tag{4.7.41}$$

wherein it follows that

$$u_\epsilon(\eta, z) \in K_\epsilon(\eta, z) \quad \Rightarrow \quad \|\eta(t)\| \leq \frac{1}{\epsilon} \sqrt{\frac{c_2}{c_1}} e^{-\frac{c_3}{2\epsilon} t} \|\eta(0)\|. \tag{4.7.42}$$

Therefore, picking $\epsilon > 0$ to be a small value increases the rate of convergence of η, i.e., increases the rate of convergence to the zero dynamics surface Z. In addition, this yields specific feedback controllers, e.g., the min-norm controller (Freeman and Kokotović, 1996),

$$m_\epsilon(\eta, z) = \operatorname{argmin}\{\|u\| : u \in K_\epsilon(\eta, z)\}. \tag{4.7.43}$$

The importance of RES-CLFs is made apparent by the following theorem, which says that *any* controller $u_\epsilon \in K_\epsilon$ results in a stable orbit for the full-order dynamics if one exists in the reduced order dynamics:

$$u_\epsilon(\eta, z) \in K_\epsilon(\eta, z) + \mathcal{O}_Z \text{ exponentially stable}$$
$$\Rightarrow \quad \mathcal{O} = \iota(\mathcal{O}_Z) \text{ exponentially stable.}$$

Or, more formally, we have the following result on RES-CLF + HZD (Ames et al., 2014a):

Theorem 3 (Control Lyapunov Functions + HZD). *Consider the hybrid control system $\mathscr{HC}$ given in (4.7.6) with any Lipschitz continuous $u_\epsilon(\eta, z) \in K_\epsilon(\eta, z)$ applied to obtain a hybrid system $\mathscr{H}$, and assume hybrid zero dynamics (4.7.34), $\Delta(S \cap Z) \subset Z$. If there exists a locally exponentially stable hybrid periodic orbit $\mathcal{O}_Z$ of the restricted hybrid system $\mathscr{H}|_Z$, then there exists an $\bar{\epsilon} > 0$ such that for all $\bar{\epsilon} > \epsilon > 0$ the hybrid periodic orbit $\mathcal{O} = \iota(\mathcal{O}_Z)$ is locally exponentially stable for the full-order hybrid system $\mathscr{H}$.*

To provide a specific example of an RES-CLF, we can utilize the constructions in Section 4.7.4.1 to obtain a specific RES-CLF. In particular, recall that the feedback linearizing controller resulted in η dynamics of the form (4.7.22), or in η notation,

$$\dot{\eta} = F\eta + G\mu. \tag{4.7.44}$$

For this linear control system, we can consider the continuous-time algebraic Riccati equations (CARE),

$$F^T P + PF - PGG^T P + Q = 0, \tag{4.7.45}$$

with solution $P = P^T > 0$. One can use P to construct a RES-CLF that can be used to exponentially stabilize the output dynamics (4.7.44) at a user defined rate of $\frac{1}{\epsilon}$. In particular, define

$$V_\epsilon(\eta) = \eta^T \underbrace{M_\epsilon P M_\epsilon}_{P_\epsilon} \eta, \quad \text{with} \quad M_\epsilon = \text{diag}(\epsilon I, I), \tag{4.7.46}$$

wherein it follows that

$$\dot{V}_\epsilon(\eta, \mu) = L_F V_\epsilon(\eta) + L_G V_\epsilon(\eta)\mu$$

with

$$L_F V_\epsilon(\eta) = \eta^T (F^T P_\epsilon + P_\epsilon F)\eta,$$

$$L_G V_\epsilon(\eta) = 2\eta^T P_\epsilon G.$$

Note that it is easy to verify that $V(\eta) = \eta^T P_\epsilon \eta$ is an RES-CLF with $c_1 = \lambda_{\min}(P)$, $c_2 = \lambda_{\max}(P)$, and $c_3 = \gamma = \frac{\lambda_{\min}(Q)}{\lambda_{\max}(P)}$. This can be seen by noting that, from (4.7.45) and the form of F and G, P_ϵ solves the CARE (Ames et al., 2014a),

$$F^T P_\epsilon + P_\epsilon F - \frac{1}{\epsilon} P_\epsilon G G^T P_\epsilon + \frac{1}{\epsilon} M_\epsilon Q M_\epsilon = 0, \tag{4.7.47}$$

and noting that $\gamma P_\epsilon \leq M_\epsilon Q M_\epsilon$ so

$$\inf_\mu \left[L_F V_\epsilon(\eta) + L_G V_\epsilon(\eta)\mu + \frac{\gamma}{\epsilon} V_\epsilon(\eta) \right] \leq \eta^T P_\epsilon G(\frac{1}{\epsilon} G^T P_\epsilon + 2\mu) \leq 0,$$

which is satisfied, for example, by $\mu(\eta) = -\frac{1}{\epsilon} G^T P_\epsilon \eta$. And, therefore, V_ϵ is an RES-CLF. We can convert this back to a control law u_ϵ via (4.7.20):

$$u_\epsilon(x) = \left(L_g L_f y(x) \right)^{-1} \left[-L_f^2 y(x) + \mu_\epsilon(\eta(x)) \right] \in K_\epsilon(\eta(x), z(x)) \tag{4.7.48}$$

for $\mu_\epsilon(\eta(x))$ satisfying

$$\dot{V}_\epsilon(\eta(x)) = L_F V_\epsilon(\eta(x)) + L_G V_\epsilon(\eta)\mu_\epsilon(\eta(x)) \leq \frac{\gamma}{\epsilon} V_\epsilon(\eta(x)) \tag{4.7.49}$$

where we converted back to the x coordinates, i.e., $\eta(x) = (y(x), \dot{y}(x))$. This gives concrete conditions that can be checked to stabilize walking gaits in the hybrid system $\mathscr{H}$ according to Theorem 3.

It is important to note that RES-CLFs can also be constructed in the case of partial hybrid zero dynamics. In this case, the linear control system for the output dynamics (4.7.30) is described by $\eta = (y_1, y_2, \dot{y}_2)$. From these dynamics, (4.7.45) can be used to construct an RES-CLF as in (4.7.46). The end result is a reformulation of Theorem 2 so that a stable periodic orbit is guaranteed for the full order dynamics for any $u_\epsilon(\eta) \in K_\epsilon(\eta)$. That is, we obtain stable walking through the entire class of controllers that satisfy the inequality constraint obtained via the CLF condition.

4.7.6 EXTENSIONS OF HYBRID ZERO DYNAMICS

The goal of this section is to consider extensions of the HZD framework with a view toward more rich application domains. In particular, we consider the HZD framework developed in this chapter in the context of optimization-based controllers via CLFs, multidomain hybrid system models, their application to powered prostheses and compliant hybrid zero dynamics.

4.7.6.1 CLF-Based QPs

The advantage of the control Lyapunov functions introduced in Section 4.7.5.3 is that they give a family of controllers which stabilize the system. That is, for any $u_\epsilon(\eta, z) \in K_\epsilon(\eta, z)$ the system has a stable periodic gait (given a stable periodic orbit in the zero dynamics); a specific example of this is given by the traditional feedback linearizing controller. The importance of CLFs goes beyond simply producing a class of controllers – it suggests an optimization-based control framework for bipedal robots and, in fact, nonlinear systems in general. This allows for these control methods to be extended to a variety of application domains from robotic walking, to prostheses to manipulation to safety-critical control methods.

To see how control Lyapunov functions yield optimization-based controllers, we can consider the set (4.7.41) giving the family of stabilizing controllers. Note that this set is affine in the control input u and, therefore, the min-norm controller (4.7.43) can be equivalently stated as a quadratic program (QP) of the form:

$$m(\eta, z) = \underset{u \in U = \mathbb{R}^m}{\operatorname{argmin}} \quad u^T u \tag{4.7.50}$$

$$\text{s.t.} \quad L_{\hat{f}} V_\epsilon(\eta, z) + L_{\hat{g}} V_\epsilon(\eta, z) u + \frac{c_3}{\epsilon} V_\epsilon(\eta) \le 0 \tag{CLF}$$

where we assume that $U = \mathbb{R}^m$ to ensure solvability of the QP. Not only is this QP guaranteed to have a solution, but the solution can be written in closed form (see Ames et al., 2014a) and is Lipschitz continuous. Moreover, one can utilize the RES-CLF given in (4.7.46) to explicitly construct the inequality constraint in this QP. Finally, because it is a QP it can be solved in real-time; in fact, the CLF-based QP has been implemented in real-time (e.g., at a 1 kHz loop rate) on MABEL (Galloway et al., 2015) and DURUS-2D (Cousineau and Ames, 2015) to achieve dynamic walking. Additionally, it was implemented at over 5 kHz as an embedded level controller on series elastic actuators in Ames and Holley (2014).

The advantage of the QP formulation of CLFs, as opposed to simply utilizing the closed form min-norm solution, is that it allows for additional constraints and objectives to be unified with the CLF. To provide a concrete example, suppose that we have torque bounds on the actuators given by a scalar $u_{\max}$ (similar ideas extend to actuators with different max torques). While one might typically simply saturate the control input, doing so prevents the controller from taking these torque saturations into account. Therefore, through the CLF-based QP framework, we can incorporate the torque bounds directly into the controller via the quadratic program:

$$u^*(\eta, z) = \underset{(\delta, \mu) \in \mathbb{R}^{n+1}}{\mathrm{argmin}} \quad u^T H(\eta, z)u + p\delta^2 \tag{4.7.51}$$

$$\text{s.t.} \quad L_{\hat{f}} V_\epsilon(\eta, z) + L_{\hat{g}} V_\epsilon(\eta, z)u + \frac{c_3}{\epsilon} V_\epsilon(\eta) \leq \delta, \quad \text{(CLF)}$$

$$u \leq u_{\max} \mathbf{1}, \quad \text{(Max Torque)}$$

$$-u \leq u_{\max} \mathbf{1} \quad \text{(Min Torque)}$$

where $H(\eta, z)$ is positive-definite and $p > 0$ is a large value that penalizes for violations of the CLF constraint. That is, we relax the CLF condition to ensure satisfaction of the physical constraints of the system. While this takes away guarantees on achieving the control objective, it will try to achieve convergence of the CLF in a pointwise optimal fashion when at all possible – the result is that the robot is able to accommodate tighter torque bounds than if one was to simply saturate the control input (see Galloway et al., 2015 for a detailed discussion, and experimental implementation). Note that as one expands the number of constraints in the QP, it is important to be aware of the impact on the resulting solvability and, as a byproduct, the continuity of the solutions to the QP; a discussion can be found in Morris et al. (2013), and conditions on continuity in Morris et al. (2015a).

Utilizing the observation that CLFs (and hence control objectives) can be represented as affine constraints in a QP results in a new paradigm for the control of walking robots. In particular, going beyond simply adding torque bounds, one can consider multiobjective controllers consisting of multiple CLF wherein each control objective results in an additional constraint in the QP (Ames and Powell, 2013); for example, in the context of unifying locomotion and manipulation objectives. Additionally, ground reaction forces on the robot also appear in an affine fashion in the dynamics; thus one can use the CLF-based QP framework in the context of force control. Finally, a recent line of work aimed at safety-critical control makes the observation that safety conditions, i.e., set invariance, can be stated in the context of *control barrier functions* which again are affine in the control input (originally formulated in Ames et al. (2014b) and studied in detail in Ames et al. (2016)); this framework has been applied in the context of robotic walking (Nguyen and Sreenath, 2016; Nguyen and Sreenath, 2015, 2016), automotive safety systems (Xu et al., 2015; Ames et al., 2016; Mehra et al., 2015) and swarm robotics (Borrmann et al., 2015; Wang et al., 2016a, 2016b). Therefore, safety conditions can be unified with control objectives, physical constraints, force objectives and safety constraints all in the context of a single optimization-based controller that can be realized in realtime on robotic systems.

4.7.6.2 Multidomain Hybrid Zero Dynamics

The analysis thus far has focused on hybrid system models of walking robots with a single continuous and discrete domain, i.e., single domain hybrid systems. Yet, in the context of advanced walking and running behaviors, it is necessary to consider multidomain hybrid systems models as introduced in Section 4.7.2.4. As indicated in this section, and as motivated by human walking, throughout the course of a step humans naturally display heel–toe behavior in their feet while locomoting (see Fig. 4.7.4). In the context of these types of walking behaviors, the end result is the multi-domain hybrid system model given in (4.7.8); specific examples of this model are shown in Fig. 4.7.4 for the bipedal robots AMBER 2 and ATRIAS (see Zhao et al., 2015a for the specific hybrid system constructions for these robots).

In the context of multidomain hybrid systems, we can extend the concept of hybrid zero dynamics. In particular, we now have a collection of continuous domains $D = \{D_v\}_{v \in V}$ on which we have associated control systems: $\dot{x} = f_v(x) + g_v(x)u_v$. Note that the domains may be of different actuation types, e.g., some may be underactuated while others may be fully actuated or over actuated. For each of these domains, we can define virtual constraints of the form (4.7.9), denoted by $y_{2,v}$; in the case of full (and over) actuation, $y_v = (y_{1,v}, y_{2,v})$ as in (4.7.13) and (4.7.14)), and in the case of over actuation, care must be taken to define constraints that result in a nonsingular decoupling matrix $A(x)$ in (4.7.29). Therefore, we can construct controllers, $u_v(x)$ for each $v \in V$ as in (4.7.20) for the underactuated domains and as in (4.7.30) for the full (and over) actuated domains.

We can consider the zero dynamic surfaces (and partial zero dynamics surfaces) denoted, for notational simplicity, uniformly by

$$Z_v = \left\{ x \in D_v \mid y_{2,v}(x) = 0 \text{ and } \dot{y}_{2,v}(x) = 0 \right\}.$$

Correspondingly, the control laws $u_v(x)$ drive the system to the surface Z_v for each $v \in V$ and, in addition, renders each of these surfaces attractive. To ensure stability of the overall dynamics, we must ensure hybrid zero dynamics for all of the discrete transitions, i.e., *multidomain hybrid zero dynamics (MDHZD)*:

$$\Delta_e(S_e \cap Z_{\text{source}(e)}) \subset Z_{\text{target}(e)}, \qquad \forall e \in E \qquad \text{(Multidomain HZD)}$$

where source(e) and target(e) are the source and target of the edge $e \in E$ of the oriented graph Γ in (4.7.8), respectively. As in Section 4.7.5, if the multidomain hybrid system (4.7.8) has MDHZD, then if there is an exponentially stable periodic orbit contained in Z_v for $v \in V$, then there exists an exponentially stable hybrid periodic in the full order dynamics when the control laws $u_v(x)$ are applied in each domain.

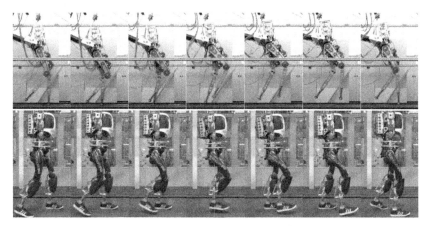

FIGURE 4.7.6 Multicontact walking and running utilizing multidomain hybrid system models realized on the DURUS-2D running robot and the humanoid robot DURUS.

This framework has been applied to numerous bipedal robots to obtain stable walking and running. Specific examples of walking robots include MABEL (Sreenath et al., 2011), AMBER 2 (Zhao et al., 2014b), ATRIAS (Hereid et al., 2014), and DURUS (Hereid et al., 2016). In the case of DURUS, due to the passive springs in the ankles, a two domain hybrid system model was considered. The end result was stable 3D robotic walking, demonstrated publicly during the DARPA Robotics Challenge, where the motion was sustained for over 5 hours with the robot traversing almost 4 km on a treadmill. Importantly, this was the most efficient walking ever realized on a bipedal humanoid robot (Reher et al., 2016a). This can be attributed to the fact that the MDHZD allows for the full dynamics of the robot to be utilized in the generation of walking gaits (through the shaping of the surfaces Z_v) and, importantly for the compliant elements in the system, e.g., springs, to be fully utilized during the walking gait. Recently, these methods were extended to yield a four-domain model of DURUS capturing the natural heel-toe behavior of the foot that humans display when locomoting; the end result was dynamic walking that is efficient and human-like (Reher et al., 2016b) (tiles of this walking gait are shown in Fig. 4.7.6). Finally, note that running motions provide natural examples of two-domain models (Chevallereau et al., 2005; Morris et al., 2006; Poulakakis and Grizzle, 2009b) – see also Fig. 4.7.8 below – and that additional domains can be introduced depending on the control action to enhance control authority over the system, as in the control of running on Thumper (Poulakakis and Grizzle, 2009a), MABEL (Sreenath et al., 2013) and the DURUS-2D runner (Ma et al., 2017) (shown in Fig. 4.7.6).

4.7.6.3 Application to Prostheses

The concepts presented throughout this chapter have natural application to powered prostheses. In particular, a prosthetic device can be simply viewed as a component of a bipedal robot and, with proper representation of the human and the interaction of the human with the device, one can generate controllers for the device via HZD-based methods (Zhao et al., 2011). The core idea to synthesizing prosthetic controllers is to first model the human and the prosthesis as two robotic systems that are coupled at the prosthetic attachment (see Fig. 4.7.7); the parameters of the "human" are taken from measurements of the human and used to generate a corresponding model, and the model of the prosthesis is then added on the affected leg to yield an overall model of the combined human–robot system. This robotic model can then be approached in the same way one would approach generating walking gaits for bipedal robots: the hybrid system model is constructed based upon the desired foot behavior, and gaits are generated through an optimization problem that enforces the HZD conditions together with physical constraints. This idea was first explored in the context of human-inspired control (Sinnet et al., 2011b), and was experimentally validated through the application on both bipedal robots (wherein one leg of the robot plays the role of the prosthesis) (Zhao et al., 2014a), followed by the evaluation with an amputee subject (Zhao et al., 2011). The advantage of the HZD-based approach to designing controllers for powered prostheses is that all of the advanced control and locomotion related concepts of this chapter can be translated to this domain. In particular, multi-domain hybrid system models of locomotion can be utilized to achieve advanced foot behaviors on the device (Zhao et al., 2016b). Additionally, CLF-based QP controllers (as in Section 4.7.6.1) can be realized on prosthetic devices in realtime through a novel model-independent variant (Zhao et al., 2015b); this allows for efficient locomotion that leverages the use of compliant elements as in AMPRO 3 shown in Fig. 4.7.6.

4.7.6.4 Compliant Hybrid Zero Dynamics

To recover part of the energy required to sustain cyclic walking or running motions in legged robots and to ensure safe interaction with the ground surface, compliant elements in the form of mechanical springs have been incorporated in the legs of many such platforms; in the context of robotic bipeds, Thumper (Hurst and Rizzi, 2008), MABEL (Grizzle et al., 2009), and ATRIAS (Hubicki et al., 2016) are just few examples of robots in this family. The role of elastic energy storage in compliant elements becomes more prominent in running motions (McMahon and Cheng, 1990; Alexander, 1990). However, the inclusion of physical springs in a robot's structure poses additional

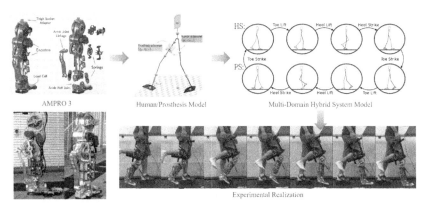

FIGURE 4.7.7 Illustration of the application of HZD methods to prostheses; in this case, in the context of the third generation AMPRO device utilized to achieve multidomain walking (Zhao et al., 2016a, 2017).

challenges to control design. More specifically, in the pursuit of closed-loop stability, the control action must actively exploit open-loop compliance instead of – as is usual in the control of flexible mechanisms – working to replace it. The concept of *compliant hybrid zero dynamics* introduced in Poulakakis (2008) extends HZD controllers so that open-loop compliance is "preserved" in the closed-loop system and determines its behavior. To avoid complexity, we first discuss the main ideas of this method in the context of a simplified hopping model – namely, the *asymmetric spring-loaded inverted pendulum (ASLIP)* (Poulakakis and Grizzle, 2007a, 2009b) – and then provide some information on its application to the control of walking and running motions in MABEL (Sreenath et al., 2011, 2013).

The Asymmetric Spring-Loaded Inverted Pendulum

The ASLIP shown in Fig. 4.7.8 was originally proposed in Poulakakis and Grizzle (2007a) as an intermediate model to bridge the gap between point-mass SLIP-like models and monopedal robots with significant torso pitch dynamics. The ASLIP includes a torso nontrivially coupled to the leg motion,[6] an issue not addressed in the widely studied SLIP, or in its straightforward extensions in which the torso COM coincides with the hip joint. As in the SLIP, the ASLIP features a massless leg and the contact between the leg end and the ground is modeled as an unactuated pin joint.

---

6. Along the same lines with the ASLIP, the Virtual Pivot Point (VPP) model was introduced in Maus et al. (2010) as a template for studying torso stabilization in running; see Subchapters 2.3 and 3.6 for more details.

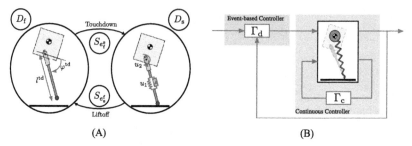

(A)   (B)

FIGURE 4.7.8 (A) The asymmetric spring-loaded inverted pendulum (ASLIP). Running comprises stance and flight phases, separated by touchdown and liftoff events. (B) The structure of the SLIP embedding controller.

The ASLIP alternates between stance and flight phases – denoted by "s" and "f", respectively – resulting in a multidomain hybrid system of the type described in Section 4.7.6.2. Let $\Gamma = (V, E)$, where $V = \{s, f\}$ and $E = \{e_s^f, e_f^s\}$, be the oriented graph that captures the contact state of the model; in this notation, e_s^f and e_f^s denote transition from stance to flight and vice versa, respectively. The model consists of two domains D_s and D_f, within which the dynamics $FG = \{(f_v, g_v)_{v \in V}\}$ of the ASLIP evolve until the state intersects the corresponding switching surface $S = \{S_e\}_{e \in E}$; at this point, a switching map $\Delta = \{\Delta_e\}_{e \in E}$ is triggered to provide initial conditions for the ensuing phase. During stance, the ASLIP is controlled by two inputs: the force u_1 acting along the leg and the torque u_2 applied at the hip; $(u_1, u_2) \in U_s$, where U_s is the set of the admissible stance inputs. As in Poulakakis and Grizzle (2007b, 2009b), the leg force u_1 is modeled as a spring in parallel with a prismatic force source. During flight, on the other hand, the assumption of a massless leg implies that the ASLIP follows a ballistic motion. Furthermore, the leg attains its desired configuration $\alpha_f = (l^{td}, \varphi^{td}) \in A_f$ in anticipation to touchdown kinematically, just like the SLIP; see also Fig. 4.7.8. The inherently hybrid nature of the dynamics of the ASLIP can then be represented by a system of the form (4.7.8) as

$$\mathcal{HC}_{\text{ASLIP}} := (\Gamma, D, U, A, S, \Delta, FG) \tag{4.7.52}$$

where $U = \{U_s, \emptyset\}$ and $A = \{\emptyset, A_f\}$ include the inputs available in continuous and in discrete time, respectively. The system (4.7.52) can be brought in the standard form of a system with impulse effects by integrating the flight phase dynamics until touchdown, thereby obtaining a map $\Delta : S_{e_s^f} \to D_s$ that takes the liftoff conditions $x_s^- \in S_{e_s^f}$ together with the desired configuration $\alpha_f \in A_f$ of the leg at touchdown to the initial conditions $x_s^+ \in D_s$ of the next stance phase. The details can be found in Poulakakis and Grizzle (2009b), and the resulting form is

$$\mathscr{HC}_{\mathrm{ASLIP}} : \begin{cases} \dot{x}_{\mathrm{s}} = f_{\mathrm{s}}(x_{\mathrm{s}}) + g_{\mathrm{s}}(x_{\mathrm{s}})u_{\mathrm{s}}, & x_{\mathrm{s}} \in D_{\mathrm{s}} \setminus S_{e_{\mathrm{s}}^{\mathrm{f}}}, \\ x_{\mathrm{s}}^{+} = \Delta\left(x_{\mathrm{s}}^{-}, \alpha_{\mathrm{f}}\right), & x_{\mathrm{s}}^{-} \in S_{e_{\mathrm{s}}^{\mathrm{f}}}, \ \alpha_{\mathrm{f}} \in A_{\mathrm{f}}. \end{cases} \tag{4.7.53}$$

It should be emphasized that the initial condition $x_{\mathrm{s}}^{+} \in D_{\mathrm{s}}$ of the ensuing stance phase does not only depend on the exit condition $x_{\mathrm{s}}^{-} \in S_{e_{\mathrm{s}}^{\mathrm{f}}}$ of the previous stance phase. It also depends on the parameter α_{f} that determines the configuration of the leg at touchdown, thereby strongly influencing the ensuing stance phase. Clearly, updating α_{f} in an event-based fashion provides a powerful control input.

Embedding the SLIP in the Dynamics of the ASLIP

As was mentioned in Chapter 3, a growing body of evidence in biomechanics indicates that, when running, diverse species tune their musculoskeletal system so that their center of mass bounces along as if it is following the dynamics of a SLIP (Holmes et al., 2006). In the light of this evidence, the SLIP is construed as a canonical model of running, and can be used as a *behavioral control target* for legged robots or robot models. In what follows, we describe a feedback control law that organizes the ASLIP so that its closed-loop dynamics is governed by the dynamics of a variant of the SLIP; namely, the *energy-stabilized SLIP (ES-SLIP)* shown in Fig. 4.7.9. The ES-SLIP is a modification of the standard SLIP that admits exponentially stable[7] hopping motions (Poulakakis and Grizzle, 2007a, 2009b). The dynamics of the ES-SLIP in closed loop with an exponentially stabilizing feedback controller – see Poulakakis and Grizzle (2007a, 2009b) for details – can be written as

$$\mathscr{H}_{\mathrm{ES-SLIP}} : \begin{cases} \dot{z} = f_{z}(z), & z \notin S_{z}, \\ z^{+} = \Delta_{z}\left(z^{-}\right), & z^{-} \in S_{z}, \end{cases} \tag{4.7.54}$$

where S_{z} corresponds to the stance-to-flight switching surface and the rest of the components of (4.7.54) are defined in a similar fashion to those in (4.7.53); see Poulakakis and Grizzle (2007a, 2009b) for details.

The goal of the SLIP embedding controller is to render *any*[8] exponentially stable periodic running orbit of the ES-SLIP exponentially stable in the ASLIP. As Fig. 4.7.8B shows, control action is distributed over continuous and discrete time as follows. The continuous-time feedback law $u_{\mathrm{s}} = \Gamma_{\mathrm{c}}(x_{\mathrm{s}})$ is employed

---

7. The standard SLIP is energy conservative and thus it cannot reject perturbations that shift the total energy of the system.

8. Provided, of course, that the physical constraints associated with ground reaction forces and actuator limitations are respected.

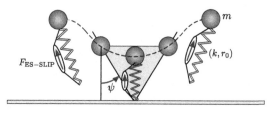

FIGURE 4.7.9   The energy-stabilized SLIP (ES-SLIP), a variant of the standard SLIP which features a prismatic leg actuator in parallel with the spring. The leg actuator can develop nonconservative forces capable or rejecting perturbations that alter the total energy of the system.

during stance with the purpose of (i) creating an *invariant* and *attractive* submanifold Z_s embedded in the stance state space, D_s, and (ii) rendering the restriction of the closed-loop stance dynamics of the ASLIP on Z_s diffeomorphic to the ES-SLIP stance dynamics; formally, $\left(f_s(x_s) + g_s(x_s)\Gamma_c(x_s) \right)\Big|_{Z_s} \cong f_z(z)$. The discrete-time feedback law $\alpha_f = \Gamma_d(x_s^-)$ is employed at transitions from stance to flight with the purpose of updating the leg configuration α_f at touchdown so that (i) Z_s is *hybrid invariant*, i.e., invariant under the closed-loop transition map $\Delta_{cl}(x_s^-) := \Delta\left(x_s^-, \Gamma_d(x_s^-)\right)$ as defined in Section 4.7.5.1, and (ii) the restricted closed-loop reset map $\Delta_{cl}|_{Z_s}$ of the ASLIP is equivalent to the transition map Δ_z of the ES-SLIP; formally, $\Delta_{cl}|_{Z_s} \cong \Delta_z$. More details on the design of the feedback laws (Γ_c, Γ_d) can be found in Poulakakis and Grizzle (2007a, 2009b).

Implications to the Control of Robots

The approach described above essentially combines the practical advantages of compliant reductive models typically used to intuitively tune empirical controllers – the SLIP is a classical example – with the analytical tractability offered by constructive feedback synthesis methods. But, how can we leverage these feedback constructions to introduce a general control synthesis framework for compliant legged robots?

Clearly, the direct implementation of the SLIP embedding controller in legged robots like Thumper and MABEL depicted in Fig. 4.7.1 is far from being a straightforward task. The primary reason is that the ASLIP is based on a number of simplifying assumptions that do not faithfully capture the structural and morphological characteristics of these robots. More specifically, the assumption of a massless leg together with the requirement that the HZD is equivalent – in a strict mathematical sense – to the SLIP severely limit the applicability of the SLIP embedding controller to Thumper and MABEL. Yet, the following lessons learnt from the SLIP embedding controller are important: (i) the HZD is an explicitly compliant system, possessing more than one degrees of freedom, thus capturing not only the progression of the leg's sweep angle – as in the classical HZD method (Westervelt et al., 2003) – but also leg compression and decompression in a way that respects compliance; (ii) control authority is available *within* the zero dynamics – which is now *controlled* – thus allowing the development of additional feedback action to realize compliance "shaping" and active force control for greater flexibility; and (iii) although keeping the torso at a constant angle – which is in fact a necessary condition for embedding the SLIP in the ASLIP (Poulakakis, 2010) – is restricting, commanding zero pitch velocity during the late stage of the stance phase ensures that the angular momentum associated with the torso is small when the system switches to flight so that excessive pitching during flight is eliminated. These three considerations, which underlie the SLIP embedding controller, can be encoded in a set of suitably parametrized outputs of the form (4.7.9) and enforced on the dynamics of Thumper and MABEL though feedback linearization as in Section 4.7.4.1; see Poulakakis (2008, Chapter VI) and Poulakakis and Grizzle (2009a) for the development of the method. Skipping details, we only note that, similar to the block diagram of Fig. 4.7.8B, the continuous-time control action introduces a set of parameters which are updated in discrete time using event-based feedback. A refined version of this method was implemented in Sreenath et al. (2011) to generate experimentally dynamically stable, fast and efficient walking motions on MABEL at top sustained speeds 1.5 m/s. Beyond walking, the notion of compliant hybrid zero dynamics is at the core of stabilizing running on MABEL (Sreenath et al., 2013). Running presents unique challenges due to the presence of substantial flight phases that limit control authority over the system. Addressing these challenges calls for active force control within the compliant HZD as detailed in Sreenath et al. (2013). This method resulted in MABEL running at an average speed of 1.95 m/s and a peak speed of 3.06 m/s.

4.7.7 SUMMARY

Most traditional legged locomotion control approaches heavily rely on heuristic methods which do not provide stability and performance guarantees, thus hindering the use of legged robots in real-life applications. The hybrid zero dynamics (HZD) method described in this chapter has been proposed as a general framework for the synthesis of feedback control laws that induce provably stable, fast, and reliable walking and running motions in legged robots. At the core of the method is the idea of encoding desired locomotion behaviors via a set of suitably parametrized virtual constraints, which effectively coordinate the higher-dimensional robot plant into a lower-dimensional hybrid subsystem – namely, the HZD – that governs the robot's locomotion behavior. This chapter briefly discussed the main concepts as well as key implementation aspects underlying the applications of the method, pointing to the relevant literature for detailed accounts. Beyond its theoretical value, perhaps the most impressive feature of the HZD method is its versatile nature. This feature supports implementation on robots with different structural and morphological characteristics, ranging from rigid walking to compliant running bipeds and to prostheses.

REFERENCES

Ackermann, M., 2007. Dynamics and Energetics of Walking with Prostheses. PhD thesis. University of Stuttgart.

Alexander, R., 1988. Elastic Mechanisms in Animal Movement. Cambridge University Press, Cambridge.

Alexander, R., 1990. Three uses for springs in legged locomotion. Int. J. Robot. Res. 9, 53–61.

Ames, A.D., 2012. First steps toward automatically generating bipedal robotic walking from human data. In: Robot Motion and Control 2011. Springer, London, pp. 89–116.

Ames, A.D., 2014. Human-inspired control of bipedal walking robots. IEEE Trans. Autom. Control 59, 1115–1130.

Ames, A.D., Cousineau, E.A., Powell, M.J., 2012a. Dynamically stable bipedal robotic walking with NAO via human-inspired hybrid zero dynamics. In: Proceedings of the 15th ACM international conference on Hybrid Systems: Computation and Control. ACM, pp. 135–144.

Ames, A.D., Galloway, K., Grizzle, J.W., 2012b. Control Lyapunov functions and hybrid zero dynamics. In: Proc. 51st IEEE Conf. Decision and Control.

Ames, A.D., Galloway, K., Sreenath, K., Grizzle, J.W., 2014a. Rapidly exponentially stabilizing control Lyapunov functions and hybrid zero dynamics. IEEE Trans. Autom. Control 59, 876–891.

Ames, A.D., Grizzle, J.W., Tabuada, P., 2014b. Control barrier function based quadratic programs with application to adaptive cruise control. In: IEEE 53rd Annual Conference on Decision and Control (CDC). IEEE, pp. 6271–6278.

Ames, A.D., Holley, J., 2014. Quadratic program based nonlinear embedded control of series elastic actuators. In: IEEE 53rd Annual Conference on Decision and Control (CDC). IEEE, pp. 6291–6298.

Ames, A.D., Powell, M., 2013. Towards the unification of locomotion and manipulation through control Lyapunov functions and quadratic programs. In: Control of Cyber-Physical Systems. Springer International Publishing, pp. 219–240.

Ames, A.D., Vasudevan, R., Bajcsy, R., 2011. Human-data based cost of bipedal robotic walking. In: Proceedings of the 14th International Conference on Hybrid Systems: Computation and Control. ACM, pp. 153–162.

Ames, A.D., Xu, X., Grizzle, J.W., Tabuada, P., 2016. Control barrier function based quadratic programs with application to automotive safety systems. arXiv:1609.06408v1.

Borrmann, U., Wang, L., Ames, A.D., Egerstedt, M., 2015. Control barrier certificates for safe swarm behavior. IFAC-PapersOnLine 48, 68–73.

Burden, S., Revzen, S., Sastry, S.S., 2011. Dimension reduction near periodic orbits of hybrid systems. In: 2011 50th IEEE Conference on Decision and Control and European Control Conference (CDC-ECC). IEEE, pp. 6116–6121.

Chevallereau, C., Abba, G., Aoustin, Y., Plestan, F., Westervelt, E.R., Canudas, C., Grizzle, J.W., 2003. RABBIT: a testbed for advanced control theory. Control Syst. Mag. 23, 57–79.

Chevallereau, C., Westervelt, E.R., Grizzle, J.W., 2005. Asymptotically stable running for a five-link, four-actuator, planar bipedal robot. Int. J. Robot. Res. 24, 431–464.

Collins, S.H., Ruina, A., 2005. A bipedal walking robot with efficient and human-like gait. In: Proceeding of 2005 International Conference on Robotics and Automation. Barcelona, Spain.

Cousineau, E., Ames, A.D., 2015. Realizing underactuated bipedal walking with torque controllers via the ideal model resolved motion method. In: 2015 IEEE International Conference on Robotics and Automation (ICRA). IEEE, pp. 5747–5753.

Freeman, R.A., Kokotović, P.V., 1996. Robust Nonlinear Control Design. Birkhäuser.

Galloway, K., Sreenath, K., Ames, A.D., Grizzle, J.W., 2015. IEEE Access 3, 323–332.

Goebel, R., Sanfelice, R., Teel, A., 2009. Hybrid dynamical systems. IEEE Control Syst. Mag. 29, 28–93.

Goldstein, H., Poole, C., Safko, J., 2002. Classical Mechanics, third edition. Addison-Wesley, San Francisco.

Goswami, A., 1999a. Foot rotation indicator (FRI) point: a new gait planning tool to evaluate postural stability of biped robots. In: International Conference on Robotics and Automation. IEEE, pp. 47–52.

Goswami, A., 1999b. Postural stability of biped robots and the foot-rotation indicator (FRI) point. Int. J. Robot. Res. 18, 523–533.

Gregg, R.D., Tilton, A.K., Candido, S., Bretl, T., Spong, M.W., 2012. Control and planning of 3-D dynamic walking with asymptotically stable gait primitives. IEEE Trans. Robot. 28, 1415–1423.

Grizzle, J.W., Abba, G., Plestan, F., 2001. Asymptotically stable walking for biped robots: analysis via systems with impulse effects. IEEE Trans. Autom. Control 46, 51–64.

Grizzle, J.W., Chevallereau, C., Ames, A.D., Sinnet, R.W., 2010. 3D bipedal robotic walking: models, feedback control, and open problems. In: IFAC Proceedings, vol. 43, pp. 505–532.

Grizzle, J.W., Chevallereau, C., Sinnet, R.W., Ames, A.D., 2014. Models, feedback control, and open problems of 3D bipedal robotic walking. Automatica 50, 1955–1988.

Grizzle, J.W., Hurst, J., Morris, B., Park, H.W., Sreenath, K., 2009. MABEL, a new robotic bipedal walker and runner. In: American Control Conference, 2009. ACC'09. IEEE, pp. 2030–2036.

Haddad, W.M., Chellaboina, V.S., Nersesov, S.G., 2006. Impulsive and Hybrid Dynamical Systems: Stability, Dissipativity, and Control. Princeton University Press, Princeton, NJ.

Hereid, A., Cousineau, E.A., Hubicki, C.M., Ames, A.D., 2016. 3D dynamic walking with underactuated humanoid robots: a direct collocation framework for optimizing hybrid zero dynamics. In: 2016 IEEE International Conference on Robotics and Automation (ICRA). IEEE, pp. 1447–1454.

Hereid, A., Kolathaya, S., Jones, M.S., Van Why, J., Hurst, J.W., Ames, A.D., 2014. Dynamic multi-domain bipedal walking with ATRIAS through slip based human-inspired control. In: Proceedings of the 17th International Conference on Hybrid Systems: Computation and Control. ACM, pp. 263–272.

Holmes, P., Full, R.J., Koditschek, D.E., Guckenheimer, J., 2006. The dynamics of legged locomotion: models, analyses, and challenges. SIAM Rev. 48, 207–304.

Hubicki, C., Grimes, J., Jones, M., Renjewski, D., Spröwitz, A., Abate, A., Hurst, J., 2016. ATRIAS: design and validation of a tether-free 3D-capable spring-mass bipedal robot. Int. J. Robot. Res. 35, 1497–1521.

Huihua, Z., Yadukumar, S.N., Ames, A.D., 2012. Bipedal robotic running with partial hybrid zero dynamics and human-inspired optimization. In: 2012 IEEE/RSJ International Conference on Intelligent Robots and Systems (IROS). IEEE, pp. 1821–1827.

Hurst, J.W., 2008. The Role and Implementation of Compliance in Legged Locomotion. PhD thesis. The Robotics Institute, Carnegie Mellon University.

Hurst, J.W., Chestnutt, J.E., Rizzi, A.A., 2007. Design and philosophy of the BiMASC, a higly dynamic biped. In: Proceedings of the IEEE International Conference of Robotics and Automation. Roma, Italy, pp. 1863–1868.

Hurst, J.W., Rizzi, A., 2008. Series compliance for an efficient running gait. IEEE Robot. Autom. Mag. 15, 42–51.

Isidori, A., 1995. Nonlinear Control Systems, third edition. Springer-Verlag, Berlin.

Kajita, S., Kanehiro, F., Kaneko, K., Fujiwara, K., Harada, K., Yokoi, K., Hirukawa, H., 2006. Biped walking pattern generator allowing auxiliary ZMP control. In: Proceedings of the 2006 IEEE/RSJ International Conference on Intelligent Robots and Systems, pp. 2993–2999.

Kajita, S., Morisawa, M., Harada, K., Kaneko, K., Kanehiro, F., Fujiwara, K., Hirukawa, H., 2003. Biped walking pattern generation by using preview control of zero-moment point. In: IEEE International Conference on Robotics and Automation. Taipei, Taiwan, vol. 2, pp. 1620–1626.

Lamperski, A., Ames, A.D., 2013. Lyapunov theory for Zeno stability. IEEE Trans. Autom. Control 58, 100–112.

Lygeros, J., Johansson, K.H., Simic, S., Zhang, J., Sastry, S., 2003. Dynamical properties of hybrid automata. IEEE Trans. Autom. Control 48, 2–17.

Ma, W., Kolathaya, S., Ambrose, E., Hubicki, C., Ames, A.D., 2017. Bipedal robotic running with DURUS-2D: bridging the gap between theory and experiment. In: Proceedings of the 20th ACM International Conference on Hybrid Systems: Computation and Control. ACM.

Maus, H.-M., Lipfert, S., Gross, M., Rummel, J., Seyfarth, A., 2010. Upright human gait did not provide a major mechanical challenge for our ancestors. Nat. Commun. 1.

McMahon, T.A., Cheng, G.C., 1990. The mechanics of running: how does stiffness couple with speed? J. Biomech. 23, 65–78. Suppl. 1.

Mehra, A., Ma, W.-L., Berg, F., Tabuada, P., Grizzle, J.W., Ames, A.D., 2015. Adaptive cruise control: experimental validation of advanced controllers on scale-model cars. In: American Control Conference (ACC). IEEE, pp. 1411–1418.

Morris, B., 2008. Stabilizing Highly Dynamic Locomotion in Planar Bipedal Robots with Dimension Reducing Control. Ph.D. thesis. University of Michigan.

Morris, B., Grizzle, J.W., 2009. Hybrid invariant manifolds in systems with impulse effects with application to periodic locomotion in bipedal robots. IEEE Trans. Autom. Control 54, 1751–1764.

Morris, B., Powell, M.J., Ames, A.D., 2013. Sufficient conditions for the lipschitz continuity of QP-based multi-objective control of humanoid robots. In: 2013 IEEE 52nd Annual Conference on Decision and Control (CDC). IEEE, pp. 2920–2926.

Morris, B., Westervelt, E.R., Chevallereau, C., Buche, G., Grizzle, J.W., 2006. Achieving bipedal running with RABBIT: six steps toward infinity. In: Mombaur, K., Dheil, M. (Eds.), Fast Motions Symposium on Biomechanics and Robotics. In: Lecture Notes in Control and Information Sciences. Springer-Verlag, Heidelberg, Germany, pp. 277–297.

Morris, B.J., Powell, M.J., Ames, A.D., 2015a. Continuity and smoothness properties of nonlinear optimization-based feedback controllers. In: 2015 IEEE 54th Annual Conference on Decision and Control (CDC). IEEE, pp. 151–158.

Motahar, M.S., Veer, S., Huang, J., Poulakakis, I., 2015b. Integrating dynamic walking and arm impedance control for cooperative transportation. In: Proceedings of the IEEE/RSJ International Conference on Intelligent Robots and Systems. Hamburg, Germany, pp. 1004–1010.

Motahar, M.S., Veer, S., Poulakakis, I., 2016. Composing limit cycles for motion planning of 3D bipedal walkers. In: Proceedings of IEEE Conference on Decision and Control, pp. 6368–6374.

Murray, R.M., Li, Z., Sastry, S.S., 1994. A Mathematical Introduction to Robotic Manipulation. CRC Press, Boca Raton.

Nguyen, Q., Hereid, A., Grizzle, J.W., Ames, A.D., Sreenath, K., 2016. 3D dynamic walking on stepping stones with control barrier functions. In: 2016 IEEE 55th Conference on Decision and Control (CDC). IEEE, pp. 827–834.

Nguyen, Q., Sreenath, K., 2015. Safety-critical control for dynamical bipedal walking with precise footstep placement. IFAC-PapersOnLine 48, 147–154.

Nguyen, Q., Sreenath, K., 2016. Optimal robust control for constrained nonlinear hybrid systems with application to bipedal locomotion. In: American Control Conference (ACC). IEEE, pp. 4807–4813.

Park, H.-W., Ramezani, A., Grizzle, J.W., 2013. A finite-state machine for accommodating un-expected large ground-height variations in bipedal robot walking. IEEE Trans. Robot. 29, 331–345.

Poulakakis, I., 2008. Stabilizing Monopedal Robot Running: Reduction-by-Feedback and Compli-ant Hybrid Zero Dynamics. PhD thesis. Department of Electrical Engineering and Computer Science, University of Michigan.

Poulakakis, I., 2010. Spring loaded inverted pendulum embedding: extensions toward the control of compliant running robots. In: Proceedings of the IEEE International Conference on Robotics and Automation, pp. 5219–5224.

Poulakakis, I., Grizzle, J., 2007a. Formal embedding of the Spring Loaded Inverted Pendulum in an Asymmetric hopper. In: Proceedings of the European Control Conference. Kos, Greece.

Poulakakis, I., Grizzle, J.W., 2007b. Monopedal running control: SLIP embedding and virtual con-straint controllers. In: Proceedings of the IEEE/RSJ International Conference of Intelligent Robots and Systems. San Diego, USA, pp. 323–330.

Poulakakis, I., Grizzle, J.W., 2009a. Modeling and control of the monopedal running robot thumper. In: Proceedings of the IEEE International Conference on Robotics and Automation. Kobe, Japan, pp. 3327–3334.

Poulakakis, I., Grizzle, J.W., 2009b. The spring loaded inverted pendulum as the hybrid zero dy-namics of an asymmetric hopper. IEEE Trans. Autom. Control 54, 1779–1793.

Powell, M.J., Hereid, A., Ames, A.D., 2013. Speed regulation in 3D robotic walking through motion transitions between human-inspired partial hybrid zero dynamics. In: 2013 IEEE International Conference on Robotics and Automation (ICRA). IEEE, pp. 4803–4810.

Powell, M.J., Zhao, H., Ames, A.D., 2012. Motion primitives for human-inspired bipedal robotic locomotion: walking and stair climbing. In: 2012 IEEE International Conference on Robotics and Automation (ICRA). IEEE, pp. 543–549.

Raibert, M.H., 1986. Legged Robots that Balance. MIT Press, Cambridge, MA.

Reher, J., Cousineau, E.A., Hereid, A., Hubicki, C.M., Ames, A.D., 2016a. Realizing dynamic and efficient bipedal locomotion on the humanoid robot DURUS. In: 2016 IEEE International Con-ference on Robotics and Automation (ICRA). IEEE, pp. 1794–1801.

Reher, J.P., Hereid, A., Kolathaya, S., Hubicki, C.M., Ames, A.D., 2016b. Algorithmic foundations of realizing multi-contact locomotion on the humanoid robot DURUS. In: Twelfth International Workshop on Algorithmic Foundations on Robotics.

Sastry, S., 1999. Nonlinear Systems: Analysis, Stability and Control. Springer-Verlag.

Sinnet, R.W., Jiang, S., Ames, A.D., 2014. A human-inspired framework for bipedal robotic walking design. Int. J. Biomechatronics Biomedical Robot. 3, 20–41.

Sinnet, R.W., Powell, M.J., Shah, R.P., Ames, A.D., 2011a. A human-inspired hybrid control ap-proach to bipedal robotic walking. IFAC Proceedings 44, 6904–6911.

Sinnet, R.W., Zhao, H., Ames, A.D., 2011b. Simulating prosthetic devices with human-inspired hybrid control. In: 2011 IEEE/RSJ International Conference on Intelligent Robots and Systems (IROS). IEEE, pp. 1723–1730.

Spong, M.W., Hutchinson, S., Vidyasagar, M., 2006. Robot Modeling and Control. John Wiley & Sons, New York.

Sreenath, K., 2011. Feedback Control of a Bipedal Walker and Runner with Compliance. PhD thesis. University of Michigan.

Sreenath, K., Park, H.-W., Poulakakis, I., Grizzle, J., 2011. A compliant hybrid zero dynamics controller for stable, efficient and fast bipedal walking on MABEL. Int. J. Robot. Res. 30, 1170–1193.

Sreenath, K., Park, H.-W., Poulakakis, I., Grizzle, J., 2013. Embedding active force control within the compliant hybrid zero dynamics to achieve stable, fast running on MABEL. Int. J. Robot. Res. 32.

van der Schaft, A., Schumacher, H., 2000. An Introduction to Hybrid Dynamical Systems. Lecture Notes in Control and Information Sciences, vol. 251. Springer-Verlag.

Vasudevan, R., Ames, A., Bajcsy, R., 2011. Using persistent homology to determine a human-data based cost for bipedal walking. In: 18th IFAC World Congress. Milano, Italy.

Vasudevan, R., Ames, A., Bajcsy, R., 2013. Persistent homology for automatic determination of human-data based cost of bipedal walking. Nonlinear Anal. Hybrid Syst. 7, 101–115.

Veer, S., Motahar, M.S., Poulakakis, I., 2015. On the adaptation of dynamic walking to persistent external forcing using hybrid zero dynamics control. In: Proceedings of the IEEE/RSJ International Conference on Intelligent Robots and Systems. Hamburg, Germany, pp. 997–1003.

Veer, S., Motahar, M.S., Poulakakis, I., 2016. Local input-to-state stability of dynamic walking under persistent external excitation using hybrid zero dynamics. In: Proceedings of the American Control Conference, pp. 4801–4806.

Veer, S., Motahar, M.S., Poulakakis, I., 2017. Almost driftless navigation of 3D limit-cycle walking bipeds. In: Proc. of IEEE/RSJ Int. Conf. on Intelligent Robots and Systems.

Vukobratović, M., Borovac, B., Potkonjak, V., 2006. ZMP: a review of some basic misunderstandings. Int. J. Humanoid Robot. 3, 153–175.

Wang, L., Ames, A., Egerstedt, M., 2016a. Safety barrier certificates for heterogeneous multi-robot systems. In: American Control Conference (ACC). IEEE, pp. 5213–5218.

Wang, L., Ames, A.D., Egerstedt, M., 2016b. Multi-objective compositions for collision-free connectivity maintenance in teams of mobile robots. In: 2016 IEEE 55th Conference on Decision and Control (CDC). IEEE, pp. 2659–2664.

Wendel, E., Ames, A.D., 2012. Rank deficiency and superstability of hybrid systems. Nonlinear Anal. Hybrid Syst. 6, 787–805.

Wendel, E.D., Ames, A.D., 2010. Rank properties of Poincaré maps for hybrid systems with applications to bipedal walking. In: Proceedings of the 13th ACM International Conference on Hybrid Systems: Computation and Control. ACM, pp. 151–160.

Westervelt, E.R., 2003. Toward a Coherent Framework for the Control of Planar Biped Locomotion. PhD thesis. University of Michigan.

Westervelt, E.R., Buche, G., Grizzle, J.W., 2004. Experimental validation of a framework for the design of controllers that induce stable walking in planar bipeds. Int. J. Robot. Res. 23, 559–582.

Westervelt, E.R., Grizzle, J.W., Chevallereau, C., Choi, J.H., Morris, B., 2007. Feedback Control of Dynamic Bipedal Robot Locomotion. Taylor & Francis/CRC Press.

Westervelt, E.R., Grizzle, J.W., Koditschek, D.E., 2003. Hybrid zero dynamics of planar biped walkers. IEEE Trans. Autom. Control 48, 42–56.

Xu, X., Tabuada, P., Grizzle, J.W., Ames, A.D., 2015. Robustness of control barrier functions for safety critical control. IFAC-PapersOnLine 48, 54–61.

Zhao, H., Ambrose, E., Ames, A.D., 2017. Preliminary results on energy efficient 3D prosthetic walking with a powered compliant transfemoral prosthesis. In: 2017 IEEE International Conference on Robotics and Automation (ICRA). IEEE.

Zhao, H., Hereid, A., Ambrose, E., Ames, A.D., 2016a. 3D multi-contact gait design for prostheses: hybrid system models, virtual constraints and two-step direct collocation. In: 2016 IEEE 55th Conference on Decision and Control (CDC). IEEE, pp. 3668–3674.

Zhao, H., Hereid, A., Ma, W.-l., Ames, A.D., 2015a. Multi-contact bipedal robotic locomotion. Robotica, 1–35.

Zhao, H., Horn, J., Reher, J., Paredes, V., Ames, A.D., 2011. First steps toward translating robotic walking to prostheses: a nonlinear optimization based control approach. Auton. Robots, 1–18.

Zhao, H., Horn, J., Reher, J., Paredes, V., Ames, A.D., 2016b. Multicontact locomotion on transfemoral prostheses via hybrid system models and optimization-based control. IEEE Trans. Autom. Sci. Eng. 13, 502–513.

Zhao, H., Kolathaya, S., Ames, A.D., 2014a. Quadratic programming and impedance control for transfemoral prosthesis. In: 2014 IEEE International Conference on Robotics and Automation (ICRA). IEEE, pp. 1341–1347.

Zhao, H., Powell, M., Ames, A., 2014b. Human-inspired motion primitives and transitions for bipedal robotic locomotion in diverse terrain. Optim. Control Appl. Methods 35, 730–755.

Zhao, H., Reher, J., Horn, J., Paredes, V., Ames, A.D., 2015b. Realization of nonlinear real-time optimization based controllers on self-contained transfemoral prosthesis. In: Proceedings of the ACM/IEEE Sixth International Conference on Cyber-Physical Systems. ACM, pp. 130–138.

Chapter 4.8

Robot Locomotion Control Based on Central Pattern Generators

Auke Ijspeert

Biorobotics Laboratory, EPFL – Ecole Polytechnique Fédérale de Lausanne, Lausanne, Switzerland

4.8.1 INTRODUCTION

The locomotion control circuits of animals, both vertebrate and invertebrate, contain networks of neuronal oscillators called *central pattern generators* (CPGs) that can produce coordinated rhythmic patterns of neural activity (Grillner, 2006), see also Subchapter 7.2. From a control point of view, CPGs can be viewed as some kind of feed-forward controllers that can produce high-dimensional rhythmic muscle activation patterns while receiving low-dimensional signals such as a desired speed of locomotion as inputs.

The biological concept of CPG can be mathematically modeled as systems of coupled nonlinear oscillators, and forms an interesting approach for controlling a large variety of robots from swimming to walking robots (see Ijspeert, 2008 for a review). The approach has been applied to several quadruped and biped robots, and is especially suited for legged robots with interesting natural dynamics due to pendulum and/or elastic components (i.e., robots that will naturally produce oscillations). For those robots, the coupling of a CPG with a mechanical body can be viewed as two systems of oscillators, one neuronal and one mechanical, coupled together. With the right design, the whole system can lead to mechanical entrainment (Taga et al., 1991, 1991), with a synchronized regime, i.e., a common resulting frequency. A key concept of CPG-based control is therefore to view locomotion as a stable limit cycle behavior of these two coupled systems, and to design the CPG controller such that the limit cycle has a large basin of attraction (in order to be robust against perturbations, e.g., from the environment) and that it can be modulated by simple control signals, e.g., for changing speed or the type of gait, for instance, with transitions between walk, trot, and gallop.

4.8.2 CENTRAL PATTERN GENERATORS IN ANIMALS

Central pattern generator networks for locomotion have been identified in many animals, both vertebrate (Grillner, 2006) and invertebrate (Getting, 1988). A key demonstration of the existence of CPGs was done by isolating CPG circuits in a petri dish and inducing rhythmic neural activity using excitatory chemicals. For instance, an isolated spinal cord of the lamprey can produce traveling waves of neural activity that resemble those produced during intact swimming. Such coordinated patterns are called *fictive locomotion* and have been observed, for instance, in lamprey (Cohen and Wallen, 1980), salamander (Delvolvé et al., 1999), and neonatal rat (Cazalets et al., 1992). These experiments using subparts of the locomotor circuits have also shown that CPGs are distributed networks made of multiple coupled local neuronal oscillators, for example, with two neural oscillators per segment in the lamprey, and specific oscillators for limb flexor and extensors in salamander (Cheng et al., 1998). As proposed by Sten Grillner, a CPG can therefore be seen as a system of coupled oscillators (that he calls unit-burst generators) with typically one or two oscillators per degree of freedom in the animal's musculoskeletal system (Grillner, 2006).

Other indirect demonstrations of the existence of CPGs have been made with paralyzed or deafferented (i.e., without sensory feedback) animals. This type of fictive locomotion has been observed in cat (Jordan et al., 1979) and monkeys (Fedirchuk et al., 1998).

An interesting property of CPGs is that they can modulate locomotion and even produce different gaits under the control of simple signals. This has been demonstrated with electrical stimulation of a specific region of the brain stem, the mesencephalic locomotor region (MLR) that directly projects to the locomotor CPGs in the spinal cord of vertebrate animals. Electrical stimulation of the MLR can indeed generate walk, trot, and gallop in a decerebrated cat (Shik et al., 1966) and transition between swimming-like and walking-like gaits in the salamander (Cabelguen et al., 2003).

There is still an ongoing debate whether human locomotion relies on CPGs. Some researchers propose that human locomotion does not use CPGs and only relies on sensory feedback and descending signals from higher part of the brain, while others propose that human locomotion uses CPGs like most vertebrate animals. See MacKay-Lyons (2002) for discussions and Danner et al. (2015) for a recent article suggesting the existence of CPGs in humans. Numerical models of biological CPGs can help investigating such questions. Taga developed a neuromechanical model of biped locomotion that combines a CPG model, reflexes, and a simple two-dimensional mechanical model of the body (Taga et al., 1991). That model, which is described in more details below, was influential as it demonstrated how mechanical entrainment could be obtained between the CPG and the body, and how the whole system could exhibit limit cycle behavior. Taga demonstrated that the model could handle perturbations and that it could receive control inputs for modulating speed, ground clearance and step length (Taga, 1998). Geyer and Herr developed a similar but purely sensory-driven model of human locomotion, with more realistic muscle models and without CPGs (Geyer and Herr, 2010). They could demonstrate that robust human-like gaits (e.g., in terms of kinematics, ground reaction forces and even EMG signals) can be produced without CPGs, solely relying on a set of reflexes for generating rhythms and stable locomotion. Dzeladini et al. (2014) extended that model by adding CPGs in the network. That article tested the hypothesis that the addition of CPG could simplify the control of speed, compared to the sensory-driven controller that needs multiple reflex gains to be retuned for reaching different speeds. It was found that a relatively large range of walking speeds could be obtained by simply modulating the frequency of the CPG, supporting the hypothesis and suggesting a useful contribution of CPGs for locomotion control. This is in agreement with the idea that body velocity is the command sent to CPG as proposed by Prochazka and colleagues (see Chapter 7, and Prochazka and Ellaway, 2012; Prochazka and Yakovenko, 2007). In Song and Geyer (2015), the 2D sensory-driven model of Geyer and Herr (2010) is extended to 3D and a supraspinal controller is added that continuously adjusts reflex gains and leg target angles. Thanks to that supraspinal controller, the 3D model can exhibit different adaptive locomotion behaviors, "including walking and running, acceleration and deceleration, slope and stair negotiation, turning, and deliberate obstacle avoidance" (Song and Geyer, 2015).

Note that Art Kuo presents an interesting analysis of the benefits of combining feedback (i.e., sensory-driven) and feedforward (CPG-based) control (Kuo, 2002). He also investigates an alternative view of the role of CPGs in biological systems: he proposes that "the neural oscillator as a filter for processing sensory information rather than as a generator of commands." In that view, the CPG helps to perform state estimation and "acts as an internal model of limb motion that predicts the state of the limb" (Kuo, 2002).

More generally, there is a large amount of work on mathematical and numerical modeling of biological CPGs, see Ijspeert (2008). As we will see next, such models can also serve as robot controllers.

4.8.3 CPGS AS ROBOT CONTROLLERS

There are many examples of legged and articulated robots controlled by CPG models for various modes of locomotion. In the next sections, we will briefly review different types of CPG implementations, different examples of robot locomotion, and different design methods.

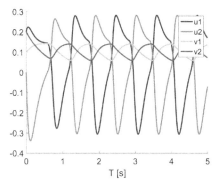

FIGURE 4.8.1 Matsuoka oscillator. Time evolution of the 4 state variables of the Matsuoka oscillator.

4.8.3.1 Different Types of Implementation

The type and complexity of CPG controllers for robot locomotion vary from detailed spiking neural network models (Lewis et al., 2005), to leaky integrator neural networks (Arena, 2000; Lu et al., 2006), to systems of coupled nonlinear oscillators (Kimura et al., 1999; Ijspeert et al., 2007). One popular approach, in between neural network and systems of coupled oscillators, is based on Matsuoka oscillators (Matsuoka, 1985, 1987). A Matsuoka oscillator is typically composed of 2 leaky-integrator neurons with mutual inhibition and a fatigue mechanism. It has 4 state variables and its dynamics is determined by differential equations that are linear except for a *max* operator that makes the whole system nonlinear and capable of producing limit cycle behavior:

$$\tau \dot{u}_1 = -u_1 - wy_2 - \beta v_1 + u_0,$$
$$\tau \dot{u}_2 = -u_2 - wy_1 - \beta v_2 + u_0,$$
$$\tau' \dot{v}_1 = -v_1 + y_1,$$
$$\tau' \dot{v}_2 = -v_2 + y_2,$$
$$y_i = \max(0, u_i) \quad (i = 1, 2),$$

where y_i is the output of the ith neuron; u_i is its inner state variable; v_i is a second state variable representing a degree of self-inhibition; u_0 is an external input, w is a connection weight, and τ and τ' are time constants. Because of its simplicity and its robust limit cycle behavior, it is probably the type of oscillator that has been mostly used to design CPG controllers for robots (Kimura et al., 1999, 2007; Williamson, 1998). (See Fig. 4.8.1.)

Other types of oscillators have been used such as Van der Pol, Hopf oscillators, (Kuramoto) phase oscillators, and Rayleigh oscillators (Ijspeert, 2008).

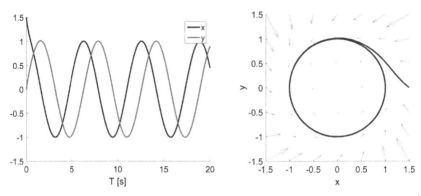

FIGURE 4.8.2   Hopf oscillator. (Left) Time evolution of the two state variables of the Hopf oscillator. (Right) Phase portrait. Notice that the limit cycle has a circular shape (harmonic oscillation).

Among these, the phase and Hopf oscillators are among the simplest possible models. A phase oscillator has only one state variable, the phase θ_i,

$$\dot{\theta}_i = \omega_i + \sum_j w_{ij} \sin(\theta_j - \theta_i - \phi_{ij}),$$

where ω_i determines the oscillation frequency, w_{ij} is the weight of a connection coming from another oscillator j, and ϕ_{ij} is a phase bias term. Such an oscillator allows one to modulate relative phases between actuated joints and can produce periodic joint angle signals through simple periodic functions such as $x = A\cos(\theta)$, where A defines the amplitude of oscillations, or more complex wave forms using arbitrary periodic functions.

The Hopf oscillator has 2 state variables. It has a harmonic limit cycle (i.e., a perfect circle in Cartesian space), and it can bifurcate between a limit cycle regime and a stable attractor point regime depending on the sign of a control parameter μ:

$$\dot{x} = \left(\mu - r^2\right)x - \omega y,$$

$$\dot{y} = \left(\mu - r^2\right)y + \omega x,$$

$$r = \sqrt{x^2 + y^2},$$

where x and y are the two state variables and ω determines the oscillation frequency. (See Fig. 4.8.2.)

Most CPG controllers for robots are implemented as systems of differential equations that are numerically solved on a microprocessor or a microcontroller on board of the robot. The computation can be distributed with different oscillators being implemented on different robot modules (Conradt and Varshavskaya,

2003; Inagaki et al., 2006; Sproewitz et al., 2008). A few CPGs have been implemented directly on a chip with digital or analog electronics (DeWeerth et al., 1997; Still and Tilden, 1998; Nakada et al., 2003; Lewis et al., 2005; Simoni and DeWeerth, 2007).

4.8.3.2 Examples of CPG Controllers

In this section, a few examples of projects will be presented in order to illustrate what type of research has been carried out in developing CPG controllers for legged robots. The list is not meant to be comprehensive by any means.

Taga's Neuromechanical Simulation of Biped Locomotion

Taga and colleagues developed a CPG controller for a two-dimensional biped simulator (Taga et al., 1991, 1993; Taga, 1998). While only in simulation, that seminal work influenced many researchers in robotics. As mentioned above, it was one of the first projects to demonstrate how mechanical entrainment between control oscillators and a body could lead to stable limit cycle behavior for locomotion. The CPG model is constructed out of coupled Matsuoka oscillators, one per joint. More specifically, the torque of each joint is computed from the difference of signals produced by the two, flexor and extensor, neurons forming the oscillator (Fig. 4.8.3). Oscillators are coupled together through neuronal connections and receive feedback from the body and environment through reflexes such as the stretch reflex. The model can generate stable walking gaits in the sagittal plane and resist to perturbations (e.g., pushes and slopes of the ground) without any parameter change. It could even switch to a running gait when excitatory input to the CPG model is increased. In a follow-up article (Taga, 1998), Taga demonstrated how the model could be extended to perform anticipatory behavior, such as stepping over an obstacle. For this, discrete signals are provided for the CPG model in order to adjust step length and ground clearance at specific moments. This demonstrates how a CPG model can be modulated by higher control centers for adaptive, visually-guided locomotion.

CPG Models for Quadruped Locomotion

Several CPG controllers for the locomotion of compliant quadruped robots have been developed by Kimura and colleagues (Kimura et al., 1999, 2001, 2007; Fukuoka et al., 2003). The CPG controllers are composed of coupled Matsuoka oscillators with a set of reflexes based on body orientation, tendon force, and contact with the floor. In Kimura et al. (2001) they investigated how sensory feedback could best be integrated into the CPG and compared two different

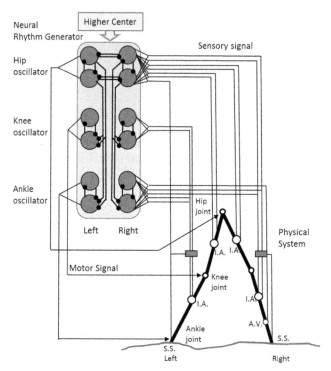

FIGURE 4.8.3 Taga's CPG-controlled bipedal model in two dimensions (adapted from Taga et al., 1993). The model is made of a 2D mechanical model of a biped, a CPG (the neural rhythm generator) and several sensory feedback loops. I.A. stands for inertial angle, A.V. for angular velocity, and S.S. for somatic sense.

implementations: one in which sensory feedback acts independently from the CPG and one in which sensory feedback is fed through the CPG network.

The second implementation led to the most stable locomotion in complex terrain. They concluded that the interaction between sensory feedback and the CPG is important for adjusting the cycle duration to the environment (e.g., when going up or down a slope) and for making sensory feedback phase-dependent. Reflexes in animals are known to be similarly phase-dependent, e.g., the stumbling correction reflex is only active during swing (Andersson et al., 1978; Forssberg, 1979; Pearson, 1995). In a later paper, the authors added different types of reflexes and obtained robust outdoors locomotion (Kimura et al., 2007).

While most CPG models rely heavily on direct interoscillator couplings for generating coordinated patterns, it is possible to obtain stable gaits and even gait transitions based only on indirect mechanical coupling, similarly to what has been observed in the stick insect (Cruse, 1990). This was demonstrated by Owaki and colleagues on a quadruped robot (Owaki et al., 2013). Each limb of the robot is controlled by a phase oscillator that receives a load-dependent feedback signal from the same limb:

$$\dot{\theta}_i = \omega - \sigma N_i \cos(\theta_i),$$

where the θ_i is the phase, ω is the intrinsic frequency of the oscillator, N_i the normal force felt by the limb, and σ is a gain that determines the strength of the sensory feedback. The feedback signal slows down or even prevents oscillations of the limb as long as there is load on the limb. This creates an indirect mechanical coupling between the phase oscillators. Interestingly stable gaits emerge from that coupling, and different gaits (e.g., trot, pace, diagonal sequence walk, lateral sequence walk) can be generated depending on the mass distribution in the robot. When the mass is placed more in the front, as in camels, or more to the rear, as in monkeys, the same gaits emerge as in their biological counterparts, respectively lateral sequence walk and diagonal sequence walk.

Other examples of quadruped robots controlled by CPGs include (Tsujita et al., 2001; Buchli et al., 2006a, 2006b; Manoonpong et al., 2007; Ajallooeian et al., 2013a).

CPG Models for Biped Locomotion

There are many examples of biped robots controlled by CPG-controllers (Aoi and Tsuchiya, 2005; Endo et al., 2005; Miyakoshi et al., 1998; Shan and Nagashima, 2002; Nakanishi et al., 2004; Righetti and Ijspeert, 2006; Zaier and Nagashima, 2006).

Aoi and Tsuchiya (2005) for instance designed a CPG controller for the HOAP-1 robot. The controller was based on phase oscillators with a phase resetting mechanism at heel strike for synchronizing the oscillators with the robot leg motions. The robot could change speed by modulating the stride length and handle perturbations such as small slopes and small pushes against the trunk. A similar approach that replaces the phase resetting mechanism (which can be abrupt) with a smoother estimation of the natural phase of the robot dynamics was proposed in Morimoto et al. (2006).

Endo and colleagues investigated how reinforcement learning could be applied to train a CPG controller for Sony's humanoid robot Qrio (Endo et al., 2008). The CPG model was based on Matsuoka oscillators, and a policy-gradient method was used to set the parameters in the model, in particular in the feedback loops, using learning iterations in simulation followed by learning iterations on the real robot. The reward function was designed to favor upright forward motion, with negative rewards when falling. Stable walking with the real robot could be obtained after a few thousands of iteration in simulation. The controllers could be improved further with online learning for 200 iterations (2.5 hours) on the real robot.

CPGs for Amphibious Locomotion

CPG controllers are well-suited for many modes of locomotion, in particular for producing coordinating patterns for swimming for instance anguilliform swimming in lamprey-like robots (Arena, 2001; Crespi and Ijspeert, 2008; Stefanini et al., 2012). They have also been used to control amphibious robots such as salamander-like robots. Salamanders can swim using anguilliform swimming gaits like eels and lampreys and walk on dry ground using walking trot gaits. Based on neurophysiological findings, we developed a CPG model for salamander that could explain the generation and transition between these two gaits (Ijspeert et al., 2007). The model is made of amplitude-controlled phase oscillators, i.e. phase oscillators with an additional state variable for amplitude, with a saturation function that determines the intrinsic frequency and amplitude of oscillation depending on an input signal. The main idea behind the model is that the bi-modal locomotion of salamander can be obtained by adding limb oscillators with lower intrinsic frequency and lower saturation frequencies to a lamprey swimming circuit made of a double chain of coupled oscillators. The model could generate realistic gaits in water and on ground, and dynamically modulate the speed, direction, and type of gait under the control of simple signals sent from a remote control. Interestingly, these gaits could be obtained in open-loop because of the intrinsic stability offered by surface swimming and walking with a sprawled posture. It is clear that sensory feedback can play an important role in modulating the locomotor patterns (Knuesel et al., 2010), and this is currently under investigation.

4.8.3.3 Design Methods for CPG Controllers

Different approaches have been taken to design CPG controllers. Many approaches are guided by the theory of dynamical systems, which can provide guidelines for instantiating parameters such as intrinsic frequencies and coupling weights in order to obtain synchronization and the generation of desired gaits (Buchli et al., 2006a; Golubitsky et al., 1998, 1999; Pham and Slotine, 2007; Seo and Slotine, 2007; Righetti and Ijspeert, 2008; Schoner et al., 1990).

Unsupervised learning approaches have also been followed. These approaches are useful when the desired behavior of the robot controller is only partially known, typically with only high-level desired characteristics such as moving fast without falling. Examples include the use of reinforcement algorithms (Matsubara et al., 2006; Nakamura et al., 2007; Ogino et al., 2004) and heuristic optimization algorithms such as Powell's method (Crespi and Ijspeert, 2008; Marbach and Ijspeert, 2005; Sproewitz et al., 2008). Most of these approaches have been applied online, i.e., directly during runtime on the robot. An alternative approach is to perform some optimization offline, using dynamic simulators of the robot, before porting them online (Van der Noot et al., 2015). Stochastic optimization algorithms such as genetic algorithms and particle swarm optimization are often used in such case (Beer and Gallagher, 1992; Sims, 1994; Gruau and Quatramaran, 1997; Ijspeert, 2001; Kamimura et al., 2003; Lewis et al., 1993; Van der Noot et al., 2015).

When a desired output pattern is available (e.g., from animal gait recordings or other insights), supervised learning approaches can also be taken. In that case, the desired periodic pattern can be used to define an explicit error function to be minimized. The learning can be done for instance using statistical learning algorithms (Ijspeert et al., 2013), gradient descent algorithms for recurrent neural networks (Pearlmutter, 1995), or instantiations of vector fields (Okada et al., 2002). With clever mappings, it is possible to generate arbitrary limit cycle shapes while keeping the underlying dynamics low-dimensional (Ajallooeian et al., 2013a, 2013b). Alternatively, arbitrary limit cycle shapes can be learned by pools of adaptive frequency oscillators (Righetti et al., 2006; Righetti and Ijspeert, 2006).

4.8.4   DISCUSSION

As discussed in this chapter, CPGs present several interesting properties as locomotion controllers for robots. They can exhibit stable limit cycle behavior. They are well-suited to be coupled to and entrained by a mechanical system. They allow for smooth modulation of locomotion (i.e., their limit cycle properties typically act as filters of abrupt input signal changes). They can be implemented in a distributed fashion (e.g., coupled oscillators on different microcontrollers). And they offer a good substrate for learning algorithms (online or offline).

These interesting properties are probably also the reasons why CPGs evolved in biological systems and why they are found in so many animals both vertebrate and invertebrate. CPGs complement feedback loops based on reflexes by adding a feedforward component to the whole locomotor circuitry. As discussed earlier this is useful for generating locomotor patterns that can then be shaped by sensory feedback, for handling noise in the sensory signals (Kuo, 2002) and for simplifying the control of speed (Dzeladini et al., 2014). Note that unlike what is sometimes believed, CPGs do not need to produce stereotyped behavior; and many biological and robotic CPGs can be extensively modulated to produce rich motor behavior (e.g., gait transitions between multiple gaits).

In robotics, CPG controllers are well-suited for fast locomotion, robust locomotion on unstructured terrains, compliant robots, modular/reconfigurable robots, and robots for which an accurate dynamical model does not exist. They also have a great potential for the control of prostheses and exoskeletons (Ronsse et al., 2011).

They are not so well-suited for accurate feet placement, accurate full-body control, and rich motor skills (e.g., rapid transitions and superimposition of different motor behaviors). Also if an accurate dynamic model of the robot and its environment exist, alternative model-based control approaches such as optimal control might be better alternatives.

Both in terms of biology and robotics, much research remains to be done to better decode the functioning of CPGs and the design of better CPG-based controllers for robots. For instance, designing a CPG-based controller for a particular robot remains a bit of an art, and more generic design methods would be useful. Also it is not yet clear how to generate the rich motor skills exhibited by animals, with combinations of discrete and rhythmic movements, rapid transients, and superimposition of different motor behaviors. One promising approach for this is the concept of a modular control architecture made of motor primitives, i.e., building blocks of motor behavior that can be combined in several ways (Thoroughman and Shadmehr, 2000; Flash and Hochner, 2005). CPGs could be viewed as one particular type of motor primitive for generating periodic behavior, which can be combined with other primitives for richer motor behavior.

4.8.5 CONCLUSION

CPG-based control is an interesting biologically-inspired control approach for robotics. It offers interesting control properties for some types of locomotion and of robots, typically fast locomotion of compliant robots. It is also a research area in which robotics can usefully collaborate with neurobiology and biomechanics in order to decode the interactions between CPGs, sensory feedback, descending modulation, and musculoskeletal properties that underlie the amazing locomotor abilities of animals.

REFERENCES

Ajallooeian, M., Gay, S., Tuleu, A., Sproewitz, A., Ijspeert, A.J., 2013a. Modular control of limit cycle locomotion over unperceived rough terrain. In: 2013 IEEE/RSJ International Conference on Intelligent Robots and Systems (IROS 2013), pp. 3390–3397.

Ajallooeian, M., van den Kieboom, J., Mukovskiy, A., Giese, M.A., Ijspeert, A.J., 2013b. A general family of morphed nonlinear phase oscillators with arbitrary limit cycle shape. Physica D: Nonlinear Phenomena 263, 41–56.

Andersson, O., Forssberg, H., Grillner, S., Lindquist, M., 1978. Phasic gain control of the transmission in cutaneous reflex pathways to motoneurones during "fictive" locomotion. Brain Res. 149 (2), 503–507.

Aoi, S., Tsuchiya, K., 2005. Locomotion control of a biped robot using nonlinear oscillators. Auton. Robots 19, 219–232.

Arena, P., 2000. The central pattern generator: a paradigm for artificial locomotion. Soft Comput. 4 (4), 251–266.

Arena, P., 2001. A mechatronic lamprey controlled by analog circuits. In: Proceedings of the 9th IEEE Mediterranean Conference on Control and Automation (MED '01).

Beer, R.D., Gallagher, J.C., 1992. Evolving dynamical neural networks for adaptive behavior. Adapt. Behav. 1 (1), 91–122.

Buchli, J., Iida, F., Ijspeert, A.J., 2006a. Finding resonance: adaptive frequency oscillators for dynamic legged locomotion. In: Proceedings of the IEEE/RSJ International Conference on Intelligent Robots and Systems (IROS 2006). IEEE, pp. 3903–3909.

Buchli, J., Righetti, L., Ijspeert, A.J., 2006b. Engineering entrainment and adaptation in limit cycle systems – from biological inspiration to applications in robotics. Biol. Cybern. 95 (6), 645–664.

Cabelguen, J.M., Bourcier-Lucas, C., Dubuc, R., 2003. Bimodal locomotion elicited by electrical stimulation of the midbrain in the salamander *Notophthalmus viridescens*. J. Neurosci. 23 (6), 2434–2439.

Cazalets, J.R., Sqalli-Houssaini, Y., Clarac, F., 1992. Activation of the central pattern generators for locomotion by serotonin and excitatory amino acids in neonatal rat. J. Physiol. 455, 187.

Cheng, J., Stein, R.B., Jovanovic, K., Yoshida, K., Bennett, D.J., Han, Y., 1998. Identification, localization, and modulation of neural networks for walking in the mudpuppy (*Necturus Maculatus*) spinal cord. J. Neurosci. 18 (11), 4295–4304.

Cohen, A.H., Wallen, P., 1980. The neural correlate of locomotion in fish. "Fictive swimming" induced in an in vitro preparation of the lamprey spinal cord. Exp. Brain Res. 41, 11–18.

Conradt, J., Varshavskaya, P., 2003. Distributed central pattern generator control for a serpentine robot. In: International Conference on Artificial Neural Networks (ICANN 2003).

Crespi, A., Ijspeert, A.J., 2008. Online optimization of swimming and crawling in an amphibious snake robot. IEEE Trans. Robot. 24 (1), 75–87.

Cruse, H., 1990. What mechanisms coordinate leg movement in walking arthropods? Trends Neurosci. 13 (1), 15–21.

Danner, S.M., Hofstoetter, U.S., Freundl, B., Binder, H., Mayr, W., Rattay, F., Minassian, K., 2015. Human spinal locomotor control is based on flexibly organized burst generators. Brain 138 (3), 577–588. https://doi.org/10.1093/brain/awu372.

Delvolvé, I., Branchereau, P., Dubuc, R., Cabelguen, J.-M., 1999. Fictive rhythmic motor patterns induced by NMDA in an in vitro brain stem-spinal cord preparation from an adult urodele. J. Neurophysiol. 82, 1074–1077.

DeWeerth, S.P., Patel, G.N., Simoni, M.F., Schimmel, D.E., Calabrese, R.L., 1997. A VLSI architecture for modeling intersegmental coordination. In: Brown, R., Ishii, A. (Eds.), 17th Conference on Advanced Research in VLSI. IEEE Computer Society, pp. 182–200.

Dzeladini, F., van den Kieboom, J., Ijspeert, A., 2014. The contribution of a central pattern generator in a reflex-based neuromuscular model. Front. Human Neurosci. 8. http://dx.doi.org/10.3389/fnhum.2014.00371.

Endo, G., Morimoto, J., Matsubara, T., Nakanishi, J., Cheng, G., 2008. Learning CPG-based biped locomotion with a policy gradient method: application to a humanoid robot. Int. J. Robot. Res. 27 (2), 213–228.

Endo, G., Nakanishi, J., Morimoto, J., Cheng, G., 2005. Experimental studies of a neural oscillator for biped locomotion with Q RIO. In: Proceedings of the 2005 IEEE International Conference on Robotics and Automation (ICRA2005). Barcelona, Spain, pp. 598–604.

Fedirchuk, B., Nielsen, J., Petersen, N., Hultborn, H., 1998. Pharmacologically evoked fictive motor patterns in the acutely spinalized marmoset monkey (*Callithrix jacchus*). Exp. Brain Res. 122 (3), 351–361. http://dx.doi.org/10.1007/s002210050523.

Flash, T., Hochner, B., 2005. Motor primitives in vertebrates and invertebrates. Curr. Opin. Neurobiol. 15 (6), 660–666.

Forssberg, H., 1979. Stumbling corrective reaction: a phase-dependent compensatory reaction during locomotion. J. Neurophysiol. 42 (4), 936–953.

Fukuoka, Y., Kimura, H., Cohen, A.H., 2003. Adaptive dynamic walking of a quadruped robot on irregular terrain based on biological concepts. Int. J. Robot. Res. 3–4, 187–202.

Getting, P.A., 1988. Comparative analysis of invertebrate central pattern generators. In: Cohen, A.H., Rossignol, S., Grillner, S. (Eds.), Neural Control of Rhythmic Movements in Vertebrates. Jon Wiley & Sons, pp. 101–127.

Geyer, H., Herr, H., 2010. A muscle-reflex model that encodes principles of legged mechanics produces human walking dynamics and muscle activities. IEEE Trans. Neural Syst. Rehabil. Eng. 18 (3), 263–273. http://dx.doi.org/10.1109/TNSRE.2010.2047592.

Golubitsky, M., Stewart, I., Buono, P., Collins, J.J., 1998. A modular network for legged locomotion. Physica D: Nonlinear Phenomena 115 (1–2), 56–72.

Golubitsky, M., Stewart, I., Buono, P.-L., Collins, J.J., 1999. Symmetry in locomotor central pattern generators and animal gaits. Nature 401, 693–695.

Grillner, S., 2006. Biological pattern generation: the cellular and computational logic of networks in motion. Neuron 52 (5), 751–766.

Gruau, F., Quatramaran, K., 1997. Cellular encoding for interactive evolutionary robotics. In: Husbands, P., Harvey, I. (Eds.), Proceedings of the Fourth European Conference on Artificial Life, ECAL97. MIT Press, pp. 368–377.

Ijspeert, A.J., 2001. A connectionist central pattern generator for the aquatic and terrestrial gaits of a simulated salamander. Biol. Cybern. 84 (5), 331–348.

Ijspeert, A.J., 2008. Central pattern generators for locomotion control in animals and robots: a review. Neural Netw. 21 (4), 642–653.

Ijspeert, A.J., Crespi, A., Ryczko, D., Cabelguen, J.-M., 2007. From swimming to walking with a salamander robot driven by a spinal cord model. Science 315 (5817), 1416–1420.

Ijspeert, A.J., Nakanishi, J., Hoffmann, H., Pastor, P., Schaal, S., 2013. Dynamical movement primitives: learning attractor models for motor behaviors. Neural Comput. 25 (2), 328–373.

Inagaki, S., Yuasa, H., Suzuki, T., Arai, T., 2006. Wave CPG model for autonomous decentralized multi-legged robot: gait generation and walking speed control. Robot. Auton. Syst. 54 (2), 118–126.

Jordan, L.M., Pratt, C.A., Menzies, J.E., 1979. Locomotion evoked by brain stem stimulation: occurrence without phasic segmental afferent input. Brain Res. 177 (1), 204–207. http://dx.doi.org/10.1016/0006-8993(79)90933-8.

Kamimura, A., Kurokawa, H., Toshida, E., Tomita, K., Murata, S., Kokaji, S., 2003. Automatic locomotion pattern generation for modular robots. In: IEEE International Conference on Robotics and Automation (ICRA2003).

Kimura, H., Akiyama, S., Sakurama, K., 1999. Realization of dynamic walking and running of the quadruped using neural oscillators. Auton. Robots 7 (3), 247–258.

Kimura, H., Fukuoka, Y., Cohen, A.H., 2007. Adaptive dynamic walking of a quadruped robot on natural ground based on biological concepts. Int. J. Robot. Res. 26 (5), 475–490.

Kimura, H., Fukuoka, Y., Konaga, K., 2001. Adaptive dynamic walking of a quadruped robot using a neural system model. Adv. Robot. 15 (8), 859–878.

Knuesel, J., Cabelguen, J.-M., Ijspeert, A., 2010. Decoding the mechanisms of gait generation and gait transition in the salamander using robots and mathematical models. In: Motor Control: Theories, Experiments, and Applications, pp. 417–451. http://dx.doi.org/10.1093/acprof:oso/9780195395273.003.0018.

Kuo, A.D., 2002. The relative roles of feedforward and feedback in the control of rhythmic movements. Mot. Control 6 (2), 129–145.

Lewis, M.A., Fagg, A.H., Bekey, G.A., 1993. Genetic algorithms for gait synthesis in a hexapod robot. In: Zheng, Y.F. (Ed.), Recent Trends in Mobile Robots. World Scientific.

Lewis, M.A., Tenore, F., Etienne-Cummings, R., 2005. CPG design using inhibitory networks. In: IEEE International Conference on Robotics and Automation (ICRA2005).

Lu, Z., Ma, B., Li, S., Wang, Y., 2006. 3D locomotion of a snake-like robot controlled by cyclic inhibitory CPG model. In: The Proceedings of the IEEE/RSJ International Conference on Intelligent Robots and Systems (IROS 2006), pp. 3897–3902.

MacKay-Lyons, M., 2002. Central pattern generation of locomotion: a review of the evidence. Phys. Ther. 82 (1), 69–83.

Manoonpong, P., Pasemann, F., Roth, H., 2007. Modular reactive neurocontrol for biologically inspired walking machines. Int. J. Robot. Res. 26 (3), 301–331.

Marbach, D., Ijspeert, A.J., 2005. Online optimization of modular robot locomotion. In: Proceedings of the IEEE Int. Conference on Mechatronics and Automation (ICMA 2005), pp. 248–253.

Matsubara, T., Morimoto, J., Nakanishi, J., Sato, M., Doya, K., 2006. Learning CPG-based biped locomotion with a policy gradient method. Robot. Auton. Syst. 54, 911–920.

Matsuoka, K., 1985. Sustained oscillations generated by mutually inhibiting neurons with adaptation. Biol. Cybern. 52, 367–376.

Matsuoka, K., 1987. Mechanisms of frequency and pattern control in the neural rhythm generators. Biol. Cybern. 56, 345–353.

Miyakoshi, S., Taga, G., Kuniyoshi, Y., Nagakubo, A., 1998. Three dimensional bipedal stepping motion using neural oscillators – towards humanoid motion in the real world. In: Proceedings of the IEEE/RSJ Int. Conference on Intelligent Robots and Systems (IROS1998), pp. 84–89.

Morimoto, J., Endo, G., Nakanishi, J., Hyon, S., Cheng, G., Bentivegna, D., Atkeson, C.G., 2006. Modulation of simple sinusoidal patterns by a coupled oscillator model for biped walking. In: Proceedings of the 2006 IEEE International Conference on Robotics and Automation (ICRA2006), pp. 1579–1584.

Nakada, K., Asai, T., Amemiya, Y., 2003. An analog CMOS central pattern generator for interlimb coordination in quadruped locomotion. IEEE Trans. Neural Netw. 14 (5), 1356–1365.

Nakamura, Y., Mori, T., Sato, M., Ishii, S., 2007. Reinforcement learning for a biped robot based on a CPG-actor-critic method. Neural Netw. 20 (6), 723–735.

Nakanishi, J., Morimoto, J., Endo, G., Cheng, G., Schaal, S., Kawato, M., 2004. Learning from demonstration and adaptation of biped locomotion. Robot. Auton. Syst. 47, 79–91.

Ogino, M., Katoh, Y., Aono, M., Asada, M., Hosoda, K., 2004. Reinforcement learning of humanoid rhythmic walking parameters based on visual information. Adv. Robot. 18 (7), 677–697.

Okada, M., Tatani, K., Nakamura, Y., 2002. Polynomial design of the nonlinear dynamics for the brain-like information processing of whole body motion. In: IEEE International Conference on Robotics and Automation (ICRA2002), pp. 1410–1415.

Owaki, D., Kano, T., Nagasawa, K., Tero, A., Ishiguro, A., 2013. Simple robot suggests physical interlimb communication is essential for quadruped walking. J. R. Soc. Interface 10 (78), 20120669. http://dx.doi.org/10.1098/rsif.2012.0669.

Pearlmutter, B.A., 1995. Gradient calculations for dynamic recurrent neural networks: a survey. IEEE Trans. Neural Netw. 6 (5), 1212–1228.

Pearson, K.G., 1995. Proprioceptive regulation of locomotion. Curr. Opin. Neurobiol. 5, 786–791.

Pham, Q.-C., Slotine, J.-J., 2007. Stable concurrent synchronization in dynamic system networks. Neural Netw. 20 (1), 62–77.

Prochazka, A., Ellaway, P., 2012. Sensory systems in the control of movement. In: Comprehensive Physiology. Retrieved from http://onlinelibrary.wiley.com/doi/10.1002/cphy.c100086/full.

Prochazka, A., Yakovenko, S., 2007. The neuromechanical tuning hypothesis. In: Cisek, P., Drew, T., Kalaska, J.F. (Eds.), Progress in Brain Research, vol. 165. Elsevier, pp. 255–265. Retrieved from http://www.sciencedirect.com/science/article/pii/S0079612306650164.

Righetti, L., Buchli, J., Ijspeert, A.J., 2006. Dynamic hebbian learning in adaptive frequency oscillators. Physica D 216 (2), 269–281.

Righetti, L., Ijspeert, A.J., 2006. Programmable central pattern generators: an application to biped locomotion control. In: Proceedings of the 2006 IEEE International Conference on Robotics and Automation (ICRA2006), pp. 1585–1590.

Righetti, L., Ijspeert, A.J., 2008. Pattern generators with sensory feedback for the control of quadruped locomotion. In: Proceedings of the 2008 IEEE International Conference on Robotics and Automation (ICRA 2008).

Ronsse, R., Lenzi, T., Vitiello, N., Koopman, B., van Asseldonk, E., Rossi, S., et al., 2011. Oscillator-based assistance of cyclical movements: model-based and model-free approaches. Med. Biol. Eng. Comput. 49 (10), 1173–1185. http://dx.doi.org/10.1007/s11517-011-0816-1.

Schoner, G., Jiang, W.Y., Kelso, J.A.S., 1990. A synergetic theory of quadrupedal gaits and gait transitions. J. Theor. Biol. 142, 359–391.

Seo, K., Slotine, J.-J.E., 2007. Models for global synchronization in CPG-based locomotion. In: Proceedings of the IEEE International Conference on Robotics and Automation (ICRA 2007), pp. 281–286.

Shan, J., Nagashima, F., 2002. Neural locomotion controller design and implementation for humanoid robot HOAP-1. In: Proceedings of the 20th Annual Conference of the Robotics Society of Japan.

Shik, M.L., Severin, F.V., Orlovsky, G.N., 1966. Control of walking by means of electrical stimulation of the mid-brain. Biophysics 11, 756–765.

Simoni, M.F., DeWeerth, S.P., 2007. Sensory feedback in a half-center oscillator model. IEEE Trans. Biomed. Eng. 54, 193–204.

Sims, K., 1994. Evolving virtual creatures. In: Computer Graphics Proceedings, Annual Conferences Series. Association for Computing Machinery, New York, pp. 15–24.

Song, S., Geyer, H., 2015. A neural circuitry that emphasizes spinal feedback generates diverse behaviours of human locomotion. J. Physiol. 593 (16), 3493–3511. http://dx.doi.org/10.1113/JP270228.

Sproewitz, A., Moeckel, R., Maye, J., Ijspeert, A.J., 2008. Learning to move in modular robots using central pattern generators and online optimization. Int. J. Robot. Res. 27 (3–4), 423–443.

Stefanini, C., Orofino, S., Manfredi, L., Mintchev, S., Marrazza, S., Assaf, T., et al., 2012. A novel autonomous, bioinspired swimming robot developed by neuroscientists and bioengineers. Bioinspir. Biomim. 7 (2), 25001. http://dx.doi.org/10.1088/1748-3182/7/2/025001.

Still, S., Tilden, M.W., 1998. Controller for a four legged walking machine. In: Smith, L.S., Hamilton, A. (Eds.), Neuromorphic Systems: Engineering Silicon from Neurobiology. World Scientific.

Taga, G., 1998. A model of the neuro-musculo-skeletal system for anticipatory adjustment of human locomotion during obstacle avoidance. Biol. Cybern. 78 (1), 9–17.

Taga, G., Miyake, Y., Yamaguchi, Y., Shimizu, H., 1993. Generation and coordination of bipedal locomotion through global entrainment. In: International Symposium on Autonomous Decentralized Systems, Proceedings. ISADS 93, pp. 199–205.

Taga, G., Yamaguchi, Y., Shimizu, H., 1991. Self-organized control of bipedal locomotion by neural oscillators in unpredictable environment. Biol. Cybern. 65, 147–159.

Thoroughman, K.A., Shadmehr, R., 2000. Learning of action through adaptive combination of motor primitives. Nature 407 (6805), 742–747. https://doi.org/10.1038/35037588.

Tsujita, K., Tsuchiya, K., Onat, A., 2001. Adaptive gait pattern control of a quadruped locomotion robot. In: IEEE International Conference on Intelligent Robots an Systems (IROS2001).

Van der Noot, N., Colasanto, L., Barrea, A., van den Kieboom, J., Ronsse, R., Ijspeert, A.J., 2015. Experimental validation of a bio-inspired controller for dynamic walking with a humanoid robot. In: IEEE, pp. 393–400. https://doi.org/10.1109/IROS.2015.7353403.

Williamson, M.M., 1998. Neural control of rhythmic arm movements. Neural Netw. 11 (7–8), 1379–1394.

Zaier, R., Nagashima, F., 2006. Motion pattern generator and reflex system for humanoid robots. In: 2006 IEEE/RSJ International Conference on Intelligent Robots and Systems (IROS2006). IEEE, pp. 840–845. https://doi.org/10.1109/IROS.2006.281734.

Chapter 5

Torque Control in Legged Locomotion*

Juanjuan Zhang*,†, Chien Chern Cheah†, and Steven H. Collins*,‡

*Department of Mechanical Engineering, Carnegie Mellon University, United States
†School of Electric and Electronic Engineering, Nanyang Technological University, Singapore
‡Robotics Institute, Carnegie Mellon University, United States

Many torque control approaches have been proposed for robotic devices used in legged locomotion, but few comparisons have been performed across controllers in the same system. In this study, we compared the torque-tracking performance of nine control strategies, including variations on classical feedback control, model-based control, adaptive control, and iterative learning. To account for interactions between patterns in desired torque and tracking performance, we tested each in combination with four high-level controllers that determined desired torque based on time, joint angle, a neuromuscular model, or electromyographic measurements. Controllers were implemented on an ankle exoskeleton with series elastic actuation driven by an off-board motor through a uni-directional Bowden cable. The exoskeleton was worn by one human subject walking on a treadmill at 1.25 m·s^{-1} for one hundred steady-state steps under each condition. We found that the combination of proportional control, damping injection and iterative learning resulted in substantially lower root-mean-squared error than other torque control approaches for all high-level controllers. With this low level torque controller, RMS errors can be as low as 1.3% of peak torque for real-time tracking, and 0.2% for the average stride. Model-free, integration-free feedback control seems to be well suited to the uncertain, changing dynamics of the human–robot system, while iterative learning is advantageous in the cyclic task of walking.

5.1 INTRODUCTION

Robotic legged locomotion, including walking robots and powered lower-limb exoskeletons and prostheses, has been an area of active research for

---

*. Supplementary document of this chapter is located at https://www.andrew.cmu.edu/user/shc17/Zhang_2016_BLL—SuppMat.pdf.

Bioinspired Legged Locomotion. http://dx.doi.org/10.1016/B978-0-12-803766-9.00007-5

decades (Cloud, 1965; Mosher, 1967; Frank, 1968). Most early walking related robots used kinematic trajectory control, an approach that persists today (Jezernik et al., 2003; Aguirre-Ollinger et al., 2007; Suzuki et al., 2007; Tsai et al., 2010). In case of exoskeletons and prostheses, however, position control strategies tend to result in less safe and less comfortable human–robot interactions since they can cause large forces to develop when human and robot motions differ (de Luca et al., 2006; Haddadin et al., 2008). Position-controlled exoskeletons have also been shown to be less effective in rehabilitation compared to traditional human-based therapies (Hidler et al., 2009) for similar reasons.

Increasingly, the control of exoskeletons and prostheses has shifted from kinematic methods to strategies that respond more fluidly to actions of the user. One reason for this shift is the concern for human safety and comfort. Another driver is our improved understanding of the natural dynamics of human motion (Mochon and McMahon, 1980; McGeer, 1990; Collins et al., 2005; Verdaasdonk et al., 2009; Ijspeert, 2014), which suggests a more dynamic approach to human–robot interactions than afforded by kinematic control.

One method for improved interaction between humans and robots is impedance manipulation (Andrews and Hogan, 1983), in which the reaction of a robot to external forces is regulated rather than the resulting position trajectory (Kazerooni et al., 2005). Whereas position control strategies typically impose high impedance to improve trajectory tracking performance, this method allows lower impedance at the robot interface and a greater influence of human actions on the resulting motions.

Direct control of interaction forces or torques can also be used to reduce human–robot interface impedance (Haddadin et al., 2008; Lasota et al., 2014). Torque control provides a simple means of manipulating the flow of energy from the exoskeleton to the human, which can be useful in biomechanics studies (Veneman et al., 2007; Sawicki and Ferris, 2009; Stienen et al., 2010; Malcolm et al., 2013; Jackson and Collins, 2015). Torque control can also be used to exploit passive dynamics or render virtual systems with alternate dynamics in humanoid robots (Pratt et al., 1997), active prostheses (Au et al., 2008; Sup et al., 2009; Caputo and Collins, 2013), and exoskeletons (Kawamoto et al., 2010; Unluhisarcikli et al., 2011; Giovacchini et al., 2014; Witte et al., 2015). In exoskeletons and prostheses, the quality of torque control is a limiting factor in precision of the applied intervention and can be the limiting factor in human–robot system performance.

Series elastic actuation, in which compliance is placed between a motor and its end-effector, can improve torque control in devices. This is especially important when human–robot interaction is involved, in which case there often exists unknown, changing human–exoskeleton interaction dynamics. Elasticity in the actuator transmission decouples motor inertia from the structure of the exoskeleton or prosthesis (Pratt and Williamson, 1995), which physically reduces interface impedance and results in smaller torques when human and robot motions unexpectedly diverge. Series elastic actuation can thereby provide improved human safety (Zinn et al., 2004) and improve torque tracking performance in the face of complex, dynamic user movements (Robinson et al., 1999). Unlike direct-drive actuators, torque output in a series elastic actuator is usually not directly related to motor torque, but instead to the position of the motor relative to the joint. Motor position is therefore better correlated to load torque, especially in the presence of transmission friction. For these reasons, series elastic actuators with a motor drive running in velocity mode typically have lower actuation impedance and smoother torque tracking with lower error (Pratt et al., 2004; Wyeth, 2006).

Bowden cable transmissions are often used in exoskeletons and prostheses to further reduce physical impedance through drive relocation. Bowden cables allow motor and gearbox elements to be placed in more desirable locations than the joint they actuate, resulting in reduced device inertia. Motors can be moved proximally on the limb or body (Collins et al., 2005; Schiele et al., 2006; Hobbelen et al., 2008; Mooney et al., 2014) or off the body altogether (Veneman et al., 2006; Caputo and Collins, 2013; Witte et al., 2015). Bowden cables are flexible, producing little interference with joint motions (Caputo and Collins, 2014), but have complex stick–slip transmission dynamics that pose additional torque control challenges (Schiele et al., 2006).

Unidirectional Bowden cables can completely isolate the human from motor inertia when desired. The capacity to become transparent, or produce zero impedance, is desirable in exoskeletons, as it is frequently useful to apply precisely zero torque to the human (Veneman et al., 2007; Kong et al., 2009; Zanotto et al., 2013; van Dijk et al., 2013). Uni-directional Bowden cables can be kept slack, preventing any torque from being transmitted regardless of human dynamics (Collins and Jackson, 2013; Witte et al., 2015; Jackson and Collins, 2015). However, allowing the transmission to become slack introduces complex dynamics and uncertainty during reengagement, as in other systems with intermittent contact, which can make torque control more difficult.

The human ankle produces more than half of the mechanical work of the lower limbs during walking (Winter, 1991) and has been a frequent target for exoskeleton and prosthesis assistance (Ferris et al., 2006) and humanoid robot control (Kim and Oh, 2004). In fact, ankle joint assistance has led to the first systems that reduce the energy cost of walking for humans (Malcolm et al., 2013), including one device that does so passively (Collins et al., 2015). Improved torque control at ankle joints of robotic legged locomotion systems would provide immediate benefits for such systems, and could also be beneficial at knee and hip joints.

Torque control is typically implemented at a low level in walking-related robot control hierarchies, with higher level controllers determining behaviors and commanding desired torques. In such schemes, desired torque is not a control objective selected in advance, but rather a mid-level signal, often with complex dynamics that reflect interactions with the human user. In this chapter, we will refer to the class of control elements that generate desired torque as *high-level controllers*, and to the elements that enforce desired torque, the torque controllers that are the primary focus of this study, as *low-level controllers*. Since the dynamics of the desired torque signal depend on the high-level control type, we expect interactions with low-level controllers that will affect torque tracking performance.

Many potential low-level control elements have been proposed for tracking torque and position in walking related robots and series elastic actuators. Prominent categories of torque control include classical feedback, model-based control, adaptive control and iterative learning.

Classical proportional-integral-derivative (PID) feedback control, and simple variations thereon, have been widely employed in exoskeletons due to their simplicity and ease of tuning. Integral control elements are used to reduce steady state errors in series elastic actuators with consistent dynamics and low impedance (Pratt et al., 2004; Wyeth, 2006; Vallery et al., 2007; Lenzi et al., 2011; van Dijk et al., 2013; Paine et al., 2014; Giovacchini et al., 2014). Integration-free proportional-derivative (PD) control is often used in high-impedance exoskeletons (Nef et al., 2007; Gupta et al., 2008; Farris et al., 2011) and in series elastic actuators with more modeling uncertainties (Kong et al., 2009; Caputo and Collins, 2014). In cases where the derivative of the error signal is noisy, damping injection, or negative feedback on a less noisy velocity in the system, can be used instead to provide similar stabilizing effects (Arimoto and Takegaki, 1981; Kelly, 1999). Gain scheduling, in which control gain values change according to system states, is sometimes used in the control of robots that interact with humans for improved safety or intervention efficiency (Cai et al., 2006; Banala et al., 2009; Zhang et al., 2013).

Model-based control elements are often used in robots to improve torque-tracking performance. Approaches typically include feed-forward terms that use inverted plant dynamics to shape impedance or torque (Pratt et al., 2004; Kazerooni et al., 2005; Fleischer et al., 2005; Aguirre-Ollinger et al., 2007; van Dijk et al., 2013; Paine et al., 2014). This approach works best with an accurate model of the system.

Adaptive control has also often been used in systems with human–robot interaction (Zhang and Cheah, 2015). One example of adaptive control that has been applied to human–robot interaction is passivity-based control. These controllers manipulate the energy balance of the system using a system model and adaptive control elements, and can improve tracking performance with provable closed-loop stability (Ortega and Spong, 1989). Passivity-based control has been proposed for series elastic actuators (Calanca and Fiorini, 2014) and used during human–robot interactions (Zhang and Cheah, 2015).

Variations on iterative learning derived from industrial robots have also been applied to robotic legged locomotion (Bae and Tomizuka, 2012; van Dijk et al., 2013). This approach improves tracking performance by exploiting the cyclic nature of gait; tracking errors from past walking steps are used to predict errors in the ensuing step, and feed-forward corrections are applied. Since corrections are based on an accumulation of past errors, this approach bears some resemblance to classical integral control, in which errors in previous steps are used to improve performance during the present step.

High-level controllers intended to assist human walking include schemes that command desired torque based on time, joint angle, neuromuscular models, and electromyographic measurements. Perhaps the simplest way to generate desired torques is as a function of time, which can be used to regulate the relative timing of robot actions as well as human actions in cases of exoskeletons and prostheses (Malcolm et al., 2013, 2015; Jackson and Collins, 2015). Another common method is to imitate observed relationships between human joint angles and joint torques (Fite and Goldfarb, 2006; Au et al., 2008; Hitt et al., 2010), which can be especially useful in regulating net joint work (Caputo and Collins, 2014). Virtual neuromuscular systems with complex internal dynamics have also been used to generate desired joint torques in assistive devices (Rosen et al., 2001; Cavallaro et al., 2005; Perry et al., 2007; Eilenberg et al., 2010; Geyer and Herr, 2010; Shultz et al., 2014; Dorn et al., 2015; Van der Noot et al., 2015). This method has demonstrated benefits in the control of adaptive prosthetic limbs (Markowitz et al., 2011). Direct neuromuscular interfaces, such as through electromyographic measurement of muscle activity, promise more intuitive control of exoskeletons by users (Fleischer et al., 2005; Ferris et al., 2006; Kawamoto et al., 2010; Loconsole et al., 2014; Huang et al., 2014; Takahashi et al., 2015). Each of these high-level control approaches may be advantageous in some assistance paradigm, and each results in desired torque signals with different dynamics.

Many approaches to torque control in robotic legged locomotion have been established, but a more complete comparison would be helpful when designing controllers for new systems. The classical feedback, model-based, adaptive and iterative learning control approaches reviewed in this section all have strengths for human–robot interaction. Several of these controllers have been tested in lower-limb exoskeletons and have shown good performance (Zoss et al., 2006; Veneman et al., 2006; Wang et al., 2013; Zanotto et al., 2013; van Dijk et al., 2013; Giovacchini et al., 2014). Comparisons across studies are made difficult, however, due to differences in protocol, performance metrics, hardware, and high-level controllers. Some results are reported for benchtop tests (Sulzer et al., 2009; Kobayashi et al., 2010; Accoto et al., 2013), which may provide more positive results than during complex interactions with humans. Some results are not reported quantitatively (Zoss et al., 2006), which makes comparisons difficult. In some cases a small number of controllers have been tested on the same hardware (van Dijk et al., 2013), but in most cases torque tracking results are provided for a single controller working with a single system. This makes comparisons across studies difficult, since some portion of the differences in performance may be due to differences in the capability of the hardware used. Similarly, comparisons have been performed with different high-level controllers, which could interact with low-level controllers and contribute to differences in performance across studies. Studies comparing a wide range of torque controllers in human-interaction protocols with quantitative performance metrics, consistent hardware setups and a variety of high-level controllers would help establish guidelines for selecting and tuning controllers for new robotic legged locomotion systems.

The aim of this chapter is to compare the tracking performance of prominent torque control methods, with multiple high-level desired torque conditions, in a single robotic legged locomotion platform during walking. Promising methods using classical feedback, model-based, adaptive and iterative learning control elements were used. Although it was impractical to test all possible control strategies, the chosen controllers span the set of candidate methods and provide a more comprehensive test than previously available. A diverse sample of high-level controllers were used to test for interactions with low-level control dynamics and provide insights into the generality of tracking results. A single exoskeleton system was used, experimentally controlling for hardware capabilities. Tests were conducted while a human wore the exoskeleton and walked on a treadmill, making results relevant to conditions with complex interactions between the robot, a human user, and the environment. We anticipate these results to help guide the selection and tuning of torque control elements in various robotic legged locomotion systems.

5.2 SYSTEM OVERVIEW

5.2.1 System Modeling

A diagram of a typical one degree-of-freedom lower-limb robot driven by a series elastic actuator through a cable with a geared motor is shown in Fig. 5.1. Based on this structure, we used the following simplified models of system components to aid in our understanding of the system, make reasonable choices for model-free control elements, and design model-based control elements.

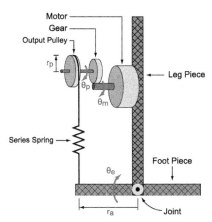

FIGURE 5.1 A schematic diagram of a one degree-of-freedom, cable-driven, robotic locomotion system with a series elastic actuator. This diagram uses an ankle device as an example. θ_m and θ_p are motor position and pulley position after gearing, respectively. θ_e is the device joint angle. R is the effective aspect ratio between motor output pulley radius r_p and device joint velocity lever arm r_a defined as $R = \dfrac{r_a}{r_p}$.

- Motor Dynamics
  Assuming armature inductance dynamics occur at a substantially higher frequency than rotor dynamics, and therefore have negligible effects, the dynamics of the motor can be written as

$$
\begin{cases}
K_a \cdot i_a(t) = I_e \cdot N \cdot \ddot{\theta}_p(t) + f_e \cdot N \cdot \dot{\theta}_p(t) + \dfrac{1}{N} \cdot \tau_o(t), \\
V_a(t) = R_a \cdot i_a(t) + K_b \cdot N \cdot \dot{\theta}_p(t),
\end{cases}
\tag{5.1}
$$

in which K_a is the motor-torque constant, i_a is the armature current, I_e is the effective moment of inertia of the motor and gear referred to the motor shaft, $N = \dot{\theta}_m/\dot{\theta}_p$ is the gear ratio, θ_m is the angular position of the motor shaft, θ_p is the angular position of the gear output shaft, f_e is the effective viscous friction coefficient of the combined motor and gear referred to the motor shaft, τ_o is output torque at the gear output pulley, V_a is the armature voltage, R_a is the armature resistance, and K_b is the motor-voltage constant.
- Transmission Model
  The pulley transmits load to the cable as

$$
\tau_o = F \cdot r_p,
\tag{5.2}
$$

in which r_p is the radius of the pulley attached to the gear output and F is the tension in the cable on the motor side. Making the simplifying assumption that there is no friction in the transmission, the torque at the device side is

$$\tau = F \cdot r_a, \tag{5.3}$$

in which r_a is the lever arm at the ankle joint. We further assume that the angular excursion of the ankle joint is small, and that the lever arm is therefore approximately constant.

- Force–Position Relationship
  Making the simplifying assumption that the cable has either spring-like compliance or negligible compliance compared to the series spring, we have

$$F = K_c \cdot (r_p \cdot \theta_p - r_a \cdot \theta_e), \tag{5.4}$$

in which K_c is the total effective stiffness of the cable transmission and series spring, and θ_p and θ_e are the pulley and device joint angles relative to a neutral position at which the cable begins to go slack.

- Torque–Angle Relationship
  Defining the gear ratio of the transmission, R, as

$$R = \frac{r_a}{r_p}, \tag{5.5}$$

the torque applied by the device can be written as

$$\begin{aligned}
\tau &= F \cdot r_a \\
&= r_p \cdot r_a \cdot K_c \left[\theta_p - \frac{r_a}{r_p} \theta_e \right] \\
&= K_t (\theta_p - \theta_e R)
\end{aligned} \tag{5.6}$$

with transmission stiffness, K_t, defined as

$$K_t = r_p \cdot r_a \cdot K_c, \tag{5.7}$$

relating torque at the device to the angles of the motor output pulley and device joint.

- Device Joint Dynamics
  Applying law of balance of angular momentum to the device joint, we have

$$\tau - \tau_h - B_e \cdot \dot{\theta}_e = I_e \cdot \ddot{\theta}_e \tag{5.8}$$

where τ_h is the torque applied to the robot by the environment, which is mostly human body in case of exoskeletons, B_e is the device joint damping coefficient, and I_e is the moment of inertia of the device.

- Motor Velocity Control Dynamics

  Motors are often operated in velocity control mode in series elastic actuators, which tends to result in lower actuation impedance and better torque tracking (Pratt et al., 2004; Wyeth, 2006). Without access to the proprietary controller used by the commercialized motor drivers, the precise relationship between desired motor velocity, $\dot{\theta}_{m,des}$, and input voltage to the motor, V_a, is unknown. However, from Eq. (5.1), we can derive the relationship between input voltage and actual motor velocity as

$$
\begin{aligned}
V_a &= \frac{R_a I_e N}{K_a}\ddot{\theta}_p + \left(\frac{R_a f_e N}{K_a} + K_b N\right)\dot{\theta}_p + \frac{R_a}{K_a}\tau_o \\
&= \frac{R_a I_e}{K_a}\ddot{\theta}_m + \left(\frac{R_a f_e}{K_a} + K_b\right)\dot{\theta}_m + \frac{R_a}{K_a}\tau_o.
\end{aligned}
\tag{5.9}
$$

When the angular acceleration is zero, this reduces to

$$
V_a = \left(\frac{R_a f_e}{K_a} + K_b\right)\dot{\theta}_m + \frac{R_a}{K_a}\tau_o.
\tag{5.10}
$$

At moderate torques and speeds, the contribution of armature resistance to voltage drop is small at moderate speeds. Neglecting this term, we have

$$
V_a = \left(\frac{R_a f_e}{K_a} + K_b\right)\dot{\theta}_m,
\tag{5.11}
$$

and input voltage and motor velocity are linearly related at low torque and high speed.

5.2.2 Potential Control Issues

While the dynamic models described by Eqs. (5.1)–(5.11) capture the basic properties of the system, they do not address its full complexity. There are additional uncertain or changing dynamics that are difficult to model, which contribute to most of the challenges of the control problem we are addressing. These additional complexities are described below.

- Bowden Cable Nonlinearities and Stiction

  Bowden cables are often used as a part of the transmission in robotic legged locomotion systems. For simplicity, we modeled the Bowden cable as a frictionless linear spring, but its stiffness is actually nonlinear and there are substantial frictional effects. The cable is stiffening, exhibiting greater local stiffness at high loads. This can be seen in the torque versus ankle angle curves generated by fixing the motor and passively flexing the device joint

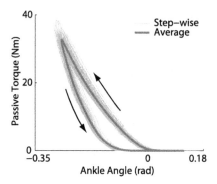

FIGURE 5.2 Torque versus exoskeleton ankle joint angle relationship with motor position fixed and the ankle joint being passively flexed for one hundred strides.

during walking (Fig. 5.2). The cable warms over the course of a many strides, which decreases its overall stiffness. It exhibits creep, which increases the slack length. If the cable is allowed to go slack, the state corresponding to reengagement is uncertain. There is substantial friction in the cable, including dissipation with characteristics of Coulomb friction, viscous damping, and stiction, some of which are visible in Fig. 5.2. The cable heats over the course of many strides, which increases overall friction. Stiction leads to sudden changes in cable force, and propagation of the slipping point along the length of the cable makes these changes unpredictable. These transmission properties are complex, nonlinear and time varying.

- Human–Robot Interaction and Human Adaptation
  In the case of exoskeletons and prostheses, the robot device works together with the human body. The device often contacts the soft tissues and muscles of the human body, often using flexible straps. This interface is complex and nonlinear, with low overall impedance. For example, muscle activity beneath the straps can substantially affect stiffness and damping at the interface. Straps may also shift on the limb, altering lever arms and engaging different tissues. Human kinematics, kinetics and underlying neural and muscular activity also vary in time and across steps. This can be seen in the variations in ankle joint angle curves over many steps, even when the motor position is fixed (Fig. 5.3).

- Delays Caused by Communication and Motor Velocity Tracking
  Delays in generating desired motor position also pose a control challenge. A portion of these delays can come from communication between subsystems. For example, in the hardware used in this study there was a 6 ms closed-loop communication delay. Another effective delay comes from accelerating the motor rotor. For the hardware used in this study, the motor

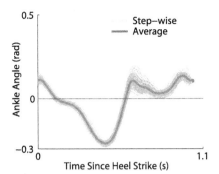

FIGURE 5.3 Variability in exoskeleton ankle joint angle trajectory during one hundred strides of walking with the motor position fixed and the ankle exoskeleton passively flexing.

velocity rise time was about 7 ms. These delays cause feedback controllers to become unstable as gains are increased, limiting closed-loop performance.

An effective low-level torque controller must accommodate these complex, non-linear, time-varying system features. As discussed in the Introduction section, various control methods have been employed in addressing this issue, which include model-free and model-based controllers that were used as feedback or feed-forward elements. These methods span the range of classical proportional-integral-derivative control, passivity-based control, model-based feed-forward compensation and iterative learning compensation. The relative performance of these control approaches will be investigated in this chapter with an experimental case study in an ankle exoskeleton testbed, which will be detailed in the next section.

5.3 A CASE STUDY WITH AN ANKLE EXOSKELETON

To investigate the relative performance of various control methods in torque tracking for robotic legged locomotion, a case study was conducted. We compared torque tracking performance for nine common torque control methods that used combinations of classical feedback control, model-based control, adaptive control and iterative learning. These included examples of model-free and model-based feedback and feed-forward control. Each low-level torque controller was tested with four high-level walking controllers that set desired torque based on time, ankle angle, a neuromuscular model, or electromyographic measurements. All controllers were implemented on a tethered ankle-foot exoskeleton with series-elastic actuation driven by a uni-directional Bowden cable, and each was tuned to minimize error. The exoskeleton was then worn by one subject who walked on a treadmill for one hundred strides at steady state under

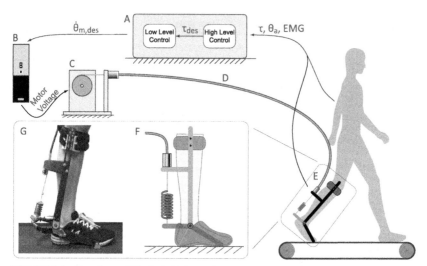

FIGURE 5.4 Experimental Testbed: (A) A real-time controller reads sensory information, computes desired torques using a high-level controller, computes desired motor velocity using a low level controller, and outputs desired motor velocity to the motor drive. (B) A dedicated motor drive. (C) An off-board geared motor and pulley. (D) A Bowden cable transmission. (E) A lightweight instrumented ankle exoskeleton. (F) An enlarged schematic of the exoskeleton. (G) A photograph of the exoskeleton.

each condition, and the root mean squared errors between desired and measured torque were calculated for each stride and for the averaged stride.

5.3.1 Exoskeleton System

We tested on a tethered ankle exoskeleton that comprised an off-board real-time control module and geared electric motor, a uni-directional Bowden cable transmission with a series spring, and an exoskeleton frame that interfaced with the human foot and shank (Fig. 5.4). This system is described in detail in Witte et al. (2015), and a summary is provided below.

We used a dedicated real-time control system (ACE1103, dSPACE Inc.) to sample sensors at 5000 Hz, filter sensor data at 200 Hz, and generate desired motor velocity commands at 500 Hz. The motor unit was composed of a low-inertia, 1.6 kW AC servo motor and a 5:1 planetary gear, with input voltage regulated by a motor driver running in velocity control mode (BSM90N-175AD, GBSM90-MRP120-5 and MFE460A010B, Baldor Electric Co.). A digital optical encoder (E4, US Digital Corp.) measured motor position. As an indication of motor module performance, the 100% rise time to peak motor velocity was 0.013 s.

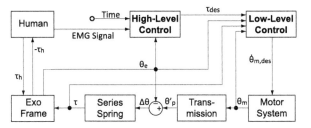

FIGURE 5.5 Flowchart of the control system. High Level Control and Low Level Control are the two blocks to be varied in this study.

A flexible uni-directional Bowden cable transmitted forces from the motor to the exoskeleton frame while minimally restricting leg motions. The cable was composed of a coiled-steel outer conduit (415310-00, Lexco Cable Mfg.) and a 0.003 m diameter Vectran® inner rope, and was 2 m in length. A series spring (DWC-148M-12, Diamond Wire Spring Co.) with an effective stiffness of 190 N·m·rad^{-1} (in terms of ankle rotation) was attached at the end of the rope to provide increased compliance.

The exoskeleton frame applied forces on the front of the human shank below the knee, beneath the heel, and on the ground beneath the toe, so as to generate an ankle plantarflexion torque in proportion to transmission force. Torque was measured using strain gauges (MMF003129, Micro-Measurements) applied in a full Wheatstone bridge on the heel lever, with 1000 Hz signal conditioning (CSG110, Futek Inc.). Joint angle was measured using a digital optical encoder (E5, US Digital Corp.). For one of the high-level controllers, we measured gastrocnemius muscle activity using a wired electromyography system (Bagnoli 4 EMG System, Delsys Inc.).

The high-level and low-level controllers, motor, transmission, exoskeleton frame and human interacted as shown in Fig. 5.5. The high-level controller used time, t, device joint angle, θ_e, or human electromyography, EMG, to determine desired torque. The low-level controller regulated torque, using desired torque, τ_{des}, measured torque, τ, motor angle, θ_m, and/or device angle to command desired motor velocity, $\dot{\theta}_{m,des}$. A hardware motor driver regulated motor velocity. Motor rotations were transmitted through a cable to one end of a series spring. Together with device rotation, this determined spring deflection, which in turn generated device torque. Exoskeleton movements resulted from the balance of torques from the series spring and from the human leg.

5.3.2 Low-Level Torque Controllers

We tested torque tracking performance with nine prominent low-level torque control methods. Low-level controllers were selected based on prominence in the literature, expected performance based on system modeling, and the results of pilot testing. They included model-free and model-based feedback and feed-forward elements. Desired torque was set with each of four high-level exoskeleton control strategies, chosen based on prominence in the literature. High-level controllers set desired torque based on time, joint angle, a neuromuscular model or electromyography.

5.3.2.1 Motor Velocity Control

All torque controllers investigated in this study included motor velocity control performed by a dedicated hardware motor controller. Series elastic actuators with a drive running in velocity mode typically have lower actuation impedance and smoother torque tracking with lower error (Pratt et al., 2004; Wyeth, 2006) than when torque is commanded to the drive. With series elastic actuation, controlling motor velocity is similar to controlling the rate of change of applied torque, since torque is approximated by the product of series stiffness and the difference between motor angle and exoskeleton joint angle (Eq. (5.6)). Desired motor velocity was calculated as

$$\dot{\theta}_{m,des} = \frac{1}{T} \cdot \Delta\theta_{m,des}$$
$$= \frac{N}{T} \cdot \Delta\theta_{p,des} \qquad (5.12)$$

where $\dot{\theta}_{m,des}$ is commanded motor velocity, T is a gain related to rise time, $\Delta\theta_{m,des}$ is desired change in motor position, N is the motor gear ratio, and $\Delta\theta_{p,des}$ is desired change in pulley position, determined by one of the low-level torque controllers described below. The value of T was tuned so as to minimize motor position rise time without causing oscillations during torque tracking.

5.3.2.2 Model-Free Feedback Control

The first group of torque controllers used model-free feedback control, comprising variations on classical proportional-integral-derivative control. Gains were tuned systematically using model-free procedures. Following tuning and pilot testing, four low-level controllers were experimentally compared, $_L1$–$_L4$.

$_L1$: Proportional Control with Damping Injection (PD*)

This controller was analogous to classical proportional-derivative control of torque, with damping injection (Arimoto and Takegaki, 1981; Kelly, 1999) on motor velocity taking the place of the derivative term:

$$\Delta\theta_{p,des} = -K_p \cdot e_\tau - K_d \cdot \dot{\theta}_p \qquad (5.13)$$

where K_p is a proportional gain, $e_\tau = \tau - \tau_{des}$ is torque error, τ is measured exoskeleton torque, τ_{des} is desired exoskeleton torque, K_d is a damping gain, and $\dot{\theta}_p$ is measured velocity of the motor pulley. In pilot testing, we found the damping term more effective than a term with the derivative of torque error; torque was measured with analog strain gauges, which included substantial noise, while pulley position was measured with a digital encoder. Damping was placed on motor pulley velocity alone rather than the relative velocity between the motor pulley and the exoskeleton joint. In pilot tests, using relative velocity was less effective, likely due to the irregular effects of stiction in the Bowden cable transmission on ankle joint velocity.

$_L2$: Proportional Control with Damping Injection and Error-Dependent Gains (PD*+EDG)

This controller was identical to $_L1$, with the exception that the proportional gain was error-dependent (Zhang et al., 2013; Zhang and Cheah, 2015), and increased with torque error:

$$K_p^* = \min\left(\left\lceil \frac{|e_\tau|}{h_\tau}\right\rceil \cdot h_k, \ K_{max}\right),$$
$$\Delta\theta_{p,des} = -K_p^* \cdot e_\tau - K_d \cdot \dot{\theta}_p$$
(5.14)

where the symbol $\lceil . \rceil$ denotes the *ceiling* operation, K_p^* is the error-dependent proportional gain, h_τ and h_k are torque error and proportional gain step sizes, and K_{max} is the maximum allowable gain. This is similar to performing proportional control on the square of the torque error, with a sign and gain adjustment. This type of gain scheduling is expected to result in slower corrections, and fewer oscillations, when torque tracking errors are small.

$_L3$: Proportional Control with Damping Injection and Previous-Error Compensation (PD* + PEC)

This controller was identical to $_L1$, except that desired torque was altered based on torque error from the previous instant in time (Gordon and Ghez, 1987; Klimchik et al., 2012) as:

$$\tau'_{des} = \tau_{des} - e_{\tau,prev},$$
$$\Delta\theta_{p,des} = -K_{pec} \cdot (\tau - \tau'_{des}) - K_d \cdot \dot{\theta}_p$$
(5.15)

where τ'_{des} is the compensated torque error and $e_{\tau,prev}$ is the torque error from the previous time step. K_{pec} is a proportional gain. This approach is expected to increase the control response to large errors. It bears some similarity to integral control, in that it includes a term on prior error, but differs in that only the prior error at the previous sampling time is used rather than the entire time history. In cases where torque error changes slowly, this approach approximates a doubling of the proportional gain.

$_L$4: Proportional-Integral Control with Damping Injection (PID*)
This controller was analogous to classical proportional-integral-derivative control, with damping injection substituted for the derivative term:

$$\Delta\theta_{p,des} = -K_p \cdot e_\tau - K_i \cdot \int_{t_0}^{t} e_\tau dt - K_d \cdot \dot{\theta}_p \qquad (5.16)$$

where K_i is the gain on the integral of torque error, t_0 is the time at which the stride began, and t is the present time. Integral control is expected to eliminate steady-state error by accumulation of control input (Pratt et al., 2004; Wyeth, 2006; Vallery et al., 2007).

5.3.2.3 Model-Based Feed-Forward Control

Many systems with series elastic actuators use an inverse dynamics model of the series spring as a feed-forward control element, typically added to a model-free feedback component. We implemented one such model-based controller in this study, $_L$5.

$_L$5: Proportional Control with Damping Injection and Model-based Compensation (PD* + M)
This controller included both the classical feedback controller of $_L$1 and a model-based feed-forward term, which was intended to anticipate changes in desired motor position due to either changes in exoskeleton joint angle or changes in desired joint torque:

$$\Delta\theta_{p,des} = -K_p \cdot e_\tau - K_d \cdot \dot{\theta}_p + (\theta_{mdl} - \theta_p),$$
$$\theta_{mdl} = \theta_e \cdot \tilde{R} - \tau_{des} \cdot \tilde{K}_t^{-1} \qquad (5.17)$$

where θ_{mdl} is a model-based motor position compensation generated from Eq. (5.6), θ_p is measured motor pulley position, θ_e is measured exoskeleton ankle joint angle, $\tilde{R}$ is the estimate of R as defined in Eq. (5.5), or the ratio of the exoskeleton lever arm to the motor pulley radius, and $\tilde{K}_t$ is an estimate of K_t, which is the total stiffness of the tether, series spring, and other structures between the motor pulley and exoskeleton joint as defined by Eq. (5.7). This is an inverse dynamics approach similar to computed torque and feedback linearization in nonlinear control. If desired torque remains constant but the exoskeleton joint moves, we expect the motor to need to move in proportion. If the joint is stationary but desired torque changes, we expect we know how much to move the motor to obtain the desired change in torque.

We also performed pilot tests with a version of this controller in which change in pulley angle, rather than absolute pulley angle, was anticipated based on the rate of change in exoskeleton angle and the rate of change in desired torque. This approach was less stable, owing to the effects of Bowden cable stiction on exoskeleton joint angle and the interplay between user behavior and desired torque through the high-level controller.

5.3.2.4 Model-Based Feedback Control

Adaptive control approaches (Slotine and Li, 1987; Slotine et al., 1991) using more complete system models have also been applied to exoskeletons. Such regimes have the capacity to exploit additional knowledge of system dynamics and allow theoretical tests of stability and performance. We developed a new adaptive controller for this system using a passivity-based approach, $_L6$.

$_L6$: Passivity-Based Adaptive Control (PAS)

Combining the dynamics of the subsystems described by Eqs. (5.1)–(5.8) so as to eliminate F and i_a, we have following dynamics of the system:

$$\ddot{\tau} + K_1 \dot{\tau} + K_2 \tau + S_\theta \dot{\theta}_e = K_V V_a + K_h \tau_h, \qquad (5.18)$$

in which τ denotes the torque transmitted to the exoskeleton from the motor, θ_e denotes the exoskeleton joint angle, V_a is the voltage applied to the armature of the motor, and τ_h denotes the torque applied to the exoskeleton by the human body. K_V, K_1, K_2 and K_h are positive gains expressed as

$$K_V = \frac{r_a r_p K_a K_c}{I_e N R_a},$$

$$K_1 = \frac{1}{I_e} \left(\frac{K_a K_b}{R_a} + f_e \right),$$

$$K_2 = \frac{r_p^2 K_c}{N^2 I_e} + \frac{r_a^2 K_c}{I_e},$$

$$K_h = \frac{r_a^2 K_c}{I_e},$$

and S_θ is a gain expressed as

$$S_\theta = \left[\frac{r_a^2 K_c}{I_e} \left(\frac{K_a K_b}{R_a} + f_e \right) - \frac{r_a^2 K_c B_e}{I_e} \right]$$

with definitions of system constants provided in Section 5.2.1. Based on the system model in Eq. (5.18), we developed a new, provably stable, adaptive controller for the system.

We define the controller as

$$\begin{aligned} V_a = &-K_p \cdot e_\tau - K_s \cdot s \\ &+ Y_d(\tau, \dot{\tau}_r, \ddot{\tau}_r, \dot{\theta}_e) \cdot \tilde{\Gamma} \\ &- K_{sw} \cdot \text{sign}(s) \end{aligned} \qquad (5.19)$$

where K_p and e_τ are as defined in $_L1$, K_s is the sliding control gain, s is the sliding vector, defined below, Y_d is a regressor, defined below, Γ and $\tilde{\Gamma}$ are the system parameter vector and its estimate, defined below, and K_{sw} is the switching term gain. The sliding vector s is defined as

$$s = \dot{\tau} - \dot{\tau}_{des} + \lambda \cdot e_\tau = \dot{\tau} - \dot{\tau}_r$$

where λ is a positive scalar and τ_r is a virtual reference torque. The regressor, Y_d, is defined as

$$Y_d(\tau, \dot{\tau}_r, \ddot{\tau}_r, \dot{\theta}_e) = [\, \tau \; \dot{\tau}_r \; \ddot{\tau}_r \; \dot{\theta}_e \,],$$

and is used to express the dynamics as a linear combination of system parameters as

$$Y_d \cdot \Gamma = \frac{1}{K_v} \cdot \ddot{\tau}_r + \frac{K_1}{K_v} \cdot \dot{\tau}_r + \frac{K_2}{K_v} \cdot \tau + \frac{S_\theta}{K_v} \cdot \dot{\theta}_e.$$

The system parameter, Γ, is defined as

$$\Gamma = K_v^{-1} \cdot [1 \; K_1 \; K_2 \; S_\theta]^\mathsf{T}.$$

With full knowledge of system parameters, or $\tilde{\Gamma} = \Gamma$, Eq. (5.19) describes a model-based computed torque controller. For practical reasons, however, it is difficult to identify the value of Γ. Therefore, an update law is added to estimate the system parameters, $\tilde{\Gamma}$, as follows:

$$\dot{\tilde{\Gamma}} = -L Y_d^\mathsf{T} s, \tag{5.20}$$

where L is a symmetric positive definite parameter adaptation gain matrix. This parameter updating process reduces the model-dependency of the controller in Eq. (5.19), because only the structure of the dynamic model is used in the construction of the controller.

The closed-loop system with the model-based adaptive controller described by Eqs. (5.19)–(5.20) and dynamics described by Eq. (5.18) is stable and the exoskeleton torque trajectory τ converges to the desired value of τ_{des}, provided that the human input torque, τ_h, the desired torque trajectory, τ_{des}, and their time derivatives, $\dot{\tau}_h$, $\dot{\tau}_{des}$, and $\ddot{\tau}_{des}$, are bounded. A proof is provided in Appendix 5.A.

In pilot tests, we found that better performance was obtained with this controller by setting the time rate of change in desired torque to zero. In practice, for most high-level controllers, the time derivative of desired torque, $\dot{\tau}_{des}$, could not be calculated in advance and contained substantial noise when approximated numerically. Noise on this signal arose from the human measurements used by the high-level controllers to calculate desired torque. We also found that the time derivative of torque error, $\dot{e}_\tau$, contained substantial noise, in part due to noise on the analog strain gauge signal and in part due to complex Bowden cable transmission dynamics. Better performance was obtained using motor output pulley velocity, $\dot{\theta}_p$, in its place. This substitution is equivalent to assuming that the characteristic time of exoskeleton joint dynamics was much larger than the characteristic time of motor dynamics (Eq. (5.6)). It is also analogous to the use of damping injection in place of derivative control in the other controllers tested. The sliding vector and regressor are therefore approximated as:

$$s \approx \dot{\theta}_p + \lambda \cdot e_\tau,$$
$$Y_d \approx [\, \tau \;-\lambda \cdot e_\tau \;-\lambda \cdot \dot{\theta}_p \; \dot{\theta}_e \,].$$

Additionally, in pilot tests we found that it was more effective to operate the dedicated motor drive in velocity control mode, rather than voltage control mode. This difference in performance is likely due to the faster control loop in the motor driver, which allowed voltage to be changed more frequently and with less delay than for the control system as a whole. Motor velocity is strongly related to applied voltage, since the two are linearly related for a given torque at steady state (Eq. (5.11)). This led to a similar formulation as for all other low-level controllers:

$$
\begin{aligned}
\Delta\theta_{p,des} = &-K_p \cdot e_\tau \;-\; K_s \cdot s \\
&+ Y_d(\tau, e_\tau, \dot{\theta}_p, \dot{\theta}_e) \cdot \tilde{\Gamma} \\
&- K_{sw} \cdot \mathrm{sign}(s).
\end{aligned}
\tag{5.21}
$$

5.3.2.5 Model-Free Feed-Forward Control

We also tested a group of controllers that use iterative learning as a feed-forward component, which were expected to improve performance by exploiting the cyclic nature of human gait.

$_L7$: Iterative Learning of Desired Motor Position (LRN)

This controller used torque error at each instant of one stride to update a feed-forward trajectory of desired motor position for each instant of the next stride. This is a variation on iterative learning, which, more generally, exploits the cyclic nature of a task to compensate complex system dynamics without an explicit model (Arimoto et al., 1984; Heinzinger et al., 1992; Van de Wijdeven et al., 2009; Schuitema et al., 2010). While walking is not as consistent as the operations of most industrial robots, it is cyclic, which was expected to afford some improvement in torque errors that occurred consistently from stride to stride.

The feed-forward trajectory of desired motor position, $\theta_{p,des}$, was calculated as

$$\theta_{p,des}(i, n+1) = \theta_{p,des}(i, n) - K_l \cdot e_\tau(i, n) \qquad (5.22)$$

where i is the time index or number of control cycles elapsed within this stride, n is this stride and $n+1$ is the next stride, and K_l is the iterative learning gain. Desired motor position was then enforced as

$$\Delta\theta_{p,des}(i, n) = \theta_{p,des}(i + D, n) - \theta_p(i, n) \qquad (5.23)$$

where D is an estimate of the delay between commanding and achieving a change in motor position. During tuning, both K_l and D were adjusted. Current torque error thereby updates desired motor position for the same time index on the next stride, while commanded motor velocity at this time index is based on a preview of desired motor position later in the same stride. Since the learned trajectory used in the present step has no dependence on the present torque, this method is feed-forward. However, present motor pulley position measurements were used in generating present motor velocity commands. This method can therefore be viewed as a feed-forward iterative learning control of torque combined with proportional feedback control of motor position.

Forgetting During Learning

To avoid divergence due to excessive accumulation of ripples during the learning process, a "forgetting" constant was introduced to Eq. (5.22) as

$$\theta_{p,des}(i, n+1) = \beta \cdot \theta_{p,des}(i, n) - K_l \cdot e_\tau(i, n) \qquad (5.24)$$

where $\beta \in [0, 1]$ is a weight on the learned trajectory. For $\beta = 1$, all learning is retained, zero steady-state offset is expected, but ripples can form if the value of D is incorrect. For $\beta < 1$, torque errors from strides before the last stride have a reduced effect on controller behavior, reducing likelihood of ripple formation, but leading to some steady-state torque offset. For $\beta = 0$, iterative learning is disabled. $L7$ then becomes proportional control based on the torque error delayed by one step, and poor torque tracking performance is expected.

Learning from Filtered Errors

Noise in the error signal leads to inappropriate updates on the learned trajectory, which can excite unstable ripple formation. This excitation can be reduced by filtering the error signal across strides:

$$e_{flt}(i, n) = (1 - \mu) \cdot e_{flt}(i, n-1) + \mu \cdot e_\tau(i, n) \qquad (5.25)$$

where e_{flt} is the filtered torque error trajectory, initially an array of zeros, used in place of e_τ in Eqs. (5.22) and (5.24), and $\mu \in [0, 1]$ is a weighting term on the learned error. For $\mu = 1$, only the error from the last stride is used to update the motor trajectory, resulting in faster convergence but larger effects of sensor noise. For $\mu < 1$, errors at this time increment from all prior strides have some effect on the motor trajectory update, resulting in slower but more stable convergence. For $\mu = 0$, torque error is not updated, and iterative learning is disabled.

$_L 8$: Iterative Learning of Desired Motor Position + Proportional-Damping Compensation (LRN + PD*)

This controller combined iterative learning with proportional-damping feedback control to compensate remaining torque errors. It is a direct superposition of controllers $_L 1$ and $_L 7$, in which the absolute desired motor position was learned as in $_L 7$ and feedback control was applied as in $_L 1$:

$$
\begin{aligned}
\theta_{p,des}^{LRN}(i, n + 1) &= \beta \cdot \theta_{p,des}^{LRN}(i, n) - K_l \cdot e_{flt}(i, n), \\
\theta_{p,des}(i, n) &= \theta_{p,des}^{LRN}(i + D, n) \\
&\quad - K_p \cdot e_\tau(i, n) - K_d \cdot \dot{\theta}_p(i, n), \\
\Delta\theta_{p,des}(i, n) &= \theta_{p,des}(i, n) - \theta_p(i, n).
\end{aligned}
\tag{5.26}
$$

Combining iterative learning with feedback control is thought to result in improved performance compared to either component used in isolation (van Dijk et al., 2013). Iterative learning is expected to generate a feed-forward trajectory that tracks torque for an average stride with zero steady-state error regardless of the complexity of the command signal required, but to be susceptible to step-by-step variability. Proportional-damping control is expected to quickly compensate for small torque errors, but to be susceptible to rapid changes in desired or measured torque.

$_L 9$: Proportional Control with Damping Injection + Iterative Learning Compensation (PD* + ΔLRN)

This controller is another combination of proportional-damping feedback and iterative learning feed-forward control elements. Unlike controllers $_L 7$ and $_L 8$, the values to be learned are desired changes in motor position instead of absolute desired positions:

$$
\begin{aligned}
\Delta\theta_{p,des}^{LRN}(i, n + 1) &= \beta \cdot \Delta\theta_{p,des}^{LRN}(i, n) - K_l \cdot e_{flt}(i, n), \\
\Delta\theta_{p,des}(i, n) &= -K_p \cdot e_\tau(i, n) - K_d \dot{\theta}_p(i, n) \\
&\quad + \Delta\theta_{p,des}^{LRN}(i + D, n).
\end{aligned}
\tag{5.27}
$$

This controller is very similar to $_L 8$, and is expected to have similar strengths and weaknesses. Differences in motor position are learned rather than absolute positions, however, which eliminates measured motor pulley position, θ_p, from the formulation. It is therefore a velocity control approach rather than a position control approach. This may affect stability, drift and the level and source of noise in the learned trajectory, which may in turn affect the allowable gains and speed of convergence. Learning desired changes in position also affects interactions between the feedback and feed-forward elements of the controller in the presence of step-by-step variability; learned changes in position add the same way regardless of present position and error, while the contribution of learned absolute positions depends upon the present motor position. Either approach can oppose feedback contributions under some conditions, but in different ways. A detailed mathematical comparison of these two approaches is provided in Appendix 5.B.

5.3.2.6 Additional Feedback Control Terms Piloted

Several control elements that seemed likely to improve performance in theory did not fare well in pilot tests. This may be due to the unique features of the control problem at hand, in particular the noisy sensory information and the complex, changing dynamics of both the Bowden cable transmission and the human. These approaches were not included in the final data collection.

One such example is the traditional derivative control element,

$$-K_d(\dot{\tau} - \dot{\tau}_{des}),$$

which involves the derivative of torque error. Analog noise in the derivative of measured torque limited the magnitude of the derivative gain that could be applied without causing oscillations. This limited the capacity of the derivative term to stabilize the system, in turn limiting the magnitude of the proportional gains that could be applied.

Using the model described by Eq. (5.6), we next approximated the derivative term as

$$-K_d\left[(\dot{\theta}_p - \dot{\theta}_e R) - \dot{\tau}_{des} \cdot \tilde{K}_t\right]$$

where the relative velocity between the motor and exoskeleton was substituted for the noisy measured torque derivative. The derivative of desired torque is also problematic, however, because it generally cannot be calculated in advance and its numerical approximation online is subject to noise from the human measurements used by the high-level controller to calculate desired torque, for example electromyographic measurements.

We next tried using just the relative velocity between the motor pulley and exoskeleton joint,

$$-K_d(\dot{\theta}_p - \dot{\theta}_e R),$$

which is equivalent to making the additional approximation that the derivative of desired torque, $\dot{\tau}_{des}$, is negligible. However, this control element was also found to be ineffective in pilot tests due to noise on the derivative of the exoskeleton joint angle, which seems to primarily arise from stiction in the Bowden cable transmission.

Finally, we arrived at the simple damping term,

$$-K_d\dot{\theta}_p,$$

which relied only upon the derivative of motor pulley position, which had little noise to amplify, and provided sufficient damping to improve stability.

We also pilot tested proportional control without a damping term, which was effective. However, the addition of some damping always allowed for higher proportional gains and improved tracking performance. Therefore, proportional control was always used together with damping injection in our tests.

5.3.3 High-Level Assistance Controllers

5.3.3.1 Stance Torque Control

During the stance period, desired exoskeleton joint torque was set according to one of four high-level assistance controllers, $H1$–$H4$, described below.

$H1$: Time Based Desired Torque Trajectory (TIME)
   This high-level controller set desired torque as a function of time. Time-based control elements are simple and easily understood, and have been incorporated into many exoskeleton systems (Fite and Goldfarb, 2006; Cain et al., 2007; Malcolm et al., 2013, 2015). We used a curve that resembled a scaled-down version of the human ankle moment during unassisted walking, calculated as:

$$\tau_{des} =$$
$$\begin{array}{llll}
0 & <t & <0.15\xi: & 0, \\[6pt]
0.15\xi & <t & <0.30\xi: & \dfrac{\tau_p}{2}\sin(\dfrac{t-0.15\xi}{0.3\xi}\pi), \\[8pt]
0.30\xi & <t & <0.45\xi: & \dfrac{-\tau_p}{4}\cos(\dfrac{t-0.3\xi}{0.15\xi}\pi)+\dfrac{3\tau_p}{4}, \\[8pt]
0.45\xi & <t & <0.60\xi: & \dfrac{\tau_p}{2}\cos(\dfrac{t-0.45\xi}{0.15\xi}\pi)+\dfrac{\tau_p}{2}, \\[8pt]
0.60\xi & \leq t & : & 0,
\end{array} \qquad (5.28)$$

   where t is the time since the current stride began, ξ is stride period and τ_p is peak torque. We used $\xi = 1.1$ s and $\tau_p = 45$ N·m in this experiment, which produced the desired torque profile shown in Fig. 5.6.

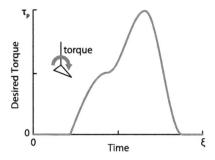

FIGURE 5.6 High-level control based on a trajectory in time.

TABLE 5.1 Angle-based control parameter values

Parameter	Value	Parameter	Value
$[\theta_0, \tau_0]$	[0.018, 0.00]	$[\theta_3, \tau_3]$	[0.00, 11.3]
$[\theta_1, \tau_1]$	[−0.122, 18.1]	$[\theta_4, \tau_4]$	[0.140, 0.00]
$[\theta_2, \tau_2]$	[−0.209, 45.2]		

$H2$: Joint Angle Based Desired Torque (ANGLE)

This high-level controller set desired torque as a function of exoskeleton ankle joint angle and phase of the gait cycle. This approach is a subset of impedance control, and is similar to setting desired torque based on a phase variable rather than clock time. Variations have been employed in many assistive devices (Au et al., 2006; Sup et al., 2009; Caputo and Collins, 2014). We used a piece-wise linear curve that resembled a scaled-down version of the human ankle moment during unassisted walking, calculated as

$$\tau_{des} = \frac{\tau_i - \tau_{i-1}}{\theta_{e,i} - \theta_{e,i-1}}(\theta_e - \theta_{e,i-1}), \quad i = \{1, 2, 3, 4\}, \tag{5.29}$$

with curve parameter values as listed in Table 5.1.

Here, (θ_i, τ_i) defines a node in torque–angle space (Fig. 5.7). The node (θ_2, τ_2) marked the transition from the dorsiflexion phase, in which ankle velocity was negative, to the plantarflexion phase, in which ankle velocity was positive. Since the exact transition point varied on each stride, we used the angle and torque at the moment of transition, (θ_2', τ_2'), when calculating desired torque in the first portion of Plantarflexion.

$H3$: Neuromuscular Model Based Desired Torque (NMM) This high-level controller set desired exoskeleton torque based on a Hill-type muscle model and a positive force feedback reflex model. The resulting dynamics produce human-like motions and muscle activation patterns in simulation

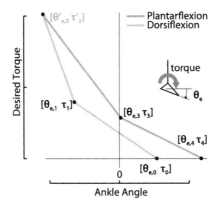

FIGURE 5.7 High-level control based on ankle joint angle.

(Song et al., 2015) and are thought to interact well with the human neuromuscular system (Eilenberg et al., 2010; Geyer and Herr, 2010; Shultz et al., 2014; Dorn et al., 2015; Van der Noot et al., 2015). Virtual muscle-tendon-unit length and velocity were set by measured exoskeleton joint angle and angular velocity. Virtual fiber length, velocity and activation were then used to determine muscle-tendon-unit force, F_{mtu}, which, after conditioning, was used to set desired exoskeleton torque. We conditioned the force signal by applying a low-pass filter with frequency ω_q, adding a small negative offset of τ_o^{nmm}, and applying a gain of K_{nmm}. Virtual muscle force was also used to drive a positive force feedback loop in which increased force led to increased muscle activation. The virtual neural system multiplied muscle force by a reflex gain, K_R, applied a time delay of D_R, added a small positive offset, $PreStim$, then applied a threshold, yielding the virtual muscle stimulation. Virtual muscle activation was driven by stimulation through first-order dynamics. A high-level schematic is provided in Fig. 5.8, high-level parameters are found in Table 5.2, and a full set of equations and parameters are available in Appendix 5.C.

$_H4$: Electromyography Based Desired Torque (EMG)

This high-level controller set desired torque in proportion to electromyographic measurements from the human gastrocnemius muscle. This approach gives the user direct neural control of the device, which is intended to make interactions more intuitive (Ferris et al., 2006; Huang et al., 2014; Takahashi et al., 2015), but can result in more complex desired torque dynamics. Electrical activity in the gastrocnemius was measured using surface electrodes and a commercial electromyography system. The signal was then high-pass filtered at a frequency of ω_{hp}, rectified, and low-pass filtered at a frequency of ω_{lp}. A small negative offset, τ_o^{emg}, was applied,

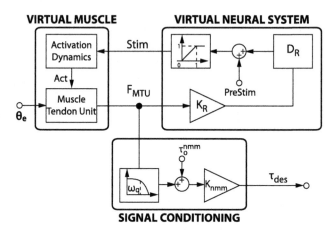

FIGURE 5.8 Neuromuscular model control schematic.

TABLE 5.2 NMM parameter values

Parameter	Value	Parameter	Value
K_R	0.002	ω_q	50 Hz
D_R	0.020 s	τ_0^{nmm}	−20
$PreStim$	0.05	K_{nmm}	0.057 N·m

which prevented desired torque generation at low levels of muscle activity. The signal was then amplified by a gain, K_{emg}, yielding desired torque. A high-level schematic is provided in Fig. 5.9, and the parameters used in this experiment can be found in Table 5.3.

5.3.3.2   Swing Control

When the foot was off the ground, motor position control was employed to allow free motion of the human ankle and maintain a small amount of slack in the cable:

$$\theta_{p,des} = \theta_e \cdot \tilde{R},$$
$$\Delta\theta_{p,des} = \theta_{p,des} - \theta_p \tag{5.30}$$

where θ_e is exoskeleton joint angle and $\tilde{R}$ is the estimated total gear ratio from motor to exoskeleton joint. Maintaining low slack in the Bowden cable reduced the time required for cable winding at the beginning of stance.

5.3.4   Experimental Methods

All experiments were conducted with one ($N = 1$) healthy adult participant (30 yrs, 56 kg, 1.65 m tall, female). Multiple participants were not warranted,

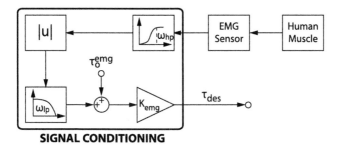

SIGNAL CONDITIONING

FIGURE 5.9 Proportional electromyography control schematic.

TABLE 5.3 EMG parameter values

Parameter	Value	Parameter	Value
F_{hp}	20 Hz	F_{lp}	6 Hz
K_{emg}	283	τ_o^{emg}	−0.008

TABLE 5.4 Low-level torque control parameter values

Parameter	Value	Parameter	Value	Parameter	Value
K_p	0.093	$\tilde{R}$	2.90	K_l	0.0077
K_d	0.010	$\tilde{K}_c$	195 N·m·rad^{-1}	D	0.022 s
K_{max}	0.15	K_s	0.005	β	1
K_{pec}	0.046	λ	0.077	μ	1
K_i	7.7e−5	L	1.0e−9I_3^*	T	0.250 s
h_τ	11.3 N·m	h_k	0.039	K_{sw}	0

* I_3 denotes a 3×3 identity matrix.

as the aim of the study was to examine torque tracking performance by the exoskeleton, not biomechanical response of the human. The participant walked on a treadmill at 1.25 m·s^{-1} with a self-selected stride period of 1.08 ± 0.06 s while wearing the exoskeleton on one leg. The participant provided written informed consent prior to participation in the study, which was conducted in accordance with a protocol approved by the Carnegie Mellon University Institutional Review Board.

Before collecting data, we tuned parameters for each combination of high- and low-level controller as the participant walked with the exoskeleton. High-level control parameters listed in the prior section were selected so as to result in peak instantaneous desired torques of approximately 45 N·m during the course of one hundred steps of walking. Low-level control parameters listed in Table 5.4 were systematically tuned with the aim of minimizing torque error.

Feedback, model and adaptive control gains in $_L1$–$_L6$ and $_L8$–$_L9$ were tuned using a variant of the Ziegler–Nichols method (Ziegler and Nichols, 1942), in which:

1. All gains (proportional, damping, integral, model, sliding, and/or adaptive) were set to zero.
2. The proportional gain was increased until significant oscillations were observed.
3. Gain value and oscillation period were then recorded and used to estimate optimal values for proportional and damping gains.
4. Fine tuning of gains for all control elements, other than iterative learning, was then performed by the experimenter.

The iterative learning gain in $_L7$ was tuned such that steady state was reached at approximately 10 strides, which led to a value of K_I that was about one tenth the tuned value of K_p. The same gains were used for iterative learning elements in controllers $_L7$–$_L9$. During tuning we found very similar suitable low-level control parameters across high-level controllers, and so used identical values within each low-level controller for consistency. Tuning was performed on a separate day from data collection. For model-based compensation, the value of $\tilde{R}$ was based on measurements of the motor output pulley radius, motor gear ratio, and exoskeleton lever arm. $\tilde{K}_c$ was estimated based on measurement of the passive relationship between exoskeleton torque and exoskeleton joint angle measured during walking experiments (Fig. 5.2).

For each high-level controller, all low-level control conditions were tested on the same day, without removal of the exoskeleton between trials. A table of condition order is presented in Supporting Materials[1] Table SI.

For each combination of low-level torque control and high-level assistance control, we collected data from 100 steady-state strides. Steady state was typically reached after about 20 strides. The subsequent 100 strides were then decomposed into individual strides, each beginning at heel strike as detected by a shoe-embedded switch. Data for an average stride were then calculated by taking the mean for each instant within the stride, in time, beginning at heel strike.

For each condition, we calculated torque error both for the set of all steady-state strides and for the average stride. We quantified torque error as the root-mean-squared error of the difference between measured and desired torque. For the set of all steady-state strides, we calculated root mean squared error for each stride individually, then calculated the mean and standard deviation. For the set of all steady-state strides, we compared means within high-level controllers and

---

1. The supporting document of this chapter is located at http://biomechatronics.cit.cmu.edu/publications/Zhang_2016_BLL—SuppMat.pdf.

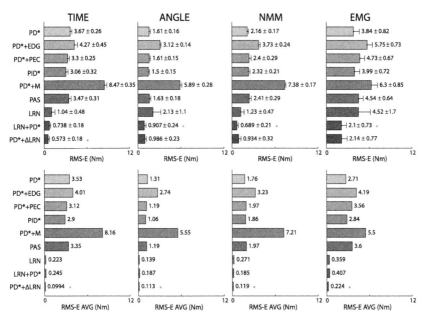

FIGURE 5.10 Root-mean-squared torque error calculated for all strides (RMS-E) and for an average stride (RMS-E AVG) across all high- and low-level control combinations.

across low-level controllers using unpaired t-tests, with a significance level of $\alpha = 0.05$.

5.3.5 Results

Means and standard deviations of stride-wise root-mean-squared torque error (RMS-E) and average-stride root-mean-squared error (RMS-E AVG) of all low- and high-level controller combinations are shown in Fig. 5.10. A complete table of p values for statistical comparisons between the RMS-E of all torque controllers are provided as Supporting Materials (Tables SII–SV). Overlapped time trajectories of desired and measured joint torques across all one hundred steady-state strides in each condition are shown in Fig. 5.11. Ankle angle trajectories in time and torque trajectories in ankle angle space are also provided for all conditions as Supporting Materials (Figs. S1–S2). Convergence plots for controllers that involved iterative learning are provided as Supporting Materials (Fig. S4).

The combination of proportional control and damping injection with iterative learning (PD* + ΔLRN or LRN + PD*) resulted in the lowest torque tracking errors for all high-level controllers, both in real-time and for average trajectories (Fig. 5.10). Of these two combinations with comparable performance, feedback control with learning compensation (PD* + ΔLRN) was

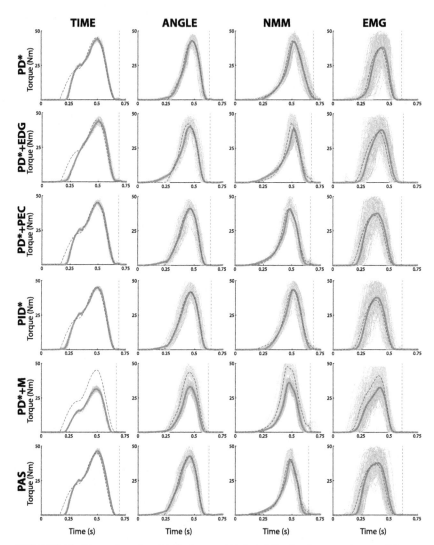

FIGURE 5.11 Time trajectories of desired torque (think pink lines) and measured torque (thin gray lines) for 100 strides of walking, and average-stride desired torque (dotted red line) and measured torque (gray line), for all combinations of controllers. (For interpretation of the references to color in this figure legend, the reader is referred to the web version of this chapter.)

simpler and converged faster. Stride-wise torque errors with PD* + ΔLRN were between 38% and 84% lower than with PD* alone ($p < 1.9 \cdot 10^{-43}$), while average-stride torque errors were between 91% and 97% lower, depending on high-level controller. Iterative learning control alone tended to result in low errors for average trajectories, but higher real-time errors than when combined

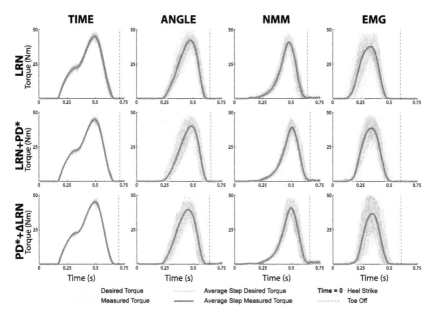

FIGURE 5.11 (*continued*)

with feedback control. Other additions to feedback control had minor effects on performance, except for model-based compensation, which increased torque error substantially. When desired torque was based on EMG, torque tracking error and variability were higher for almost all torque controllers. Values for the PD* + ΔLRN controller, including errors as a percentage of the maximum of the average desired torque, are provided in Table 5.5. The contributions of each component of the PD* + ΔLRN controller to desired motor displacement, and their evolution in time, are depicted in Fig. 5.12.

There were some interactions between high-level control type and low-level torque control performance. With Angle and EMG based high-level controllers, pure feedback control was more effective than pure iterative learning control, while for Time and NMM based controllers this trend was reversed. With Time-based desired torque, all controllers that did not have a learning component had poor tracking at the onset of desired torque, including a delay and overshoot, that comprised a large portion of the total torque error (Fig. 5.11). The addition of iterative learning to PD* control led to the greatest reductions in torque errors when desired torque was based on Time. An integral term (PID*) provided a small improvement in performance over PD* control for Time and Angle based controllers. With Time-based high-level controllers, passivity (PAS) and previous-error compensation (PD* + PEC) provided a small benefit as well.

TABLE 5.5 Tracking errors with PD* + ΔLRN torque control for all four high level controllers

	RMSE	% τ_{max}	RMSE-A	% τ_{max}
Time	0.57 ± 0.18 Nm	**1.3%**	0.10 Nm	**0.2%**
Angle	0.99 ± 0.23 Nm	2.5%	0.11 Nm	0.3%
NMM	0.93 ± 0.32 Nm	2.3%	0.12 Nm	0.3%
EMG	2.14 ± 0.77 Nm	5.9%	0.22 Nm	0.6%

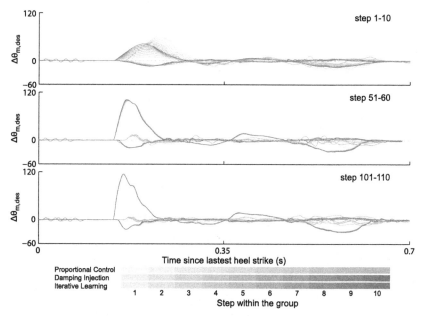

FIGURE 5.12 The contributions of each component of the PD* + ΔLRN controller at steps 1–10, steps 51–60 and steps 101–110. In the first step proportional control dominates, and at steady state the learned component dominates. Data shown are from the Time-based high-level controller. Plots for all high-level conditions are available as Supplementary Materials (Fig. S5).

5.4 DISCUSSION

In this study, we investigated the effectiveness of several prominent torque control techniques in robotic legged locomotion, implemented on a tethered ankle exoskeleton, with unidirectional series-elastic actuation, during human walking, with a variety of high-level assistance controllers. We found that model-free proportional control with damping injection compensated by iterative learning (Fig. 5.13) resulted in the lowest torque errors for all high-level controllers, both in real-time and for an averaged trajectory. This controller resulted in improved normalized torque tracking errors compared to prior torque control techniques.

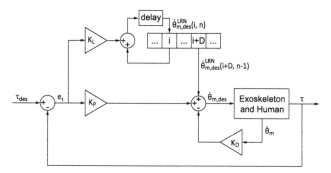

FIGURE 5.13 A block diagram of controller $_L 9$, PD* + ΔLRN.

5.4.1 Proportional-Learning-Damping Control

The most successful controller identified in this study has features that are analogous to those of classical proportional-integral-derivative control: a proportional term provides tracking during transients, iterative learning eliminates steady-state cyclic errors, and damping injection provides stability. We might therefore label the approach "proportional-learning-damping" control.

Each component of the proportional-learning-damping controller contributes to overall commands in different ways across the learning process. During the first few walking steps, proportional control is the primary contributor, moderated by damping injection, while the learned trajectory remains near its initial value (Fig. 5.12, steps 1–10). At steady state, inputs are primarily the result of learned trajectories, which anticipate and override damping injection, while proportional control compensates for step-by-step variations in required input (Fig. 5.12, steps 91–100). This results in strong performance during transients and exceptional performance at steady state.

Designing and tuning the proportional-learning-damping controller in the form of Eq. (5.27) is straightforward. First, the proportional gain on torque error K_p is slowly increased until some overshoot and oscillations are observed. Next, the damping gain on motor velocity K_d is increased until high-frequency motor oscillations are observed, and the gain is backed off of this limit. The proportional gain is then re-tuned such that it is as high as possible without resulting in oscillations in torque error. Next, the learning gain K_l is set to a value of one tenth that of the proportional gain and fine-tuned until convergence occurs within the desired time, in this case about ten strides. Finally, a parameter sweep is performed on the delay parameter D used to preview learned desired motor position. The effects of learning are sensitive to this choice; without a delay value very close to optimal, ripples in the learned desired motor position will form and grow. In those cases, non-unity values of the forgetting and error filtering terms, i.e., $\beta \neq 1$, $\mu \neq 1$, are required to stabilize the system. With the correct choice of delay, however, ripples did not form during at least one thousand strides in our experiments, even without forgetting or error filtering terms. We expect that a similar tuning process would be effective for a wide variety of lower-limb robotic systems used in legged locomotion.

This approach builds on the strengths of torque control techniques implemented in prior lower-limb exoskeletons and other legged locomotion robots. Feedback control terms similar to those tested in this study have been used in Bowden-cable driven hip–knee exoskeletons (Veneman et al., 2006), hip–knee exoskeletons with collocated drives (Zanotto et al., 2013) and mobile hip exoskeletons (Giovacchini et al., 2014). Effective joint position tracking has been achieved in a knee exoskeleton using an iterative learning approach analogous to that tested in this study (Bae and Tomizuka, 2012). Improvements in torque tracking have been achieved in a Bowden-cable driven hip–knee exoskeleton using a lower-dimensional "kernel-based" version of the iterative learning approach tested in this study (van Dijk et al., 2013). The proportional-learning-damping controller identified in this study incorporates the most effective permutations of these previously-identified control concepts.

Comparisons to prior torque tracking results can be complicated by differences in hardware. For example, the present system has higher-power off-board motors than most exoskeletons, and the unidirectional Bowden cable can go slack during the swing phase, eliminating the need for active control to achieve transparency, i.e., the ability to apply zero impedance. On the other hand, some prior exoskeletons have estimated joint torques using simplified system models rather than direct measurement, which can result in the appearance of low-error torque tracking despite substantial unmeasured torque errors. Nevertheless, the proportional-learning-damping controller identified in this study achieved the lowest torque errors as a percentage of desired torque of any exoskeleton to date.

It is likely that iterative learning of the form used here or in other studies, would improve torque tracking during most human locomotor activities, since even irregular gaits exhibit some degree of repeatability. The case of the EMG-based high-level controller provides an insight into such scenarios, because the EMG signal contains substantial noise and is highly variable from step to step (Fig. 5.11; Fig. S2). Despite these irregularities, the addition of a learning component reduced torque tracking errors by 45% compared to feedback control alone in the EMG condition.

Further improvements in torque tracking for some high-level controllers might have been possible with alternate phase variables. Iterative learning resulted in the greatest improvements in torque tracking for the Time-based high-level controller, presumably because motor position adjustments were also learned in time. Learning as a function of ankle angle in the Angle condition, for example, might have resulted in greater improvements. On the other hand, time provides a unique and monotonically increasing phase variable, with consistent indexing across steps, capable of capturing control inputs with very high levels of complexity. This topic merits further exploration.

Although this case study was conducted on an ankle exoskeleton, the system characteristics that lead to the effectiveness of the proportional-damping-learning controller during torque tracking also apply to other robotic legged locomotion systems such as lower-limb prostheses and walking robots. The complex, uncertain and changing dynamics introduced by transmissions, gait variations, or human–robot interactions make model-based and continuous-time integral control actions ineffective. The cyclic behavior of walking leads to improved performance with the addition of an iterative learning element. The presence of delays leads to a benefit from a predictive feed-forward element, in this case a learned compensation. These factors are discussed in the following section.

5.4.2   Benefits of Additional Control Elements

There appeared to be some interactions between high-level and low-level control elements, which might provide insights into strategies for circumstances that were not tested in this experiment.

5.4.2.1   Continuous-Time Integration

When proportional-damping control was augmented by an integral term (PID*), previous-error compensation (PD* + PEC), or passivity-based adaptation (PAS), torque error with Time-based high-level control was slightly improved. However, these continuous-time integral components showed no effect or negative effects in tracking performance for other high-level controllers. We can identify two factors that may explain the ineffectiveness of integral control in these cases. One factor is integral windup; the rapid changes in set point over the course of the stride could lead to either excessive or insufficient error accumulation. Another factor is the nonlinear, changing, delayed dynamics of the system; the Bowden cable has nonlinear stiffness, both the transmission and human change in time, and there are delays in the control loop and in motor actuation. All of these factors are known to limit integral control performance due to the linear accumulation of torque errors that are not linearly comparable (Sung and Lee, 1996). In the case of the Time-based high-level controller, continuous-time integration may have been more effective due to the consistency of the desired torque in time, which may have translated into more constant torque errors than with other high-level controllers.

The apparent success of integral terms in other robotic legged locomotion systems with Bowden-cable transmissions may be due to differences in torque sensing, cable composition or activities tested. For example, controllers of the Lopes system have typically included an integral term on torque error (Veneman et al., 2006; Vallery et al., 2007; van Dijk et al., 2013). Differences might relate to torque sensing. In the present study, torque was measured at the joint using strain gauges. In Lopes, torque has been estimated from relative angle of the motor and joint or from series spring deflection, which might result in a more linear relationship between motor angle input and apparent joint torque. Differences might also relate to hardware. The Bowden cables in Lopes have features that could make them more consistent with a simple spring model, such as stiffer, pre-stretched steel cables and bidirectional drives that do not allow slack. Differences might also relate to the characteristic behavior of the joint being assisted. Lopes assists the hip and knee, which have relatively smooth, continuous patterns of joint torque. In this study, we assisted the ankle joint, which typically involves sharper changes in dynamics. For example, the foot intermittently contacts the ground, discontinuously changing both the impedance of the ankle joint and the magnitude of desired torques. These changes connote rapid changes in set point for the controller, which leads to windup of integral control elements, among other challenges. The proportional-learning-damping controller developed here may therefore be expected to provide strong torque tracking performance under a larger range of conditions.

5.4.2.2 Model-Based Control Elements

Although model-based control elements show promise in simulation and in theory, these generally worsened or had no effect on tracking performance in our experiment. One might expect that better performance could have been obtained with a more accurate estimate of model parameters in the PD* + M controller. However, in exploratory tests we found that the best performance was obtained by driving model-based contributions to zero. One reason for the ineffectiveness of model-based compensation may have been the nonlinear stiffness of the Bowden cable, which we modeled as a linear spring, and slow changes in cable stiffness and length due to heating over the course of each trial. Another reason may have been the exclusion of friction and stiction. However, in pilot tests using model-based compensation that included these elements, we found them to make the controller less robust; in each case the effects were highly sensitive to choice of parameter value and torque tracking error was not reduced. Sensitivity to model errors seems to be a fundamental issue in implementing this type of inverse-model control in devices with Bowden cable transmissions.

The passivity-based controller (PAS) fared slightly better, perhaps due to its adaptive nature, but still did not yield substantial benefits. One factor that may have limited its effectiveness is input mismatch; the controller was designed with motor voltage as input (Eq. (5.18)), but implemented using motor velocity commands instead. While these terms are closely related (Eq. (5.11)), it is possible that an alternate mode of motor control, or an alternate formulation, could have led to improved results. Another factor that may have limited performance was the inclusion of a term akin to continuous-time integration, which is subject to windup as discussed above. The primary limitations likely stem from the reliance on any explicit model, however, since the dynamics of this human–robot system are highly complex and time-varying.

Some prior controllers have used model-based control elements, apparently to good effect. For example, Zanotto et al. (2013) used continuous-time integration and model-based terms that were not found to improve performance in the present study. This may be because the Bowden cable transmission used here relocated heavy actuators off of the leg, making gravity compensation unnecessary, but had more unmodeled dynamics than a gear train, making friction compensation less effective. The proportional-learning-damping controller described here resulted in lower torque tracking error than in prior systems, suggesting improved performance might be achieved even in systems for which model-based control has been effective. The proportional-learning-damping controller described here resulted in lower torque tracking error than in prior systems, suggesting improved performance might be achieved even in systems for which model-based control has been effective.

5.4.2.3 Gain Scheduling, Optimal Control, and Learning

Error-dependent gains (PD* + EDG) did not provide benefits for any high-level controller. Lower gains when torque errors were low seem to have led to larger errors at other times, since the set point changed rapidly and there were substantial execution delays. Gain scheduling methods that instead use optimal control might improve torque tracking for this system, but such feedback control techniques would still be limited by communication and actuation delays. By contrast, iterative learning realizes another form of optimal control, but uses a feed-forward approach to overcome delays. The iterative learning controller developed here is a variation on one-dimensional root finding using Newton's method. The problem is to find the desired motor position ($_L7$ and $_L8$) or displacement ($_L9$) for zero torque error, i.e., to solve the equation

$$e_\tau(\Delta\theta_{m,des}) = 0. \tag{5.31}$$

The solutions are approximated in an iterative manner by

$$\Delta\theta_{m,des}(n+1) = \Delta\theta_{m,des}(n) - K_l \cdot e_\tau(n), \tag{5.32}$$

which can be rewritten as

$$\Delta\theta_{m,des}(n+1) = \Delta\theta_{m,des}(n) - \frac{e_\tau(\Delta\theta_{m,des}(n))}{K_l^{-1}}. \tag{5.33}$$

This demonstrates a variation of Newton's method to solve $e_\tau = 0$ with the estimate for the derivative of e_τ fixed as $e_\tau' = K_l^{-1}$. Unlike gain scheduling in feedback control, this optimal control approach addresses control delays through the combination of a feed-forward term and a delay-compensating prediction term. Therefore, even with an optimized gain schedule, PD* + EDG probably would not out-perform PD* + ΔLRN.

5.4.3   Factors Limiting Interpretation

5.4.3.1   High-Level Controllers

It is difficult to make comparisons across high-level controllers for the same low-level torque controller because tracking difficulty may vary with desired torque pattern. For example, we can imagine a Time-based trajectory with step changes in desired torque for which precise tracking would be infeasible. Similarly, the ease of tracking Angle-based desired torques likely depends on the similarities between the target torque-angle curve and the passive relationship arising from series elasticity. EMG-based desired torques generally seem to be difficult to track, given the unpredictable signals that directly drive desired torque, but increasing filter frequency could make this task easier. We did not test multiple values for high-level parameters in this experiment, which is an area for future work.

5.4.3.2   Interactions with Human Response

We designed this experiment with the implicit expectation that low-level torque control would not significantly affect human response to high-level assistance modes, but this does not appear to have been the case. Changes in the patterns of desired torque (Fig. 5.11) and joint kinematics (Supporting Materials, Section II) across torque controllers within the same high-level controller reveal an interaction effect. For example, we found more variability in joint kinematics with PD* torque control than with LRN torque control when desired torque was generated on the basis of Time. In this case, differences seem to be related to the smoothness of the measured torque generated by the two controllers; the subject reported that the PD* controller had uncomfortable oscillations, leading to compensatory activity, while the LRN controller did not. As another example, we found more variability in desired torque with NMM-based assistance than Angle-based assistance using PD* torque control, but an opposite trend using LRN torque control. This appears to be the result of complex, multi-time-scale, dynamic interactions between continuous behavior of the torque controller, within-stride variations by the human, high-level control responses, and human adaptation over multiple strides. These effects may also be important in selecting and tuning an exoskeleton or prosthesis torque controller.

5.4.3.3 Hardware Dependence

Robot hardware, particularly series compliance, also interacted with the quality of torque tracking. We performed pilot tests with no series spring, other than the Bowden cable, and found significant increases in torque error and subject discomfort for all control combinations. We also tried more compliant series springs, and found small increases in torque tracking error. When series stiffness is too high, we expect small position changes by the human to result in large, undesirable, changes in torque, and when it is too low we expect motor dynamics to limit performance (Pratt and Williamson, 1995). Interactions between series elasticity, low-level torque control, high-level assistance, and assisted task should be explored in the future.

Some of the control elements found to be ineffective in this system might be more effective in devices with different hardware or task characteristics. For example, there are many examples of torque controllers incorporating model-based terms, often accompanied by hardware that lends itself more readily to modeling (e.g., Zanotto et al., 2013). Experimental comparisons of torque tracking performance with and without model-based terms in these systems would lend further insights into their potential contributions to effective torque tracking in exoskeletons and other robotic legged locomotion systems, and might provide a useful point of contrast to the present work. Future work could also address the effects of hardware changes as an additional dimension in the space sampled here. Measuring performance for each combination of a set of high-level controllers, low-level controllers and hardware setups would provide the best insights into interactions between these features. As the dynamical contributions of nonlinear time-varying elements, intermittent contact or forceful human interactions increase, continuous-time integral terms, gain scheduling, and model-based compensations are expected to become less effective, as observed with the unidirectional Bowden-cable-driven ankle-foot exoskeleton used in this study. The present system therefore represents a more challenging case for torque control, suggesting that the proportional-learning-damping controller identified here would perform even better in a system with simpler, more consistent dynamics.

The uni-directional Bowden cable used in this study made torque control more challenging, but is not the dominant factor in the observed patterns of torque error. The onset of applied torque tended to lag that of desired torque, particularly with feedback controllers and TIME based desired torque. This pattern might suggest that slack in the Bowden cable was the primary cause of torque error, but this is not the case. All high-level controllers set desired torque to zero during the beginning of stance, from about 0 to 0.25 s, consistent with typical human ankle torque patterns. All low-level controllers acted to track desired torque throughout stance, quickly eliminating slack from the previous swing phase early in the period of zero desired torque. The Bowden cable was therefore not slack upon the onset of desired torque. This is evident from the pattern of torque with feedback control and ANGLE based desired torque, which led desired torque beginning at about 0.10 s. Patterns in torque error across controllers are better-explained by a combination of rapid changes in desired torque, rapid movements of the human, and electromechanical delays in applying desired changes in motor position. These issues are common to most lower-limb exoskeletons, prostheses and walking robots, especially systems using series-elastic actuation.

5.4.4 Implications for Control of Future Systems

The insights gained from this study are expected to help guide the design of torque controllers for systems with similar traits, particularly complex, changing dynamics and cyclic motions, such as lower-limb exoskeletons, active lower-limb prostheses and walking robots. Based on the present results, there is reason to expect that the combination of feedback control and feed-forward iterative learning, without continuous-time integration or model-based compensation, will provide strong torque tracking performance in any such system. Other control elements might further improve performance for some systems and control objectives. For example, if system dynamics are relatively constant and easy to identify, model-based compensation might be useful. If the measured and desired torque both change slowly, continuous-time integral control may also lead to some improvements. In any case, proportional-learning-damping control is expected to provide good baseline torque tracking.

5.5 CONCLUSIONS

We performed a systematic comparison of torque control techniques in a robotic legged locomotion system under realistic operating conditions, and found that the combination of proportional control, damping injection and iterative learning resulted in smaller torque errors relative to peak torque than any other approach tested or previously demonstrated. Designing this proportional-learning-damping controller was straightforward, requiring sequential tuning of only four parameters. Our results generally support such an approach for any torque-controlled lower-limb exoskeleton, prosthesis and walking robot. The complex interactions between device hardware, torque control, assistance control, task goals and human behavior, in case of exoskeletons and prostheses, remain a rich area for future research.

ACKNOWLEDGEMENTS

The authors thank Rachel Jackson and Kirby Witte for hardware assistance. This material is based upon work supported by the National Science Foundation under Grant No. IIS-1355716 and by the Singapore Economic Development Board NTU-CMU Dual PhD Scholarship.

APPENDIX 5.A   STABILITY AND CONVERGENCE OF THE PASSIVITY BASED CONTROLLER

5.A.1   Passivity

Substituting Eq. (5.19) into Eq. (5.18), we have the closed-loop equation:

$$\frac{1}{K_v}\dot{s} + \frac{K_1}{K_v}s + K_p e_\tau + K_s s + Y_d(\tau, \dot{\tau}_r, \ddot{\tau}_r, \dot{\theta}_e)\Delta\Gamma = \frac{K_h}{K_v}\tau_h, \qquad (5.34)$$

in which $\Delta\Gamma = \Gamma - \tilde{\Gamma}$.

Let the output be $y = s$. Multiplying both sides of Eq. (5.34) by the output and then integrating it, we have

$$\int_0^t \frac{K_h}{K_v}s(\varsigma)\tau_h(\varsigma)d\varsigma$$

$$= \int_0^t [\frac{1}{K_v}s(\varsigma)\dot{s}(\varsigma) + \frac{K_1}{K_v}s^2(\varsigma) + K_p\dot{e}_\tau(\varsigma)e_\tau(\varsigma) + K_p\lambda e_\tau^2(\varsigma)$$

$$+ K_s s^2(\varsigma) + s(\varsigma)Y_d(\varsigma)\Delta\Gamma(\varsigma)]d\varsigma$$

$$= \underbrace{P(t) - P(0)}_{\text{Stored Energy Change}} + \underbrace{\int_0^t W(\varsigma)d\varsigma}_{\text{Dissipated Energy}},$$

in which the stored system energy is

$$P = \frac{1}{2K_v}s^2 + \frac{1}{2}K_p e_\tau^2 + \frac{1}{2}\Delta\Gamma^\intercal L^{-1}\Delta\Gamma \geq 0 \qquad (5.35)$$

and the energy dissipating rate is

$$W = (\frac{K_1}{K_v} + K_s)s^2 + K_p\lambda e_\tau^2 \geq 0. \qquad (5.36)$$

Since both V and W are nonnegative, we can conclude the passivity of output y for the system (5.34).

With no disturbance from human, i.e., $\tau_h = 0$, we have

$$\dot{P} = -W.$$

5.A.2 Convergence

To compensate for nonzero human loading, $\tau_h \neq 0$, an additional switching term is added to the controller

$$V_a = -K_p e_\tau - K_s s + Y_d(\tau, \dot{\tau}_r, \ddot{\tau}_r, \dot{\theta}_e)\tilde{\Gamma} - K_{sw}\mathrm{sign}(s) \qquad (5.37)$$

where K_{sw} is the compensation gain. With Eq. (5.37), the closed-loop system equation becomes

$$\frac{1}{K_v}\dot{s} + \frac{K_1}{K_v}s + K_p e_\tau + K_s s + Y_d(\tau, \dot{\tau}_r, \ddot{\tau}_r, \dot{\theta}_e)\Delta\Gamma$$
$$- \frac{K_h}{K_v}\tau_h + K_{sw}\mathrm{sign}(s) = 0. \qquad (5.38)$$

The integral of the product of $y = s$ with the above equation is

$$\int_0^t [\frac{1}{K_v}s(\varsigma)\dot{s}(\varsigma) + \frac{K_1}{K_v}s^2(\varsigma) + K_p\dot{e}_\tau(\varsigma)e_\tau(\varsigma) + K_p\lambda e_\tau^2(\varsigma)$$

$$+ K_s s^2(\varsigma) + s(\varsigma)Y_d(\varsigma)\Delta\Gamma(\varsigma) - s\frac{K_h}{K_v}\tau_h + sK_{sw}\mathrm{sign}(s)]d\varsigma$$

$$= \underbrace{P(t) - P(0)}_{\text{Stored Energy Change}} + \underbrace{\int_0^t W(\varsigma)d\varsigma}_{\text{Dissipated Energy}}$$

$$+ \int_0^t [-s\frac{K_h}{K_v}\tau_h + sK_{sw}\mathrm{sign}(s)]d\varsigma$$

$$= 0,$$

i.e.,

$$\dot{P} = -W + s\frac{K_h}{K_v}\tau_h - sK_{sw}\mathrm{sign}(s). \qquad (5.39)$$

Note that

$$s\frac{K_h}{K_v}\tau_h - sK_{sw}\mathrm{sign}(s) = s\frac{K_h}{K_v}\tau_h - K_{sw}|s|$$

$$\leq \frac{K_h}{K_v}|s|k_0 - K_{sw}|s|$$

where k_0 is a finite nonnegative number defined by the boundedness of the human loading,

$$|\tau_h| \leq k_0.$$

By choosing a K_{sw} such that

$$K_{sw} \geq \frac{K_h}{K_v} k_0,$$

we have

$$s \frac{K_h}{K_v} \tau_h - s K_{sw} \text{sign}(s) \leq 0. \tag{5.40}$$

It is already established that

$$W \geq 0,$$

therefore, it is proved that

$$\dot{P} \leq 0.$$

Based on the nonnegativeness of P and nonpositiveness of $\dot{P}$, we have

$$P(t) \leq P(0),$$

i.e., $P(t)$ is bounded. Since P is quadratic in s, e_τ and $\Delta\Gamma$, these three terms are bounded.

Eq. (5.39) means

$$P(t) - P(0) = -\int_0^t W(\varsigma)d\varsigma + \int_0^t \left[s(\varsigma) \frac{K_h}{K_v} \tau_h(\varsigma) - K_{sw}|s(\varsigma)| \right] d\varsigma.$$

$$\tag{5.41}$$

Combining Eq. (5.41) with Eq. (5.40) and (5.36), we have

$$P(t) - P(0) \leq -\int_0^t W(\varsigma)d\varsigma \leq 0. \tag{5.42}$$

By definition, Eqs. (5.36) and (5.42) lead to

$$s, e_\tau \in L_2(0, \infty). \tag{5.43}$$

Based on the definition of s, $\dot{e}_\tau$ is also bounded, and hence so is $\dot{\tau}$ since $\dot{\tau}_{des}$ is bounded. The boundedness of $\dot{e}_\tau$ proves the uniform continuity of e_τ.

Now due to the boundedness of $\dot{\tau}_{des}$ and $\ddot{\tau}_{des}$, $\dot{\tau}_r$ and $\ddot{\tau}_r$ are also bounded, which then lead to the boundedness of $Y_d(\tau, \dot{\tau}_r, \ddot{\tau}_r, \dot{\theta}_e)$. Based on Eq. (5.38), $\dot{s}$ is then bounded, which then proves the uniform continuity of s.

Combining the conclusions of uniform continuity, boundedness and Eq. (5.43), we have (Desoer and Vidyasagar, 1975; Arimoto, 1996):

$$s \to 0, \quad \text{as } t \to \infty,$$
$$e_\tau \to 0, \quad \text{as } t \to \infty.$$

By the definition of s and e_τ, we then can conclude

$$\tau \to \tau_{des}, \quad \text{as } t \to \infty,$$
$$\dot{\tau} \to \dot{\tau}_{des}, \quad \text{as } t \to \infty,$$

i.e., both the actual ankle torque and its changing rate converge to the desired values.

APPENDIX 5.B   PD* + ΔLRN VERSUS LRN + PD*

For controller $_L9$, i.e., PD* + ΔLRN, as described in Eq. (5.27), the total desired motor displacement at a certain time stamp t within step n, with a forgetting factor $\beta = 1$ and a filtering factor $\mu = 1$, is

$$\Delta\theta_{p,des}(n)\Big|_t = \underbrace{\Delta\theta^{LRN}_{p,des}(0)\Big|_t - K_I \sum_{m=1}^{n-1} e_\tau(m)\Big|_t}_{\text{Step-wise Integral Control}} \tag{5.44}$$

$$\underbrace{-K_p e_\tau(n)\Big|_t}_{\text{Proportional Control}} \underbrace{-K_d \dot{\theta}_p(n)\Big|_t}_{\text{Damping Injection}},$$

in which the operator

$$*(n)\Big|_t$$

denotes the variable $*$ at time t within step n, i.e., the time lapsed from the latest heel strike. The term

$$\Delta\theta^{LRN}_{p,des}(0)\Big|_t$$

denotes the initial value of the desired motor displacement to be learned. Iterative learning of desired motor displacement realizes a stepwise integral control action. Considering Eq. (5.44), it can be seen that $_L9$ is analogous to traditional PID control; damping injection improves system stability in the manner of derivative control, and iterative learning takes eliminates steady-state errors across steps similar to integral control.

For controller $_L8$, LRN + PD*, combining Eqs. (5.26) and (5.12), we have

$$
\begin{aligned}
\dot{\theta}_{p,des}(n)\Big|_t &= \frac{1}{T}\Delta\theta_{p,des}(n)\Big|_t \\
&= \frac{1}{T}\Bigg[\theta_{p,des}^{LRN}(0)\Big|_t - K_I\sum_{m=1}^{n-1}e_\tau(m)\Big|_t - \theta_p(n)\Big|_t \\
&\quad - K_p e_\tau(n)\Big|_t - K_d\dot{\theta}_p(n)\Big|_t\Bigg]
\end{aligned}
\tag{5.45}
$$

when $\beta = 1$ and $\mu = 1$.

Assuming perfect motor velocity tracking, namely

$$
\dot{\theta}_p(n)\Big|_t = \dot{\theta}_{p,des}(n)\Big|_t,
\tag{5.46}
$$

Eq. (5.26) becomes

$$
\begin{aligned}
\dot{\theta}_p(n)\Big|_t &= \frac{1}{T}\Bigg[\theta_{p,des}^{LRN}(0)\Big|_t - K_I\sum_{m=1}^{n-1}e_\tau(m)\Big|_t - \theta_p(n)\Big|_t \\
&\quad - K_p e_\tau(n)\Big|_t - K_d\dot{\theta}_p(n)\Big|_t\Bigg]
\end{aligned}
\tag{5.47}
$$

which can be written as

$$
\begin{aligned}
\dot{\theta}_p(n)\Big|_t &= -\frac{1}{T + K_d}\theta_p(n)\Big|_t \\
&\quad + \frac{1}{T + K_d}\Bigg[\theta_{p,des}^{LRN}(0)\Big|_t - K_I\sum_{m=1}^{n-1}e_\tau(m)\Big|_t \\
&\quad - K_p e_\tau(n)\Big|_t\Bigg].
\end{aligned}
\tag{5.48}
$$

Assuming the same ankle kinematics for each step and the same motor position at each heel strike,

$$
\theta_p(n)\Big|_0 = \theta_p\Big|_0,
$$

the dynamics described in Eq. (5.48) can be treated as a linear-time-invariant system starting from the latest heel strike in the format of

$$
\dot{x} = Ax + Bu
\tag{5.49}
$$

where x is $\theta_p(n)$, A is $-\dfrac{1}{T+K_d}$, and Bu is

$$\frac{1}{T+K_d}\left[\left.\theta_{p,des}^{LRN}(0)\right|_t - K_I\sum_{m=1}^{n-1}\left.e_\tau(m)\right|_t - \left.K_pe_\tau(n)\right|_t\right].$$

Therefore, the solution of LTI system as described in Eq. (5.48) is

$$
\begin{aligned}
\left.\theta_p(n)\right|_t = &\left.\theta_p\right|_0 \exp(-\frac{t}{T+K_d})\\
&+\frac{1}{T+K_d}\int_0^t \exp(-\frac{t-\varsigma}{T+K_d})\left[\left.\theta_{p,des}^{LRN}(0)\right|_t\right.\\
&\left.- K_I\sum_{m=1}^{n-1}\left.e_\tau(m)\right|_\varsigma - \left.K_pe_\tau(n)\right|_\varsigma\right]d\varsigma,
\end{aligned}
\tag{5.50}
$$

in which t denotes the time lapsed since the last heel strike and $\varsigma \in [0, t]$ is a variable tracing t.

Therefore, the control input of $_L8$ can be expressed as:

$$
\begin{aligned}
\left.\Delta\theta_{p,des}(n)\right|_t = &\underbrace{\left.\theta_{p,des}^{LRN}(0)\right|_t - K_I\sum_{m=1}^{n-1}\left.e_\tau(m)\right|_t}_{\text{Step-wise Integral Control}}\ \underbrace{\left.-K_pe_\tau(n)\right|_t}_{\text{Proportional Control}}\ \underbrace{\left.-K_d\dot\theta_p(n)\right|_t}_{\text{Damping Injection}}\\
&-\left.\theta_p\right|_0 \exp(-\frac{t}{T+K_d}) - \frac{1}{T+K_d}\int_0^t \exp(-\frac{t-\varsigma}{T+K_d})\\
&\times\left[\theta_{p,des}^{LRN}(0) - K_I\sum_{m=1}^{n-1}\left.e_\tau(m)\right|_\varsigma - \left.K_pe_\tau(n)\right|_\varsigma\right]d\varsigma.
\end{aligned}
\tag{5.51}
$$

It can be seen that the LRN + PD* controller differs from the PD* + ΔLRN controller in that it has additional exponential and low-pass filtered continuous-time integral terms. These arise from the presence of $-\theta_p(n)$ in the control input.

This difference can also be illustrated in the frequency domain. The convolution term

$$\int_0^t \exp(-\frac{t-\varsigma}{T+K_d})\left[\left.\theta_{p,des}^{LRN}(0)\right|_t - K_I\sum_{m=1}^{n-1}\left.e_\tau(m)\right|_\varsigma - \left.K_pe_\tau(n)\right|_\varsigma\right]d\varsigma \tag{5.52}$$

translates to

$$\frac{1}{s + \dfrac{1}{T + K_d}} \left[\left. \theta^{LRN}_{p,des}(0) \right|_s - K_l \left. \sum_{m=1}^{n-1} e_\tau(m) \right|_s - K_p \left. e_\tau(n) \right|_s \right] \qquad (5.53)$$

in the frequency domain. The low-pass filter term

$$\frac{1}{s + \dfrac{1}{T + K_d}}$$

realizes constant scaling at low frequency and integration at high frequency. Each evaluation of the term to be filtered,

$$\left. \theta^{LRN}_{p,des}(0) \right|_t - K_l \left. \sum_{m=1}^{n-1} e_\tau(m) \right|_t - K_p \left. e_\tau(n) \right|_t ,$$

is generated independently at the sampling frequency. Therefore, the term is expected to have high frequency. A continuous-time integration effect is therefore included in the control input with the LRN + PD* controller.

In our experimental comparison, we found that continuous-time integral action is not desirable due to the linear accumulation of terms that are not linearly correlated. Therefore, PD* + ΔLRN is preferred due to its simplicity and lack of continuous-time integration.

APPENDIX 5.C   NEUROMUSCULAR REFLEX MODEL

The neuromuscular model included a Hill-type muscle-tendon-unit (MTU) (Fig. 5.14), with total length, l_{mtu}, equal to the length of the series elastic element, l_{se}, plus that of the contractile element, l_{ce}. The length of the parallel elastic element, l_{pe}, was identical to l_{ce}. The total force produced by the MTU, F_{mtu}, was equal to the force on the series elastic element, F_{se}, and also equal to the sum of the contractile element force, F_{ce}, and the parallel elastic element force, F_{pe} (which was usually zero). The contractile element force was expressed as

$$F_{ce} = f_L(l_{ce}) \cdot f_V(\dot{l}_{ce}) \cdot Act \cdot F_{max},$$

in which $f_L(l_{ce})$ and $f_V(\dot{l}_{ce})$ are force scaling factors that reflect the force–length and force–velocity properties of muscle, Act is the muscle activation state, and F_{max} is the maximum force that can be produced by the muscle. The

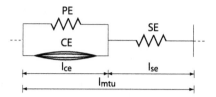

FIGURE 5.14 The Hill-type muscle-tendon-unit model includes a contractile element (CE), a parallel elastic element (PE) and series elastic element (SE).

force–length and force–velocity relationships were:

$$f_L(l_{ce}) = \exp[-\frac{(l_{ce} - l_{ce,opt})^2}{(W \cdot l_{ce,opt})^2}],$$

and

$$f_V(\dot{l}_{ce}) = \begin{cases} A\dfrac{v_{max} + F_v^{max} \cdot \dot{l}_{ce}}{v_{max} \cdot A + \dot{l}_{ce}}, & \dot{l}_{ce} < 0, \\[3mm] \dfrac{v_{max} \cdot A + F_v^{max} \cdot (A+1) \cdot (F_v^{max} - 1)}{v_{max} \cdot A + \dot{l}_{ce} \cdot (A+1) \cdot (F_v^{max} - 1)}, & \dot{l}_{ce} \geq 0, \end{cases}$$

in which $l_{ce,opt}$ is the optimal contractile element length, W is a force–length constant, $\dot{l}_{ce}$ is the contractile element velocity, A is a force–velocity constant, v_{max} is the maximum contraction velocity, and F_v^{max} is the maximum eccentric (lengthening) muscle scaling factor.

Series and parallel elastic element forces were determined as:

$$F_{se} = F_{max} \cdot \frac{1}{(u_{max} \cdot l_{se,sl})^2} \cdot \max(\delta l_{se}, 0)^2$$

and

$$F_{pe} = \begin{cases} k_1 \cdot \delta l_{pe} + F_{max} \cdot \dfrac{k_{pe}}{l_{ce,opt}^2} \cdot \delta l_{pe}^2, & \delta l_{pe} > 0, \\[3mm] k_1 \cdot \delta l_{pe}, & \delta l_{pe} \leq 0, \end{cases}$$

where $\delta l_{pe} = l_{ce} - 1.5 \cdot l_{ce,opt}$ and $\delta l_{se} = l_{se} - l_{se,sl}$, and $l_{se,sl}$ is the slack length of the series elastic element, u_{max} is an elastic element curve parameter and k_1 is the subslack elastic element stiffness of the parallel elastic element.

The virtual muscle had activation dynamics as:

$$Act(t = 0) = PreAct,$$

$$\dot{Act} = (Stim - Act) \cdot \left(\frac{Stim}{t_a} - \frac{1 - Stim}{t_d}\right)$$

TABLE 5.6 Muscle model parameter values used in NMM

Parameter	Value	Parameter	Value
$l_{ce,opt}$	0.055 m	$l_{se,sl}$	0.245 m
$l_{mtu,0}$	0.284 m	F_{max}	1000 N
A	0.25	W	0.56
$F_{v,max}$	1.8	u_{max}	0.04
v_{max}	0.55 m/s	k_1	1
t_a	0.02 s	t_d	0.05 s
$PreAct$	0.05		

where $Stim$ is neural stimulation from the positive reflex mechanism of the virtual neural system, t_a and t_d are muscle activation and deactivation time constants, respectively, and the initial activation value $A(t = 0)$ is defined by $PreAct$.

The parameter values used in this study are listed in Table 5.6.

REFERENCES

Accoto, D., Carpino, G., Sergi, F., Tagliamonte, N.L., Zollo, L., Guglielmelli, E., 2013. Design and characterization of a novel high-power series elastic actuator for a lower limb robotic orthosis. Int. J. Adv. Robot. Syst. 10, 359.

Aguirre-Ollinger, G., Colgate, J.E., Peshkin, M.A., Goswami, A., 2007. Active-impedance control of a lower-limb assistive exoskeleton. In: Proceedings of the IEEE International Conference on Rehabilitation Robotics (ICORR). IEEE, pp. 188–195.

Andrews, J.R., Hogan, N., 1983. Impedance control as a framework for implementing obstacle avoidance in a manipulator. In: Control of Manufacturing Processes and Robotic Systems. ASME, pp. 243–251.

Arimoto, S., 1996. Control Theory of Nonlinear Mechanical Systems. Oxford University Press, Inc.

Arimoto, S., Takegaki, M., 1981. A new feedback method for dynamic control of manipulators. J. Dyn. Syst. Meas. Control 102, 119–125.

Arimoto, S., Kawamura, S., Miyazaki, F., 1984. Bettering operation of robots by learning. J. Robot. Syst. 1 (2), 123–140.

Au, S.K., Dilworth, P., Herr, H., 2006. An ankle-foot emulation system for the study of human walking biomechanics. In: Proceedings of the IEEE International Conference on Robotics and Automation (ICRA). IEEE, pp. 2939–2945.

Au, S., Berniker, M., Herr, H., 2008. Powered ankle-foot prosthesis to assist level-ground and stair-descent gaits. Neural Netw. 21 (4), 654–666.

Bae, J., Tomizuka, M., 2012. A gait rehabilitation strategy inspired by an iterative learning algorithm. Mechatronics 22 (2), 213–221.

Banala, S.K., Kim, S.H., Agrawal, S.K., Scholz, J.P., 2009. Robot assisted gait training with active leg exoskeleton (ALEX). IEEE Trans. Neural Syst. Rehabil. Eng. 17 (1), 2–8.

Cai, L.L., Fong, A.J., Otoshi, C.K., Liang, Y., Burdick, J.W., Roy, R.R., Edgerton, V.R., 2006. Implications of assist-as-needed robotic step training after a complete spinal cord injury on intrinsic strategies of motor learning. J. Neurosci. 26 (41), 10564–10568.

Cain, S.M., Gordon, K.E., Ferris, D.P., 2007. Locomotor adaptation to a powered ankle-foot orthosis depends on control method. J. NeuroEng. Rehabil. 4.

Calanca, A., Fiorini, P., 2014. Human-adaptive control of series elastic actuators. Robotica 32 (08), 1301–1316.

Caputo, J.M., Collins, S.H., 2013. An experimental robotic testbed for accelerated development of ankle prostheses. In: Proceedings of the IEEE International Conference on Robotics and Automation (ICRA). IEEE, pp. 2645–2650.

Caputo, J.M., Collins, S.H., 2014. Prosthetic ankle push-off work reduces metabolic rate but not collision work in non-amputee walking. Sci. Rep. 4.

Cavallaro, E., Rosen, J., Perry, J.C., Burns, S., Hannaford, B., 2005. Hill-based model as a myoprocessor for a neural controlled powered exoskeleton arm – parameters optimization. In: Proceedings of the IEEE International Conference on Robotics and Automation (ICRA). Barcelona, Spain.

Cloud, W., 1965. Man amplifiers: machines that let you carry a ton. Pop. Sci. 187 (5), 70–73.

Collins, S.H., Jackson, R.W., 2013. Inducing self-selected human engagement in robotic locomotion training. In: Proceedings of the IEEE International Conference on Rehabilitation Robotics (ICORR). IEEE, pp. 1–6.

Collins, S., Ruina, A., Tedrake, R., Wisse, M., 2005. Efficient bipedal robots based on passive-dynamic walkers. Science 307 (5712), 1082–1085.

Collins, S.H., Wiggin, M.B., Sawicki, G.S., 2015. Reducing the energy cost of human walking using an unpowered exoskeleton. Nature 522, 212–215.

de Luca, A., Albu-Schaffer, A., Haddadin, S., Hirzinger, G., 2006. Collision detection and safe reaction with the DLR-III lightweight manipulator arm. In: Proceedings of the IEEE/RSJ International Conference on Intelligent Robots and Systems (IROS). IEEE, pp. 1623–1630.

Desoer, C.A., Vidyasagar, M., 1975. Feedback Systems: Input–Output Properties, vol. 55. SIAM.

Dorn, T.W., Wang, J.M., Hicks, J.L., Delp, S.L., 2015. Predictive simulation generates human adaptations during loaded and inclined walking. PLoS ONE 10 (4).

Eilenberg, M.F., Geyer, H., Herr, H., 2010. Control of a powered ankle–foot prosthesis based on a neuromuscular model. IEEE Trans. Neural Syst. Rehabil. Eng. 18 (2), 164–173.

Farris, R.J., Quintero, H., Goldfarb, M., et al., 2011. Preliminary evaluation of a powered lower limb orthosis to aid walking in paraplegic individuals. IEEE Trans. Neural Syst. Rehabil. Eng. 19 (6), 652–659.

Ferris, D.P., Gordon, K.E., Sawicki, G.S., Peethambaran, A., 2006. An improved powered ankle–foot orthosis using proportional myoelectric control. Gait Posture 23 (4), 425–428.

Fite, K.B., Goldfarb, M., 2006. Design and energetic characterization of a proportional-injector monopropellant-powered actuator. IEEE/ASME Trans. Mechatron. 11 (2), 196–204.

Fleischer, C., Reinicke, C., Hommel, G., 2005. Predicting the intended motion with EMG signals for an exoskeleton orthosis controller. In: Proceedings of the IEEE/RSJ International Conference on Intelligent Robots and Systems (IROS). IEEE, pp. 2029–2034.

Frank, A.A., 1968. Automatic Control Systems for Legged Locomotion Machines. DTIC Document, Tech. Rep.

Geyer, H., Herr, H., 2010. A muscle-reflex model that encodes principles of legged mechanics produces human walking dynamics and muscle activities. IEEE Trans. Neural Syst. Rehabil. Eng. 18 (3), 263–273.

Giovacchini, F., Vannetti, F., Fantozzi, M., Cempini, M., Cortese, M., Parri, A., Yan, T., Lefeber, D., Vitiello, N., 2014. A light-weight active orthosis for hip movement assistance. Robot. Auton. Syst (73), 123–134.

Gordon, J., Ghez, C., 1987. Trajectory control in targeted force impulses. Exp. Brain Res. 67 (2), 253–269.

Gupta, A., O'Malley, M.K., Patoglu, V., Burgar, C., 2008. Design, control and performance of RiceWrist: a force feedback wrist exoskeleton for rehabilitation and training. Int. J. Robot. Res. 27 (2), 233–251.

Haddadin, S., Albu-Schaffer, A., De Luca, A., Hirzinger, G., 2008. Collision detection and reaction: a contribution to safe physical human–robot interaction. In: Proceedings of the IEEE/RSJ International Conference on Intelligent Robots and Systems (IROS). IEEE, pp. 3356–3363.

Heinzinger, G., Fenwick, D., Paden, B., Miyazaki, F., 1992. Stability of learning control with disturbances and uncertain initial conditions. IEEE Trans. Autom. Control 37 (1), 110–114.

Hidler, J., Nichols, D., Pelliccio, M., Brady, K., Campbell, D.D., Kahn, J.H., Hornby, T.G., 2009. Multicenter randomized clinical trial evaluating the effectiveness of the Lokomat in subacute stroke. Neurorehabil. Neural Repair 23 (1), 5–13.

Hitt, J.K., Sugar, T.G., Holgate, M., Bellman, R., 2010. An active foot-ankle prosthesis with biomechanical energy regeneration. J. Med. Devices 4 (1), 011003.

Hobbelen, D., de Boer, T., Wisse, M., 2008. System overview of bipedal robots flame and tulip: Tailor-made for limit cycle walking. In: Proceedings of the IEEE/RSJ International Conference on Intelligent Robots and Systems (IROS). IEEE, pp. 2486–2491.

Huang, S., Wensman, J.P., Ferris, D.P., 2014. An experimental powered lower limb prosthesis using proportional myoelectric control. J. Med. Devices 8 (2), 024501.

Ijspeert, A.J., 2014. Biorobotics: using robots to emulate and investigate agile locomotion. Science 346, 196–203.

Jackson, R.J., Collins, S.H., 2015. An experimental comparison of the relative benefits of work and torque assistance in ankle exoskeletons. J. Appl. Physiol. 119 (5), 541–557.

Jezernik, S., Colombo, G., Keller, T., Frueh, H., Morari, M., 2003. Robotic orthosis lokomat: a rehabilitation and research tool. Neuromodulation 6 (2), 108–115.

Kawamoto, H., Taal, S., Niniss, H., Hayashi, T., Kamibayashi, K., Eguchi, K., Sankai, Y., 2010. Voluntary motion support control of robot suit HAL triggered by bioelectrical signal for hemiplegia. In: Proceedings of the IEEE International Conference of the Engineering in Medicine and Biology Society. IEEE, pp. 462–466.

Kazerooni, H., Racine, J.-L., Huang, L., Steger, R., 2005. On the control of the Berkeley lower extremity exoskeleton (BLEEX). In: Proceedings of the IEEE International Conference on Robotics and Automation (ICRA). IEEE, pp. 4353–4360.

Kelly, R., 1999. Regulation of manipulators in generic task space: an energy shaping plus damping injection approach. IEEE Trans. Robot. Autom. 15 (2), 381–386.

Kim, J.H., Oh, J.H., 2004. Walking control of the humanoid platform khr-1 based on torque feedback control. In: Proceedings of the IEEE International Conference on Robotics and Automation (ICRA), pp. 623–628.

Klimchik, A., Pashkevich, A., Chablat, D., Hovland, G., 2012. Compensation of compliance errors in parallel manipulators composed of non-perfect kinematic chains. In: Latest Advances in Robot Kinematics. Springer, pp. 51–58.

Kobayashi, T., Akazawa, Y., Naito, H., Tanaka, M., Hutchins, S.W., et al., 2010. Design of an automated device to measure sagittal plane stiffness of an articulated ankle–foot orthosis. Prosthet. Orthot. Int. 34 (4), 439–448.

Kong, K., Bae, J., Tomizuka, M., 2009. Control of rotary series elastic actuator for ideal force-mode actuation in human–robot interaction applications. IEEE/ASME Trans. Mechatron. 14 (1), 105–118.

Lasota, P.A., Rossano, G.F., Shah, J.A., 2014. Toward safe close-proximity human–robot interaction with standard industrial robots. In: 2014 IEEE International Conference on Automation Science and Engineering (CASE). IEEE, pp. 339–344.

Lenzi, T., Vitiello, N., De Rossi, S.M.M., Roccella, S., Vecchi, F., Carrozza, M.C., 2011. Neuroexos: a variable impedance powered elbow exoskeleton. In: Proceedings of the IEEE International Conference on Robotics and Automation (ICRA). IEEE, pp. 1419–1426.

Loconsole, C., Dettori, S., Frisoli, A., Avizzano, C.A., Bergamasco, M., 2014. An EMG-based approach for on-line predicted torque control in robotic-assisted rehabilitation. In: 2014 IEEE, Haptics Symposium (HAPTICS). IEEE, pp. 181–186.

Malcolm, P., Derave, W., Galle, S., De Clercq, D., 2013. A simple exoskeleton that assists plantarflexion can reduce the metabolic cost of human walking. PLoS ONE 8 (2), e56137.

Malcolm, P., Quesada, R.E., Caputo, J.M., Collins, S.H., 2015. The influence of push-off timing in a robotic ankle–foot prostheses on the energetics and mechanics of walking. J. NeuroEng. Rehabil. 12 (21), 1–14.

Markowitz, J., Krishnaswamy, P., Eilenberg, M.F., Endo, K., Barnhart, C., Herr, H., 2011. Speed adaptation in a powered transtibial prosthesis controlled with a neuromuscular model. Philos. Trans. R. Soc. Lond. B, Biol. Sci. 366 (1570), 1621–1631.

McGeer, T., 1990. Passive dynamic walking. Int. J. Robot. Res. 9 (2), 62–82.

Mochon, S., McMahon, T., 1980. Ballistic walking: an improved model. Math. Biosci. 52, 241–260.

Mooney, L.M., Rouse, E.J., Herr, H.M., 2014. Autonomous exoskeleton reduces metabolic cost of human walking during load carriage. J. NeuroEng. Rehabil. 11 (80), 0003-11.

Mosher, R.S., 1967. Handyman to Hardiman. SAE Technical Paper, Tech. Rep.

Nef, T., Mihelj, M., Kiefer, G., Perndl, C., Muller, R., Riener, R., 2007. ARMin-exoskeleton for arm therapy in stroke patients. In: Proceedings of the IEEE International Conference on Rehabilitation Robotics (ICORR). IEEE, pp. 68–74.

Ortega, R., Spong, M.W., 1989. Adaptive motion control of rigid robots: a tutorial. Automatica 25 (6), 877–888.

Paine, N., Oh, S., Sentis, L., 2014. Design and control considerations for high-performance series elastic actuators. IEEE/ASME Trans. Mechatron. 19 (3), 1080–1091.

Perry, J.C., Rosen, J., Burns, S., 2007. Upper-limb powered exoskeleton design. IEEE/ASME Trans. Mechatron. 12 (4), 408–417.

Pratt, G.A., Williamson, M.M., 1995. Series elastic actuators. In: Proceedings of the IEEE/RSJ International Conference on Intelligent Robots and Systems (IROS), vol. 1. IEEE, pp. 399–406.

Pratt, J., Dilworth, P., Pratt, G., 1997. Virtual model control of a bipedal walking robot. In: 1997 IEEE International Conference on Robotics and Automation, Proceedings, vol. 1. IEEE, pp. 193–198.

Pratt, G.A., Willisson, P., Bolton, C., Hofman, A., 2004. Late motor processing in low-impedance robots: impedance control of series-elastic actuators. In: Proceedings of the American Control Conference, vol. 4. IEEE, pp. 3245–3251.

Robinson, D.W., Pratt, J.E., Paluska, D.J., Pratt, G.A., 1999. Series elastic actuator development for a biomimetic walking robot. In: 1999 IEEE/ASME International Conference on Advanced Intelligent Mechatronics, Proceedings. IEEE, pp. 561–568.

Rosen, J., Brand, M., Fuchs, M.B., Arcan, M., 2001. A myosignal-based powered exoskeleton system. IEEE Trans. Syst. Man Cybern., Part A, Syst. Hum. 31 (3), 210–222.

Sawicki, G.S., Ferris, D.P., 2009. Powered ankle exoskeletons reveal the metabolic cost of plantar flexor mechanical work during walking with longer steps at constant step frequency. J. Exp. Biol. 212 (1), 21–31.

Schiele, A., Letier, P., van der Linde, R., Van der Helm, F., 2006. Bowden cable actuator for force-feedback exoskeletons. In: Proceedings of the IEEE/RSJ International Conference on Intelligent Robots and Systems (IROS). IEEE, pp. 3599–3604.

Schuitema, E., Wisse, M., Ramakers, T., Jonker, P., 2010. The design of LEO: a 2D bipedal walking robot for online autonomous reinforcement learning. In: Proceedings of the IEEE/RSJ International Conference on Intelligent Robots and Systems (IROS). IEEE, pp. 3238–3243.

Shultz, A.H., Mitchell, J.E., Truex, D., Lawson, B.E., Ledoux, E., Goldfarb, M., 2014. A walking controller for a powered ankle prosthesis. In: Engineering in Medicine and Biology Society (EMBC), 2014 36th Annual International Conference of the IEEE. IEEE, pp. 6203–6206.

Slotine, J.-J.E., Li, W., 1987. On the adaptive control of robot manipulators. Int. J. Robot. Res. 6 (3), 49–59.

Slotine, J.-J.E., Li, W., et al., 1991. Applied Nonlinear Control. Prentice-Hall, Englewood Cliffs, NJ.

Song, S., Kim, J., Yamane, K., 2015. Development of a bipedal robot that walks like an animation character. In: Proceedings of the IEEE International Conference on Robotics and Automation (ICRA), pp. 3596–3602.

Stienen, A.H., Hekman, E.E., ter Braak, H., Aalsma, A.M., van der Helm, F.C., van der Kooij, H., 2010. Design of a rotational hydroelastic actuator for a powered exoskeleton for upper limb rehabilitation. IEEE Trans. Biomed. Eng. 57 (3), 728–735.

Sulzer, J.S., Roiz, R., Peshkin, M., Patton, J.L., et al., 2009. A highly backdrivable, lightweight knee actuator for investigating gait in stroke. IEEE Trans. Robot. 25 (3), 539–548.

Sung, S.W., Lee, I.-B., 1996. Limitations and countermeasures of PID controllers. Ind. Eng. Chem. Res. 35 (8), 2596–2610.

Sup, F., Varol, H.A., Mitchell, J., Withrow, T.J., Goldfarb, M., 2009. Preliminary evaluations of a self-contained anthropomorphic transfemoral prosthesis. IEEE/ASME Trans. Mechatron. 14 (6), 667–676.

Suzuki, K., Mito, G., Kawamoto, H., Hasegawa, Y., Sankai, Y., 2007. Intention-based walking support for paraplegia patients with robot suit HAL. Adv. Robot. 21 (12), 1441–1469.

Takahashi, K.Z., Lewek, M.D., Sawicki, G.S., 2015. A neuromechanics-based powered ankle exoskeleton to assist walking post-stroke: a feasibility study. J. NeuroEng. Rehabil. 12 (1), 23.

Tsai, B.-C., Wang, W.-W., Hsu, L.-C., Fu, L.-C., Lai, J.-S., 2010. An articulated rehabilitation robot for upper limb physiotherapy and training. In: Proceedings of the IEEE/RSJ International Conference on Intelligent Robots and Systems (IROS). IEEE, pp. 1470–1475.

Unluhisarcikli, O., Pietrusinski, M., Weinberg, B., Bonato, P., Mavroidis, C., 2011. Design and control of a robotic lower extremity exoskeleton for gait rehabilitation. In: Proceedings of the IEEE/RSJ International Conference on Intelligent Robots and Systems (IROS). IEEE, pp. 4893–4898.

Vallery, H., Ekkelenkamp, R., Van Der Kooij, H., Buss, M., 2007. Passive and accurate torque control of series elastic actuators. In: Proceedings of the IEEE/RSJ International Conference on Intelligent Robots and Systems (IROS). IEEE, pp. 3534–3538.

Van de Wijdeven, J., Donkers, T., Bosgra, O., 2009. Iterative learning control for uncertain systems: robust monotonic convergence analysis. Automatica 45 (10), 2383–2391.

Van der Noot, N., Ijspeert, A.J., Ronsse, R., 2015. Biped gait controller for large speed variations, combining reflexes and a central pattern generator in a neuromuscular model. In: Proceedings of the IEEE International Conference on Robotics and Automation (ICRA). IEEE, pp. 6267–6274.

van Dijk, W., Van Der Kooij, H., Koopman, B., van Asseldonk, E., 2013. Improving the transparency of a rehabilitation robot by exploiting the cyclic behaviour of walking. In: Proceedings of the IEEE International Conference on Rehabilitation Robotics (ICORR). IEEE, pp. 1–8.

Veneman, J.F., Ekkelenkamp, R., Kruidhof, R., van der Helm, F.C., van der Kooij, H., 2006. A series elastic- and Bowden-cable-based actuation system for use as torque actuator in exoskeleton-type robots. Int. J. Robot. Res. 25 (3), 261–281.

Veneman, J.F., Kruidhof, R., Hekman, E.E., Ekkelenkamp, R., Van Asseldonk, E.H., Van Der Kooij, H., 2007. Design and evaluation of the lopes exoskeleton robot for interactive gait rehabilitation. IEEE Trans. Neural Syst. Rehabil. Eng. 15 (3), 379–386.

Verdaasdonk, B.W., Koopman, H.F., van der Helm, F.C., 2009. Energy efficient walking with central pattern generators: from passive dynamic walking to biologically inspired control. Biol. Cybern. 101, 49–61.

Wang, S., Meijneke, C., Van Der Kooij, H., 2013. Modeling, design, and optimization of mindwalker series elastic joint. In: Proceedings of the IEEE International Conference on Rehabilitation Robotics (ICORR). IEEE, pp. 1–8.

Winter, D.A., 1991. The Biomechanics and Motor Control of Hhuman Gait: Normal, Elderly and Pathological, 2nd ed.. University of Waterloo Press, Waterloo.

Witte, K.A., Zhang, J., Jackson, R.W., Collins, S.H., 2015. Design of two lightweight, high-bandwidth torque-controlled ankle exoskeletons. In: Proceedings of the IEEE International Conference on Robotics and Automation (ICRA). Seattle, WA.

Wyeth, G., 2006. Control issues for velocity sourced series elastic actuators. In: Proceedings of the Australasian Conference on Robotics and Automation 2006. Australian Robotics and Automation Association Inc.

Zanotto, D., Lenzi, T., Stegall, P., Agrawal, S.K., 2013. Improving transparency of powered exoskeletons using force/torque sensors on the supporting cuffs. In: Proceedings of the IEEE International Conference on Rehabilitation Robotics (ICORR). IEEE, pp. 1–6.

Zhang, J., Cheah, C.C., 2015. Passivity and stability of human–robot interaction control for upper-limb rehabilitation robots. IEEE Trans. Robot. 31, 233–245.

Zhang, J., Cheah, C.C., Collins, S.H., 2013. Stable human–robot interaction control for upper-limb rehabilitation robotics. In: Proceedings of the IEEE International Conference on Robotics and Automation (ICRA). IEEE, pp. 2201–2206.

Ziegler, J.G., Nichols, N.B., 1942. Optimum settings for automatic controllers. Trans. Am. Soc. Mech. Eng. 64 (11).

Zinn, M., Khatib, O., Roth, B., Salisbury, J.K., 2004. Playing it safe [human-friendly robots]. IEEE Robot. Autom. Mag. 11 (2), 12–21.

Zoss, A.B., Kazerooni, H., Chu, A., 2006. Biomechanical design of the Berkeley lower extremity exoskeleton (BLEEX). IEEE/ASME Trans. Mechatron. 11 (2), 128–138.

Chapter 6

Neuromuscular Models for Locomotion

Arthur Prochazka, Simon Gosgnach, Charles Capaday, and
Hartmut Geyer

Nature has solved the problem of controlling legged locomotion many thousands of times in animals exhibiting an enormous variety of neuromechanical structures. Control systems features that are common to nearly all species include feedback control of limb displacement and force and feedforward generation of motor patterns by neural networks within the central nervous system (Central Pattern Generators: CPGs). Here we review the components of locomotor control systems, with a focus on mammalian animals including humans. We then propose a generalized model of reflex control and discuss the mathematical functions that describe the properties of the components of a spinal stretch reflex model.

Bioinspired Legged Locomotion. http://dx.doi.org/10.1016/B978-0-12-803766-9.00008-7
401

Chapter 6.1

Introduction: Feedforward vs Feedback in Neural Control: Central Pattern Generators (CPGs) Versus Reflexive Control

Arthur Prochazka* and Hartmut Geyer†

*Neuroscience and Mental Health Institute, University of Alberta, Edmonton, AB, Canada
†Robotics Institute, Carnegie Mellon University, Pittsburgh, PA, United States

The control of animal locomotion by the nervous system has been studied for many years. Early work in mammals explored the reflexive control of simple movements, including locomotion, after surgical removal of the cerebrum and transection of the spinal cord, abolishing all descending neural input from the brain, the brainstem and the spinal cord above the transection (Flourens, 1824; Freusberg, 1874; Goltz, 1869; Magnus, 1909; Sherrington, 1910). In the spinally-transected animals, dropping one hind-limb from a flexed position could initiate a sequence of alternating, locomotor-like contractions of the flexor and extensor muscles of the hind-limbs (Freusberg, 1874). The rhythmical sequence could be halted simply by constraining a limb to a fixed position (Freusberg, 1874; Sherrington, 1910). These findings led Freusberg to suggest that locomotion was the result of a sequence of spinal reflexes, the sensory input signaling the end of the swing phase, triggering the onset of the stance phase, and vice versa. The notion of chains of reflexes was not new. It had been proposed a decade earlier by the Russian neurophysiologist Ivan Sechenov to be the general mechanism underlying the control of all movement (Sechenov, 1863).

The observation that locomotion can be achieved in animals without active brain coordination may come as a surprise at first. However, passive legged robots have been built that have neither actuators nor controllers but can walk down a shallow ramp, switching between swing and stance simply as an outcome of their mechanics (McGeer, 1990; Collins et al., 2005). A similar phenomenon has been demonstrated for running machines (Owaki et al., 2010).

The idea that animal locomotion was controlled entirely by a chain of reflexes was seriously challenged when T. Graham Brown discovered that locomotor-like rhythms in cats with transected spinal cords, could still occur even after the sensory nerve roots entering the spinal cord were cut (Brown, 1911). This led him to propose the existence of an "intrinsic factor" in the spinal cord which could generate a locomotor rhythm without descending control from the brain and also without sensory input. In 1975 Sten Grillner and Peter Zangger renamed this mechanism "central pattern generation" (Grillner and Zangger, 1975). The existence of central pattern generators (CPGs) that can generate rhythmical movements in the absence of sensory input had also been suggested from work in locusts (Wilson, 1961) and has since been demonstrated in many other vertebrate and invertebrate species.

In this chapter we will review some of the research that has explored how mechanical structure, sensory input and CPGs interact in different animals and under different circumstances to generate locomotion. We will then discuss various neuromechanical models of locomotion and conclude with comments on the relevance of an understanding of biological locomotor control to control in legged robots.

REFERENCES

Brown, T.G., 1911. The intrinsic factors in the act of progression in the mammal. Proc. R. Soc. Lond., Ser. B 84, 308–319.

Collins, S., Ruina, A., Tedrake, R., Wisse, M., 2005. Efficient bipedal robots based on passive-dynamic walkers. Science 307, 1082–1085.

Flourens, P., 1824. Recherches expérimentales sur les propriétés et les fonctions du système nerveux dans les animaux vertébrés. Crevot, Paris.

Freusberg, A., 1874. Reflexbewegungen beim Hunde. Pflügers Arch. 9, 358–391.

Goltz, F.L., 1869. Beitraege zur Lehre von den Functionen der Nervencentren des Frosches. Hirschwald, Berlin.

Grillner, S., Zangger, P., 1975. How detailed is the central pattern generation for locomotion? Brain Res. 88, 367–371.

Magnus, R., 1909. Zur Regelung der Bewegungen durch das Zentralnervensystem. Mitteilung I. Pflügers Arch. Gesamte Physiol. Menschen Tiere 130, 219–252.

McGeer, T., 1990. Passive bipedal running. Proc. R. Soc. Lond. B, Biol. Sci. 240, 107–134.

Owaki, D., Koyama, M., Yamaguchi, S., Kubo, S., Ishiguro, A., 2010. A two-dimensional passive dynamic running biped with knees. In: IEEE Int. Conf. Robot. Autom. IEEE, pp. 5237–5242.

Sechenov, I.M., 1863. Reflexes of the Brain (Refleksy Golovnogo Mozga). In: Subkov, A.A. (Ed.), I.M. Sechenov, Selected Works. State Publishing House, Moscow, pp. 264–322.

Sherrington, C.S., 1910. Flexion-reflex of the limb, crossed extension-reflex, and reflex stepping and standing. J. Physiol. (Lond.) 40, 28–121.

Wilson, D.M., 1961. The central control of flight in a locust. J. Exp. Biol. 38, 471–490.

Chapter 6.2

Locomotor Central Pattern Generators

Simon Gosgnach and Arthur Prochazka

Neuroscience and Mental Health Institute, University of Alberta, Edmonton, AB, Canada

There is now overwhelming evidence for the existence of locomotor CPGs in invertebrates and nonprimate mammals. Most of the mammalian evidence is based on neuronal firing patterns recorded during "fictive locomotion" elicited by electrical stimulation in the midbrain of decerebrated, paralyzed animals (Grillner and Zangger, 1974), or by chemical or electrical stimuli in isolated spinal cord preparations (Ayers et al., 1983). As will be seen in this section, it has been posited that the mammalian locomotor CPG has separate components controlling the timing of the locomotor rhythm and the selection of α-motoneuron pool activation patterns (Perret et al., 1988; Orsal et al., 1990; McCrea and Rybak, 2007; Rybak et al., 2006). Genetic techniques have identified candidate interneuronal populations that contribute to these separate components. Attempts have been made recently to incorporate these interneurons into a model of left–right coordination by the locomotor CPG (Shevtsova et al., 2015). The existence of a locomotor CPG in humans is less certain, though there is increasing evidence to support this idea (Calancie et al., 1994; Danner et al., 2015; Dimitrijevic et al., 1998).

6.2.1 NEURONAL NETWORKS THAT MAKE UP THE LOCOMOTOR CPG

Walking is a complex task that requires precise coordination of dozens of muscles. As discussed above, while the first theories to account for the neural control of locomotion suggested that propriospinal reflexes were responsible for the rhythmic, repetitive alternation of the hindlimbs (Sherrington, 1910), it has now been accepted that in many animals, intrinsic neural networks (CPGs) located in the spinal cord of mammals, are responsible for controlling the timing and activation of α-motoneurons and muscles in an appropriate sequence (for a review, see Kiehn, 2006). As mentioned above, this was initially proposed by Graham Brown (1911, 1914) who found that in cats, rabbits, and guinea pigs, after spinal transection and deafferentation that abolished descending and sensory input to the lumbosacral spinal cord, the animals' hindlimbs could still display alternating locomotor-like movements. To account for this, Graham Brown suggested that for each limb, the spinal cord contained a pair of mutually inhibitory neural centers, later called "half-centers." The half-centers were mutually and reciprocally inhibitory such that activity in the extensor half-center activated extensor α-motoneurons while inhibiting the flexor half-center and flexor α-motoneurons. This inhibition waned, so that after a short time the flexor half-center began activating flexor α-motoneurons while inhibiting the extensor half-center and extensor α-motoneurons. Brown also proposed that activity of the half-centers could be modulated by sensory input.

6.2.2 IN VIVO PREPARATIONS USED TO STUDY THE LOCOMOTOR CPG

In spite of the compelling evidence and hypothesis of Graham Brown, the idea that sensory-evoked reflexes were solely responsible for generating locomotor activity continued to be strongly supported up to the 1960s and remains influential in certain types of locomotor models to this day (Song and Geyer, 2015). Around this time a series of experiments by Andres Lundberg's group (Lundberg, 1967; Jankowska et al., 1967) incorporated a preparation in which a decerebrate cat was spinalized and L-DOPA and 5-HTP were applied intravenously. In the presence of these pharmacological agents, stimulation of high-threshold cutaneous or muscle afferents which evoke the flexion reflex under normal conditions (i.e., flexor reflex afferents – FRAs) suppressed this reflex and instead evoked rhythmic, alternating activity in flexor and extensor α-motoneurons ipsilateral to the stimulated afferents, a locomotor-like pattern (Jankowska et al., 1967). The Lundberg group used intracellular techniques to record from interneurons in the ventromedial aspect of the lumbar spinal cord and were able to identify individual cells that received input from either ipsilateral or contralateral FRAs as well as input from descending systems thought to be involved in locomotor initiation. These experiments provided experimental evidence to support the half-center architecture of the locomotor CPG originally proposed by Brown (1911, 1914). They are also of historical significance since they involved the first direct recordings from interneuronal components presumed to be part of the mammalian locomotor CPG.

Shortly after these experiments, studies taking place in Moscow were successful in identifying a small area within the midbrain which, when electrically stimulated, was able to reliably evoke locomotion in the premammillary-transected cat. This region became known as the mesencephalic locomotor region (MLR) and was shown to have discrete control over locomotor activity, in that increases or decreases of stimulation strength were able to modulate locomotor speed accordingly (Shik et al., 1966, 1969).

While initial experiments using MLR stimulation were performed in decerebrate cats walking on a treadmill, much of the work in the subsequent two decades was aimed at elucidating the structure and function of the loco-motor CPG. This work was mainly done in the fictive locomotor preparation. For these experiments adult cats are decerebrated and the neuraxis is severed in the midbrain to abolish descending inhibitory drive. After a laminectomy exposes the spinal cord for intracellular recording, the animals are fixed in a frame and paralyzed with curare. Hindlimb nerves are cut and placed on bipolar electrodes for recording of efferent activity. This also provides the opportu-nity to stimulate afferents in order to investigate the effect of their activity on the locomotor pattern. Locomotor activity in this preparation was termed fictive locomotion since the hindlimbs do not move; rather the electroneuro-gram activity that is evoked in the motor axons innervating the hindlimbs is recorded, providing a read-out of the neural correlates of locomotor activity. By incorporating intracellular recording of neurons in the spinal cord during MLR-evoked fictive locomotor activity, this preparation enabled identification of a number of the interneuronal components of the locomotor CPG. Identified cells were typically grouped together based on how they responded to affer-ent input during fictive locomotion. In addition to the interneurons that receive input from FRAs described above (Jankowska et al., 1967), there are interneu-rons in the intermediate nucleus of the caudal spinal cord that receive input from extensor afferents and are only active during the extensor phase of loco-motion (Angel et al., 2005). Another population of interneurons was shown to be preferentially active during the flexor phase of locomotion and receive exci-tation primarily from group II afferents (Edgley and Jankowska, 1987). Finally, a population of cholinergic cells was identified that are primarily active dur-ing extension and project axons to the contralateral spinal cord (Huang et al., 2000).

6.2.3 IN VITRO PREPARATIONS USED TO STUDY THE LOCOMOTOR CPG

While the in vivo cat preparation provided insight into the general location of the locomotor CPG as well as the manner in which it was activated, this work was expanded upon greatly following the development of an in vitro neonatal rodent preparation (Kudo and Yamada, 1987; Smith et al., 1988). Here the spinal cord is dissected out of a newborn (typically postnatal day 0–4) rodent, placed in a recording chamber and perfused with oxygenated artificial cerebrospinal fluid to maintain viability. Suction electrodes are used to record electroneuro-grams (ENGs) from ventral roots in the lumbar spinal cord. Initial experiments demonstrated that the second lumbar ventral root (i.e., L2) is almost entirely made up of axons from flexor α-motoneurons while the 5th lumbar ventral root (i.e., L5) consists primarily of extensor α-motoneurons. Locomotor activity can be evoked in this preparation via either electrical stimulation (of the brainstem or dorsal roots) or application of various substances including NMDA and 5-HT to the perfusate. In this preparation an alternation of ENG activity recorded in flexor and extensor-related ventral roots on both sides of the spinal cord is observed. Two separate groups performed a series of experiments in which various lesions and segmentations were made to the isolated spinal cord to determine the location of the hindlimb locomotor CPG (Cowley and Schmidt, 1997; Kjaerulff and Kiehn, 1996). Both groups concluded that the CPG is distributed throughout the lower thoracic and lumbar segments of the spinal cord with a rostral–caudal gradient such that the rostral segments were more able to generate fast and regular rhythmic activity than the caudal segments and they tended to entrain the rhythm when both segments were connected. Furthermore, they demonstrated that the rhythm-generating components were located ventromedially, in regions roughly corresponding to lamina VII, VIII, and X of the spinal cord.

Because the spinal cord could be isolated and pharmacological agents applied directly to it (rather than systemically in the adult cat preparations), interneurons that comprised the locomotor CPG were more accessible for whole cell recording and their morphology and connectivity could be more easily investigated with neuroanatomical tracers. It was also easier to perform post hoc immuno-histochemical staining. Several interneuronal populations with homogeneous characteristics were identified using this preparation (Butt et al., 2002; Butt and Kiehn, 2003). In spite of these advantages, little progress was made deciphering the manner in which these cell populations were interconnected to generate locomotor activity. The slow headway is attributable to the plethora of interneurons in the mammalian spinal cord and the extent to which those with heterogeneous functions are intermingled.

6.2.4 IMPLEMENTATION OF MOLECULAR GENETIC TECHNIQUES TO STUDY THE LOCOMOTOR CPG

Over the past 15 years, molecular techniques have complemented traditional anatomical and electrophysiological approaches in identifying components of the locomotor CPG and providing insight into its network structure. These experiments have shown that the developing neural tube in the embryonic mouse can be divided into ten distinct populations of spinal interneurons (dI1–dI6, V0–V3) based on transcription factor expression (see Goulding, 2009, for a review; Fig. 6.2.1 – schematic of interneuronal populations). These populations can first be identified around embryonic day 10 (E10). By E13, they begin to migrate towards their settling position which they reach by E16, where they remain. Since gene and transcription-factor expression dictate neuronal charac-

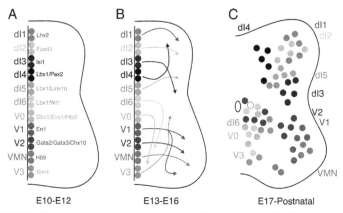

FIGURE 6.2.1 Genetic characterization of the mammalian spinal cord. (A) Transcription factors expressed postmitotically between approximately embryonic day 10–12 (E10–E12) allow spinal cord neurons to be divided into 11 genetically-distinct populations. (B) From E13 to E16 each population migrates towards its settling position. (C) Just before birth, populations reach the positions in the spinal cord where they remain throughout adulthood.

teristics such as cell fate, channel composition, axonal projection patterns, and neurotransmitter phenotype, it was initially postulated that populations of neurons with a similar genetic background will have similar characteristics and a similar function during locomotor activity.

The genetic characterization of interneurons in the spinal cord enables the use of powerful molecular tools to silence or ablate entire neuronal populations and pinpoint their specific function during locomotion. Alternatively, transcription factors can be used to drive expression of reporter proteins such as green or red fluorescent protein (i.e., GFP or tdTomato). This allows each population to be visualized in live or fixed tissue with a fluorescent microscope and renders anatomical and electrophysiological approaches much more efficient. Validation of this multidisciplinary approach has come from a number of studies which have characterized many of the neuronal populations that originate in the ventral neural tube and defined their function during locomotor activity (Garcia-Campmany et al., 2010).

Initially, all of the components of the locomotor CPG were predicted to be derived from the four populations of interneurons that originated from the ventral half of the neural tube during development – the V0, V1, V2, and V3 populations. Anatomical and electrophysiological experiments were performed on each of these populations to identify their neurotransmitter phenotype, axonal projection pattern and intrinsic properties. In addition, the use of molecular techniques enabled the selective ablation or silencing of each population in order to assess fictive locomotion activity in their absence. The molecular work has demonstrated that each of these four "parent" populations of cells located in the ventral neural tube can be subdivided into two or more subpopulations, each of which expresses the same transcription factor as its parent but differs in downstream expression (Alaynick et al., 2011). Recent work has indicated that there is substantial diversity in the properties, connectivity, and locomotor function of these genetically-related subpopulations.

V0 cells project commissural axons. The V0 population has been divided into 3 subpopulations: $V0_D$ and $V0_V$ cells that cross to the opposite (contralateral) side of the spinal cord (Lanuza et al., 2004) and coordinate left–right limb alternation at slow and fast locomotor cadences, respectively (Talpalar et al., 2013), and $V0_{C/G}$ cells that project to the same (ipsilateral) side of the cord and regulate α-motoneuronal output (Zagoraiou et al., 2009).

V1 cells have been divided into a subset comprising Renshaw cells and Ia inhibitory interneurons (Sapir et al., 2004), and while these cells are not presumed to be a key component of the locomotor CPG (Noga et al., 1987) elimination of the entire V1 population drastically reduces the cadence of locomotor activity (Gosgnach et al., 2006). A second subset of V1 cells is involved in coordinating the ipsilateral alternation of activity in flexor and extensor α-motoneuron pools during stepping (Britz et al., 2015).

V2 interneurons are extremely diverse. Broadly they can be subdivided into a V2a subset which is excitatory and projects ipsilaterally. Some of these cells make synaptic contact with V0v interneurons and affect left–right alternation at fast cadences (Crone et al., 2008, 2009). A further subset has all the characteristics of locomotor rhythm generation but deleting this subset does not abolish the locomotor rhythm, indicating that these cells are part of a larger rhythm-generating network comprised of several cell types (Dougherty et al., 2013; Hagglund et al., 2013). V2b interneurons are inhibitory, and work together with V1 cells to coordinate ipsilateral flexor and extensor α-motoneurons during locomotion (Britz et al., 2015). V2c cells form a small, inhibitory, Sox1-expressing population that has not been well characterized, nor has its function during locomotion been established.

V3 interneurons project to contralateral α-motoneurons and regulate their output (Zhang et al., 2008). They are the least well characterized of the ventrally-derived interneuronal populations and have yet to be subdivided into subpopulations based on function or transcription factor expression. However, it seems likely that there are several subsets since it has recently been shown that there are three distinct regions in the postnatal spinal cord in which V3 interneurons are clustered, and cells in each position have a unique set of electrophysiological characteristics (Borowska et al., 2015).

6.2.5 NETWORK MODELS OF THE LOCOMOTOR CPG

Over the past century several network models have been proposed to account for the connectivity and mechanism of function of the locomotor CPG. The first of these, the half-center model described above postulated that stepping in each limb is controlled by a separate CPG which contains two groups of excitatory interneurons with mutual inhibitory connections that project to either flexor or extensor α-motoneurons (see Fig. 6.2.2A, half-center schematic). While many of the central tenets of this model still hold up after a century of testing with experimental data, the observation that activity of many flexor and extensor α-motoneurons is not strictly reciprocal (i.e., certain α-motoneuronal pools are active during both flexion and extension phases and the onset and cessation of activity of certain α-motoneuronal pools vary in their timing) were inconsistent with the hypothesis. This resulted in the development of the "unit burst generator" hypothesis (Grillner and Zangger, 1975) which was proposed after it was demonstrated that the more complex activity patterns of flexor and extensor α-motoneurons was maintained following complete bilateral de-afferentation of the hindlimbs. The unit burst generator hypothesis held that, rather than a single half-center arrangement overseeing activity in each hindlimb, there were half-centers around each joint that were connected to one another (see Fig. 6.2.2B, unit burst generator schematic). Activity in these individual "units" was linked but each could be individually controlled by supraspinal inputs to produce different patterns of locomotor output as required by the environment (Grillner, 1981).

An issue with both the half-center and unit burst generator models is that they were constructed such that the excitatory interneurons generating the locomotor rhythm were directly connected to the α-motoneurons generating locomotion. This makes it difficult to imagine how independent control of amplitude and duration of individual motor pool activity is achieved. A number of groups (Burke, 2001; Kriellaars et al., 1994; Orsal et al., 1990; Perret et al., 1988) have suggested that, rather than direct connectivity between a rhythm generating module and α-motoneurons, the CPG is a multilayered entity with an upper layer acting as a clock and setting the timing of activity. The clock connects to a lower layer which controls the firing of α-motoneurons. Compelling evidence for this multilayered arrangement comes from studies investigating the timing of missing bursts of activity in α-motoneuronal pools (called "deletions"). Deletions occur relatively frequently during fictive locomotion (Lafreniere-Roula and McCrea, 2005). Analysis indicated that in most cases the burst following a deletion typically occurred at a time point expected from the prior rhythm. These were termed nonresetting deletions since the locomotor rhythm had not been reset. A much less common occurrence was a deletion in

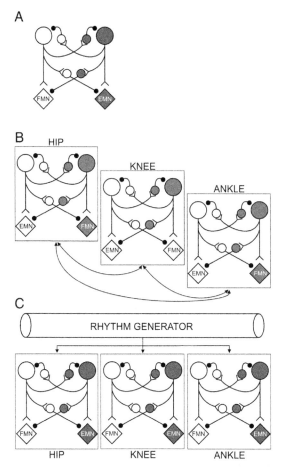

FIGURE 6.2.2 Schematics of network models of the locomotor CPG. (A) In the half-center model alternation of flexor and extensor α-motoneurons on one side of the body is controlled by half-centers (large circles), which activate α-motoneurons (diamonds) and interneurons (small circles) which inhibit the antagonistic half-center and associated α-motoneurons. (B) The unit burst generator (UBG) model comprises neurons interconnected in a similar arrangement as the half-center model but differs in that there is an independent UBG that controls flexor–extensor alternation at each joint. (C) The two-layer model is unique in that the functions of locomotor rhythm generation and regulation of α-motoneuronal activity are carried out by distinct neuronal populations. In this arrangement the rhythm generating layer acts as a clock and activates the pattern forming layer at each joint which is responsible for regulating α-motoneuronal activity.

which activity in the affected α-motoneuronal pools occurred at a random time point after the "missed burst", an event known as a resetting deletion. From these data the authors proposed a two-layer model with the upper layer comprising the rhythm generator (RG) module which feeds out to the lower layer, the pattern forming (PF) layer, which in turn activates α-motoneurons. Nonresetting deletions were proposed to occur due to aberrant activity in the PF layer rather than in the RG module. Conversely, resetting deletions were ascribed to the RG layer since both timing and α-motoneuronal activity were affected.

The structure of the network model constructed in light of these data (Fig. 6.2.2C, three-layer half-center model) includes a flexor and extensor RG module in each limb which is activated via intrinsic cellular properties (persistent sodium channels). These RG modules act as traditional half-centers made up of excitatory interneurons which mutually excite each other, and also excite a module of inhibitory interneurons responsible for reciprocal inhibition of the antagonistic half-center. Cells of the PF layer receive input from their respective RG layer and excite agonist α-motoneurons as well as inhibitory interneurons which inhibit activity in antagonist α-motoneurons. A number of simulations have been run using this model (Rybak et al., 2006) and it has been shown to accurately reproduce many of the experimental phenomena observed in the in vivo cat fictive locomotor preparation.

The initial three-layer model of the locomotor CPG was constructed to account primarily for in vivo cat fictive locomotor data involving deletions. Recently, attempts have been made to incorporate the genetically-defined interneuronal populations described above into an updated version of this integrated model (Shevtsova et al., 2015). Since the interneuronal populations involved in coordinating left/right alternation have been so well defined, the model specifically includes them and attempts to explain how they may interact with each other as well as with the RG layer (Shevtsova et al., 2015).

The general structure of the integrated model includes a RG module for both flexors and extensors on the right and left sides. The V2a, V0$_D$, and V3 interneuronal populations, each receive direct excitatory input from the RG layer of the model. The V0$_D$, V0$_V$, and V3 cells synapse back onto the RG layer in the contralateral spinal cord with the V0$_V$ and V3 cells making excitatory connections and the V0$_D$ cells making inhibitory connections. The experimental data generated in the intact in vitro fictive locomotor preparation, as well as the preparation in which the V0$_D$, V0$_V$, or V2a cells are inactivated, were reproduced in this computational model.

While constructing computational models can be useful for generating and testing hypotheses in regards to network structure of the locomotor CPG, this specific model may be premature given the preliminary nature of the connectivity of the various cells types. First, while it has been shown that V2a cells project to the V0$_V$ subpopulation, so far only a handful of V0$_V$ cells has been shown to receive glutamatergic V2a-derived input. Furthermore, the full extent of axonal projections of V2a cells has not been investigated. While some axons from V2a cells likely project to V0$_V$ cells, there is a good chance they also project to other cell types and regions of the CNS given that V2a cells have been shown to have monosynaptic contacts with α-motoneurons and that virtually all interneuronal populations investigated have a promiscuous axonal projection pattern with individual interneurons projecting multiple axons. In addition, the model proposes that each of the V2a, $V0_V$, and $V0_D$ cells projects onto the RG layer of the locomotor CPG. While this may be the case it is a premature claim given that each of these cell populations has been shown to make mono- or di-synaptic connections with α-motoneurons and there are no data indicating that they project to rhythm-generating components of the locomotor CPG. Finally, the claim that the V3 population is responsible for coordinating left/right activity during fast locomotion (i.e., gallop) is made without any experimental evidence and the fact that the locomotor phenotype expressed in the absence of this population (Zhang et al., 2008) cannot be replicated with this model makes it unlikely that these cells play the proposed role.

Despite the preliminary nature of this model, the idea that three commissural pathways exist and coordinate left–right activity is a logical one. In the model, $V0_D$ cell activity predominated in low-speed left–right alternating gait (walk), $V0_V$ cells at higher speed alternating gait (trot), and V3 cells took over when the hindlimbs were activated synchronously (gallop). The key experiment required to test this hypothesis is to examine the activity of each cell population at various locomotor cadences in order to determine when they are preferentially active.

6.2.6 CPG CONTROL OF LOCOMOTOR PHASE DURATIONS

In most animals, step cycle duration varies mainly as a result of changes in the duration of the extensor phases with swing phase durations varying much less (Goslow et al., 1973; Halbertsma, 1983). However, in fictive locomotion in decerebrate cats in which the locomotor rhythm is largely generated by the CPG, flexor phase durations were found to vary just as much, if not more, than extensor phase durations (Yakovenko et al., 2005). In a given sequence of step cycles the phase (flexion of extension) that varied more was termed the "dominant" phase. It was proposed that the locomotor CPG is not inherently extensor- or flexor-dominant, but that this depends on the level of descending drive received by the flexion and extension half-centers of the CPG. The half-center receiving the lower level of drive would take longer to reach switching threshold and therefore it would generate longer and more adjustable phase durations.

Neurons in which persistent inward currents (PICs) have been activated show an inverse relationship between PIC level and sensitivity to synaptic inputs (Lee et al., 2003; Li et al., 2004). This raises the possibility that interneurons in the extensor timing element may receive less PIC-generating input and therefore they are not only set to have longer phase durations, but they are also more sensitive to descending synaptic commands for higher or lower durations. As we will see below, in models of locomotion controlled by finite state rules, extensor-dominant phase-duration characteristics were associated with the most stable gait, which suggests that the descending drives to the extensor and flexor half-centers to produce locomotion are tuned to the biomechanical requirements and that this may in a sense be hard-wired into the locomotor CPG.

REFERENCES

Alaynick, W.A., Jessell, T.M., Pfaff, S.L., 2011. SnapShot: spinal cord development. Cell 146. 178–178.e1.

Angel, M.J., Jankowska, E., McCrea, D.A., 2005. Candidate interneurones mediating group I disynaptic EPSPs in extensor motoneurones during fictive locomotion in the cat. J. Physiol. 563, 597–610.

Ayers, J., Carpenter, G.A., Currie, S., Kinch, J., 1983. Which behavior does the lamprey central motor program mediate? Science 221, 1312–1314.

Borowska, J., Jones, C.T., Deska-Gauthier, D., Zhang, Y., 2015. V3 interneuron subpopulations in the mouse spinal cord undergo distinctive postnatal maturation processes. Neuroscience 295, 221–228.

Britz, O., Zhang, J., Grossmann, K.S., Dyck, J., Kim, J.C., Dymecki, S., Gosgnach, S., Goulding, M., 2015. A genetically defined asymmetry underlies the inhibitory control of flexor–extensor locomotor movements. eLife 4.

Brown, T.G., 1911. The intrinsic factors in the act of progression in the mammal. Proc. R. Soc. Lond., Ser. B 84, 308–319.

Brown, T.G., 1914. On the nature of the fundamental activity of the nervous centres; together with an analysis of the conditioning of rhythmic activity in progression, and a theory of the evolution of function in the nervous system. J. Physiol. 48, 18–46.

Burke, R.E., 2001. The central pattern generator for locomotion in mammals. Adv. Neurol. 87, 11–24.

Butt, S.J., Kiehn, O., 2003. Functional identification of interneurons responsible for left-right coordination of hindlimbs in mammals. Neuron 38, 953–963.

Butt, S.J., Harris-Warrick, R.M., Kiehn, O., 2002. Firing properties of identified interneuron populations in the mammalian hindlimb central pattern generator. J. Neurosci. 22, 9961–9971.

Calancie, B., Needham-Shropshire, B., Jacobs, P., Willer, K., Zych, G., Green, B.A., 1994. Involuntary stepping after chronic spinal cord injury. Evidence for a central rhythm generator for locomotion in man. Brain 117, 1143–1159.

Cowley, K.C., Schmidt, B.J., 1997. Regional distribution of the locomotor pattern-generating network in the neonatal rat spinal cord. J. Neurophysiol. 77, 247–259.

Crone, S.A., Quinlan, K.A., Zagoraiou, L., Droho, S., Restrepo, C.E., Lundfald, L., Endo, T., Setlak, J., Jessell, T.M., Kiehn, O., Sharma, K., 2008. Genetic ablation of V2a ipsilateral interneurons disrupts left-right locomotor coordination in mammalian spinal cord. Neuron 60, 70–83.

Crone, S.A., Zhong, G., Harris-Warrick, R., Sharma, K., 2009. In mice lacking V2a interneurons, gait depends on speed of locomotion. J. Neurosci. 29, 7098–7109.

Danner, S.M., Hofstoetter, U.S., Freundl, B., Binder, H., Mayr, W., Rattay, F., Minassian, K., 2015. Human spinal locomotor control is based on flexibly organized burst generators. Brain 138, 577–588.

Dimitrijevic, M.R., Gerasimenko, Y., Pinter, M.M., 1998. Evidence for a spinal central pattern generator in humans. Ann. N.Y. Acad. Sci. 860, 360–376.

Dougherty, K.J., Zagoraiou, L., Satoh, D., Rozani, I., Doobar, S., Arber, S., Jessell, T.M., Kiehn, O., 2013. Locomotor rhythm generation linked to the output of spinal shox2 excitatory interneurons. Neuron 80, 920–933.

Edgley, S.A., Jankowska, E., 1987. Field potentials generated by group II muscle afferents in the middle lumbar segments of the cat spinal cord. J. Physiol. 385, 393–413.

Garcia-Campmany, L., Stam, F.J., Goulding, M., 2010. From circuits to behaviour: motor networks in vertebrates. Curr. Opin. Neurobiol. 20, 116–125.

Gosgnach, S., Lanuza, G.M., Butt, S.J., Saueressig, H., Zhang, Y., Velasquez, T., Riethmacher, D., Callaway, E.M., Kiehn, O., Goulding, M., 2006. V1 spinal neurons regulate the speed of vertebrate locomotor outputs. Nature 440, 215–219.

Goslow Jr., G.E., Reinking, R.M., Stuart, D.G., 1973. The cat step cycle: hind limb joint angles and muscle lengths during unrestrained locomotion. J. Morphol. 141, 1–41.

Goulding, M., 2009. Circuits controlling vertebrate locomotion: moving in a new direction. Nat. Rev. Neurosci. 10, 507–518.

Grillner, S., 1981. Control of locomotion in bipeds, tetrapods, and fish. In: Handbook of Physiology. The Nervous System. American Physiological Society, Bethesda, pp. 1179–1236.

Grillner, S., Zangger, P., 1974. Locomotor movements generated by the deafferented spinal cord. Acta Physiol. Scand. 91, 38–39.

Grillner, S., Zangger, P., 1975. How detailed is the central pattern generation for locomotion? Brain Res. 88, 367–371.

Hagglund, M., Dougherty, K.J., Borgius, L., Itohara, S., Iwasato, T., Kiehn, O., 2013. Optogenetic dissection reveals multiple rhythmogenic modules underlying locomotion. Proc. Natl. Acad. Sci. USA 110, 11589–11594.

Halbertsma, J.M., 1983. The stride cycle of the cat: the modelling of locomotion by computerized analysis of automatic recordings. Acta Physiol. Scand., Suppl. 521, 1–75.

Huang, A., Noga, B.R., Carr, P.A., Fedirchuk, B., Jordan, L.M., 2000. Spinal cholinergic neurons activated during locomotion: localization and electrophysiological characterization. J. Neurophysiol. 83, 3537–3547.

Jankowska, E., Jukes, M.G., Lund, S., Lundberg, A., 1967. The effect of DOPA on the spinal cord. 5. Reciprocal organization of pathways transmitting excitatory action to alpha motoneurones of flexors and extensors. Acta Physiol. Scand. 70, 369–388.

Kiehn, O., 2006. Locomotor circuits in the mammalian spinal cord. Annu. Rev. Neurosci. 29, 279–306.

Kjaerulff, O., Kiehn, O., 1996. Distribution of networks generating and coordinating locomotor activity in the neonatal rat spinal cord in vitro: a lesion study. J. Neurosci. 16, 5777–5794.

Kriellaars, D.J., Brownstone, R.M., Noga, B.R., Jordan, L.M., 1994. Mechanical entrainment of fictive locomotion in the decerebrate cat. J. Neurophysiol. 71, 2074–2086.

Kudo, N., Yamada, T., 1987. N-methyl-D,L-aspartate-induced locomotor activity in a spinal cord-hindlimb muscles preparation of the newborn rat studied in vitro. Neurosci. Lett. 75, 43–48.

Lafreniere-Roula, M., McCrea, D.A., 2005. Deletions of rhythmic motoneuron activity during fictive locomotion and scratch provide clues to the organization of the mammalian central pattern generator. J. Neurophysiol. 94, 1120–1132.

Lanuza, G.M., Gosgnach, S., Pierani, A., Jessell, T.M., Goulding, M., 2004. Genetic identification of spinal interneurons that coordinate left-right locomotor activity necessary for walking movements. Neuron 42, 375–386.

Lee, R.H., Kuo, J.J., Jiang, M.C., Heckman, C.J., 2003. Influence of active dendritic currents on input-output processing in spinal motoneurons in vivo. J. Neurophysiol. 89, 27–39.

Li, Y., Gorassini, M.A., Bennett, D.J., 2004. Role of persistent sodium and calcium currents in motoneuron firing and spasticity in chronic spinal rats. J. Neurophysiol. 91, 767–783.

Lundberg, A., 1967. The supraspinal control of transmission in spinal reflex pathways. Electroencephalogr. Clin. Neurophysiol. Suppl, 35–46.

McCrea, D.A., Rybak, I.A., 2007. Modeling the mammalian locomotor CPG: insights from mistakes and perturbations. Prog. Brain Res. 165, 235–253.

Noga, B.R., Shefchyk, S.J., Jamal, J., Jordan, L.M., 1987. The role of Renshaw cells in locomotion: antagonism of their excitation from motor axon collaterals with intravenous mecamylamine. Exp. Brain Res. 66, 99–105.

Orsal, D., Cabelguen, J.M., Perret, C., 1990. Interlimb coordination during fictive locomotion in the thalamic cat. Exp. Brain Res. 82, 536–546.

Perret, C., Cabelguen, J.M., Orsal, D., 1988. Analysis of the pattern of activity in "Knee Flexor" motoneurons during locomotion in the cat. In: Gurfinkel, V.S., Ioffe, M.E., Massion, J., Roll, J.P. (Eds.), Stance and Motion: Facts and Concepts. Plenum Press, New York, pp. 133–141.

Rybak, I.A., Shevtsova, N.A., Lafreniere-Roula, M., McCrea, D.A., 2006. Modelling spinal circuitry involved in locomotor pattern generation: insights from deletions during fictive locomotion. J. Physiol. 577, 617–639.

Sapir, T., Geiman, E.J., Wang, Z., Velasquez, T., Mitsui, S., Yoshihara, Y., Frank, E., Alvarez, F.J., Goulding, M., 2004. Pax6 and engrailed 1 regulate two distinct aspects of renshaw cell development. J. Neurosci. 24, 1255–1264.

Sherrington, C.S., 1910. Flexion-reflex of the limb, crossed extension-reflex, and reflex stepping and standing. J. Physiol. (Lond.) 40, 28–121.

Shevtsova, N.A., Talpalar, A.E., Markin, S.N., Harris-Warrick, R.M., Kiehn, O., Rybak, I.A., 2015. Organization of left-right coordination of neuronal activity in the mammalian spinal cord: insights from computational modelling. J. Physiol. 593, 2403–2426.

Shik, M.L., Severin, F.V., Orlovsky, G.N., 1966. Control of walking and running by means of electrical stimulation of the mid-brain. Biophysics 11, 756–765.

Shik, M.L., Severin, F.V., Orlovsky, G.N., 1969. Control of walking and running by means of electrical stimulation of the mesencephalon. Electroencephalogr. Clin. Neurophysiol. 26, 549.

Smith, J.C., Feldman, J.L., Schmidt, B.J., 1988. Neural mechanisms generating locomotion studied in mammalian brain stem-spinal cord in vitro. FASEB J. 2, 2283–2288.

Song, S., Geyer, H., 2015. A neural circuitry that emphasizes spinal feedback generates diverse behaviours of human locomotion. J. Physiol. 593, 3493–3511.

Talpalar, A.E., Bouvier, J., Borgius, L., Fortin, G., Pierani, A., Kiehn, O., 2013. Dual-mode operation of neuronal networks involved in left-right alternation. Nature 500, 85–88.

Yakovenko, S., McCrea, D.A., Stecina, K., Prochazka, A., 2005. Control of locomotor cycle durations. J. Neurophysiol. 94, 1057–1065.

Zagoraiou, L., Akay, T., Martin, J.F., Brownstone, R.M., Jessell, T.M., Miles, G.B., 2009. A cluster of cholinergic premotor interneurons modulates mouse locomotor activity. Neuron 64, 645–662.

Zhang, Y., Narayan, S., Geiman, E., Lanuza, G.M., Velasquez, T., Shanks, B., Akay, T., Dyck, J., Pearson, K., Gosgnach, S., Fan, C.M., Goulding, M., 2008. V3 spinal neurons establish a robust and balanced locomotor rhythm during walking. Neuron 60, 84–96.

Chapter 6.3

Corticospinal Control of Human Walking

Charles Capaday

Universitätsmedizin Göttingen, Institute for Neurorehabilitation Systems, Georg-August University Göttingen, Göttingen, Germany

Human walking has four main gait characteristics: (1) humans walk erect on two legs, (2) at the moment of contact with the ground the leg is almost fully extended, (3) the foot strikes the ground heel first (plantigrade gait), and (4) during the late swing phase, the body's center of gravity (COG) is outside the base of support. By contrast, the COG of bipedal walking robots, such as Mark Tilden's Robosapien and Honda's more complex Asimo, is always within the base of support. As a consequence of the straight legged nature of human gait, there is a mixed activation of extensor and flexor muscles at heel contact and the activities of the various leg extensors are not in phase. Ankle extensor activity is delayed, occurring after heel contact when activity in most other leg extensors has ceased (Capaday, 2002). In other mammals, such as cats, the activities of leg extensors are in phase when the foot makes contact with the ground, digits first (digitigrade gait). Alexander (1992) suggested that the straight-legged characteristic of human walking minimizes muscular activity by using the legs like struts. Birds walk on two legs, but in a squatted position. Penguins walk with a more erect posture than other birds, but they still walk in a squatted position and, like other birds, walk on their toes. Thus, with the exception of a similar gait adopted occasionally by some monkeys and apes, an erect, bipedal, plantigrade gait pattern is unique to humans and its neural control needs to be understood on its own terms (Capaday, 2002). Here I review in a critical manner studies on the role of the motor cortex (MCx) during human walking and some aspects of spinal reflex mechanisms as they relate to MCx control.

It may seem surprising to suggest a role for the MCx in a seemingly automatic task such as walking, but there are good reasons for this. The MCx not only issues voluntary motor commands, but it also mediates reflex-like responses to stretch of upper limb muscles (Matthews et al., 1990; Capaday et al., 1991) and integrated responses such as contact placing (Amassian et al., 1979). Its importance increases with phylogenetic order, as judged from the motor deficits that result from lesions of the corticospinal tract (CST) (Passingham et al., 1983). For example, damage to the MCx disrupts swallowing, a seemingly automatic and unconscious task (Hamdy and Rothwell, 1998). Likewise, damage to the MCx or CST results in severe walking deficits in humans and macaques, the most conspicuous being foot drop (e.g., Knutsson and Richards, 1979; Jagiella and Sung, 1989; Nathan, 1994; Courtine et al., 2005). Moreover, locomotor recovery of persons with incomplete spinal cord injury (SCI) is associated with improved corticospinal transmission assessed with transcranial magnetic stimulation (TMS) methods (Thomas and Gorassini, 2005). However, it is not clear from these clinical observations what aspect(s) of walking the MCx controls. Nor does a parallel improvement of corticospinal transmission and locomotion necessarily prove that the MCx has a direct role in controlling human walking. It may simply be that other descending tracts recover with a similar time course. This also applies to ascending, or sensory tracts. Human walking requires critical coordination of the upper body (head, arms and trunk (HAT)), with leg movements. Spinal lesions interrupt sensory inflow from the legs to supraspinal centers involved in balancing the HAT, depriving these centers of the required feedback.

In the cat, evidence from single-unit MCx recordings, intracortical microstimulation and deficits following lesions of ascending and descending spinal cord tracts, suggests that the MCx may be involved in the transition from the stance to swing phases of the step cycle (Armstrong and Drew, 1984a, 1984b; Jiang and Drew, 1996; Rho et al., 1999). However, no single observation directly proves this. For example, the peak firing rate of different MCx neurons occurs at widely different times during the step cycle (Armstrong and Drew, 1984a, 1984b). Additionally, sensory inputs may also modulate the firing rate of MCx neurons, making interpretation of their activity ambiguous. What is clear is that the MCx can initiate voluntary corrective adjustments, such as stepping over a suddenly-appearing obstacle (Drew, 1988). In a major series of studies on methods to restore walking deficits after a spinal cord lesion in rats, it was shown that behavioral therapies which encourage supraspinally-mediated movements result in a cortex-dependent recovery of locomotor capacity (van den Brand et al., 2012). Strong evidence was provided for MCx involvement in initiating and sustaining locomotion, as well as in corrective movements. However, in rodents the MCx is not essential for locomotion (Courtine et al., 2007). Thus, the results of van den Brand et al. (2012) show that the MCx can affect control actions normally mediated by other neural systems, a finding of potential clinical value. Let us now consider studies on the role of the MCx during human walking.

6.3.1 FORWARD WALKING

In a study that tackled this directly, the MCx leg area was activated by TMS at various phases of the step cycle (Capaday et al., 1999). Input–output curves of motor-evoked potentials (MEPs) in the ankle extensor soleus and the ankle flexor tibialis anterior (TA) were measured (Devanne et al., 1997). TMS during the stance phase elicited MEPs in both muscles. In 4 of 6 subjects, TA MEPs were larger than those of soleus throughout stance. This is surprising, since soleus is active during the stance phase, but the TA remains active only at the onset of stance. In contrast, TA MEPs were not elicited when soleus was activated voluntarily. Additionally, soleus MEPs were reduced by ∼30% during the stance phase compared to those during voluntary contractions at matched background electromyographic (EMG) levels. No comparable reduction of TA MEPs was observed. Finally, TMS of the MCx at various phases of the step cycle did not alter the timing of the next step, indicating that the MCx was not part of the neural system controlling the timing of step cycles, nor did it have access to putative spinal timing circuits. It was concluded that during locomotion, the corticospinal system taken as a whole (MCx circuitry and spinal relays of the corticospinal pathway) affects spinal circuits controlling the ankle flexor TA more than those controlling the ankle extensor soleus, but during voluntary contractions requiring attention, it affects both equally (Capaday et al., 1999).

It had been suggested that TA MEPs are enhanced at the transition from stance to swing (Schubert et al., 1997) but this was not seen in the Capaday et al. study (Capaday et al., 1999), though at the onset of a voluntary reaction time (RT) task, TA MEPs do increase substantially, prior to any measurable change in background EMG (Schneider et al., 2004; Davey et al., 1998; MacKinnon and Rothwell, 2000). It was suggested that MEPs may depend more on α-motoneuron activity than on activity in MCx (Schneider et al., 2004), which has important methodological implications to be discussed below.

In another study, subthreshold TMS of MCx was found to suppress muscle activity during walking (Petersen et al., 2001). It was proposed that this was due to intracortical inhibition, and that the result supported the idea that MCx was directly involved in activating both TA and soleus (Petersen et al., 2001). However, subthreshold TMS of MCx only suppresses voluntarily-generated EMG activity in about 10% of trials and the effect is weak. It is therefore not a reliable indicator of the proposed involvement of MCx. More importantly, H-reflexes elicited at the time of maximal reduction of voluntary soleus EMG activity by sub-threshold TMS are reduced relative to control H-reflexes (Fig. 6.3.1). This suggests that TMS that is subthreshold for activating α-motoneurons activates spinal interneurons, possibly Ia-interneurons which inhibit α-motoneurons. The important point is that the inhibition is at the spinal rather than at the cortical level.

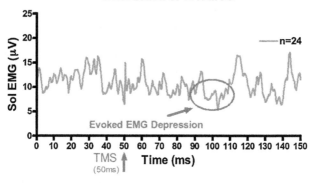

Control TMS at 34% During Soleus Isometric Contraction of 10% MVC

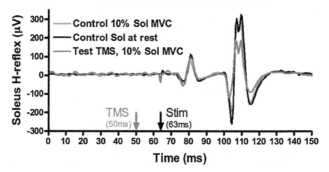

TMS Induced H-reflex Inhibition

FIGURE 6.3.1 A single subthreshold TMS pulse produces a small depression, or inhibition, of the ongoing EMG. When the H-reflex is timed to arrive at the time of maximal inhibition it is markedly reduced compared to its value at rest. This shows that the cortical stimulus evokes inhibition in the spinal cord.

6.3.2 BACKWARD WALKING

Lacquaniti et al. (1999) proposed that backward walking is controlled at the kinematic level by the time-reversed motor program of forward walking (Lacquaniti et al., 1999). Interestingly, the modulation pattern of the soleus H-reflex is not a time-reversed version of the pattern during forward walking. In forward walking, the soleus H-reflex increases progressively during the stance phase nearly in parallel with soleus EMG levels (Capaday and Stein, 1986; Crenna and Frigo, 1987; Ethier et al., 2003). It is abruptly reduced just before swing and remains essentially shut off throughout the swing phase and early stance while TA is active (Ethier et al., 2003). The modulation pattern of the H-reflex during forward walking thus follows the classic pattern of reciprocal inhibition between antagonistic muscles (Lavoie et al., 1997). But when untrained subjects walked backward on a treadmill the modulation pattern was very different. There was a marked increase of the soleus H-reflex in mid-swing, well before soleus EMG activity started and toe contact occurred (Schneider et al., 2000). It was suggested that this was associated with reduced confidence due to uncertainties of balance and timing of toe contact. In support of this idea, when subjects held onto handrails, the high-amplitude H-reflex in mid-swing was no longer present (Schneider and Capaday, 2003). This was also the case after ten days of training without handrail support. During the training period the maximal H-reflex shifted progressively from mid-swing to early stance, suggesting that the reflex activity was anticipatory and gradually declined as subjects gained confidence. The reflex changes were not due to changes in ankle muscle activity or leg kinematics, indicating that they were adaptations in the motor program controlling backward walking.

Because backward walking on a treadmill appears to require greater conscious control, it seemed reasonable to ask whether the MCx might be involved. Specifically, it was posited that in untrained subjects, CST activity during mid-swing depolarizes soleus α-motoneurons subliminally and thus brings them closer to threshold, explaining the unexpectedly high amplitude H-reflex (Ung et al., 2005). To test this hypothesis, TMS was applied to the leg area of the MCx during backward walking. MEPs were recorded from soleus and TA in untrained subjects at different phases of the step cycle. It was reasoned that if soleus MEPs could be elicited in mid-swing while soleus was inactive, this would be strong evidence for increased postsynaptic excitability of soleus α-motoneurons. In the event, despite the presence of an unexpectedly large H-reflex in mid-swing, no soleus MEPs were observed at that time. Rather, they were in phase with soleus EMG activity (Fig. 6.3.2). During backward walking soleus MEPs increased less rapidly as a function of voluntary EMG activity than they did in voluntary contractions. Furthermore, a conditioning stimulus to the MCx facilitated the soleus H-reflex at rest and during voluntary plantarflexion, but not in the mid-swing phase of backward walking. As mentioned above, with daily training, the maximal H-reflex shifted progressively from mid-swing to early stance, and its amplitude was considerably reduced compared with its value on the first day. By contrast, no changes were observed in the timing or amplitude of soleus MEPs with training (Fig. 6.3.2).

Taken together, these observations make it unlikely that the MCx is involved in the control of the H-reflex during the backward step cycle of untrained subjects, nor in its progressive adaptation with training. Instead, the large amplitude of the H-reflex in untrained subjects in backward walking, and its adaptation with training, may be due to control of presynaptic inhibition of Ia-afferents by other descending tracts.

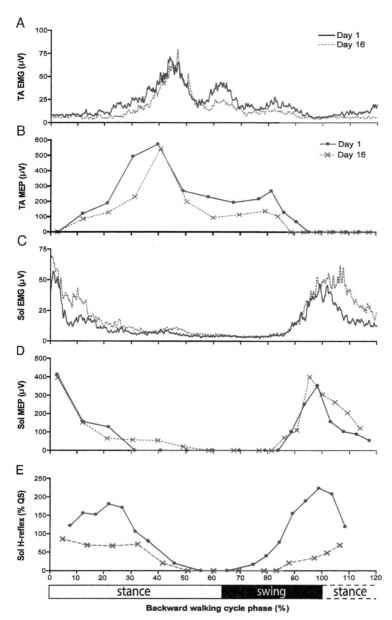

FIGURE 6.3.2 Comparisons of soleus H-reflex and MEP modulation patterns during backward walking on day 1 and day 16 of training. On day 1, the soleus H-reflex began to increase in the mid-swing phase prior to soleus EMG onset, whereas the soleus MEP increased at soleus EMG onset. At day 16, the soleus H-reflex had decreased throughout the step cycle, most markedly in mid-swing and early stance. In contrast, there was little difference between soleus and TA MEPs on days 1 and 16.

6.3.3 COMMENTS ON THE ROLE OF MOTOR CORTEX IN FORWARD AND BACKWARD WALKING

Let us first consider the detailed study of Courtine et al. (2005) on the effects of unilateral transections at the thoracic level of the CST in macaques (Courtine et al., 2005). A week after transection the animals regained some stepping capacity, with notable leg drag. However, the monkeys were unsuccessful at retrieving an item with the affected hindlimb. Thus a group of muscles could be activated during locomotion, but not voluntarily. Some 12 weeks later the spatial and temporal characteristics of the kinematics and EMG activation patterns during walking returned to near control values and yet the ability of the animals to retrieve an item with the affected hindlimb remained significantly impaired. Clearly, the CST lesion had a differential effect on stepping and voluntary activation of the affected hindlimb. If the MCx were directly and equally involved in the two tasks, one would expect that muscles activated in one task would also be equally activated in the other. After spinal cord injury, Sherrington (1947) noted *"In the monkey and in man spinal shock is not only peculiarly intense but peculiarly long lasting"* (Sherrington, 1947). The mechanisms of spinal shock are poorly understood, but we know that many spinal neurons, including interneurons, undergo what Sherrington termed *"isolation dystrophy"*, they degenerate and die. We also know that many different central and peripheral inputs converge on common interneurons. Furthermore, studies on fictive locomotion in cats have shown that pattern generating and reflex circuits may be intertwined (McCrea, 2001). Thus, gait deficits after CST lesions may be the result of interneuronal dysfunctions, as well as other yet unknown mechanisms. Put simply, the effects of CST lesions do not necessarily reveal the function of the CST per se. The same logic applies to hemiplegic gait after a stroke; the resulting gait deficits do not necessarily prove that cortical systems drive human walking.

Let us now consider the TMS studies. No facilitation of TA MEPs was found at the transition from stance to swing, by contrast to the large facilitation observed some 12 ms before the onset of voluntary TA EMG activity. It may be argued that at the transition from stance to swing, TA α-motoneurons are repolarizing from a state of hyperpolarization and that this is different from a voluntary dorsiflexion, where the α-motoneurons may be either at a resting membrane potential, or even subliminally depolarized. However, at or very near the onset of activity these considerations do not apply in either case. It therefore seems unlikely that the human MCx is involved in either triggering or driving flexor muscle activity during walking. During backward walking, whilst a prominent soleus H-reflex can be elicited in mid-swing, it is not possible to evoke an MEP in soleus, or to facilitate the H-reflex with MCx TMS. Thus two principal observations argue against the direct involvement of the MCx during either forward or backward walking. Regarding the enhanced TA MEPs during the stance phase of walking, in this phase the flexion reflex is enhanced relative to that during standing. Additionally, it is significantly enhanced during swing relative to voluntary leg and ankle flexion at matched levels of EMG activity. Presumably therefore, TMS of the MCx during walking activates interneurons in the spinal circuit mediating the flexion reflex which are in a greater excitable state.

The last point that requires explanation is the apparent reduction of CST input to soleus α-motoneurons during the stance phase of gait. As discussed in Subchapter 6.4, a significant portion of ankle extensor muscle activity may be due to inputs from muscle spindle and tendon organ afferents, mediated in part by spinal interneurons. CST inputs to these interneurons may be partially occluded by the afferent input. Another possibility is that the enhanced excitability of the flexion reflex network inhibits interneurons that transmit part of the CST input to soleus α-motoneurons. This issue requires further investigation.

6.3.4 CONCLUSIONS ON CORTICOSPINAL CONTROL

The human MCx may well act to voluntarily initiate or stop walking (Jiang et al., 2015). However, the locomotor drive is likely to be mediated by brainstem nuclei that have been shown in numerous studies to be capable of initiating and sustaining locomotion (e.g., Steeves and Jordan, 1980; Noga et al., 1991; Cowley et al., 2008. Damage to the MCx would then lead to impairments in the initiation of gait and also of voluntary gait modifications, such as changing direction. These effects must be distinguished from direct phasic drive of locomotor muscles once walking has started. On balance, the evidence presented above does not support the notion that the MCx is involved in this direct control. In stating this I do not claim that the MCx has no role in controlling walking that is in progress. Perhaps it is involved in balancing the HAT and along with other descending systems, in maintaining excitability and balance within brainstem and spinal circuits. The present conclusions are based on noninvasive neurophysiological experiments. They stand to be corrected by more probing methods that may be developed in the future. Finally, human walking involves more than just placing one foot after the other on the ground. The body must remain balanced and for most of the time it is balanced on only one leg. This is the main difficulty of human walking and it requires the integrative action of many ascending and descending neural systems.

REFERENCES

Alexander, M.R., 1992. The Human Machine. Columbia University Press, NY.

Amassian, V.E., Eberle, L., Rudell, A., 1979. Changes in forelimb trajectory during maturation of contact placing in kittens [proceedings]. J. Physiol. 289, 54P.

Armstrong, D.M., Drew, T., 1984a. Discharges of pyramidal tract and other motor cortical neurones during locomotion in the cat. J. Physiol. 346, 471–495.

Armstrong, D.M., Drew, T., 1984b. Locomotor-related neuronal discharges in cat motor cortex compared with peripheral receptive fields and evoked movements. J. Physiol. 346, 497–517.

Capaday, C., 2002. The special nature of human walking and its neural control. Trends Neurosci. 25, 370–376.

Capaday, C., Stein, R.B., 1986. Amplitude modulation of the soleus H-reflex in the human during walking and standing. J. Neurosci. 6, 1308–1313.

Capaday, C., Forget, R., Fraser, R., Lamarre, Y., 1991. Evidence for a contribution of the motor cortex to the long-latency stretch reflex of the human thumb. J. Physiol. 440, 243–255.

Capaday, C., Lavoie, B.A., Barbeau, H., Schneider, C., Bonnard, M., 1999. Studies on the corticospinal control of human walking. I. Responses to focal transcranial magnetic stimulation of the motor cortex. J. Neurophysiol. 81, 129–139.

Courtine, G., Roy, R.R., Raven, J., Hodgson, J., McKay, H., Yang, H., Zhong, H., Tuszynski, M.H., Edgerton, V.R., 2005. Performance of locomotion and foot grasping following a unilateral thoracic corticospinal tract lesion in monkeys (Macaca mulatta). Brain 128, 2338–2358.

Courtine, G., Bunge, M.B., Fawcett, J.W., Grossman, R.G., Kaas, J.H., Lemon, R., Maier, I., Martin, J., Nudo, R.J., Ramon-Cueto, A., Rouiller, E.M., Schnell, L., Wannier, T., Schwab, M.E., Edgerton, V.R., 2007. Can experiments in nonhuman primates expedite the translation of treatments for spinal cord injury in humans? Nat. Med. 13, 561–566.

Cowley, K.C., Zaporozhets, E., Schmidt, B.J., 2008. Propriospinal neurons are sufficient for bulbospinal transmission of the locomotor command signal in the neonatal rat spinal cord. J. Physiol. 586, 1623–1635.

Crenna, P., Frigo, C., 1987. Excitability of the soleus H-reflex arc during walking and stepping in man. Exp. Brain Res. 66, 49–60.

Davey, N.J., Rawlinson, S.R., Maskill, D.W., Ellaway, P.H., 1998. Facilitation of a hand muscle response to stimulation of the motor cortex preceding a simple reaction task. Mot. Control 2, 241–250.

Devanne, H., Lavoie, B.A., Capaday, C., 1997. Input–output properties and gain changes in the human corticospinal pathway. Exp. Brain Res. 114, 329–338.

Drew, T., 1988. Motor cortical cell discharge during voluntary gait modification. Brain Res. 457, 181–187.

Ethier, C., Imbeault, M.-A., Ung, V., Capaday, C., 2003. On the soleus H-reflex modulation pattern during walking. Exp. Brain Res. 151, 420–425.

Hamdy, S., Rothwell, J.C., 1998. Gut feelings about recovery after stroke: the organization and reorganization of human swallowing motor cortex. Trends Neurosci. 21, 278–282.

Jagiella, W.M., Sung, J.H., 1989. Bilateral infarction of the medullary pyramids in humans. Neurology 39, 21–24.

Jiang, W., Drew, T., 1996. Effects of bilateral lesions of the dorsolateral funiculi and dorsal columns at the level of the low thoracic spinal cord on the control of locomotion in the adult cat. I. Treadmill walking. J. Neurophysiol. 76, 849–866.

Jiang, N., Gizzi, L., Mrachacz-Kersting, N., Dremstrup, K., Farina, D., 2015. A brain-computer interface for single-trial detection of gait initiation from movement related cortical potentials. Clin. Neurophysiol. 126, 154–159.

Knutsson, E., Richards, C., 1979. Different types of disturbed motor control in gait of hemiparetic patients. Brain 102, 405–430.

Lacquaniti, F., Grasso, R., Zago, M., 1999. Motor patterns in walking. News Physiol. Sci. 14, 168–174.

Lavoie, B.A., Devanne, H., Capaday, C., 1997. Differential control of reciprocal inhibition during walking versus postural and voluntary motor tasks in humans. J. Neurophysiol. 78, 429–438.

MacKinnon, C.D., Rothwell, J.C., 2000. Time-varying changes in corticospinal excitability accompanying the triphasic EMG pattern in humans. J. Physiol. 528, 633–645.

Matthews, P.B., Farmer, S.F., Ingram, D.A., 1990. On the localization of the stretch reflex of intrinsic hand muscles in a patient with mirror movements. J. Physiol. (Lond.) 428, 561–577.

McCrea, D.A., 2001. Spinal circuitry of sensorimotor control of locomotion. J. Physiol. 533, 41–50.

Nathan, P.W., 1994. Effects on movement of surgical incisions into the human spinal cord. Brain 117 (Pt 2), 337–346.

Noga, B.R., Kriellaars, D.J., Jordan, L.M., 1991. The effect of selective brainstem or spinal cord lesions on treadmill locomotion evoked by stimulation of the mesencephalic or pontomedullary locomotor regions. J. Neurosci. 11, 1691–1700.

Passingham, R.E., Perry, V.H., Wilkinson, F., 1983. The long-term effects of removal of sensorimotor cortex in infant and adult rhesus monkeys. Brain 106 (Pt 3), 675–705.

Petersen, N.T., Butler, J.E., Marchand-Pauvert, V., Fisher, R., Ledebt, A., Pyndt, H.S., Hansen, N.L., Nielsen, J.B., 2001. Suppression of EMG activity by transcranial magnetic stimulation in human subjects during walking. J. Physiol. 537, 651–656.

Rho, M.J., Lavoie, S., Drew, T., 1999. Effects of red nucleus microstimulation on the locomotor pattern and timing in the intact cat: a comparison with the motor cortex. J. Neurophysiol. 81, 2297–2315.

Schneider, C., Capaday, C., 2003. Progressive adaptation of the soleus H-reflex with daily training at walking backward. J. Neurophysiol. 89, 648–656.

Schneider, C., Lavoie, B.A., Capaday, C., 2000. On the origin of the soleus H-reflex modulation pattern during human walking and its task-dependent differences. J. Neurophysiol. 83, 2881–2890.

Schneider, C., Lavoie, B.A., Barbeau, H., Capaday, C., 2004. Timing of cortical excitability changes during the reaction time of movements superimposed on tonic motor activity. J. Appl. Physiol. 97, 2220–2227.

Schubert, M., Curt, A., Jensen, L., Dietz, V., 1997. Corticospinal input in human gait: modulation of magnetically evoked motor responses. Exp. Brain Res. 115, 234–246.

Sherrington, C.S., 1947. Integrative Action of the Nervous System. Cambridge University Press, Cambridge, p. 433. Originally published in 1906 by Yale University Press, New Haven (CT).

Steeves, J.D., Jordan, L.M., 1980. Localization of a descending pathway in the spinal cord which is necessary for controlled treadmill locomotion. Neurosci. Lett. 20, 283–288.

Thomas, S.L., Gorassini, M.A., 2005. Increases in corticospinal tract function by treadmill training after incomplete spinal cord injury. J. Neurophysiol. 94, 2844–2855.

Ung, R.V., Imbeault, M.A., Ethier, C., Brizzi, L., Capaday, C., 2005. On the potential role of the corticospinal tract in the control and progressive adaptation of the soleus h-reflex during backward walking. J. Neurophysiol. 94, 1133–1142.

van den Brand, R., Heutschi, J., Barraud, Q., DiGiovanna, J., Bartholdi, K., Huerlimann, M., Friedli, L., Vollenweider, I., Moraud, E.M., Duis, S., Dominici, N., Micera, S., Musienko, P., Courtine, G., 2012. Restoring voluntary control of locomotion after paralyzing spinal cord injury. Science 336, 1182–1185.

Chapter 6.4

Feedback Control: Interaction Between Centrally Generated Commands and Sensory Input

Arthur Prochazka

Neuroscience and Mental Health Institute, University of Alberta, Edmonton, AB, Canada

6.4.1 LOCOMOTOR CONTROL

In order to be useful, locomotion must be controlled so as to achieve the desired bodily movement and to cope with changes in terrain. The main controlling inputs to the mammalian locomotor CPG arise from supraspinal centers, notably the cerebral cortex and midbrain locomotor region (MLR) and from sensory inputs to the spinal cord. In the intact animal the cortical drive most likely depends on visual and auditory inputs that provide the brain with information about the surroundings, terrain and the locations of external objects. It has been suggested that the basal ganglia send commands to the MLR to initiate locomotion for the purposes of exploration and the hypothalamus does the same to satisfy basic drives such as hunger (Jordan, 1998). Cadence and left–right coordination of hindlimb movements was shown nearly 50 years ago to change as the intensity of MLR stimulation in the high decerebrate cat was increased (Shik et al., 1966). This suggested that the velocity of bodily movement might be the main controlled variable in locomotion (Prochazka and Ellaway, 2012).

6.4.2 EFFECT ON LOCOMOTION OF SENSORY LOSS

Locomotion is possible in the absence of input from proprioceptors of the legs, but at least initially, it is irregular and uncoordinated. This holds true in vertebrates and invertebrates alike (Giuliani and Smith, 1987; Bassler, 1983, 1993). Intense training has been shown to restore unobstructed over-ground locomotion after chemical de-afferentation, but adaptive responses are not restored (Pearson et al., 2003). In humans who have lost limb proprioception, locomotion is severely impaired and requires conscious attention (Cole and Sedgwick, 1992). If neck proprioception is also lost, locomotion becomes virtually impossible (Lajoie et al., 1996). Given the profound effects of sensory loss on the control of locomotion, the question naturally arises, how do the three main neural mechanisms, supraspinal control, CPG control and sensory control, interact?

6.4.3 CENTRALLY-GENERATED COMMANDS VERSUS SENSORY-DOMINATED CONTROL

Even to this day, theories on the control of animal locomotion range from those in which neural activation patterns generated within the central nervous system play the major role, to those in which locomotion is entirely controlled by spinally-mediated reflexes (Song and Geyer, 2015). Between these extremes lie models in which centrally-generated patterns are modulated, fine-tuned or even overridden, in response to sensory input signaling specific biomechanical events (Prochazka, 1993).

Let us now briefly review the types of sensory input that play a role in the neural control of locomotion.

6.4.4 SENSORY INPUTS

Most of the mechanoreceptors in the limbs of mammals are cutaneous or hair follicle receptors. These receptors are sporadically active during the locomotor step cycle. For example, hair follicle receptors and cutaneous receptors in the footpad fire transiently at the moment of ground contact at the onset of the stance phase. Hair follicle receptors covering the limb fire unpredictably in response to surface airflow (Prochazka, 1996). Cutaneous afferents generally do not have strong direct reflex actions on α-motoneurons, but they can initiate more global motor responses such as the corrective response to a trip (Prochazka et al., 1978) and transitions between the stance and swing phases of the locomotor step cycle (Rossignol et al., 2006). Most of the *continuous* sensory input during movement is provided by the proprioceptive afferents: Golgi tendon organs (TOs) and muscle spindles.

Broadly speaking, TOs signal muscle force and muscle spindles signal muscle length and velocity. Few of the group Ib afferents that arise from TOs are active in a resting muscle (Houk et al., 1971). They start firing when a specific force threshold is reached. This threshold varies widely between individual Ib afferents. They are more sensitive to variations in active force generated by motor units whose muscle fibers insert into the musculotendinous capsule of the receptor (Jansen and Rudjord, 1964). With the recruitment of each such motor unit, there is a stepwise increase in the firing rate of a Ib afferent, but these steps are smoothed out when the firing rates of ensembles of Ib afferents are summed, as they would be in the spinal cord as a result of the convergence of synaptic input from ensembles of Ib afferents onto spinal neurons (Crago et al., 1982). The summed Ib firing rate saturates at high force levels even though most of the contributing Ib afferents respond fairly linearly (Crago et al., 1982). This is because more Ib afferents are recruited at low forces than at high forces, leading to a power-law relationship between ensemble Ib firing rate and force (Lin and Crago, 2002; Mileusnic and Loeb, 2009).

Muscle spindle group Ia and II afferents respond both to length changes of their parent muscle and to the activity of fusimotor efferents emanating from the spinal cord (γ-motoneurons and β-motoneurons, the latter being α-motoneurons which innervate intrafusal as well as extrafusal muscle). The responses of spindle afferents to length changes both in the absence and presence of fusimotor action have been modeled in numerous studies of varying complexity spanning 50 years. The reader is referred to a recent review which provides the mathematical details of the main models, along with a discussion of their pros and cons (Prochazka, 2015). In the simpler spindle models, fusimotor action is represented by two parameters: gain and offset. The more complex models have up to 30 parameters representing fusimotor action, intrafusal mechanics, sensory adaptation, and so on. Because γ-motoneurons are small, very few researchers have managed to record from them in normally behaving animals. This is possible in decerebrate cats (Durbaba et al., 2003; Ellaway et al., 2002; Taylor et al., 2006) but in this case, descending drive is absent or abnormal, and there is strong nociceptive input from open surgical wounds, both of which affect γ-motoneurons. Because there is still no clear consensus on how fusimotor activity is controlled during locomotion and because fusimotor action plays a big role in the more complex models, it is unclear whether these models provide better predictions of muscle spindle afferent activity during normal locomotion than the simpler models.

Muscle spindle models have only been compared head-to-head in a single study of afferent activity recorded during normal locomotion in cats (Prochazka and Gorassini, 1998a, 1998b). Interestingly, in some muscles, more than 80% of the variance in spindle afferent firing was accounted for by the simpler models, presumably because fusimotor action did not vary much in these muscles during locomotion. In other muscles such as the ankle extensors, the predictions of the models were less satisfactory, even after presumed alpha-linked fusimotor action was added.

Because roboticists who design walking machines have been inspired just as much, if not more, by locomotor control in invertebrates versus that in vertebrates, it is worth pointing out that there are remarkable analogs of vertebrate muscle spindles and TOs in invertebrates. Crustacean thoracico-coxal muscle receptor organs have response properties similar to those of mammalian muscle spindles (Bassler, 1983, 1993). They have two types of sensory afferents, T and S, which are analogous to spindle Ia and II afferents (Bush, 1981). They transmit their nonspiking signals to the CNS electrotonically, but in all other respects their responses to length changes and their efferent control by Rml and Rm2 motoneurons (equivalent to γ and β fusimotor neurons) are astonishingly similar to those of muscle spindles. Campaniform sensilla in the external cuticle of many invertebrates show response characteristics comparable to mammalian TOs. Similar response characteristics have also been reported for other classes of invertebrate proprioceptors including locust forewing stretch receptors (Pearson and Ramirez, 1990) and cockroach femoral tactile spines (Pringle and Wilson, 1952).

6.4.5 STRETCH REFLEXES AND "PREFLEXES": DISPLACEMENT AND FORCE FEEDBACK

Muscles of the legs generate the forces that support the mass of the body during locomotion. When muscles are activated at a constant level by neural input, they respond to stretch like damped springs (Houk and Henneman, 1967a; Partridge, 1967). The tendons in series with the muscle fibers are viscoelastic and contribute to this response. Resistance to stretch can be viewed as a form of displacement feedback control, in this case resulting purely from mechanical properties. The term "preflex", has been coined to describe such feedback control (Loeb et al., 1999).

Feedback from sensory receptors signaling muscle displacement and force reflexly adjusts the neural input to muscles, the simplest case being the monosynaptic stretch reflex, whereby muscle spindle afferents responding to stretch activate α-motoneurons that innervate the receptor-bearing muscles, resisting the stretch. This is also equivalent to displacement feedback. It was first thought that TOs had high thresholds and served only as overload sensors. In fact as mentioned above, ensembles of TO afferents signal active muscle force over the whole physiological range (Houk and Henneman, 1967b; Prochazka and Wand, 1980; Stephens et al., 1975). The available evidence from animal experiments suggests that in quiet stance, TOs reflexly inhibit homonymous α-motoneurons (Eccles et al., 1957) which is equivalent to negative force feedback. However, during locomotion, this reflex action switches to excitation, equivalent to *positive* force feedback (Conway et al., 1987). Interestingly, the same type of reflex reversal occurs in stick insects (Hellekes et al., 2012).

Positive feedback with an open loop gain greater than unity causes instability in man-made control systems and is therefore avoided in the design of servo control systems. In animals, the open loop gain of force feedback mediated by TO afferents presumably remains below unity during normal locomotion, partly because muscles produce less force when they shorten, which automatically reduces the loop gain (Donelan and Pearson, 2004; Prochazka et al., 1997). The open-loop gain may transiently exceed unity in bouncing gaits as animal systems switch reflex gains between different phases of the locomotor step cycle (Geyer et al., 2003). For instance, positive force feedback gains of the leg antigravity muscles that are larger than unity in the stance phase approach zero in the subsequent swing phase. Phase-dependent modulation of reflex gains has been demonstrated not only in quadruped mammals (Forssberg et al., 1975; Prochazka et al., 1978; Wand et al., 1980), but also in crustaceans (DiCaprio and Clarac, 1981), insects (Bässler, 1986), and humans (Capaday and Stein, 1986; Stein and Capaday, 1988). It is evidently a common and widespread mechanism of sensorimotor gain control in many species (Prochazka, 1989).

In humans, it was proposed that the short-latency monosynaptic stretch reflex mediated by spindle Ia afferents was the dominant contributor to stretch reflexes during locomotion (Capaday and Stein, 1986; Capaday, 2001, 2002). Sinkjaer and colleagues challenged this, instead suggesting that spindle group II afferents were the main contributors (Grey et al., 2001; Sinkjaer et al., 2000). This was in turn challenged in experiments which implicated short-latency positive force feedback mediated by TO afferents (af Klint et al., 2010; Grey et al., 2007). In a study in decerebrated cats, TO input alone was estimated to contribute up to 30% of extensor muscle activation during locomotion, with very little attributed to muscle spindle input (Donelan et al., 2009; Donelan and Pearson, 2004; Stein et al., 2000).

6.4.6 ROLE OF SENSORY INPUT IN PHASE-SWITCHING

Whereas stretch reflexes provide continuous feedback control of muscle activation, the alternating transitions between flexion and extension in the locomotor step cycle are discontinuous, switch-like events. Since the observations of Freusberg nearly 150 years ago (Freusberg, 1874), sensory input has been implicated in these transitions, but as we have seen, the locomotor CPG can also switch between flexion and extension in the absence of sensory input.

Sensory-dominated control of phase-switching is exemplified by work done in the 1960s on an above-knee prosthesis (Tomovic and McGhee, 1966). This led to the concept of a "cybernetic actuator", a system that produced continuous, alternating motions from sensory inputs that had a finite number of states. Signals from sensors monitoring joint angles and ground contact were compared to a set of threshold values corresponding to specific moments in the step cycle. Tomovic and his colleagues acknowledged the differences between what they called artificial reflex control and neural control in animals (Popovic et al., 1991). Nonetheless, in 1990, Holk Cruse formulated a set of sensory-mediated rules that neatly described the control of phase transitions and interlimb coupling in invertebrates (Cruse, 1990; Cruse et al., 1998). The "Cruse Rules" have since been used in the control systems of robotic walking machines. Contrary to the idea of CPGs, Cruse and his colleagues concluded: "Our investigations of the stick insect Carausius morosus indicate that these animals gain their adaptivity and flexibility mainly from the extremely decentralized organization of the control system that generates the leg movements. Neither the movement of a single leg nor the coordination of all six legs appears to be centrally pre-programmed. Thus, instead of using a single, central controller with global knowledge, each leg appears to possess its own controller with only procedural knowledge for the generation of the leg's movement."

REFERENCES

af Klint, R., Mazzaro, N., Nielsen, J.B., Sinkjaer, T., Grey, M.J., 2010. Load rather than length sensitive feedback contributes to soleus muscle activity during human treadmill walking. J. Neurophysiol. 103, 2747–2756.

Bassler, U., 1983. Neural Basis of Elementary Behavior in Stick Insects. Springer, Berlin, p. 169.

Bässler, U., 1986. Afferent control of walking movements in the stick insectCuniculina impigra II. Reflex reversal and the release of the swing phase in the restrained foreleg. J. Comp. Physiol. 158, 351–362.

Bassler, U., 1993. The femur-tibia control system of stick insects – a model system for the study of the neural basis of joint control. Brains Res. Rev. 18, 207–226.

Bush, B.M.H., 1981. Non-impulsive stretch receptors in crustaceans. In: Roberts, A., Bush, B.M.H. (Eds.), Neurones Without Impulses. Cambridge University Press, Cambridge, pp. 147–176.

Capaday, C., 2001. Force-feedback during human walking. Trends Neurosci. 24: 10.

Capaday, C., 2002. The special nature of human walking and its neural control. Trends Neurosci. 25, 370–376.

Capaday, C., Stein, R.B., 1986. Amplitude modulation of the soleus H-reflex in the human during walking and standing. J. Neurosci. 6, 1308–1313.

Cole, J.D., Sedgwick, E.M., 1992. The perceptions of force and of movement in a man without large myelinated sensory afferents below the neck. J. Physiol. 449, 503–515.

Conway, B.A., Hultborn, H., Kiehn, O., 1987. Proprioceptive input resets central locomotor rhythm in the spinal cat. Exp. Brain Res. 68, 643–656.

Crago, P.E., Houk, J.C., Rymer, W.Z., 1982. Sampling of total muscle force by tendon organs. J. Neurophysiol. 47, 1069–1083.

Cruse, H., 1990. What mechanisms coordinate leg movement in walking arthropods? Trends Neurosci. 13, 15–21.

Cruse, H., Dean, J., Kindermann, T., Schmitz, J., Schumm, M., 1998. Simulation of complex movements using artificial neural networks. Z. Naturforsch., C 53, 628–638.

DiCaprio, R.A., Clarac, F., 1981. Reversal of a walking leg reflex elicited by a muscle receptor. J. Exp. Biol. 90, 197–203.

Donelan, J.M., Pearson, K.G., 2004. Contribution of force feedback to ankle extensor activity in decerebrate walking cats. J. Neurophysiol. 92, 2093–2104.

Donelan, J.M., McVea, D.A., Pearson, K.G., 2009. Force regulation of ankle extensor muscle activity in freely walking cats. J. Neurophysiol. 101, 360–371.

Durbaba, R., Taylor, A., Ellaway, P.H., Rawlinson, S., 2003. The influence of bag2 and chain intrafusal muscle fibers on secondary spindle afferents in the cat. J. Physiol. 550, 263–278.

Eccles, J.C., Eccles, R.M., Lundberg, A., 1957. Synaptic actions on motoneurones caused by impulses in Golgi tendon organ afferents. J. Physiol. 138, 227–252.

Ellaway, P., Taylor, A., Durbaba, R., Rawlinson, S., 2002. Role of the fusimotor system in locomotion. Adv. Exp. Med. Biol. 508, 335–342.

Forssberg, H., Grillner, S., Rossignol, S., 1975. Phase dependent reflex reversal during walking in chronic spinal cats. Brain Res. 85, 103–107.

Freusberg, A., 1874. Reflexbewegungen beim Hunde. Pflügers Arch. 9, 358–391.

Geyer, H., Seyfarth, A., Blickhan, R., 2003. Positive force feedback in bouncing gaits? Proc. R. Soc. Lond. B, Biol. Sci. 270, 2173–2183.

Giuliani, C.A., Smith, J.L., 1987. Stepping behaviors in chronic spinal cats with one hindlimb deafferented. J. Neurosci. 7, 2537–2546.

Grey, M.J., Ladouceur, M., Andersen, J.B., Nielsen, J.B., Sinkjaer, T., 2001. Group II muscle afferents probably contribute to the medium latency soleus stretch reflex during walking in humans. J. Physiol. 534, 925–933.

Grey, M.J., Nielsen, J.B., Mazzaro, N., Sinkjaer, T., 2007. Positive force feedback in human walking. J. Physiol. 581, 99–105.

Hellekes, K., Blincow, E., Hoffmann, J., Buschges, A., 2012. Control of reflex reversal in stick insect walking: effects of intersegmental signals, changes in direction, and optomotor-induced turning. J. Neurophysiol. 107, 239–249.

Houk, J., Henneman, E., 1967a. Feedback control of skeletal muscles. Brain Res. 5, 433–451.

Houk, J., Henneman, E., 1967b. Responses of Golgi tendon organs to active contractions of the soleus muscle of the cat. J. Neurophysiol. 30, 466–481.

Houk, J.C., Singer, J.J., Henneman, E., 1971. Adequate stimulus for tendon organs with observations on mechanics of ankle joint. J. Neurophysiol. 34, 1051–1065.

Jansen, J.K., Rudjord, T., 1964. On the silent period and golgi tendon organs of the soleus muscle of the cat. Acta Physiol. Scand. 62, 364–379.

Jordan, L.M., 1998. Initiation of locomotion in mammals. In: Kiehn, O., Harris-Warrick, R.M., Jordan, L.M., Hultborn, H., Kudo, N.N.Y. (Eds.), Neuronal Mechanisms for Generating Locomotor Activity. Academy of Sciences, New York, pp. 83–93.

Lajoie, Y., Teasdale, N., Cole, J.D., Burnett, M., Bard, C., Fleury, M., Forget, R., Paillard, J., Lamarre, Y., 1996. Gait of a deafferented subject without large myelinated sensory fibers below the neck. Neurology 47, 109–115.

Lin, C.C., Crago, P.E., 2002. Neural and mechanical contributions to the stretch reflex: a model synthesis. Ann. Biomed. Eng. 30, 54–67.

Loeb, G.E., Brown, I.E., Cheng, E.J., 1999. A hierarchical foundation for models of sensorimotor control. Exp. Brain Res. 126, 1–18.

Mileusnic, M.P., Loeb, G.E., 2009. Force estimation from ensembles of Golgi tendon organs. J. Neural Eng. 6, 036001.

Partridge, L.D., 1967. Intrinsic feedback factors producing inertial compensation in muscle. Biophys. J. 7, 853–863.

Pearson, K.G., Ramirez, J.M., 1990. Influence of input from the forewing stretch receptors on motoneurones in flying locusts. J. Exp. Biol. 151, 317–340.

Pearson, K.G., Misiaszek, J.E., Hulliger, M., 2003. Chemical ablation of sensory afferents in the walking system of the cat abolishes the capacity for functional recovery after peripheral nerve lesions. Exp. Brain Res. 150, 50–60.

Popovic, D., Tomovic, R., Tepavac, D., 1991. Control aspects of active above-knee prosthesis. Int. J. Man-Mach. Stud. 35, 751–767.

Pringle, J.W.S., Wilson, V.J., 1952. The response of a sense organ to a harmonic stimulus. J. Exp. Biol. 29, 220–234.

Prochazka, A., 1989. Sensorimotor gain control: a basic strategy of motor systems? Prog. Neurobiol. 33, 281–307.

Prochazka, A., 1993. Comparison of natural and artificial control of movement. IEEE Trans. Rehabil. Eng. 1, 7–17.

Prochazka, A., 1996. Proprioceptive feedback and movement regulation. In: Rowell, L., Sheperd, J.T. (Eds.), Handbook of Physiology, Section 12. Exercise: Regulation and Integration of Multiple Systems. American Physiological Society, New York, pp. 89–127.

Prochazka, A., 2015. Proprioceptor models. In: Jung, R., Jaeger, D., Tresch, M. (Eds.), Encyclopedia of Neurocomputation. Springer-Verlag, New York, NY, pp. 2501–2518.

Prochazka, A., Ellaway, P.H., 2012. Sensory systems in the control of movement. In: Baldwin, K., Edgerton, V.R., Wagner, P. (Eds.), Comprehensive Physiology, Supplement 29: Handbook of Physiology, Exercise: Regulation and Integration of Multiple Systems. Wiley, in conjunction with the American Physiological Society, New York, pp. 2615–2627.

Prochazka, A., Gorassini, M., 1998a. Ensemble firing of muscle afferents recorded during normal locomotion in cats. J. Physiol. 507, 293–304.

Prochazka, A., Gorassini, M., 1998b. Models of ensemble firing of muscle spindle afferents recorded during normal locomotion in cats. J. Physiol. 507, 277–291.

Prochazka, A., Wand, P., 1980. Tendon organ discharge during voluntary movements in cats. J. Physiol. 303, 385–390.

Prochazka, A., Sontag, K.H., Wand, P., 1978. Motor reactions to perturbations of gait: proprioceptive and somesthetic involvement. Neurosci. Lett. 7, 35–39.

Prochazka, A., Gillard, D., Bennett, D.J., 1997. Implications of positive feedback in the control of movement. J. Neurophysiol. 77, 3237–3251.

Rossignol, S., Dubuc, R., Gossard, J.P., 2006. Dynamic sensorimotor interactions in locomotion. Physiol. Rev. 86, 89–154.

Shik, M.L., Severin, F.V., Orlovsky, G.N., 1966. Control of walking and running by means of electrical stimulation of the mid-brain. Biophysics 11, 756–765.

Sinkjaer, T., Andersen, J.B., Ladouceur, M., Christensen, L.O., Nielsen, J.B., 2000. Major role for sensory feedback in soleus EMG activity in the stance phase of walking in man. J. Physiol. 523, 817–827.

Song, S., Geyer, H., 2015. A neural circuitry that emphasizes spinal feedback generates diverse behaviours of human locomotion. J. Physiol. 593, 3493–3511.

Stein, R.B., Capaday, C., 1988. The modulation of human reflexes during functional motor tasks. Trends Neurosci. 11, 328–332.

Stein, R.B., Misiaszek, J.E., Pearson, K.G., 2000. Functional role of muscle reflexes for force generation in the decerebrate walking cat. J. Physiol. 525, 781–791.

Stephens, J.A., Reinking, R.M., Stuart, D.G., 1975. Tendon organs of cat medial gastrocnemius: responses to active and passive forces as a function of muscle length. J. Neurophysiol. 38, 1217–1231.

Taylor, A., Durbaba, R., Ellaway, P.H., Rawlinson, S., 2006. Static and dynamic gamma-motor output to ankle flexor muscles during locomotion in the decerebrate cat. J. Physiol. 571, 711–723.

Tomovic, R., McGhee, R., 1966. A finite state approach to the synthesis of control systems. IEEE Trans. Hum. Factors Electron. 7, 122–128.

Wand, P., Prochazka, A., Sontag, K.H., 1980. Neuromuscular responses to gait perturbations in freely moving cats. Exp. Brain Res. 38, 109–114.

Chapter 6.5

Neuromechanical Control Models

Arthur Prochazka[‡] and Hartmut Geyer[§]

[‡]*Neuroscience and Mental Health Institute, University of Alberta, Edmonton, AB, Canada*
[§]*Robotics Institute, Carnegie Mellon University, Pittsburgh, PA, United States*

In the last few years the interaction between biomechanical properties, sensory inputs and the locomotor CPG has been explored in neuromechanical models. For instance, Yakovenko and colleagues (2004) developed a planar model comprising a pair of hind limbs attached to a horizontal torso supported at the front by a frictionless wheel (Fig. 6.5.1A). In this model, each hindlimb comprised four rigid-body segments (thigh, shank, foot, and toes) actuated by six muscle–tendon units. The CPG activation patterns of the muscle tendon units were based on EMG recordings in normal locomotion in cats (Fig. 6.5.1B). The model assumed that proprioceptive reflexes mediated by TO and muscle spindle input contributed up to 30% to the muscle activations (Prochazka et al., 2002). In addition, the timing of step cycle phase transitions set by the CPG patterns could be overridden by the following finite-state rules (Prochazka, 1993):

1. Stance to swing transition: IF stance AND ipsilateral hip is extended AND contralateral leg is loaded THEN swing;
2. Swing to stance transition: IF swing AND ipsilateral hip is flexed AND ipsilateral knee is extended THEN stance.

Interestingly, the kinematic effect on locomotion of completely removing the stretch reflex components was surprisingly modest, except when the amplitudes of the CPG profiles were set too low to maintain load-bearing. It had been noted previously that in the cat there are relatively long delays (20 to 40 ms) before the reflexly-evoked EMG response to ground contact develops (Gorassini et al., 1994; Gritsenko et al., 2001). When the delay in the development of muscle force is added (50 to 100 ms), the overall delay in the force contribution of stretch reflexes (70 to 140 ms) becomes a significant part of the entire duration of the stance phase ($\sim$550 ms in slow walking, $\sim$150 ms in a gallop). Along the same lines, in humans, the stiffness of electrically-activated muscles in the absence of stretch reflexes was compared to that of voluntarily-activated muscles in the presence of stretch reflexes (Sinkjaer et al., 1988). At medium forces, stretch reflexes increased stiffness by up to 60%, but at low and high forces the reflex contributions dropped to zero. Furthermore these contributions did not develop fully until 200 ms after stretch onset.

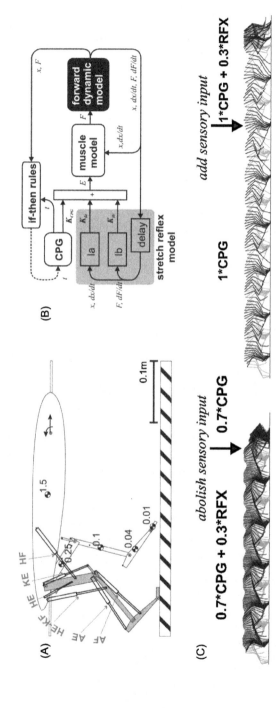

FIGURE 6.5.1 Neuromuscular model of cat hind limb locomotion: (A) musculoskeletal structure; (B) control overview; (C) traces of hindlimb motion with different contributions by CPG and proprioceptive reflexes. Without sensory input, the model falls (left). Added sensory input stabilizes gait (right).

When the If–Then rules were added to the basic CPG activation profiles, the stability of the model in over-ground locomotion and in dealing with variations in the terrain was significantly enhanced (Fig. 6.5.1C). The key to the improvement was that the duration of each CPG-generated step cycle was automatically adjusted to the prevailing kinematic state. Similar conclusions were reached in a later neuromechanical study in which locomotion was generated entirely by If–Then rules, in the absence of a modulated CPG pattern (Ekeberg and Pearson, 2005).

In finite-state control, sensory thresholds, actuator states and state-transition rules are rigidly defined. There is no provision for the input and output states to be described in probabilistic terms or for sensory inputs to be weighted such that the sums of weighted inputs fire state transitions as opposed to every sensory input having to exceed its specified threshold (Cruse, 1990). Over the years, finite-state systems have been "softened" to incorporate such probabilistic features (e.g., fuzzy controllers, Kalman filters, and Hidden Markov models) that can cope with uncertain sensory inputs and uncertain motor outputs in a probabilistic manner (Prochazka, 1996).

6.5.1 ARE EXTENSOR-DOMINATED PHASE DURATIONS OBLIGATORY FOR BIOMECHANICAL REASONS?

The spinal locomotor CPG is effectively blind to the unfolding kinematics. We saw that in fictive locomotion in decerebrate MLR-stimulated cats the spinal CPG could generate cycles ranging from extensor- to flexor-dominant, depending on the balance of descending drives to the half-centers. It was therefore interesting to discover in the above neuromechanical model that stable locomotion was associated with phase durations that conformed to the extensor-dominated pattern. This suggests that the biomechanics of stable locomotion require extensor-dominant phase-duration characteristics. It would also suggest that to harmonize with the kinematics and therefore the sensory input, the CPG oscillator should not only have an extensor-dominant phase-duration characteristic, but its operating points on this characteristic should be matched as closely as possible to forthcoming biomechanical requirements by drive descending from supraspinal areas. In other words, phase-duration characteristics are dictated by biomechanical attributes and the nature of the motor task (Prochazka and Yakovenko, 2007a, 2007b).

6.5.2 NEUROMECHANICAL ENTRAINMENT IN HUMAN MODELS OF LOCOMOTION

The interdependence between neural control and biomechanics has been stressed even more in models of human locomotion. Early models by Taga and colleagues (Taga, 1995a, 1995b; Taga et al., 1991) were based on the assumption that the entire process of human locomotion results from mutual entrainment between the oscillatory, pendulum-like body mechanics and the oscillatory stimulation by neural CPGs. Corresponding biped models showed stable and robust locomotion without having to resort to motion trajectories, and adaptation to changing slopes and load conditions. Hase and Yamazaki showed that this entrainment approach generalizes to 3D neuromuscular models with many CPGs and muscles (Hase and Yamazaki, 2002).

Other models included reflexive entrainment. For example, Ogihara and Yamazaki (2001) found in a modeling study that human locomotion could be generated with a neural controller architecture in which CPG input was required only for some muscles. The mutual entrainment between the CPG and the mechanical system in this model was achieved mostly through reciprocal reflex inhibition of antagonistic leg muscles, which provided an alternative source of rhythmic control. Several other neuromechanical models of human locomotion have been proposed whose control did not include a CPG layer. For instance, in testing the equilibrium point hypothesis, Gunther and Ruder (2003) found that only two sets of reference muscle lengths were needed in a model driven by stretch reflexes to entrain human walking in different gravitational environments. In other recent models the essential mechanical functions of legged locomotion were achieved for a range of human locomotion behaviors with entrainment only by reflexive feedback control (Geyer and Herr, 2010; Song and Geyer, 2015; Wang et al., 2012).

6.5.3 ALTERNATIVE ROLES OF CPGS IN THE LIMB CONTROLLER

Given the various entrainment options proposed, there is a renewed interest in the theoretical exploration of alternative roles of CPGs. Possible alternatives are to consider CPGs as observers of reflex output rather than as generators of limb motion (Kuo, 2002), and as sole providers of an internal clock driving muscle output frequency (Dzeladini et al., 2014). In the latter study, which was based on a neuromechanical model of human locomotion, it was concluded that a locomotor CPG could function as an internal drive and speed control mechanism of a primarily reflex-based control network.

In contrast to neuromechanical models of cat locomotion, the experimental validation of human models has been pursued less intensively, which makes it difficult to decide between the different control ideas. For several human models, the predicted leg kinematics, kinetics, and muscle activations patterns have been compared to the ones observed in human gait and found to be more or less in agreement. More intensive comparisons will be necessary to really test candidate models. For instance, the reaction to mechanical or electrical perturbations could be simulated and could help to refute some control hypotheses while supporting others.

6.5.4 INSPIRATION FOR CONTROL IN ROBOTICS

The control systems in neuromechanical models differ from those prevalent in legged robots (compare with Subchapters 4.1–4.7), in which the control structure is derived in a principled way from the dynamical model of the entire robot. By contrast, the control system in neuromechanical models is based largely on the known neuromuscular and biomechanical properties of animals and theories concerning how these interact to generate legged locomotion. Such a heuristic strategy is at odds with a rigorous mathematical approach, but, on the other hand, it can inspire new ideas for overcoming control problems in legged robots.

One example concerns leg prostheses. The control of robotic leg prostheses cannot be derived with a model-based approach, as the states of the human in this human–robot system are unknown. This decentralized control problem is currently being solved using joint impedance control. In this approach, the torque–angle relationships observed for individual leg joints in human locomotion are reproduced in motorized prostheses to provide amputees with an artificial limb behavior appropriate for normal human locomotion (Sup et al., 2009). While this works for walking over level ground and slopes, impedance control does not react well to unexpected pushes, trips, or slips. Neuromechanical models with their decentralized control architecture have inspired alternative control schemes in this application. For instance, reflex-like controllers have been implemented in robotic ankle and knee prosthesis prototypes with the goals of providing natural joint behavior and improving balance recovery after gait disturbances (Eilenberg et al., 2010; Markowitz et al., 2011; Thatte and Geyer, 2015).

Another example is the exploration of neuromechanical entrainment in legged robots (compare also with Subchapter 4.8). Quadrupedal robots like Tekken (Kimura et al., 2007) and Cheetah-cub (Sproewitz et al., 2013) use a control architecture which directly mimics that of neuromechanical models. In particular, a CPG network generates motor commands in these robots. In the case of Cheetah-cub, the network runs open-loop without sensory feedback, and the neuromechanical entrainment is sufficient to produce self-stabilizing locomotion over a large range of speeds. For Tekken, the network is augmented by reflexes that influence joint torques either directly or by modulating the duration of step-cycle phases generated by a CPG. Outdoor experiments with this quadruped robot demonstrate the adaptiveness of entrainment to unexpected disturbances on unstructured natural ground, a task that still presents a challenge for many model-based control approaches.

Similar explorations of entrainment have been performed with humanoid robots. One example is the work of Nassour et al. (2014) who studied walking of a NAO humanoid driven by a multilayer CPG network similar to the one proposed by Rybak and colleagues for cat locomotion (Rybak et al., 2006). In general, however, simulation and experimental studies with humanoid robots emphasize the importance of reflexes, as dynamic balance and its control by feedback becomes a major part of bipedal locomotion. For instance, for stabilizing the gait of a 26 degree of freedom humanoid, Huang and Nakamura required specific reflexes related to the zero moment point, the landing phase, and body posture (Huang and Nakamura, 2005). Zaier and Kanda utilized similar reflexes for the stabilization of the Fujitsu HOAP-3 against sudden and unexpected obstacles (Zaier and Kanda, 2008). These reflexes represent control concepts derived from the traditional robotics literature on humanoid locomotion. Thus, the exploration of neuromechanical control ideas in humanoid robots may in return inspire neural control architectures not previously considered.

6.5.5 MODELING THE MAMMALIAN LOCOMOTOR SYSTEM

Fig. 6.5.2 provides a schematic of a locomotor control system in mammals that includes many of the control elements discussed in this chapter. Supraspinal areas initiate locomotion and set and adjust the desired body velocity according to cognitively generated goals, exteroceptive inputs such as vision, and proprioceptive input from bodily mechanoreceptors. The velocity command is mediated by the midbrain locomotor region. The two components of the spinal CPG receive this command and translate it into corresponding extensor and flexor phase durations and activity levels of α-motoneuronal pools, which in turn activate muscles. The intrinsic stiffness of the active muscles provides non-neural local negative displacement feedback ("preflexes"). Input from sensory receptors mediates both negative displacement feedback and positive force feedback via spinal reflex pathways. The sensory input also reaches the timing elements of the CPG where switching from one phase of the step cycle to the next can be influenced and possibly overridden. The sensory input also feeds back to some of the brain areas responsible for generating the overall commands.

Fig. 6.5.3 shows a schematic in which the preflex and reflex loops controlling a single muscle, and the presumed positive force feedback from Golgi tendon organs appear in more detail. The problem faced by neuromechanical modelers is to choose from the large number of mathematical equations in the literature that model the various components of this system (α-motoneurons, muscles and muscle receptors, labeled A–J in the figure and discussed in the Appendix). In

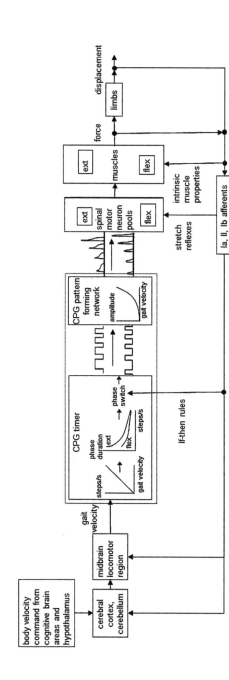

FIGURE 6.5.2 Schematic of a locomotor control system in mammals. Supraspinal areas initiate locomotion and set and adjust desired body velocity. The midbrain locomotor region translates this into a velocity command to the spinal CPG. The two components of the CPG translate the velocity command into phase durations and activity levels of flexor and extensor muscles. The intrinsic stiffness of the active muscles provides immediate negative displacement feedback. Sensory input mediates negative displacement feedback and positive force feedback via spinal reflex pathways, fine-tuning of phase durations via the CPG timing elements and higher-level decisions in the brain areas responsible for generating the overall commands.

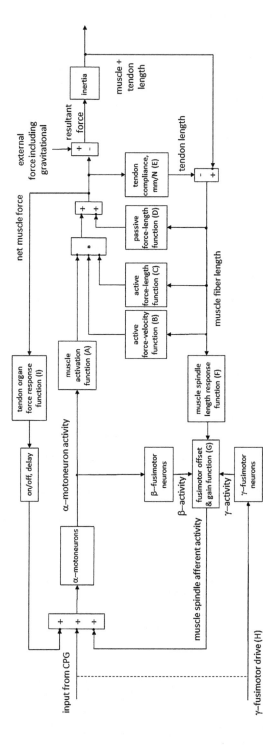

FIGURE 6.5.3 Schematic of the reflex part of the locomotor control system in mammals. For a detailed discussion of all the components of this schematic, see Appendix.

a recent paper, author AP reviewed and compared numerous models of muscle, skin and joint receptors. In principle, the more complex and recent the model, the more accurately it should predict the afferent responses. However, complexity is generally associated with increased numbers of free parameters (up to 20 in one spindle model), all of which have to be chosen according to a quite limited database of relevant recordings in animals and humans.

To paraphrase from the above review, "Regarding the accuracy that is actually required of a model, this depends on the task at hand. For example, the gain of displacement and force feedback in spinal stretch reflexes is probably quite low. Open-loop gains of spindle feedback of more than 1 or 2 tend to cause instability because of the sluggish responses of load-moving muscles (Prochazka et al., 1997a, 1997b). Because of the low gain, such reflexes contribute quite modestly to load compensation during locomotion (Prochazka et al., 2002; Prochazka and Yakovenko, 2002) and therefore the accuracy of the simpler models may suffice (i.e., a higher accuracy would make very little difference to the performance of the neuromechanical system). On the other hand, the timing of locomotor phase transitions (from swing to stance and back) relies on accurate information on the position and velocity of a limb (Markin et al., 2010; Prochazka and Yakovenko, 2007b), so here a higher accuracy of ensemble afferent input might be desirable."

The same comments regarding the trade-off between simplicity and accuracy apply to models predicting the nonlinear dependence of muscle force on muscle length, velocity, and excitation–contraction coupling (activation). For example, force–velocity models range from a simple pair of hyperbolic and exponential functions (see Appendix), to a sigmoid, or to more complex, structurally-based models (Cheng et al., 2000). Finally, in a neuromechanical model of locomotion, much depends on the biomechanical details: the skeleton, number and attachments of muscles, limb segment dimensions, and so on. Fortunately, software platforms such as Matlab Simulink, and Working Model 2D enable graphics-based construction of the physical model and the use of block diagrams such as that in Fig. 6.5.3 to specify neuromuscular properties, neuronal command signals and local feedback control for each muscle.

6.5.6   EXPLICIT EXAMPLE OF NEUROMECHANICAL MODEL OF HUMAN LOCOMOTION

As described earlier in this section, human neuromechanical models emphasize the entrainment of locomotion by reflexes. Fig. 6.5.4 provides an explicit example from Song and Geyer (2015). The model does not have a locomotor CPG that enforces explicit relationships between body velocity, step cycles, and muscle activation amplitudes of the type shown in Fig. 6.5.2. Instead, the

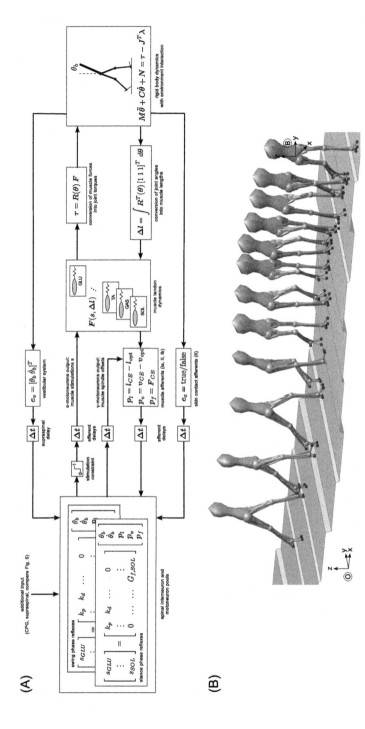

FIGURE 6.5.4 Explicit example of a neuromechanical model of human gait. The model emphasizes locomotion entrainment by reflexes. (A) Schematic of model implementation from neural control (left) to muscle dynamics (middle) to rigid body mechanics (right). (B) Snapshots of the model walking at normal speed over ground with unexpected height changes. Modified from Fig. 1 of Song and Geyer (2015).

model relies on sensory input mediated by reflexes with step cycles and muscle activation profiles appropriate for a given body velocity emerging from the interaction between sensory feedback and mechanical environment. The general model implementation is outlined in the top panel and consists of three layers: the reflex control of the spinal α-motoneuronal pools (left), the muscle dynamics (middle), and the rigid body mechanics (right). The rigid body mechanics interact with the muscle dynamics through geometric conversion of joint angles (Θ) into muscle length changes (Δl) and of muscle forces ($\mathbf{F}$) into joint torques (τ). The instantaneous moment arms of each muscle are captured in the matrix $\mathbf{R}(\Theta)$. The matrix is diagonal except for a few off-diagonal terms accounting for the biarticular nature of some leg muscles.

The length changes Δl together with the muscle stimulations $\mathbf{s}$ form the input for the computation of the muscle tendon dynamics $\mathbf{F}(\mathbf{s}, \Delta l)$, which were represented by Hill-type muscle models. More details on these models can be found in Subchapter 8.1 of this book. Major leg muscles that are often represented in neuromechanical models of human gait include the monoarticular soleus (SOL), tibialis anterior (TA), biceps femoris short head (BFsH), vastus (VAS), gluteus (GLU) and combined hip flexors (HFL), as well as the biarticular gastrocnemius (GAS), hamstrings (HAM), and rectus femoris (RF). Besides the forces $\mathbf{F}$, the contractile elements (CE) of the muscle tendon dynamics also generate proprioceptive signals from the muscle spindles ($\mathbf{p}_l$ and $\mathbf{p}_v$) and the Golgi tendon organs ($\mathbf{p}_f$) carrying information about the muscle length, velocity, and force. Although more complex models of these sensory organs exist (see Appendix), they are reduced to proportional signals with offsets for the length and velocity in this particular model.

The reflex control layer receives a range of sensory inputs and generates the muscle stimulations (α-motoneurons) and fusimotor drives (γ-motoneurons). The sensory inputs include the proprioceptive signals from the muscle dynamics and exteroceptive signals from the rigid body mechanics ($\mathbf{e}_v$ and $\mathbf{e}_c$). The latter represent the vestibular system providing information about the upper body orientation (Θ_b) and the mechanoreceptors providing information about the environment interaction (contact detection and sometimes ground reaction forces). Note that in neuromechanical gait models the sensory organs for exteroception are generally modeled with less detail than proprioceptors. The sensory pathways as well as the motor pathways interfacing the spinal α-motoneuronal pools and the mechanical layers are time delayed (Δt), mimicking the signal transmission delays in the sensory and motor axons.

The synaptic interconnections between sensory inputs and motor outputs that form the reflex control of the different muscles in the spinal α-motoneuronal pools are based on ten functional control modules which embed key functions of legged systems including compliant leg behavior and trunk balance in stance, and ground clearance and leg placement in swing. The actual reflex pathways used to embed these modules were handcrafted based on prior work that investigated each module separately.

In the figure, the synaptic interconnections are represented by matrix multiplications. While this linear representation is accurate for many of the modeled reflexes, the model has more complex interconnections as well. For instance, some reflexes use multiplication of several inputs similar to presynaptic inhibition; others include nonlinear effects such as the switching between stance and swing reflex connections due to input from the mechanoreceptors, e_c. Finally, this reflex control network may be further modulated by additional input from central pattern generators (compare with Fig. 6.5.2) and voluntary supraspinal control such as a desired leg placement angle in swing.

The bottom panel of Fig. 6.5.4 shows snapshots of the model walking overground with unexpected height changes of up to 10 cm. The red cylinders indicate the location of the leg muscles used in this model with their color saturation signaling the level of muscle activation. The spinal reflex control is sufficient to generate robust walking and other locomotor behaviors.

These behaviors were identified by optimization of the model's 82 control parameters. The optimization criteria were selected for stable and energy efficient gait at the target velocity. For more detailed explanations about the model and its implementation, see Song and Geyer (2015).

6.5.7   CONCLUDING REMARKS

In this chapter we have tried to describe the present state of neurophysiological knowledge regarding the control of legged locomotion in animals, with a focus on mammalian systems. There are still many uncertainties and gaps in our knowledge, some of which are quite basic, for example, whether human locomotion depends mainly on a spinal CPG or on chains of reflexes. Given the complexity and fluidity of the phasic activations of dozens of muscles during the relatively simple activity of walking in humans, it is no wonder that humanoid robots still walk "robotically." A biomimetic approach to robot design is quite likely to change this situation. In return, the evaluation of hypotheses about the control of the mammalian system in legged robots is likely to alter our understanding of animal and human locomotion. Neuromuscular control models will play an important role for this interplay, as they bridge between the two domains.

REFERENCES

Cheng, E.J., Brown, I.E., Loeb, G.E., 2000. Virtual muscle: a computational approach to understanding the effects of muscle properties on motor control. J. Neurosci. Methods 101, 117–130.

Cruse, H., 1990. What mechanisms coordinate leg movement in walking arthropods? Trends Neurosci. 13, 15–21.

Dzeladini, F., van den Kieboom, J., Ijspeert, A., 2014. The contribution of a central pattern generator in a reflex-based neuromuscular model. Front. Human Neurosci. 8, 371.

Eilenberg, M.F., Geyer, H., Herr, H., 2010. Control of a powered ankle-foot prosthesis based on a neuromuscular model. IEEE Trans. Neural Syst. Rehabil. Eng. 18, 164–173.

Ekeberg, O., Pearson, K., 2005. Computer simulation of stepping in the hind legs of the cat: an examination of mechanisms regulating the stance-to-swing transition. J. Neurophysiol. 94, 4256–4268.

Geyer, H., Herr, H., 2010. A muscle-reflex model that encodes principles of legged mechanics produces human walking dynamics and muscle activities. IEEE Trans. Neural Syst. Rehabil. Eng. 18, 263–273.

Gorassini, M.A., Prochazka, A., Hiebert, G.W., Gauthier, M.J., 1994. Corrective responses to loss of ground support during walking. I. Intact cats. J. Neurophysiol. 71, 603–610.

Gritsenko, V., Mushahwar, V., Prochazka, A., 2001. Adaptive changes in locomotor control after partial denervation of triceps surae muscles in the cat. J. Physiol. 533, 299–311.

Gunther, M., Ruder, H., 2003. Synthesis of two-dimensional human walking: a test of the lambda-model. Biol. Cybern. 89, 89–106.

Hase, K., Yamazaki, N., 2002. Computer simulation study of human locomotion with a three-dimensional entire-body neuro-musculo-skeletal model. I. Acquisition of normal walking. JSME Int. J. 45, 1040–1050.

Huang, Q., Nakamura, Y., 2005. Sensory reflex control for humanoid walking. IEEE Trans. Robot. 21, 977–984.

Kimura, H., Fukuoka, Y., Cohen, A.H., 2007. Biologically inspired adaptive walking of a quadruped robot. Philos. Trans. R. Soc., Math. Phys. Eng. Sci. 365, 153–170.

Kuo, A.D., 2002. The relative roles of feedforward and feedback in the control of rhythmic movements. Mot. Control 6, 129–145.

Markin, S.N., Klishko, A.N., Shevtsova, N.A., Lemay, M.A., Prilutsky, B.I., Rybak, I.A., 2010. Afferent control of locomotor CPG: insights from a simple neuromechanical model. Ann. N.Y. Acad. Sci. 1198, 21–34.

Markowitz, J., Krishnaswamy, P., Eilenberg, M.F., Endo, K., Barnhart, C., Herr, H., 2011. Speed adaptation in a powered transtibial prosthesis controlled with a neuromuscular model. Philos. Trans. R. Soc. Lond. B, Biol. Sci. 366, 1621–1631.

Nassour, J., Henaff, P., Benouezdou, F., Cheng, G., 2014. Multi-layered multi-pattern CPG for adaptive locomotion of humanoid robots. Biol. Cybern. 108, 291–303.

Ogihara, N., Yamazaki, N., 2001. Generation of human bipedal locomotion by a bio-mimetic neuro-musculo-skeletal model. Biol. Cybern. 84, 1–11.

Prochazka, A., 1993. Comparison of natural and artificial control of movement. IEEE Trans. Rehabil. Eng. 1, 7–17.

Prochazka, A., 1996. The fuzzy logic of visuomotor control. Can. J. Physiol. Pharm. 74, 456–462.

Prochazka, A., Yakovenko, S., 2002. Locomotor control: from spring-like reactions of muscles to neural prediction. In: Nelson, R.J. (Ed.), The Somatosensory System: Deciphering The Brain's Own Body Image. CRC Press, Boca Raton, pp. 141–181.

Prochazka, A., Yakovenko, S., 2007a. The neuromechanical tuning hypothesis. In: Cisek, P., Drew, T., Kalaska, J. (Eds.), Progress in Brain Research, Computational Neuroscience: Theoretical Insights into Brain Function. Elsevier, NY, pp. 255–265.

Prochazka, A., Yakovenko, S., 2007b. Predictive and reactive tuning of the locomotor CPG. Integr. Comp. Biol. 47, 474–481.

Prochazka, A., Gillard, D., Bennett, D.J., 1997a. Implications of positive feedback in the control of movement. J. Neurophysiol. 77, 3237–3251.

Prochazka, A., Gillard, D., Bennett, D.J., 1997b. Positive force feedback control of muscles. J. Neurophysiol. 77, 3226–3236.

Prochazka, A., Gritsenko, V., Yakovenko, S., 2002. Sensory control of locomotion: reflexes versus higher-level control. Adv. Exp. Med. Biol. 508, 357–367.

Rybak, I.A., Shevtsova, N.A., Lafreniere-Roula, M., McCrea, D.A., 2006. Modelling spinal circuitry involved in locomotor pattern generation: insights from deletions during fictive locomotion. J. Physiol. 577, 617–639.

Sinkjaer, T., Toft, E., Andreassen, S., Hornemann, B.C., 1988. Muscle stiffness in human ankle dorsiflexors: intrinsic and reflex components. J. Neurophysiol. 60, 1110–1121.

Song, S., Geyer, H., 2015. A neural circuitry that emphasizes spinal feedback generates diverse behaviours of human locomotion. J. Physiol. 593, 3493–3511.

Sproewitz, A., Tuleu, A., Vespignani, M., Ajallooeian, M., Badri, E., Ijspeert, A.J., 2013. Towards dynamic trot gait locomotion: design, control, and experiments with Cheetah-cub, a compliant quadruped robot. Int. J. Robot. Res. 32, 932–950.

Sup, F., Varol, H.A., Mitchell, J., Withrow, T.J., Goldfarb, M., 2009. Self-contained powered knee and ankle prosthesis: initial evaluation on a transfemoral amputee. IEEE Eng. Med. Biol. Mag. 2009, 638–644.

Taga, G., 1995a. A model of the neuro-musculo-skeletal system for human locomotion. I. Emergence of basic gait. Biol. Cybern. 73, 97–111.

Taga, G., 1995b. A model of the neuro-musculo-skeletal system for human locomotion. II Real-time adaptability under various constraints. Biol. Cybern. 73, 113–121.

Taga, G., Yamaguchi, Y., Shimizu, H., 1991. Self-organized control of bipedal locomotion by neural oscillators in unpredictable environment. Biol. Cybern. 65, 147–159.

Thatte, N., Geyer, H., 2015. Toward balance recovery with reg prostheses using neuromuscular model control. IEEE Trans. Biomed. Eng.

Wang, J.M., Hamner, S.R., Delp, S.L., Koltun, V., 2012. Optimizing Locomotion Controllers Using Biologically-Based Actuators and Objectives. ACM Trans. Graph. 31, 1–27.

Yakovenko, S., Gritsenko, V., Prochazka, A., 2004. Contribution of stretch reflexes to locomotor control: a modeling study. Biol. Cybern. 90, 146–155.

Zaier, R., Kanda, S., 2008. Adaptive locomotion controller and reflex system for humanoid robots. In: IEEE International Conference on Intelligent Robots and Systems. Nice, France, pp. 2492–2497.

Chapter 6.6

Appendix

Arthur Prochazka[¶] and Hartmut Geyer[‖]

[¶]*Neuroscience and Mental Health Institute, University of Alberta, Edmonton, AB, Canada*
[‖]*Robotics Institute, Carnegie Mellon University, Pittsburgh, PA, United States*

Reflex model of neuromuscular control. Here we provide and discuss the mathematical functions for the components in the reflex model presented in Subchapter 6.6. These include functions that describe muscle intrinsic properties (excitation–contraction coupling, force–velocity and orce–length curves) and sensory receptor models (muscle spindle responses to length changes and fusimotor drive, Golgi tendon organ responses to changes in muscle force).

The mathematical functions that have been used by modelers to specify the properties of elements A to J in Fig. 6.5.3 are discussed below.

6.6.1 MUSCLE ACTIVATION FUNCTION

$$F = \frac{da(t)}{dt} + \frac{1}{\tau_{act}} \cdot \left(\frac{\tau_{act}}{\tau_{deact}} + \left[1 - \frac{\tau_{act}}{\tau_{deact}} \right] \cdot u(t) \right) \cdot a(t) = \frac{1}{\tau_{act}} \cdot u(t)$$

where F is muscle force, $u(t)$ is the neural input signal (range 0–1), $a(t)$ the state variable associated with muscle activation, $\rightarrow_{act}$ is the time constant when $u(t) = 1$ and $\rightarrow_{deact}$ is the time constant when $u(t) = 1$ (Zajac, 1989).

6.6.2 FORCE–VELOCITY FUNCTION

The most widely used is a combination of a hyperbola for shortening (Hill, 1938) and an exponential for lengthening (Winters, 1990).

For shortening,

$$(F + a) = \frac{b \cdot (F_{max} + a)}{b - V}$$

where F is muscle tensile force, V is shortening velocity, F_{max} is isometric tensile force, a specifies a force asymptote (at $V = \infty$), b specifies a velocity asymptote (at $F = \infty$). Since the force asymptote is negative (compressive, not tensile) and the velocity asymptote is positive (lengthening, not shortening), this function is only valid when F is positive and V is negative. Force is zero at V_{max}, the maximal possible shortening velocity, which is typically 5 rest lengths/s for mammalian muscle.

For lengthening,

$$F = \left(F_{max} + c \cdot \left(1 - e^{-kV}\right)\right)$$

where k sets the curvature. This function is only valid for V positive.

The following function generates a similarly-shaped relationship and requires less computation:

$$F = F_{max}\left(\left(\left(1 - e^{-aV}\right)/\left(1 - e^{bV}\right)\right) + 1\right)$$

where a and b are constants that set the curvatures for shortening and lengthening, respectively (Winters, 1990).

6.6.3   FORCE–LENGTH (LENGTH–TENSION) FUNCTION

The force generated by individual sarcomeres is approximately related to sarcomere length by an inverted-U function and this is sometimes used to describe the length–tension curve of the whole muscle. However, in practice, the length–tension curve of whole muscle at very low velocities is virtually monotonic, with some hysteresis-like properties (Gillard et al., 2000). Arguably, over most of the physiological operating range it is sufficient to model the length–tension function of the muscle fiber component by Hooke's law for a spring:

$$dF = k \cdot dl$$

where dF is the change in force, k is a constant and dl is muscle fiber displacement.

Another version of this simple model is to divide the function into three sections, a spring at short muscle fiber lengths, a plateau at medium lengths, and a negative spring function at long lengths (Zajac, 1989).

6.6.4   PASSIVE STIFFNESS

due to connective tissue can be modeled by a spring or exponential that starts generating force at a long muscle length (Zajac, 1989). This compensates for the decline in sarcomere force at very long muscle lengths.

6.6.5   TENDON COMPLIANCE

may also be modeled by Hooke's law, with the inclusion of a small viscous component if deemed necessary.

6.6.6 MUSCLE SPINDLE LENGTH RESPONSE FUNCTION

Numerous models have been developed for muscle spindle group Ia and II afferent responses to dynamic changes in muscle length. Many of these models have been reviewed by one of the authors recently (Prochazka, 2015). The functions below are illustrative of the types of models that have been used in neuromechanical simulations. For comments on the parameters, please refer to the review, which is available on request.

Models comprising transfer functions (Poppele and Bowman, 1970):
Spindle group Ia,

$$r(s) = \frac{kxs(s+0.44)(s+11.3)(s+44)}{(s+0.04)(s+0.82)}$$

where r is firing rate, k is a constant, x is displacement;
Spindle group II,

$$r(s) = \frac{k_2xs(s+0.44)(s+11.3)}{(s+0.82)};$$

Models with power-law relationships:
Spindle group Ia and II (Houk et al., 1981),

$$r(t) - r_0 = k(x - x_1)v^n$$

where r is firing rate, k and n are constants, x is displacement;
Spindle group Ia (Prochazka, 1999),

$$r(t) = av^n + bx(t) + c + d(a(t))$$

where r is firing rate, a, c, d, n are constants, x is displacement, $\alpha(t)$ is α-motoneuronal output, assuming that a component of Ia firing is due to biasing by α-linked γ_s action (see main text and 6.6.8 below).

For more complex, structurally-based models, refer to Mileusnic et al. (2006).

6.6.7 FUSIMOTOR OFFSET AND GAIN FUNCTION

There are few models of the effects of fusimotor drive on spindle afferent firing. The following is a transfer function modeling the dynamics of γ_s and γ_d action, assuming concurrent length changes (Chen and Poppele, 1978):

$$r(s) = \frac{1}{(s+8.8)(s+50.2)}$$

where $r(s)$ is spindle group Ia firing rate in the frequency domain.

A more complex structurally-based model has been developed that includes biasing action (Mileusnic et al., 2006).

6.6.8 γ-FUSIMOTOR DRIVE

This variable is the most uncertain part of the reflex model, as discussed in the main body of the chapter. One option is to assume that γ_s action is mirrored by α-motoneuronal activity (see above), but there is much controversy as to the extent of such coactivation versus independent control of fusimotor activity that could be task- or context-related (fusimotor set) or phasically modulated in some way related to the intended kinematics for example (Ellaway et al., 2015; Taylor et al., 2006).

6.6.9 GOLGI TENDON ORGAN MODEL

The most widely used model is a transfer function developed for single Ib afferents (Houk and Simon, 1967), namely

$$r(s) = \frac{kF(s+0.15)(s+1.45)(s+16)}{(s+0.2)(s+2)(s+37)}$$

where $r(s)$ is Ib firing rate in the frequency domain F is muscle force and k is a constant.

Another version of this model has a logarithmic static response term (Lin and Crago, 2002):

$$r(s) = \frac{G_g \ln(\frac{F}{G_f}+1)(s+0.15)(s+1.45)}{(s+0.2)(s+2)}$$

where $r(s)$ is Ib firing rate in the frequency domain, F is muscle force, G_g and G_f are gains.

A more complex, structurally-based model has been developed (Mileusnic et al., 2006).

ACKNOWLEDGMENTS

AP and SG were funded by the Canadian Institute of Health Research (CIHR). CC was funded by CIHR and the Fonds de Recherche du Québec – Santé (FRQS).

REFERENCES

Chen, W.J., Poppele, R.E., 1978. Small-signal analysis of response of mammalian muscle spindles with fusimotor stimulation and a comparison with large-signal responses. J. Neurophysiol. 41, 15–27.

Ellaway, P.H., Taylor, A., Durbaba, R., 2015. Muscle spindle and fusimotor activity in locomotion. J. Anat. 227 (2), 157–166.

Gillard, D.M., Yakovenko, S., Cameron, T., Prochazka, A., 2000. Isometric muscle length–tension curves do not predict angle–torque curves of human wrist in continuous active movements. J. Biomech. 33, 1341–1348.

Hill, A.V., 1938. The heat of shortening and the dynamic constants of muscle. Proc. R. Soc. Lond. B 126, 136–195.

Houk, J., Simon, W., 1967. Responses of Golgi tendon organs to forces applied to muscle tendon. J. Neurophysiol. 30, 1466–1481.

Houk, J.C., Rymer, W.Z., Crago, P.E., 1981. Dependence of dynamic response of spindle receptors on muscle length and velocity. J. Neurophysiol. 46, 143–166.

Lin, C.C., Crago, P.E., 2002. Neural and mechanical contributions to the stretch reflex: a model synthesis. Ann. Biomed. Eng. 30, 54–67.

Mileusnic, M.P., Brown, I.E., Lan, N., Loeb, G.E., 2006. Mathematical models of proprioceptors. I. Control and transduction in the muscle spindle. J. Neurophysiol. 96, 1772–1788.

Poppele, R.E., Bowman, R.J., 1970. Quantitative description of linear behavior of mammalian muscle spindles. J. Neurophysiol. 33, 59–72.

Prochazka, A., 1999. Quantifying proprioception. Prog. Brain Res. 123, 133–142.

Prochazka, A., 2015. Proprioceptor models. In: Jung, R., Jaeger, D., Tresch, M. (Eds.), Encyclopedia of Neurocomputation. Springer-Verlag, New York, NY, pp. 2501–2518.

Taylor, A., Durbaba, R., Ellaway, P.H., Rawlinson, S., 2006. Static and dynamic gamma-motor output to ankle flexor muscles during locomotion in the decerebrate cat. J. Physiol. 571, 711–723.

Winters, J.M., 1990. Hill-based muscle models: a systems engineering perspective. In: Winters, J.M., Woo, S.L.-Y. (Eds.), Multiple Muscle Systems: Biomechanics and Movement Organization. Springer, NY, pp. 69–93.

Zajac, F.E., 1989. Muscle and tendon: properties, models, scaling, and application to biomechanics and motor control. Crit. Rev. Biomed. Eng. 17, 359–411.

Part III

Implementation

Chapter 7

Legged Robots with Bioinspired Morphology

Ioannis Poulakakis, Madhusudhan Venkadesan, Shreyas Mandre,
Mahesh M. Bandi, Jonathan E. Clark, Koh Hosoda, Maarten Weckx,
Bram Vanderborght, and Maziar A. Sharbafi

This chapter discusses how biologically inspired principles and mechanisms can be transferred in the engineering domain with the purpose of designing legged robots capable of reproducing animal locomotion behaviors. A bioinspired design approach is described, which relies on quantifying the basic principles underlying a desired locomotion behavior through simple, yet predictive, mathematical models. The chapter begins by examining the human foot and exploring its implications toward the design of the next generation of prosthetic and orthotic devices. Then, a detailed discussion of various approaches for designing robot legs is provided. The chapter concludes with a description of a number of bipedal and quadrupedal robots as instances of a bio-inpired approach to design. Throughout the chapter, the emphasis is on realizing natural-like locomotion behaviors on engineered systems without necessarily copying the exact mechanisms by which animals produce these behaviors in nature.

PREFACE

During the long history of evolution, animals have developed a variety of mechanisms to survive in challenging habitats. Virtually every organ, muscle, or nerve of their bodies has been called into being in the struggle to live in dynamically changing and uncertain environments and act in compelling circumstances. As a result, an inexhaustible set of working design ideas can be found in nature, offering valuable inspiration to engineers in their quest to build machines, systems, and algorithms that address societal needs. It is tempting to just copy the mechanisms by which a behavior of interest is realized in the biological world. However, such an approach quickly arrives at an impasse since biological solutions do not often translate directly into viable engineering designs. Furthermore, the complexity of biological systems is usually overwhelming as

Bioinspired Legged Locomotion. http://dx.doi.org/10.1016/B978-0-12-803766-9.00010-5

457

these systems may support many vital functions beyond those of interest to an engineer; hence, reproducing the underlying mechanisms can be an extremely challenging – if not impossible – task that may even be unnecessary. Finally, attempting to engineer systems that mimic exactly the form and structure of their biological counterparts alleviates the possibility to improve. This part of the book provides examples of a *bioinspired* design approach for robotic systems that relies on extracting *principles* relevant to a task of interest – in our case, legged locomotion – and then adapting these principles and incorporating them to engineering designs through the means technology provides.

Over the past few years, there has been an explosion of legged robots that are inspired by the way animals move. Whether legs push against a solid substrate to create crawling, walking, running, or climbing gaits, these truly exceptional machines offer the potential of unprecedented mobility that can match – or, in certain cases, surpass – that of their counterparts in nature. The complexity of the mechanisms underlying the capacity of biological legged systems for generating and supporting their movement can be daunting. Nonetheless, on a macroscopic level, the principles governing legged locomotion can be understood through the introduction of archetypical reductive models that are composed by idealized mechanical elements, such as springs, dampers, and inertias. In previous parts of this book we have seen examples of models of this kind. Beyond their usefulness in analyzing the mechanics of locomotion, these models can provide design guidelines for the hardware realization of robots that exhibit the model's behaviors, doing so without delving into the fine structural and morphological details of the mechanisms by which this behavior is achieved in the biological world. In other words, the principles that characterize legged locomotion behaviors can be distilled in simple idealized mechanical models, which can then be adapted to build machines that capture the functions of interest of a biological system without necessarily reproducing its form. Of course, passing from a model that encodes the fundamental principles characterizing a locomotion behavior to a robotic system that actually realizes this behavior entails a number of nontrivial design decisions. In the following chapters, we will discuss a bioinspired approach to designing legged robots that begins with forming a hypothesis about locomotion through biological observation, and proceeds with testing this hypothesis through modeling, hardware realization, and experimental evaluation and assessment.

This part of the book is composed of four chapters that address various aspects of incorporating biological principles in legged robots. We begin with Subchapter 7.1 contributed by Madhusudhan Venkadesan, Shreyas Mandre and Mahesh M. Bandi. This chapter discusses the mechanical functions of the human foot, which can act as an active suspension system for protection against collisions, as a regenerative brake for elastic energy storage, as a mechanism for modulating stiffness and damping to enhance gait stability, and finally as a sensing unit for providing essential feedback for motion control. Emphasis is placed on the derivation of simple mechanical models that encapsulate the functional principles of the foot, focusing on the relationship between structure and function. The next chapter, Subchapter 7.2, is contributed by Jonathan Clark and it deals with leg design. For the purpose of locomotion, each leg can act as a strut, a spring, a damper and an actuator, keeping the body from falling while providing means for regulating the energy associated with the locomotion task. The chapter begins with a brief description of the functions of a leg in a robot, and proceeds with leg design aspects, focusing on different actuation and transmission strategies. Then, it examines how inspiration from biology can determine the morphology and dynamics of robotic legs, focusing again on the role the simple mathematical models as means to capture and quantify biological observations. The chapter concludes with multiuse leg designs for the development of robots capable of multimodal locomotion; that is, robots which can combine walking and running behaviors with climbing of vertical surfaces.

The final two chapters of this part discuss bipedal and quadrupedal robot design. Subchapter 7.3 deals with bioinspired bipedal robots, and is contributed by Koh Hosoda, Maarten Weckx, Bram Vanderborght, Ioannis Poulakakis, and Maziar A. Sharbafi. Bipedal robots are inspired by the human morphology and have the potential to play an important role for automating tasks in typical human-centric or natural environments. The chapter begins with a brief overview of different approaches to designing robotic bipeds, including two extreme approaches in terms of actuation: passive bipedal walkers and humanoids. It is discussed that, although passive bipedal walkers are much simpler than humanoids, they are capable of reproducing the dynamics of human walking more faithfully. The chapter concludes with a brief discussion of bi-articular muscles – that is, muscle units that span more than one joints – as an actuation architecture for designing legged robots. The final chapter, Subchapter 7.4, discusses quadrupedal robots and is contributed by Ioannis Poulakakis. A brief overview of common quadrupedal gaits is provided first, and then the discussion focuses on the role of the torso in dynamic quadrupedal locomotion. Although biological observations and mathematical models suggest that torso flexibility appears to be advantageous – especially at high locomotion speeds – the vast majority of successful robotic quadrupeds feature rigid, nondeformable torsos. This is primarily due to the fact that the specifics of the mechanical realization of a flexible torso significantly increase the complexity of the platform, to the point that any performance enhancement may be overwhelmed by the cost associated with the added complexity. To summarize, as was mentioned above, nature uses its own means – shaped through millions of year of evolution – to realize locomotion in animals. On the other hand, incorporating the underlying principles to robotic designs is subject to the conditions of operation of the mechanical, actuation and control components of the system. An engineer that seeks inspiration from the biological world to reproduce animal locomotion behaviors on robot platforms should always make decisions that respect the limitations and haness the capabilities of the structural elements of the proposed design.

Chapter 7.1

Biological Feet: Evolution, Mechanics and Applications

Madhusudhan Venkadesan*, Shreyas Mandre[†], and Mahesh M. Bandi[‡]
*Department of Mechanical Engineering & Materials Science, Yale University, New Haven, CT, United States
[†]School of Engineering, Brown University, Providence, RI, United States
[‡]Collective Interactions Unit, OIST Graduate University, Tancha, Okinawa, Japan

The foot is an active suspension system that mitigates potentially injurious collisions, acts like a regenerative brake by storing elastic energy, modulates stiffness and damping to aid in stability, and provides essential sensory feedback. In this unit, we examine the function and evolutionary history of human feet from the perspective of mechanics. The emphasis is on simple mathematical models that provide insight into the relationship between structure and function. We conclude with implications for diagnosis or treatment of foot disabilities.

7.1.1 OVERVIEW

The foot lies at the interface between the ground and our body. The spring-like elastic interface provided by the foot is something that we pay little attention to, unless something goes wrong. Even mild disability of the foot can severely impact our mobility, and thereby the entire quality of our life. Interest in the human foot is also deeply rooted in the scientific endeavor to understand the evolutionary origins and implications of bipedal locomotion (Lieberman, 2012). It is therefore not surprising that scientific, medical and engineering interest in the foot dates back at least one century (Keith, 1894; Morton, 1922; Dunn, 1928; Elftman and Manter, 1935; Jones, 1941; Hicks, 1955; Ker et al., 1987; Harcourt-Smith and Aiello, 2004; Lee et al., 2005; Fey et al., 2012; Kelly et al., 2014; Zelik et al., 2014). What is perhaps more striking is that the mechanics of the foot has remained an active area of research despite continuous research interest for over a century. Our feet comprise over 50 bones, almost a quarter of all the bones in our body. With so many bones comes an astounding complexity in the number of joints, ligaments, and muscles that comprise the foot. Even counting the number of joints, ligaments, or muscles remains mired in debate (e.g., Taniguchi et al., 2003). If two distinct bones are tightly held together and move together, is that counted as a joint or not? Is a fan-shaped ligament or muscle counted as one ligament or more than one, and how does that depend on

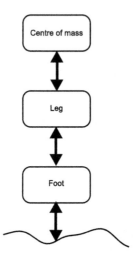

FIGURE 7.1.1　The interaction of our body with the ground is mediated by the mechanical properties of the foot. These mechanical properties arise from the interaction between the geometric structure of the foot, the skeletal elements such as the bones and ligaments, and the neurally controlled activity of the muscles.

load transmission? Is such reductionist accounting of the elements of the foot even necessary to understand its overall function? The foot remains an active area of research partly because of this structural complexity, and the importance of its function for healthy living.

The goal of this chapter is to present a mechano-centric perspective on how the foot functions, a perspective that we consider to be of great importance in the study of the foot or any anatomical organ. This unit does not attempt to present the foot's complexity in its entirety, and it does not claim to be the definitive word on its mechanics. Such a goal deserves more than a single unit, and certainly requires a substantial amount of new research. This unit therefore adopts the viewpoint of a *mathematical mechanician*, one who wants to understand a mechanism in terms of its relationships between forces and displacements using simple mathematical models. From that perspective, despite its enormous complexity, the foot can be treated as an elastic body that transforms applied loads into displacements, and transmits forces from the ground to the body according to the laws of mechanics; see Fig. 7.1.1. We bring to bear the mathematical machinery of mechanics for such an analysis of the foot. One of the primary motivations in undertaking such an analysis is to glean underlying principles that have the potential to guide the design of feet for applications in robotics and prosthetics.

After a brief overview of the anatomy and evolution of the human foot in Section 7.1.2, we motivate the functional role of the foot using a cost–benefit

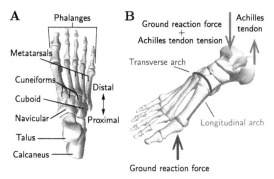

FIGURE 7.1.2   Anatomy and external forces on the human foot. (A) The skeletal anatomy of the foot, with the *proximal* and *distal* ends identified with the toes and the heel, respectively. (B) The arches of the foot, and external loads on it. The human foot has a doubly arched structure in the mid-foot region; the longitudinal and the transverse arches. This figure depicts the typical external loads when pushing-off against the ground using the ball of the foot.

analysis in Section 7.1.3. This discussion is rooted in the evolutionary story of the human foot. In Section 7.1.4, we treat the foot as a mechanical filter, one that transforms applied forces and displacements in time. The presentation is organized as models of increasing mathematical sophistication, but not of increasing complexity (measured by the number of free parameters),[1] and is restricted to consideration of either point-like or rigid feet. The spatial extent and flexibility of the foot are explicitly considered in Section 7.1.5. Section 7.1.6 concludes this unit with a summary and brief comments about the state of the art in robotic and prosthetic feet.

7.1.2   THE HUMAN FOOT

7.1.2.1   Anatomy

The feet and hands share striking muscular and skeletal similarities, including a large number of homologous bones that have shared relative arrangements, evolutionary history, and developmental origins (Rolian et al., 2010). The foot, like the hand, not only comprises a large number of bones (over 25 per foot), but also consists of a large number of muscles and sensory organs. We will revisit this comparison with the hand later, with an emphasis on their differences. For now, we use homologies between the foot and the hand to help orient the reader to the skeleton of the foot.

The skeleton of the foot comprises its bones (Fig. 7.1.2A), and the elastic ligaments that interconnect and hold the bones in place. The toes are at the *distal* end, and the calcaneus, which forms the heel, is at the *proximal* end. The

---

1. Retaining as few free parameters as possible is a central goal of effective mathematical models.

foot articulates with the shank of the leg through the ankle. The ankle is a complex joint that consists of multiple articular joints, with the talus at its core. At the least, the ankle comprises the tibiotalar, the subtalar, and the talonavicular joints. The talus articulates with the rest of the leg through the tibiotalar joint, and also forms joints with the calcaneus (subtalar joint) and the navicular (talonavicular joint). Further distally, and articulating with the navicular are a complex set of bones that are collectively called the tarsal bones (the cuboid and the cuneiforms). The tarsals are homologous to the carpal bones that form the wrist. The tarsals in turn articulate with the metatarsals at their distal end. The metacarpals of the hand, forming the palm, are homologous to the metatarsals. The metatarsal heads are commonly referred to as the balls of the foot, and is the location under the foot where you push off when walking or running. The knuckles in the hand are homologous to the metatarsal heads.

External loads on the foot, when pushing off against the ground with the balls of the foot are as depicted in Fig. 7.1.2B. The combined effect of the loads from the leg, through the tibiotalar joint, and the Achilles tendon, is approximated as a net force and a net moment at the tibiotalar joint or the calcaneus.

7.1.2.2 Evolution

The current form of the human foot evolved from an ancestral morphology that resembled our present day hands, rather than the feet (Morton, 1922, 1924a; Harcourt-Smith and Aiello, 2004; Lieberman, 2012) (Fig. 7.1.3). The toes were long, and the big toe was pointed away from the rest of the foot, termed as an abducted hallux. The tarsal and metatarsals bones were arranged such that the foot was flat, lacking the arches that are present in modern human feet. These hand-like ancestral feet are also present in extant non-human primates such as chimpanzees and gorillas. The human lineage is thought to have diverged from chimpanzees at least 7–8 million years ago (Bramble and Lieberman, 2004; Langergraber et al., 2012), and from gorillas at least 8–10 million years ago (Bramble and Lieberman, 2004; Katoh et al., 2016). Although fossil remains have been found from close to the time when the split between the human and chimpanzee lineages happened, none of those findings contain samples of the foot (*Sahelanthropus tchadensis*, Brunet et al., 2005; *Orrorin Tugenensis*, Senut et al., 2001). The earliest intact archaeological evidence from feet after the split from chimpanzees dates back to the ~ 4.4 million year old fossil of *Ardipithecus ramidus* (White et al., 1994; Lovejoy et al., 2009). This foot is different from the human foot in having a highly abducted hallux (a big toe that points sideways), flat feet, and many other morphological differences. Feet that are strikingly similar to modern human feet appear in the genus *Homo*, around 1.8 million years ago (Day and Napier, 1964). A cursory examination

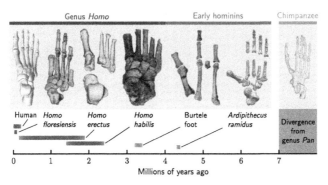

FIGURE 7.1.3   Partial sample of fossils representing evolution of the human foot, adapted from Lieberman (2012). Humans, the sole surviving member of the genus *Homo*, diverged from chimpanzees over 7 million years ago. Chimpanzees, members of the genus *Pan*, are phylogenetically the closest relatives to humans among extant species. As a result, *Pan* is often used as a hypothetical model for the last common ancestor between humans and chimpanzees. The foot of the chimpanzee more closely resembles our hands than our feet. It is a flexible and prehensile appendage. The feet of early hominins shared several features with the chimpanzees. More human-like features appear only as recently as ∼ 2 million years ago. Note that images of these skeletons are scaled to show detail, and are therefore not to scale. The human foot is larger in size than all the other feet shown here.

of these feet shows that older feet more closely resembled that of a chimpanzee than human. There is still some debate on whether the last common ancestor between humans and chimpanzees resembled the chimpanzee or not, and this debate impacts the choice of the chimpanzee as a model organism to study human evolution. This debate notwithstanding, it is clear that human feet look and behave very differently from the chimpanzee or other primates. Human feet have short stubby toes that are all aligned to point forward, pronounced arches in the longitudinal and transverse directions, and a pronounced calcaneus (heel bone). Chimpanzees and gorillas have long toes, an abducted hallux, highly curved phalanxes (digits of the foot), flat feet without arches, and a relatively smaller calcaneus.

The obvious difference in the functionality between human and the chimpanzee or gorilla foot is bipedal locomotion. We regularly walk over ground with our feet, and without using our hands or knuckles. In walking, we repeatedly push against the ground using the balls of the foot and apply forces that exceed body-weight. In running, this force can easily exceed twice our body weight. Despite such high loads, the foot retains its form during propulsion, remaining relatively undeformed. In contrast, a knuckle walking chimpanzee applies lesser forces than its body weight on its foot, but leads to severe deformation of its foot at the mid-foot location as seen in Fig. 7.1.4. Humans who are flatfooted also occasionally exhibit similar severe bending of the mid-foot

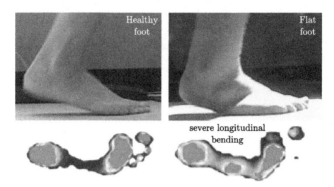

FIGURE 7.1.4 Figure showing differing flexibilities of human feet; modified from DeSilva et al. (2015). Flat feet, with a weak or absent arches, are a consistent skeletal correlate of a flexible mid-foot, the so-called mid-foot break. Such flexibility is associated with symptoms of pain, which can sometimes be debilitating.

(DeSilva et al., 2015), commonly referred to as the midtarsal break. Besides being painful, such loss of stiffness may have other consequences to our locomotion. We will now examine the benefits of having a stiff foot, and some of the "costs" associated with maintaining such a structure. Such an analysis will have implications for how we think about trade-offs in designing a foot, as well as for understanding the evolution of bipedal locomotion in humans.

7.1.3 COST–BENEFIT ANALYSIS OF THE HUMAN FOOT

7.1.3.1 Costs

During locomotion, most animals actively swing their leg because of overall energetic benefits (Kuo, 2007). Because mass at the extremities is more expensive during active leg swing than being closer to the center of mass, having a foot places a disproportionate energetic cost on locomotion.[2] The torque required to accelerate a mass is proportional to its moment of inertia with respect to the actuated joint. In case of a foot of mass m_{foot}, the torque M_{hip} required to swing the leg of length ℓ_{leg} back and forth in locomotion is

$$M_{hip} \propto m_{foot}\ell_{leg}^2. \tag{7.1.1}$$

Metabolic power consumption increases with increasing muscle force generation. Therefore, an increase in foot mass without a change in the gait will incur

2. This additional cost is, however, offset by the energetic benefits of pushing off using a stiff foot (Kuo et al., 2005).

higher metabolic cost. This expectation is reflected in experimental measurements of power consumption as a function of adding distal mass (Myers and Steudel, 1985). Metabolic power consumption, as measured by oxygen consumption, increases by as much as 20% by distributing 5% of the body weight at the ankles, instead of at the waist. Similarly, for quadrupeds, adding an additional mass of 3.5% of body mass, distributed distally versus close to the center of mass led to an increase in metabolic power consumption of up to 20% (Steudel, 1990). In fact swinging the leg is an appreciable cost of human locomotion, up to 30% (Doke et al., 2005; Marsh et al., 2004). Additionally, a majority of the metabolic energy consumption in legged locomotion is to make up for the losses from ground collisions, and adding distal mass exacerbates collisional losses (Ruina et al., 2005). Therefore, be it for reasons of collisions, or active leg swinging, a heavier foot always costs more metabolic energy for the animal.

The costs of having a foot are not all in the energetics. There is a significant risk of injury because the initial collision of the foot with the ground depends on its mass. Heavier feet cause greater impacts and could be injurious (Lieberman et al., 2010). Adding extra material, to cushion the impact or to strengthen the foot, could in fact worsen the problem of collisions. The choice of materials with high strength, low stiffness (Young's modulus) and low density is one way to deal with foot collisions, and is the topic of active research in the shoe industry. Whether biological materials are somehow "optimal" with respect to these considerations is an open research question.

A more general consequence of a heavier foot is that it imposes a lower bound on the impedance of the leg. Mass, damping, and stiffness parameterize the linear, second-order response of any mechanical system; the moment of inertia is the quantity analogous to mass for rotational motion. An increase in the mass of the foot disproportionately increases the moment of inertia of the leg. The impedance of a mechanical system, which affects its frequency dependent response to external forcing, is a function of the mass, damping and stiffness parameters (Hogan, 1984). Although stiffness and damping are under neural control, little can be done by the neural control system to alleviate the impedance arising from the moment of inertia of the leg. In theory, exact feedback compensation can be used to mimic even zero mass. However, such neural feedback-dependent strategies are fundamentally and unavoidably prone to limitations that arise from sensory or neural time-delays. The time-delay determines the highest frequency at which the neural system has control over the impedance of the foot, and for humans, this frequency is certainly less than 5 Hz (Venkadesan et al., 2007). As a result, increased distal mass at the foot limits the ability of the nervous system to control the leg. This has implications for actively controlling foot placement to deal with terrain irregularities, and for the control of foot–ground collisions.

Every terrestrial vertebrate has a foot despite all the costs mentioned above. We argue that the benefits of a foot, in the form of an active mechanical filter, outweighs the costs. Such an argument is difficult to prove, if not impossible, in the context of evolutionary biology. Therefore, we will treat this argument as a conjecture that motivates the analyses presented here.

7.1.3.2 Benefits

Modulating the stiffness of the interface between the ground and the body has implications for stability, injury, and energy consumption during locomotion. It is essential for the foot to remain soft when it collides with the ground during vigorous activities like running or landing from a jump. A soft foot absorbs some of the kinetic energy of the body, and reduces the severity of the collision by spreading out the collision over a longer duration of time. This is much like a shock absorber in a car.

On the other hand, the foot should be stiff during propulsion in order to convert the mechanical energy output of the ankle into useful mechanical work that propels the center of mass of the body. In walking, the mechanical energy gained by the center of mass during push-off is almost entirely attributed to the ankle joint (Kuo et al., 2005; Collins et al., 2015). A stiff foot is essential to convert this energy output of the ankle joint into kinetic energy of the body. Importantly, a stiffer foot allows the leg to impulsively push on the ground close to push-off. It is well known that an impulsive push-off late in stance is important for reducing the energetic cost of locomotion, and to enhance stability (Kuo et al., 2005; Ruina et al., 2005; Kuo, 2007; Srinivasan, 2011).

The ability to modulate the stiffness of the foot is like an adaptive suspension system of some cars. While a passive suspension system can be optimized for one function or the other, say collision mitigation or sharp turns, an adaptive system can achieve multiple goals. Such an elastic mechanical response of the foot is sometimes characterized as a temporal filter. The term *filter* is used in the general sense of a dynamical system that transforms inputs into outputs, and typically modeled using ordinary differential equations. In addition to its ability to filter inputs over time, the foot also has a nonzero spatial extent and behaves as a spatial filter. Some of the beneficial mechanical properties of the foot are by virtue of having a finite spatial extent. As we will see later in some detail, the foot is able to function in mitigating the initial collision during agile activities like running, and is also able to smooth out roughness of the terrain. Mathematical models of the foot as a temporal filter typically use ordinary differential

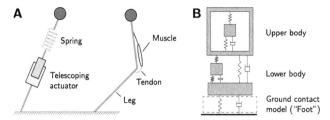

FIGURE 7.1.5 Point-like models of the foot. Reproduced, with modifications, from Srinivasan (2011) and Liu and Nigg (2000). (A) A prevalent model of the foot, such as those in Srinivasan (2011), uses a rigid and point-like mass for the foot. Some models, such as that shown on the left use a massless foot. Others assume a point mass at the foot, and additionally need to assume specific collision laws to model the foot's collision with the ground. (B) These point-like models of the foot are sometimes expanded to include additional parameters that capture the mass, stiffness and damping (Liu and Nigg, 2000). In the example shown here, the colliding mass of the foot is ignored, but its stiffness and damping are modeled by a viscoelastic "ground contact model." The ground contact model lumps together the foot and the ground. On nearly rigid ground, the ground contact model reflects the viscoelastic properties of the foot.

equations to characterize the foot, and models as a spatial filter often rely on partial differential equations.

There are also benefits of having a foot for purposes of neural control. For example, the base of the foot is made of a specialized type of skin called glaborous skin (Kandel, 2013). Such skin is superficially characterized by the presence of ridge patterns, such as fingerprints in the hand. Underneath these ridges are high densities of specialized mechanical receptors called Pacinian corpuscles that are sensitively tuned to vibrations and play a central role in detecting slip. Such specialized receptors on the foot pad may play an important role in maintaining stability, although this has not yet been investigated in any depth.

7.1.4 TEMPORAL FILTERING

Temporal filters are dynamical systems that transform inputs to outputs. Such models are prevalent in models of electrical circuits and in rigid body mechanics. The spatial extent of the foot is ignored, and the inputs that act at specific points on the foot are transformed by its mechanical response.

7.1.4.1 Point-Like Foot

One approximation of the foot is to treat it as a point-like object that is connected to the rest of the body by springs and dampers (Fig. 7.1.5). This allows the characterization of the temporal dynamics of the foot without having to deal with the complexities that arise from its spatial extent. Such mathematical approximations are routine in the analysis of system dynamics, where the foot is assigned input and output "ports" that approximate its interaction with the ground and the body. Physically, this is equivalent to approximating the distribution of ground reaction forces acting on the foot with an effective force or force/moment pair acting at a single point; likewise for the effect of the leg and body on the foot. A useful idealization of such an approximation is the spring-legged inverted pendulum (SLIP) model of walking and running (Fig. 7.1.5A). In such a model, the foot and leg are lumped together into a single spring, either constant or time-varying (Seyfarth et al., 2002; Geyer et al., 2006; Maus et al., 2015). These models have proven useful in characterizing whole-body gait dynamics in running. However, by lumping the temporal dynamics of the foot together with the rest of the leg, it is no longer possible to delineate the effect of the foot.

Others have developed more complex models (Fig. 7.1.5B), with several internal degrees of freedom for the body, including the foot (Thompson and Raibert, 1990; McMahon and Cheng, 1990; Liu and Nigg, 2000; Chi and Schmitt, 2005; Zadpoor and Nikooyan, 2006). In these idealizations, the foot is treated as a separate mass which is connected to the leg and to the ground. The contact with the ground is in some cases approximated by a rigid collision. The alternative is to place a massless spring–damper between the foot and the ground. In all these cases, the leg and foot are approximated as a one-dimensional telescoping actuator with specified viscoelastic responses.

7.1.4.2 Spatially Extended and Rigid Foot

An alternative to the one-dimensional or point-like (zero-dimensional) models of the foot is to explicitly consider its spatial extent. One such simple model is to regard the leg and foot as rigid bars, connected with a simple hinge joint. Such a model has been previously used to understand the consequences of different running styles (forefoot versus heel strikes) on the ground collision (Lieberman et al., 2010). We present this model in some detail as an instructive example of the utility and limitations of such rigid-body models of the foot.

We model the ground collision of the foot during running as a rigid inelastic collision of an L-shaped object. Two variants of this model are considered; one with an ankle-like frictionless hinge at the corner (Fig. 7.1.6A), and another with an infinitely stiff ankle (Fig. 7.1.6B). A *rigid* collision refers to the assumption that the collision is instantaneous, and thus gives rise to infinitely large instantaneous forces over an infinitesimal time period so that the net impulse due to the force is finite. Because the duration of the collision is nearly zero, the configuration of the system remains constant over the period of the collision. The term inelastic refers to the assumption that there is no rebound. Under these assumptions, the only forces that affect the collision are infinite forces. This is

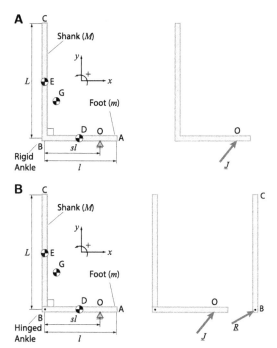

FIGURE 7.1.6 The shank of the leg and the foot modeled as an L-shaped object. The foot is assumed to first touch down and collide at the point O, which is located between the metatarsal heads (A) and the heel (B). The location of the collision is parametrized by the strike index $0 \le s \le 1$, so that the distance from the ankle to the point of collision is sl. The collision occurs at B occurs when $s = 0$, and at A when $s = 1$. (A) To model a stiff ankle, we consider an L-shaped bar with a rigid (infinitely stiff) hinge at the ankle. The free body diagram on the right is used to solve for the ground reaction impulse due to the collision. (B) Modeling the ankle as a frictionless hinge, located at the heel B, considers the effect of a soft ankle. The two free body diagrams on the right are used in solving the collision equations. All forces are depicted as general vectors with positive components in the chosen coordinate system, so that a negative component would act along the negative axis.

because the mathematical idealization of rigidity reduces the time duration of the collision to zero. Therefore, finite forces such as those arising from gravity or elastic joints will not affect the momentum of the body; momentum is proportional to the impulse, i.e., force multiplied by time. The two variants therefore represent the two extremes of ankle joint stiffness, infinite and zero, respectively.

Condition I: Infinitely Stiff Ankle Joint We consider the case of an L-shaped bar falling vertically downward with velocity $v_G^- = -v^- \hat{y}$ and no rotational velocity, i.e., $\mathbf{\Omega}^- = 0$. Because of the collision, there is an abrupt jump in the angular and linear velocities of the L-shaped bar to give v_G^+ and $\mathbf{\Omega}^+ = \omega^+ \hat{k}$.

Fig. 7.1.6A shows a free-body-diagram of the L-shaped bar with the foot in contact with the ground at a point O (center of pressure) that is located between the tip of the foot and the ankle. Because the only external force-impulse on the L-bar is applied at O, no external torque-impulses act on it about the same point. Therefore, the angular momentum $H_{/O}$ about O is the same before and after the collision. In terms of the angular velocity vector $\mathbf{\Omega}$, the velocity of the center of mass G is given by

$$v_G = v_O + \mathbf{\Omega} \times r_{G/O}, \qquad (7.1.2)$$

where $v_O^- = -v^- \hat{y}$ and $v_O^+ = 0$. Then, the angular impulse–momentum balance is expressed as

$$H_{/O}^+ = H_{/O}^-, \qquad (7.1.3)$$

$$H_{/O}^- = (m + M)r_{G/O} \times v_G^-, \qquad (7.1.4)$$

$$H_{/O}^+ = (m + M)r_{G/O} \times v_G^+ + I_{/G}\mathbf{\Omega}^+. \qquad (7.1.5)$$

The mass of the foot is m, the mass of the shank is M, and $\mathbf{I}_{/G}$ is the moment of inertia matrix of the foot plus shank about the center of mass G, where only the principal inertia I_{zz} parallel to the z-axis is relevant for this planar object. Because the problem is planar, the only nonzero component of the angular momentum vectors is the z-component, yielding one equation with one unknown, ω^+, which we can solve. We find v_G^+ using Eq. (7.1.2) and linear impulse–momentum balance for the L-shaped bar gives us the impulse J at the contact point O,

$$J = (m + M)(v_G^+ - v_G^-). \qquad (7.1.6)$$

Referring back to the point-like models of the foot, this collisional impulse can be thought of as arising from a collision of a point with an effective mass M_{eff}, which is given by

$$M_{\text{eff}} = \left| \frac{J \cdot \hat{y}}{v^-} \right|. \qquad (7.1.7)$$

An L-shaped bar with an infinitely stiff ankle strikes the ground at some intermediate point on the foot of length ℓ, parameterized by $0 \le s \le 1$, where $s = 0$ is a heel-strike and $s = 1$ is a forefoot-strike. The effective mass for a foot is given by

$$M_{\text{eff}} = \frac{m\ell^2(m + 4M) + 4ML^2(m + M)}{12(m + M)\ell^2 s^2 + 4m\ell^2(1 - 3s) + 4ML^2}. \qquad (7.1.8)$$

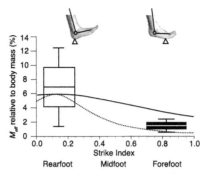

FIGURE 7.1.7 Comparison of the foot–ground collision between heel-strike runners versus forefoot strike runners, quantified using the effective mass M_{eff}; modified from Lieberman et al. (2010). The solid line is the predicted M_{eff} for an infinitely stiff ankle (condition I), and the dotted line shows the predicted M_{eff} for condition II with an infinitely compliant ankle. The infinitely compliant ankle effectively decouples the foot mass from the rest of the foot during forefoot strikes, thereby reducing the collisional impulse over and above the reduction because of the landing location.

L is the length of the shank. For an average adult human, $m = 1.4\%$ of body mass, $M = 4.5\%$ of body mass, and $L = 1.53\ell$ (Dempster, 1955). With these numbers, we calculate the effective mass of the foot as a function of which point makes contact with the ground at collision, as shown by the solid line in Fig. 7.1.7.

Condition II: Infinitely Compliant Ankle Joint We next derive M_{eff} as a function of the strike index s for conditions in which the ankle is modeled as a frictionless hinge, i.e., the shank and foot are modeled as a double-pendulum with its joint flexed to $90°$ when it collides with the ground. This is analogous to an infinitely compliant ankle. Fig. 7.1.6B shows the free-body diagram for this posture of the double pendulum. Like earlier, we assume that just before impact the entire object is moving downward with speed v^- (i.e., $v_D^- = v_E^- = v_D^- = -v\hat{y}$) and no angular velocity (i.e., $\mathbf{\Omega}_m^- = \mathbf{\Omega}_M^- = 0$). Assuming a rigid plastic collision, we use angular impulse–momentum balance of the entire double pendulum about the collision point O, and of the shank segment (of mass M) about the hinge point B. The angular momentum vectors about O and B are given by

$$\boldsymbol{H}_{m+M/O} = (m\boldsymbol{r}_{D/O} \times \boldsymbol{v}_D + \mathbf{I}_{m/D}\mathbf{\Omega}_m) + (M\boldsymbol{r}_{E/O} \times \boldsymbol{v}_E + \mathbf{I}_{M/E}\mathbf{\Omega}_M) \tag{7.1.9}$$

$$\boldsymbol{H}_{M/B} = M\boldsymbol{r}_{E/B} \times \boldsymbol{v}_E + \mathbf{I}_{M/E}\mathbf{\Omega}_M, \tag{7.1.10}$$

where

$$\boldsymbol{v}_D = \boldsymbol{v}_O + \mathbf{\Omega}_m \times \boldsymbol{r}_{D/O}, \tag{7.1.11}$$

$$v_E = v_O + \mathbf{\Omega}_m \times r_{B/O} + \mathbf{\Omega}_M \times r_{E/B}. \qquad (7.1.12)$$

The subscripts m, M, and $m + M$ refer to the two segments separately or together, and the only relevant element of the matrix $\mathbf{I}$ is the principal moment of inertia I_{zz} parallel to the z-axis. Because there are no external torques on the body about the points O and B, the angular impulse–momentum equations are

$$H^+_{m+M/O} = H^-_{m+M/O}, \qquad (7.1.13)$$

$$H^+_{M/B} = H^-_{M/B}. \qquad (7.1.14)$$

Solving these equations as before, we find that the effective mass with a soft ankle is given by

$$M_{\text{eff}} = \frac{m(m + 4M)}{12(m + M)s^2 + 4m(1 - 3s)}. \qquad (7.1.15)$$

The predicted effective mass of a foot collision is shown by the dotted line in Fig. 7.1.7, and compared against data from both heel-strike and forefoot-strike runners. This simple model resembles data, and also establishes bounds for the lowest collision that is possible, given that the foot has a nonzero mass.

Unlike point-like models of the foot, introducing the spatial extent allows us to relate the foot mass to the anatomy. Importantly, we have shown that the mass in point-like models of the foot is not a constant, and depends on the landing posture. However, all the models presented thus far suffer from the drawback that they are unable to predict details of the ground reaction force, because of the rigid collision assumption. Dealing with realistic collisions remains an open topic of research in mechanics.

7.1.4.3   Conclusion: Temporal Filtering

The clear advantage of these lumped-parameter models of the foot as a temporal filter is their simplicity. However, such models are unable to predict the ground reaction force. Although it appears that the model of a foot as a point with a spring between the foot and the ground indeed predict the forces, that is not the case. This is because the predictions are a direct consequence of choice of parameters values for the spring, damper and mass constants. As we saw with the L-shaped foot model, even the mass of the foot that affects the dynamics may be a consequence of factors not considered by the point-like foot models. Therefore, while such temporal filtering models are of value in mathematically assessing the consequences of parameter choices, they remain limited in their ability to predict the foot–ground interaction forces.

7.1.5 SPATIAL FILTERING

In order to address the limitations of the earlier (temporal) models of the foot that we have seen, it becomes necessary to explicitly consider not only its spatial extent, but also its flexibility. This brings us to the realm of elasticity theory. A soft and spatially extended elastic foot introduces interaction between the application of forces at different locations. Moreover, the spatial geometry becomes closely coupled to its elastic properties, thereby introducing an additional layer of complexity. Unlike work on the temporal filtering characteristics, there has been little research into the spatially extended and elastic characterization of the foot. We present some initial forays towards this direction in this section.

7.1.5.1 Smoothing Over Rough Terrains

The elasticity of the foot smooths out the influence of uneven terrain on the body dynamics. The area of contact between the foot and the ground is determined by the shape of the foot and the ground. If a rigid foot were to make contact with a rigid ground, the forces and torques transmitted to the ankle will depend on the precise shapes of both. However, the foot is not rigid and deforms in response to the uneven terrain. The change in character of the transmitted force resulting from the interaction between a flexible foot and uneven ground, which we term *spatial filtering*, has been poorly studied. Here we outline some simple principles and the associated reduced models that may serve to guide more detailed investigations.

Spatial filtering by the feet occurs by two mechanisms; first because of the padding soft tissue provides, and second because of the flexibility provided by the numerous skeletal joints in the foot. The soft padding filters ground unevenness on the scale of the padding thickness, which is approximately 1 cm. The skeletal joint flexibility deforms the foot on the scale of its width, and therefore filters ground unevenness on that scale, which is approximately 10 cm.

Spatial Filtering by Soft Padding The soft padding around the skeleton provides a spatial filtering mechanism for smoothing stress exerted by a terrain with roughness scale of about 1 cm. This mechanism is depicted in Fig. 7.1.8A. The contact stress on the foot due to the ground reaction is the ratio of the applied load and the area of contact. The area of contact depends on the scale of the ground roughness, while the applied load is comparable to body weight. For rough ground, contact is made on a small fraction of the foot area, as shown in Fig. 7.1.8A, and the contact stress is high. The stress diffuses and is spread over a larger area as the padding deforms and the area of contact itself changes owing the elasticity of the padding. Hertz contact theory may be used to estimate the stresses resulting from the roughness with radius of curvature R_{rough}.

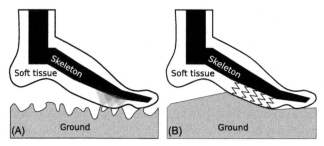

FIGURE 7.1.8   The spatial filtering because of soft tissue in the foot. (A) Role of soft padding tissue of a foot on rough ground. For contact with ground surface with a roughness length scale shorter than the soft tissue padding, the padding diffuses the applied ground reaction stress as it is transmitted to the skeleton. (B) Role of soft padding tissue of a foot on smooth ground. The ground surface has a length scale longer than the thickness of the soft tissue. The soft tissue acts as a Winkler foundation, i.e., a bed of springs, as it transmits the ground reaction force to the skeleton.

For a normal displacement δ_{pad}, and radius of contact $r_{pad} = \sqrt{R_{rough}\delta_{pad}}$ (where equality stands for "scales as" in this estimate), the strain is δ_{pad}/r_{pad} and the stress is $E_{pad}\delta_{pad}/r_{pad} = E_{pad}\sqrt{\delta_{pad}/R_{rough}}$. The force supported by the contact area is $f_{pad} = E_{pad}\delta_{pad}^{3/2}R_{rough}^{1/2}$. For a given force, the normal displacement is $\delta_{pad} = (f_{pad}/E_{pad})^{2/3}R_{rough}^{-1/3}$ and the resulting stress in the pad is $E_{pad}^{2/3}f_{pad}^{1/3}R_{rough}^{-2/3}$. This estimate shows that the maximum stress in the pad scales as $E_{pad}^{2/3}$, and therefore softer pads result is reduced stress for the same supported force because of increase in the area of contact. This mechanism is therefore effective in distributing the contact force over a greater area and in reducing the maximum stress experienced by the tissue when the ground roughness scale is smaller than the padding thickness.

When the unevenness scale is longer than its thickness, the soft padding also provides a simple compliance to the surface of the foot. This possibility is shown in Fig. 7.1.8B. The soft padding acts as a Winkler foundation, or in other words, an array of independent springs. The characteristic stiffness of these springs, defined as the ratio of surface pressure to normal displacement, scales as E_{pad}/t_{pad}, where E_{pad} is the elastic modulus of the soft tissue and t_{pad} is the padding thickness. The presence of the Winkler foundation implies that contact with uneven ground happens over an extended area rather than at a point, which reduces the maximum stress experienced by the contacting surfaces. Since the characteristic stress experienced by the foot $p_{foot} \approx mg/A_{contact}$, where mg is body weight and A_{foot} is foot area, the characteristic compression of the soft padding is $\delta_{pad} = mgt_{pad}/E_{pad}A_{foot}$. This type of deformation accounts for roughness of the ground with an amplitude smaller than δ_{pad}, so that elastic deformation of the pad can accommodate unevenness.

Spatial Filtering by Skeletal Joint Flexibility The flexibility arising in the foot due to the presence of numerous skeletal joints also filters out ground unevenness on a scale of foot width and with amplitude greater than δ_{pad}. In this case, the foot may be considered to be akin to an elastic plate and its response may be approximated as a bending-torsion deformation. The details underlying the bending rigidity of the foot will be considered in Section 7.1.5.2. The whole foot, and not just its bottom surface, conforms to features on the ground using this flexibility. As described in Section 7.1.3, a foot which is soft in bending can act as a shock absorber and as a spatial filter, but it needs to be stiff enough to transmit ground reaction forces to the body without deforming too much. This trade-off is an important principle underlying the design of feet.

The consequences of a stiff human foot are understood by juxtaposing them against those of arboreal primates and human hands. Human hands and primate feet are general purpose grasping appendages. An important function of human hands is to grasp objects of varying sizes. Similarly, the feet of arboreal primates are used for hanging on tree limbs of varying sizes. This involves conforming the appendage by changing its curvature to the shape of the object being grasped. In order for the muscles to not fight against the skeletal structure, the appendages should be soft under bending so that the action of muscles can deform them to the required shape. Indeed, feet of primates such as the chimpanzee are significantly softer than humans, as seen from cadaveric bending tests (Ker et al., 1987; Bennett et al., 1989) and from severe mid-foot deformation during walking (D'Août et al., 2002; Thompson et al., 2015). Although human feet have soft padding that can accommodate small amounts of ground unevenness (e.g., small gravel), the overall foot is not soft enough for use in grasping. This makes human feet well suited for propulsion, but poor for grasping. Such trade-offs in function, arising from natural selection, were probably driven by the improved energetic advantage of human-like walking gaits at the expense of adaptations that benefit an arboreal lifestyle (Jungers, 1988; Wood and Collard, 1999; Bramble and Lieberman, 2004).

7.1.5.2   Effect of the Foot Arches on Stiffness

Recall from the section on anatomy and evolution that unlike other primates, human feet have a pronounced arched morphology. Two specific arches are anatomically easily identifiable, the longitudinal and transverse (Fig. 7.1.2B). The longitudinal arch is oriented from the heel to the toe, and the transverse arch is approximately orthogonal to that direction. It is a long-standing hypothesis that the longitudinal arch underlies the higher stiffness of the human foot (Morton, 1924b; Ker et al., 1987; Williams and McClay, 2000), although its importance in the context of walking is contested (Bennett et al., 1989; Bram-

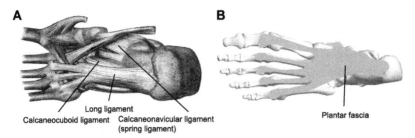

A

B

Long ligament
Calcaneocuboid ligament   Calcaneonavicular ligament
(spring ligament)

Plantar fascia

FIGURE 7.1.9 Soft skeletal tissues on the plantar (sole-side) of the human foot. (A) Three prominent longitudinally oriented ligaments in the mid-foot region are shown. The attachment sites for the calcaneocuboid and calcaneonavicular ligaments are evident from their names. The long plantar ligament, the longest ligament in the midfoot region and one of the strongest (stiffest) in the foot, attaches to the calcaneus at the proximal end. It splits into branches at the distal end, attaching to cuboid and to the proximal heads of the 2nd, 3rd, and 4th metatarsals. Image adapted from Gray's Anatomy. (B) The plantar fascia are a band of tissue (aponeurosis), stretching from the calcaneus to the phalanges.

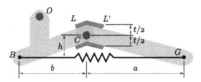

FIGURE 7.1.10 A model of the foot to compare the relative contributions of the plantar fascia and plantar ligaments to the stiffness of the longitudinal arch.

ble and Lieberman, 2004). In this section we first show that the plantar fascia contributes to foot stiffness as much as the longitudinally oriented ligaments holding up the bones. We then present an analysis to show that the contribution from the transverse arch most likely overshadows that of the longitudinal ligaments, and therefore the plantar fascia. These mathematical arguments are supported by data (Ker et al., 1987; Bennett et al., 1989; Huang et al., 1993) from cadaveric dissection experiments, which show that the plantar fascia contribute to only about 27% of the total foot stiffness.

Longitudinal Arch The longitudinal arch is spanned by a band of elastic tissue called the plantar fascia (Fig. 7.1.9B). In terms of material properties, the plantar fascia are distinct from both ligaments and tendons, and more closely resemble aponeuroses that form the bridge from tendon to muscle. If the foot is deformed by an external load, such as during push-off, the plantar fascia will be stretched. To understand the relative contribution of the plantar fascia to overall foot stiffness, we compare against the other mechanisms stiffening the foot through the model shown in Fig. 7.1.10. In this model, contact with the ground is made only at the forefoot G. The base of the heel B is supported by the net forces and torques at the point O, which represent the joint contact forces at the talotibial joint (ankle), and also the forces exerted by the calf muscles and the Achilles tendon. The foot in this model consists of the tibiotalar joint (point O), the talus and the calcaneus assumed rigid (OBC), and the metatarsals (CG). Changing the angle BCG represents the mid-foot bending longitudinally. The plantar fascia is represented by the spring BG, and the ligaments stiffening the midtarsal joints that maintain the longitudinal arch are schematically shown as the red segment LL'. The longitudinally oriented ligaments (Fig. 7.1.9A), which run parallel to the plantar fascia, comprise of several *plantar* ligaments; the long plantar ligament (from the calcaneus to the 2nd, 3rd, and 4th metatarsals), the plantar calcaneocuboid ligament (short plantar ligament), the plantar calcaneonavicular ligament (spring ligament), and the plantar tarsometatarsal ligaments. These plantar ligaments are much shorter in length and narrower in width than the plantar fascia, and therefore stiffer than the plantar fascia. Together, the plantar fascia and the plantar ligaments contribute to the longitudinal bending stiffness of the foot.

We consider the foot to deform in longitudinal bending by increasing the angle BCG in response to an applied ground reaction force at G. The stiffness of the foot to bending (increasing the angle BCG) that arises from the plantar fascia stretching is $k_{\mathrm{BG}}h^2/a^2$, where k_{BG} is the spring constant of the plantar fascia, h is the moment arm of the plantar fascia about the point of rotation C, and a is the moment arm of the vertical force at G about point C. This spring constant may be estimated from the properties of the plantar fascia as $k_{\mathrm{BG}} \approx E_{\mathrm{pf}}A_{\mathrm{pf}}/L_{\mathrm{pf}}$, where E_{pf} is the Young's modulus, A_{pf} is the approximate cross sectional area, and L_{pf} is the length of the plantar fascia. Note that because we seek an estimate, approximate representative values for the properties of the plantar fascia suffice.

The ligaments maintaining the longitudinal arch of the foot also contribute to the stiffness of the foot in longitudinal bending. This contribution to the stiffness may be estimated as $Nk_{\mathrm{LL'}}t^2/a^2$, where N is the number of these ligaments, $k_{\mathrm{LL'}}$ is the spring constant for the ligaments, and t is the moment arm of the ligaments about point of rotation C. The moment arm of these ligaments is approximated by the thickness of the tarsometatarsal joints, and for purposes of this estimation, we assume these ligaments to be similar in length and cross-sectional area. The properties of the ligaments in turn determine the spring stiffness; we estimate $k_{\mathrm{LL'}} \approx E_{\mathrm{lig}}A_{\mathrm{lig}}/L_{\mathrm{lig}}$.

Our claim that the plantar fascia contributes to the mid-foot stiffness as much as the longitudinal plantar ligaments is equivalent mathematically to the statement that $k_{\mathrm{BG}}h^2/a^2 \approx Nk_{\mathrm{LL'}}t^2/a^2$. Substituting k_{BG} and $k_{\mathrm{LL'}}$ in terms of the properties of plantar fascia and the ligaments respectively, this condition can be translated to $E_{pf}A_{pf}/L_{pf}h^2 = NE_{lig}A_{lig}/L_{lig}t^2$. The ratio of the two sides scales as

$$\rho = \frac{E_{pf}}{E_{lig}} \frac{A_{pf}}{A_{lig}} \frac{L_{lig}}{L_{pf}} \left(\frac{h}{t}\right)^2 \frac{1}{N}. \tag{7.1.16}$$

Since both the tissues have similar modulus and similar cross-sectional areas, the ratio simplifies to

$$\rho = \frac{L_{lig}}{L_{pf}} \left(\frac{h}{t}\right)^2. \tag{7.1.17}$$

Using anatomical values for these lengths and heights (Gomberg, 1981), we find $L_{pf} \approx 4L_{lig}$ and $h \approx 2t$. Therefore,

$$\rho \approx 1.$$

Transverse Arch The role of the transverse arch in the foot's stiffness has only recently been studied (Dias et al., 2015; Yawar et al., 2016), and has traditionally not been typically considered as important as the longitudinal arch. This recent study (Dias et al., 2015) shows that the contribution of the transverse arch is several fold higher than the longitudinal arch. The mechanics underlying this is demonstrated from casual observation of thin elastic structures, like a dollar bill. Curling the dollar bill transversely to the bending direction dramatically increases its stiffness (Fig. 7.1.11A). This principle impacts many things in our everyday life, from inserting currency into vending machines, eating pizza, to the design of measuring tape. This is because curvature in the transverse direction couples longitudinal bending to transverse stretching, and thin elastic structures are much harder to stretch in-plane than to bend out-of-plane. In the context of the foot, this is evident in the form of splaying of the distal metatarsal heads if you try to bear your weight using the balls of your feet, and is demonstrated by a foot-mimic with tunable curvature (Fig. 7.1.11). This in turn stretches several short transversely oriented ligaments such as the intermetatarsal ligaments. To make this heuristic argument more rigorous, we model the foot as idealized by the schematics in Figs. 7.1.11B and 7.1.12A. The structure consists of three rigid triangular and rectangular elements analogous to the tarsal and metatarsal bones respectively. The tarsals are assumed to be held rigidly in place by the rest of the foot, the clamped boundary condition model for the proximal end of the foot. The metatarsals are connected to the tarsal bones through a flexible joint, but the permissible motion of the metatarsals is along directions misaligned by angle

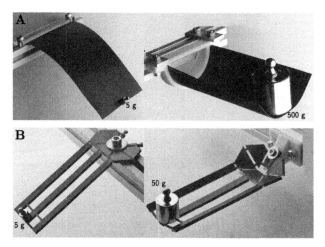

FIGURE 7.1.11 Effect of the transverse arch, demonstrated using (A) a continuum and (B) a discrete mechanical structure. Introducing transverse curvature dramatically increases the longitudinal bending stiffness in both cases. The underlying mechanics in both cases is a coupling of out-of-plane bending with in-plane stretching of transversely oriented elastic elements because of curvature.

θ due to the nature of the tarsometatarsal joint, assigning the structure a transverse radius of curvature of $R_T = \ell_0/\theta$. A displacement δ of the metatarsal heads in the normal direction is accompanied by a stretching of the distance between them as $\ell = \ell_0 + \delta \tan \theta$ and an energy storage of $E_s = k_s \delta^2 \tan^2 \theta$ in the two ligaments (not shown) with spring constant k_s each spanning the distal metatarsal heads. This energy is in addition to approximately $E_b = 3k_0(\delta/L)^2/2$ stored in the tarsometatarsal joint modeled here as three independent torsional springs with spring constant k_0. The restoring force exerted by the structure $F = d(E_s + E_b)/d\delta$, predicts an effective stiffness to bending to be $k_b = F/\delta = 2k_s \tan^2 \theta + 3k_0/L^2$. For small curvatures, $\tan \theta \approx \theta$, and the stiffness may be approximated as

$$k_b = \frac{2k_s \ell_0^2}{R_T^2} + \frac{3k_0}{L^2} \quad \text{or} \quad \frac{k_b}{k_{b,0}} = \left(\frac{R_0}{R_T}\right)^2 + 1, \qquad (7.1.18)$$

$$\text{where} \quad k_{b,0} = \frac{3k_0}{L^2} \quad \text{and} \quad R_0 = \left(\frac{2k_s}{3k_0}\right)^{1/2} \ell_0 L. \qquad (7.1.19)$$

Fig. 7.1.12B compares the experimentally measured stiffness of the discrete structure shown in Fig. 7.1.11B with the predictions of Eq. (7.1.19). Only two parameters are needed to accurately describe the stiffness dependence on the transverse arch radius, viz. $k_{b,0}$ and R_0. The former governs the zero curvature stiffness, whereas the latter governs the scale for the radius of transverse curvature below which stiffening is predicted. For the experimental fit shown, the

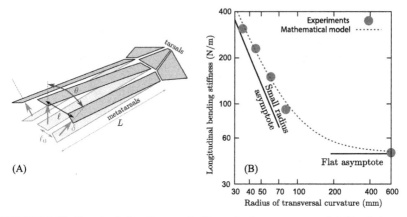

FIGURE 7.1.12 Curvature induced increased stiffness of a discrete structure mimicking the human foot.

value of $k_{b,0}$ was independently experimentally measured, and R_0 is a fitting parameter. The exact fit is less important than the agreement of the curvature dependence, i.e., the slope of the line passing through the data points for small radii of curvature.

If $R_0 \gg R_T$, then the stiffness is dominated by the transverse curvature induced bending-stretching coupling. If $R_0 \gg R_T$ for the foot, reinforcing the transversally oriented ligament at the distal metatarsal head using external elastic tape increases R_0^2, and therefore proportionally increases the stiffness of the foot.

The analyses in the previous section, of the longitudinal arch and the plantar fascia, showed that the effect of the plantar fascia is at most comparable to that of the mid-foot ligaments. On the other hand, the transverse curvature appears to show a dramatic increase in stiffness, at least for the physical model. Importantly, even slight transverse curvature appears to increase the bending stiffness of the physical foot by nearly 100%. It remains to be shown where the transverse curvature of the human foot falls. Nevertheless, this analysis presents an exciting method to tune the stiffness of a foot in the robotic or prosthetic context. Instead of using variable stiffness motors within the foot, modulating the transverse curvature can lead to stiffening, without adding any additional mass to the foot. In theory, the actuation necessary for modulating the transverse curvature could reside outside the foot, in the shank of the leg, thereby further reducing foot mass. Practical implementations of such a tunable stiffness design remains to be developed.

7.1.6 CONCLUSION

The foot plays a central role in mediating the interaction of the body with the ground. Its importance in human evolution is well known, and yet, its mechanics in only recently being understood better. The current state of the art in robotic feet is relatively primitive, and often consist of partial mimics of biological feet (Au et al., 2007), or simply a rigid foot (Collins et al., 2005; Caputo and Collins, 2014; Ananthanarayanan et al., 2012; Collins et al., 2015). Mimics of biological feet are faced with the daunting complexity of the foot, and therefore choose to mimic some elements and not others. Such mimics have been limited to introducing a longitudinal arch, but not the other elements that we examined in this unit. Such design choices are presently not informed by the mechanics presented in this unit, and it is in this context that a mathematical mechanician, one who spans the areas of biomechanics, applied mathematics, and mechanisms design, plays an important role. By distilling the complexity underlying biological feet into unifying principles, it is possible to create control and design guidelines. For example, controlling the landing location to be at the distal end of the foot is a significant method to ameliorate the severity of ground collisions. Once the foot is in contact with the ground, its stiffness is best modulated through the transverse arch. Such findings are made possible only through a systematic study of biological feet. Future robotic and prosthetic feet may benefit from incorporating the principle of curvature-induced stiffening in their control and design.[3]

REFERENCES

Ananthanarayanan, A., Azadi, M., Kim, S., 2012. Towards a bio-inspired leg design for high-speed running. Bioinspir. Biomim. 7, 046005.

Au, S.K., Herr, H., Weber, J., Martinez-Villalpando, E.C., 2007. Powered ankle-foot prosthesis for the improvement of amputee ambulation. In: Conf. Proc. IEEE Eng. Med. Biol. Soc., vol. 2007, pp. 3020–3026.

Bennett, M.B., Ker, R.F., Alexander, R.M., 1989. Elastic strain energy storage in the feet of running monkeys. J. Zool. 217, 469–475.

Bramble, D.M., Lieberman, D.E., 2004. Endurance running and the evolution of Homo. Nature 432, 345–352.

Brunet, M., Guy, F., Pilbeam, D., Lieberman, D.E., Likius, A., Mackaye, H.T., Ponce de León, M.S., Zollikofer, C.P.E., Vignaud, P., 2005. New material of the earliest hominid from the Upper Miocene of Chad. Nature 434, 752–755.

Caputo, J.M., Collins, S.H., 2014. A universal ankle–foot prosthesis emulator for human locomotion experiments. J. Biomech. Eng. 136, 035002.

Chi, K.-J., Schmitt, D., 2005. Mechanical energy and effective foot mass during impact loading of walking and running. J. Biomech. 38, 1387–1395.

---

3. The goal in robotic or prosthetic feet should be to encapsulate the functional principle, and not the detailed morphology seen in biology.

Collins, S., Ruina, A., Tedrake, R., Wisse, M., 2005. Efficient bipedal robots based on passive-dynamic walkers. Science 307, 1082–1085.

Collins, S.H., Wiggin, M.B., Sawicki, G.S., 2015. Reducing the energy cost of human walking using an unpowered exoskeleton. Nature 522, 212–215.

D'Août, K., Aerts, P., De Clercq, D., De Meester, K., Van Elsacker, L., 2002. Segment and joint angles of hind limb during bipedal and quadrupedal walking of the bonobo (Pan paniscus). Am. J. Phys. Anthropol. 119, 37–51.

Day, M.H., Napier, J.R., 1964. Fossil foot bones. Nature 201, 969.

Dempster, W.T., 1955. Space Requirements of the Seated Operator, Geometrical, Kinematic, and Mechanical Aspects of the Body with Special Reference to the Limbs. Technical report, DTIC Document.

DeSilva, J.M., Bonne-Annee, R., Swanson, Z., Gill, C.M., Sobel, M., Uy, J., Gill, S.V., 2015. Midtarsal break variation in modern humans: functional causes, skeletal correlates, and paleontological implications. Am. J. Phys. Anthropol. 156, 543–552.

Dias, M.A., Singh, D.K., Bandi, M.M., Venkadesan, M., Mandre, S., 2015. Role of the tranverse arch in stiffness of the human foot. In: APS Meeting Abstracts, vol. 1, p. 47003.

Doke, J., Donelan, J.M., Kuo, A.D., 2005. Mechanics and energetics of swinging the human leg. J. Exp. Biol. 208, 439–445.

Dunn, H.L., 1928. Arch mechanics of the normal adult foot. Am. J. Hyg. 8, 410–447.

Elftman, H., Manter, J., 1935. The evolution of the human foot, with especial reference to the joints. J. Anat. 70, 56–67.

Fey, N.P., Klute, G.K., Neptune, R.R., 2012. Optimization of prosthetic foot stiffness to reduce metabolic cost and intact knee loading during below-knee amputee walking: a theoretical study. J. Biomech. Eng. 134, 111005.

Geyer, H., Seyfarth, A., Blickhan, R., 2006. Compliant leg behaviour explains basic dynamics of walking and running. Proc. Biol. Sci. 273, 2861–2867.

Gomberg, D.N., 1981. Form and Function of the Hominoid Foot. PhD thesis. University of Massachusetts.

Harcourt-Smith, W.E.H., Aiello, L.C., 2004. Fossils, feet and the evolution of human bipedal locomotion. J. Anat. 204, 403–416.

Hicks, J.H., 1955. The foot as a support. In: Cells Tissues Organs.

Hogan, N., 1984. Adaptive control of mechanical impedance by coactivation of antagonist muscles. IEEE Trans. Autom. Control 29, 681–690.

Huang, C.-K., Kitaoka, H.B., An, K.-N., Chao, E.Y.S., 1993. Biomechanical evaluation of longitudinal arch stability. Foot Ankle Int. 14, 353–357.

Jones, R.L., 1941. The human foot. An experimental study of its mechanics, and the role of its muscles and ligaments in the support of the arch. Am. J. Anat. 68, 1–39.

Jungers, W.L., 1988. Relative joint size and hominoid locomotor adaptations with implications for the evolution of hominid bipedalism. J. Hum. Evol. 17, 247–265.

Kandel, E.R., 2013. Principles of Neural Science, 5th edition. McGraw-Hill Education/Medical. October 2012.

Katoh, S., Beyene, Y., Itaya, T., Hyodo, H., Hyodo, M., Yagi, K., Gouzu, C., Wolde, G., Hart, W.K., Ambrose, S.H., Nakaya, H., Bernor, R.L., Boisserie, J.-R., Bibi, F., Saegusa, H., Sasaki, T., Sano, K., Asfaw, B., Suwa, G., 2016. New geological and palaeontological age constraint for the gorilla–human lineage split. Nature 530, 215–218.

Keith, A., 1894. The ligaments of the catarrhine monkeys, with references to corresponding structures in man. J. Anat. Physiol. 28, 149–168.

Kelly, L.A., Cresswell, A.G., Racinais, S., Whiteley, R., Lichtwark, G., 2014. Intrinsic foot muscles have the capacity to control deformation of the longitudinal arch. J. R. Soc. Interface 11, 2013118.

Ker, R.F., Bennett, M.B., Bibby, S.R., Kester, R.C., Alexander, R.M., 1987. The spring in the arch of the human foot. Nature 325, 147–149.

Kuo, A.D., 2007. The six determinants of gait and the inverted pendulum analogy: a dynamic walking perspective. Hum. Mov. Sci. 26, 617–656.

Kuo, A.D., Donelan, J.M., Ruina, A., 2005. Energetic consequences of walking like an inverted pendulum: step-to-step transitions. Exerc. Sport Sci. Rev. 33, 88–97.

Langergraber, K.E., Prüfer, K., Rowney, C., Boesch, C., Crockford, C., Fawcett, K., Inoue, E., Inoue-Muruyama, M., Mitani, J.C., Muller, M.N., Robbins, M.M., Schubert, G., Stoinski, T.S., Viola, B., Watts, D., Wittig, R.M., Wrangham, R.W., Zuberbühler, K., Pääbo, S., Vigilant, L., 2012. Generation times in wild chimpanzees and gorillas suggest earlier divergence times in great ape and human evolution. Proc. Natl. Acad. Sci. USA 109, 15716–15721.

Lee, M.S., Vanore, J.V., Thomas, J.L., Catanzariti, A.R., Kogler, G., Kravitz, S.R., Miller, S.J., Gassen, S.C., 2005. Diagnosis and treatment of adult flatfoot. J. Foot Ankle Surg. 44, 78–113.

Lieberman, D.E., 2012. Human evolution: those feet in ancient times. Nature 483, 550–551.

Lieberman, D.E., Venkadesan, M., Werbel, W.A., Daoud, A.I., D'Andrea, S., Davis, I.S., Mang'eni, R.O., Pitsiladis, Y., 2010. Foot strike patterns and collision forces in habitually barefoot versus shod runners. Nature 463, 531–535.

Liu, W., Nigg, B.M., 2000. A mechanical model to determine the influence of masses and mass distribution on the impact force during running. J. Biomech. 33, 219–224.

Lovejoy, C.O., Latimer, B., Suwa, G., Asfaw, B., White, T.D., 2009. Combining prehension and propulsion: the foot of Ardipithecus ramidus. Science 326, 72e1–72e8.

Marsh, R.L., Ellerby, D.J., Carr, J.A., Henry, H.T., Buchanan, C.I., 2004. Partitioning the energetics of walking and running: swinging the limbs is expensive. Science 303, 80–83.

Maus, H.-M., Revzen, S., Guckenheimer, J., Ludwig, C., Reger, J., Seyfarth, A., 2015. Constructing predictive models of human running. J. R. Soc. Interface 12.

McMahon, T.A., Cheng, G.C., 1990. The mechanics of running: how does stiffness couple with speed?. J. Biomech. 23 (Suppl. 1), 65–78.

Morton, D.J., 1922. Evolution of the human foot. Am. J. Phys. Anthropol. 5, 305–347.

Morton, D.J., 1924a. Evolution of the human foot II. Am. J. Phys. Anthropol. 7, 1–52.

Morton, D.J., 1924b. Evolution of the longitudinal arch of the human foot. J. Bone Jt. Surg. 6, 56–90.

Myers, M.J., Steudel, K., 1985. Effect of limb mass and its distribution on the energetic cost of running. J. Exp. Biol. 116, 363–373.

Rolian, C., Lieberman, D.E., Hallgrímsson, B., 2010. The coevolution of human hands and feet. Evolution 64, 1558–1568.

Ruina, A., Bertram, J.E.A., Srinivasan, M., 2005. A collisional model of the energetic cost of support work qualitatively explains leg sequencing in walking and galloping, pseudo-elastic leg behavior in running and the walk-to-run transition. J. Theor. Biol. 237, 170–192.

Senut, B., Pickford, M., Gommery, D., Mein, P., Cheboi, K., Coppens, Y., 2001. First hominid from the Miocene (Lukeino formation, Kenya). C. R. Acad. Sci., Ser. 2, Earth Planet. Sci. 332, 137–144.

Seyfarth, A., Geyer, H., Gunther, M., Blickhan, R., 2002. A movement criterion for running. J. Biomech. 35, 649–655.

Srinivasan, M., 2011. Fifteen observations on the structure of energy-minimizing gaits in many simple biped models. J. R. Soc. Interface 8, 74–98.

Steudel, K., 1990. The work and energetic cost of locomotion. I. The effects of limb mass distribution in quadrupeds. J. Exp. Biol. 154, 273–285.

Taniguchi, A., Tanaka, Y., Takakura, Y., Kadono, K., Maeda, M., Yamamoto, H., 2003. Anatomy of the spring ligament. J. Bone Jt. Surg., Am. Vol. 85-A, 2174–2178.

Thompson, C.M., Raibert, M.H., 1990. Passive dynamic running. In: Experimental Robotics I. Springer, pp. 74–83.

Thompson, N.E., Demes, B., O'Neill, M.C., Holowka, N.B., Larson, S.G., 2015. Surprising trunk rotational capabilities in chimpanzees and implications for bipedal walking proficiency in early hominins. Nat. Commun. 6.

Venkadesan, M., Guckenheimer, J., Valero-Cuevas, F.J., 2007. Manipulating the edge of instability. J. Biomech. 40, 1653–1661.

White, T.D., Suwa, G., Asfaw, B., 1994. Australopithecus ramidus, a new species of early hominid from Aramis, Ethiopia. Nature 371, 306–312.

Williams, D.S., McClay, I.S., 2000. Measurements used to characterize the foot and the medial longitudinal arch: reliability and validity. Phys. Ther. 80, 864–871.

Wood, B., Collard, M., 1999. The human genus. Science 284, 65–71.

Yawar, A., Lugo-Bolanos, M.F., Mandre, S., Venkadesan, M., 2016. Stiffness of the arched human foot. In: Proceedings of the 2016 SIAM Annual Meeting, PP1, p. 67.

Zadpoor, A.A., Nikooyan, A.A., 2006. A mechanical model to determine the influence of masses and mass distribution on the impact force during running – a discussion. J. Biomech. 39, 388–389.

Zelik, K.E., La Scaleia, V., Ivanenko, Y.P., Lacquaniti, F., 2014. Coordination of intrinsic and extrinsic foot muscles during walking. Eur. J. Appl. Physiol. 115, 691–701.

Chapter 7.2

Bioinspired Leg Design

Jonathan E. Clark

Department of Mechanical Engineering, FAMU/FSU College of Engineering, Tallahassee, FL, United States

In this section we address the question of how to utilize our understanding of the dynamics of legged locomotion to design and build robot limbs. We begin with a review of the basic functional roles of legs, and then discuss how available materials and actuators can be used to achieve these functions. The interplay between leg form and function is considered as we review some common designs for walking and running robots. We conclude with a discussion on design philosophy as we move toward building robots capable of the fast and agile motions that we currently only see exhibited by animals.

7.2.1   FUNCTIONS OF A LEG IN A ROBOT

7.2.1.1   Four Basic Functions

Essential to the operation of any legged robot is the structure and design of the leg itself. Although leg design in biology and in robotics have taken a great variety of forms there are a few basic functions that each leg fulfills, these include the roles of a: strut, spring, damper, and actuator.

Whether the body is moving or it is stationary, each leg provides a role as a strut, a structural element keeping the body from contacting the ground. In vertebrates the bones and in invertebrates the chitinous exoskeleton provides structural support that withstands the compressive forces generated by gravity and the reaction forces from the ground. These forces peak during stance (often at the midpoint of stance) and provide a functional lower limit on the mechanical strength of the leg.

A second role that legs fulfill during locomotion is that of a spring – a mechanism that deflects during loading to store and return energy during locomotion. Used in both walking and running systems, it appears that springs benefit locomotion in at least three distinct ways (Galloway et al., 2013):

Physical Robustness Leg springs act as low-pass filters on the impact forces from ground contact, reducing the shock experience by the robot's body, significantly increasing the overall system's physical robustness.

Energetic Efficiency Springs act in concert with the rhythmically excited actuators so that the system behaves as a tuned harmonic system, increasing the efficiency of locomotion.

Dynamic Stability Properly designed spring elements alter the dynamics of the overall mechanical plant, thereby contributing to the overall stability of the robot against perturbative forces.

A third role of legs is that of an energy dissipative element – a damper. The ability to selectively remove energy either through passive means such as friction, or actively by doing negative work with actuators, is an essential tool in creating stable gaits and recovering from perturbations that add energy to the gait, such as going down a step. Whether included as inherent structural element or included as needed through reactive or higher-level commands, the ability to slow down the system and remove energy during a step is key to maintaining balance in difficult environments.

The fourth role of legs is to provide actuation or energy to propel the animal or robot forward. The active generation of force or torque through muscles in animals and motors, pneumatics, hydraulics, or active materials in robots allows the necessary accelerations to induce and maintain any motion.

7.2.1.2 Obstacle Clearance and Foot Scuffing

In addition to serving these four basic functional roles while they are contacting the ground, each leg also needs to be able to move in a way that allows the robot to clear obstacles and avoid foot scuffing during the swing phase in anticipation of touchdown.

Most animal and robot legs undergo a combination of flight and stance phases within each step. The flight phase repositions the leg for stance and needs to move the foot in a motion high enough to overcome terrain variations. This provides some limitations on the possible kinematics of legs. The three link serial kinematic chain favored by most animals nicely allows for the retraction of the foot to a position relatively near the body when obstacles are present. While effective, many innovative robotic designs have found alternative kinematics or other means to accomplish this essential task.

Another design consideration that follows from the cyclic succession of stance and flight phases is the resulting impact forces that occur during touchdown. Depending on the effective inertia of the leg at impact these forces can be considerable. Even when these forces are mitigated, the legs need to be able to support the increased ground reaction forces resulting from lower duty cycles. In running gaits robot legs typically need to be able to support at least three times the body weight. As discussed in other chapters these intermittent contacts also sometimes result in control complexity as controllers switch from position control in flight to force control during stance.

7.2.1.3 Material and Manufacturing Considerations

While each of these basic roles are filled by legs in both biological and synthetic systems, the manner in which they are manifested vary greatly. In particular, it is worth remembering that the materials available for robots are fundamentally different than those used in animals. This results in dramatically different design decisions. The good news for roboticists is that there is a great variety of materials and manufacturing methods at our disposal. The bad news, on the other hand, is that the properties and limitations of each of these methods differ from those of the muscles, tendons, bones, etc., that make up the structure of our biological precedents. This limits our ability to directly mimic the elegant designs used in nature and forces us to search for new and effective ways to combine our available tools with our fundamental understanding of how locomotion works.

In particular, it should be noted that the primary structural materials used in robots vary greatly depending on operational environment and scale. For example, it is not uncommon to see robots being built from paper at small scales (Birkmeyer et al., 2009), even thought it would fail if it were to operate in a wet environment or if the same design were built at a larger scale.

A major limitation constraining the design of robots is in our current manufacturing processes. Although new techniques are being developed to overcome this, our ability to fabricate anisotropic, small scale, and spatially varying material properties limits our capacity to duplicate the design sophistication found in animals. Recently there have been significant advances in, and use of, 3D printing techniques. These include the use of stronger materials and the ability to print with multiple materials simultaneously. Although this is providing additional options to designers, most of these machines are only able to print plastics, restricting their use to smaller robot parts. For structural parts and larger robots where scale dictates the use of metals, traditional machining and welding processes (and their limitations) still dominate. A similar set of trade-offs associated with complexity and scale hold for actuators, which are discussed in greater detail in the next section.

7.2.2 ACTUATION STRATEGIES

Just as the materials that engineers use to build their robots differ from biological systems, so do the elements of actuation that are available to the designer. A nice overview of how muscles work and their limitations is given in Subchapter 8.1. It is worth noting here that while biological muscle generates force by contracting muscle fibers, mechanical actuators can create force either linearly (by pushing or pulling) or rotationally. Furthermore, rotational actuators can have a much larger range of motion than is found in animals. Less obvious, but equally important, are the differences in how force generation changes with position and velocity. For biological tissue there is a well understood range of positions (near the center of the range of motion) and velocities (speeds near zero) where maximum force can be generated. These trade-offs also exist for our mechanical compliments, but the details differ from actuator to actuator.

In this section we will consider first the most common actuation techniques: fluidic and DC motor based systems. We will then briefly review some of the more commonly used active (or "smart") materials that have found use in robots – mostly at small scales. Following this is a description of some of the techniques or transmission strategies used to transfer the force from the actuators to the joints of the limbs. Lastly we will briefly review recent robotic efforts to introduce actuators that emulate the ability of natural systems to vary the mechanical compliance of their limbs.

The approximate specific power, efficiency, and scale at which these classes of actuators are commonly used is shown in the table below.

Actuator type	Specific power	Max efficiency	Common scale
Hydraulic	10^5 W/kg	98%	>50 kg
Pneumatic	10^4 W/kg	40%	500 g–50 kg
Electromechanical	10^2 W/kg	92%	10 g–50 kg
Smart materials	10^4 W/kg	99%	<10 g

7.2.2.1   Pneumatics, Hydraulics, and DC Motors

Pneumatics, hydraulics, and DC motors are three of the most common actuation schemes used in legged robots. Pneumatics are advantageous due their light weight and built-in compliance. They do, however, require tubing and a source of compressed air – often placed off-board. Consequently, they are typically used in smaller robots in the lab, or on robots that weight tens of kilograms. Hydraulics are much more powerful, but are also heavier, messier, and also require a compressor or off-board storage tank. These usually drive prismatic pistons in the largest of legged robots. Electric motors are probably the most common option due to their mechanical simplicity, ease of use, and low cost. Their continuous rotation allows (at least theoretically) infinite range of motion. Thus, unlike muscles, their force generation is not usually position dependent. It is, however, still velocity dependent. When prismatic motion is desired with motors some type of transmission mechanism in needed; a few common options are

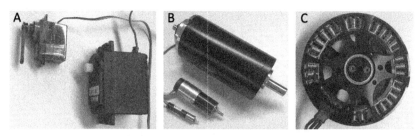

FIGURE 7.2.1   Common classes of DC motors: (A) servo, (B) brushed, and (C) hollow core brushless.

described in Section 7.2.2.3. By design, most motors operate at speeds higher than needed for leg motions and therefore utilize gearboxes to increase their available torque generation. While planetary gearboxes are quite compact, there is still an additional layer of complexity and energetic losses. The use of gear reduction also adds reflected inertia to the limbs. Fig. 7.2.1 shows three common types of DC motors used in robots. Self-contained servo motors (Fig. 7.2.1A) are simple to use, requiring only a PWM signal to specify position or speed and are frequently used in smaller platforms. Brushed DC motors (Fig. 7.2.1B) are probably the most common, with these feedback control is usually enabled via an attached quadrature shaft encoder. Brushless motors can be stronger, but required dedicated control electronics. Recently, hollow core brushless motors (Fig. 7.2.1C) have become popular due to their high power density.

7.2.2.2   Active Materials

Especially at smaller scales where DC motors become inefficient (often when they are less than 10 g), alternative actuation schemes, usually involving active materials, become commonplace. Piezo-electric actuators can operate at high frequency, have a high-power-to-weight ratio and are quite mechanically robust. They, however, can only produce a very limited range of motion. To compensate for that they are often connected in series or combined with a mechanism to extend their stroke length. Electro-active polymers (EAP) provide a softer and more flexible actuator alternative. These are also driven by high-voltage electrical supply but they can undergo much larger deflections. They, however, are usually only capable of moderate force generation. Several configurations such as discs or rolls have been implemented in robotic designs. Smart memory alloys (SMA), most commonly nickel–titanium (NiTi), or nitenol, are very light-weight actuators capable of generating large forces. Thermal heating of the material (often in wire or spring form) causes compression and, depending on the geometry, bending of the wire. The "memory" effect of material is sometimes used in medical stints, but is rarely utilized in robot legs. The primary

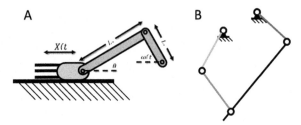

FIGURE 7.2.2 Classes of common actuation schemes: (A) four-bar crank–rocker mechanism, (B) schematic of a 5-bar pantograph leg.

limitation of this actuation scheme is in the low-frequency of operation due to limits on thermal cooling.

7.2.2.3 Transmission Strategies

Rather than using tendons to compliantly anchor muscles to the skeletal (or exoskeletal) system, robots rely on a number of mechanisms to connect their actuators and structural members. For motors, these mechanisms convert high-speed continuous rotation into motions that are more useful for locomotion. Most commonly the first step in the transmission is a gear box which increases the torque, but does so at a cost to speed, back-drivability, and efficiency and increases the reflected inertia of the system. Fig. 7.2.2 shows a few examples of linkages that have been used as transmission mechanisms in legged robots.

Fig. 7.2.2A shows a special case of four-bar linkage (a crank–slider) that is often used to transform rotation of the coupler link (a) into linear motion of the slider $(x(t))$. This particular diagram is utilized in iSprawl robot to drive the three push–pull cables coming out of the left hand side of the figure (Kim et al., 2006). Flexible cables are commonly used to apply force to a distal limb from a proximally located motor. This arrangement allows for light-weight limbs, especially near the distal segments where the velocities are the highest.

The schematic of a mechanism shown in Fig. 7.2.2B is a special class of 5-bar known as a pantograph (used, for example, by Waldron, 1986). The linkage allows input from two actuators located, for example, on the body of the robot (the motors indicated at the top of figure) to control the planar motion of the distal end. When arranged properly the vertical and horizontal motion of the foot can be decoupled, thereby allowing each motor to primarily control one direction.

A hybrid linkage design was implement on the RiSE robot (Spenko et al., 2008). This mechanism converts the motion of two motors via a differential to control both the foot trajectory specified by the four bar linkage (when moving at the same speed) and to control the abduction–adduction of the leg (when moving at different speeds).

7.2.2.4 Variable Stiffness Mechanisms

As animals traverse obstacles or rough terrain they often alter the effective compliance or stiffness of their legs (Ferris et al., 1998). For robots, stiffness control has been traditionally implemented via adjusting the controller gaits at the joint or motor level (Semini et al., 2013; Hyun et al., 2014).

For high speed and dynamic running behaviors characterized by significant impacts and unpredictable timing, the power limitations and bandwidth delays inherent in motor control have led to the development and adoption of passively compliant legs (e.g., Altendorfer et al., 2001; Poulakakis et al., 2005). Using physical springs gives high-bandwidth response and allows energy to be stored and returned, but effectively modulating or varying the natural frequency of these limbs is more challenging.

Several different actuators or leg designs with variable passive compliance have been developed over the years (Wolf et al., 2015). Some of these include: mechanical stiffness control, antagonistic nonlinear springs, and structure-controlled stiffness (Vanderborght et al., 2009). More details of these methods, in particular how they have been applied in bipedal robots, can be found in Subchapter 8.2. Some notable examples include the PPAM (pleated pneumatic artificial muscles) (Van Ham et al., 2006), which uses a pair of opposing pleated membranes that when pressurized with air contract longitudinally, and the biped with mechanically adjustable series compliance (BiMASC) which uses antagonistic aligned nonlinear springs to adjust the stiffness of the leg. Another example of mechanical stiffness control is provided by MACCEPA (mechanically adjustable compliance and controllable equilibrium position actuator) (Beyl et al., 2006), where each joint's stiffness is controlled by a pair of servo motors. One controls the equilibrium, or set point, while the other pretensions the spring.

A third approach to achieving variable mechanical stiffness, known as structure-controlled, alters the effective length or compliance of an elastic element, such as a cantilevered beam or helical spring. The active length is modulated by a small, dedicated actuator. Several groups have developed structure-controlled stiffness mechanisms (Morita and Sugano, 1995; Hollander et al., 2005; Tabata et al., 1999). Galloway et al. (2013) have used this approach on the C-shaped compliant legs of the hexapod RHex to show that varying the stiffness of the legs can improve locomotive efficiency by over 40%.

7.2.3 BIO-INSPIRATION: MORPHOLOGY

The actuation and transmission systems described in Section 7.2.2 have been utilized and combined in a number of robot morphologies. Many of these designs have been inspired by animals, at least in terms of the kinematic characteristics of their legs. In this section we review a number of these designs that have been used in walking robots. Most of these robots move slowly and carefully, concentrating more on leg coordination and foot placement than center of mass dynamics. In Section 7.2.4 the discussion turns to robot designs that focus more on bio-like dynamic function, rather than form.

The adaptive suspension vehicle (ASV) was one of the first walking robots that could traverse large outdoor obstacles. Even though it was one of the earliest autonomous walking robots, it is still one of the largest, weighing in at over 2,600 kg and about 5 m long. Its legs were powered by hydraulics and utilized a 5-bar pantograph mechanism to decouple the vertical and horizontal plane motions (Waldron, 1986). Another early legged robot, but built on a smaller scale, was Boadecia which was pneumatically powered and borrowed its limb design from the cockroach leg morphology. While it also used a pantograph mechanism in the legs, it – like the cockroach – featured significant leg differentiation, with the front legs only having 2 DOF, while the middle and hind legs had 3 DOF. In addition, its feet had overlapping workspaces to increase the maximum stride length (Binnard, 1995). Also biologically-inspired in terms of leg design, The NU Lobster Robot used SMA wires to actuate its eight legs. Designed to operate underwater and in surf regions, the legs were designed to be lightweight, waterproof, and very stable (Ayers and Witting, 2007).

In what is now a common design, the CWRU robot II used a stick insect-like 3-DOF serial chain leg with point feet. The proximal actuators controlled the forward motion and the distal mass was kept extremely low to decrease motor loading (Espenschied et al., 1996). StarlETH uses compliant series elastic legs to improve the efficiency and robustness of the classic 3-DOF leg design (Hutter et al., 2012). With this design it is able to move faster than most other robots with 3-DOF legs and can negotiate a range of environmental obstacles. More recently, JPL's Robosimian which competed in the 2015 DARPA Robotic Challenge has extremely strong 7-DOF limbs. Each L-shaped leg segment is designed with the same motor and gearbox. The modular nature of the limbs allows for a variety of configurations, but the high distal mass makes rapid maneuvers difficult. This ape-like robot excels at moving in very complex environments, such as vehicle egress maneuvers (Hebert et al., 2015).

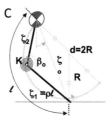

FIGURE 7.2.3 Dynamic running robots with 1 active DOF legs: (A) iSprawl, (B) RHex. (C) Model of the effective dynamics of compliant legs with rolling contact.

7.2.4 BIO-INSPIRATION: DYNAMICS

Another class of robots have been designed to capture the dynamics of animals' motion, rather than their kinematic structure. While some of these leg designs do not look like animal legs, they have been shown to accurately capture the center of mass dynamics of running in animals of different structure and morphology, effectively realizing spring-loaded inverted pendulum (SLIP) and occasional lateral-leg spring (LLS) running.

7.2.4.1 Single Active DOF Legs

The earliest dynamic running robots utilized simple leg designs featuring a single active DOF. Successful examples of this approach include the RHex (Saranli et al., 2001) (Fig. 7.2.3B) and iSprawl (Kim et al., 2006) (Fig. 7.2.3A) robots. With only one actuator powering the leg, these devices had to choose whether to generate force in a torsional or prismatic direction. The RHex family of robots is an example of using a single motor at the hip to generate sufficient torques to accelerate the body. The compliant legs compress, store, and then release the their energy in second half of stance to generate a forward and upward acceleration. On the other hand, the Sprawl family of robots use prismatic actuators in their legs to generate force, and relies on springs at the hips to redirect the ground reaction forces into forward velocity.

It has been traditional in analytical studies to use the compliant, prismatic legs characteristic of the SLIP model (Blickhan, 1989). However, these models, due to their simplicity, abstract away many of the physical features that contribute to the performance of legs such as the ones used on RHex or in human prostheses. The dynamic effects of the C-shape of the leg, including: the change in rest length, stiffness, and rolling contact point, all contribute to improved running. Recently, several researchers have attempted to characterize running with half-circle legs using a modified version of SLIP model (Sayginer, 2010; Huang et al., 2014; Jun and Clark, 2012). Jun and Clark's torque-driven and damped half-circle-leg (TD-HCL) model (Fig. 7.2.3C), for example, has shown that curved feet promote better disturbance rejection and faster running (Jun and Clark, 2012).

Since, as Asano and Luo have shown, that a rolling foot can act in a manner similar to an ankle joint for walking (Asano and Luo, 2006, 2007), it will be interesting to see if compliant monolithic legs with rolling contact can have similar dynamics to the three link articulated limbs employed by animals.

7.2.4.2 2+ Active DOF Legs

Despite their success in achieving fast running motions, 1 DOF leg designs are greatly constrained in their range of motion and the types of complex behaviors that the legs can perform. To overcome these limitations a number of researchers have developed running robots with more complex designs. These higher DOF designs, however, still have to deal with the limited power density and the need to understand how to coordinate the active DOFs to generate the desired (usually SLIP-like) COM motions.

The most common morphologies for higher DOF legs are built around using revolute actuators at the hips and then a selection of either prismatic legs or adding a second active (rotary) joint at the knee. This allows a direct mapping to a SLIP-like running model, and was the approach used by Marc Raibert in the first dynamic running robots. These first hoppers where relied on the air in the pneumatic pistons in their legs to store and return energy for each hop. The inclusion of revolute joints in the knees, as done with, the KOLT robot (Nichol et al., 2004), allows for greater obstacle clearance and has been shown to improve stability (Rummel and Seyfarth, 2008). For KOLT, the primary energy addition was done by prestretching an elastic member during flight and releasing it during stance.

BigDog and most other robots built by Boston Dynamics utilize an onboard hydraulic system to provide the necessary power for running. The use of hydraulics allowed them to implement a light-weight, serial-chain 3 DOF leg design. Although noisy, the power provided by the hydraulic system enabled subsequent robots such as Wildcat to achieve impressive running speeds (Raibert et al., 2008). The MIT Cheetah robot, on the other hands, is smaller and quieter due to its use of high-power density hollow-core DC motors. The leg designs combine the range of motion of a serial chain morphology with a 5-bar linkage in the upper limb. This allows for both the "hip" and the "knee" motors to be placed in the body, dramatically reducing the inertia of the leg. This combined with compliant tendons connected to the lower limbs allows the robot to reach high speeds and to leap over nearly hip-height obstacles (Park et al., 2015).

While robots that feature legs designs with both prismatic and revolute degrees of freedom have been successful, the specific advantages afforded by each method of actuation are not well understood. Some preliminary investigations have been undertaken to compare these approaches. These have included simulation studies comparing strictly prismatic or strictly torsional actuation of modified SLIP like runners (Larson and Seipel, 2012) or studies that have utilized multivariate optimization to examine hybrid systems (Remy et al., 2012; Görner and Albu-Schäffer, 2013). Some of these have been able to accurately reproduce several different animal-like gaits, just by altering the actuation scheme or optimization criteria. In a study that explicitly looked at the relative effort of prismatic and radial action in a SLIP-like model with damping, Miller et al. (2014) found for certain types of running that instead of an even 50/50 split, the optimal distribution of work between prismatic and torsional is about 70/30, suggesting that designs that feature nonuniform actuator power distribution may be preferable.

This brief survey has highlighted some of the successes in developing walking and running with 1 and 2-DOF legs. The discussion of human-like leg designs with three or more active degrees of freedom is not covered here, but is treated in some detail in Subchapter 7.3.

7.2.4.3 Climbing and Other Uses of Legs

In addition to walking and running over flat surfaces, animals are able to use their legs to negotiate a wide variety of terrains. They use their limbs to jump, climb, or even swim. The next subchapter includes a discussion of jumping robots in the context of bi-articular muscles. Here we look particularly at climbing. Climbing not only allows scaling of vertical objects, but also enables transitions to other modes of locomotion such as gliding or flapping flight (Paskins et al., 2007; Byrnes et al., 2008; McGuire and Dudley, 2005; Sato et al., 2009).

The unique requirements of attaching to the surface and moving up in the face of gravity have affected the limb designs for climbing robots, some of which are more animal-like than others. For examples, the Ninja robot uses suction to attach to smooth vertical surfaces such as glass (Nagabuko and Hirose, 1994). Precise orientation and location of the footpads is important, so they utilize a parallel mechanism in each leg. The DIG robot, on the other hand, is more animal like and uses simple claws and directional in-pulling forces between the feet to ensure attachment. A simple 3 DOF serial chain morphology and distal claws allow it to climb prepared surfaces (such as wire mesh or screens) even when inverted. The RiSE robot (Spenko et al., 2008) extends the use of directional adhesion to a wider range of surfaces by using arrays of microspines at each foot. It uses a hybrid 2-DOF four-bar/differential leg mechanism to climb stucco, brick, and cinderblock walls. The four-bar linkage chosen allows for straight line motion of the foot during stance, and trace a curved path during the flight-phase for recirculation. Motion of the whole arm toward or away from the wall is controlled by the relative speed of differential.

The Capucin robot climbs much like a human rock-climber, exploiting hand-holds on the surface to propel itself upwards on flat surfaces (Zhang and Latombe, 2013). The simplified attachment and need for a large workspace lead to the use of a 2-DOF, planar serial-chain leg design. The Stickybot robot (Kim et al., 2008) uses gecko-inspired dry adhesives to climb smooth hard surfaces such as glass. It uses a servo-driven parallel four-bar mechanism in the legs to provide the force to move the robot up the wall. Since it climbs smooth surfaces, the alternating legs can be synchronized and it does not have to lift the feet very far off the wall. The underactuated four-toed feet hyperextend off of the wall, allowing for detachment at the end of the stroke and then to compliantly conform to the wall to ensure adhesion at touchdown.

These robots highlight several different attachment mechanisms and legs design for climbing walls. In each case, however, the climbing speed is slow and the power requirements for accelerating the robot are minimal. Much like for running robots, the leg designs that have worked for rapid climbers are markedly different. The models and mechanisms that work for these robots are described in the following section.

Dynamic Climbing

While the SLIP model has been effective in describing running behaviors that are orthogonal to gravity, moving into the face of gravity requires a different mechanism and model. To address this, the Full–Goldman (FG) model of climbing has been developed by Goldman et al. (2006) based on the dynamic climbing behavior of geckos and cockroaches. This model – or template in the terminology of Full and Koditschek (1999) – captures the fundamental pendular motion and forces exhibited by these animals as they rapidly scale vertical surfaces. Other models, such as brachiation (Bertram et al., 1999; Parsons and Taylor, 1977) or body-mass oscillation (Provancher et al., 2011), have also been used to design robots for arboreal or scansorial settings.

The FG Model of climbing (see Fig. 7.2.4A) represents the body as a point mass, with massless springs as arms. Each arm is prismatically actuated, and pulls the body upwards in an alternating fashion. The body can swing during each stance phase as the contact point is modeled as a revolute joint. A spring in series with the actuator in each arm stores and returns the energy in each stride and reduces the peak wall-reaction forces from each step. A schematic, including the resulting COM trajectory, is given in Fig. 7.2.4A.

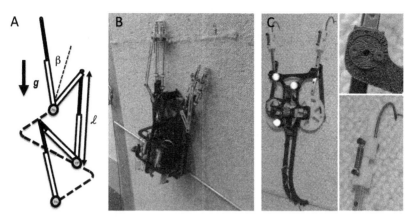

FIGURE 7.2.4 (A) Schematic depicting the Full–Goldman template for dynamic climbing. (B) DynoClimber, the first robotic platform shown to anchor the FG template and (C) BOB.

The first platform to instantiate the FG template, DynoClimber, can climb at speeds up to 0.66 m/s (Lynch et al., 2012). The design of this 2 kg platform, shown in Fig. 7.2.4B, demonstrates a close adherence the template morphology. The design features single DOF legs that act prismatically via a crank–slider mechanism. A spring in parallel with the linkage allows energy to be stored during the flight phase and released during stance to aid the motor. Key to instantiating the FG-template is the fact that the arms pull inwards on each stroke. The same mechanism is used on a dynamically similar but smaller (200 g) version of the bipedal climber, BOB (Dickson et al., 2015) which is driven by a single motor. BOB has been shown to climb at up to 40 cm/s and using microspines (Kim et al., 2005) can climb surfaces such as stone aggrogate, brick, and cinderblocks. Additional actuators, however, are required for steering, climbing on curved surfaces, or climbing downwards (Miller et al., 2015). For both robots, the horizontal bar at the bottom acts as passive rear legs to stabilize rolling motions not captured in the FG-template.

7.2.4.4 Multiuse Leg Designs

Most of the dynamic legged robots examined thus far are only capable of high-speed motion in a single mode of operation. Animals, on the other hand, such as the gecko *Hemidactylus garnotii* are able to nimbly switch, for example, between climbing and running on level ground without losing speed (Autumn et al., 2006). While the animal maintains the same kinematic motions (Zaaf et al., 2001) the force generation is switched from pushing laterally during running to pulling inwards during climbing (Autumn et al., 2006). Similar adaptations are observed for the cockroach *Blaberus discoidalis* (Kram et al., 1997; Goldman et al., 2006). In order to develop robots capable of achieving this kind of agile, multimodal locomotion, the limbs need to be designed in a way that allows for both the range of motion and force generation for both modes of locomotion. Combining these is particularly difficult in light of the limitations in power density of current actuators.

Our experience with single-domain robotic limb designs suggest that these robot legs will require sufficient degrees of freedom, tuned compliant elements, and properly allocated actuators. For example, a robot that can both run and climb dynamically should be able to instantiate multiple templates. Although structurally the SLIP and the FG templates are very similar (point-mass body, simple spring legs), the dynamic constraints are distinct in at least two ways. First, the direction of force generation is switched from pushing outward to pulling inwards. Second, the magnitude of the effective limb compliance changes significantly. In addition, the design and placement of the feet is different for each mode.

The challenges could, in theory, be met simply by adding enough actuators. In practice, however, this leads to a heavier payload and greater demands on each actuator. It also significantly complicates the control problem.

Control vs. Design-Based Approaches

Attempts to resolve these constraints have generally fallen into one of two approaches, either (1) a control-based (CB) approach or (2) a design-based (DB) approach.

In a control-based approach, the platform is designed for maximum flexibility. Serial-chain leg morphologies that maximize workspace are often chosen. In order to simplify the design, robots such as Robosimian (Hebert et al., 2015) and ANYmal (Hutter et al., 2016) even use the same motors to drive each DOF in the leg. The first challenge here is to design a control strategy to coordinate the limbs to create functional and adaptive gaits (Hyun et al., 2014; Semini et al., 2013). Techniques such as hybrid-zero dynamics for asymmetric SLIP runners (Poulakakis and Grizzle, 2009) or virtual chassis for snake robots (Rollinson et al., 2012) are used to reduce or map the high DOF system to a low level template for control purposes. The second challenge for CB robots is to provide sufficient power density, as their motors tend to be heavily geared and each motor is not being used to full capacity in each mode.

The design-based approach, on the other hand, explicitly attempts to encode the template dynamics into the structure of the body and legs. Robots such as iSprawl (Cham et al., 2002), DASH (Birkmeyer et al., 2009), and Dyno-Climber (Lynch et al., 2012) each have leg structures with passive dynamics tuned to running or climbing in the desired domain. While these light-weight

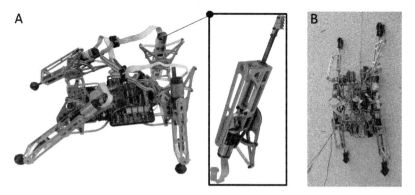

FIGURE 7.2.5 SCARAB, a quadrupedal running and climbing platform. (A) Running mode, (B) close up of the leg design, and (C) in climbing mode.

robots are fast (sometimes exceeding the speed of their biological exemplar), they lack the flexibility to adapt their motions as they change domains.

There have been some efforts to build legged robots capable of moving in multiple modalities. The RiSE robot uses a combination of four-bar linkages, a differential hip, and bi-directional springs (Saunders et al., 2006) to enable both walking on the ground and the scaling of vertical surfaces. Weight and power limitations, however, prevented it from moving dynamically without further specialization (i.e., RiSE v3) (Haynes et al., 2009).

The quadrupedal platform, SCARAB, shown in Fig. 7.2.5, is an example of following the iterative, DB-approach to achieve dynamic operation in multiple modalities. SCARAB's legs combine the crank–rocker mechanism used on iSprawl with hip joints that can rotate (Miller et al., 2015). As shown in Fig. 7.2.5 each leg features a pair of springs: one that is compressed in running and a softer one that is compressed during climbing. This configuration, combined with the ability to adjust the leg angles and phasing of the legs, allows for rapid motion in both horizontal and vertical domains. An assessment of the velocity profiles of SCARAB suggests that the robot is capturing the biologically relevant SLIP, LLS, and FG dynamics. Overcoming large obstacles and transitioning between running and climbing, however, is still a challenge.

7.2.5 SUMMARY AND FUTURE DIRECTIONS

While there are a wide variety of robotic leg designs, all of them provide a few basic functions. Each leg acts as a strut, spring, damper, and actuator. Whether actuated by pneumatics, hydraulics, DC motors, or smart materials, robots use their legs to propel themselves forward during stance and raise the leg for retraction during the swing phase. Many of the differences in leg design revolve around the number and location of the active degrees of freedom. Kinematic linkages have often been used to simplify the gait development process and this in turn has lead to fast robots with a minimal number of actuators. Other designs have used a more mammalian-like serial-chain leg morphology, allowing for a larger workspace and dexterous foot positioning, but this makes achieving desired whole-body dynamics more challenging.

In recent years, promising advances have been made in combining high-DOF morphologies with powerful, proximally located actuators to anchor the reduced order dynamic models that have guided the design of high-speed runners. As we progress in our ability to combine the advantages of control-based add design-based robot limbs we will enable a new generation of dynamic multimodal legged robots capable of rapid motion in difficult, unstructured environments.

REFERENCES

Altendorfer, R., Moore, N., Komsuoglu, H., Buehler, M., Brown Jr., H.B., McMordie, D., Saranli, U., Full, R., Koditschek, D.E., 2001. RHex: a biologically inspired hexapod runner. Auton. Robots 11, 207–213.

Asano, F., Luo, Z., 2006. On energy-efficient and high-speed dynamic biped locomotion with semi-circular feet. In: IEEE/RSJ International Conference on Intelligent Robots and Systems.

Asano, F., Luo, Z., 2007. The effect of semicircular feet on energy dissipation by heel-strike in dynamic biped locomotion. In: Proceedings of the IEEE International Conference on Robotics and Automation.

Autumn, K., Hsieh, S.T., Dudek, D.M., Chen, J., Chitaphan, C., Full, R.J., 2006. Dynamics of geckos running vertically. J. Exp. Biol. 209, 260–272.

Ayers, J., Witting, J., 2007. Biomimetic approaches to the control of underwater walking machines. Philos. Trans. R. Soc., Math. Phys. Eng. Sci. 365, 273–295.

Bertram, J.E.A., Ruina, A., Cannon, C.E., Chang, Y.H., Coleman, M.J., 1999. A point-mass model of gibbon locomotion. J. Exp. Biol. 202, 2609–2617.

Beyl, P., Vanderborght, B., Van Ham, R., Van Damme, M., Versluys, R., Lefeber, D., 2006. Compliant actuation in New Robotic Applications. In: NCTAM06 – 7th National Congress on Theoretical and Applied Mechanics.

Binnard, M.B., 1995. Design of a Small Pneumatic Walking Robot. Massachusetts Institute of Technology.

Birkmeyer, P., Peterson, K., Fearing, R.S., 2009. DASH: a dynamic 16g hexapedal robot. In: IEEE/RSJ International Conference on Intelligent Robots and Systems (IROS). St. Louis, MO, pp. 2683–2689.

Blickhan, R., 1989. The spring-mass model for running and hopping. J. Biomech. 22, 1217–1227.

Byrnes, G., Lim, N.T.-L., Spence, A.J., 2008. Take-off and landing kinetics of a free-ranging gliding mammal, the Malayan colugo (*Galeopterus variegatus*). Proc. R. Soc. Lond. B, Biol. Sci. 275, 1007–1013.

Cham, J.G., Bailey, S.A., Clark, J.E., Full, R.J., Cutkosky, M.R., 2002. Fast and robust: hexapedal robots via shape deposition manufacturing. Int. J. Robot. Res. 21, 869–882.

Dickson, J.D., Patel, J., Clark, J.E., 2015. Towards maneuverability in plane with a dynamic climbing platform. In: Proceedings of the IEEE International Conference on Robotics and Automation (ICRA). Karlsruhe, Germany, pp. 1355–1361.

Espenschied, K.S., Quinn, R.D., Beer, R.D., Chiel, H., 1996. Biologically based distributed control and local reflexes improve rough terrain locomotion in a hexapod robot. Robot. Auton. Syst. 18, 59–64.

Ferris, D.P., Louie, M., Farley, C.T., 1998. Running in the real world: adjusting leg stiffness for different surfaces. Proc. R. Soc. Lond. B, Biol. Sci. 265, 989–994.

Full, R., Koditschek, D.E., 1999. Templates and anchors: neuromechanical hypotheses of legged locomotion on land. J. Exp. Biol. 202, 3325–3332.

Galloway, K.C., Clark, J.E., Koditschek, D.E., 2013. Variable stiffness legs for robust, efficient, and stable dynamic running. J. Mech. Robot. 5, 011009.

Goldman, D.I., Chen, T.S., Dudek, D.M., Full, R.J., 2006. Dynamics of rapid vertical climbing in cockroaches reveals a template. J. Exp. Biol. 209, 2990–3000.

Görner, M., Albu-Schäffer, A., 2013. A robust sagittal plan hexapedal running model with serial elastic actuation and simple periodic feedforward control. In: IEEE/RSJ International Conference on Intelligent Robots and Systems (IROS). Tokyo, Japan, pp. 5586–5592.

Haynes, G.C., Khripin, A., Lynch, G., Amory, J., Saunders, A., Rizzi, A.A., Koditschek, D.E., 2009. Rapid pole climbing with a quadrupedal robot. In: Proceedings of the IEEE International Conference on Robotics and Automation (ICRA). Kobe, Japan, pp. 2767–2772.

Hebert, P., Bajracharya, M., Ma, J., Hudson, N., Aydemir, A., Reid, J., Bergh, C., Borders, J., Frost, M., Hagman, M., et al., 2015. Mobile manipulation and mobility as manipulation – design and algorithms of RoboSimian. J. Field Robot. 32, 255–274.

Hollander, K., Sugar, T., Herring, D., 2005. Adjustable robotic tendon using a 'Jack Spring'TM. In: IEEE International Conference on Rehabilitation Robotics.

Huang, K.-J., Huang, C.-K., Lin, P.-C., 2014. A simple running model with rolling contact and its role as a template for dynamic locomotion on a hexapod robot. Bioinspir. Biomim. 9, 046004.

Hutter, M., Gehring, C., Bloesch, M., Hoepflinger, M.A., Remy, C.D., Siegwart, R., 2012. StarlETH: a compliant quadrupedal robot for fast, efficient, and versatile locomotion. In: International Conference on Climbing and Walking Robots.

Hutter, M., Gehring, C., Jud, D., Lauber, A., Bellicoso, C.D., Tsounis, V., Hwangbo, J., Bodie, K., Frankhauser, P., Bloesch, M., Deithelm, R., Bachmann, S., Melzer, A., Hoepfinger, M., 2016. ANYmal – a highly mobile and dynamic quadrupedal robot. In: IEEE International Conference on Intelligent Robotics and Systems (IROS), pp. 9–14.

Hyun, D.J., Seok, S., Lee, J., Kim, S., 2014. High speed trot-running: implementation of a hierarchical controller using proprioceptive impedance control on the MIT Cheetah. Int. J. Robot. Res. 33, 1417–1445.

Jun, J.Y., Clark, J.E., 2012. A reduced-order dynamical model for running with curved legs. In: Proceedings of the 2012 IEEE International Conference on Robotics and Automation (ICRA). IEEE, pp. 2351–2357.

Kim, S., Asbeck, A.T., Cutkosky, M.R., Provancher, W.R., 2005. SpinybotII: climbing hard walls with compliant microspines. In: IEEE International Conference on Advanced Robotics (ICAR). Seattle, WA, pp. 601–606.

Kim, S., Clark, J.E., Cutkosky, M.R., 2006. iSprawl: design and tuning for high-speed autonomous open-loop running. Int. J. Robot. Res. 25, 903–912.

Kim, S., Spenko, M., Tujillo, S., Heyneman, B., Santos, D., Cutkosky, M.R., 2008. Smooth vertical surface climbing with directional adhesion. IEEE Trans. Robot. 24, 65–74.

Kram, R., Wong, B., Full, R.J., 1997. Three-dimensional kinematics and limb kinetic energy of running cockroaches. J. Exp. Biol. 200, 1919–1929.

Larson, P., Seipel, J., 2012. A spring-loaded inverted pendulum locomotion model with radial forcing. In: ASME International Design Engineering Technical Conferences & Computers and Information in Engineering Conference (IDETC/CIE), vol. 4. Chicago, IL, pp. 877–883.

Lynch, G.A., Clark, J.E., Lin, P.C., Koditschek, D.E., 2012. A bioinspired dynamical vertical climbing robot. Int. J. Robot. Res. 31, 974–996.

McGuire, J.A., Dudley, R., 2005. The cost of living large: comparative gliding performance in flying lizards (Agamidae: Draco). Am. Nat. 166, 93–106.

Miller, B.D., Brown, J.M., Clark, J.E., 2014. On prismatic and torsional actuation for running legged robots. In: Experimental Robotics. Springer, pp. 17–31.

Miller, B.D., Rivera, P.R., Dickson, J.D., Clark, J.E., 2015. Running up a wall: the role and challenges of dynamic climbing in enhancing multi-modal legged systems. Bioinspir. Biomim. 10, 025005.

Morita, T., Sugano, S., 1995. Design and development of a new robotic joint using a mechanical impedance adjuster. In: Proceedings of the IEEE International Conference on Robotics and Automation, vol. 3, pp. 2469–2475.

Nagabuko, A., Hirose, S., 1994. Walking and running of the quadruped wall-climbing robot. In: Proceedings of the IEEE International Conference on Robotics and Automation (ICRA), vol. 2. San Diego, CA, pp. 1005–1012.

Nichol, J.G., Singh, S., Waldron, K., Palmer, L., Orin, D., 2004. System design of a quadrupedal galloping machines. Int. J. Robot. Res. 23, 1013–1027.

Park, H.-W., Wensing, P.M., Kim, S., 2015. Online planning for autonomous running jumps over obstacles in high-speed quadrupeds. In: Proceedings of Robotics: Science and Systems. Rome, Italy.

Parsons, P.E., Taylor, C.R., 1977. Energetics of brachiation versus walking: a comparison of a suspended and an inverted pendulum mechanism. Physiol. Zool. 50, 182–188.

Paskins, K.E., Bowyer, A., Megill, W.M., Scheibe, J.S., 2007. Take-off and landing forces and the evolution of controlled gliding in northern flying squirrels *Glaucomys sabrinus*. J. Exp. Biol. 210, 1413–1423.

Poulakakis, I., Grizzle, J.W., 2009. The spring loaded inverted pendulum as the hybrid zero dynamics of an asymmetric hopper. IEEE Trans. Autom. Control 54, 1779–1793.

Poulakakis, I., Smith, J., Buehler, M., 2005. Modeling and experiments of untethered quadrupedal running with a bounding gait: the Scout II robots. Int. J. Robot. Res. 24, 239–256.

Provancher, W.R., Jensen-Segal, S.I., Fehlberg, M.A., 2011. ROCR: an energy-efficient dynamic wall-climbing robot. IEEE/ASME Trans. Mechatron. 16, 897–906.

Raibert, M., Blankespoor, K., Nelson, G., Playter, R., the BigDog Team, 2008. BigDog, the rough-terrain quadruped robot. In: Proceedings of the 17th IFAC World Congress. Seoul, Korea, pp. 10822–10825.

Remy, C.D., Buffinton, K., Siegwart, R., 2012. Comparison of cost functions for electrically driven running robots. In: Proceedings of the IEEE International Conference on Robotics and Automation (ICRA). Saint Paul, MN, pp. 2343–2350.

Rollinson, D., Buchan, A., Choset, H., 2012. Virtual chassis for snake robots: definition and applications. Adv. Robot. 26, 1–22.

Rummel, J., Seyfarth, A., 2008. Stable running with segmented legs. Int. J. Robot. Res. 27, 919–934.

Saranli, U., Buehler, M., Koditschek, D.E., 2001. RHex: a simple and highly mobile hexapod robot. Int. J. Robot. Res. 20, 616–631.

Sato, K., Sakamoto, K.Q., Watanuki, Y., Takahashi, A., Katsumata, B., Bost, C.-A., Weimerskirch, H., 2009. Scaling of soaring searbirds and implications for flight abilities of giant pterosaurs. PLoS ONE 4. e5400(6).

Saunders, A., Goldman, D.I., Full, R.J., Buehler, M., 2006. The RiSE climbing robot: body and leg design. In: Unmanned Systems Technology VIII. Orlando, FL. In: Proc. SPIE, vol. 6230, 623017.

Sayginer, E., 2010. Modelling the Effects of Half Circular Compliant Legs on the Kinematics and Dynamics of a Legged Robot. PhD thesis.

Semini, C., Barasuol, V., Boaventura, T., Frigerio, M., Buchli, J., 2013. Is active impedance the key to a breakthough for legged robots?. In: International Symposium of Robotics Research (ISRR).

Spenko, M.J., Haynes, G.C., Saunders, J.A., Cutkosky, M.R., Rizzi, A.A., Full, R.J., Koditschek, D.E., 2008. Biologically inspired climbing with a hexapedal robot. J. Field Robot. 25, 223–242.

Tabata, O., Konishi, S., Cusin, P., Ito, Y., Kawai, F., Hirai, S., Kawamura, S., 1999. Microfabricated tunable stiffness device. In: Proc. of the 13th Annual Int. Conf. on Micro Electro Mechanical Systems.

Van Ham, R., Van Damme, M., Vanderborght, B., Verrelst, B., Lefeber, D., 2006. MACCEPA: the mechanically adjustable compliance and controllable equilibrium position actuator. In: Proceedings of CLAWAR, pp. 196–203.

Vanderborght, B., Van Ham, R., Lefeber, D., Sugar, T., Hollander, K., 2009. Comparison of mechanical design and energy consumption of adaptable, passive-compliant actuators. Int. J. Robot. Res. 28, 90–103.

Waldron, K., 1986. Force and motion management in legged locomotion. IEEE J. Robot. Autom. RA/2, 214–220.

Wolf, S., Grioli, G., Friedl, W., Grebenstein, M., Hoeppner, H., Burdet, E., Caldwell, D., Bicchi, A., Stramigioli, S., Vanderborght, B., 2015 Variable stiffness actuators: review on design and components.

Zaaf, A., Van Damme, R., Herrel, A., Aerts, P., 2001. Limb joint kinematics during vertical climbing and level running in a specialist climber: *Gekko gecko* Linneus, 1758 (Lacertilia : Gekkonidae). Belg. J. Zool. 131, 173–182.

Zhang, R., Latombe, J.-C., 2013. Capuchin: a free-climbing robot. Int. J. Adv. Robot. Syst. 10, 194.

Chapter 7.3

Human-Inspired Bipeds

Koh Hosoda[§], Maarten Weckx[¶], Maziar A. Sharbafi[‖], and
Ioannis Poulakakis[**]

[§]*Department of System Innovation, Graduate School of Engineering Science, Osaka University,
Toyonaka, Japan*
[¶]*Department of Mechanical Engineering, Vrije Universiteit Brussel, Brussels, Belgium*
[‖]*School of Electrical and Computer Engineering, College of Engineering, University of Tehran,
Iran*
[**]*Department of Mechanical Engineering, University of Delaware, Newark, DE, United States*

To some extend, almost all bipedal robots that have been developed thus far are
inspired by the human figure. Indeed, robots in this family are designed with
the objective to reproduce certain features of human locomotion that are important for automating tasks in typical human-centric or natural environments. The
approaches adopted by researchers in this field to incorporate such features in
a bipedal robot's behavioral repertoire vary significantly, resulting in systems
with drastically different morphological and geometric characteristics, and actuation and control architectures. These differences manifest in the resulting
locomotion behaviors, with important consequences in mobility, versatility and
energy efficiency. For example, humanoid robots – perhaps the most visible to
the general public bipedal robots – are versatile platforms capable of accomplishing a diverse array of locomotion and manipulation tasks at the expense of
energy economy. On the other hand, passive dynamic walkers are champions
in energy efficiency but their locomotion behaviors are limited; the remarkable
elegance and energy efficiency of these systems comes at the cost of poor ability
in achieving tasks such as climbing stairs, turning or running.

This chapter begins with a brief overview of some of the different approaches
that have been adopted for the mechanical design, actuator architecture and control of bipedal robots. Our objective in this part is to provide a glimpse to the
different levels of biological inspiration that underlie the development of bipedal
robots. It is deduced that creating robots that share the same morphological
characteristics with their natural models – in this case, humans – does not necessarily imply that the properties of the resulting robot locomotion behaviors
can faithfully capture those of the corresponding natural model. As we will see,
passive walkers are more "human-like" than humanoid robots, despite the fact
that humanoids are more anthropomorphic machines. Throughout this brief review, attention is placed on the actuation approaches that are common in bipedal

robots – i.e., electrical, hydraulic, and pneumatic – to motivate the discussion in the second part of the chapter, which focuses on the role of biarticular muscle units in powering locomotion on bipedal robots. The use of biarticular muscles – that is, muscles that span more than one joints – serves as a concrete example of a bioinspired robot architecture, and it represents a departure from the classical robotics actuation paradigm, in which each joint is actuated by one actuator. The implications of this novel actuation approach are discussed in more detail.

7.3.1 MIMICKING THE HUMAN FIGURE

The history of designing and building modern bipedal robots can be traced back in the 1960s. One of the earliest bipeds is WAP-1 shown in Fig. 7.3.1, an anthropomorphic, pneumatically actuated "pedipulator" developed by Prof. Kato of Waseda University in 1969 (Lim, 2007). This robot was able to walk slowly along a straight line, and was constrained to move within the sagittal plane. The motivation underlying the development of WAP-1 has been to study and understand the biomechanics of human walking, and to emulate such motions in an engineered system. Due to the strong desire to mimic human movement, the robot was actuated by artificial pneumatically-driven muscles. Following WAP-1, a series of bipedal robots has been designed by the same group, leading to WAP-3, a three-dimensional bipedal robot driven by pulse width modulation (PWM)-controlled bag-like pneumatic actuators. The aforementioned robots were controlled on the basis of a "teaching-playback paradigm" (Ito and Tsutomu, 1983), and had no balancing capabilities. To provide such capabilities, more involved feedback control schemes had to be implemented, and pneumatic actuation was dropped in favor of more powerful hydraulic actuator units, which are capable of delivering control commands to the robot's joints fast, with minimal response time. As a result, more powerful control schemes could be implemented based on a stability criterion that aims at maintaining balance by keeping the center of pressure (COP) within the support plane. However, the adoption of hydraulic actuators introduced dynamics that are markedly different from the dynamics of the human-like pneumatic artificial muscles, thereby shifting the focus from robots that mimic the structural properties of human actuation units to robots that mimic function in terms of balancing and walking motions. Yet, it is important to point out that one of the earliest bipedal machines used pneumatic actuation.

Based on this work, Prof. Kato pointed out three research approaches for designing and constructing bipedal robots (Kato, 1983). One approach emphasizes mechanical design and dynamics, resulting in motions that are largely based on the system's capabilities with minimal – or, in some cases, without any – control influence. A second approach is based on imitating the biomechanics of human

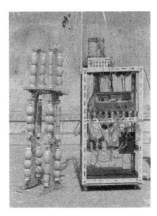

FIGURE 7.3.1   The bipedal robot WAP-1, which was actuated by artificial pneumatically-driven muscles (Lim, 2007) (courtesy of Atsuo Takanishi Lab., Waseda University).

motion by introducing robots that closely reproduce not only the morphology but also the musculoskeletal structure of a human. Finally, a third approach relies more on feedback control theory coupled with sufficient actuation authority to ensure high-performance trajectory tracking that replicates the kinematics of human motion. This categorization provides a basis for understanding the different levels at which inspiration from human structure, morphology and dynamics can be applied to bipedal robot design. We will briefly review these approaches, but in a different order – starting from feedback-oriented methods, continuing with mechanical design and dynamics methods, and finally concluding with biomechanical approaches – to make the connection with the second part of the chapter that deals with biarticular actuation.

7.3.1.1   Early Control-Based Approaches

One of the earliest bipedal robots capable of dynamic walking motions was developed by Miura and Shimoyama (1984). As shown in Fig. 7.3.2, the stilt biped BIPER had small feet and heavy torso, and was modeled as an inverted pendulum. Based on linearizing the equations of motion around a desired trajectory, a linear feedback controller has been designed to stabilize the system. BIPER was driven by electric motors, controlling the torque applied at each joint of the robot. This robot did not share the geometric characteristics of the legs and torso of a human. In fact, its legs had no knee and ankle joints, and the dynamics was simplified so that the controller implemented based on the linearized equations was effective. However, following this work, the inverted pendulum model and its extensions have become standard in capturing the dynamics of balancing in bipedal walking machines.

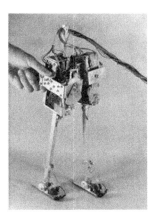

FIGURE 7.3.2 The bipedal robot BIPER closely resembles an inverted pendulum (Miura and Shimoyama, 1984).

Following Miura and Shimoyama (1984), a number of researchers developed bipedal robots with more human-like morphologies than BIPER and used the inverted pendulum paradigm for designing control laws. The majority of these robots were driven by electric motors controlled by computers with small time delays. In general, though, the ability of electric motors to generate torque is speed-dependent; typically, the magnitude of the torque that can be delivered to the motorshaft is small and the rotating speed is much faster than that required in bipedal robot applications. Therefore, it is common to use transmission mechanisms – often, gearboxes – to reduce the rotating speed and increase the output torque. The introduction of gearboxes in the drive train, introduces dynamic effects that significantly affect the behavior of the robot. In particular, the gear reducers tend to increase the friction of the joint and to add reflected inertia, which, depending on the gear reduction ratio, can be significant. Hence, joints that are driven by combined motor–gearbox actuation units can be difficult to move from the output side – that is, the side of the gearshaft – resulting in systems that lack backdrivability.

The lack of backdrivability can be advantageous when one is interested in precise control of each joint individually. Indeed, any disturbances that are developed at the gearbox output – either due to the motion of other joints and links of the robot or due to its interaction with the environment – cannot significantly affect the torque applied by the motor. As a result, each joint can be controlled at a satisfactory degree without the need to take into account the whole dynamics of the robot's structure, and precise motion control can be achieved using only local feedback (Sciavicco and Siciliano, 1996). The externally applied forces are simply treated as disturbances that can be sufficiently attenuated by the non-backdrivable nature of the combined motor–gearbox unit. While this property can be helpful in simplifying feedback design, robots the lack backdrivability *cannot* produce highly dynamic movements, for such systems are very "stiff" and insensitive to interaction forces. As an example, consider a running biped. Following the impact of a leg with the ground, the corresponding knee joint would have to bend as the center of mass of the system lowers and decelerates. A non-backdrivable knee joint would not be able to realize this motion naturally, and explicit control action would be required, which – beyond the energy cost that it entails – is subjected to the bandwidth limitations of the motors involved, especially at the touchdown instant when impulsive forces are developed.

We have seen thus far one of the great challenges in designing human-inspired bipedal robots, which, as a matter of fact, permeates the development of any biologically inspired robot. Although, we can reproduce the geometry of the human figure at a satisfactory degree, and we can achieve high-precision control of the joint motion according to desired human-like trajectories, it is extremely hard to capture the *dynamic* characteristics of human movement. This is due to the fact that common engineering materials, actuation modules and controller design approaches are characterized by limitations and capabilities that are markedly different from those of bones, ligaments, tendons, muscles, and the neural circuitry that controls the ensemble. Clearly, the challenge in reproducing human-like locomotion behaviors on bipedal robots lies on the fact that the technological means available to the designer are fundamentally different from those used in animals, thereby resulting in dramatically different design decisions.

7.3.1.2 Morphologically Inspired Bipeds and Quasistatic Balancing

As we have seen, robots that are driven by combinations of motors and gearboxes do not exhibit significant joint back-drivability, making it possible to control each joint individually to track desired trajectories. In this case, the behavior of the robot largely depends on the nature of the trajectories that are imposed on its joints, and – owing to the availability of fast local feedback controllers – can be easily altered by changing these desired trajectories. There are several approaches on how to design desired trajectories for realizing walking on bipedal robots. Perhaps one of the simplest ways is to record – for example, by using a motion capture system – human joint trajectories during walking and enforce properly scaled versions of them on the robot joints. However, in most cases, this method would fail to realize stable walking motions. Indeed, as was mentioned above, the robot's dynamics is significantly different from the dynamics of a human during walking, even when proper scaling is used to match the geometric characteristics of the human figure. As a result, imposing human-like trajectories on bipedal robots does not automatically guarantee stable operation, unless special care is taken; see Ames (2014), for example.

Clearly, one of the principal problems of designing functional walking machines is locomotion stability. To avoid complexity and ensure that the robot will not fall, many developers of bipedal robots – particularly humanoids – have adopted a simple notion of gait stability. Indeed, for many of these robots, the stabilization algorithm boils down to maintaining the center of pressure of the ground reaction forces of the stance foot strictly within the convex hull of the foot. Combined with simple pendulum-like models of human locomotion, the ZMP stability criterion can be used to generate desired trajectories for highly complex bipedal robots. For example, if the torso is relatively large, the dynamics of the robot can be approximated by an inverted pendulum with a prismatic knee joint, called the linear inverted pendulum model (LIP) (Kajita and Tani, 1991). One can then design the desired trajectory of the center of gravity (COG) of the robot based on high-level motion planning objectives and then use the LIP model to compute the corresponding evolution of the ZMP. Walking can then be realized by the robot though the use of sensory feedback to track the calculated ZMP trajectory (Kajita et al., 2002). Alternatively, one can utilize a ZMP-based walking pattern generation to first design the ZMP trajectory and compute the corresponding COG trajectory. The desired trajectories for the robot's joints can then be computed so that the COG trajectory is realized (Kajita et al., 2003). Depending on the available sensory information about the ZMP, preview control can be implemented to modify the COG trajectory as in Kajita et al. (2003).

No matter how the COG or ZMP trajectories are generated, the resulting walking motions are often flat-footed and distinctly not human like. This is clearly visible in humanoid robots that typically employ controllers like the ones described above. Humanoids are complex, high-degree-of-freedom prototypes, developed as part of an effort to create robots that will be able to serve humans – or even directly replace them – in tasks that may be dull or dangerous. As such, humanoids involve a very broad-ranging development effort that includes machine vision, portable power sources, artificial intelligence, force sensing, durability, packaging, etc. Upright, stable bipedal locomotion is only one piece of the overall effort, and – largely for reasons of expediency – the designers of these robots have adopted ZMP-based notions of gait stability, that do not faithfully capture human walking. Clearly, although these robots are inspired by human morphology, they are limited in their ability to reproduce the natural dynamics of human locomotion.

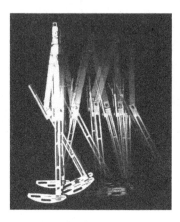

FIGURE 7.3.3   McGeer's passive dynamic walker.

7.3.1.3   Passive Walking and Dynamic Balancing

At the opposite end of the complexity spectrum – compared to humanoid robots – are the passive dynamic walkers. In an exercise of "creative neglect" the designers of these systems effectively seek to minimize the number of joints, links, actuators and sensors that are required to realize a specific locomotion task. This area was inspired by the groundbreaking work of Tad McGeer, who in the late 1980s and early 1990s, analyzed and built planar bipeds which could walk stably (in the sense of possessing an exponentially stable periodic orbit) down a slight incline with no sensing or actuation whatsoever (McGeer, 1990a,b). Although the original passive dynamic walker does not have any torso, to some extent its morphology is human like, as can be seen in Fig. 7.3.3. Its legs feature a passive (unactuated) knee joint and a round-shape foot, and move freely in the sagittal plane as rigid-body pendula under the influence of gravity. McGeer's insight has been that if the inertia and geometric properties of the system are tuned just right, stable periodic walking motions can be generated without the need of any feedback control law – this is in stark contrast with the feedback-based walking machines described above. Passive dynamic walking is purely the outcome of the interplay between gravity and the geometric and inertia properties of the robot.

Merely powered by gravity, McGeer's original passive biped was capable of walking downhill. Subsequent research efforts in this area, focused on developing (nearly) passive dynamic walkers that are able to walk on flat ground by supplying the system with just enough power (Collins et al., 2005). An example of such systems is the pneumatically actuated robot Mike shown in Fig. 7.3.4, which was able to walk stably on flat ground with a simple and intuitive control strategy (Wisse et al., 2005; Wisse and van der Linde, 2007). Inspired by the

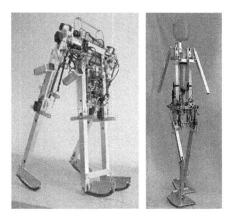

FIGURE 7.3.4 The (nearly) passive dynamic bipedal robots Mike (left) and Denise (right) (Wisse and van der Linde, 2007).

rimless wheel paradigm of walking (Wisse et al., 2005), the objective of Mike's controller was to avoid falling down in the forward direction by rapidly placing the swing leg at a proper angle in front of the stance leg. Albeit simple, this control strategy was very effective in increasing the basin of attraction of passive walking motions with only the minimum amount of energy input required. Essentially, Mike was mostly relying on its natural dynamics, continuously falling on one leg and recapturing itself by the other in a manner that closely resembles the dynamics of human walking. Indeed, this cyclic energy exchange between kinetic and gravitational potential energies is known to be of fundamental importance in explaining the economy of walking in humans (Margaria, 1976). Following Mike – which was a sagittal plane walker – the three-dimensional passive dynamic walking robot Denise was developed; see Fig. 7.3.4. Denise was able to avoid sideways falling through suitable design of the shape of the foot sole, much like a tumble doll. Mike and Denise were part of a series of robots that have been developed in an increasing level of design complexity with the objective to identify what each increase in complexity can contribute to bipedal walking, in terms of versatility – such as flat ground versus inclines, or spatial versus planar walking – and enhanced stability – such as the ability to tolerate deviations in the walking surface without falling (Wisse and van der Linde, 2007).

As can be seen from the examples of passive dynamic walkers presented in Figs. 7.3.3 and 7.3.4, one common feature of these robots is their circular foot design. It turns out that this shape is advantageous in at least two ways; first, by generating a propulsive force as the foot rolls over the ground when the robot falls forward, and second by promoting energy efficiency. It is very interesting to note that the trajectory of the COP of a human with respect to a coordinate sys-

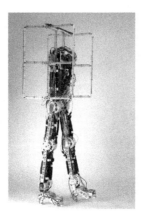

FIGURE 7.3.5   The bipedal robot Pneumat-BB, which based on the design of passive dynamic walkers.

tem fixed at the ankle joint also features a circular shape (Hansen et al., 2004). Actually, it is more than just a coincidence that humans utilize their – much more complex as we have seen in Subchapter 7.1 – feet to generate propulsive forces in a similar manner as the simple bipeds discussed here. In fact, the circular foot shape of passive walkers can be conceived as a simple design solution capable of capturing aspects of the dynamics of human walking. An alternative foot design has been proposed in Narioka et al. (2013), which investigated the effect of the circular foot shape and deduced that similar propulsion forces can be generated by a flat foot combined with a compliant ankle joint. To experimentally validate this idea, the bipedal robot Pneumat-BB shown in Fig. 7.3.5 has been designed. Similarly to the passive dynamic walkers – although Pneumat-BB is sufficiently actuated – this robot takes advantage of its passive dynamics to generate robust walking motions. However, in Pneumat-BB, ankle compliance was realized by artificial pneumatic muscles; note that an electric motor emulating compliance could also be used to actuate the ankle joint.

In a way, although McGeer's original passive dynamic walker and other robots in that category feature much fewer degrees of freedom and actuators than more complex bipeds, they do capture certain key aspects of the dynamics of human walking in a surprisingly – given their simplicity – good way. In these highly underactuated machines, the emphasis is on achieving energy-efficient walking through the effective use of the "natural dynamics" of the system. Note also that taking advantage of the natural dynamics of the bipedal plant in a suitable way can drastically simplify the control problem as well. This does *not* mean that feedback control is not used. For example, implementation of the leg recirculation strategy to enhance sagittal plane stability in Mike requires event-based triggering of the hip actuators (Wisse and van der Linde, 2007). Feedback control laws are present in a more subtle way as well; that is, by embedding them in the mechanical design of the system. For example, the hip bisecting mechanism used to stabilize the bipedal robot Max mechanically imposes a holonomic constraint that reduces the problem of walking stability to one that can be addressed by simple leg recirculation control (Wisse and van der Linde, 2007). Such "mechanical" solutions combined with minimal feedback control laws are consistent with the design philosophy of passive or nearly passive bipedal walkers. The flip side of the coin is that passive dynamic walkers exhibit a very limited notion of locomotion. The remarkable elegance and economy of these machines comes at the cost of their limited ability to achieve tasks other than walking at a fixed speed. On the other hand, the impressive versatility demonstrated by humanoid robots comes at the cost of increased power consumption, heavy actuators, and expensive electronics. It is therefore natural to ask how the efficiency and elegance of the minimalist walkers can be combined with the versatility of humanoid robots. To address this challenge, novel design solutions and feedback laws must be developed that work in concert with – and not against – the natural dynamics of the system in realizing the intended locomotion behaviors. This is at the core of current research efforts in bioinspired legged locomotion.

7.3.2 HUMAN-INSPIRED MUSCULOSKELETAL BIPEDS

In the previous section, we briefly examined just a few design approaches for developing human-inspired bipedal robots capable of implementing walking motions. We have seen that, although the underlying mechanism may be drastically different from the structure, actuation and control of the corresponding biological system, reliable locomotion behaviors that capture certain human-like features can indeed be realized. In this section, we turn our attention to the discussion of a novel actuator architecture that is inspired by biological systems and it differs from the classical approach in robotic design; namely, biarticular actuator units.

7.3.2.1 Biarticular Muscles: Biomechanics and Inspiration

In robotics, actuation is typically introduced in the robot's mechanical structure so that each motor is devoted to the control of only one joint, the joint it actuates. In the context of the musculoskeletal system, muscles that cross only one joint are called *monoarticular*. In addition to monoarticular muscles, the musculoskeletal system is supplied with muscles that cross two joints; these muscles are termed *biarticular*. Focusing on the human leg, the rectus femoris (RF), hamstrings (HAMS), and gastrocnemius (GAS) are examples of biarticular muscles, each actuating two joints: RF flexes the hip and extends the knee

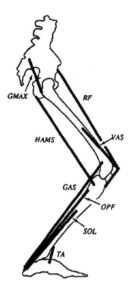

FIGURE 7.3.6   Schematic representation of the musculoskeletal model of the human leg and its primary muscles (Pandy et al., 1990). Symbols appearing in the diagram are: gluteus maximus (GMAX), hamstring muscles (HAMS), rectus femoris (RF), vastus lateralis (VAS), gastrocnemius (GAS), other plantar flexors (OPF), soleus (SOL), and tibialis anterior (TA).

joint, HAMS extends the hip and flexes the knee joint, while GAS flexes the knee and extends and ankle joint. Measurements of the cross sectional area of various human muscles indicate that monoarticular extensor[4] muscles produce more force than biarticular ones. The purpose of this section is to provide an overview of the functions of biarticular muscles in human locomotion, and prepare the ground for a discussion regarding the application of biarticular elements in robotic devices. Since we are interested in legged robots, our focus will be on biarticular muscle groups located in the legs, such as those mentioned above. Restricting attention to actions that cannot be performed by an alternative pair of monoarticular muscles, the following functions of biarticular muscles have been proposed in the literature (van Ingen Schenau et al., 1990).

Coupling Joint Movements

Biarticular muscles couple the joints they cross in a way that can facilitate more proximally located – and more powerful – monoarticular muscles to apply indirect actions on joints they do not directly affect. Consider, for example, the RF, displayed in Fig. 7.3.6. If, the hip is extended by its monoarticular exten-

---

4. In general, extensor muscles are used to support the body weight against gravity, while flexor muscles are used to lift the limbs. Flexors, therefore, are generally much smaller and generate significantly less force than extensors.

sors – e.g., by the gluteus maximus (GMAX), one of the largest and strongest muscles in the human body – while the RF retains its length, the knee must be extended as well, thus allowing the proximally located GMAX to indirectly affect knee extension. In a completely analogous way, knee extension can be coupled with plantar flexion through the biarticular GAS, eventually allowing GMAX to support plantar flexion as the leg extends through the RF and GAS. This type of coupling is characterized by limited contraction of the biarticular muscles involved and is known as *tendinous action* due to the fact that, under these conditions, the corresponding biarticular muscles can be largely regarded as tendons. Coupling multiple joints through the tendinous action of the corresponding biarticular muscles can offer several advantages by

- Enabling the transmission of the work of proximally located powerful muscles to an extremity;
- Reducing the mass of more distant segments by placing larger and heavier muscles closer to the trunk;
- Facilitating the coordinated control of multi-joint movement.

Clearly, these advantages are of interest to the design of robots, as we will see below.

Low Contraction Velocity

During the simultaneous movement of two adjacent joints crossed by a biarticular muscle – such "concurrent" movements cause the origins and insertions of the muscle to move in the same direction – the corresponding muscle operates at a lower shortening velocity than that of the monoarticular muscles involved. For example, during simultaneous hip and knee extension the biarticular HAMS have a lower shortening velocity than the monoarticular hip extensors, and the biarticular RF has a lower shortening velocity than the knee extensors. As a result, the biarticular muscles operate in a more favorable region of their force-velocity characteristic compared to the case where origin and insertion do not move in the same direction.

Transport of Energy

As was mentioned above, one of the advantages of the tendinous action associated with biarticular muscles is the transport of energy from proximally located monoarticular muscles to the more distal joints. This action turns out to be more prominent in explosive movements, such as vertical jumping. In van Ingen Schenau et al. (1987), kinematic data and muscle activation patterns from ten experienced jumpers during counter movement jumps were analyzed. The results demonstrate the role of the biarticular GAS in supporting plantar flexion prior

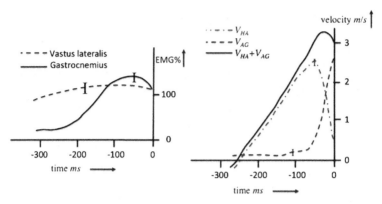

FIGURE 7.3.7   Recorded EMG patterns of the vastus lateralis and mean of both heads of the gastrocnemius (l.) and measured velocity differences during jumps (van Ingen Schenau et al., 1987). Time is expressed in ms preceding liftoff.

to push off by transporting energy from the knee extensor – namely the monoarticular vastus lateralis (VAS) – to the ankle joint. In more detail, joint position measurements during a jump were used to calculate the difference between the velocities of the hip and ankle joints, V_{HA}, as well as the velocities of these two joints with respect to the ground, V_{HG} and V_{AG}, respectively. In addition to the kinematic data, ground reaction forces were measured and electromyographic (EMG) patterns of the muscles involved were recorded. The average recorded EMG patterns and calculated velocity differences are displayed in Fig. 7.3.7, indicating that the GAS compensates for the inevitable decrease of the velocity difference between the hip and the ankle joints, V_{HA}, at the end of push-off. To provide some intuition, note that the maximum V_{HA} is reached prior to full knee extension, at a mean knee angle equal to 132°, due to geometric and anatomic constraints (van Ingen Schenau, 1989).

- Geometric constraints. As the knee approaches its fully extended configuration, the velocity difference V_{HA} decreases to zero. Clearly, at this configuration the hip and ankle joints move with the same velocity, and thus it is impossible to maintain an increasing V_{HA} up to full knee extension. Furthermore, converting knee angular velocity to translational velocity difference V_{HA} becomes less effective at large knee angles that correspond to more straight leg configurations.
- Anatomic constraints. To avoid hyperextension of the knee, the knee angular velocity should be decelerated to zero at full extension.

Given these constraints, the decrease of V_{HA} toward the end of the contact phase cannot be avoided. However, the biarticular GAS compensates for this decrease by opposing knee extension and promoting plantar flexion. It allows the knee

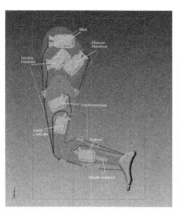

FIGURE 7.3.8 Actuator architecture of Achilles (left) and Achilles performing a squat experiment (right) (Klein et al., 2008).

extensors to further contract and deliver power, which is *not* used to further increase V_{HA} through additional knee extension but for plantar flexion. Plantar flexion accelerates V_{AG} and thus contributes to V_{HG} as well. Note that if only monoarticular muscles were employed, the plantar flexors would have to contract rapidly, and the further extension of the knee would be useless, resulting only to power dissipated into heat. This is a typical example of how biarticular muscles can be recruited to transport energy to more distal joints. A similar mechanism takes place in the upper leg, with the biarticular muscle RF. Several simulations have also proven the importance of biarticular muscles in explosive movements, including maximum height human jumping (Pandy et al., 1990) and running (Jacobs et al., 1993).

7.3.2.2 Applications to Robotics

To better understand the coordination of mono- and biarticular muscles and to test the idea of proximo-distal energy transfer in a robotic context, the anthropomorphic leg of Fig. 7.3.8 has been constructed (Klein et al., 2008). The joints of the leg – which is part of the bipedal robot Achilles – are actuated by a combination of high-performance modular motors connected in series with Kevlar straps, designed to mimic the action of a muscle. The muscle units incorporated in Achilles's leg are shown in Fig. 7.3.8, among which the RF and the GAS are biarticular. The leg terminates at a passive toe and all the distances and properties of the limb segments closely follow those of the human leg. A number of experiments were performed to access the contribution of the soleus (SOL) and GAS to the ankle power, and to analyze the effect of the activation timing of the SOL and GAS on peak power production. In these experiments, the leg

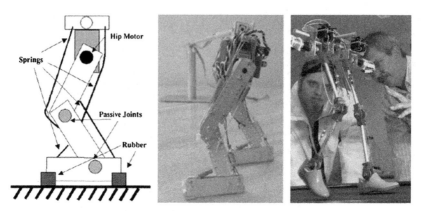

FIGURE 7.3.9 The structure (left) and realization (middle) of JenaWalker I, and JenaWalker II (right).

was commanded to lift itself up on its toe from a squat position, as shown in Fig. 7.3.8. Measurements of the forces developed in the SOL and GAS were used to calculate the work performed at the ankle when either or both muscle units were activated. This analysis showed that the work at the ankle is increased when both muscles are active, and that GAS delivers more power than the SOL, demonstrating the effect of energy transport from the knee to the ankle when the GAS was active. By varying the delay between the activation of the SOL and the GAS, it was also found that the highest peak in power output at the ankle occurred when the activation of GAS preceded that of SOL by 350 ms. This slight delay between the activation of the SOL and GAS and its effect in optimizing energy transfer at the ankle is consistent with measurements of muscle activation in humans (Winter, 1990).

The function of biarticular muscles as power transfer mechanisms through tendinous action can be captured in a minimalistic way by substituting these muscles with biarticular tension springs[5] as in the JenaWalker I, which is shown in Fig. 7.3.9; see Iida et al. (2009) for details. The JenaWalker I represents the mechanical realization of a spring–mass model that was proposed in Iida et al. (2008) to study the salient features of the human musculoskeletal system in a template setting. Each leg consists of three segments connected with passive knee and ankle joints via four linear tension springs, as displayed in Fig. 7.3.9. The springs represent the monoarticular tibialis anterior and biarticular rectus femoris, biceps femoris, and gastrocnemius. Each leg is actuated by a single motor, which is located at the corresponding hip joint and is controlled to track

---

5. Note that in animals such as the horse, a number of the biarticular muscles exhibit only limited shortening capacity. Thus, they can be regarded largely as tendons (van Ingen Schenau et al., 1990), justifying the simplification that they can be represented as tension springs.

a sinusoidal trajectory. Simulation and experimental results show that the basic motor oscillation signals excite the whole-body dynamics of the robot, which converges to planar periodic gait cycles that correspond to human-like walking and running motions. Interestingly, no sensory feedback is required, and the resulting gaits are purely the outcome of the interaction between the robot's "musculoskeletal" system and the environment.

A number of additions and improvements to JenaWalker I led to JenaWalker II (Seyfarth et al., 2009), which is shown in Fig. 7.3.9. JenaWalker II retains the biarticular compliant elements of its predecessor. However, additional servomotors have been introduced above the hip with the purpose of tuning the rest length of the biarticular springs – namely those representing the rectus femoris, biceps femoris and gastrocnemius muscles – thus allowing for offline and online postural adjustments of the legs. The stride frequency and step length can be adapted at a given speed by appropriately changing the frequency and amplitude of the sinusoidal signal driving the hip motors. Due to torque limitations of the servo motors, only jogging with almost straight knees has been implemented by tuning the gastrocnemius so that an extended foot position was obtained.

Although based on vastly simplifying assumptions regarding the structure of the human musculoskeletal system, the JenaWalkers I and II showed how biarticular muscles – reduced to simple compliant elements – could be implemented in a robotic context to realize walking and jogging motions under the condition that actuation is provided only by proximal motors located at the hip. However, as was mentioned above, other muscle units – such as the SOL, for example – play an important role in generating locomotion power, which cannot be captured by JenaWalkers I and II. Aiming at a bipedal robot that reproduces the human musculoskeletal system more faithfully by combining the biarticular elements of JenaWalkers I and II with additional monoarticular units, the robot BioBiped I has been designed (Radkhah and von Stryk, 2011a,b); see also Fig. 7.3.10. The leg structure of BioBiped I is shown in Fig. 7.3.10, and it features three biarticular elements per leg – as in the JenaWalkers these represent the gastrocnemius, rectus femoris and biceps femoris muscles – and five monoarticular structures. The biarticular elements and the monoarticular tibialis anterior are passive, while the soleus and vasti lateralis are active, realized by series elastic actuators (SEAs) (Radkhah and von Stryk, 2011a). The hips roll and pitch degrees of freedom are driven by bidirectional SEAs or bionic drives with fixed elastic elements but adjustable quasistiffness through active compliance. The purpose of these actuators is to emulate the function of the GMAX and the iliopsoas (ILIO) muscles; more details on the bidirectional SEAs can be found in Chapter 8. The trunk is allowed to freely lean forward and backward. All in all, the robot has nine degrees of freedom and nine motors. Finally, in

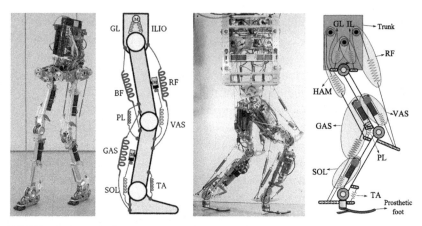

FIGURE 7.3.10 (Left) The BioBiped I, and (middle-left) its leg actuation architecture; gray and purple indicating active and passive components, respectively (Radkhah and von Stryk, 2011a). (Middle-right) BioBiped III with active biarticular muscles (right) and its actuation architecture.

contrast to the JenaWalkers, which did not require any sensory feedback, Bio-Biped features an inertial measurement unit, encoders at every joint, as well as force contact sensors at the heel and ball (Radkhah et al., 2010). In preparation to running, vertical hopping motions were realized on BioBiped I by suitably constraining the robot's pelvis motion along the vertical hopping direction. The resulting hopping motions feature both alternating and synchronous leg movements, with flight phase durations up to 200 ms with a duty factor of about 0.5 and ground clearance of up to 5 cm (Maufroy et al., 2011). In BioBiped III as the latest version of BioBiped series robots the hip lateral DoF is removed and three SEAs are added for biarticular muscles (RF, BF, and GA) to focus on their contributions in stance, swing and balance control in sagittal plane (Sharbafi et al., 2016).

The aforementioned bipeds are actuated by electrical motors and feature compliant elements to reproduce certain structural characteristics of the human musculoskeletal system, such as biarticular muscles. In parallel to the development of these electrically actuated systems, a number of bipedal robots driven by artificial muscles has been introduced. Although, the majority of these robots utilized antagonistic artificial muscle pairs to drive joints, initially they did not include biarticular muscles. One of the earliest robots that uses pneumatically driven artificial muscles in both monoarticular and biarticular configurations was the monopod Que-Kaku-K (an air-leg in Japanese), which is depicted in Fig. 7.3.11 and detailed in Hosoda et al. (2010). As shown in Fig. 7.3.11, this robot closely follows the musculoskeletal structure of the human leg in the sagittal plane, featuring 9 muscles organized in 5 pairs. In accordance to the dis-

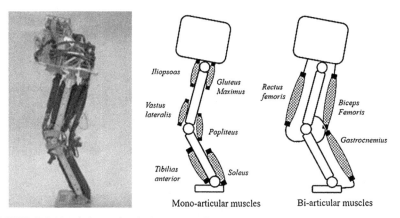

Mono-articular muscles   Bi-articular muscles

FIGURE 7.3.11 A human-inspired monopod Que-Kaku-K. It has 6 monoarticular muscles (3 pairs) and 3 biarticular muscles (Hosoda et al., 2010).

cussion in Section 7.3.2.1, the biarticular muscles in Que-Kaku-K were mainly used for power transfer, while the more powerful monoarticular muscles – especially the antigravity ones – generated the power necessary to lift the body up. A number of hopping experiments have been performed by changing the tensile forces developed by the biarticular GAS and investigating its influence on the direction of the hoping motion. The controller design does not rely on any given desired trajectories; instead, only the balance of the biarticular muscles has been altered to realize changes in the hopping direction, keeping the control of the antigravity muscles unchanged for propulsion. It is important to mention that the proposed controller does not utilize intense feedback, thus the computational cost for realizing hopping is small. Hence, although pneumatic artificial muscles have low bandwidth capabilities, fast hopping motions have been realized in Que-Kaku-K without the need of feedback control. An extension of Que-Kaku-K is Pneumat-BB (Narioka et al., 2013), a planar bipedal walker whose muscles are all actuated by pneumatics. As was mentioned above, antigravity muscles are those primarily responsible for propulsion, but the antagonistic muscles also play an important role in changing joint compliance. As in passive bipedal walkers, Pneumat-BB has been designed so that its natural dynamics is suitably tuned for walking motions. As a result, in closed loop with a simple control law, the robot was able to realize reliable and efficient limit-cycle walking gaits on a treadmill for more than 15 minutes.

To further investigate the influence of joint compliance on locomotion, the bipedal robot Pneumat-BR shown in Fig. 7.3.12 (left) has been designed. Each joint of Pneumat-BR is driven by a pair of muscles and thus the robot's joint compliance can be easily modified. As a result, a wider range of behaviors can be realized. Indeed, Pneumat-BR is a 3D biped capable of walking, jumping

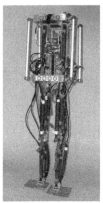

FIGURE 7.3.12 Pneumat-BR, a biped runner (left) and Pneumat-BS, a human-like compliant biped (right).

and running motions through suitably tuned joint compliance. Since this robot is not confined to move in the sagittal plane, a stretch reflex controller in Soleus has been incorporated to maintain frontal balance (Rosendo et al., 2015). Note that the musculoskeletal structure of the robot enables us to evaluate the efficacy of such bioinspired control designs, rendering such robots an excellent dynamic model of a human. Along the same philosophy with Pneumat-BR, the biped Pneumat-BS shown in Fig. 7.3.12 (right) has been constructed. Up to date, Pneumat-BS is perhaps the closest robot to the musculoskeletal structure of a human (Ogawa et al., 2011). It has a total of 40 pneumatic artificial muscles, and, owing to its leg actuation structure, is also capable of 3D locomotion through biarticular muscles that extend form the pelvis to the knee joint and contribute to knee flexion, hip flexion, hip external rotation, and hip abduction. The complexity of Pneumat-BS calls for more advanced control strategies than those derived on basis of simplified functional locomotion models like the SLIP, which cannot fully capture the fundamental behaviors of the robot. Despite this complexity however, the close resemblance of the musculoskeletal structure of this robot with that of the human legs makes it an ideal platform for studying the biomechanics of human locomotion in a more controlled setting.

7.3.3 CONCLUSIONS

This section provided a brief overview of a different design philosophies for human-inspired bipedal robots. Inspired by human locomotion, the objective of these bipedal machines is to produce dependable walking motions in typical human-centric and natural environments. Through this overview, we emphasized the fact that reproducing the morphology of the human figure on a robot is far from being a sufficient condition for capturing the dynamics of human walking. For example, despite their mechanical simplicity, passive dynamic walkers can capture the energy transformations underlying human walking motions more accurate than more anthropomorphic robot designs. Furthermore, decisions on the actuation architecture and on the control system are fundamental to how the robot interacts with its environment, and thus significantly affect the locomotion behaviors it can generate. One recent approach to actuation design which is inspired by biological systems relies on the use of biarticular actuator units – namely actuators that span more than one joints – thus marking a point of departure to traditional robot actuation design. The use of biarticular muscles in robot design allows proximal placement of heavy motors and enables the transmission of work to the extremities, all the while facilitating multi-joint coordination. Both electrically and pneumatically actuated bipeds have been constructed that employ biarticular muscle units, with encouraging results in reproducing aspects of human locomotion dynamics.

REFERENCES

Ames, A.D., 2014. Human-inspired control of bipedal walking robots. IEEE Trans. Autom. Control 59, 1115–1130.

Collins, S., Ruina, A., Tedrake, R., Wisse, M., 2005. Efficient bipedal robots based on passive-dynamic walkers. Science 307, 1082–1085.

Hansen, A.H., Childress, D.S., Knox, E.H., 2004. Roll-over shapes of human locomotor systems: effects of walking speed. Clin. Biomech. 19, 407–414.

Hosoda, K., Sakaguchi, Y., Takayama, H., Takuma, T., 2010. Pneumatic-driven jumping robot with anthropomorphic muscular skeleton structure. Auton. Robots 28, 307–316.

Iida, F., Rummel, J., Seyfarth, A., 2008. Bipedal walking and running with spring-like biarticular muscles. J. Biomech. 41, 656–667.

Iida, F., Minekawa, Y., Rummel, J., Seyfarth, A., 2009. Toward a human-like biped robot with compliant legs. Robot. Auton. Syst. 57, 139–144.

Ito, M., Tsutomu, M., 1983. Current state of biped robots. J. Robotics Soc. Jpn. 1, 44–52.

Jacobs, R., Bobbert, M., van Ingen Schenau, G.J., 1993. Function of mono- and biarticular muscles in running. Med. Sci. Sports Exerc. 25, 1163–1173.

Kajita, S., Tani, K., 1991. Study of dynamic biped locomotion on rugged terrain-derivation and application of the linear inverted pendulum mode. In: Proceedings of the 1991 IEEE International Conference on Robotics and Automation. IEEE, pp. 1405–1411.

Kajita, S., Kanehiro, F., Kaneko, K., Fujiwara, K., Yokoi, K., Hirukawa, H., 2002. A realtime pattern generator for biped walking. In: Proceedings of the IEEE International Conference on Robotics and Automation, ICRA'02, vol. 1. IEEE, pp. 31–37.

Kajita, S., Kanehiro, F., Kaneko, K., Fujiwara, K., Harada, K., Yokoi, K., Hirukawa, H., 2003. Biped walking pattern generation by using preview control of zero-moment point. In: Proceedings of the IEEE International Conference on Robotics and Automation, 2003, ICRA'03, vol. 2. IEEE, pp. 1620–1626.

Kato, I., 1983. Bipedal walking robots. J. Robotics Soc. Jpn. 1, 164–166.

Klein, T., Pham, T., Lewis, M., 2008. On the design of walking machines using biarticulate actuators. Adv. Mob. Robot. 1, 229–237.

Lim, H.-o., Takanishi, A., 2007. Biped walking robots created at Waseda University: WL and WABIAN family. Philos. Trans. R. Soc., Math. Phys. Eng. Sci. 365, 49–64.

Margaria, R., 1976. Biomechanics and Energetics of Muscular Exercise. Clarendon Press, Oxford.

Maufroy, C., Maus, H.-M., Radkhah, K., Scholz, D., von Stryk, O., Seyfarth, A., 2011. Dynamic leg function of the biobiped humanoid robot. In: International Symposium on Adaptive Motions of Animals and Machines.

McGeer, T., 1990a. Passive dynamic walking. Int. J. Robot. Res. 9, 62–82.

McGeer, T., 1990b. Passive walking with knees. In: Proceedings of the IEEE International Conference on Robotics and Automation, vol. 3. Cincinnati, OH, pp. 1640–1645.

Miura, H., Shimoyama, I., 1984. Dynamic walk of a biped. Int. J. Robot. Res. 3, 60–74.

Narioka, K., Homma, T., Hosoda, K., 2013. Roll-over shapes of musculoskeletal biped walker. Automatisierungstechnik 61, 4–14.

Ogawa, K., Narioka, K., Hosoda, K., 2011. Development of whole-body humanoid Pneumat-BS with pneumatic musculoskeletal system. In: Proceedings of the IEEE/RSJ International Conference on Intelligent Robots and Systems.

Pandy, M., Zajac, F., Sim, E., Levine, W., 1990. An optimal control model for maximum-height human jumping. J. Biomech. 23.

Radkhah, K., von Stryk, O., 2011a. Actuation requirements for hopping and running of the musculoskeletal robot BioBiped 1. In: Proceedings of the IEEE/RSJ International Conference on Intelligent Robots and Systems.

Radkhah, K., von Stryk, O., 2011b. Concept and design of the BioBiped robot for human-like walking and running. Int. J. Humanoid Robot. 48, 439–458.

Radkhah, K., Scholz, D., von Stryk, O., Maus, M., Seyfarth, A., 2010. Towards human-like bipedal locomotion with three-segmented elastic legs. In: Proceedings of the joint International Symposium on Robotics and the German Conference on Robotics, pp. 696–703.

Rosendo, A., Liu, X., Shimizu, M., Hosoda, K., 2015. Stretch reflex improves rolling stability during hopping of a decerebrate biped system. Bioinspir. Biomim. 10, 016008.

Sciavicco, L., Siciliano, B., 1996. Modeling and Control of Robot Manipulators. McGraw-Hill.

Seyfarth, A., Iida, F., Tausch, R., Stelzer, M., von Stryk, O., Karguth, A., 2009. Towards bipedal jogging as a natural result of optimizing walking speed for passively compliant three-segmented legs. Int. J. Robot. Res. 28, 257–265.

Sharbafi, M.A., Rode, C., Kurowski, S., Scholz, D., Möckel, R., Radkhah, K., Zhao, G., Rashty, A.M., von Stryk, O., Seyfarth, A., 2016. A new biarticular actuator design facilitates control of leg function in BioBiped3. Bioinspir. Biomim. 11, 046003.

van Ingen Schenau, G.J., 1989. From rotation to translation: constraints on multi-joint movements and the unique action of bi-articular muscles. Hum. Mov. Sci. 8, 301–337.

van Ingen Schenau, J.G., Bobbert, M., Rozendal, R., 1987. The unique action of bi-articular muscles in complex movements. J. Anat. 155, 1–5.

van Ingen Schenau, G.J., Bobbert, M., Rozendal, R., 1990. The unique action of bi-articular muscles in leg extensions. In: Winter, J., Woo, S.L. (Eds.), Multiple Muscle Systems. Springer-Verlag, New York, pp. 639–652.

Winter, D.A., 1990. Biomechanics and Motor Control of Human Movement. John Wiley & Sons, New York.

Wisse, M., van der Linde, R.Q., 2007. Delft Pneumatic Bipeds. Springer Tracts in Advanced Robotics, vol. 34. Springer-Verlag, Berlin, Heidelberg.

Wisse, M., Schwab, A.L., van der Linde, R.Q., van der Helm, F.C.T., 2005. How to keep from falling forward: elementary swing leg action for passive dynamic walkers. IEEE Trans. Robot. 21, 393–401.

Chapter 7.4

Bioinspired Robotic Quadrupeds

Ioannis Poulakakis

Department of Mechanical Engineering, University of Delaware, Newark, DE, United States

Our focus in this chapter will be on dynamically-moving quadrupedal robots. Compared to bipedal and hexapedal robots, robotic quadrupeds offer a good tradeoff among (i) stability, (ii) load-carrying capacity, and (iii) mechanical complexity. Such properties render machines of this kind an attractive alternative to conventional vehicles in many real-world applications that require enhanced mobility and versatility. This has been recognized by the robotics community, and sophisticated quadrupeds have been constructed; see, for instance (Raibert et al., 2008; Wooden et al., 2010) for the recently constructed, highly mobile, BigDog and also (Raibert, 1990; Poulakakis et al., 2005; Nichol et al., 2004) for earlier robot designs. The objective of this section is to provide a brief description of the role of the torso in dynamic quadrupedal locomotion, and to offer an account on how biological observations in the context of quadrupedal animals provided inspiration for the design of robotic quadrupeds. It is important to emphasize upfront that our use of the term "bio-inspiration" does *not* suggest an effort to copy the exact mechanisms by which animals achieve a desired behavior in their natural environments. Nature uses its own means – shaped through millions of year of evolution – to realize a wealth of locomotion behaviors depending on the circumstances animals face in their everyday lives. As was discussed in the previous chapters, more often than not the mechanisms available to roboticists in their effort to create machines capable of realizing animal-like behaviors differ substantially from those employed by nature. Hence, in this chapter "bio-inspiration" refers to extracting the principles underlying a behavior of interest – in our case, a desired quadrupedal gait – and to exploiting the properties of the means technology provides to reproduce this behavior on a suitably designed robot.

7.4.1 PRELIMINARIES ON GAITS

The term *gait* simply means the way animals – or, in our case, bioinspired robots – move. The study of gaits has a long and rich history, and gait descriptions have been available since the late 18th century with the work of Goiffon and Vincent (1779), who studied the gaits of horses by recording the ring pattern of a bell system attached to their hooves. A detailed account of the history of gait studies would lead us too far astray; yet, it is important for our subsequent discussion of quadrupedal robots to provide a brief description of some common quadrupedal gaits that are also employed by running robots. More detailed descriptions of gaits can be found in the relevant literature. The interested reader is referred to the pioneering work of Hildebrand (1965, 1977), as well as to Gambaryan (1974, Chapter 2) and to Bertram (2016, Chapter 2) for a more recent account; mathematical treatments are also available in McGhee (1968) and Collins and Stewart (1993).

When a legged animal or robot moves, its legs exhibit a progressive and retrogressive motion with respect to the body. According to Muybridge (1957), a *step* is an act that involves the motion of a leg as it goes through its regular functions in the course of supporting and propelling the body; that is, a leg contacts the ground to provide support and generate propulsion, then lifts off the ground and swings forward in preparation for the next step. A *stride* on the other hand is a combination of actions that involves all legs, moving either alone or in association with other legs. It begins and ends with two consecutive footfalls of some reference leg, and it includes the contact of all other legs in between (Bertram, 2016, Chapter 2). Essentially, the term stride is used to denote the fundamental repeating pattern within a regular gait.

Depending on the movement of the legs within a stride, a wide variety of gaits can be observed. Gaits in which the footfalls of the left and right legs in a pair – front or hind – are equally spaced in time are called "symmetrical"; in these gaits, the legs in a pair contact the ground half a stride out of phase with each other. According to this definition, the walk, the trot and the pace presented in Fig. 7.4.1 are examples of symmetrical gaits. As can be seen in Fig. 7.4.1A, in the walk, a front leg contacts the ground after the hind leg of the same side (lateral sequence) has touched down, forming a sequence of independent footfalls that enhances stability. On the other hand, legs on the same – in the pace – or opposite – in the trot – sides of the body swing more or less in unison, resulting in a sequence of coupled footholds; see Figs. 7.4.1B and C, respectively. The coordinated use of legs on the same side in pacing helps avoid interference between the front and hind legs in animals with long limbs. Coupling diagonal legs in trotting causes the line of support formed by the feet in contact with the ground to pass near the center of mass, facilitating animals with splayed leg postures or wide bodies. Hildebrand (1965) devised an effective system for quantifying and classifying symmetrical quadrupedal gaits on the basis of a minimal set of variables; namely, the stride duration, the duty factor and the relative phase. The *duty factor* of a foot is the fraction of the stride duration over which that foot is in contact with the ground. Typical "walks" have duty factors more than 50%, implying that there is always at least one leg providing support, as in the walking gait of Fig. 7.4.1. Typical "runs", on the other hand, have duty factors less

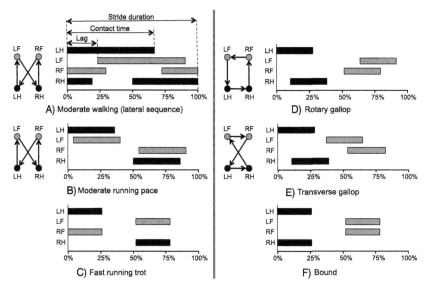

FIGURE 7.4.1 Examples of symmetrical (left) and asymmetrical (right) gaits. The horizontal bars indicate the period over which the corresponding leg is in contact with the ground. "L" stands for left, "R" for right, "H" for hind, and "F" for front. The arrows in the footfall patterns on the left of the gait diagrams show the succession of the legs touching the ground; e.g., starting with the "LH" leg in the rotary gallop the "RH" touches the ground next followed by the "RF" and then the "LF" legs in a sequence that rotates around the torso's center, explaining the term "rotary gallop." Footfall patterns for the trot and bound gaits are not shown since the legs forming a pair touch the ground in unison, and cannot be distinguished.

than 50%, indicating that there are periods in running when all the legs are in the air (Collins and Stewart, 1993); examples include the running pacing and trotting gaits of Fig. 7.4.1. In symmetrical gaits, knowledge of the stride duration and the duty factor of one of the legs in a pair – front or hind – completely determines touchdown and liftoff of both legs within that pair. Yet, a number of different gaits emerge depending on the temporal relationship between the front and hind leg pairs. To capture this effect, the *relative phase* defined as the fraction of the stride duration from the footfall of a rear leg until the front leg on the same side of the body strikes the ground was proposed in Hildebrand (1965).

Contrary to symmetrical gaits, gaits in which the footfalls of the legs forming a pair – front or hind – are unevenly spaced in time are called "asymmetrical" (Hildebrand, 1977). In this case, we distinguish between the legs in a pair, calling the leg that contacts the ground first[6] *trailing* and the leg that touches down next *leading*. Examples of asymmetrical gaits include the transverse and

---

6. The terminology emphasizes spatial position over temporal succession, and may appear confusing since the leading leg contacts the ground after the trailing leg.

rotary variations of the gallop and the bound; see Fig. 7.4.1. In the rotary gallop, which is typical of the Cheetah, the leading legs of both the front and hind pairs lie on opposite sides of the body, and the succession of footfalls "rotates" around the body, as shown in Fig. 7.4.1. Conversely, in the transverse gallop, which is characteristic of the Horse, the leading legs of both the front and rear pairs lie on the same side. A limiting case in which there is no phase lag between the legs within a pair – and thus no distinction to trailing and leading legs is needed – corresponds to the bounding gait of Fig. 7.4.1, variations of which are observed in smaller quadrupeds. Note that the bound is left–right symmetric, although it is classified as asymmetrical in the sense of Hildebrand (1977) due to the fact that the touchdown events of the legs forming a pair are not equally spaced in time. As expected, more parameters than those used to quantify symmetrical gaits are needed in the description and classification of asymmetrical gaits; see Hildebrand (1977) for details. Asymmetrical gaits are typically employed at high speeds and are characterized by the emergence of prolonged suspension phases, in which all the legs are off the ground so that more distance can be covered in a singe stride. Suspension phases typically occur after the front legs take off the ground. These are often referred to as *gathered* suspension phases due to the configuration of the legs that are collected toward the center of mass of the body; see Fig. 7.4.3 below. On certain occasions, a second suspension phase may emerge after the hind legs take off the ground; this phase is referred to as an *extended* flight phase due to the fact that the legs are stretched out.

Animals employ different gaits depending on desired rate of travel, energy efficiency, stability and maneuverability. In transitioning between gaits, irregular stepping patterns emerge, which are difficult to describe using Hildebrand's approaches (Hildebrand, 1965, 1977). An alternative approach proposed in Abourachid (2003) describes the gait as a succession of cycles, each composed by sequences of contacts starting with the trailing front leg and propagating in an anterior–posterior fashion to the hind legs. The method does not rely on the existence of clearly defined temporal boundaries of the gait pattern, and it can be used to describe symmetrical and asymmetrical gaits (Abourachid, 2003) as well as irregular gaits patterns (Abourachid et al., 2007) in a concise way. Although the aforementioned gait description methods have been quite useful in classifying and comparing gaits, they do not by themselves offer mechanical explanations of the underlying force generation and energy transformation mechanisms that characterize the emergence of an observed gait behavior[7].

---

7. For example, as was mentioned above, contrary to typical "runs", "walks" have duty factors more than 50%, and thus walking and running could be distinguished merely on the basis of the duty factor. While this distinction may hold most of the times and for most animals, McMahon and Cheng (1990) pointed out in the context of human running that there are cases where it fails. A better criterion for distinguishing walking and running in humans – and one that reflects the underlying

Similarly to their counterparts in the animal world, robotic quadrupeds have demonstrated several different gaits. Many research efforts have focused on statically-stable robots, which avoid tipping by keeping the torso's center of mass over the polygon formed by the feet in contact with the ground; detailed accounts on statically-stable quadrupeds can be found in Song and Waldron (1989); Gonzalez de Santos et al. (2006). In this section, we focus on balancing quadrupedal robots that can tolerate departures from static equilibrium, thereby relaxing the rules on how the legs can be used for support (Raibert, 1986). The most common quadrupedal gaits implemented in robots are the pace, the trot and the bound, footfall patterns of which are depicted in Fig. 7.4.1. Characteristic of these gaits is that pairs of legs – lateral legs in pacing, diagonal legs in trotting and front and hind legs in bounding – are coordinated to work together. Synergies of this kind significantly simplify controller design and implementation by effectively reducing the dimensionality of the system. In fact, the first dynamically balancing quadrupedal robots designed by Raibert and his collaborators in the mid-1980s realized paired gaits (Raibert, 1986). In Raibert's highly influential work, the collective action of a pair of legs that are used in unison is represented by a single *virtual* leg, effectively mapping paired quadrupedal gaits to equivalent bipedal gaits, which can be further simplified to one-foot hopping gaits. This reduction drastically simplifies the control task by allowing controllers for single-leg hoppers to be applied in the context of multi-legged systems, as will be discussed below and is detailed in Raibert (1986, Chapter 4) and in Raibert (1990). On the other hand, only a few robotic implementations of the rotary and transverse gallop gaits can be found in the relevant literature – see the early work in Smith and Poulakakis (2004), for example – reflecting the difficulty associated with the purely three-dimensional nature of these gaits.

7.4.2   THE ROLE OF THE TORSO: OBSERVATIONS FROM BIOLOGY

The torso (or trunk), incorporating the vertebral column and the muscle groups that actuate it, is the main locomotion organ of vertebrates; see Fischer and Witte (2007) and references therein for an evolutionary perspective and Witte et al. (2000, 2004) for potential uses in robotic legged locomotion. In our brief discussion of certain morphological characteristics of four-legged animals that can be of interest to the design of quadrupedal robots, we will distinguish amphibians and sauropsides – particularly, quadrupedal reptiles such as lizards and crocodiles, for example – from mammals. The source of this distinction lies on

---

mechanics and energetics – is that at mid-stride the center of mass of the body is at its highest position in walking while it falls at its lowest position in running (McMahon and Cheng, 1990).

FIGURE 7.4.2 Sprawled-posture lizards and diagonal trotting motions.

the movement of the torso, and on the arrangement of the legs and their connection to the torso. As we will see both configurations inspired the development of machines with different properties.

As they emerged from the aquatic medium, the first terrestrial vertebrates inherited from their fish ancestors the *lateral* undulating bending motions of their bodies. To convert these torso oscillations to propulsive forces on the ground, the legs of sprawled-postured reptilian quadrupeds are arranged so that the proximal limbs (upper arm, thigh) are *horizontally* attached to the torso, while the distal parts (lower arm, shank) are vertically oriented, ending at attachment appendages at the points of contact with the substrate. This configuration favors locomotion patterns in which lateral torso bending motions are synchronized with a diagonally symmetric sequence of limb movements as Fig. 7.4.2 indicates; such patterns proved to be beneficial both in terms of increasing step length and of ensuring stability of the resulting gaits, and are characteristic of slow reptilian quadrupedal locomotion. Evolutionary pressures for faster movement than what the relatively slow, diagonally-symmetric walking gaits could provide resulted in changes in the locomotion rhythm and in differentiation of leg function with the hindlegs providing most of the propulsive forces. As a result, trot-like gaits with predominance of the rear legs emerged, which at extremely high-speed locomotion give rise to bipedal running.[8] It is interesting to mention that the differential leg function of sprawled-postured reptilian quadrupeds – such as geckos – resembles that of sprawled-posture hexapods – such as cockroaches – more than it resembles that of upright-postured quadrupeds (Chen et al., 2006). In particular, sprawled-postured quadrupeds generate substantial lateral ground reaction forces directed toward the midline of the animal with magnitudes that exceed those of the fore-aft forces, contrary to upright-postured quadrupeds in which the ground reaction forces are primarily concentrated in the fore-aft direction.

By way of contrast to reptilian quadrupeds, four-legged mammals employ *vertical* torso bending movements with their legs arranged vertically, in planes

---

8. As mentioned in Gambaryan (1974) the straddle configuration of the proximal parts of the legs render lateral support phases prone to tipping instability, a fact that does not favor acceleration through normal walking gaits where lateral support phases dominate.

parallel to the sagittal plane. Sagittal-plane bending motions of the spine result in flexions and extensions of the body axis, which become more pronounced in carnivores and small therian mammals, and – at high running speeds – they are associated with the regular use of asymmetrical gaits, such as the galloping gait described in Section 7.4.1. A number of hypotheses have been put forward concerning the contribution of the sagittal bending movements of the torso to locomotion performance. In running quadrupedal mammals these movements are believed to contribute to (i) faster traveling speeds; (ii) reduced metabolic cost; and (iii) improved gait stability. In more detail:

- *Faster traveling speeds.* Running speed is the product of stride length and stride frequency. Early studies (Hildebrand, 1959, 1961) on the cheetah (*Acinonyx jubatus*) indicate how torso flexibility contributes to longer strides and faster stride rates, resulting in the spectacular speeds achieved by these animals. While a detailed account can be found in Hildebrand (1959, 1961), it worth mentioning here that flexion and extension of the torso drastically increases the swing of the legs; see Fig. 7.4.3. As a result, more distance can be covered during the suspension phases, which, in the cheetah, span a substantial proportion of the gait cycle, thus dramatically increasing stride length. Remarkably, as noted in Hildebrand (1959), relative to the shoulder height, the length of the cheetah's stride is more than twice that of the horse.[9] Another point of interest regards the contribution of the muscles of the back to advancing the legs more rapidly on the recovery stroke. As pointed in Hildebrand (1959), two independent groups of muscles – one located in the back and the other inserted in the limbs – accelerate the segments of the legs as they "unfold" during the swing phase. Compared to the case where only one muscle group – the intrinsic limb muscles – recirculates the legs, recruiting two muscle groups effectively accelerates the swing of the leg by sequentially adding the relative velocities of its segments (Hildebrand, 1959).
- *Reduced metabolic cost.* It has been hypothesized that elastic structures located in the back of galloping mammals can store and return in elastic recoil part of the internal kinetic energy[10] that is required to recirculate the legs; see Alexander (1988b) and Alexander (2003, pp. 125–128) for details. The

---

9. Note that similar flexion/extension oscillations of the spine are employed by small ancestral mammals to increase the spacial gain per movement cycle; for instance, the quadrupedal mammal Pika (*Ochotona rufescens*) obtains up to 50% of the spacial gain by bending its torso for more than 40° (Fischer, 1998; Witte et al., 2004).

10. The difference between the role of compliance in the torso and that of the legs should be emphasized: the springs in the legs are responsible for saving energy that is used for propulsion – that is, *external* kinetic energy – while the springs in the torso are responsible for saving energy that is used to move parts of the body relative to itself – that is, *internal* kinetic energy; see Alexander (2003) for details.

FIGURE 7.4.3   Cheetah galloping. The length of the cheetah's torso in its maximally flexed configuration is approximately 67% of its length when fully extended. The pronounced flexion and extension of the torso during the gathered and the extended flight phases contribute to the legendary speeds achieved by cheetahs.

aponeurosis of the principal extensor muscle in the back and the lumbar part of the vertebral column have been proposed as sites for elastic energy storage (Alexander et al., 1985). Energy storage in the back may explain the economy of asymmetrical gaits – such as galloping – that are employed at speeds that are higher than those typically realized by symmetrical gaits. Indeed, as the speed increases, larger fluctuations in the internal kinetic energy are observed due to the fact that the legs swing back and forth at faster rates. Switching to asymmetrical gaits at higher speeds further amplifies the fluctuations of the internal kinetic energy due to the oscillations of the torso. Hence, if no elastic energy could be stored in the back, the emergence of asymmetrical gaits could only be justified at extreme speeds, where energy economy is not a concern. However, such gaits are not only observed at the highest running speeds, but also at more modest speeds. A plausible explanation of this fact is that the increased internal kinetic energy fluctuations are balanced – at least in part – by elastic energy storage in the back that facilitates the forward and backward swinging of the legs. Because this energy storage mechanism is available only in asymmetrical gaits – through the coordination of the legs with the movement of the torso – such motions become energetically more attractive than symmetrical gaits at relatively higher speeds (Alexander, 1988b).

- *Improved gait stability.* The coordinated motion of the torso and legs in asymmetrical gaits has been suggested as a means for enhancing stability in high-speed locomotion. Recent studies of the kinematics of the spinal motion in small mammals implementing asymmetrical gaits in Hackert (2002); Hackert et al. (2006); Schilling and Hackert (2006) reveal that sagittal spine movements may improve gait stability through the effective implementation of self-stabilizing mechanisms that rely on swing leg retraction (Seyfarth et al., 2002, 2003). In particular, using high-speed cineradiography to study the intervertebral joint movements, it was observed in Schilling and Hackert (2006) that the maximum flexion of the spine occurs during the last third of the swing phase implying that spinal extension begins prior to touchdown, thereby resulting in a retraction of the pelvis and hence of the rear legs –

either the trailing or leading leg of the rear pair in galloping or both the rear legs in the half bound – as they strike the ground. Leg retraction during the later stages of the swing phase, in anticipation of touchdown, is known to significantly improve stability, as was observed based on numerical return maps studies of conservative spring–mass models like the spring-loaded inverted pendulum (SLIP) by Seyfarth et al. (2003). Note also that the sagittal plane movements of the torso in asymmetrical gaits induces cyclic horizontal displacements of the center of mass (COM), which can be interpreted as ways to adjust the optimal leg angle of attack, a mechanism that also contributes to self-stabilization (Hackert, 2002; Hackert et al., 2006).

In general, both the reptilian and mammalian morphologies have inspired the development of successful robotic quadrupeds. In what follows we will focus exclusively in models and robots inspired by the vertical torso and leg arrangement of mammals; for robots inspired by the reptilian morphology see Ijspeert et al. (2007).

7.4.3 MODELING: TEMPLATE CANDIDATES FOR QUADRUPEDAL LOCOMOTION

As we have seen in previous chapters, simple *point-mass*, pendulum-based models, such as the spring-loaded inverted pendulum (SLIP), have been instrumental in uncovering basic principles of legged locomotion. Clearly though, distinct morphological characteristics of quadrupedal runners indicate that the standard SLIP (Blickhan, 1989; Full and Koditschek, 1999) and its immediate extensions (Hyon and Emura, 2004; Cherouvim and Papadopoulos, 2005; Ghigliazza et al., 2005), cannot capture certain key aspects of quadrupedal running;[11] namely, leg sequencing and coordination through the torso's movement. Broadly speaking, two general classes of models have been proposed in the relevant literature to describe the dynamics of running quadrupeds in a template setting; namely, simple *spring–mass models* and *collisional models*. Collisional models have been proposed recently to provide insight in leg sequencing and its implications to the energetics of quadrupedal running gaits; more information about models in this family can be found in the recent book of Bertram (2016). Spring–mass models on the other hand have been found useful in designing controllers for quadrupedal robots, and will be discussed below in more detail.

---

11. This is particularly true in the gallop and bound gaits (Blickhan and Full, 1993). For example, to realize bounding, Raibert et al. (1986) observed that the legs cannot be organized to form a single virtual leg that places the effective point of support close to the torso's COM; see also Murphy (1985) and Raibert (1986, p. 193). As a result, Raibert's original three-part controller that regulates the system's high-level behavior through the notion of the virtual leg had to be modified to control bounding.

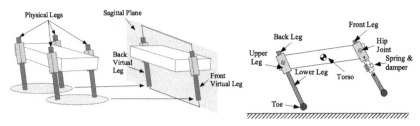

FIGURE 7.4.4 Reducing the dynamics of bounding to a planar spring–mass model. The front and rear virtual legs of the model represent the collective effect of the front and rear physical leg pairs, respectively.

7.4.3.1   Spring–Mass Models for Quadrupedal Locomotion

To keep the development simple, we restrict our attention to the bounding gait; see Fig. 7.4.1. As was mentioned in Section 7.4.1, bounding is an asymmetrical gait, and it can be thought of as a "limiting" case of galloping, where the front and rear physical legs touch and lift off the ground in unison. This assumption greatly simplifies analysis. Indeed, it implies that the essentials of the motion take place in the sagittal plane, and thus the fundamental aspects of the locomotion behavior of the system can be captured by a planar model, as indicated in Fig. 7.4.4.

Simple spring–mass models[12] similar to the one shown in Fig. 7.4.4 have been proposed in the relevant literature to study various aspects of quadrupedal running. An early example can be found in Nanua (1992) and in Nanua and Waldron (1995), with the purpose of studying the energetics of trotting, bounding and galloping. In a way analogous to Fig. 7.4.4, this model was composed of a rigid torso and prismatic, massless, springy legs. Focusing on stability in the sagittal-plane, a similar model was used in Poulakakis et al. (2003, 2006) to explain the success of minimalistic controllers on realizing bounding gaits on the Scout II quadrupedal robot (Poulakakis et al., 2005). Echoing the self-stability of the SLIP (Seyfarth et al., 2002; Ghigliazza et al., 2005), it was found in Poulakakis et al. (2003, 2006) that, for suitable parameters and initial conditions, the model was able to reject perturbations passively, provided – of course – that these perturbations do not alter the total energy of the system. At first sight, it may appear surprising that an activity so apparently complex as bounding can be stably realized in a simple model without employing any control law specifically designed for that purpose. In a way entirely analogous to self-stability

---

12. Note that, at the opposite end of modeling complexity, more detailed quadrupedal models have also appeared in the relevant literature. These models provide a detailed biological account of quadrupedal gaits (Herr and McMahon, 2001), or are grounded to the morphology of specific platform designs (Ananthanarayanan et al., 2012). Due to space limitations, this section does not discuss such higher-dimensional models.

of passive dynamic walkers (McGeer, 1990; Collins et al., 2005) and of the SLIP (Seyfarth et al., 2002; Ghigliazza et al., 2005), this observation indicates that quadrupedal running may simply be a natural mode of a properly tuned mechanical system,[13] thereby explaining the success of simple control laws in exciting and sustaining running motions in quadrupeds like Scout II (Poulakakis et al., 2005). Loosely speaking, these results are in agreement with experimental observations in biology (Kubow and Full, 1999), suggesting that *mechanical feedback* is important in simplifying neural control, and it becomes dominant at top locomotion speeds that challenge the ability of the nervous system to respond on time, as happens in the rapid running cockroach *Blaberous discoidalis* (Full et al., 1998), for example. information

Guided by the structural form of the majority of existing robotic quadrupeds, many of the relevant modeling efforts – including all the models mentioned above – focus predominantly on systems with rigid, nondeformable torsos. On the other hand, as was discussed in Section 7.4.2, torso flexibility may enhance locomotion performance in a number of ways. To investigate the influence of spinal flexion and extension on quadrupedal running, reduced-order, sagittal-plane models with segmented torsos have been introduced. In these models, the torso consists of two segments connected via a spinal joint, as shown in Fig. 7.4.5. An early example of such models can be found in Nanua (1992), which considered a passive flexible spinal joint and massless springy legs. However, the additional degree of freedom in the torso rendered the computation of periodic motions difficult, and lead to the conclusion that torso flexibility without actuation may make the realization of running motions overly complex. Subsequent efforts in Culha and Saranli (2011) and Pouya et al. (2012) focused on actuated spinal joints, while Deng et al. (2012) considered a quasipassive case in which an otherwise passive spinal joint can be "locked" when it reaches its maximum flexion and extension. The possibility of generating bounding in a completely passive setting has been investigated in Seipel (2011) using a model corresponding to the particular geometry of two spring loaded inverted pendula connected through a rotational spring. A similar model was used in Haynes et al. (2012b) which focused only on the stance dynamics without considering cyclic motions and in Satzinger and Byl (2013) which provided preliminary results toward the control of bounding with a passive flexible spine.

To investigate the implications of torso flexibility on quadrupedal running in a unified fashion, a hierarchy of planar models with increasing complexity has been proposed in Cao (2015). These models are presented in Fig. 7.4.5, and

---

13. More generally, how to optimally exploit the inherent properties of the mechanical system in robot design and control has also been examined in the relevant literature; see Remy et al. (2010); Remy (2011), for example.

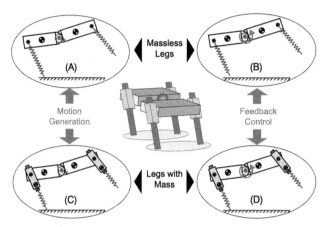

FIGURE 7.4.5 A hierarchy of sagittal-plane models for quadrupedal bounding with a flexible torso. (Center) A three-dimensional virtual prototype. (A) A passive model with massless springy legs used to generate self-stable motions; (B) The template in (A) with an input toque at the torso joint for feedback coordination. (C) The template in (A) with nontrivial leg mass and hip actuation to capture energy efficiency; (D) The template in (C) with a control input at the torso joint for feedback coordination.

they all include compliant legs and a segmented flexible torso. At one end of the complexity spectrum, the conservative model of Fig. 7.4.5A has been proposed in Cao and Poulakakis (2012) to study conditions under which bounding can be generated passively, as the outcome of a properly tuned compliant mechanical system. Furthermore, for suitable torso and leg stiffness values, the existence of passively *stable* cyclic bounding motions has been established in Cao and Poulakakis (2013a). It worth noting that the emergence of self-stability in the context of the model of Fig. 7.4.5A is not as straightforward as it may seem given the self-stable motions in the SLIP (Seyfarth et al., 2002; Ghigliazza et al., 2005) and in the rigid-torso bounding model of Poulakakis et al. (2003, 2006). This is due to the additional degree of freedom in the torso of the models of Fig. 7.4.5, which introduces a large number of spurious solutions that do not correspond to physically relevant bounding motions. At the next level of the modeling hierarchy, an active torque component is introduced in parallel to the torso spring of the model in Fig. 7.4.5A, resulting in model Fig. 7.4.5B. The purpose of this input is to actively coordinate the torso's flexion and extension oscillations in response to the motion of the legs, thereby enhancing the stability and robustness of passively generated bounding gaits (Cao and Poulakakis, 2013a,b). With this being the only input available, it was shown in Cao and Poulakakis (2013a,b) that significantly large disturbances can be rejected with minimal control effort, further supporting the idea that a properly tuned mechanical system is essential in simplifying feedback control.

In the spirit of the SLIP, the models in Figs. 7.4.5A and 7.4.5B feature massless springy legs. As a result, they cannot capture adequately well the effect of torso flexibility on gait energetics, for they do not address the energy that is required to recirculate the legs during flight in anticipation to touchdown. This cost contributes significantly to the total cost of transport, and should be incorporated in any model that intends to capture the contribution of torso flexibility in the economy of running. To address this need the model of Fig. 7.4.5C has been proposed, which differs from the models of Figs. 7.4.5A and 7.4.5B in that it includes nontrivial leg mass. Comparisons of this model with rigid-torso models in terms of the mechanical cost of transport documented in Cao and Poulakakis (2014, 2015) reveal that torso compliance promotes locomotion efficiency, but only at speeds that are sufficiently high. Furthermore, by considering nonideal torque generating and compliant elements with biologically reasonable efficiency values, Cao and Poulakakis (2015) showed that the flexible-torso model of Fig. 7.4.5C can predict the metabolic cost of transport for different animals, as this cost estimated using measurements of oxygen consumption. The hierarchy of models depicted in Figs. 7.4.5 has been found effective in unifying a number of observations regarding the stability and energetics of quadrupedal running gaits in the presence of torso flexibility, and extensions of these models have been used to study feedback control strategies for gait transitions as in Cao et al. (2015); Cao and Poulakakis (2016); see also Cao (2015). Next, we study the model of Fig. 7.4.5A in more detail, as an example of a template for bounding quadrupeds with compliant torso and legs.

7.4.3.2 A Passive Template Candidate for Bounding With a Flexible Torso

We consider the model of Fig. 7.4.5A, which features a segmented torso with two identical rigid bodies connected via a rotational spinal joint; see Koutsoukis and Papadopoulos (2015) for the case of a prismatic joint at the torso. To simplify the development, we assume that the torque produced by the torsional spring follows Hooke's law

$$\tau_{\text{torso}} = k_{\text{torso}}(\theta_{\text{a}} - \theta_{\text{p}}) \tag{7.4.1}$$

where θ_{a}, θ_{p} are the pitch angles of the two segments, as shown in Fig. 7.4.6, and k_{torso} is its stiffness. As in the SLIP, the mass of each legs is assumed to be negligible, and the legs are represented by prismatic springs with stiffness k_{leg} and natural (uncompressed) length l_0. The interaction between the toe and the ground is modeled as an unactuated, frictionless pin joint. The flexible-torso model of Fig. 7.4.6A can be considered as an extension of the standard

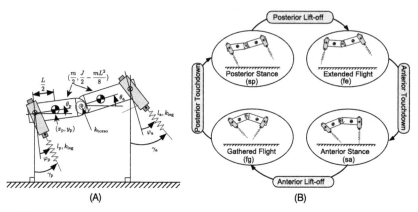

FIGURE 7.4.6   (A) A flexible-torso model for studying bounding. (B) Bounding phases and events.

SLIP, suitable for studying torso–leg coordination in quadrupedal running with a bounding gait. The bounding gait considered here is shown in Fig. 7.4.6B. Depending on the state of the legs – whether on the ground or in the air – and the configuration of the torso, the bounding cycle can be divided into four phases: the posterior stance phase, "sp"; the anterior stance phase, "sa"; the extended flight phase, "fe"; and, the gathered flight phase, "fg." A mathematical model for bounding will thus be composed by the continuous phases mentioned above separated by discrete events. To derive such model – which falls in the category of multiphase hybrid systems – we proceed as follows.

Continuous-Time Dynamics   In each stance phase $i \in \{sp, sa\}$, the configuration space Q_i can be parameterized by the length of the leg in contact with the ground, the corresponding leg angle relative to the torso, and the pitch angles of the torso's segments. Hence, with reference to Fig. 7.4.6A, we have[14]

$$q_{sp} = (l_p, \varphi_p, \theta_p, \theta_a)' \quad \text{and} \quad q_{sa} = (l_a, \varphi_a, \theta_p, \theta_a)' \quad (7.4.2)$$

for the stance-posterior and the stance-anterior phases, respectively. During the flight phases, the configuration space Q_i, $i \in \{fe, fg\}$ can be parameterized by the Cartesian coordinates of the COM of the posterior part of the torso and the corresponding pitch angles. Hence, with the definitions of Fig. 7.4.6A,

$$q_i := (x_p, y_p, \theta_p, \theta_a)' \quad (7.4.3)$$

---

14. Notation: To avoid cluttering, we denote the transpose of a matrix A by A' instead of the commonly used symbol A^{T}.

for $i \in \{$fe, fg$\}$. Defining $x_i = (q_i', \dot{q}_i')$, the dynamics can be written in state-space form as

$$\frac{d}{dt}\begin{bmatrix} q_i \\ \dot{q}_i \end{bmatrix} = \begin{bmatrix} \dot{q}_i \\ -D_i(q_i)^{-1} \left(C_i(q_i, \dot{q}_i)\dot{q}_i + G_i(q_i) \right) \end{bmatrix} \Leftrightarrow \dot{x}_i = f_i(x_i) \ , \quad (7.4.4)$$

where, for each $i \in \{$sp, sa, fe, fg$\}$, $D_i(q_i)$ is the mass matrix, and $C_i(q_i, \dot{q}_i)\dot{q}_i$, $G_i(q_i)$ are vectors containing the centrifugal and Coriolis forces and the gravitational forces, respectively.

The physical parameters in (7.4.4) include the inertia $\{m, J\}$, geometric $\{l_0, L\}$, and stiffness $\{k_{\text{torso}}, k_{\text{leg}}\}$ properties of the model, and their number can be reduced by writing the equations of motion (7.4.4) in non-dimensional form using dimensional analysis. Let τ be the characteristic time scale defined by

$$\tau := \sqrt{l_0/g} \ , \quad (7.4.5)$$

where l_0 is the nominal leg length and g is the gravitational acceleration. Then, the configuration variables (7.4.2)–(7.4.3) and their derivatives with respect to time become

$$\zeta^* := \frac{\zeta}{l_0}, \quad \dot{\zeta}^* := \frac{\tau\dot{\zeta}}{l_0}, \quad \ddot{\zeta}^* := \frac{\tau^2\ddot{\zeta}}{l_0} \ , \quad (7.4.6)$$

for $\zeta \in \{x_{\text{p}}, y_{\text{p}}, x_{\text{com}}, y_{\text{com}}, l_{\text{p}}, l_{\text{a}}\}$ and

$$\psi^* := \psi, \quad \dot{\psi}^* := \tau\dot{\psi}, \quad \ddot{\psi}^* := \tau^2\ddot{\psi} \ , \quad (7.4.7)$$

for $\psi \in \{\varphi_{\text{p}}, \varphi_{\text{a}}, \theta_{\text{p}}, \theta_{\text{a}}, \theta\}$. In (7.4.6)–(7.4.7), the superscript "*" denotes a dimensionless quantity. Substitution of (7.4.6) and (7.4.7) into (7.4.4) results in the dimensionless form of the continuous-time dynamics (7.4.4),

$$\frac{d}{d\tau}x_i^* = f_i^*(x_i^*), \quad (7.4.8)$$

where $x_i^* := ((q_i^*)', (\dot{q}_i^*)')'$ and $i \in \{$sp, sa, fe, fg$\}$. In this dimensionless setting, the parameters of the model are reduced to the following four dimensionless quantities

$$I := \frac{J}{mL^2}, \quad d := \frac{L}{l_0}, \quad \kappa_{\text{torso}} := \frac{k_{\text{torso}}}{mgl_0}, \quad \kappa_{\text{leg}} := \frac{k_{\text{leg}}l_0}{mg}, \quad (7.4.9)$$

corresponding to the dimensionless moment of inertia I, the relative hip-to-COM distance d, and the relative torso κ_{torso} and leg κ_{leg} stiffness. It is emphasized that (7.4.8) does not depend on the choice of units.

Event-Based Transitions The continuous-time phases are separated by the event-based transitions; namely, the touchdown and liftoff events shown in Fig. 7.4.6B. Touchdown occurs when the height of the toe of either the posterior or the anterior leg becomes zero. To realize this condition, we assume that the legs touch the ground at their nominal (uncompressed) length l_0 and that the corresponding touchdown angles are γ_p^{td*} for the posterior and γ_a^{td*} for the anterior leg; the values of these angles will be specified below. Hence, the threshold functions that signify the touchdown of the posterior and the anterior legs are given by

$$H_{fg \to sp}(x_{fg}^*, \alpha^*) = y_p^* - \frac{d}{2}\sin\theta_p^* - \cos\gamma_p^{td*} \quad \text{and}$$

$$H_{fe \to sa}(x_{fe}^*, \alpha^*) = y_p^* + \frac{d}{2}\sin\theta_p^* + d\sin\theta_a^* - \cos\gamma_a^{td*}, \qquad (7.4.10)$$

respectively, where $\alpha^* = (\gamma_p^{td*}, \gamma_a^{td*})'$. To model liftoff, it is assumed that the contact between the ground and the toe of the stance leg is broken when the acceleration of the stance leg end is positive – that is, directed upwards – and the ground force becomes zero. Due to the assumption of massless legs, the stance-to-flight condition can be simplified so that liftoff occurs when the stance leg, as it extends, achieves its natural length l_0. In our dimensionless setting, the zeroing of the threshold functions

$$H_{sa \to fe}(x_{sa}^*) = l_a^* - 1, \quad \dot{l}_a^* > 0 \quad \text{and} \quad H_{sp \to fg}(x_{sp}^*) = l_p^* - 1, \quad \dot{l}_p^* > 0, \tag{7.4.11}$$

signifies transition from the stance-anterior and stance-posterior to the subsequent flight phases, respectively.

Existence and Stability of Cyclic Bounding Gaits To study the existence and stability of bounding gaits according to the phase sequence of Fig. 7.4.6B, the method of Poincaré is used (Guckenheimer and Holmes, 1996). The Poincaré section is taken at the apex height of the extended flight, when the vertical velocity of the torso joint is zero, i.e.,

$$S_{apex}^* := \left\{ x_{fe}^* \in T Q_{fe}^* \mid \dot{y}_p^* + d/2\dot{\theta}_p^* \cos\theta_p^* = 0, \ \theta_a^* > 0 \right\}. \tag{7.4.12}$$

As in Altendorfer et al. (2004) and Poulakakis et al. (2006), since we are interested in periodic bounding gaits, in constructing the Poincaré map we will not consider the horizontal coordinate x_p^* that increases monotonically with time. A further dimensional reduction – inherent to Poincaré's method (Guckenheimer and Holmes, 1996) – can be achieved by projecting out $\dot{y}_p^*$, which always satisfies the condition defining $\mathcal{S}^*_{\text{apex}}$ in (7.4.12). Thus, the state vector is reduced to "locomotion-relevant" variables

$$z^*_{\text{fe}} := (y^*_{\text{p}}, \theta^*_{\text{p}}, \theta^*_{\text{a}}, \dot{x}^*_{\text{p}}, \dot{\theta}^*_{\text{p}}, \dot{\theta}^*_{\text{a}})'. \tag{7.4.13}$$

The Poincaré map $\mathcal{P}^*$ can then be constructed numerically by starting with the states $z^*_{\text{fe}}[k]$ at the kth apex height as initial conditions and integrating forward the dynamics (7.4.8) for all the phases according to the sequence of Fig. 7.4.6B and until the next apex height event occurs. This process results in the nonlinear discrete-time control system

$$z^*_{\text{fe}}[k+1] = \mathcal{P}^* \left(z^*_{\text{fe}}[k], \alpha^*[k] \right). \tag{7.4.14}$$

It worth emphasizing that, despite the fact that the touchdown angles $\alpha^* = (\gamma^{\text{td}^*}_{\text{p}}, \gamma^{\text{td}^*}_{\text{a}})'$ are not part of the state vector (7.4.13), they directly affect the value of $\mathcal{P}^*$. The appearance of $\alpha^* = (\gamma^{\text{td}^*}_{\text{p}}, \gamma^{\text{td}^*}_{\text{a}})'$ in (7.4.14) is a consequence of the dependence of the threshold functions (7.4.10) on the values of the touchdown angles. It is apparent from (7.4.14) that the touchdown angles are (kinematic) inputs available for "cheap" control, since, in general, it is relatively easy to place the legs at their target angles during the flight phases.

With (7.4.14), periodic bounding motions can be computed by solving

$$\bar{z}^*_{\text{fe}} = \mathcal{P}^* \left(\bar{z}^*_{\text{fe}}, \bar{\alpha}^* \right), \tag{7.4.15}$$

to compute a fixed point $\bar{z}^*_{\text{fe}}$ at a given value $\bar{\alpha}^*$. Finally, to analyze the local stability properties of bounding, we linearize (7.4.14) at a fixed point $(\bar{z}^*_{\text{fe}}, \bar{\alpha}^*)$ obtaining the companion system

$$\Delta z^*_{\text{fe}}[k+1] = A \Delta z^*_{\text{fe}}[k] + B \Delta \alpha^*[k], \tag{7.4.16}$$

where $\Delta z^*_{\text{fe}} = z^*_{\text{fe}} - \bar{z}^*_{\text{fe}}$, and $\Delta \alpha^* = \alpha^* - \bar{\alpha}^*$, and $A = \left. \dfrac{\partial \mathcal{P}^*}{\partial z^*_{\text{fe}}} \right|_{z^*_{\text{fe}} = \bar{z}^*_{\text{fe}}, \alpha^* = \bar{\alpha}^*}$, and $B = \left. \dfrac{\partial \mathcal{P}^*}{\partial \alpha^*} \right|_{z^*_{\text{fe}} = \bar{z}^*_{\text{fe}}, \alpha^* = \bar{\alpha}^*}$. When the eigenvalues of A are all within the unit disc, the corresponding fixed point is locally exponentially stable.[15]

Using the aforementioned procedure, a large number of fixed points can be computed and their local stability properties can be evaluated. Fig. 7.4.7A presents snapshots of the bounding motion associated with one such fixed point, and Fig. 7.4.7B shows the corresponding evolution of the relative pitch angle

---

15. Note that one of the eigenvalues is always equal to 1 due to the conservative nature of the system. If all the remaining eigenvalues are within the unit circle, the system is exponentially stable within a constant total energy level.

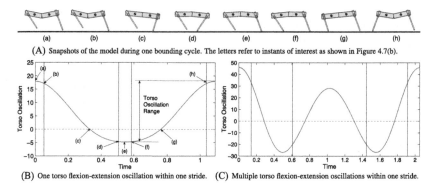

(A) Snapshots of the model during one bounding cycle. The letters refer to instants of interest as shown in Figure 4.7(b).

(B) One torso flexion-extension oscillation within one stride.   (C) Multiple torso flexion-extension oscillations within one stride.

FIGURE 7.4.7 Evolution of the torso bending angle computed as $\theta_a^* - \theta_p^*$ for fixed point with one (Fig. 7.4.7B) and multiple (Fig. 7.4.7C) torso flexion-extension oscillations. The labels in Fig. 7.4.7B correspond to the sequence of phases in Fig. 7.4.7A. From (a) to (h): apex height, anterior leg touchdown, torso flat, anterior leg liftoff, minimum torso bending, posterior leg touchdown, torso flat and posterior leg liftoff.

$\theta_a^* - \theta_p^*$. It can be seen that the anterior and posterior leg stance phases effectively "translate" the configuration of the torso from convex to concave and vice versa in order to prepare the system for the gathered and extended flight phases, respectively. Note, however, that computing passively generated bounding motions that correspond to physically realistic torso bending oscillations – like those in Fig. 7.4.7B – is not a straightforward task. This difficulty has been pointed in previous work by Nanua (1992) and Deng et al. (2012), and is attributed to the sensitive dependence of the motion on the combination between the torso stiffness and the leg stiffness. Even when fixed points can be computed – which is not always the case – they may correspond to spurious motions in which the torso exhibits multiple oscillations within a single stride. Fig. 7.4.7C presents an instance of such motions when the leg stiffness and the torso stiffness are not properly tuned.

Let us consider the effect of the relative torso and leg stiffnesses on the system's motion in more detail, for these parameters are of key importance to the leg–torso coordination. Fig. 7.4.8 shows how the spectral radius $\rho(A) := \max_i |\lambda_i|$ of the matrix A in (7.4.16) changes as a function of the pair $(\kappa_{leg}, \kappa_{torso})$ defined in (7.4.9) keeping the rest of the (dimensionless) parameters constant. The gray area in Fig. 7.4.8 corresponds to periodic motions with torso bending movements similar to those of Fig. 7.4.7C for which multiple torso flexions and extensions exist within one stride. These types of periodic behaviors appear for small values of leg stiffness. Clearly, a softer leg requires a relatively longer time period to go through a complete compression and decompression cycle during stance, thereby allowing the torso to oscillate multiple times within one stride, as in Fig. 7.4.7C. An interesting observation from Fig. 7.4.8 is that

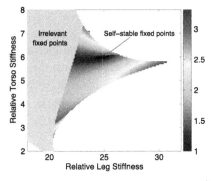

FIGURE 7.4.8 Fixed points computed for the same total energy ($E_t^* = 7.95$), average speed ($Fr = 2.41$), and hopping height ($y_{tj}^* = 0.82$) and for different values of dimensionless leg and torso stiffness. The color code corresponds to the values of the spectral radius of A.

the range of values of the relative leg stiffness over which bounding gaits – not necessarily stable ones – can be generated passively depends strongly on the torso stiffness. Fig. 7.4.8 also illustrates that self-stable bounding motions emerge for particular combinations of the relative torso and leg stiffness. These motions correspond to a small fraction (less than 1%) of the bounding gaits that can be generated passively. Note that self-stability in the presence of a flexible segmented torso does not immediately follow the existence of such self-stable motions in quadrupedal models with rigid torso (Poulakakis et al., 2006; Chatzakos and Papadopoulos, 2009). The reason is that torso bending movements may cause divergent behavior when they are not properly coordinated with the hybrid oscillations of the legs. While, in the rigid-torso case, the inertia properties of the torso – captured by the dimensionless moment of inertia of the torso – dominate self-stability (Poulakakis et al., 2006; Chatzakos and Papadopoulos, 2009), in the flexible-torso case, the combination of the stiffness properties of the legs and the torso appears to be the dominant factor. These observations reflect the significance of suitably tuning the properties of passive elements used in the system so that the desired behavior is generated. Below we will see that the use of passive compliant elements has a number of benefits when it comes to robot design, but it does restrict the versatility of a robot by imposing constraints on how these elements are selected and inserted in the robotic platform.

7.4.4 QUADRUPEDAL ROBOT DESIGN: RIGID OR FLEXIBLE TORSOS?

In striking contrast to their counterparts in nature – which owe much of their remarkable locomotion performance to their flexible torsos and limbs – the vast majority of running robotic quadrupeds incorporate rigid, nondeformable torsos.

Yet, as we will see in this section, some of these robots demonstrated impressive rough terrain mobility (Raibert et al., 2008) and high-speed running (Boston Dynamics, 2013), at a level that is comparable to those of running animals. The performance of these robots brings up a classical dilemma regarding the design and control of bioinspired machines. That is, to what extent should animal morphological characteristics be reproduced in robotic platforms to enable high-performance, natural-like behaviors in these robots?

7.4.4.1   Robots With Rigid Torso

Raibert and his collaborators were the first to report on actively-balancing legged robots, including quadrupeds (Raibert, 1986). Their quadrupedal robot depicted in Fig. 7.4.9A comprises a nondeformable torso and four legs, each having three hydraulically actuated DOFs. Two actuators are located at the hip and they are responsible for placing the leg fore-aft (protraction–retraction) and sideways (abduction–adduction). One actuator is placed inside the leg, and it acts in series with an air spring to maintain the resonant bouncing motion and to shorten the leg after liftoff to ensure adequate toe clearance. Based on the concept of a *virtual leg*, dynamic gaits that employ legs in pairs – namely, trotting, pacing and bounding – have been realized experimentally by mapping them into virtual bipedal gaits (Raibert, 1990). The underlying control algorithm is organized on two levels. On the low level, control action is responsible for synchronizing the physical legs that form a pair, so that their combined action can be represented by an equivalent virtual leg. On the high level, the control system manipulates the virtual legs to regulate task-level variables such as forward velocity, hopping height and torso attitude; details on the control strategy can be found in Raibert (1986, 1990).

In high performance, power autonomous running robots, deciding how to distribute control authority is extremely important. This is not only for energy efficiency, but also for reasons associated with the fact that corrective action in these robots must be developed over time intervals that are sufficiently small to challenge the ability of the control system to react on time. It is thus desirable that the controller works in concert with the *natural dynamics* of the system to generate and sustain the desired motion. For example, to stabilize bounding one might expect that significant control effort should be devoted to regulating the oscillatory pitching motion of the torso. It turns out, however, that this is *not* necessarily true. It depends on the geometric and inertia properties of the torso in a way that is captured by the dimensionless moment of inertia $I = J/mL^2$, where J is the moment of inertia of the torso, m is its mass and L is half the distance between the hips; see also (7.4.9). As was observed in Murphy and Raibert (1985) and further supported theoretically in Berkemeier (1998),

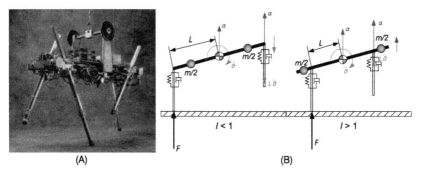

FIGURE 7.4.9 (A) Raibert's quadruped. (B) Bounding and the dimensionless moment of inertia, which effectively describes the "resistance" to rotational versus translational motion due to mass distribution. The distance at which the point masses $m/2$ are located corresponds to the radius of gyration. When the point masses are located between the hips so that $I < 1$, the ground force F transferred through the leg spring at the back hip tends to rotate the torso clockwise more than it pushes it upwards, thereby favoring bounding (Murphy and Raibert, 1985).

if $I < 1$ the natural dynamics favors bounding, and the torso pitch oscillations can be passively stabilized without the need of a feedback control component specifically dedicated to it; Fig. 7.4.9B provides an intuitive explanation of the stabilizing effect of the torso.

Motivated by the need to explore the limits of the mechanical system in stabilizing highly complex dynamic running motions, the Scout II quadruped shown in Fig. 7.4.10 has been introduced (Papadopoulos and Buehler, 2000; Talebi et al., 2001; Poulakakis et al., 2005). Scout II demonstrated efficient bounding gaits using only *one* actuator per leg located at the hip. In more detail, each of Scout's legs consists of a lower and an upper segment connected via a passive, prismatic spring to form a compliant, *unactuated* knee joint, as shown in Fig. 7.4.10. It was found that simple control laws that excite the robot's natural dynamics by merely placing the legs at desired touchdown angles during flight and sweeping them backwards to propel the robot during stance – see Fig. 7.4.11 – are sufficient to generate robust and highly efficient bounding gaits at top speeds 1.2 m/s (Papadopoulos and Buehler, 2000; Talebi et al., 2001; Poulakakis et al., 2005). It worth mentioning that the controller of Fig. 7.4.11 does *not* require any task-level feedback, and the resulting bounding motion is purely due to the interaction of the natural dynamics of the system and its environment.

Scout II is an example of a minimalistic approach to robot design. Beyond mechanical simplicity, the benefits of this approach lie in exposing the key elements that underlie specific locomotion behaviors, and in forcing controllers to exploit the natural dynamics of the system in realizing these motions. Despite its simple minimally-actuated structure, Scout II demonstrated various

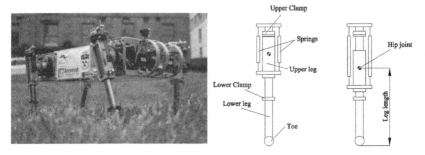

FIGURE 7.4.10 The Scout II quadruped and its leg design.

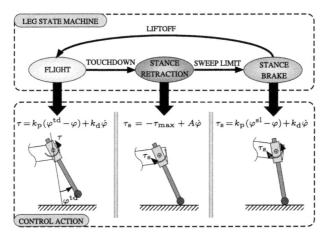

FIGURE 7.4.11 Phases and corresponding control action during bounding with Scout II.

behaviors beyond bounding, including dynamic compliant walking (de Lasa and Buehler, 2001) and step climbing (Talebi, 2000). Similar design and control ideas have subsequently been implemented to generate bounding in a modified (one-actuator-per-leg) version of the SONY AIBO dog (Yamamoto et al., 2001), and in the hexapedal RHex (Campbell and Buehler, 2003). On the other hand, analogously to passive dynamic walkers (McGeer, 1990; Collins et al., 2005), robots like Scout II are often limited in the behaviors they can generate, and modifications to the mechanical platform may be inevitable for realizing different gaits. For example, to generate a walking trot, Scout's legs had to be modified through the addition of a rotational knee joint (Hawker and Buehler, 2000).

Generally, incorporating additional features in a robotic platform – e.g., extra degrees of freedom, actuators or compliant elements – comes at the cost of increasing the complexity of the system. It is thus desirable that this additional complexity contributes to extending the operational capabilities of the

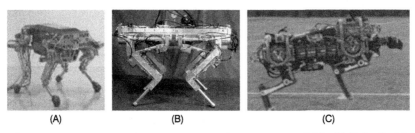

| (A) | (B) | (C) |

FIGURE 7.4.12 Examples of quadrupeds with rigid torso. (A) StarlETH; (B) HyQ; (C) MIT Cheetah v.2.

robot. For example, the ability to actively shorten the swing leg during the early stages of its flight is essential for ensuring sufficient toe clearance, which is necessary, particularly in rough terrain locomotion. Thus, unlike Scout II, the majority of quadrupedal robots incorporate actuators dedicated specifically to shortening the swing leg. Examples include the Patrush (Kimura et al., 1999) and Tekken (Fukuoka et al., 2004) quadrupeds, which, owing to their controller design and compliant structure demonstrated some degree of adaptability to natural terrain (Kimura et al., 2007). The addition of an actuated abduction/adduction degree of freedom further enhances mobility by providing the robot with three dimensional capabilities as it allows placing the legs sideways. As was mentioned above, Raibert's quadruped features legs with actuated abduction/adduction. Other platforms include KOLT (Nichol et al., 2004), which experimentally realized planar pronking and trotting gaits through a differential leg thrust controller (Estremera and Waldron, 2008). The fully-autonomous quadrupedal robot StarlETH shown in Fig. 7.4.12A also features three actuated degrees of freedom per leg (Hutter et al., 2012, 2013). The robot is driven by 12 (three per leg) precisely torque controlled Series Elastic Actuators (SEAs), whose springs essentially act as "mechanical capacitors" that store elastic energy during stance (Remy et al., 2012). StarlETH demonstrated a variety of static and dynamic walking and running gaits under different terrain conditions using a cascade control architecture organized into a motion generator and a motion controller (Gehring et al., 2013, 2014).

Regardless of the leg actuation scheme, a common design choice in all the aforementioned quadrupeds is the use of *passive* elastic energy storage elements – in the form of mechanical or pneumatic springs – intended to recover part of the negative work performed during the early stages of the support phase, where the body lowers and decelerates. Elastic energy storage in compliant elements is of central importance in explaining the mechanics and efficiency of running in animals (McMahon, 1984; Alexander, 1988a). Indeed, ground reaction force profiles and metabolic energy activity measured in experiments with diverse animals cannot explain the high efficiency of running without assuming the existence of elastic energy storage mechanisms (McMahon, 1984; Alexander, 1988a). However, inserting passive compliant elements in a robot's structure entails some critical design decisions, for such elements may limit the system's locomotion capabilities. In more detail, springs introduce low-frequency resonant modes, which have to be tuned according to a specific task. For example, to maximize energy efficiency at a given running speed, the stiffness of the spring needs to be well tuned with the natural dynamics of the system so that the rate at which energy is stored in the spring and returned to the system matches the stride frequency. As a result, extending the range of the achievable running speeds without compromising energy economy requires stiffness adjustment capabilities (Hurst and Rizzi, 2008), which can increase the complexity of the design. Furthermore, even in the case of SEAs where stiffer springs are typically used, the mere existence of the spring reduces the bandwidth of the actuator, thus limiting the use of the system in tasks the require precise movements. Finally, from a control design perspective, if compliance is added in series with actuators, it increases the degree of underactuation, and thus it may reduce the authority of the controller over the system. For example, in compliant leg designs where the spring is inserted in series with a motor – typically the knee motor – the interaction forces between the robot and its environment cannot be regulated in a direct way. In general, as we have pointed out in Section 4.7.6.4, the addition of explicit compliant elements poses strict requirements on controller design so that control action works in concert with the physical springs (Poulakakis and Grizzle, 2007; Poulakakis, 2008).

If energy efficiency were not a concern, the complexity introduced by passively compliant legs could be eliminated. In this case, active regulation of the forces developed at the port of interaction of the robot with its environment could be achieved by exploiting the hip and knee actuators so that each leg behaves as a *virtual* impedance element. Subject to actuator bandwidth limitations, this approach enhances control authority over the system, and it offers additional flexibility since the parameters of the virtual impedance – e.g., the stiffness and damping coefficients – can be adjusted in real time by the control software. As a result, platform versatility is significantly enhanced. An example of a quadrupedal robot that can actively regulate leg impedance is the torque-controlled Hydraulic Quadruped, HyQ (Semini et al., 2011, 2015) depicted in Fig. 7.4.12B. HyQ does not have any physical springs and dampers; the required impedance of the leg is tuned by the control software and implemented through high-performance hydraulic servovalves that enable joint-level torque and position control with excellent tracking capabilities (Boaventura et al., 2012). Experimental studies in Semini et al. (2015) indicate that active compliance can emulate passive elements in the dynamic range relevant to legged locomotion tasks. As a result, this approach can combine the advantages of variable stiffness actuators without the complexity associated with them, and has the potential to enable legged robots to execute a range of difference tasks that require different interaction dynamics with the environment.

We have seen that active impedance control may remove the constraints associated with the use of passive elements at the cost of increasing the energy requirements. But are these two highly desirable requirements mutually exclusive? The electrically actuated MIT Cheetah robot (Hyun et al., 2014) shown in Fig. 7.4.12C proves that this is not the case, and that high locomotion efficiency can be combined with enhanced control authority over the system. Through careful analysis of the energy flow in dynamic legged robots, a set of design principles has been proposed and implemented at the MIT Cheetah robot (Seok et al., 2015). In more detail, three principal sources of energy dissipation have been identified and for each source a solution has been suggested, as follows.

- Heat dissipation losses. Fast, dynamic legged locomotion requires the development of large forces, which entail energy losses in the form of heat at the actuators. For example, when electromagnetic motors are used, the demand for high torque requires larger currents and the energy dissipated due to Joule heating increases. To cope with these losses, the use of high torque-density motors – here, the term refers to high mass-specific continuous torque τ/m – is suggested in Seok et al. (2015) as a means to decrease the current required to provide the necessary torques for the locomotion task. Note also that legged locomotion is an inherently dissipative process and there are inevitable periods where negative work is performed. In the absence of elastic members, the actuators would have to supply the energy lost when negative work is performed. As an alternative solution to mechanical springs for storing energy, the MIT Cheetah incorporates electrical regeneration implemented through a motor driver designed to act like a three-phase generator when the motors perform negative work.

- Transmission losses. These correspond to the energy lost as the actuator torques are transmitted to the joints. To minimize these losses, a low impedance (back-drivable) transmission is adopted, and is realized in the MIT Cheetah through the selection of a single stage, low gear ratio transmission. This design choice results in low reflected inertia and damping. Beyond decreasing the friction losses, keeping the mechanical impedance of the transmission small facilitates proprioceptive force control.

- Interaction losses. These include the energy lost at the impacts of the legs with the ground; that is, at the interface between the robot and its environment. To cope with these losses, a low inertia bioinspired leg design has been proposed and implemented through a composite bone-like structure (Ananthanarayanan et al., 2012), while all actuation has been placed in the torso. The leg architecture is also discussed Subchapter 7.2.

The MIT Cheetah employs programmable leg impedance and high bandwidth control *without* series compliance. High-speed bounding gaits at 6.4 m/s have been realized with a sequentially organized controller that addresses the complexity of the platform (Park et al., 2017). At the core of the controller process lies an impulse-based gait design module, which relies on impulse scaling (Park and Kim, 2015) to provide feed-forward force profiles for a wide range of running speeds. This gait design module is followed by a gait stabilization module and an additional control layer that incorporates implementation specifics; see Park et al. (2017) for details.

7.4.4.2 Robots With Segmented Torsos

Despite consensus that torso flexibility improves performance in quadrupedal mammals, only a few quadrupedal robots incorporate a flexible torso and are capable of animal-like locomotion performance. This is primarily due to the complexity associated with the design and control of systems with segmented torsos in view of the additional degrees of freedom associated with the torso, their mechanical realization and feedback control.

Quadrupedal robots with articulated torsos appeared in the mid-1990s with the work of Leeser at the MIT Leg Lab (Leeser, 1996). Following a minimalistic design approach, the objective of this work was to explore the role of torso articulation in quadrupedal running using the minimum hardware necessary. As a result, the robot was planar – essentially the sagittal-plane half of a quadruped, as shown in Fig. 7.4.13 – constrained to move in the plane by a planarizing mechanism. The articulated torso of the robot had three segments so that both standing and traveling wave behavior could be exhibited, and the corresponding joints were actuated. Finally, as in other MIT Leg Lab robots, this planar quadruped featured hydraulically actuated telescopic legs with air springs located in series with the actuators. This simple arrangement is capable of capturing the dominant leg–torso behavior found in bounding, which as was mentioned in Section 7.4.1 can be considered as a limiting case of gallop; see Fig. 7.4.1. Experiments with the planar quadrupedal design of Fig. 7.4.13 suggest that thrusting with the back can be used to amplify the thrust provided by the legs. Furthermore, the motion of the back effectively modulates the energy of the leg springs and the impedance characteristics of the legs, which has been found useful in developing bounding controllers for the robot.

Contemporary examples of quadrupedal robots with articulated torsos include Canid (Haynes et al., 2012a; Pusey et al., 2013), Bobcat (Khoramshahi et al., 2013; Spröwitz et al., 2013), and the flexible-torso version of the MIT

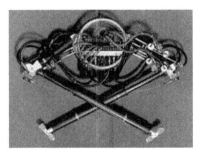

FIGURE 7.4.13 The planar quadruped, one of the earliest robotic quadrupeds with a segmented torso.

Cheetah (Folkertsma et al., 2012). Let us consider these robots in more detail. Bobcat is depicted in Fig. 7.4.14A, and is a relatively small-sized quadruped composed of springy kneed legs and a segmented torso that features a spinal joint actuated by a single RC servo motor. The torso can be reconfigured by locking a pin at the spinal joint so that no relative motion is permitted between the two segments. This simple modification allows for direct comparisons between segmented and rigid torso configurations. A large number of bounding gaits with different spinal joint activities has been generated on Bobcat using a CPG-based controller. Careful analysis of the resulting data revealed that active spinal joint movements tend to increase the maximum achievable speed and that larger amplitudes of the corresponding oscillation contribute directly to higher speeds (Khoramshahit et al., 2013). Furthermore, it was observed that active spinal oscillations resulted in less foot sliding and better directional accuracy, indicating that a segmented torso can enhance gait stability. On the other hand, the realization of faster gaits through spine actuation results at a higher cost of transport, as may have been expected due to the additional actuated degree of freedom in the spinal joint (Khoramshahit et al., 2013). It worth mentioning however that comparisons between bounding gaits generated with fixed and actively controlled spinal joint showed that the average of the horizontal impulse per stride in the latter configuration has been three times lower than that corresponding to fixed spine gaits. Given that an active spinal joint can result in faster gaits, it has been deduced that suitably coordinated torso oscillations allow to redirect the impulse into forward motion, implying better mechanical energy management.[16]

The Canid platform shown in Fig. 7.4.14B has been developed as part of an effort to realize the advantages of elastic elements in the torso in generating

16. Note that this improvement in the mechanical cost of transport did not significantly reduce the electrical (total) cost of transport in Bobcat, which – in general – highly depends on the hardware realization of the platform.

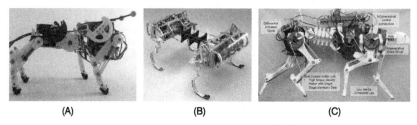

(A)   (B)   (C)

FIGURE 7.4.14 Examples of quadrupeds with segmented torsos: (A) Bobcat; (B) Canid; (C) MIT Cheetah.

highly dynamic locomotion (Haynes et al., 2012a; Pusey et al., 2013). Following the minimalistic design philosophy underlying the Scout II (Poulakakis et al., 2005) and RHex (Saranli et al., 2001) robots, Canid comprises four compliant semicircular (C-shaped) legs with one actuator per leg located at the hip joint. Its segmented torso differs from Bobcat's in that it incorporates an elastic energy storage element – a leaf spring – placed in parallel with the torso actuators (Haynes et al., 2012a; Pusey et al., 2013). The goal of Canid's spine design has been to experimentally assess the contribution of torso compliance in high-power thrusting and energy harvesting, and thus in promoting gait speed and efficiency. An index capturing the robot's body energy has been proposed as an empirically accessible measure of the robot's readiness to perform work rapidly (Pusey et al., 2013). Coupled with an open-loop control strategy that implements predetermined actuator set points, Canid has been capable of highly-energetic, power autonomous leaping behaviors. However, the lack of exact coordination between the posterior legs when they touch the ground resulted in the excitation of undesirable oscillations in the highly elastic spine that severely disturb the robot's motion to the point that steady-state bounding could not be realized. Clearly, the complexity added by the elastic nature of the spine, and the corresponding difficulty in achieving the desired level of coordination between the dorsoventral oscillations of the torso and the movement of the legs – which are also highly compliant structures – call for more powerful closed-loop control laws. Transforming Canid's highly-energetic leaping behaviors into steady-state locomotion is at the core of current research efforts with this platform.

The early design of the MIT Cheetah depicted in Fig. 7.4.14 is one more example of a quadrupedal robot that exhibits torso flexibility. Its spine consists of several hard plastic vertebrae alternated with softer polyurethane intervertebral discs. In contrast to Canid and Bobcat, which incorporate direct torso actuation – either with or without springs – the motion of the torso of the MIT Cheetah is mechanically coupled with its rear legs through a differential mechanism (Folkertsma et al., 2012). Inspired by the coordinated motion between the torso and the legs of a running cheetah – see Fig. 7.4.3 – this mechanism effectively "slaves" the spine so that it flexes or extends in response to the rear legs moving in phase forward (protraction) or backward (retraction), respectively; that is, in-phase rear leg protraction causes spinal flexion while in-phase rear leg retraction causes spinal extension. Furthermore, the differential mechanism ensures that no significant spinal flexion and extension oscillations are produced when the left and right legs of the rear pair move out of phase. Essentially, this mechanically imposed coordination ensures that spinal flexion and extension occurs during bounding and that such oscillations are not present in walking or trotting gaits. Using a programmable impedance controller that recruits the leg motors to actively introduce and tune virtual leg springs, the MIT Cheetah has been able to trot at high speeds. However, since the two rear legs move out of phase in trotting, the flexible torso remained in its neutral configuration and did not contribute to the robot's motion during these experiments (Kim, 2013). Despite the fact that torso bending capabilities were present in this early design of the MIT Cheetah, these capabilities were not exploited in experiments to enhance the robot's performance in terms of gait speed. In fact, subsequent research on the MIT Cheetah resulted in substituting the earlier flexible torso design with a rigid torso – as in the robots of Section 7.4.4.1 – and high-speed bounding motions have been realized in the more recent rigid-torso version of the platform depicted in Fig. 7.4.12C. These results demonstrate that torso flexibility may not be necessary for high-performance quadrupedal robot running.

To summarize the discussion above, the introduction of torso flexibility undoubtedly complicates the mechanical design of the robot. It also complicates the design of the control system, which needs to regulate the additional degrees of freedom to produce reliable locomotion behaviors that realize the potential advantages of the flexible torso. Bobcat, Canid and the flexible-torso version of the MIT Cheetah represent early steps toward the experimental realization of highly dynamic quadrupeds that mimic their biological counterparts in that they feature – *and exploit* – torso flexibility. Along this direction, Boston Dynamics released a video of their hydraulically actuated Cheetah robot galloping on a treadmill with the help of a support mechanism at the record speed of 29 mph (Boston Dynamics, 2013). While this result demonstrates the potential of realizing fast quadrupedal running motions in the presence of a segmented torso, only limited information on how torso bending movements actually affect locomotion is available in the context of this platform. The extend to which torso flexibility enables performance enhancement that is significant enough to overcome the cost associated with the added complexity still remains a mystery.

7.4.5 CONCLUSIONS

This chapter provides a brief overview of a few topics that are relevant to the design of bioinspired robotic quadrupeds, with particular emphasis on the role of the torso. A number of studies in biology and biomechanics suggest that torso movements contribute significantly in locomotion performance. As a concrete example, we have focused on sagittal-plane torso bending motions, which are believed to be of fundamental importance in achieving faster running speeds, in reducing metabolic cost and in enhancing gait stability in quadrupedal mammals when they run at high speeds. A hierarchy of simple mathematical models has been proposed to capture and quantify some of these biological observations. By way of contrast, however, the vast majority of robotic quadrupeds feature rigid, nondeformable torsos. We have examined in this chapter a few of these quadrupeds with varying levels of mechanical complexity and control authority, and with different actuation architectures. These robots constitute representative examples of design philosophies that emphasize different – but not mutually exclusive – objectives, such as fast locomotion, energy efficiency and platform versatility, and have been highly successful in realizing dynamic gaits. On the other hand, much fewer experimental results are available in the context of quadrupedal robots with flexible torsos. Although, torso flexibility appears to be advantageous – especially at high speeds – the specifics of the mechanical realization of a flexible torso undoubtedly increase the complexity of the platform. Realizing the expected performance enhancement is the subject of ongoing research efforts.

ACKNOWLEDGEMENTS

Authors Madhusudhan Venkadesan, Mahesh M. Bandi, and Shreyas Mandre would like to thank the generous funding support from the Human Frontier Science Program. We also thank Marcelo Dias, Dhiraj Singh, Maria Lugo-Bolanos, and Kenneth Meacham III for their assistance in preparing some of the figures. Images of the foot skeleton were produced using Wolfram Research, Inc., Mathematica, Version 11.0, Champaign, IL (2016).

REFERENCES

Abourachid, A., 2003. A new way of analyzing symmetrica and asymmetrical gaits in quadrupeds. C. R. Biol. 367.

Abourachid, A., Herbin, M., Hackert, R., Maes, L., Martin, V., 2007. Experimental study of coordination patterns during unsteady locomotion in mammals. J. Exp. Biol. 210.

Alexander, R., 1988a. Elastic Mechanisms in Animal Movement. Cambridge University Press, Cambridge.

Alexander, R., 1988b. Why mammals gallop. Am. Zool. 28, 237–245.

Alexander, R., 2003. Principles of Animal Locomotion. Princeton University Press, Princeton, NJ.

Alexander, R., Dimery, N., Ker, R., 1985. Elastic structures in the back and their role in galloping in some mammals. J. Zool. 207, 467–482.

Altendorfer, R., Koditschek, D.E., Holmes, P., 2004. Stability analysis of legged locomotion models by symmetry-factored return maps. Int. J. Robot. Res. 23, 979–999.

Ananthanarayanan, A., Azadi, M., Kim, S., 2012. Towards a bio-inspired leg design for high-speed running. Bioinspir. Biomim. 7, 046005.

Berkemeier, M., 1998. Modeling the dynamics of quadrupedal running. Int. J. Robot. Res. 19, 971–985.

Bertram, J.E. (Ed.), 2016. Understanding Mammalian Locomotion: Concepts and Applications. Wiley.

Blickhan, R., 1989. The spring-mass model for running and hopping. J. Biomech. 22, 1217–1227.

Blickhan, R., Full, R.J., 1993. Similarity in multi-legged locomotion: bouncing like a monopode. J. Comp. Physiol., A Sens. Neural Behav. Physiol. 173, 509–517.

Boaventura, T., Semini, C., Buchli, J., Frigerio, M., Focchi, M., Caldwell, D.G., 2012. Dynamic torque control of a hydraulic quadruped robot. In: Proceedings of the IEEE International Conference on Robotics and Automation, pp. 1889–1894.

Boston Dynamics, 2012. CHEETAH – fastest legged robot, March 15, 2013.

Campbell, D., Buehler, M., 2003. Preliminary bounding experiments in a dynamic hexapod. In: Siciliano, B., Dario, P. (Eds.), Experimental Robotics VIII. In: Springer Tracts in Advanced Robotics, vol. 5. Springer-Verlag, Berlin, pp. 612–621.

Cao, Q., 2015. A Modeling Hierarchy of Quadrupedal Running with Torso Compliance. PhD thesis. University of Delaware.

Cao, Q., Poulakakis, I., 2012. Passive quadrupedal bounding with a segmented flexible torso. In: Proceedings of the IEEE/RSJ International Conference on Intelligent Robots and Systems, pp. 2484–2489.

Cao, Q., Poulakakis, I., 2013a. Passive stability and control of quadrupedal bounding with a flexible torso. In: Proceedings of the IEEE/RSJ International Conference on Intelligent Robots and Systems, pp. 6037–6043.

Cao, Q., Poulakakis, I., 2013b. Quadrupedal bounding with a segmented flexible torso: passive stability and feedback control. Bioinspir. Biomim. 8, 046007.

Cao, Q., Poulakakis, I., 2014. On the energetics of quadrupedal bounding with and without torso compliance. In: Proceedings of the IEEE/RSJ International Conference on Intelligent Robots and Systems, pp. 4901–4906.

Cao, Q., Poulakakis, I., 2015. On the energetics of quadrupedal running: predicting the metabolic cost of transport via a flexible-torso model. Bioinspir. Biomim. 10, 056008.

Cao, Q., Poulakakis, I., 2016. Quadrupedal running with a flexible torso: control and speed transitions with sums-of-squares verification. Artif. Life Robot. 21, 384–392.

Cao, Q., Van Rijn, A.T., Poulakakis, I., 2015. On the control of gait transitions in quadrupedal running. In: Proceedings of the IEEE/RSJ International Conference on Intelligent Robots and Systems, pp. 5136–5141.

Chatzakos, P., Papadopoulos, E., 2009. Self-stabilizing quadrupedal running by mechanical design. Appl. Bionics Biomech. 6, 73–85.

Chen, J.J., Peattie, A.M., Autumn, K., Full, R.J., 2006. Differential leg function in a sprawled-posture-quadrupedal trotter. J. Exp. Biol. 209, 249–259.

Cherouvim, N., Papadopoulos, E., 2005. Single actuator control analysis of a planar 3DOF hopping robots. In: Thrun, S., Sukhatme, G., Schaal, S. (Eds.), Robotics: Science and Systems I. MIT Press, pp. 145–152.

Collins, J.J., Stewart, I.N., 1993. Coupled nonlinear oscillators and the symmetries of animal gaits. J. Nonlinear Sci. 3, 349–392.

Collins, S., Ruina, A., Tedrake, R., Wisse, M., 2005. Efficient bipedal robots based on passive-dynamic walkers. Science 307, 1082–1085.

Culha, U., Saranli, U., 2011. Quadrupedal bounding with an actuated spinal joint. In: Proceedings of the IEEE International Conference on Robotics and Automation, pp. 1392–1397.

Deng, Q., Wang, S., Xu, W., Mo, J., Liang, Q., 2012. Quasi passive bounding of a quadruped model with articulated spine. Mech. Mach. Theory 52, 232–242.

de Lasa, M., Buehler, M., 2001. Dynamic compliant quadruped walking. In: Proceedings of the IEEE International Conference on Robotics and Automation, vol. 3, pp. 3153–3158.

Estremera, J., Waldron, K., 2008. Thrust control, stabilization and energetics of a quadruped running robots. Int. J. Robot. Res. 27, 1135.

Fischer, M.S., Lehmann, R., 1998. Application of cineradiography for the metric and kinematic study of in-phase gaits during locomotion of the pika (Ochotona rufescens, Mammalia: Lagomorpha). Zoology 101, 148–173.

Fischer, M.S., Witte, H., 2007. Legs evolved only at the end!. Philos. Trans. R. Soc., Math. Phys. Eng. Sci. 367, 185–198.

Folkertsma, G.A., Kim, S., Stramigioli, S., 2012. Parallel stiffness in a bounding quadruped with flexible spine. In: Proceedings of the IEEE/RSJ International Conference on Intelligent Robots and Systems, pp. 2210–2215.

Fukuoka, Y., Kimura, H., Cohen, A.H., 2004. Adaptive dynamic walking of a quadruped robot on irregular terrain based on biological concepts. Int. J. Robot. Res. 23, 1059–1073.

Full, R., Koditschek, D.E., 1999. Templates and anchors: neuromechanical hypotheses of legged locomotion on land. J. Exp. Biol. 202, 3325–3332.

Full, R.J., Autumn, K., Chung, J.I., Ahn, A.N., 1998. Rapid negotiation of rough terrain by the death-head cockroach. Am. Zool. 38, 81A.

Gambaryan, P.P., 1974. How Mammals Run: Anatomical Adaptations. Wiley, New York.

Gehring, C., Coros, S., Hutter, M., Bloesch, M., Hoepflinger, M.A., Siegwart, R., 2013. Control of dynamic gaits for a quadrupedal robot. In: Proceedings of the IEEE International Conference on Robotics and Automation, pp. 3272–3277.

Gehring, C., Coros, S., Hutter, M., Bloesch, M., Hoepflinger, M.A., Siegwart, R., 2014. Toward automatic discovery of agile gaits for quadrupedal robots. In: Proceedings of the IEEE International Conference on Robotics and Automation, pp. 4243–4248.

Ghigliazza, R.M., Altendorfer, R., Holmes, P., Koditschek, D.E., 2005. A simply stabilized running model. SIAM Rev. 47, 519–549.

Goiffon, M., Vincent, M., 1779. Mémoire artificielle des principes rélatifs à la fidèle représetation des animaux tant en peinture, qu'en sculpture. I partie concernant le cheval (Memoir on the artificial principles of faithful representation of animals both in painting and sculpture. Part I concerning the horse) Alfort (Reference from P.P. Gambaryan, How Mammals Run: Anatomical Adaptations, New York: Wiley, 1974.).

Gonzalez de Santos, P., Garcia, E., Estremera, J., 2006. Quadrupedal Locomotion: An Introduction to the Control of Four-legged Robots. Springer.

Guckenheimer, J., Holmes, P., 1996. Nonlinear Oscillations, Dynamical Systems, and Bifurcations of Vector Fields. Applied Mathematical Sciences, vol. 42. Springer-Verlag, New York.

Hackert, R., 2002. Dynamics of the Pikas' Half-Bound – Spinal Flexion Contributes to Dynamic Stability. Aspects of Sagittal Bending of the Spine in Small Mammals. PhD thesis. Friedrich Schiller University of Jena.

Hackert, R., Schilling, N., Fischer, M.S., 2006. Mechanical self-stabilization, a working hypothesis for the study of the evolution of body proportions in terrestrial mammals?. C. R. Palevol 5, 541–549.

Hawker, G., Buehler, M., 2000. Quadruped trotting with passive knees: design, control, and experiments. In: Proceedings of the IEEE International Conference on Robotics and Automation, vol. 3, pp. 3046–3051.

Haynes, G.C., Pusey, J., Knopf, R., Johnson, A.M., Koditschek, D.E., 2012a. Laboratory on legs: an architecture for adjustable morphology with legged robots. In: Proceedings of the SPIE Defense, Security, and Sensing Conference, Unmanned Systems Technology XIV, vol. 8387.

Haynes, G.C., Pusey, J., Knopf, R., Koditschek, D.E., 2012b. Dynamic bounding with a passive compliant spine. In: Proceedings of the Dynamic Walking Conference.

Herr, H.M., McMahon, T.A., 2001. A galloping horse model. Int. J. Robot. Res. 20, 26–37.

Hildebrand, M., 1959. Motions of the running cheetah and horse. J. Mammal. 40, 481–495.

Hildebrand, M., 1961. Further studies on locomotion of the cheetah. J. Mammal. 42, 84–91.

Hildebrand, M., 1965. Symmetrical gaits of horses. Science 150, 701.

Hildebrand, M., 1977. Analysis of asymetrical gaits. J. Mammal. 58, 131–156.

Hurst, J.W., Rizzi, A., 2008. Series compliance for an efficient running gait. IEEE Robot. Autom. Mag. 15, 42–51.

Hutter, M., Gehring, C., Bloesch, M., Hoepflinger, M.A., Remy, C.D., Siegwart, R., 2012. StarlETH: a compliant quadrupedal robot for fast, efficient, and versatile locomotion. In: International Conference on Climbing and Walking Robots.

Hutter, M., Remy, C.D., Hoepflinger, M.A., Siegwart, R., 2013. Efficient and versatile locomotion with highly compliant legs. IEEE/ASME Trans. Mechatron.

Hyon, S.-H., Emura, T., 2004. Energy-preserving control of a passive one-legged running robots. Adv. Robot. 18, 357–381.

Hyun, D.J., Seok, S., Lee, J., Kim, S., 2014. High speed trot-running: implementation of a hierarchical controller using proprioceptive impedance control on the MIT Cheetah. Int. J. Robot. Res. 33, 1417–1445.

Ijspeert, A.J., Crespi, A., Ryczko, D., Cabelguen, J.-M., 2007. From swimming to walking with a salamander robot driven by a spinal cord model. Science 315, 1416–1420.

Khoramshahit, M., Spröwizt, A., Tuleut, A., Ahmadabadi, M.N., Ijspeert, A.J., 2013. Benefits of an active spine supported bounding locomotion with a small compliant quadruped robot. In: Proceedings of the IEEE International Conference on Robotics and Automation, pp. 3329–3334.

Kim, S., Cheetah-inspired quadruped, March 15, 2013.

Kimura, H., Akiyama, S., Sakurama, K., 1999. Realization of dynamic walking and running of the quadruped using neural oscillators. Auton. Robots 7, 247–258.

Kimura, H., Fukuoka, Y., Cohen, A.H., 2007. Adaptive dynamic walking of a quadruped robot on natural ground based on biological concepts. Int. J. Robot. Res. 26, 475–490.

Koutsoukis, K., Papadopoulos, E., 2015. On passive quadrupedal bounding with flexible linear torso. Int. J. Robot. 4, 1–8.

Kubow, T.M., Full, R.J., 1999. The role of the mechanical system in control: a hypothesis of self-stabilization in hexapedal runners. Philos. Trans. R. Soc. Lond. B, Biol. Sci. 354, 849–862.

Leeser, K.F., 1996. Locomotion Experiments on a Planar Quadruped Robot with Articulated Spine. Master's thesis. Massachusetts Institute of Technology.

McGeer, T., 1990. Passive dynamic walking. Int. J. Robot. Res. 9, 62–82.

McGhee, R.B., 1968. Some finite state aspects of legged locomotion. Math. Biosci. 2, 67–84.

McMahon, T.A., 1984. Muscles, Reflexes, and Locomotion. Princeton University Press.

McMahon, T.A., Cheng, G.C., 1990. The mechanics of running: how does stiffness couple with speed?. J. Biomech. 23 (Suppl. 1), 65–78.

Murphy, K.N., 1985. Trotting and Bounding in a Simple Planar Model. Technical report. Carnegie Mellon University, The Robotics Institute.

Murphy, K.N., Raibert, M.H., 1985. Trotting and bounding in a planar two-legged models. In: Morecki, K.K.A., Bianchi, G. (Eds.), 5th Symposium on Theory and Practice of Robots and Manipulators. MIT Press, MA, pp. 411–420.

Muybridge, E., 1957. Animals in Motion. Dover.

Nanua, P., 1992. Dynamics of a Galloping Quadruped. PhD thesis. Ohio State University.

Nanua, P., Waldron, K., 1995. Energy comparison between trot, bound and gallop using a simple model. J. Biomech. Eng. 117, 466–473.

Nichol, J.G., Singh, S., Waldron, K., Palmer, L., Orin, D., 2004. System design of a quadrupedal galloping machines. Int. J. Robot. Res. 23, 1013–1027.

Papadopoulos, D., Buehler, M., 2000. Stable running in a quadruped robot with compliant legs. In: Proceedings of the IEEE International Conference on Robotics and Automation. San Francisco, CA, pp. 444–449.

Park, H.-W., Kim, S., 2015. Quadrupedal galloping control for a wide range of speed via vertical impulse scaling. Bioinspir. Biomim. 10, 025003.

Park, H.-W., Wensing, P., Kim, S., 2017. High-speed bounding with the MIT Cheetah 2: control design and experiments. Int. J. Robot. Res. 36, 167–192.

Poulakakis, I., 2008. Stabilizing Monopedal Robot Running: Reduction-by-Feedback and Compliant Hybrid Zero Dynamics. PhD thesis. Department of Electrical Engineering and Computer Science, University of Michigan.

Poulakakis, I., Grizzle, J.W., 2007. Monopedal running control: SLIP embedding and virtual constraint controllers. In: Proceedings of the IEEE/RSJ International Conference of Intelligent Robots and Systems. San Diego, USA, pp. 323–330.

Poulakakis, I., Papadopoulos, E., Buehler, M., 2003. On the stable passive dynamics of quadrupedal running. In: Proceedings of the IEEE International Conference on Robotics and Automation, vol. 1, pp. 1368–1373.

Poulakakis, I., Smith, J., Buehler, M., 2005. Modeling and experiments of untethered quadrupedal running with a bounding gait: the Scout II robots. Int. J. Robot. Res. 24, 239–256.

Poulakakis, I., Papadopoulos, E.G., Buehler, M., 2006. On the stability of the passive dynamics of quadrupedal running with a bounding gait. Int. J. Robot. Res. 25, 669–687.

Pouya, S., Khodabakhsh, M., Moeckel, R., Ijspeert, A.J., 2012. Role of spine compliance and actuation in the bounding performance of quadruped robots. In: Proceedings of the Dynamic Walking Conference.

Pusey, J.L., Duperret, J.M., Haynes, G.C., Knopf, R., Koditschek, D.E., 2013. Free-standing leaping experiments with a power-autonomous elastic-spined quadruped. In: SPIE Defense, Security, and Sensing, 87410W.

Raibert, M.H., 1986. Legged Robots that Balance. MIT Press, Cambridge, MA.

Raibert, M.H., 1990. Trotting, pacing and bounding by a quadruped robot. J. Biomech. 23 (Suppl. 1), 79–98.

Raibert, M.H., Chepponis, M., Brown, J.H., 1986. Running on four legs as though they were one. IEEE Trans. Robot. Autom. 2, 70–82.

Raibert, M., Blankespoor, K., Nelson, G., Playter, R., the BigDog Team, 2008. BigDog, the rough-terrain quadruped robot. In: Proceedings of the 17th IFAC World Congress. Seoul, Korea, pp. 10822–10825.

Remy, C.D., 2011. Optimal Exploitation of Natural Dynamics in Legged Locomotion. PhD thesis. Eidgenössische Technische Hochschule Zürich.

Remy, C., Buffinton, K., Siegwart, R., 2010. Stability analysis of passive dynamic walking of quadrupeds. Int. J. Robot. Res. 29, 1173–1185.

Remy, C.D., Hutter, M., Hoepflinger, M., Bloesch, M., Gehring, C., Siegwart, R., 2012. Vierbeinige Laufroboter mit weicher und steifer Aktuierung (Quadrupedal robots with stiff and compliant actuation). Automatisierungstechnik 60, 682–691.

Saranli, U., Buehler, M., Koditschek, D.E., 2001. RHex: a simple and highly mobile hexapod robot. Int. J. Robot. Res. 20, 616–631.

Satzinger, B., Byl, K., 2013. Control of planar bounding quadruped with passive flexible spine. In: International Symposium of Adaptive Motion in Animals and Machines.

Schilling, N., Hackert, R., 2006. Sagittal spine movements of small therian mammals during asymetrical gaits. J. Exp. Biol. 209, 3925–3939.

Seipel, J.E., 2011. Analytic-holistic two-segment model of quadruped back-bending in the sagittal plane. In: Proceedings of ASME Mechanisms and Robotics Conference, vol. 6, pp. 855–861.

Semini, C., Tsagarakis, N.G., Guglielmino, E., Focchi, M., Cannella, F., Caldwell, D.G., 2011. Design of HyQ – a hydraulically and electrically actuated quadruped robot. Proc. Inst. Mech. Eng., Part I, J. Syst. Control Eng. 225, 831–849.

Semini, C., Barasuol, V., Boaventura, T., Frigerio, M., Focchi, M., Caldwell, D.G., Buchli, J., 2015. Towards versatile legged robots through active impedance control. Int. J. Robot. Res. 34, 1003–1020.

Seok, S., Wang, A., Chuah, M.Y., Hyun, D.J., Lee, J., Otten, D.M., Lang, J., Kim, S., 2015. Design principles for energy efficient legged locomotion and implementation on the MIT Cheetah. IEEE/ASME Trans. Mechatron. 20, 1117–1129.

Seyfarth, A., Geyer, H., Gunther, M., Blickhan, R., 2002. A movement criterion for running. J. Biomech. 35, 649–655.

Seyfarth, A., Geyer, H., Herr, H., 2003. Swing leg retraction: a simple control model for stable running. J. Exp. Biol. 206, 2547–2555.

Smith, J.A., Poulakakis, I., 2004. Rotary gallop in the untethered quadrupedal robot Scout II. In: Proceedings of the IEEE/RSJ International Conference on Intelligent Robots and Systems, vol. 3, pp. 2556–2561.

Song, S.-M., Waldron, K.J., 1989. Machines that Walk: The Adaptive Suspension Vehicle. MIT Press, Cambridge, MA.

Spröwitz, A., Badri, E., Khoramshahi, M., Tuleu, A., Ijspeert, A.J., 2013. Use your spine! Effect of active spine movements on horizontal impulse and cost of transport in a bounding, quadruped robot. In: Dynamic Walking.

Talebi, S., 2000. Legged Locomotion in Rough Terrain. Master's thesis. McGill University.

Talebi, S., Poulakakis, I., Papadopoulos, E., Buehler, M., 2001. Quadruped robot running with a bounding gait. In: Rus, D., Singh, S. (Eds.), Experimental Robotics VII. In: Lecture Notes in Control and Information Sciences Series. Springer-Verlag, pp. 281–290.

Witte, H., Hackert, R., Ilg, W., Biltzinger, J., Schilling, N., Biedermann, F., Jergas, M., Preuschoft, H., Fischer, M.S., 2000. Quadrupedal mammals as paragons for walking machines. In: International Symposium on Adaptive Motion of Animals and Machines (AMAM). Montreal, Canada.

Witte, H., Hoffmann, H., Hackert, R., Schilling, C., Fischer, M.S., Preuschoft, H., 2004. Biomimetic robotics should be based on functional morphology. J. Anat. 204, 331–342.

Wooden, D., Malchano, M., Blankespoor, K., Howardy, A., Rizzi, A., Raibert, M., 2010. Autonomous navigation for BigDog. In: Proceedings of the IEEE International Conference on Robotics and Automation. Anchorage, AK, pp. 4736–4741.

Yamamoto, Y., Fujita, M., De Lasa, M., Talebi, S., Jewell, D., Playter, R., Raibert, M., 2001. Development of dynamic locomotion for the entertainment robot – teaching a new dog old tricks. In: Proceedings of the International Conference on Climbing and Walking Robots. Karlsruhe, Germany, pp. 695–702.

Chapter 8

Actuation in Legged Locomotion

Koh Hosoda, Christian Rode, Tobias Siebert, Bram Vanderborght, Maarten Weckx, and D. Lefeber

How the actuation provides energy to the body governs resultant locomotion. This chapter describes actuation in legged locomotion, both in natural and artificial, for understanding how the muscle-like actuation is advantageous. It begins with understanding mechanism of the natural biological muscles, which is energy-efficient, effective, and redundant. Then, the joint stiffness and compliance by electric motors is carefully investigated. Finally, artificial muscles in current Robotics are described to mimic properties of natural muscles.

Muscles are the main source for biological agents to provide energy for locomotion. Therefore, to understand locomotion of biological agent, the characteristics of muscles must be revealed in order to reproduce the patterns and forces by artificial agents. In this chapter, we will elucidate the features and characteristics of biological muscles and their structure consisting of muscles and bones. Robotic counterparts will be described to design and build artificial agents.

A living muscle behaves not only as a passive element, but also as a combination of actuation and passive elements (Abbott and Aubert, 1952), and exerts complicated dynamics. The complex characteristics of the muscle will be discussed in Subchapter 8.1. In particular, we will discuss muscle's morphological function, which contributes to the generation of the animal's adaptive locomotion. For example, the compliance of the muscle plays an important role in the so-called preflex (Loeb, 1995), i.e., the production of rapid reaction to impact with the ground (Alexander, 2003). The physical compliance can cope with the collision without any time delay. The impact energy can be conserved in the compliance, and utilized for next step, which is crucial for the animal's rapid and robust motion. When we design an artificial agent that emulates the animal's behavior, such compliance must be taken into account.

Electric motors are widely used for actuating artificial agents; however, intrinsic compliance is not inherent in the motor. Because normally an electric motor itself does not have enough torque to drive a joint directly, a gear is used

Bioinspired Legged Locomotion. http://dx.doi.org/10.1016/B978-0-12-803766-9.00011-7

for multiplying it. As a result, the rotational speed will decrease, always keeping the power constant. Now the problem is not speed, but friction and inertia if large gearbox ratios are employed. Because of the friction in the gear transmission, the joint driven by the electric motor tends to be nonbackdrivable: it requires large torques to turn the joint, which results in high stiffness values. Because of its importance for adaptive locomotion it is beneficial to add compliance to the motors. This can be done in series or in parallel, as we will clarify in more detail below. Fixed compliance devices are relatively easy to be realized, and variable impedance actuators can also be designed to change characteristics depending on the current state of the environment. In some situations, only a simple locking principle would work. Subchapter 8.2 will be devoted to these issues.

Finally, in Subchapter 8.3, the engineering counterpart of living muscles will be discussed. The point is, how to determine the key functions of the muscles or rather the entire musculoskeletal system so that engineering alternatives can be found to realize animal-like locomotion. One of the questions is at what level we want to imitate the animal's behavior. Do we simply want to reproduce the joint motion, or are we also interested in replicating the underlying mechanisms of how the movements come about? In order to achieve artificial legged locomotion, depending on our degree of mimicry, we have to carefully observe and measure the behavior of the animals so that we can recreate their movements, forces, and compliance. We have to select the level of biomimicry very carefully considering the currently available technology. Robots can be driven using electric motors, hydraulics, pneumatics, artificial pneumatic muscles, etc., all of which instantiate a different degree of biological realism.

Chapter 8.1

Muscle-Like Actuation for Locomotion

Christian Rode* and Tobias Siebert[†]

*Department of Motion Science, Friedrich-Schiller-Universität Jena, Jena, Germany [†]Institute of Sport and Motion Science, University of Stuttgart, Stuttgart, Germany

8.1.1 FUNDAMENTAL PHENOMENOLOGICAL MUSCLE MECHANICS

In humans, about 600 skeletal muscles enable complex movement tasks like locomotion or laughing. Unlike pneumatic actuators used increasingly in prosthetics and robotics (e.g., Ferris et al., 2005), skeletal muscles exhibit volume constancy during contraction (Swammerdam, 1737). Regardless of their shapes and sizes, they share the same typical hierarchical structure (Fig. 8.1.1). The muscle belly consists of hundreds of fascicles (muscle fiber bundles) that are visible by the naked eye in cooked meat. Thousands of muscle fibers are arranged in parallel within each fascicle, and each fiber in turn consists of about thousand myofibrils in parallel that contain sets of highly ordered, longitudinally arranged long, slender filaments. The whole muscle belly, fascicles, and fibers are surrounded by epimysium, perimysium, and endomysium, respectively, connective tissue sheets mainly formed by collagen fibers.

When viewed under the microscope with polarized light, the typical striation pattern of skeletal muscle is recognizable on the level of muscle fibers and myofibrils. This pattern results from the serial repetition of half-sarcomeres (Fig. 8.1.1), the basic contractile units of the muscle (Campbell et al., 2011). The dark stripe, with its higher refractive index, is the A-band formed by the (thick) myosin filaments, while the light, isotropic I-band contains the (thin) actin filaments. Half-sarcomeres are bordered by the Z-discs, a thin meshed filament structure (Knappeis and Carlsen, 1962) running through the middle of the I-bands, and the M-lines, which run through the middle of the A-band (Fig. 8.1.1). The half-sarcomere mainly consists of the actin and myosin filaments, and the titin molecules. Active muscle force is produced by the interaction of myosin heads – which are distributed along the myosin filament – with the actin filaments (Fig. 8.1.1).

The signal for the contraction to begin is the sudden release of calcium ions from the sarcoplasmic reticulum into the cytoplasm induced by action potentials transferred from the motoneuron to the muscle fiber (for a detailed review of the excitation–contraction coupling, see MacIntosh et al., 2006). The regulatory protein, Troponin C, captures calcium ions and undergoes a conformational

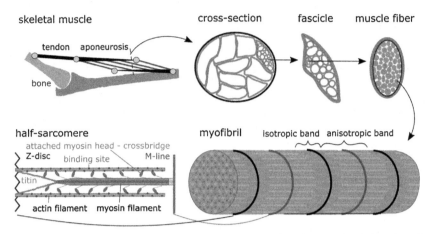

FIGURE 8.1.1   Muscle structure. Ratio of myosin to actin to titin filaments is 1:2:4. Actin and myosin filaments slide relative to each other due to forces generated by cross-bridges. For further explanations, see text.

change that moves the tropomyosin molecule away from the actin filament. Thereby, binding sites on the actin filament are exposed onto which the myosin heads can attach (Fig. 8.1.1). The myosin head binds to actin, performs a power stroke generating about 2–10 pN force (Huxley, 2000), dissociates from actin, and enters a state of pretension prior to the next cycle at the cost of one adenosine triphosphate (ATP). The action of myriads of cross-bridges causes the muscle to contract.

Muscle force depends on the number and size of recruited muscle fibers and the individual motoneuron firing rates. The most important force regulation mechanism, called the "size principle", states that motor units are recruited in order of increasing size (Henneman et al., 1965). If low force is required, small motor units comprising a comparably low number of slow twitch fibers are recruited. With increasing force, larger motor units (activating fast twitch muscle fibers) are progressively recruited. The second mechanism is the dependency of motor unit force on the stimulation frequency. Single action potentials produce a short twitch, because ATP driven pumps transport calcium ions back to the sarcoplasmic reticulum. For instance, in rat muscle *M. triceps brachii*, force is completely abolished between single twitches when electrically stimulated at 12 Hz. As the stimulus frequency increases, twitches overlap until smooth tetanic contractions are achieved at 120 Hz.

Maximal tensions in mammalian skeletal muscles are in the range of 15 to 30 N per cm^2 physiological cross-sectional area (CSA, Nelson et al., 2004; Siebert et al., 2015). Many human muscles are fusiform, i.e., fibers lie in the force axis, and their length approximates muscle belly length. Their CSA equals

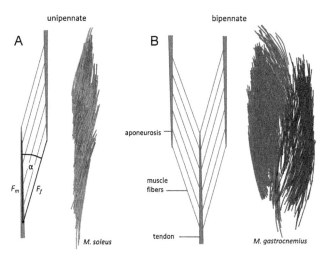

FIGURE 8.1.2 Two different arrangements of muscle fibers. Schematic of a unipennate (A) and bipennate (B) muscle architecture. Much more complex 3D fascicle traces (colored lines) determined by manual digitization are given exemplarily for unipennate rabbit *M. soleus* (green) and bipennate *M. gastrocnemius* (*caput mediale*, dark red; *caput laterale*, light red). Data were provided by Kay Leichsenring and Carolin Wick. F_f and F_m are muscle fiber force and muscle force, respectively; α, pennation angle. (For interpretation of the references to color in this figure legend, the reader is referred to the web version of this chapter.)

the anatomical cross-sectional area perpendicular to the force axis. CSA increases with pennation angle α with respect to the force axis (Fig. 8.1.2) because pennation leads to smaller lengths of muscle fibers and thus to more fibers per muscle volume. This effect typically overcompensates the loss in force due to misalignment of fibers and force axis. However, due to shorter fibers, pennate muscles have a smaller working range as well as lower absolute shortening velocities compared with fusiform muscles.

Much more complex 3D muscle architectures exist featuring different muscle compartments (De Ruiter et al., 1995; English and Letbetter, 1982), tube- and sail-like inner aponeurosis structures (Böl et al., 2015), or in series arrangement of muscle fibers within muscles (Loeb et al., 1987). Considering that muscles are incompressible three-dimensional force generators and that most muscles are packed within other muscles, bones and connective tissue, one can speculate that the 3D muscle architecture has many more functions than longitudinal force generation like, e.g., stabilization of the spine by transversal back muscles' forces (Morlock et al., 1999; Rupp et al., 2015) or minimization of mutual transversal loading within muscle groups during active muscle deformation. Thus, models of muscle packages are required to understand the complexity of

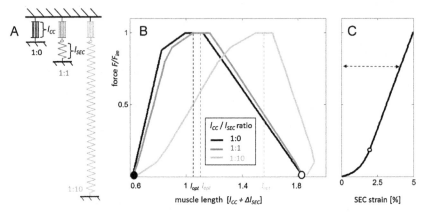

FIGURE 8.1.3 Influence of the ratio of fiber (l_{CC}) to serial elastic component length (l_{SEC}) on the active force–length relation. (A) Three muscles with l_{CC}/l_{SEC} ratios 1:0 (black, corresponding to no or rigid SEC), 1:1 (dark gray), 1:10 (light gray), respectively. (B) Active force–length (fiber length + SEC elongation) relations. (C) Typical nonlinear force-strain relation of an aponeurosis / tendon, here used for the SEC. The transition between the nonlinear foot region and the linear region is marked by a circle. Tendon strain at maximal isometric force is 5%.

3D muscle architecture as well as transverse interaction of muscles with each other (Siebert et al., 2016; Yucesoy et al., 2003).

Typically, muscles exert forces via aponeuroses and tendons to the bone (Figs. 8.1.1, 8.1.2B). In the simplest model, this structure is described as a force generating contractile component (CC) that comprises the fibers, and a serial elastic component (SEC) that lumps the effects of tendon and aponeurosis. Skeletal muscles exhibit vastly different ratios of CC to SEC length. For example, *M. plantaris* of wallabies has short contractile fibers and a long tendon (ratio = 1/14.8, Biewener et al., 2004), while *M. pectoralis* of pigeons consists of long fibers and a short tendon (ratio about 3:1, Biewener, 1998). Moreover, tendon compliance decreases during growth by a factor of 2.5 (Nakagawa et al., 1996) and varies dependent on the muscle function (Bogumill, 2002; Siebert et al., 2015). In Fig. 8.1.3, we illustrate the influence of the CC to SEC length ratio on the isometric force that the muscle can produce at a certain length (length here includes length of the CC and SEC elongation). Elasticity results in a rightward shift of the optimal length associated with the maximum active isometric force. For a more detailed treatment of how the fiber to SEC length ratio impacts contraction dynamics see Morl et al. (2016).

Some muscles resist passive stretching more than others, and it can be relevant to include a parallel elastic component (PEC, lumping effects of, e.g., connective tissue and titin) in the muscle model. For example, when calculating the active muscle force from measurements of total force (obtained typically

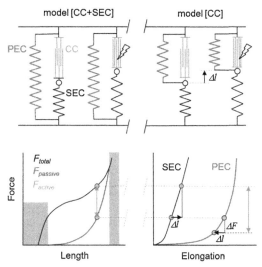

FIGURE 8.1.4 Effect of two different arrangements of the parallel elastic component (PEC) on estimated active muscle force (F_{active}). When stimulated, muscle force increases from passive force ($F_{passive}$) to total force (F_{total}) during an isometric contraction of the muscle (bottom left). In contrast to model [CC+SEC] (top left), the length of the PEC decreases with contractile component (CC) shortening by Δl in model [CC] (right column). Hence, $F_{passive}$ decreases by ΔF during electrical stimulation, and the estimated active force is higher by ΔF than the active force estimated with model [CC+SEC] (Rode et al., 2009b). For this estimation, the constitutive laws of the series elastic component (SEC) and the PEC (bottom right) must be known.

under supramaximal electrical stimulation of the muscle) and passive muscle force, it is relevant to consider the passive forces in accordance with the structural situation in the muscle (Fig. 8.1.4, right column). Only then, the obtained active force is consistent with the theoretical sarcomere force–length relationship (Gordon et al., 1966; Rode et al., 2009b).

Now we go into more detail with the dynamic contractile properties of muscle fibers. It is well known that the force a muscle can produce depends on the contraction velocity. Hill (1938) measured the relation between muscle force (F) and velocity (v_{CC}) for concentric contractions of frog *M. sartorius*. The resulting hyperbolic force–velocity relation (Fig. 8.1.5A) is a fundamental property of muscles and can be described with the following equation:

$$f_v(v_{CC}) = \frac{F(v_{CC})}{F_{im}} = \frac{1 \frac{v_{CC}}{v_{CCmax}}}{1 + \frac{v_{CC}}{curv \cdot v_{CCmax}}} \quad v_{CC} > 0, \tag{8.1.1}$$

where F_{im} is the maximum isometric force, F/F_{im} is the normalized muscle force, and $v_{CCmax} > 0$ is the maximum shortening velocity. The parameter $curv = a/F_{im} = b/v_{CCmax}$ (damping increases with decreasing $curv$; $-a$ and

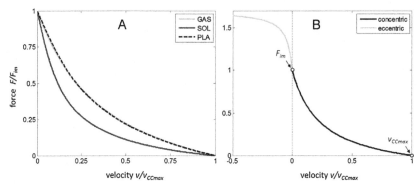

FIGURE 8.1.5 Force–velocity relation of muscle normalized to maximum isometric force (F_{im}) and maximum shortening velocity (v_{CCmax}). (A) Concentric force–velocity relation of slow twitch fibered rabbit *M. soleus* (SOL, black line), and mainly fast twitch fibered *M. gastrocnemius* (GAS, gray line) and *M. plantaris* (PLA, black broken line). (B) Eccentric and concentric force–velocity relation of rat *M. gastrocnemius medialis* (parameters from Till et al., 2008).

$-b$ describe the force and velocity asymptotes, respectively; see Hill, 1938) is an inverse measure of the relation's curvature.

Slow muscles (ST – slow twitch), e.g., the rabbit *M. soleus* (Fig. 8.1.5A), exhibit a larger curvature of the force–velocity relation corresponding to a smaller *curv* value of 0.15 than fast muscles (FT – fast twitch), as, e.g., *M. plantaris* (*curv* = 0.41) and *M. gastrocnemius* (*curv* = 0.47, Close, 1972; Siebert et al., 2015). The maximum shortening velocity is mainly determined by the fiber type (Jones et al., 2004). FT fibers are able to shorten two to three times faster than ST fibers. For human muscles, maximum shortening velocities are about two and six fiber lengths/s [l_0/s] for ST and FT muscles, respectively (Faulkner et al., 1986). By comparison, mouse, rat, and rabbit muscles are much faster (ST, 6–13 l_0/s; FT, 9–24 l_0/s; Close, 1964, 1965; Luff, 1981; Ranatunga and Thomas, 1990; Siebert et al., 2015). With decreasing mass of the animal $v_{max} \sim m^{-ks}$ increases, where ks is the scaling coefficient ranging from 0.125 to 0.33 (McMahon, 1984; Rome et al., 1990).

For eccentric contractions ($v < 0$), the force–velocity relationship (Fig. 8.1.5B, gray line) is commonly also modeled as a hyperbolic relationship (e.g., van Soest and Bobbert, 1993; Fig. 8.1.5B). For increasing eccentric speed muscle forces exceed F_{im} reaching an eccentric force limit in the range of 1.3 to 1.8 F_{im} (Curtin and Edman, 1994; Katz, 1939; Rijkelijkhuizen et al., 2003). However, as at least two different mechanisms contribute to the eccentric force generation (Pinniger et al., 2006), it is difficult to extract the eccentric force–velocity relation from experimental data. As a consequence, there is few data concerning the eccentric force–velocity relation, and the variation of the determined eccentric force–velocity parameters is much higher (Nigg and

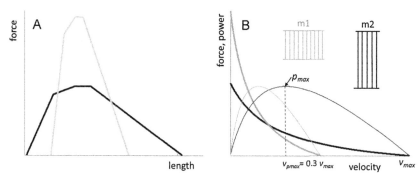

FIGURE 8.1.6 Influence of two different muscle fiber configurations (m1 [gray] vs. m2 [black]) on the force–length (A), force–velocity (B, thick lines), and power–velocity relation (B, thin lines). Power = force · velocity. The velocity axis corresponds to muscle shortening. Muscles m1 and m2 share the same volume, and m2 has half the physiological cross-sectional area but twice the fiber length of m1. Thus, m2 generates half the maximum isometric force, has a larger working range, and twice the maximum shortening velocity (v_{max}) of m1. For the used typical curvature parameter ($curv = 0.25$), maximum power (p_{max}) is generated at $0.3\ v_{max}$.

Herzog, 2007). Considering the role of titin during eccentric contraction (see Section 8.1.2), Till et al. (2008) suggested a model based method to determine realistic parameters for the eccentric force–velocity relation (Fig. 8.1.5B).

Muscle growth is related to an increase in muscle mass. As the number of muscle fibers remains constant after birth (Mozdziak et al., 2000), growth is related to an increase in fiber diameter (hypertrophy) or length increase of muscle fibers. Hypertrophy takes place by synthesis of new myofibrils in parallel in a muscle cell. Change in fiber length is accompanied by a change in the number of sarcomeres in series within a fiber. An increase in the number of sarcomeres in series can be induced by eccentric muscle contractions (Proske and Morgan, 2001) or surgical lengthening of a muscle by bone lengthening (Boakes et al., 2007). In contrast, fixation of a muscle at short length results in a reduction of the sarcomere number in series (Tabary et al., 1972). Furthermore, the muscle fiber length and thus the number of sarcomeres in series seems to depend on the typical working range at which the muscle is used in daily life (Herzog et al., 1991). The influence of an increase in CSA or in fiber length on the force–length, force–velocity, and power–velocity relation is illustrated in Fig. 8.1.6.

8.1.2   ACTIVE AND SEMIACTIVE MECHANISMS OF FORCE-PRODUCTION

This section will review the classical theories of contraction and recent modifications of these theories that alter our view of muscle fiber structure and

A

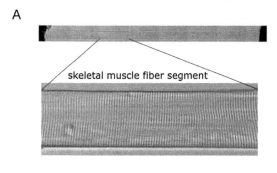

skeletal muscle fiber segment

B

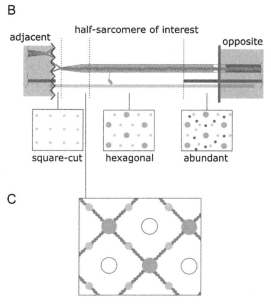

FIGURE 8.1.7 Muscle fiber structure. (A) Cross-striation of muscle fibers under polarized light (photo provided by Andre Tomalka). (B) Schematic of a half-sarcomere that is bounded by the Z-disc (left, black zigzag line) and the M-line (right, blue vertical line). At optimal length, it spans about 1 μm in width and length. The lateral spacing of filaments depends on muscle length because of isovolumetricity of the intact muscle and is of the order of 10 nm. Green and red indicate opposed polarity of the respective filaments. Cross-sections below show known (from left to right: Knappeis and Carlsen, 1962; Huxley HE, 1957; Trombitas and Tigyi-Sebes, 1985) different actin (small circles) and myosin filament (big circles) arrangements. (C) Proposed myofilament and titin (orange) arrangement near the Z-disc (Rode et al., 2016). (For interpretation of the references to color in this figure legend, the reader is referred to the web version of this chapter.)

functioning. For this, we go a bit deeper into structure of the fibers before introducing the classical and recent theories.

As stated in Section 8.1.1, the typical striation pattern on the level of muscle fibers (Fig. 8.1.7A) and myofibrils results from a serial arrangement of half-sarcomeres (Fig. 8.1.7B). Within these half-sarcomeres, giant titin molecules run through the myosin filaments and connect to adjacent Z-discs, and the distance from the tips of the myosin filaments to the Z-discs is called titin's free molecular spring region (Fig. 8.1.7B). Myosin filaments are cross-linked in the center of the anisotropic band at the M-line. Cross-sections of vertebrate myofibrils reveal hexagonal actin filament lattice with myosin filaments centered in the hexagons (Fig. 8.1.7B, middle cross-section, myosin to actin filament ratio 1:2) in typical overlap regions (e.g., Huxley HE, 1957), and square-cut actin filament lattice near the Z-disc (Fig. 8.1.7B, left cross-section), where α-actinins cross-link actin filaments to form the Z-disc (e.g., Knappeis and Carlsen, 1962). Because two titin molecules connect to each actin filament at the Z-disc (Zou et al., 2006), four titin molecules must run through each myosin filament. A novel, suggested myofilament and titin arrangement near the Z-disc for short muscle lengths (Rode et al., 2016) is illustrated in Fig. 8.1.7C.

In 1954, two research groups (Huxley and Hanson, 1954; Huxley and Niedergerke, 1954) independently concluded that the myofilaments slide relative to each other during fiber contraction (now known as sliding filament theory) and – considering the linear slope of the isometric force–length relationship (FLR) at long muscle lengths (Fig. 8.1.8) – they suggested that independent force generators along the myosin filaments (the myosin heads) lead to fiber contraction (cf. Fig. 8.1.1). Huxley HE (1957) identified 15 to 20 of such possible force generators spaced at 40 nm along a half-myosin filament projecting towards each of six surrounding actin filaments. At the same time, Huxley AF (1957) formulated mathematically the cross-bridge theory that assumes cyclical binding of myosin heads to actin filaments and the generation of a pulling force by these so-called cross-bridges. By adjusting the attachment and detachment rates of the myosin heads, he could explain the phenomenological force–velocity relationship (see Section 8.1.1, Fig. 8.1.6) and the rate of energy liberation during concentric muscle contractions known at that time (Hill, 1938). A geometric model of filament overlap (Gordon et al., 1966) could predict straightforwardly parts of the FLR, in particular the descending limb, the plateau, and the point of slope change on the ascending limb of the force–length relationship (where myosin filaments reach the Z-disc). The shallow part of the ascending limb is explained with inhibition of cross-bridges due to actin filaments of the opposite half-sarcomere entering the half-sarcomere of interest (Fig. 8.1.7B, right cross-section), and the steep part of the ascending limb is typically explained with compression or folding of myosin filaments at the Z-disc. To date, the sliding filament and cross-bridge theories shape our understanding of muscle fiber structure and function.

Interestingly, even before the sliding filament and cross-bridge theories were proposed, experimental results seemingly contradicted these theories. In 1940,

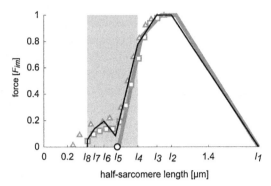

half-sarcomere length [μm]

FIGURE 8.1.8 Active isometric force–half-sarcomere length relation of frog muscle fiber. Force is given as a fraction of maximum isometric force F_{im} (data by Gordon et al., 1966, is approximated as straight lines; data by Ramsey and Street, 1940, is shown for two specimens as squares and triangles, respectively). The foot region at short lengths can be explained by a geometric model of filament overlap (black line) incorporating myosin filament sliding through the Z-disc (Rode et al., 2016). The gray area marks the range where myosin filaments are proposed to have entered the adjacent half-sarcomere instead of being compressed or folded as classically assumed. Following this assumption, specific filament overlap changes occur at lengths l_1 to l_8. The length of classical zero force is marked by a circle.

Ramsey and Street reported a pronounced foot region of the FLR for extended stimulus duration (Fig. 8.1.8). Furthermore, fibers did not restore their resting length after cessation of activation. These observations cannot be explained with compression of filaments or with the recently proposed decrease of active force due to increasing filament spacing (Williams et al., 2013). Moreover, in 1952, Abbott and Aubert showed that the force in an isometric phase after active muscle shortening/lengthening was lower/higher than the active force in an isometric contraction at the same final length (Fig. 8.1.9). These effects are called force depression/force enhancement, and are summarized as contraction history effects. These effects were reported for muscles (Abbott and Aubert, 1952; Siebert et al., 2015), fibers (e.g., Edman, 1975; Edman et al., 1982), and myofibrils (Joumaa et al., 2008; Joumaa and Herzog, 2010) and they range up to 20% of maximum isometric force for force depression and up to 200% for force enhancement (Leonard and Herzog, 2010).

To explain the behavior of fibers at short muscle lengths, Rode et al. (2016) proposed that myosin filaments are not compressed at the Z-discs but that they slide through them (Fig. 8.1.10). From available micromechanical data they estimated that the myosin filaments are too stiff to be folded or even compressed by the muscular force. Moreover, myosin filament sliding through the Z-discs would explain the basic structure of the Z-disc. By conversion from hexagonal lattice in regular overlap to tetragonal lattice at the Z-disc, the actin filament lattice can accommodate twice the number of myosin filaments at the Z-disc,

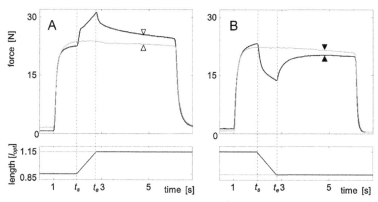

FIGURE 8.1.9 Force enhancement (A) and force depression (B) experiments of rabbit *M. soleus*. Upper figure and lower figure represent force–time and corresponding muscle length–time traces, respectively. Force enhancement (difference between white triangles) and force depression (difference between black triangles) are the force differences between ramp experiments (black) and isometric reference contractions (gray) determined two seconds after the end of the ramp (t_e), shown exemplarily for a ramp velocity of 0.35 optimal muscle lengths per second.

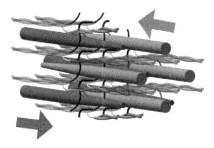

FIGURE 8.1.10 Proposed myosin filament sliding through the Z-disc and proposed myofilament arrangement. Myosin filaments (thick) of adjacent half-sarcomeres (shown in green and red, respectively) slide through the Z-disc formed by actin filaments (thin) cross-linked with α-actinins (black and gray). Titin and other molecules are omitted. (For interpretation of the references to color in this figure legend, the reader is referred to the web version of this chapter.)

which is required for an orderly sliding process. Further, this mechanism suggests a specific arrangement of titin molecules that could support the transition to a checkered pattern of myosin filaments within the actin filament lattice near the Z-disc (Fig. 8.1.7C).

A geometric model of filament overlap accounting for myosin filament sliding through the Z-disc (Rode et al., 2016) predicted the strikingly linear steep part of the FLR's ascending limb and the occurrence of a foot region (Fig. 8.1.8). Moreover, the model also reproduced the steep, linear decline of maximal contraction velocity of the sarcomere from the point below the length where myosin filaments reach the Z-disc and the constant contraction velocity above this length

(Edman, 1979). Hence, the whole force–length relationship is in accordance with the sliding-filament and cross-bridge theories, and the "strange behavior" of fibers reported by Ramsey and Street (1940) could be explained with the assumption that myosin filaments slide through the Z-disc.

Another challenge to the classic theories of contraction arises from the observed history effects. Numerous experiments have been performed to assess dependencies of history effects on contraction distance, velocity, force, or activation (e.g., Abbott and Aubert, 1952; Marechal and Plaghki, 1979; Edman, 1975; Edman et al., 1982; Herzog and Leonard, 1997). For example, force enhancement depends linearly on stretch distance and is independent of stretch velocity, while there is no unanimous opinion for the dependence of force depression on contraction distance and velocity. Although some history effects can be reproduced with multiple sarcomere models and the assumption of strong and weak half-sarcomeres (Morgan, 1990), evidence adds up that strongly suggests a very important activation-dependent non-cross-bridge contribution to muscle force production and a minor effect of contraction conditions on the cross-bridges themselves. The non-cross-bridge contribution to muscle force production is probably due to titin (Noble, 1992; Rode et al., 2009a; Nishikawa et al., 2012). We will briefly explain the concept and explain current models of this mechanism. Please, see Herzog et al. (2008), Campbell and Campbell (2011), Edman (2012), or Siebert and Rode (2014) for a review of the dependencies of history effects on contraction conditions and suggested mechanisms related to force enhancement and force depression.

In the presence of calcium, titin can attach to the actin filament (Fig. 8.1.11; e.g., Bianco et al., 2007). This would reduce its free molecular spring length and would lead to vastly increased titin forces during stretch. Also, such a connection would induce decreased forces during shortening (Rode et al., 2009a, 2009b). Actin filament winding upon activation as suggested in Nishikawa et al. (2012) would facilitate binding of titin's PEVK-region to actin. The principle of so far published molecular models to explain force enhancement is similar. However, only the "sticky-spring" model (Rode et al., 2009a) accounts for limited forces of actin–titin bindings (Bianco et al., 2007), and is formulated mathematically to be applicable to muscle shortening.

The physiological role of such semiactive titin behavior is not clear. On the one hand, such attachment and forces enable stable half-sarcomere operation when working on the descending limb of the total force–length relationship (Heidlauf et al., 2016). On the other hand, this behavior may simplify neural control as the muscle reacts like a linear spring during stretch (Tomalka et al., 2017). Further, it remains to be investigated to what extent the energy can be recovered after active stretch. The molecular mechanisms proposed so far would

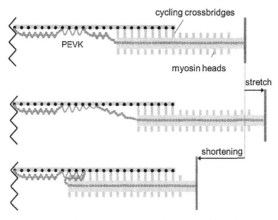

FIGURE 8.1.11 Schematic of a half-sarcomere stretching (middle) and shortening (bottom) from some initial length (top). When activated, titin's PEVK-region attaches to myosin binding sites on actin. During stretch, the number of cycling cross-bridges decreases; however, the force in titin increases as its PEVK region is stretched along the actin filament. During shortening, first passive forces decrease and then even become negative, counteracting the cross-bridge forces. Further, some cross-bridges are inhibited during shortening.

create hysteresis and thus loss of a part of the stored energy. Moreover, the mechanism would be counterproductive in muscles producing positive work during principal daily activities. Indeed, there is some evidence that muscles vary in the development of history effects (Pinniger et al., 2006; Stienen et al., 1992; Siebert et al., 2015) and that muscles are possibly able to adapt their actin–titin binding according to their use (Lindstedt et al., 2002). McBride et al. (2003) observed differential expression of titin isoforms in weight- and powerlifters in comparison to untrained non-athletic individuals. Beside training induced changes in fiber type, muscle strength and size (Hortobagyi et al., 1996; LaStayo et al., 2000), adaptation of actin–titin interaction might be a further mechanism to adjust our neuromuscular system to specific task dependent movement requirements.

8.1.3 HOW HUMAN MUSCLES WORK AS ACTUATORS IN LOCOMOTION

Muscles are versatile actuators. During locomotion, muscles can work as *motors* to inject energy, as *brakes*, e.g., to dissipate energy for control of inertial movements (Biewener and Roberts, 2000; Ahn and Full, 2002) or to prevent potentially dangerous high frequency force oscillations in tendons, e.g., in the horse (Wilson et al., 2001), and as *springs* to increase the efficiency of locomotion (Biewener and Roberts, 2000; Roberts et al., 1997). The tendon introduces an internal degree of freedom to the muscle. Tendon stiffness, which is mainly determined by the muscle fiber–tendon length ratio, governs the amount of interplay between the fibers and the tendon. This important parameter may be tuned to the main task of the muscle (Mörl et al., 2016). For example, this ratio should be low (i.e., the tendon should be long and compliant) for muscles working primarily as springs, and large for muscles working primarily as motors. However, the ability of muscles to be activated at any time during a certain movement allows for more complicated tasks and even task switching during one motion (like the vastii in the long jump, see next paragraph) or in different types of locomotion. Last but not least, muscles provide wobbling mass that can significantly reduce impacts in jumping or running (Gruber et al., 1998; Günther et al., 2003).

The force that muscles produce is not only regulated by the level of activation. Muscles are special actuators with built-in properties (cf. Sections 8.1.1, 8.1.2) that make them respond instantaneously to perturbations of, e.g., position. These instantaneous responses, e.g., due to their force–length or force–velocity relationships, have been termed *preflexes* (Loeb et al., 1999; see Subchapter 6.5), and they can reduce the control effort (i.e., they can reduce the information [bit] that has to be processed in order to generate a movement; see Häufle et al., 2014). One figurative example of the way how muscle properties can affect our motion without control is the operation of the vastii in the long jump. During knee flexion, the vastii work eccentrically, exploiting the ability to generate very high eccentric forces due to the force–velocity relationship and storing energy in the tendon (Seyfarth et al., 2000). In the subsequent phase, where the knee extends, the muscle can still contract eccentrically in the beginning and then concentrically at a low velocity because the tendon accounts for a large part of the total muscle length change during knee extension. This leads to higher muscle force during knee extension as a consequence of the force–velocity relationship compared with a muscle without tendon. Today, it seems that further mechanisms like the semiactive titin (see Section 8.1.2), or the organization of muscles in muscle packages (Siebert et al., 2015; Reinhardt et al., 2016) also influence the contraction dynamics and the force that is exerted to the locomotor system. The functional role of these properties is not completely understood. However, inclusion of the semiactive titin spring in parallel to the contractile component, enables the contractile component itself to react like a spring during eccentric contractions, and thus to save and release energy. Additionally, inclusion of semiactive titin reduces half-sarcomere length inhomogeneity and thus contributes to stability on the muscle level, too (Heidlauf et al., submitted). Understanding the muscle as a three-dimensional structure surrounded by other tissues (e.g., neighboring muscles or bones) and generating forces not only in longitudinal direction, transversal muscle forces

(Siebert et al., 2016) may have functional relevance in locomotion. For example, work performed transversally on adjacent muscles during muscle deformation may be saved and released subsequently. In addition to energy savings in series (Biewener and Baudinette, 1995) and parallel elastic structures (Rode et al., 2009a), this may represent an additional way of recovering energy during cyclical locomotion. Further, transversal forces may be functionally relevant regarding, for example, stabilization of the spine by back muscles (Morlock et al., 1999). Models often concentrate on classic force–length and force–velocity relationships and demonstrate their beneficial effects in achieving a desired motion. For example, the force–velocity relationship stabilizes hopping in the sense of a preflex that is tunable by activity (Häufle et al., 2010).

In an example of a postural task, we will show how the force–length and the force–velocity relationships can simplify the task of, e.g., balancing a tray during walking. Despite the perturbations induced by stepping, the tray should be held still. We use a Hill-type model with *constant* activation and force–length and force–velocity relations ignoring history effects and the tendon. In our simple model of the mechanical systems shown in Fig. 8.1.12A, gravitational force and inertia acting on the mass are counterbalanced via a lever arm by the muscular force (Fig. 8.1.12B). Activation of the muscle can be tuned to achieve a stable equilibrium on the ascending limb of the force–length relationship due to its positive stiffness in this range. In this case, small perturbations can be negotiated without any neural response or reflexes changing the activation of the muscle. The system can be overdamped (no oscillations after perturbation, which may be desired in the task of balancing a tray) or underdamped (oscillations after perturbation). The specific dynamic response to the perturbation is determined by muscle properties and the gearing ratio of mass lever arm b divided by muscle lever arm a.

Interestingly, the maximum isometric force, that is associated with the most common adaptation of muscle, a change in cross-sectional area, does not necessarily influence the response of the system to a perturbation because activation is tuned to achieve equilibrium, i.e., if, for instance, the muscle is stronger, less activation is needed. Only if at the same time the proportion of fast and slow muscle fibers changes, the curvature of the force–velocity relation may change and this would influence the response of the system to a perturbation. Other adaptations of the muscle such as an increase in sarcomeres in series decreases the stiffness of the muscle, but the associated increase in maximal contraction velocity overcompensates this effect leading to a more underdamped behavior of the system. A further adaptation that is only accessible to evolution or to surgery in some cases of spasms is related to the gearing ratio. The surface in Fig. 8.1.12C shows the relative force value where critical damping occurs. If the

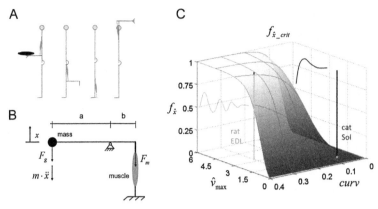

FIGURE 8.1.12   The influence of muscle preflexes on a postural task. (A) Similar postural tasks. (B) Model of postural tasks. Muscle force F_m counteracts gravitational force F_g and inertia $m \cdot \ddot{x}$ via a gearing ratio $G = b/a$. Equilibrium is achieved by tuning muscle activation. (C) Surface of static muscle force (normalized by maximum isometric force F_{im}) that corresponds to critical damping. Muscle force is $F_m = Act \cdot F_{im} \cdot f_l \cdot f_v$, with Act the activation, f_l and f_v the force–length and force–velocity relationships. $f_{\hat{x}} = 1 - \hat{x}^2$, $f_{\hat{v}}$ (a hyperbola of curvature $curv$, cf. Section 8.1.1; smaller values of $curv$ correspond to stronger damping), and $\hat{v}_{max}$ are the dimensionless scaled versions of f_l, f_v, and v_{max} (the maximum contraction velocity of the muscle), respectively. To determine the critical values $f_{\hat{x}\_crit}$, the system is linearized at the equilibrium point, and the eigenvalues of the Jacobian matrix are investigated. Increasing width of f_l decreases $\hat{v}_{max}$. Muscles with slow fibers and lower $curv$ values like the soleus of the cat tend to overdamp the system while muscles with fast fibers and higher $curv$ values tend to underdamp the system.

equilibrium point is at a higher or a lower value, then the system is overdamped or underdamped, respectively.

Muscles with their preflexes and sensors of fiber length, velocity, and force, the vestibular and visual sensors, and the morphology of the biological system can be considered at the bottom of an essentially hierarchical, biologically relevant model of sensorimotor control, where the spinal cord is in the middle, and the brain at the top (Loeb et al., 1999). Muscle activation is generated via motor commands from the spinal cord or from the brain (mediated by interneurons in the spinal cord). Reflex feedback loops are associated with information processing solely at the level of the spinal cord, and central pattern generators residing likewise in the spinal cord are thought to provide rhythmic activity (as the outcome of a dynamical system of neuron pools) to leg muscles in locomotion. Initiation of motor programs or intermittent tuning of parameters of central pattern generators as a consequence of afferent signal processing is associated with the brain (see Subchapter 6.2 for a discussion of central pattern generators). Independent of the level of control (high or low), feedback control is characterized by continual processing of afferent information (e.g., reflexes), and feedforward control is adapted intermittently as a consequence of afferent information (Riemann and Lephart, 2002).

The biological control problem in moving animals differs remarkably from the technical control problem of industry robots (Loeb et al., 1999): in contrast to motors, muscles react slowly towards changes in neural stimulus, but, due to preflexes, quickly when kinematic conditions change. Instead of routing feedback signals to individual actuators, signals from a large number of slow and noisy sensors converge with signals from many command centers before they are routed to motoneurons. Biological performance of muscles is often suboptimal, but adequate in a range of circumstances, instead of optimal with respect to a single criterion for nominal conditions (Loeb, 2012). Also, biological systems seem to employ types of feedback that are uncommon in technical applications. Geyer et al. (2003) demonstrated with a simple model of hopping that a positive feedback of force or length could stabilize hopping.

Today many aspects of biological control of human locomotion remain to be investigated. For example, it is still disputed whether central pattern generators shown to exist in many animals (Delcomyn, 1980) exist in humans, too. Geyer and Herr (2010) published a model of human walking using a seven-link model with segment mass that, aside from negative feedback trunk control and some special events, was driven exclusively by reflexes with no need of a central pattern generator. Also, the tuning or interdependencies of different types of control remain elusive. Häufle et al. (2012) could show that a combination of positive feedback (reflexes) and feedforward control (central pattern generator signals) improved stability of a simple hopping model. Despite the apparent complexity of the human locomotor system that evolved for millions of years, it seems possible to identify important mechanisms at different structural levels that enhance stability or locomotion efficiency. Still, the understanding of the interplay of these mechanisms is far from being completely understood. Thus, we can expect a series of discoveries in this area in the next years.

8.1.4 REDUNDANCY OF THE ACTUATION SYSTEM – FUNCTIONALLY RESOLVED?

First we clarify what redundancy in the actuation system means. Joints in the musculoskeletal system are not ideal pivot joints and may function differently in different configurations. For example, the human knee can bend and extend, and its ability to rotate increases when it is flexed. Therefore, several monoarticular muscles spanning one joint (e.g., the vastii) may contribute differently to extension and rotation, and they would not necessarily be redundant but have distinct functional meaning. However, in this section we concentrate on the sagittal plane and simplify joints to pivot joints. In this circumstance, redundancy describes the fact that monoarticular muscles exist for each joint of the human leg and that they are complemented by biarticular muscles, for example, the rectus femoris (RF) and the group of the hamstrings (HAM).

Biarticular muscles can fulfill a number of functions as discussed in earlier sections of the book (see, e.g., Section 7.3.2) that also illustrate their potential use for robotics. Among them is the coordination of joint movements (tendinous, or ligamentous action; see Cleland, 1867) and associated with this is the ability to transfer energy from one joint to the other (e.g., Bobbert et al., 1987; van Ingen Schenau, 1989). With that they can synchronize the motion of adjacent joints without the need for sensory feedback and high bandwidth actuators (e.g., Niiyama et al., 2007; Hosoda et al., 2010; Scholz et al., 2012). Moreover, they can resolve the singularity of the knee in explosive motions (e.g., van Soest et al., 1993), and they can decrease the amount of required energy when creating torques about two different joints, even more so in animals during leg extension due to a more favorable working condition (decreased muscle contraction velocity associated with higher force). Biarticular muscles might also facilitate the control of the swing phase (Dean and Kuo, 2009) that is associated with opposite torques in adjacent joints (DeVita and Hortobagyi, 2000) corresponding to activity in swing phase initiation (RF) and swing leg retraction (HAM; Prilutsky et al., 1998). Here, we want to put forth some ideas and provide evidence that there are further important functions that might help to explain and resolve the redundancy in the human actuation system.

Let us consider the concept that the leg function can be decomposed into two sagittal-plane components, the axial leg function (mainly associated with body support and bouncing) and the perpendicular leg function (mainly associated with control of upper body orientation, Fig. 8.1.13, left). Both components can contribute to fore-aft acceleration. The perpendicular leg function can for example be exploited for this by leaning the trunk forward to increase walking or running speed, or especially in amputees wearing passive leg prostheses using the positive hip work (Czerniecki et al., 1991). From standing to walking, to running, and hopping, the axial function increasingly dominates (Maykranz et al., 2013). If the central nervous system would be capable of accessing these functions conveniently, this might reduce the effort of learning new movements and varying movement conditions (e.g., locomotion speed), that is, the balance problem might be solved in a very elegant way.

Considering complex dynamics of segmented legs, it seems not trivial to access these leg functions by the central nervous system. For example, during stance (with low inertial effects) the monoarticular hip extensors create force along the shank (Hof, 2001). Hence, they contribute to axial *and* perpendicular leg function, and this contribution varies nonlinearly with the knee angle. Muscle dynamics (see Sections 8.1.1, 8.1.2) further complicate the scenario. In con-

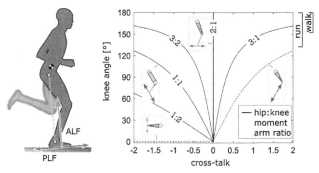

FIGURE 8.1.13 Axial (ALF) and perpendicular leg functions (PLF). The contour lines show hip to knee moment arm ratios of biarticular thigh muscles depending on the knee angle (180° is straight) and on the ratio of axial and perpendicular leg force (cross-talk) that biarticular muscles produce at the distal end of a two-segmented leg. The more the leg is extended, the more the cross talk increases for moment arm ratios deviating from 2:1. By symmetry, the gastrocnemius also produces purely perpendicular force for ankle to knee moment arm ratio of 2:1; however, in contrast to the thigh muscles, its effect on angular momentum is low.

trast, the biarticular HAM, that likewise extend the hip, would simultaneously exert a knee bending torque that would reduce their axial force contribution. For equal thigh and shank lengths (as approximated in humans), the hip to knee ratio of muscle moment arms of 2:1 would allow the HAM and RF to produce only perpendicular ground reaction force contributions independent of the knee angle (Fig. 8.1.13, right). For deviations from this ratio, the coupling of axial and perpendicular leg forces increases, especially for extended leg configurations as commonly used in human locomotion (Fig. 8.1.13, right).

The conceptual morphology (equal thigh and shank lengths, 2:1 hip to knee biarticular thigh muscle moment arms) was tested for its explanatory power for muscle EMG (electromyography) activity (Tokur et al., 2015). Subjects were exposed to an external force applied at different positions of the body in the sagittal plane, and were instructed to hold their position. Biarticular muscles consistently increased in EMG accordingly, whereas the monoarticular muscles, which might, for example, compensate for muscle moment arms of biarticular muscles deviating from the conceptual 2:1 morphology, did not (Fig. 8.1.14). Also, in perturbed walking experiments, HAM immediately controlled angular momentum (Pijnappels et al., 2005). This illustrates what is inherent to its incremental principle: the access to the perpendicular leg function still works even if biarticular muscles are used for the support of the axial leg function at the same time. These results suggest the relevance of the concept for balance in humans.

To test the relevance of these theoretical findings for robotic applications, control of ground reaction force direction was implemented in the humanoid legged robot BioBiped3 (Sharbafi et al., 2016) using monoarticular versus biar-

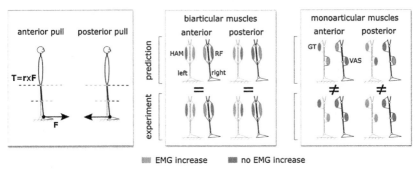

FIGURE 8.1.14 Static perturbation experiment. Subjects stand on one leg; external force **F** is applied at a distance **r** to the joints of the free leg. Signed magnitudes of static torques **T** applied at knee and hip joints, drawn in a color corresponding to the leg (broken lines). Statistically significant EMG increase (orange) is found in accordance with the predictions for the biarticular muscles (middle), but not for the monoarticular ones (right). (For interpretation of the references to color in this figure legend, the reader is referred to the web version of this chapter.)

ticular actuators in static (standing with different knee angles) and in dynamic (squatting) conditions. In both cases, biarticular muscles outperformed monoarticular ones by not only tracking the GRF direction better but also by producing less variations in the GRF magnitude and knee angle while reducing energy consumption.

To further test whether the conceptual morphology could facilitate balance, we implemented this design in a 7-link rigid-body model (trunk and three-segmented legs; see Lakatos et al., 2014) with series elastic actuators (biarticular thigh and shank actuators, monoarticular knee actuator). With the elastic decoupling of axial and perpendicular leg function, human sagittal-plane running was decomposed into a set of tasks which could be directly addressed. The benefit of this architecture for keeping balance was demonstrated with a simple control scheme enabling bipedal running of the model. On a side note, combining a spring loaded inverted pendulum model for the stance leg and a double pendulum for the swing leg, Sharbafi et al. (2017) have shown that the human swing leg motion could be reconstructed by just setting appropriate rest lengths and stiffness of biarticular springs.

The conceptual leg architecture follows simply from the idea that balance control can be facilitated by direct access to axial and perpendicular leg function. Perpendicular leg function requires opposite joint torques (flexion and extension) within adjacent joints. These requirements predict the existence of biarticular muscles, biarticular muscle moment arm ratios of 2:1 for hip to knee and ankle to knee, and equal segment thigh and shank segment lengths. Humans approximate this architecture to some extent (existence of biarticular muscles, nearly equal thigh and shank lengths, approximate 2:1 ratio for gastrocnemius and HAM, but rather 4:3 for RF in extended leg configurations, Winter, 2005). Notably, the biarticular antagonist for the gastrocnemius is missing, reflecting an asymmetry introduced by the flat foot. These predictions together with experimental and simulation results suggest that the human leg architecture is adapted to simplify balance.

Seemingly, the advantages of biarticular muscles add up and may be used and exploited in future prosthetic, orthotic, or robotic designs.

REFERENCES

Abbott, B.C., Aubert, X.M., 1952. The force exerted by active striated muscle during and after change of length. J. Physiol. 117, 77–86.

Ahn, A.N., Full, R.J.A., 2002. A motor and a brake: two leg extensor muscles acting at the same joint manage energy differently in a running insect. J. Exp. Biol. 205, 379–389.

Bianco, P., Nagy, A., Kengyel, A., Szatmári, D., Mártonfalvi, Z., Huber, T., Kellermayer, M.S., 2007. Interaction forces between F-actin and titin PEVK domain measured with optical tweezers. Biophys. J. 93 (6), 2102–2109.

Biewener, A., Baudinette, R., 1995. In vivo muscle force and elastic energy storage during steady-speed hopping of tammar wallabies (Macropus eugenii). J. Exp. Biol. 198 (9), 1829–1841.

Biewener, A.A., 1998. Muscle function in vivo: a comparison of muscles used for elastic energy storage savings versus muscles used to generate power. Am. Zool. 38, 703–717.

Biewener, A.A., Roberts, T.J., 2000. Muscle and tendon contributions to force, work, and elastic energy savings: a comparative perspective. Exerc. Sport Sci. Rev. 28, 99–107.

Biewener, A.A., McGowan, C., Card, G.M., Baudinette, R.V., 2004. Dynamics of leg muscle function in tammar wallabies (*M. eugenii*) during level versus incline hopping. J. Exp. Biol. 207, 211–223.

Boakes, J.L., Foran, J., Ward, S.R., Lieber, R.L., 2007. Muscle adaptation by serial sarcomere addition 1 year after femoral lengthening. Clin. Orthop. Relat. Res. 456, 250–253.

Bobbert, M.F., Hoek, E., van Ingen Schenau, G.J., Sargeant, A.J., Schreurs, A.W., 1987. A model to demonstrate the power transporting role of biarticular muscles. J. Physiol. 387, 24.

Bogumill, G.P., 2002. Functional anatomy of the flexor tendon system of the hand. J. Hand Surg. 7 (1), 33–46.

Böl, M., Leichsenring, K., Ernst, M., Wick, C., Blickhan, R., Siebert, T., 2015. Novel microstructural findings in *M. plantaris* and their impact during active and passive loading at the macro level. J. Mech. Behav. Biomed. Mater. 51, 25–39.

Campbell, S.G., Campbell, K.S., 2011. Mechanisms of residual force enhancement in skeletal muscle: insights from experiments and mathematical models. Biophys. Rev. 3, 199–207.

Campbell, S.G., Hatfield, P.C., Campbell, K.S., 2011. A mathematical model of muscle containing heterogeneous half-sarcomeres exhibits residual force enhancement. PLoS Comput. Biol. 7, e1002156.

Cleland, J., 1867. On the actions of muscles passing over more than one joint. J. Anat. 1, 85.

Close, R., 1964. Dynamic properties of fast and slow skeletal muscles of the rat during development. J. Physiol. 173, 74–95.

Close, R., 1965. Force: velocity properties of mouse muscles. Nature 206, 718–719.

Close, R.I., 1972. Dynamic properties of mammalian skeletal muscles. Physiol. Rev. 52, 129–197.

Curtin, N.A., Edman, K.A., 1994. Force–velocity relation for frog muscle fibres: effects of moderate fatigue and of intracellular acidification. J. Physiol. 475, 483–494.

Czerniecki, J.M., Gitter, A., Munro, C., 1991. Joint moment and muscle power output characteristics of below knee amputees during running: the influence of energy storing prosthetic feet. J. Biomech. 24, 63–75.

De Ruiter, C.J., De Haan, A., Sargeant, A.J., 1995. Physiological characteristics of two extreme muscle compartments in gastrocnemius medialis of the anaesthetized rat. Acta Physiol. Scand. 153, 313–324.

Dean, J.C., Kuo, A.D., 2009. Elastic coupling of limb joints enables faster bipedal walking. J. R. Soc. Interface 6, 561–573.

Delcomyn, F., 1980. Science 210 (4469), 492–498.

DeVita, P., Hortobagyi, T., 2000. Age causes a redistribution of joint torques and powers during gait. J. Appl. Physiol. 88, 1804–1811.

Edman, K.A.P., 1975. Mechanical deactivation induced by active shortening in isolated muscle fibres of the frog. J. Physiol. 246, 255–275.

Edman, K.A.P., 1979. The velocity of unloaded shortening and its relation to sarcomere length and isometric force in vertebrate muscle fibres. J. Physiol. 291, 143–159.

Edman, K.A.P., 2012. Residual force enhancement after stretch in striated muscle. A consequence of increased myofilament overlap? J. Physiol. 590, 1339–1345.

Edman, K.A.P., Elzinga, G., Noble, M.I., 1982. Residual force enhancement after stretch of contracting frog single muscle fibers. J. Gen. Physiol. 80, 769–784.

English, A.W., Letbetter, W.D., 1982. Anatomy and innervation patterns of cat lateral gastrocnemius and plantaris muscles. Am. J. Anat. 164, 67–77.

Faulkner, J.A., Claflin, D.R., McCully, K.K., 1986. Power output of fast and slow fibers from human skeletal muscles. In: Jones, N.L., McCartney, N., McComas, A.J. (Eds.), Human Muscle Power. Human Kinetics, Champaign.

Ferris, D.P., Czerniecki, J.M., Hannaford, B., 2005. An ankle-foot orthosis powered by artificial pneumatic muscles. J. Appl. Biomech. 21, 189–197.

Geyer, H., Herr, H., 2010. A muscle-reflex model that encodes principles of legged mechanics produces human walking dynamics and muscle activities. IEEE Trans. Neural Syst. Rehabil. 18, 263–273.

Geyer, H., Seyfarth, A., Blickhan, R., 2003. Positive force feedback in bouncing gaits? Proc. R. Soc. Ser. B 270 (1529), 2173–2183.

Gordon, A.M., Huxley, A.F., Julian, F.J., 1966. The variation in isometric tension with sarcomere length in vertebrate muscle fibres. J. Physiol. 184, 170–192.

Gruber, K., Ruder, H., Denoth, J., Schneider, K., 1998. A comparative study of impact dynamics: wobbling mass model versus rigid body models. J. Biomech. 31, 439–444.

Günther, M., Sholukha, V.A., Kessler, D., Wank, V., Blickhan, R., 2003. Dealing with skin motion and wobbling masses in inverse dynamics. J. Mech. Med. Biol. 3 (3), 309–335.

Haeufle, D.F.B., Grimmer, S., Seyfarth, A., 2010. The role of intrinsic muscle properties for stable hopping – stability is achieved by the force–velocity relation. Bioinspir. Biomim. 5, 016004.

Haeufle, D.F.B., Grimmer, S., Kalveram, K.T., Seyfarth, A., 2012. Integration of intrinsic muscle properties, feed-forward and feedback signals for generating and stabilizing hopping. J. R. Soc. Interface 9, 1458–1469.

Haeufle, D.F.B., Günther, M., Wunner, G., Schmitt, S., 2014. Quantifying control effort of biological and technical movements: an information-entropy-based approach. Phys. Rev. E 89, 012716.

Heidlauf, T., Klotz, T., Rode, C., Siebert, T., Röhrle, O., 2016. A continuum-mechanical skeletal muscle model including actin–titin interaction predicts stable contractions on the descending limb of the force–length relation. PLoS Comput. Biol. http://dx.doi.org/10.1371/journal.pcbi.1005773. in press.

Henneman, E., Somjen, G., Carpenter, D.O., 1965. Functional significance of cell size in spinal motor neurons. J. Neurophysiol. 28, 560–580.

Herzog, W., Leonard, T.R., 1997. Depression of cat soleus-forces following isokinetic shortening. J. Biomech. 30, 865–872.

Herzog, W., Guimaraes, A.C., Anton, M.G., Carter-Erdman, K.A., 1991. Moment–length relations of rectus femoris muscles of speed skaters/cyclists and runners. Med. Sci. Sports Exerc. 23, 1289–1296.

Herzog, W., Leonard, T.R., Joumaa, V., Mehta, A., 2008. Mysteries of muscle contraction. J. Appl. Biomech. 24, 1–13.

Hill, A.V., 1938. The heat of shortening and the dynamic constants of muscle. Proc. R. Soc. Lond. B, Biol. Sci. 126, 136–195.

Hof, A., 2001. The force resulting from the action of mono- and biarticular muscles in a limb. J. Biomech. 34, 1085–1089.

Hortobagyi, T., Hill, J.P., Houmard, J.A., Fraser, D.D., Lambert, N.J., Israel, R.G., 1996. Adaptive responses to muscle lengthening and shortening in humans. J. Appl. Physiol. 80, 765–772.

Hosoda, K., Sakaguchi, Y., Takayama, H., Takuma, T., 2010. Pneumatic-driven jumping robot with anthropomorphic muscular skeleton structure. Auton. Robots 28, 307–316.

Huxley, A.F., 1957. Muscle structure and theories of contraction. Prog. Biophys. Biophys. Chem. 7, 255–318.

Huxley, H.E., 1957. The double array of filaments in cross-striated muscle. J. Biophys. Biochem. Cytol. 3, 631–648.

Huxley, A.F., 2000. Cross-bridge action: present views, prospects, and unknowns. J. Biomech. 33, 1189–1195.

Huxley, H.E., Hanson, J., 1954. Changes in the crossstriations of muscle during contraction and stretch and their structural interpretation. Nature 173, 973–976.

Huxley, A.F., Niedergerke, R., 1954. Structural changes in muscle during contraction: interference microscopy of living muscle fibres. Nature 173, 971–973.

Jones, D., Round, J., De Haan, A., 2004. Skeletal Muscle: From Molecules to Movement. Churchill, Livingstone, Edinburgh.

Joumaa, V., Herzog, W., 2010. Force depression in single myofibrils. J. Appl. Physiol. 108, 356–362.

Joumaa, V., Leonard, T.R., Herzog, W., 2008. Residual force enhancement in myofibrils and sarcomeres. Proc. - Royal Soc., Biol. Sci. 275, 1411–1419.

Katz, B., 1939. The relation between force and speed in muscular contraction. J. Physiol. 96, 45–64.

Knappeis, G.G., Carlsen, F., 1962. The ultrastructure of the Z disc in skeletal muscle. J. Cell Biol. 13, 323–335.

Lakatos, D., Rode, C., Seyfarth, A., Albu-Schäfer, A., 2014. Design and control of compliantly actuated bipedal running robots: concepts to exploit natural system dynamics. In: IEEE/RAS International Conference on Humanoid Robots (Humanoids). International Society for Optics and Photonics, pp. 930–937.

LaStayo, P.C., Pierotti, D.J., Pifer, J., Hoppeler, H., Lindstedt, S.L., 2000. Eccentric ergometry: increases in locomotor muscle size and strength at low training intensities. Am. J. Physiol., Regul. Integr. Comp. Physiol. 278, R1282–R1288.

Leonard, T.R., Herzog, W., 2010. Regulation of muscle force in the absence of actin–myosin-based cross-bridge interaction. Am. J. Physiol., Cell Physiol. 299, C14–C20.

Lindstedt, S.L., Reich, T.E., Keim, P., LaStayo, P.C., 2002. Do muscles function as adaptable locomotor springs? J. Exp. Biol. 205, 2211–2216.

Loeb, G.E., 2012. Biol. Cybern., 1–9.

Loeb, G.E., Pratt, C.A., Chanaud, C.M., Richmond, F.J., 1987. Distribution and innervation of short, interdigitated muscle fibers in parallel-fibered muscles of the cat hindlimb. J. Morphol. 191, 1–15.

Loeb, G.E., Brown, I.E., Cheng, E.J., 1999. A hierarchical foundation for models of sensorimotor control. Exp. Brain Res. 126, 1–18.

Luff, A.R., 1981. Dynamic properties of the inferior rectus, extensor digitorum longus, diaphragm and soleus muscles of the mouse. J. Physiol. 313, 161–171.

MacIntosh, B.R., Gardiner, P.F., McComas, A.J., 2006. Skeletal Muscle: Form and Function, 2nd edition. Human Kinetics, Champaign.

Marechal, G., Plaghki, L., 1979. The deficit of the isometric tetanic tension redeveloped after a release of frog muscle at a constant velocity. J. Gen. Physiol. 73, 453–467.

Maykranz, D., Grimmer, S., Seyfarth, A., 2013. A work-loop method for characterizing leg function during sagittal plane movements. J. Appl. Biomech. 29, 616–621.

McBride, J.M., Triplett-McBride, T., Davie, A.J., Abernethy, P.J., Newton, R.U., 2003. Characteristics of titin in strength and power athletes. Eur. J. Appl. Physiol. 88, 553–557.

McMahon, T.A., 1984. Muscles, Reflexes and Locomotion. Princeton University Press, Princeton.

Morgan, D.L., 1990. New insights into the behavior of muscle during active lengthening. Biophys. J. 57, 209–221.

Morl, F., Siebert, T., Haufle, D., 2016. Contraction dynamics and function of the muscle-tendon complex depend on the muscle fibre-tendon length ratio: a simulation study. Biomech. Model. Mechanobiol. 15, 245–258.

Morlock, M.M., Bonin, V., Hansen, I., Schneider, E., Wolter, D., 1999. Die Rolle der Muskulatur bei bandscheibenbedingten Erkrankungen der Wirbelsäule. In: Radant, S., Grieshaber, R., Schneider, W. (Eds.), Prävention von arbeitsbedingten Gesundheitsgefahren und Erkrankungen des Stütz- und Bewegungssystems. monade Verlag, Leipzig, pp. 68–89.

Mozdziak, P.E., Pulvermacher, P.M., Schultz, E., 2000. Unloading of juvenile muscle results in a reduced muscle size 9 wk after reloading. J. Appl. Physiol. 1985 (88), 158–164.

Nakagawa, Y., Hayashi, K., Yamamoto, N., Nagashima, K., 1996. Age-related changes in biomechanical properties of the Achilles tendon in rabbits. Eur. J. Appl. Physiol. Occup. Physiol. 73, 7–10.

Nelson, F.E., Gabaldon, A.M., Roberts, T.J., 2004. Force–velocity properties of two avian hindlimb muscles. Comp. Biochem. Physiol., Part A, Mol. Integr. Physiol. 137, 711–721.

Nigg, B.M., Herzog, W., 2007. Biomechanics of the Musculo-Skeletal System, 3rd edition. John Wiley & Sons, Chichester.

Niiyama, R., Nagakubo, A., Kuniyoshi, Y., 2007. Mowgli: a bipedal jumping and landing robot with an artificial musculoskeletal system. In: 2007 IEEE International Conference on Robotics and Automation. IEEE, pp. 2546–2551.

Nishikawa, K.C., Monroy, J.A., Uyeno, T.E., Yeo, S.H., Pai, D.K., Lindstedt, S.L., 2012. Is titin a 'winding filament'? A new twist on muscle contraction. Proc. R. Soc. Lond. B, Biol. Sci. 279 (1730), 981–990.

Noble, M.I., 1992. Enhancement of mechanical performance of striated muscle by stretch during contraction. Exp. Physiol. 77 (4), 539–552.

Pijnappels, M., Bobbert, M.F., van Dieen, J.H., 2005. How early reactions in the support limb contribute to balance recovery after tripping. J. Biomech. 38, 627–634.

Pinniger, G.J., Ranatunga, K.W., Offer, G.W., 2006. Crossbridge and non-crossbridge contributions to tension in lengthening rat muscle: force-induced reversal of the power stroke. J. Physiol. 573, 627–643.

Prilutsky, B.I., Gregor, R.J., Ryan, M.M., 1998. Coordination of two-joint rectus femoris and hamstrings during the swing phase of human walking and running. Exp. Brain Res. 120, 479–486.

Proske, U., Morgan, D.L., 2001. Muscle damage from eccentric exercise: mechanism, mechanical signs, adaptation and clinical applications. J. Physiol. 537 (2), 333–345.

Ramsey, R.W., Street, S.F., 1940. The isometric length tension diagram of isolated skeletal muscle fibers of the frog. J. Cell. Compar. Physiol. 15, 11–34.

Ranatunga, K.W., Thomas, P.E., 1990. Correlation between shortening velocity, force–velocity relation and histochemical fibre-type composition in rat muscles. J. Muscle Res. Cell Motil. 11, 240–250.

Reinhardt, L., Siebert, T., Leichsenring, K., Blickhan, R., Böl, M., 2016. Intermuscular pressure between synergistic muscles correlates with muscle force. J. Exp. Biol.. http://dx.doi.org/10.1242/jeb.135566.

Riemann, B.L., Lephart, S.M., 2002. The sensorimotor system, part I: the physiologic basis of functional joint stability. J. Athl. Train. 37, 71.

Rijkelijkhuizen, J.M., de Ruiter, C.J., Huijing, P.A., de Haan, A., 2003. Force/velocity curves of fast oxidative and fast glycolytic parts of rat medial gastrocnemius muscle vary for concentric but not eccentric activity. Pflügers Arch. 446, 497–503.

Roberts, T.J., Marsh, R.L., Weyand, P.G., Taylor, C.R., 1997. Muscular force in running turkeys: the economy of minimizing work. Science 275, 1113–1115.

Rode, C., Siebert, T., Blickhan, R., 2009a. Titin-induced force enhancement and force depression: a 'sticky-spring' mechanism in muscle contractions? J. Theor. Biol. 259 (2), 350–360.

Rode, C., Siebert, T., Herzog, W., Blickhan, R., 2009b. The effects of parallel and series elastic components on the active cat soleus force–length relationship. J. Mech. Med. Biol. 9, 105–122.

Rode, C., Siebert, T., Tomalka, A., Blickhan, R., 2016. Myosin filament sliding through the Z-disc relates striated muscle fibre structure to function. Proc. R. Soc. B 283, 20153030.

Rome, L.C., Sosnicki, A.A., Goble, D.O., 1990. Maximum velocity of shortening of three fibre types from horse soleus muscle: implications for scaling with body size. J. Physiol. 431, 173–185.

Rupp, T.K., Ehlers, W., Karajan, N., Gunther, M., Schmitt, S., 2015. A forward dynamics simulation of human lumbar spine flexion predicting the load sharing of intervertebral discs, ligaments, and muscles. Biomech. Model. Mechanobiol. 14, 1081–1105.

Scholz, D., Maufroy, C., Kurowski, S., von Radkhah, K., Stryk, O., Seyfarth, A., 2012. Simulation and experimental evaluation of the contribution of biarticular gastrocnemius structure to joint synchronization in human-inspired three-segmented elastic legs. In: 3rd International Conference on Simulation, Modeling and Programming for Autonomous Robots (SIMPAR), pp. 251–260.

Seyfarth, A., Blickhan, R., Van Leeuwen, J.L., 2000. Optimum take-off techniques and muscle design for long jump. J. Exp. Biol. 203 (4), 741–750.

Sharbafi, M., Rode, C., Kurowski, S., Scholz, D., Möckel, R., Radkhah, K., Zhao, G., Rashty, A., von Stryk, O., Seyfarth, A., 2016. A new biarticular actuator design facilitates control of leg function in BioBiped3. Bioinspir. Biomim. 11, 046003.

Sharbafi, M.A., Rashty, A.M.N., Rode, C., Seyfarth, A., 2017. Reconstruction of human swing leg motion with passive biarticular muscle models. Human Movement Science 52, 96–107. https://doi.org/10.1016/j.humov.2017.01.008.

Siebert, T., Rode, C., 2014. Computational modeling of muscle biomechanics. In: Computational Modelling of Biomechanics and Biotribology in the Musculoskeletal System: Biomaterials and Tissues, pp. 173–243.

Siebert, T., Leichsenring, K., Rode, C., Wick, C., Stutzig, N., Schubert, H., Blickhan, R., Bol, M., 2015. Three-dimensional muscle architecture and comprehensive dynamic properties of rabbit gastrocnemius, plantaris and soleus: input for simulation studies. PLoS ONE 10, e0130985.

Siebert, T., Rode, C., Till, O., Stutzig, N., Blickhan, R., 2016. Force reduction induced by unidirectional transversal muscle loading is independent of local pressure. J. Biomech. 49, 1156–1161.

Stienen, G.J., Versteeg, P.G., Papp, Z., Elzinga, G., 1992. Mechanical properties of skinned rabbit psoas and soleus muscle fibres during lengthening: effects of phosphate and Ca^{2+}. J. Physiol. 451, 503–523.

Swammerdam, J., 1737. Biblia Naturae. Isaak Severinus, Boudewyn and Peter van der Aa, Leiden.

Tabary, J.C., Tabary, C., Tardieu, C., Tardieu, G., Goldspink, G., 1972. Physiological and structural changes in the cat's soleus muscle due to immobilization at different lengths by plaster casts. J. Physiol. 224, 231–244.

Till, O., Siebert, T., Rode, C., Blickhan, R., 2008. Characterization of isovelocity extension of activated muscle: a Hill-type model for eccentric contractions and a method for parameter determination. J. Theor. Biol. 255, 176–187.

Tokur, D., Rode, C., Hoitz, F., Seyfarth, A., 2015. Response of leg muscles (mono- vs. bi-articular) to horizontal perturbations in the sagittal plane. In: 25th Congress of the International Society of Biomechanics (ISB2015), pp. 923–925.

Tomalka, A., Rode, C., Schumacher, J., Siebert, T., 2017. The active force–length relationship is invisible during extensive eccentric contractions in skinned skeletal muscle fibres. Proc. R. Soc. B 284, 20162497.

Trombitas, K., Tigyi-Sebes, A., 1985. How actin filament polarity affects crossbridge force in doubly overlapped insect muscle. J. Muscle Res. Cell Motil. 6, 447–459.

van Ingen Schenau, G., 1989. From rotation to translation: constraints on multi-joint movements and the unique action of bi-articular muscles. Hum. Mov. Sci. 8, 301–337.

van Soest, A.J., Bobbert, M.F., 1993. The contribution of muscle properties in the control of explosive movements. Biol. Cybern. 69, 195–204.

van Soest, A.J., Schwab, A.L., Bobbert, M.F., van Ingen Schenau, G.J., 1993. The influence of the biarticularity of the gastrocnemius muscle on vertical-jumping achievement. J. Biomech. 26, 1–8.

Williams, C.D., Salcedo, M.K., Irving, T.C., Regnier, M., Daniel, T.L., 2013. The length–tension curve in muscle depends on lattice spacing. Proc. R. Soc. B 280, 20130697.

Wilson, A.M., McGuigan, M.P., Su, A., van den Bogert, A.J., 2001. Horses damp the spring in their step. Nature 414, 895–899.

Winter, D.A., 2005. BioMechanics and Motor Control of Human Movement, 3rd edition. John Wiley & Sons, New Jersey, USA.

Yucesoy, C.A., Koopman, B.H., Baan, G.C., Grootenboer, H.J., Huijing, P.A., 2003. Extramuscular myofascial force transmission: experiments and finite element modeling. Arch. Physiol. Biochem. 111, 377–388.

Zou, P., Pinotsis, N., Lange, S., Song, Y.H., Popov, A., Mavridis, I., Mayans, O.M., Gautel, M., Wilmanns, M., 2006. Palindromic assembly of the giant muscle protein titin in the sarcomeric Z-disk. Nature 439, 229–233.

Chapter 8.2

From Stiff to Compliant Actuation

Maarten Weckx, Bram Vanderborght, and Dirk Lefeber

Department of Mechanical Engineering, Vrije Universiteit Brussel, Brussels, Belgium

This chapter discusses the evolution from using stiff servomotors in humanoid walking to the inclusion of virtual compliance and eventually physical compliance. Firstly, the addition of compliant elements in series with a motor and possible drive train will be discussed. This is followed by a discussion on the expansion towards variable physical compliance and the categorization regarding variable stiffness actuators (VSAs). Subsequently, the possibility of adding parallel compliant elements to an actuator scheme is discussed. The chapter is finalized with a discussion on the use of locking mechanisms in actuator schemes and the implementation of compliant actuators in multi-DoF joints of humanoids.

8.2.1 STIFF SERVOMOTOR

Most of the robotic systems consist of traditional stiff servomotors. They consist of an electric motor connected with a high-gear transmission to reduce the rotational speed and increase the torque. The motors are controlled by a feedback loop, often with a high gain PD, based on the measurement of motor position. Their ideal working principle is that the desired position is quickly reached and maintained, regardless of the external forces exerted on the actuator; within the force limits of the device (Hogan, 1985). The aim for each joint is to be as stiff as possible or to approach infinite impedance. The motor is consequently a position source and the behavior is excellent when a desired trajectory must be tracked with a high bandwidth and with high accuracy. These properties are advantageous for industrial robots where precise and fast position control must be achieved, but due to their dynamical properties they cannot match the requirements of robots that need to interact with humans and with an unknown, dynamic environment. The dynamic properties of the motor, for instance, mass, inertia, and stiffness, heavily influence the control of the entire robotic system. The actuators cannot achieve comparable motion, safety, and energy efficiency of a human. Position control in any task in which a robot interacts with the environment is not a properly posed problem because the controller is dependent on parameters that cannot represent the interaction. As a result, industrial robots with servomotors are often placed in cages since the robot cannot "feel" an inter-action with the dynamic environment and will deploy its full power to continue the position commands.

Advantages of servomotors are the facts that it is compact, easy to fit in a design and easy to control with a PD controller. Main advantage is the excellent behavior when a desired trajectory must be tracked with a high bandwidth and with high accuracy. A servomotor has several disadvantages too:

- Shocks introduce large forces in the mechanism since impedance is infinite;
- The servomotors are not safe for human–robot interaction or for interaction with any dynamic environment in general;
- Since no compliant element is implemented in the hardware, a servomotor cannot store and release energy. As a result they cannot be loaded for explo-sive motions and they have to deliver energy for both positive and negative powers.

The HRP-3 (Matsui et al., 2005) and KHR-3 (HUBO) (Park et al., 2006) hu-manoids use local high gain PD position controllers of which the latter's tracking controller runs at 1000 Hz. RABBIT uses a similar PD control to ensure the tracking of trajectories (Sabourin et al., 2006). Mahru-III uses servo controlled DC motors, but incorporates active impedance control for landing and uneven terrain walking (Kwon et al., 2007).

8.2.2 STIFF SERVOMOTOR WITH ACTIVE COMPLIANCE

Position control in a task in which a robot interacts with the environment is not a properly posed problem, as previously mentioned. The controller is namely dependent on parameters that are out of the control potential. On the other hand, controlling the impedance and the so-called equilibrium position is a well-posed problem. Within certain boundaries, this can be done without knowledge of the environment. Compliant actuators allow for deviations from their equilibrium position, depending on the applied external force. The equilibrium position of a compliant actuator is defined as the position of the actuator where the actuator generates zero force or zero torque. This concept is specifically introduced for compliant actuators, since it does not exist for stiff actuators (Van Ham et al., 2009).

This trend was started by Hogan in 1984 (Hogan, 1985) with his work on impedance control which was of particular interest for robotics that had to han-dle objects in a dynamic environment. Since the compliance is implemented in the controller software and not in the hardware it is referred to as active compli-ance. The actively controlled compliance will mimic the behavior of a spring. Based on the measurements of the external force or torque, a certain deviation is calculated and set by the stiff actuator. A deviation from the desired position is thus allowed and artificially implemented by means of a desired and controlled compliance. This type of compliant actuator requires an actuator, a sensor, and a controller which are all fast enough for the application. The main disadvantage of active compliant actuators is the continuous energy dissipation, due to the inherent continuous control and the absence of a physical elastic element, in which energy can be stored and subsequently released. The limited bandwidth of the controller results in the disability of the actuator system to absorb shocks. Exploiting the natural dynamics of the actuated system in an energy efficient way, as is done for passive dynamic walkers, is not possible since energy is required to move. Furthermore, the controller is quite complex and the dynamics of the system need to be known very well. An advantage of active compliance is that the controller can make the quasistiffness online adaptable in a theoretical infinite range and with infinite speed. In a similar way as stiffness is implemented, the controller can also incorporate damping into the system, resulting in an active impedance actuator (VIA).

This radical new concept of active compliance opened a vast amount of research directions which led for example only recently to the commercialization of the DLR lightweight arm by KUKA. We can state that the technology is mature and the commercial version is used in many applications (Albu-Schaffer et al., 2008). TORO, which evolved from the DLR-Biped (Ott et al., 2012), is a humanoid with impedance-controlled legs driven by stiff electrical drives (Englsberger et al., 2014). SARCOS's CB (Herzog et al., 2014) and Petman (Nelson et al., 2012) have torque-controlled hydraulic actuators. The humanoid Mahru-III is driven by stiff servomotors and uses an impedance controller for landing and walking on uneven terrain (Kwon et al., 2007). The planar 4-DoF biped Amber uses torque-controlled DC motors and is able to walk over inclined surfaces and uneven terrain (Yadukumar et al., 2012). HyQ is an example of a hydraulic quadruped that is torque-controlled (Boaventura et al., 2012).

An active compliant actuator is equal to a servomotor concerning its hardware and thus also has the advantage of being compact and easy to fit in a design. The impedance control provides a certain level of safety for human-robot interaction, nonetheless due to the limited bandwidth of the controller no shocks can be absorbed, e.g., hitting the system with a baseball bat will not be handled by the control and probably the system will simply break. It also requires torque sensors, which can be quite expensive. Furthermore, an active compliant actuator has the disadvantage, like servomotors, that no energy can be stored in the actuation system.

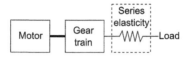

FIGURE 8.2.1   Scheme of the SEA.

8.2.3   SERIES ELASTIC ACTUATOR (SEA)

A logic extension of Hogan's work on implementing impedance by control was to introduce intrinsic compliant elements to robotic hardware. This started with using hydraulic springs in hopping robots (Raibert, 1986). Next, Pratt and Williamson introduced the series elastic actuator (SEA) (Pratt and Williamson, 1995) which consists of a servomotor with gear train in series with a spring as shown in the scheme in Fig. 8.2.1. The introduction of spring elements in robotic actuation was inspired by biology, for example, the work of Alexander (1988) and Geyer et al. (2006) explains the importance of elasticity in animal and human locomotion. The equilibrium position is the only controllable parameter, hence at least one motor is required. The introduction of a passive elastic element has several advantages with respect to a stiff servomotor:

- The passive elastic element can store and subsequently release energy. As a result, the peak power and energy requirements for a certain application can be lowered;
- Due to the passive compliance, SEAs allows certain deviations from the equilibrium position which makes them safer for human–robot interaction;
- The passive elastic element has an infinite bandwidth for shock absorption.

Of course, there are some disadvantages too:

- The system becomes more complex since the elastic element needs to be added;
- The system might store large amounts of energy in the elastic element that can potentially create an unsafe situation when released quickly (Van Damme et al., 2009).

When the impedance becomes zero, an extreme case arises where the actuator becomes a force/torque source like gravity. Examples are constant torque springs and direct-drive motors. The latter are used in the MIT Cheetah (Folkertsma et al., 2012) and in a leg consisting of two direct-drive dc motors coupled to a symmetric five bar linkage (Kenneally and Koditschek, 2015).

The planar biped Flame possesses bidirectional SEAs in the hip and knee joints. Whereas unidirectional SEAs with passive return springs are used in the ankles. The motor for the ankle actuators are placed in the upper body for a more favorable mass distribution in the limbs and are connected to the springs in the ankles by means of Bowden cables (Hobbelen et al., 2008). The planar BioBiped has bidirectional SEAs in the hip joints. These consist of a belt–pulley transmission of which the driving pulley is connected to the joint via small compression springs in an inner ring aligned around the joint. Unidirectional SEAs are used in the other leg joints, which consist of DC motors that wind up a cable on a pulley to pull a spring that is attached to a lever (Radkhah et al., 2012). The preceding JenaWalker I and II used the same actuation concept (Seyfarth et al., 2009). The sagittal knee joints and frontal ankle joints of COMAN are driven by SEAs that centralize all components in a compact module and utilize small compression springs as well (Li et al., 2013). The NASA-JRC Valkyrie humanoid uses a mix of rotational SEAs with custom-made torsion springs and ball–screw driven SEAs with compression springs placed between the DC motor and the ground, as opposed to between the DC motor and the load (Paine et al., 2014, 2015). Herbert's SEAs are based on ball–screw driven belt–pulley transmissions. The inside of the output pulley of those transmissions is machined to become a torsion spring. Thus, the output pulleys serve as the compliant element and are directly connected to the joint axes (Pierce and Cheng, 2014). Samsung's Roboray uses similar ball–screw driven belt–pulley transmissions, but with elastic wires substituting normal timing belts and acting as the compliant elements (Kim et al., 2012). The planar running robot ScarlETH with electrical SEAs employs compliant with high damping position control during flight phase and torque control during stance phase (Hutter et al., 2013).

8.2.3.1 Example: SEA in an Ankle Prosthesis

Since the passive compliant element, typically a spring, in an SEA can store and release energy, the peak power (PP) and energy requirements (ER) for the motor in certain applications can be lowered by storing negative work and releasing this energy during periods of positive work. In order to clearly indicate this, an example comparing the use of an SEA to a stiff servomotor in an ankle prosthesis is simulated. The simulations are inspired by previous work by Grimmer and Seyfarth (2011) and have been published by Mathijssen et al. regarding the potential of the series–parallel elastic actuator (SPEA) in an ankle prosthesis (Mathijssen et al., 2013).

The requirements of a sound human ankle (torque, speed, and power) are taken from Winter (1984) and summarized in Fig. 8.2.2. In case the ankle prosthesis is actuated by a stiff servomotor, the requirements for this motor would logically be identical to Winter's data.

The power curve in Fig. 8.2.2 clearly shows a high positive peak which is required to generate the push off. During the first 40% of the gait cycle, the ankle produces negative power. The total required energy (by taking the integral

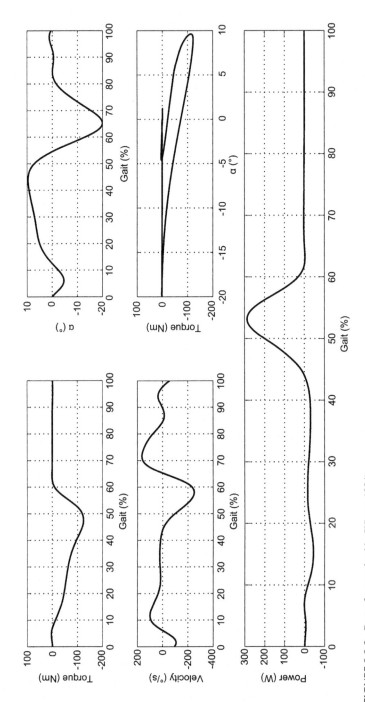

FIGURE 8.2.2 Data of a sound ankle (Winter, 1984).

of the power curve) is 35 J, of which 25 J are required to deliver the positive power and 9 J to deliver the negative power.

Regarding energy requirements, an "ideal actuator" should store the 9 J required to deliver the negative work and release this during periods of positive work. As a result, theoretically this ideal actuator only requires 16 J.

An SEA can reduce the peak power by storing and releasing energy so that the motor speed can be reduced. It is important to note, however, that the torque requirements can never be changed. Of course the torque requirements can be scaled down by means of a gearbox, but the shape of the torque curve cannot be changed. Due to the possibility of reshaping the speed curve, the power curve can be reshaped as well and as such the peak power can potentially be lowered.

Fig. 8.2.3 summarizes the results of a simulation regarding the actuation of an ankle prosthesis with a stiff servomotor or an SEA. Main goal is to show that the peak power and energy requirements can be lowered, and that for both a minimum of the spring stiffness in the SEA can be found. The upper left plot indicated that indeed a minimum can be found regarding the peak power. When the stiffness is smaller, the peak power raises since this corresponds to high speeds since large spring extensions are required. When the stiffness is bigger, the peak power increases too, with an asymptote towards infinity. This is logical, since an infinite stiffness corresponds to a stiff servomotor and thus the peak power goes toward 280 W which was the peak power of Winter's data in Fig. 8.2.2.

Fig. 8.2.3 also clearly indicates that the motor torque requirements do not change, regardless of the change in stiffness. The speed requirements on the other hand do clearly change and as a result the peak power can be reduced (dashed curves) and the energy requirements too (dashed–dotted curves).

8.2.4 VARIABLE STIFFNESS ACTUATOR (VSA)

The SEA discussed previously is a passive compliant actuator with fixed stiffness. The most basic scheme consists of only one motor that enables to change the equilibrium position of the actuator. A variable stiffness actuator (VSA) is a passive compliant actuator that has a controllable stiffness. As a result, at least two motors are required for controlling both the equilibrium position and the stiffness. In the literature they have been referred to with varying terminology, e.g., actuators with passive variable stiffness, controllable stiffness or adjustable compliance. In 2009 Van Ham et al. (2009) categorized the VSAs in four groups according to the mechanism that is used to change the stiffness. This reasoning is further extended by Vanderborght et al. (2013). The new categorization is summarized in Fig. 8.2.4. In the following subsections these categories will be briefly discussed. For designing an application like a biped, Grioli et al. (2015)

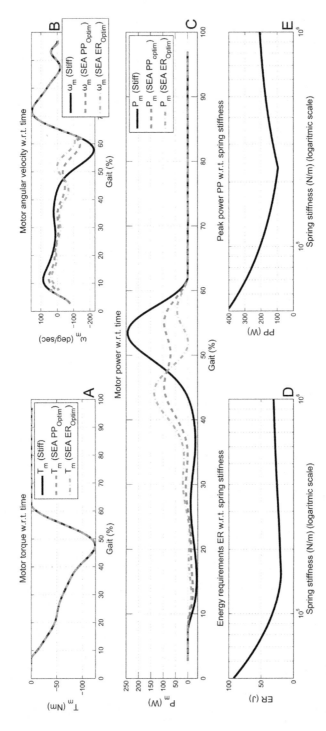

FIGURE 8.2.3 Different minima for the spring stiffness in an SEA can be found to either minimize the PP or the ER (Mathijssen et al., 2013).

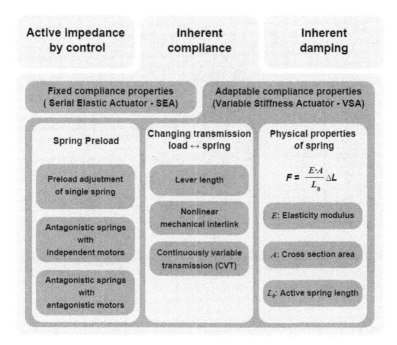

FIGURE 8.2.4 Categorization of VSAs (Vanderborght et al., 2013).

provide design procedures and data presentation of how a generic VSA could be organized so as to minimize the engineer's effort in choosing the actuator type and size that would best fit the application needs.

8.2.4.1 Spring Preload

A way of changing the stiffness of a VSA is by adjusting the preload of a spring. This can be done in three ways. An example of each of the three is presented in Fig. 8.2.5.

The first mechanism, shown in Fig. 8.2.5A, utilizes an antagonistic setup of actuators, whether they are artificial muscles or SEAs. At least two nonlinear springs are required in such a setup to obtain an actuator with variable compliance (Van Ham et al., 2006). In general, this is a disadvantage. Opposing movement of both motors changes the preload of the spring and changes the stiffness. Movement in the same direction, on the other hand, changes the equilibrium position. The main disadvantage of this working principle is that both motors need to work synchronously to control the equilibrium position or the stiffness separately. This means that the motors cannot be separately dimensioned for the equilibrium position control task and the stiffness modulation

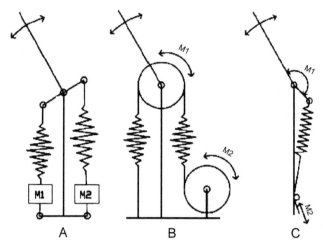

FIGURE 8.2.5 Three different setups for a VSA with spring preload mechanism. Antagonistic springs with antagonistic motors in (A). Antagonistic springs with independent motors in (B). Preload adjustment of single spring in (C) (Vanderborght et al., 2013).

task. Examples are the VSA-II (Schiavi et al., 2008), the VSA-cube (Catalano et al., 2011), and the BAVS (Friedl et al., 2011). The following bipeds are examples of the implementation of this type of antagonistic actuators: Pneumat-BB using McKibben muscles (Narioka et al., 2012) and Lucy using pleated pneumatic artificial muscles (Vanderborght et al., 2008).

A second mechanism, shown in Fig. 8.2.5B, uses a different arrangement of motors in an antagonistic spring setup to at least partially decouple the equilibrium and stiffness control. In this case one of the motors is dedicated to changing the springs' preload. Examples are the quasiantagonistic joint (Eiberger et al., 2010) and the AMASC (Hurst et al., 2004). The compliant asymmetric antagonistic actuator, on the other hand, has a dedicated motor for driving the joint and a second, smaller motor dedicated to store potential energy in a rubber-type elastic cord for release in the joint (Roozing et al., 2015).

A third mechanism, shown in Fig. 8.2.5C, does not require an antagonistic setup and uses only one spring. Again the preload of the spring is adjusted to change the stiffness of the VSA. The MACCEPA (Van Ham et al., 2007) is an example of a VSA that is based on this mechanism. It consists of a lever arm that controls the equilibrium position. The lever arm is connected to a fixed pivot point on the output link by means of a spring. When the lever arm and the output link are aligned, the exerted spring force is aligned with output link and no torque is exerted. When a deviation angle between the lever arm and the output link exists, the exerted spring force possesses a perpendicular component relative to the output link and a torque is exerted on the output.

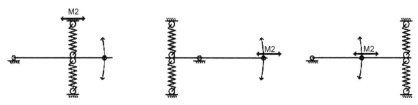

FIGURE 8.2.6 Three possibilities of controllable transmission ratio: changing force, changing spring, and changing pivot point (Vanderborght et al., 2013).

While the antagonistic setups require nonlinear springs or designs that transform linear springs into nonlinear elastic mechanisms, the MACCEPA just requires a linear spring. Its torque–angle characteristic is quasilinear. Furthermore, both equilibrium position and stiffness are controlled independently. The MACCEPA 2.0 (Vanderborght et al., 2011) uses a cam shaped lever arm allowing to shape the torque–angle relation according to the application. The MACCEPA has been implemented in three bipeds (Mao et al., 2007; Huang et al., 2013; Weckx et al., 2014; Rodriguez Cianca et al., 2015).

8.2.4.2 Changing Transmission Between Load and Spring

The stiffness can also be controlled by controlling the configuration of a lever mechanism. This can be done in three ways as shown in Fig. 8.2.6: changing the force, changing the spring, and changing the pivot point. Several actuators described in the literature use one of these mechanisms. In the automatic rigidity/compliance switching VSA (Cui et al., 2014) and the HVSA (Kim and Song, 2012) the force and pivot point are kept fixed, while the spring point is changed to change the stiffness. In the AWAS-II (Jafari et al., 2012) and the vsaUT-II (Groothuis et al., 2014) the force and the spring point are kept fixed but instead the pivot point is changed. The biped walkers Blue and miniBlue are actuated with VSAs that vary the spring point in order to change the joint stiffness (Enoch and Vijayakumar, 2015).

8.2.4.3 Changing the Physical Properties of the Spring

The variations in stiffness of the spring are obtained by changing the effective physical structure of the spring. This is best explained via the standard elastic law in Vanderborght et al. (2013):

$$F = \frac{EA}{L_0}\Delta L = K\Delta L \qquad (8.2.1)$$

where F is the force, E the material modulus, A the cross-sectional area, L the effective beam length, and ΔL is the extension, respectively. The material modulus can logically not be changed by the structure. But F, A, and L can be changed structurally and can thus be used to change the stiffness. Many designs are discussed in literature. Changing the cross-sectional area can, for example, be done by rotating a beam with an aspect ratio that differs from one. Another way to change the stiffness is to vary the effective length of the spring. For example, by changing the number of active coils of a spring, as done in the Jack spring actuator (Hollander et al., 2005). The arched flexure VSA changes the engaged length of a nonuniform cantilever beam (Schimmels and Garces, 2015). Similarly the engaged length of a custom spline shaft torsion spring is varied in Schuy et al. (2013). A different example is the EPAIA that uses a hydraulic torsion spring of which the stiffness can be changed by adjusting the internal pressure (Misgeld et al., 2014).

8.2.5 PARALLEL STIFFNESS

This section will discuss the use of additional passive parallel elements that span a joint. Previously some examples were given to demonstrate that the use of an elastic element in series with the motor can reduce peak power (PP) and the energy requirements (ER) of the motor. As shown in Fig. 8.2.3, this can be done by reshaping the velocity curve of the motor. The torque curve of the motor, however, cannot be reshaped, only rescaled by changing the gear ratio. This, however, can be done by adding an elastic element in parallel to the actuator, which is often called load cancellation. The most common implementation of a load cancellation mechanism is in an everyday lamp where the springs counteract gravity in each joint and divert the gravity load on each link towards the base. Modern excavators are also equipped with a piston parallel to their boom assembly for gravity compensation. The gravity balancers are also referred to in literature as static balancers. Just Herder finished his PhD on statically balanced spring mechanism at the Delft Technical University (Herder, 1998). Static balancers are interesting for human friendly robots since they remove significant loads from the system by static balancing of gravity forces. The personal robot PR-1 from Willow Garage, for example, uses a gravity balancer system based on the work of Herder. Because the joint motors do not need to handle gravity loads, they can be low-torque and have small and efficient gear reduction which makes the PR-1 strong, yet light, and safe. Mettin et al. (2010) studied the use of springs in parallel to a direct drive that drives a planar underactuated double pendulum and called this a parallel elastic actuator (PEA). In this article, the author follows the intuitive idea of having spring-like elements generating most of the nominal torque of the target motion, while the control efforts of the original actuators are spent mainly in achieving feedback stabilization and uncertainty compensation. Mettin also showed that the passive parallel spring mechanisms have to be configured for specific sets of trajectories and they are particularly useful for repetitive or cyclic robot motions. It is indeed not possible to find spring configurations with only passive parallel springs that give arbitrary qualitative and quantitative torque functions, which means they are not always an option for complementary actuation.

In Wang et al. (2011) the potential of an SEA and a PEA in the hip, knee, and ankle of an exoskeleton were studied. In their exoskeleton, four degrees of freedom (DOFs) were actuated: the hip flexion/extension (F/E), hip abduction/adduction (A/A), knee flexion/extension (F/E), and ankle dorsi/plantar flexion (D/PF). Theoretical optimization results (Fig. 8.2.7) show that adding parallel springs can reduce the peak torque by 66%, 53%, and 48% for hip flexion/extension (F/E), hip abduction/adduction (A/A), and ankle dorsi/plantar flexion (D/PF), respectively, and the RMS power by 50%, 45%, and 61%, respectively. Adding a spring in series (forming a series elastic actuator, SEA) reduces the peak power by 79% for ankle D/PF and by 60% for hip A/A. An SEA does not reduce the peak power demand at other joints. Remarkable is the fact that the SEA and PEA do not improve the motor requirements at the knee. This is because the quasistiffness changes dramatically after heel-off, which means the motor has to produce large torques during swing phase if a PEA with fixed spring stiffness is added. For the ankle, a unidirectional PEA is proposed.

Au et al. solved this issue by implementing a unidirectional parallel spring in his prosthesis (Au et al., 2009). Like this, the required powers and torques in the stance phase are reduced while in the swing phase the parallel spring does not counteract the desired motion.

For the other joints (flexion/extension hip and knee, and inversion/eversion hip), similar results can be found. This is indicated in Fig. 8.2.8. The motor torque and power clearly decreases. For the frontal hip DOF the results can probably be improved a lot if the parallel spring would be decoupled during swing phase (since here the torque graph was originally zero).

Haeufle et al. developed a PEA that uses a clutch to engage and disengage a spring in parallel to a stiff servomotor. One-legged periodic hopping experiments with the PEA driving the knee joint showed a reduction of 80% in energy consumption and a reduction of 66% in peak torque requirements of the motor. During normal locomotion the clutch is ideally engaged at touchdown and disengaged at takeoff (Haeufle et al., 2012). The inclusion of parallel springs into the stiff servomotor driven biped walker STEPPR was investigated in Mazumdar et al. (2015). The biped has 3-DoF hips, 1-DoF knees, and 2-DoF ankles. The investigation was done for three gait types:

1. Human gait data, the "golden standard" for humanoids
2. Dynamic human-like robot walking simulation gaits

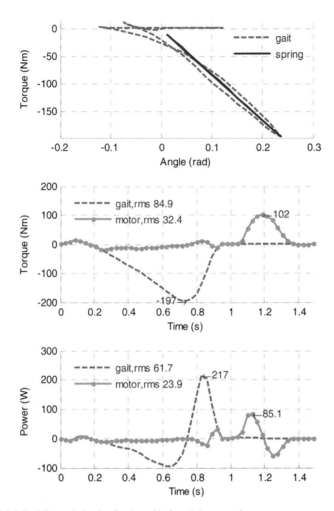

FIGURE 8.2.7 PEA optimization for the ankle for minimum peak power.

3. More static gaits, i.e., crouched gaits, that were typical for early humanoids

The authors concluded, based on simulations, that for all three gaits the sagittal ankle joint and the frontal hip joint benefit from a parallel spring. Under the condition that the parallel spring in the ankle is only engaged during stance phase. They furthermore conclude that the sagittal hip and knee joints only benefit from a parallel spring when gait type 1 applies. Subsequently, the frontal hip spring for gait type 3 and the sagittal ankle spring for gait type 2 were experimentally validated on a dedicated test bench. This resulted in the addition of the investigated parallel springs to the STEPPR biped. The inclusion of PEAs was also investigated with MIT's quadruped Cheetah, which also showed improvements

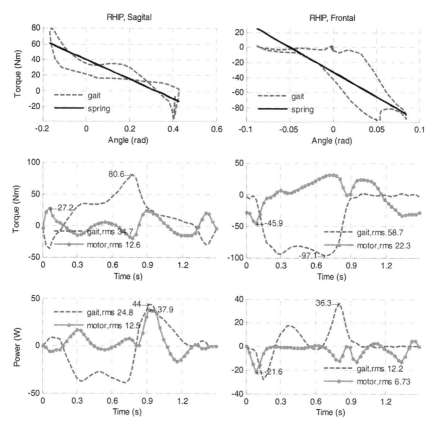

FIGURE 8.2.8 Spring stiffness optimization of the PEA for frontal and sagittal hip movement during normal walking.

in power consumption and the need for switchable parallel springs in certain joints (Folkertsma et al., 2012). One important disadvantage of a spring in parallel with a stiff servomotor is the fact that the PEA as a whole is still a stiff actuator. This is logical since the overall stiffness of two elements in parallel is the sum of the stiffness of each element. And since the servomotor is infinitely stiff, the overall stiffness of the PEA is as well. A high backdriveability is therefore desired when using a PEA. This can be achieved by low gear ratios for the motor, as is done in MIT's Cheetah and the STEPPR biped.

Eslamy et al. presented an extensive simulation study of different actuation schemes for ankle actuation using human gait measurements of walking and running at different velocities (0.5, 1, 1.6, 2.1, and 2.6 m/s). In the study they combined a stiff servomotor and an SEA with either a parallel spring or a unidirectional parallel spring. An SEA without any parallel spring was also included in the study. For every combination they subsequently optimized the spring stiffness and in case of a unidirectional parallel spring the engagement angle per measured gait type. The optimizations were done once by minimize the peak power requirements of the motor and once by minimizing the energy requirements of the motor. Finally, per optimization criteria they compared the peak power and energy requirements of the motors in every combination for the different speeds of walking and running. They conclude that the best selection of combination depends on the type of gait and the velocity (Eslamy et al., 2012).

8.2.6 LOCKING MECHANISMS

The two main reasons for using locking devices in robotic systems are energy management and reconfiguration. The most common use of locking devices in robotics is related to decreasing energy consumption. For instance, in the fields of mobile robotics, prosthetics, and exoskeletons energy consumption is paramount for performance. The advantage of the evolution towards compliant actuation in robotics is the possibility of storing and subsequently releasing energy from springs. As previously mentioned, this can lower the energy consumption of the actuators. The springs, however, are noncontrollable energy buffers. A solution to control the release of energy from springs is the use of continuous variable transmissions. These devices are, however, still in the development phase and are not readily available for use in robotic applications. Another solution is using locking devices to control the release of energy. A second common reason for using locking devices is to reconfigure a robotic system. Such systems consist of multiple modules that can be connected and disconnected to form different configurations that perform different tasks. Those modules are connected and disconnected using locking devices of various designs. An ideal locking mechanism possesses following properties: adjustable locking positions, unlocking while under load, low energy consumption, lockable in any position, compact, lightweight, short switching time, inexpensive, and high locking force. Locking devices can be categorized in three main categories. Mechanical lockers lock and unlock based on the position of a mechanical component, e.g., wedges or paws. Friction-based lockers engage or disengage two friction surfaces to lock or unlock a joint. Singularity lockers exploit singular positions of the device to lock or unlock a joint. Singularities of a mechanism result in infinite transfer ratios and in such positions the device can lock and unlock the joint with an infinitely high and low force, respectively. All categories can be implemented in an active or a passive way. Active lockers require an actuator to control the locking timing, position, or force. Passive devices require no additional actuator, control, or electronics, but rather lock and unlock based on for instance the joint's position or direction of velocity. In the following paragraph some examples of locking devices in legged robotics will be discussed. For a full review on all types of lockers in the main categories and their implementation in robotic systems, please, refer to Plooij et al. (2015).

Usage of a latch to lock the knees of a biped during specific walking phases can be found in two examples (Collins and Ruina, 2005; Wisse et al., 2007). The bipedal running robot Phides, on the other hand, uses a latch to attach a parallel spring to the knees during the stance phase and detach it during flight phase (Karssen and Wisse, 2012). Multiple passive latches are used in a knee–ankle prosthesis to control the energetic coupling between the ankle and knee joint during locomotion (Unal et al., 2010). An active latch is used in the weight acceptance mechanism of a knee–ankle prosthesis to engage a spring parallel to the knee joint during the stance phase (Flynn et al., 2015). In the AMP-Foot 1 a ratchet is used to change the internal configuration of the foot between the loading and push-off phase (Brackx et al., 2013). The energy storage device of a spherical hopping robot uses a passive ratchet (Li et al., 2009). A clutch is used to engage and disengage a parallel spring in a knee exoskeleton (Elliott et al., 2013) or to switch between ground mode and flight mode of robots (Kossett and Papanikolopoulos, 2011). Recently a mutilated gear mechanism was used as a cam-based locker to lock parallel springs of a compliant actuator (Mathijssen et al., 2015). Electromagnetic brakes were used to lock the joints of a bipedal walker during stand-still to effectively decrease the energy consumption (Sugahara et al., 2002). A strategy that is also implemented in a knee prosthesis that engages a clutch to bypass the required reaction torque required by the motor in an SEA when the spring releases energy (Rouse et al., 2014). Overrunning clutches have been used in orthotics and prosthetics to engage support springs (Shamaei et al., 2013), energy harvesting (Li et al., 2008), and control the energy storage and release of springs (Collins and Kuo, 2010). Examples of singularity lockers can be found in a bipedal walker to lock its knees (Van Oort et al., 2011), in transfemoral prostheses to lock a spring during loading by a smaller motor (Cherelle et al., 2014), and for engaging a spring in the weight acceptance mechanism in the knee of the Cyberlegs alpha prototype (Flynn et al., 2015).

8.2.7 MULTI-DOF JOINTS

Since humans posses multiple degrees of freedom joints, bio-inspired robotic applications often require multiple degrees of freedom actuators with variable stiffness. The three-dimensional shoulder and wrist joints of the human arm provide a large workspace and dexterity to the arms. During human locomotion, 23% of the total hip work is done in the frontal plane, while 74% of the total hip work is done in the sagittal plane (Eng and Winter, 1995). The hip work in the frontal plane is done to support the body weight when the body mass is shifted laterally and to counteract a destabilizing gravitational moment in order to keep the pelvis and trunk erect (MacKinnon and Winter, 1993). One of the strategies the human body uses to cope with perturbations of the center of mass, particularly in the frontal plane, is widening the stance (Goodworth and Peterka, 2010) by utilizing the frontal degrees of freedom in hips and ankles. These degrees of freedom, furthermore, allow for locomotion over surfaces inclined in the frontal plane and are used to adapt the gait when walking over irregular surfaces (Manz et al., 2003). Because of the importance and use of multiple degrees of freedom joints in the human body, bio-inspired robotic applications often require multiple degrees of freedom actuators with variable stiffness. In cases where multiple degrees of freedom are wanted in one joint, however, cascades of single-degree of freedom actuators are usually used. The VSA-CubeBot is an example in which a robot's multiple degrees of freedom joints are formed by directly connecting variable stiffness actuators in series (Catalano et al., 2011). Another way to achieve this is by moving some parts of the actuator away from the joint, e.g., the motors, and using transmissions to drive the joint and/or change its stiffness (Hobbelen et al., 2008; Gallego et al., 2010). The research on multiple degrees of freedom actuators with variable stiffness is still limited. One of the few examples is the multiple degrees of freedom actuator with variable stiffness based on two antagonistic setups of ANLES actuators. The ANLES actuator drives a linear torsion spring that wraps around a nonuniform cylinder. As a result, the engaged part of the torsion spring varies, which yields an overall nonlinear elastic behavior. It is used in a 3-DoF wrist joint (Koganezawa and Yamashita, 2010). This actuator has, however, the same disadvantages as other antagonistic setups previously mentioned. Another example, but with no possibility of stiffness modulation, couples two custom-made spring elements to a spherical mechanism and is also used as a 2-DoF wrist joint (Chu et al., 2014). Different from variable stiffness actuators is the two-degree of freedom variable damping actuator, using electro-rheological fluids, proposed by Sakaguchi and Furusho (1999). Furthermore, there are few examples of spherical electric motors without any physical compliance, but with the possible option of being used as direct drives with active compliance through control (Wang et al., 2003; Rossini et al., 2013; Kim et al., 2015). A 2-DoF ankle prosthesis is described in Ficanha and Rastgaar and couples the movement of two DC motors that drive a Cardan joint through a nylon cable. The nylon cable provides some compliance, but no variable stiffness can be obtained. A novel variable stiffness actuator, based on the MACCEPA principle, to be specifically used in a 2-DoF joint has been presented in Weckx et al. (2014). This actuator has been used for the joints of a biped (Rodriguez Cianca et al., 2015). The biped TORO possesses 2-DoF Cardan joints in its ankles. Two direct drives with active impedance control drive the Cardan joint. The frontal motor is incorporated within the joint, while the sagittal motor is placed in the shank and connected to the joint by means of a parallel bar mechanism (Englsberger et al., 2014). This arrangement of actuators can also be found in the ankles of COMAN, with the exception that COMAN is driven by compact SEAs (Li et al., 2013). The biped Valkyrie also possesses 2-DoF Cardan joints in its ankles that are driven by two linear SEA's, which are both placed more towards the biped's calves (Paine et al., 2015).

REFERENCES

Albu-Schaffer, A., Eiberger, O., Grebenstein, M., Haddadin, S., Ott, C., Wimbock, T., Wolf, S., Hirzinger, G., 2008. IEEE J. Robot. Autom. 15, 20–30.

Alexander, R.M., 1988. Elastic Mechanism in Animal Movement. Cambridge University Press.

Au, S., Weber, J., Herr, H., 2009. IEEE Trans. Robot. 25, 51–66.

Boaventura, T., Semini, C., Buchli, J., Frigerio, M., Focchi, M., Caldwell, D., 2012. In: Proc. of the 2012 IEEE International Conference on Robotics and Automation, pp. 1889–1894.

Brackx, B., Van Damme, M., Matthys, A., Vanderborght, B., Lefeber, D., 2013. Int. J. Adv. Robot. Syst. 10.

Catalano, M., Grioli, G., Garabini, M., Bonomo, F., Mancinit, M., Tsagarakis, N., Bicchi, A., 2011. In: Proceedings of the 2011 IEEE International Conference on Robotics and Automation (ICRA). IEEE, pp. 5090–5095.

Cherelle, P., Grosu, V., Matthys, A., Vanderborght, B., Lefeber, D., 2014. Design and validation of the ankle mimicking prosthetic (AMP-)foot 2.0. IEEE Trans. Neural Syst. Rehabil. Eng. 22, 138–148. https://doi.org/10.1109/TNSRE.2013.2282416.

Chu, C.-Y., Xu, J.-Y., Lan, C.-C., 2014. In: Proceedings of 2014 IEEE International Conference on Robotics & Automation (ICRA), pp. 6156–6161.

Collins, S.H., Kuo, A.D., 2010. PLoS ONE 5.

Collins, S., Ruina, A., 2005. Robotics and automation. In: Proceedings of the 2005 IEEE International Conference on, pp. 1983–1988.

Cui, Z., Yang, H., Qian, D., Cui, Y., Peng, Y., 2014. In: Proceedings of the 2014 IEEE International Conference on Robotics and Biomimetics, pp. 326–331.

Eiberger, O., Haddadin, S., Weis, M., Albu-Schaeffer, A., Hirzinger, G., 2010. In: 2010 IEEE International Conference on Robotics and Automation (ICRA), pp. 1687–1694.

Elliott, G., Sawicki, G.S., Marecki, A., Herr, H., 2013. In: 2013 IEEE International Conference on Rehabilitation Robotics (ICORR), pp. 1–6.

Eng, J., Winter, D., 1995. J. Biomech. 28, 753–758.

Englsberger, J., Werner, A., Ott, C., Henze, B., 2014. In: 2014 14th IEEE-RAS International Conference on Humanoid Robots (Humanoids), pp. 916–923.

Enoch, A., Vijayakumar, S., 2015. J. Mech. Robot. 8, 1–14.

Eslamy, M., Grimmer, M., Seyfarth, A., 2012. In: 2012 IEEE International Conference on Robotics and Biomimetics (ROBIO), pp. 2406–2412.

Ficanha, E., Rastgaar, M. In: 5th IEEE RAS & EMBS International Conference on Biomedical Robotics and Biomechatronics, pp. 1033–1038.

Flynn, L., Geeroms, J., Jimenez-Fabian, R., Vanderborght, B., Vitiello, N., Lefeber, D., 2015. Robot. Auton. Syst. 73, 4–15.

Folkertsma, G., Kim, S., Stramigioli, S., 2012. In: 2012 IEEE/RSJ International Conference on Intelligent Robots and Systems (IROS), pp. 2210–2215.

Friedl, W., Hoppner, H., Petit, F., Hirzinger, G., 2011. In: 2011 IEEE IRSJ International Conference on Intelligent Robots and Systems, pp. 1836–1842.

Gallego, J., Forner-Cordero, A., Moreno, J., Montellano, A., Turewska, E., Pons, J., 2010. In: 2010 3rd IEEE/RAS and EMBS International Conference on Biomedical Robotics and Biomechanics (BioRob). IEEE, pp. 734–739.

Geyer, H., Seyfarth, A., Blickhan, R., 2006. Proc. R. Soc. Lond. B, Biol. Sci. 273, 2861–2867.

Goodworth, A., Peterka, R., 2010. J. Neurophysiol. 2, 1103–1118.

Grimmer, M., Seyfarth, A., 2011. In: 2011 IEEE International Conference on Robotics and Automation (ICRA), pp. 1439–1444.

Grioli, G., Wolf, S., Garabini, M., Catalano, M., Burdet, E., Caldwell, D., Carloni, R., Friedl, W., Grebenstein, M., Laffranchi, M., Lefeber, D., Stramigioli, S., Tsagarakis, N., Van Damme, M., Vanderborght, B., Albu-Schaeffer, A., Bicchi, A., 2015. Int. J. Robot. Res. 34, 727–743.

Groothuis, S., Rusticelli, G., Zucchelli, A., Stramigioli, S., Carloni, R., 2014. IEEE/ASME Trans. Mechatron. 19, 589–597.

Haeufle, D., Taylor, M., Schmitt, D., Geyer, H., 2012. In: 2012 4th IEEE RAS & EMBS International Conference on Biomedical Robotics and Biomechatronics (BioRob), pp. 1614–1619.

Herder, J., 1998. Mech. Mach. Theory 33, 151–161.

Herzog, A., Righetti, L., Grimminger, F., Pastor, P., Schaal, S., 2014. In: Proc. of the 2014 IEEE/RSJ International Conference on Intelligent Robots and Systems, pp. 981–988.

Hobbelen, D., De Boer, T., Wisse, M., 2008. In: IEEE/RSJ International Conference on Intelligent Robots and Systems, pp. 2486–2491.

Hogan, N., 1985. J. Dyn. Syst. Meas. Control 107, 1–24.

Hollander, K., Sugar, T., Herring, D., 2005. In: 9th International Conference on Rehabilitation Robotics, ICORR 2005, pp. 113–118.

Huang, Y., Vanderborght, B., Van Ham, R., Wang, Q., Van Damme, M., Guangming, X., Lefeber, D., 2013. In: IEEE/ASME Trans. Mechatron., vol. 18. Vrije Universiteit Brussel, pp. 598–611.

Hurst, J., Chestnutt, J., Rizzi, A., 2004. In: 2004 IEEE International Conference on Robotics and Automation, pp. 4662–4667.

Hutter, M., Remy, C.D., Hoepflinger, M., Siegwart, R., 2013. IEEE/ASME Trans. Mechatron. 18, 449–458.

Jafari, A., Tsagarakis, N., Sardellitti, I., Caldwell, D., 2012. IEEE/ASME Trans. Mechatron., 1–9.

Karssen, D., Wisse, M., 2012. Dynamic Walking Conference.

Kenneally, G., Koditschek, D., 2015. In: 2015 IEEE/RSJ International Conference on Intelligent Robots and Systems (IROS), pp. 5712–5718.

Kim, B.-S., Song, J.-B., 2012. In: IEEE Trans. Robot., pp. 1145–1151.

Kim, J., Lee, Y., Kwon, S., Seo, K.H., Lee, H., Roh, K., 2012. In: 2012 IEEE/RSJ International Conference on Intelligent Robots and Systems (IROS), pp. 4000–4005.

Kim, H.Y., Kim, H., Gweon, D.-G., Jeong, J., 2015. IEEE/ASME Trans. Mechatron. 20, 532–540.

Koganezawa, K., Yamashita, H., 2010. In: 2010 IEEE International Symposium on Industrial Electronics (ISIE), pp. 1973–1979.

Kossett, A., Papanikolopoulos, N., 2011. In: 2011 IEEE International Conference on Robotics and Automation (ICRA), pp. 4595–4600.

Kwon, W., Kim, H., Park, J., Roh, C., Lee, J., Park, J., Kim, W.-K., Roh, K., 2007. In: 2007 7th IEEE–RAS International Conference on Humanoid Robots, pp. 583–588.

Li, Q., Naing, V., Hoffer, J.A., Weber, D.J., Kuo, A.D., Donelan, J.M., 2008. In: 2008 IEEE International Conference on Robotics and Automation, pp. 3672–3677.

Li, B., Deng, Q., Liu, Z., 2009. In: 2009 IEEE International Conference on Robotics and Biomimetics (ROBIO), pp. 402–407.

Li, Z., Tsagarakis, N., Caldwell, D., 2013. Auton. Robots 35, 1–14.

MacKinnon, C., Winter, D., 1993. J. Biomech. 26, 633–644.

Manz, H., Lord, S., Fitzpatrick, R., 2003. Gait Posture 18, 35–46.

Mao, Y., Wang, J., Jia, P., Li, S., Qiu, Z., Zhang, L., Han, Z., 2007. In: The IEEE International Conference on Robotics and Automation. IEEE, pp. 3609–3614.

Mathijssen, G., Cherelle, P., Lefeber, D., Vanderborght, B., 2013. Actuators 2, 59–73.

Mathijssen, G., Lefeber, D., Vanderborght, B., 2015. IEEE/ASME Trans. Mechatron. 20, 594–602.

Matsui, T., Hirukawa, H., Ishikawa, Y., Yamasaki, S., Kagami, F., Kanehiro, F., Saito, H., Inamura, T., 2005. In: IEEE International Conference on Embedded and Real-Time Computing Systems and Applications (RTCSA 2005), pp. 205–210.

Mazumdar, A., Spencer, S., Salton, J., Hobart, C., Love, J., Dullea, K., Kuehl, M., Blada, T., Quigley, M., Smith, J., Bertrand, S., Wu, T., Pratt, J., Buerger, S., 2015. In: 2015 IEEE International Conference on Robotics and Automation (ICRA), pp. 835–841.

Mettin, U., La Hera, P., Freidovich, L., Shiriaev, A., 2010. Int. J. Robot. Res. 29, 1186–1198.

Misgeld, B.J., Stille, J., Pomprapa, A., Leonhardt, S., 2014. World Congress 19, 6599–6605.

Narioka, K., Homma, T., Hosoda, K., 2012. In: 2012 12th IEEE–RAS International Conference on Humanoid Robots (Humanoids), pp. 15–20.

Nelson, G., Saunders, A., Neville, N., Swilling, B., Bondaryk, J., Billings, D., Lee, C., Playter, R., Raibert, M., 2012. J. Robotics Soc. Jpn. 30, 372–377.

Ott, C., Eiberger, O., Englsberger, J., Roa, M., Albu-Schaeffer, A., 2012. J. Robotics Soc. Jpn. 30, 378–382.

Paine, N., Oh, S., Sentis, L., 2014. IEEE/ASME Trans. Mechatron. 19, 1080–1092.

Paine, N., Mehling, J., Holley, J., Radford, N., Johnson, G., Fok, C.-L., Sentis, L., 2015. J. Field Robot. 32, 378–396.

Park, I.-W., Kim, J., Oh, J.-H., 2006. In: IEEE International Conference on Robotics and Automation (ICRA 2006), pp. 1231–1236.

Pierce, B., Cheng, G., 2014. In: 2014 14th IEEE-RAS International Conference on Humanoid Robots (Humanoids), pp. 7–12.

Plooij, M., Mathijssen, G., Cherelle, P., Lefeber, D., Vanderborght, B., 2015. IEEE Robot. Autom. Mag. 22, 106–117.

Pratt, G.A., Williamson, M.M., 1995. In: Human Robot Interaction and Cooperative Robots, vol. 1. IEEE/RSJ, pp. 399–406.

Radkhah, K., Lens, T., von Stryk, O., 2012. In: 2012 IEEE/RSJ International Conference on Intelligent Robots and Systems, pp. 4243–4250.

Raibert, M.H., 1986. Legged Robots That Balance, vol. 3. MIT Press, Cambridge.

Rodriguez Cianca, D., Weckx, M., Torricelli, D., Gonzalez, J., Lefeber, D., Pons, J., 2015. In: 2015 15th IEEE–RAS International Conference on Humanoid Robots (Humanoids).

Roozing, W., Li, Z., Medrano-Cerda, G., Caldwell, D., Tsagarakis, N., 2015. IEEE/ASME Trans. Mechatron. 21, 1080–1091.

Rossini, L., Chetelat, O., Onillon, E., Perriard, Y., 2013. IEEE/ASME Trans. Mechatron. 18, 1006–1018.

Rouse, E., Mooney, L., Herr, H., 2014. Int. J. Robot. Res. 33, 1611–1625.

Sabourin, C., Bruneau, O., Buche, G., 2006. Int. J. Robot. Res., 843–860.

Sakaguchi, M., Furusho, J., 1999. J. Intell. Mater. Syst. Struct. 10, 666–670.

Schiavi, R., Grioli, G., Sen, S., Bicchi, A., 2008. In: 2008 IEEE International Conference on Robotics and Automation, pp. 2171–2176.

Schimmels, J., Garces, D., 2015. In: 2015 IEEE International Conference on Robotics and Automation (ICRA), pp. 220–225.

Schuy, J., Beckerle, P., Faber, J., Wojtusch, J., Rinderknecht, S., von Stryk, O., 2013. In: 2013 IEEE/ASME International Conference on Advanced Intelligent Mechatronics (AIM), pp. 1786–1791.

Seyfarth, A., Iida, F., Tausch, R., Stelzer, M., Von Stryk, O., Karguth, A., 2009. Int. J. Robot. Res. 28, 257–265.

Shamaei, K., Napolitano, P.C., Dollar, A.M., 2013. In: 2013 IEEE International Conference on Rehabilitation Robotics (ICORR), pp. 1–6.

Sugahara, Y., Endo, T., Lim, H.O., Takanishi, A., 2002. In: IEEE/RSJ International Conference on Intelligent Robots and Systems, vol. 3, pp. 2658–2663.

Unal, R., Behrens, S., Carloni, R., Hekman, E., Stramigioli, S., Koopman, H., 2010. In: 2010 3rd IEEE RAS and EMBS International Conference on Biomedical Robotics and Biomechatronics (BioRob), pp. 191–196.

Van Damme, M., Vanderborght, B., Verrelst, B., Van Ham, R., Daerden, F., Lefeber, D., 2009. Int. J. Robot. Res. 28, 266–284.

Van Ham, R., Vanderborght, B., Van Damme, M., Verrelst, B., Lefeber, D., 2006. In: 2006 Proceedings of the IEEE International Conference on Robotics and Automation, pp. 2195–2200.

Van Ham, R., Vanderborght, B., Van Damme, M., Verrelst, B., Lefeber, D., 2007. Robot. Auton. Syst. 55, 761–768.

Van Ham, R., Sugar, T., Vanderborght, B., Hollander, K., Lefeber, D., 2009. IEEE Robot. Autom. Mag. 16, 81–94.

Van Oort, G., Carloni, R., Borgerink, D.J., Stramigioli, S., 2011. In: 2011 IEEE International Conference on Robotics and Automation (ICRA), pp. 2003–2008.

Vanderborght, B., Van Ham, R., Verrelst, B., Van Damme, M., Lefeber, D., 2008. Adv. Robot. 22, 1027–1051.

Vanderborght, B., Tsagarakis, N., Van Ham, R., Thorson, I., Caldwell, D., 2011. Auton. Robots 31, 55–65.

Vanderborght, B., Albu-Schaeffer, A., Bicchi, A., Caldwell, D., Tsagarakis, N., Van Damme, M., Lefeber, D., Van Ham, R., Burdet, E., Carloni, R., Catalano, M., et al., 2013. Robot. Auton. Syst. 61, 1601–1614.

Wang, W., Wang, J., Jewell, G., Howe, D., 2003. Mechatronics 8, 457–468.

Wang, S., van Dijk, W., van der Kooij, H., 2011. In: Proceedings of the 2011 IEEE International Conference on Rehabilitation Robotics.

Weckx, M., Van Ham, R., Cuypers, H., Jim'enez-Fabi'an, R., Torricelli, D., Pons, J., Vanderborght, B., Lefeber, D., 2014. In: 2014 14th IEEE–RAS International Conference on Humanoid Robots (Humanoids), pp. 33–38.

Winter, D., 1984. Crit. Rev. Biomed. Eng. 9, 287–314.

Wisse, M., Feliksdal, G., Van Frankkenhuyzen, J., Moyer, B., 2007. IEEE Robot. Autom. Mag. 14, 52–62.

Yadukumar, S., Pasupuleti, M., Ames, A., 2012. In: 2012 IEEE/RSJ International Conference on Intelligent Robots and Systems (IROS), pp. 2478–2483.

Chapter 8.3

Actuators in Robotics as Artificial Muscles

Koh Hosoda

Department of System Innovation, Graduate School of Engineering Science, Osaka University, Toyonaka, Japan

What is required for substituting the function of the biological muscles with artificial ones? If we suppose the biological muscles are the actuators driving joint motion through the muscular-tendon structure, the artificial muscle should exert a force to the link through tendons. The simplest engineering solution is using motors driving links though wires as we reviewed in the previous section. In this section, we begin with more muscle-like actuators used for humanoid robots.

In this section, we will focus on muscle-like actuators for realizing locomotion, electric-motor based actuators, hydraulic and pneumatic actuators, and their hybrid actuators. They can generate large force enough for driving humanoid robots. We also have some artificial muscles that can only generate small forces: e.g., EAP (electro-active polymer) (Bar-Cohen, 2005), SMA (Shape Memory Alloy) (Cho et al., 2007), and piezo-electric actuators (Ueda, 2012). They can also realize locomotion in appropriate scale according to ability of each actuator. But, since we are focusing on animal behavior like a human, we will ignore small scales in this section.

8.3.1 MUSCLE-LIKE ACTUATORS DRIVEN BY ELECTRIC ROTATIONAL MOTORS

In the previous Subchapter 8.2, we mainly reviewed rotational actuators driving tendon wires. They typically use a pulley on the motor shaft to wind up the wires. Using a pulley enables us to keep moment arm constant, which in general simplifies control (a moment arm is the length between the joint axis and the line of force acting on that joint). However, to emulate the mechanism underlying the behavior of humans and animals more closely, such actuators are needed that can generate longitudinal forces when they contract, similar to biological muscles. In biology, as a muscle contracts, its moment arm changes according to the angle of the joint. Furthermore, some biological muscles not only exert force between two bones (links), but among three or more bones (links), which are called biarticular muscles or triarticular muscles.

Some muscle-like actuators steered by electric motors have been proposed and muscular-skeletal humanoid robots have been developed which still utilize

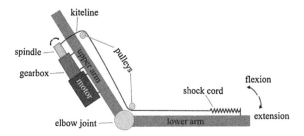

FIGURE 8.3.1 A muscle-like actuation driven by an electric motor.

FIGURE 8.3.2 Torso of an anthropomorphic robot Roboy. It has several artificial muscle actuators inside (courtesy of Rolf Pfeifer).

them. The motor of such an actuator winds up the wire so that it can generate a longitudinal force and displacement. It eventually includes a series elastic element so that it can emulate the dynamics of the biological counterpart at least to some extent.

The muscular-skeletal humanoid robot CRONOS was developed by Holland and Knight in 2006 (Holland and Knight, 2006) at the University of Essex in the UK. It is driven by muscle-like electric actuators and has a similar muscular-skeletal structure as a human. The sketch of the actuator is shown in Fig. 8.3.1. The actuator has a series elastic element in the form of a shock cord, and a motor winds up the kiteline on the spindle, the joint angle changes. Since the force generated by the actuator is not very large, the robot CRONOS and its offspring ECCE robot could only move their upper torsos, but they did not have legs.

A similar actuation principle is implemented in Roboy, a "boy robot" developed by Pfeifer and his team at the University of Zurich (Fig. 8.3.2). The structure of the actuator is similar to that of CRONOS. Springs are adopted instead of shock cords for providing series elasticity. The robot legs are too weak

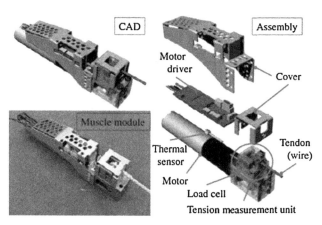

FIGURE 8.3.3   The artificial muscle unit of JSK laboratory, the University of Tokyo (courtesy of Yuki Asano).

to support the upper torso since the actuators cannot generate sufficient speed and force.

JSK laboratory of the University of Tokyo is continuously developing muscular-skeletal humanoid robots driven by artificial muscle actuators. Their actuators also consist of electric motors and wires (Fig. 8.3.3) (Asano et al., 2015). The wire is wound by the motor connected to a load cell that can be used to measure tension. It can generate a longitudinal force of up to 583 [N], continuously up to 338 [N]. The actuator unit also has a thermal sensor to monitor the temperature of the motor, which is very important to prevent over-heating, and a hall sensor is used to measure motor rotation angle, from which the wire length can be calculated. Monitoring such states of the artificial muscle unit is a crucial problem.

Applying some conventional scheme, the rotational angle of the motor can be precisely controlled. Therefore, the length and tensile force of the wire can be easily controlled based on the sensor measurement. On the other hand, this method has a trade-off between winding speed and the generated torque. If more speed is needed, the generated torque will be smaller. If we increase the diameter of the winding pulley, winding speed will increase, but the generated force will be smaller. For example, JSK' s artificial muscle unit can generate 338 [N], but it is not sufficient for walking of a human-size humanoid robot. As a result, the walking motion of these robots is very slow or almost impossible.

8.3.2   LINEAR ACTUATORS WITHOUT SLACK

The biological muscle is driving bones with tendons: it can only contract but not expand since the tendon slacks when it expands, which is why there is always

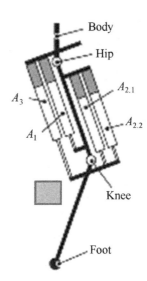

FIGURE 8.3.4   Bidirectional electromagnetic muscle-like actuator developed by Nakata et al., and jumping monopod driven by the actuator (courtesy of Yoshihiko Nakata) (Nakata et al., 2012; Ryu et al., 2016).

a flexor and an extensor. On the other hand, there are a number of linear actuators without slack, e.g., magnetic linear actuators and hydraulic and pneumatic cylinders. These actuators can generate forces in two directions.

Nakata and his research team in Osaka University developed a 3-phase direct-drive synchronous linear motor (Fig. 8.3.4) (Nakata et al., 2012; Ryu et al., 2016) based on electromagnetic effect, which can generate a force of 5.7 [N] when applied current is 1 [A]. They built a jumping monopod by using the actuators as mono- and biarticular muscles (Fig. 8.3.4). Note that the actuator is bidirectional, therefore, one joint can be driven by one actuator. Actuator A1 drives the hip joint, and A3 is corresponding to a biarticular muscle between the knee and the hip. They needed a large torque for the knee joint for jumping, so they applied two actuators for the knee. The force of the actuator is controlled by the current; therefore, it can be precisely controlled. Since the shaft is supported by a magnetic force, we have to apply feedback control based on the sensor measurement for realizing series elasticity (no physical elasticity exists in the actuator). The actuator is stiff in the radial direction and it exerts a reaction moment when it is bent, which is very different from a biological muscle.

Hydraulic and pneumatic cylinders are also linear actuators without slack. These actuators can generate larger forces than the electromagnetic ones, and are used to build whole body humanoid robots. For example, Cheng and his research team in ATR, Japan, developed a whole-body humanoid robot CB-I, which employed hydraulics (Cheng, 2007). It could walk with its two legs. A US company, Boston Dynamics (http://www.bostondynamics.com/) also developed several whole-body humanoid robots, e.g., PETMAN and ATLAS based on hydraulic actuators. Because they are capable of producing large power, and are precisely controllable, they can be used to implement balance and gait control. ATLAS serves as a common platform for DRC (DARPA Robotics Challenge) competition. HyQ, a quadruped robot developed by Semini et al. at Italian Institute of Technology (Semini et al., 2011).

Pneumatic cylinders are used to build humanoid robots as well, but their power is less than that of hydraulics. They are naturally compliant since the working fluid, air, is compressive, their energy efficiency is less than that of hydraulic ones. Asada and his research team at Osaka University developed a large baby robot CB2 (Ishiguro et al., 2011) and a small baby humanoid robot Affetto (Ishihara and Asada, 2015), which were driven by linear and rotational air cylinders. Both baby robots did not have enough power to support their weight, therefore could not move on their legs.

In sum, these slackless actuators can be controlled precisely and are energy efficient, which is why they are ideally suited for realizing walking of humanoid robots. However, we need further discussion on whether they are good candidates for emulating biological muscles. As mentioned above, biological muscles can only generate a contraction force, but they cannot actively expand. Let us now think about whether these characteristics are fundamental qualities of muscles or not.

8.3.3 PNEUMATIC ARTIFICIAL MUSCLES

The McKibben pneumatic artificial muscle, one of the most famous artificial muscles, was originally patented in the 1950s by Gayload (1958). Its structure is very simple (see Fig. 8.3.5). The tube is covered with a sleeve with a rhomboidal mesh that can transform the growing air pressure inside the tube into a longitudinal force and displacement. Since the artificial muscle is driven by compressive fluid, air, it is naturally compliant, and when there is no tension, it will slack.

McKibben artificial muscle can generate a large force, large enough for driving a same-size robot as a human. Quality of end-caps on both ends of the actuator determines applicable highest pressure, and as a result, a maximum generated force. The maximum force is also dependent on the diameter of the actuator. Typically, it can exert around 800 [N] when 7 [MPa] compressed air is applied. The diameter also determines the volume of the muscle, and as a result, the time to fill the actuator with the air. In conclusion, it is a trade-off between

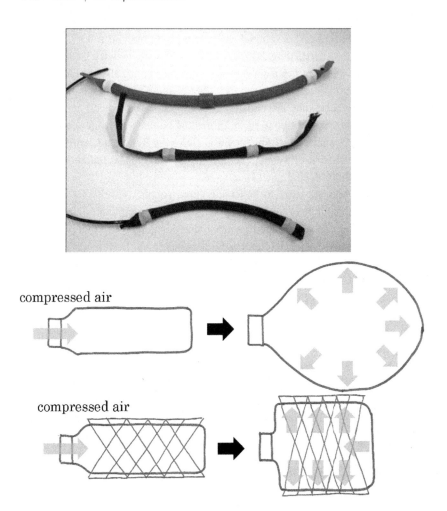

compressed air

compressed air

FIGURE 8.3.5 McKibben pneumatic artificial muscles.

the generated force and response time. Human's Achilles tendon can convey a tensile force around 2000 [N], therefore, even the artificial muscle is not strong enough for realizing human-compatible locomotion.

By using McKibben artificial muscles, human-like muscular–skeletal structures can be realized and their functions can be investigated in a constructivist approach that we could call "understanding by building." Some robots driven by McKibben artificial muscles have already been introduced in Chapter 5.

The characteristics of McKibben pneumatic artificial muscles have been carefully investigated (Chou and Hannaford, 1996; Klute and Hannaford, 2000). If we close the valve and keep the amount of air constant, they act like linear

FIGURE 8.3.6 Pleated pneumatic artificial muscle (PPAM) developed by Vanderborght et al. in Free University of Brussels.

springs within a certain range of the applied force. They are subject to hysteresis resulting from the friction between the inner tube and the sleeve. The static relation among the inner pressure, its length, and tensile force is theoretically calculated in Chou and Hannaford (1996) and is extended to nonlinear models (Klute and Hannaford, 2000; Sugimoto et al., 2013). In short, one out of three variables is determined when two of them are given, e.g., the length is determined when the pressure and the tensile force are given. However, in reality, depending on how the artificial muscles are produced, their characteristics will dramatically change. For example, if we apply pretension to the inner tube, we can change the spring constant of the actuator; if we reduce the size of the both end-caps, we can increase the contraction ratio; if we put some particles in the inner tube, we can reduce the air consumption. Since the actuator is made from soft material, it is very difficult to control the properties of the muscles. In this sense, the McKibben pneumatic actuator is not a great tool in terms of traditional engineering.

There are several variations of pneumatic artificial muscles: FESTO, a German pneumatic device company developed FESTO fluidic muscles. A flexible tube contains reinforced fibers in the form of a rhomboidal mesh. This structure is free from friction between the tube and the fiber, which drastically reduces hysteresis. The end-caps are well-engineered so that the muscle can be applied relatively large pressure, up around 1 [MPa], which enables the muscle generate a large force.

The pleated pneumatic artificial muscle (PPAM) was developed to overcome dry friction and material deformation, which is present in the widely used McKibben muscle. The essence of the PPAM is its pleated membrane structure which enables the muscle to work at low pressures and at large contractions (see Fig. 8.3.6) (Villegas et al., 2012). Another advantage is that it generates higher forces at lower pressures and no threshold of pressure is needed. With the PPAM a biped robot Lucy and a knee exoskeleton Knexo (Vanderborght et al., 2008;

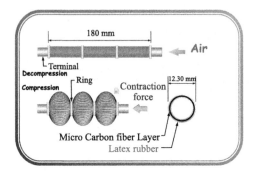

FIGURE 8.3.7 A straight-fiber-type artificial muscle developed by Tomori et al. (2013)

Beyl et al., 2011) is powered. The group also experiments with self-healing materials so cuts and damages can be healed (Terryn et al., 2015).

Nakamura and his research team at Chuo University developed what they call a straight-fiber-type artificial muscle (Fig. 8.3.7) (Tomori et al., 2013). This artificial muscle is also a tube reinforced by a fiber, and several rings are put on the tube so that they increase contraction ratio and a tensile force. The contraction ratio is 1.5 times more than a normal McKibben artificial muscle, and the tensile force is approximately 3 times more, respectively.

8.3.4 ARTIFICIAL MUSCLE EMULATING DYNAMICS OF BIOLOGICAL MUSCLE

There is a trial to produce an artificial muscle precisely emulating the dynamics of the biological muscle (Klute, 2000). As stated above, force–length properties of the McKibben pneumatic artificial muscle are similar to those of the biological muscle. If it is used under the isometric condition, therefore, it is a good model of the biological muscle. On the other hand, its dynamic properties, force–velocity properties are very different. Klute et al. put a hydraulic damper parallel to the pneumatic artificial muscles so that they intended to consist of an actuator with similar dynamic properties as a biological muscle. They also put springs in serial to the pneumatic actuator to emulate the tendon dynamics.

They developed an artificial muscle–tendon system shown in Fig. 8.3.8, and tried to reproduce the same property as the *triceps surae*. While the force–velocity relationship of the system can be qualitatively similar to that of the *triceps surae*, the actuator cannot be stretched beyond its resting length, and the force–velocity curve has a concave shape while the curve of the biological muscle has a convex one. It is very challenging to reproduce the dynamics of the biological muscle by the artificial one.

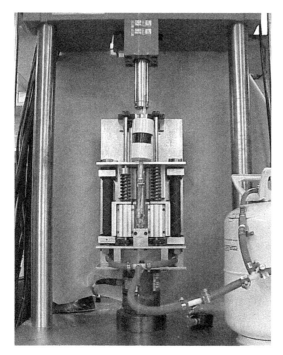

FIGURE 8.3.8 An artificial muscle–tendon system to emulate the dynamics of the biological muscle (Klute, 2000).

REFERENCES

Alexander, R.McN., 2003. Principles of Animal Locomotion. Princeton University Press.

Asano, Y., et al., 2015. A sensor-driver integrated muscle module with high-tension measurability and flexibility for tendon-driven robots. In: 2015 IEEE/RSJ International Conference on Intelligent Robots and Systems (IROS).

Bar-Cohen, Y., 2005. Biomimetics: Biologically Inspired Technologies. CRC Press.

Beyl, P., Knaepen, K., Duerinck, S., Van Damme, M., Vanderborght, B., Meeusen, R., Lefeber, D., 2011. Safe and compliant guidance by a powered knee exoskeleton for robot-assisted rehabilitation of gait. Adv. Robot. 25 (5), 513–535.

Cheng, G., Hyon, S., Morimoto, J., Ude, A., Hale, J.G., Colvin, G., Scroggin, W., Jacobsen, S.C., 2007. CB: a humanoid research platform for exploring neuroscience. J. Adv. Robot. 21 (10), 1097–1114.

Cho, K.J., Rosmarin, J., Asada, H., 2007. SBC hand: a lightweight robotic hand with an SMA actuator array implementing C-segmentation. In: 2007 IEEE International Conference on Robotics and Automation, pp. 921–926.

Chou, C.-P., Hannaford, B., 1996. Measurement and modeling of McKibben pneumatic artificial muscles. IEEE Trans. Robot. Autom. 12 (1), 90–102.

Gayload, R.H. Fluid actuated motor system and stroking device. U.S. Patent 2,884,126.

Holland, O., Knight, R., 2006. The anthropomimetic principle. In: Burn, J., Wilson, M. (Eds.), Proceedings of the AISB06 Symposium on Biologically Inspired Robotics.

Ishiguro, H., Minato, T., Yoshikawa, Y., Asada, M., 2011. Humanoid platforms for cognitive developmental robotics. Int. J. Humanoid Robot. 8 (3), 391–418.

Ishihara, H., Asada, M., 2015. Design of 22-DOF pneumatically actuated upper body for child android 'Affetto'. Adv. Robot. 29 (18), 1151–1163. Online publication date: 17 Sept. 2015.

Klute, G.K., Hannaford, B., 2000. Accounting for elastic energy storage in McKibben artificial muscle actuators. J. Dyn. Syst. Meas. Control 122 (2), 386–388.

Loeb, G.E., 1995. Control implications of musculoskeletal mechanics. In: Proceedings of 17th International Conference of the Engineering in Medicine and Biology Society 2, pp. 1393–1394.

Nakata, Y., Ide, A., Hirata, K., Ishiguro, H., 2012. Hopping of a monopedal robot with a biarticular muscle driven by electromagnetic linear actuators. In: IEEE International Conference on Robotics and Automation 2012, WeC210.3. St. Paul, USA.

Ryu, H., Nakata, Y., Nakamura, Y., Ishiguro, H., 2016. Adaptive whole-body dynamics: an actuator network system for orchestrating multijoint movements. IEEE Robotics Automation Magazine 23 (3), 85–92.

Semini, C., Tsagarakis, N.G., Guglielmino, E., Focchi, M., Cannella, F., Caldwell, D.G., 2011. Design of HyQ – a hydraulically and electrically actuated quadruped robot. Proc. Inst. Mech. Eng., Part I, J. Syst. Control Eng., 831–849.

Sugimoto, Y., et al., 2013. Static and dynamic properties of McKibben pneumatic actuator for self-stability of legged-robot motion. Adv. Robot. 27 (6), 469–480.

Terryn, S., Mathijssen, G., Brancart, J., Lefeber, D., Van Assche, G., Vanderborght, B., 2015. Development of a self-healing soft pneumatic actuator: a first concept. Bioinspir. Biomim. 10 (4), 046007.

Tomori, H., Nagai, S., Majima, T., Nakamura, T. Variable impedance control with an artificial muscle manipulator using instantaneous force and MR Brake. In: Proceedings of 2013 IEEE/RSJ International Conference on Intelligent Robots and Systems (IROS), 2013, pp. 5396–5403.

Ueda, J., Asada, H. Broadcast control for a large array of stochastically controlled piezoelectric actuators. In: Steager, E., Julius, A., Kim, M.J. (Eds.), Microbiorobotics: Biologically Inspired Microscale Robotic Systems. Elsevier. ISBN 978-1-4557-7891-1, 2012, pp. 87–114. Chapter 4.

Vanderborght, B., Van Ham, R., Verrelst, B., Van Damme, M., Lefeber, D., 2008. Overview of the Lucy project: dynamic stabilization of a biped powered by pneumatic artificial muscles. Adv. Robot. 22 (10), 1027–1051.

Villegas, D., Van Damme, M., Vanderborght, B., Beyl, P., Lefeber, D., 2012. Third-generation pleated pneumatic artificial muscles for robotic applications: development and comparison with McKibben muscle. Adv. Robot. 26 (11–12), 1205–1227.

Chapter 9

Conclusion

Maziar A. Sharbafi, David Lee, Thomas G. Sugar, Jeffrey Ward, Kevin W. Hollander, Koh Hosoda, and André Seyfarth

As the conclusion of the book, state-of-the-art research regarding legged locomotion and the application to daily activities are presented in this chapter. By comparing engineered legged systems with animals we address how far we are from nature. The recent technologies on assistive devices demonstrate the applicability of the scientific methods to fill this gap by developing products usable in human life. An overview of current research on legged locomotion presented in this chapter depicts the roadmap for future research.

Within this book we presented concepts on legged locomotion related to both biological and engineered systems. Currently, the biologically motivated principles do still not sufficiently explain many details on the organization of the locomotor systems in humans and animals. At the same time, for some of the already established principles, a direct transfer into robotic or other engineered platforms is still missing. Most of the bioinspired hardware systems focus on the implementation of biological design principles regarding the mechanical design. Only a few approaches exist which also implement biological actuator design and control.

Rather than "just" using the biological solutions as templates for technical locomotor systems for gait generation, nature could also be understood as a valuable source of a large variety of approaches, which may inspire novel design and control approaches in engineering. Sometimes, limitations in biology (e.g., due to evolutionary constraints) could even be bypassed in engineered systems, as shown in modern prosthetic leg technologies, which are at the transition to outperform their biological counterparts (e.g., in the case of the amputee long jumper Markus Rehm, who won the German competition in 2014).

In this conclusion chapter, we highlight different aspects regarding the state-of-the-art in this field of research and the potential progress that can be achieved. First, a comparison of biological and engineered locomotor systems is given, illustrating the gaps between them. Then, recent products are reviewed regarding their applicability for daily life. Finally, some recent and ongoing research

Bioinspired Legged Locomotion. http://dx.doi.org/10.1016/B978-0-12-803766-9.00012-9
623

projects are presented to provide a perspective on the future of legged locomotion science and technology. By identifying the limits in the current state-of-the-art, we point to the challenges for the future research on bioinspired legged locomotion.

Chapter 9.1

Versatility, Robustness and Economy

David Lee

School of Life Sciences, University of Nevada, Las Vegas, United States

This book has considered the principles, models, and application of legged locomotion in robotic and biological systems. It is clear from observation of legged
robots, whether in YouTube clips or in structured competitions such as the Darpa
Robotics Challenge (DRC), that the performance of biological systems exceeds
that of existing robots. However, this can only be quantified when a common
set of metrics is applied to animals and machines. Ideally, such metrics would
represent economy, robustness to perturbation, and versatility of the locomotor system. In order to improve the performance of legged robots, it is useful
to first consider the available "parts", working constraints, and developmental
processes of biological systems – as well as the fundamental principles of dynamics and control that we have been able to learn from nature. Competitions,
such as DARPA challenges and the W Prize, can act as an impetus for robotic
advancements that achieve survival within a defined environment, much as natural selection acts on biological systems.

It is readily apparent that natural selection produces better locomotor performance than that achieved by existing robotic systems. However, evolution does
not produce optimal organisms – only designs that are good enough for survival
in their environment, for competition with other individuals of the same species,
and for reproduction to pass their genes on to the next generation. So why do engineered systems fall short of our own locomotor capabilities? The answers are
many and, like our understanding of biological systems, incomplete. Nonetheless, some basic features of biological systems juxtaposed with functional levels
of the locomotor system defined in Fig. 2.0.1 could help inform engineered systems (Table 9.1.1).

Such biological observations may provide engineers with a different perspective and perhaps inspire new approaches to developing robots. Nature produces organisms capable of locomotion in their respective environments but is
also faced with many constraints including available materials and actuators,
as well as constructional and developmental processes inherited from their ancestors. Engineers, on the other hand, have a broad range of options, including
synthetic materials and actuators, and they can move between options quickly
with a broad range of constructional and design processes at their disposal. Despite this flexibility, we struggle to produce robots capable of locomotion in the

TABLE 9.1.1 Biological observations with reference to functional levels of locomotor systems from Fig. 2.0.1

• Nature produces integrated systems in which the parts work together in every stage of development – in other words, the developmental process is itself under selection.	**[Processes]**
• Living organisms can remodel their musculoskeletal and motor control systems throughout development and, to a lesser extent, in adulthood.	
• Biological systems learn about their own dynamics and store this information in the central nervous system.	**High Level Control**
• Nerve conduction velocities by action potentials are arbitrarily slow (10–100 ms^{-1}) compared with electrical signals and are directly proportional to axon diameter. Animals with longer legs have longer conduction times.	**Low Level Control**
• For a given motor unit, a single motor neuron may innervate fewer than ten or thousands of muscle fibers.	
• In vertebrates, motor units are incrementally recruited, from small units with slow-twitch muscle fibers to large units with fast-twitch muscle fibers.	
• Muscle is the only actuator used in animal locomotion. Its efficiency is about 25% but its force density is high across a range of velocities, declining rapidly from negative to zero velocity, then hyperbolically with increasing positive velocity.	**Actuators**
• A muscle can be "geared" in favor of force or velocity by changing the number of sarcomeres for a given muscle length (invertebrates only), myofibril type, and whole muscle architecture – i.e., parallel or oblique attachments muscles to their tendons of origin and insertion.	
• Muscle force is greatest at intermediate lengths, declining gradually as sarcomeres lengthen and declining more abruptly due to extreme shortening.	
• Muscle force is history dependent and is enhanced by active preloading, as in the eccentric countermovement of a jump.	
• Animals are limited to relatively few available biomaterials, tissues, and constructional processes.	**Biomechanics**
• Muscle–tendon systems act through skeletal transmissions that, together with ligaments, constrain the translation and rotation of joints.	
• Synovial joints of animals achieve lower friction than any engineered joint.	
• The mechanical advantage of muscles with respect to reaction force from the environment is increased in larger mammals and birds primarily by adopting a straighter limb posture.	
• Chemical energy (ATP) is the only energy source available to animals and its sustained rate of use is limited by oxygen delivery.	**[Energetics]**

real world. These observations underscore the key idea, emphasized throughout this text, that advances in robotic locomotion will depend upon knowledge of fundamental principles used in biological systems.

The performance discrepancy between biological and robotic systems is obvious and unchallenged, yet there are few consistent metrics to quantify this claim. Coarse parameters such as speed and leg kinematics are often reported, however, metrics of versatility (the ability to accomplish varied tasks), robustness (the rejection of perturbations), and economy (the mechanical work or total energy used to move a system's weight a unit distance) of legged robots in comparison to animals are varied and inconsistent across the literature. The limited examples discussed in this conclusion consider economy in terms mechanical and total cost of transport, and robustness in terms of the depth of vertical substrate drop that can be sustained during locomotion. Unless otherwise noted throughout this section, biological data are compared with experimental results from actual rather than simulated robots.

Learning underlying biological dynamics and control strategies, instead of simply the effects of those strategies, is critical to bridging the gap between the capabilities of biological and robotic systems. Given the vast resources available to human designers, such an approach will allow scientists and engineers to not only meet but exceed the locomotor performance of biological systems. For example, MIT Cheetah uses electric motors capable of converting about 60% of the (negative) work done on the motors into electrical energy, thereby achieving a total cost of transport of 0.50, which is slightly less than that predicted for a running animal of similar size (Seok et al., 2015). Using regenerative motors instead of physical springs, this robot captures the compliant leg function and energy return typical of mammals and birds larger than about 5 kg, while avoiding the constraints imposed by physical springs discussed in Chapter 7. Small animals such as kangaroo rats use muscular actuators in series with relatively stiff tendons that store minimal elastic strain energy during hopping (Biewener et al., 1981). Like MIT Cheetah, these animals avoid the constraints of using physical springs, so a small hopping robot using regenerative motors in place of muscles should realize even greater improvement over biological economy.

Robustness of robots is sometimes quantified by the vertical step-down that can be negotiated without falling. As discussed in Chapter 4, passive dynamic walkers can step down only about 1–2% of their leg length (Wisse et al., 2005). In contrast, running guinea fowl are robust enough to traverse unseen vertical drops 30% of their leg length (Daley et al., 2006). Despite dynamic differences between walking and running, this highlights at least an order of magnitude difference in robustness between passive dynamic robots and birds. Whereas crouched limb posture and muscle dynamics provide intrinsic robustness to birds (Daley and Biewener, 2006; Daley et al., 2009), rigid-legged passive dynamic walkers rely heavily on their controllers to avoid falling. In contrast, using compliant legs controlled by a simple spring–mass model, the Atrias biped has been simulated to run bipedally across obstacles 25% of its leg length but this has not yet been realized in the actual robot (Wu and Geyer, 2014). The Atrias monopod, however, has been shown to hop vertically into and out of a gravel-filled pit 17% of its leg length (Hubicki et al., 2016). As informed by key experiments on guinea fowl, compliant legs and actuators with appropriate dynamics, as discussed in Chapter 8, are improving the robustness of legged robots without increasing the complexity of their control. Animals remain substantially more robust and capable of rough terrain locomotion and, like some robots; they also plan their foot placements on steps and in steep or broken terrain.

Competitions that seek to overcome the performance disparities between robotic and animal locomotion, such as the DRC, allow direct comparison to humans or other animals performing the same task. Staging such an event provides a publicly visible assay of the state-of-the-art in robotics. From a technical perspective, the tasks achieved by humanoid robots are staggering; however, a casual observer might scoff at the several minutes needed for the winning robot to step out of a vehicle. The versatility required to perform a multitude of different tasks is key to success in such competitions, as well as in natural and human environments – and this is something we take for granted as integrated biological systems.

The "W Prize" represents an alternative type of robotics competition, which focuses on achieving highly economical real-world locomotion of legged robots. Competitors must complete a 10 km course with occasional obstacles, achieving an average speed no less than 1 ms^{-1} and total cost of transport no greater than 0.10. This economy is unachievable by humans because our total metabolic cost of transport assessed by oxygen consumption is 0.41 during walking at 1 ms^{-1} and increases by about 20% on uneven surfaces (Voloshina et al., 2013). If we ignore the oxygen consumed and consider our mechanical cost of transport of about 0.06 during walking at 1 ms^{-1}, our whole system efficiency would have to be 61%, compared with 15% based upon center of mass work and the actual metabolic cost of walking. We can also add the work done by the legs against one another using the individual limbs method (Donelan et al., 2002), which increases the mechanical cost of transport by about one-third and requires an even greater whole-system efficiency of 82%. Given that muscles are only 25% efficient, the cocontraction of antagonistic muscle pairs and only modest elastic energy return during walking, it is clear that robots will be the only contenders for the W Prize.

The previous discussion highlights the conundrum of whole system efficiency, which rewards the mechanical cost that we usually seek to minimize (e.g., Srinivasan and Ruina, 2006). For example, as shown in Section 2.2.1, human runners triple their mechanical cost of transport (Lee et al., 2013) during running compared with walking, yet their total metabolic cost of transport increases by less than one-fourth, based upon data from Rubenson et al. (2007) and Voloshina et al. (2013). This suggests perversely that running is about three-times more efficient than walking. Rather than elevate their mechanical cost, animals and robots really seek to reduce their total cost of transport – thereby achieving meaningful efficiency.

The world's most economical bipedal robot, Cornell Ranger, set the endurance record of 65 km for legged robots using a single battery charge (Bhounsule et al., 2014). This robot has a leg length of a tall human and achieves a mechanical cost of transport of 0.08 (including both positive and negative work) – but at a speed of only 0.6 ms^{-1} – whereas the W Prize requires a minimum average speed of 1 ms^{-1}. Cornell Ranger's total (electrical) cost of transport is between 0.19 and 0.30, depending upon the controller used, so this robot is capable of walking on level ground with a total cost of transport one-half that of humans, albeit at only one-half of our typical walking speed. Cornell Ranger would need to cut its total cost of transport in half and nearly double its walking speed to compete for the W Prize. Such improvements are made doubly hard by an inverse relationship between power and speed in rigid-legged bipeds (Garcia et al., 1998). Despite its long legs, Cornell Ranger's body mass of 9.91 kg is nearly an order of magnitude less than that of a human. Using data compiled by Rubenson et al. (2007), total cost of transport for walking mammals scales as body mass to the -0.31 power, just as it does for running mammals and birds (Taylor et al., 1982). Applying this exponent and extrapolating, the total cost of transport for a more human-like 80 kg Cornell Ranger is a convenient 0.10. In comparison, the 62 kg ATRIAS biped walks at 0.9 ms^{-1} with a total cost of transport of 1.13 (Hubicki et al., 2016) – eleven-fold too high for the W Prize – and has a modeled mechanical cost of transport of 0.19 (including both positive and negative work) at 1 ms^{-1} (Ramezani et al., 2014). As we have seen, the MIT Cheetah achieves high economy while trotting ten-times faster than Cornell Ranger walks, yet its total cost of transport is still five-fold too high for the W Prize. Boston Dynamic's Atlas biped is often used by DRC teams for its astounding real-world versatility, yet it pays a steep price for these capabilities with a total cost of transport 50-fold too high for the W Prize.

Although it's our most economical contender, Cornell Ranger could not even begin to negotiate the obstacles on the W-Prize course because its underactuated, passive dynamic design trades versatility for walking economy. Assessing the ability of our best robots to compete for the W Prize highlights the ongoing struggle to achieve real-world locomotion capabilities with high economy. One might ask if the W Prize is achievable by any robot of human scale and the answer is a qualified yes – well-placed springs and regeneration by motors, as well as careful attention to dynamics and control will probably be needed in combination to reach a total cost of transport less than 0.10 during level walking at 1 ms^{-1}. Despite nature's seamless integration of economy and versatility, surmounting the obstacles on the W Prize course presents formidable challenges in actuation and control. Considering economy and versatility together in a single design concept sets a high bar for legged robotics but might also yield unexpected synergy. The W Prize is just one concept and we need many more like it – for example, another award could emphasize speed and terrain-adaptability at a total cost of transport no greater than one.

Thinking of the minimum criteria for awards and challenges conjures up the process of natural selection acting upon a legged animal's abilities to evade predators, capture prey, compete, forage, and migrate – using the least energy possible. Like animals, robots just need to be good enough, for example, to achieve a set of metrics, to complete specified tasks, or to serve as experimental platforms for testing ideas. Unlike nature, we set the criteria that determine whether a robot is good enough. Matching human motion during a particular task, for example, is unlikely to result in robust systems, but requiring a robot to move quickly through a field of cobbles would likely result in robust systems. Choosing the right performance criteria for a specific technical challenge will motivate designers to create innovative solutions that can match or surpass certain aspects of biological systems. The collective knowledge from many such efforts may someday result in robotic "all-rounders" that can keep up with (or be worn by) organisms in the real world.

REFERENCES

Bhounsule, P.A., Cortell, J., Grewal, A., Hendriksen, B., Karssen, J.D., Paul, C., Ruina, A., 2014. Low-bandwidth reflex-based control for lower power walking: 65 km on a single battery charge. Int. J. Robot. Res. 33 (10), 1305–1321.

Biewener, A., Alexander, R., Heglund, N.C., 1981. Elastic energy storage in the hopping of kangaroo rats (Dipodomys spectabilis). J. Zool. 195 (3), 369–383.

Daley, M.A., Biewener, A.A., 2006. Running over rough terrain reveals limb control for intrinsic stability. Proc. Natl. Acad. Sci. 103 (42), 15681–15686.

Daley, M.A., Usherwood, J.R., Felix, G., Biewener, A.A., 2006. Running over rough terrain: Guinea fowl maintain dynamic stability despite a large unexpected change in substrate height. J. Exp. Biol. 209 (1), 171–187.

Daley, M.A., Voloshina, A., Biewener, A.A., 2009. The role of intrinsic muscle mechanics in the neuromuscular control of stable running in the Guinea fowl. J. Physiol. 587 (11), 2693–2707.

Donelan, J.M., Kram, R., Kuo, A.D., 2002. Simultaneous positive and negative external mechanical work in human walking. J. Biomech. 35 (1), 117–124.

Garcia, M., Chatterjee, A., Ruina, A., 1998. Speed, efficiency, and stability of small-slope 2d passive dynamic bipedal walking. In: Proceedings. 1998 IEEE International Conference on Robotics and Automation, 1998, vol. 3. IEEE, pp. 2351–2356.

Hubicki, C., Grimes, J., Jones, M., Renjewski, D., Spröwitz, A., Abate, A., Hurst, J., 2016. ATRIAS: design and validation of a tether-free 3D-capable spring-mass bipedal robot. Int. J. Robot. Res. 35 (12), 1497–1521.

Lee, D.V., Comanescu, T.N., Butcher, M.T., Bertram, J.E., 2013. A comparative collision-based analysis of human gait. Proc. R. Soc. Lond. B, Biol. Sci. 280 (1771), 20131779.

Ramezani, A., Hurst, J.W., Hamed, K.A., Grizzle, J.W., 2014. Performance analysis and feedback control of ATRIAS, a three-dimensional bipedal robot. J. Dyn. Syst. Meas. Control 136 (2), 021012.

Rubenson, J., Heliams, D.B., Maloney, S.K., Withers, P.C., Lloyd, D.G., Fournier, P.A., 2007. Reappraisal of the comparative cost of human locomotion using gait-specific allometric analyses. J. Exp. Biol. 210 (20), 3513–3524.

Seok, S., Wang, A., Chuah, M.Y.M., Hyun, D.J., Lee, J., Otten, D.M., Kim, S., 2015. Design principles for energy-efficient legged locomotion and implementation on the MIT Cheetah Robot. IEEE/ASME Trans. Mechatron. 20 (3), 1117–1129.

Srinivasan, M., Ruina, A., 2006. Computer optimization of a minimal biped model discovers walking and running. Nature 439 (7072), 72–75.

Taylor, C.R., Heglund, N.C., Maloiy, G.M., 1982. Energetics and mechanics of terrestrial locomotion. I. Metabolic energy consumption as a function of speed and body size in birds and mammals. J. Exp. Biol. 97 (1), 1–21.

Voloshina, A.S., Kuo, A.D., Daley, M.A., Ferris, D.P., 2013. Biomechanics and energetics of walking on uneven terrain. J. Exp. Biol. 216 (21), 3963–3970.

Wisse, M., Schwab, A.L., van der Linde, R.Q., van der Helm, F.C., 2005. How to keep from falling forward: elementary swing leg action for passive dynamic walkers. IEEE Trans. Robot. 21 (3), 393–401.

Wu, A., Geyer, H., 2014. Highly robust running of articulated bipeds in unobserved terrain. In: 2014 IEEE/RSJ International Conference on Intelligent Robots and Systems. IEEE, pp. 2558–2565.

Chapter 9.2

Application in Daily Life (Assistive Systems)

Thomas G. Sugar[*], Jeffrey Ward[†], and Kevin W. Hollander[†]

[*]*Fulton Schools of Engineering, The Polytechnic School, Arizona State University, AZ, USA*
[†]*SpringActive, Inc.*

Legged robotic systems are allowing users to walk and move around the world: powered rehabilitation robots mounted on treadmills assist in stroke and spinal cord rehabilitation, people with spinal cord injuries are able to stand up and walk in clinics with wearable robots, passive and powered ankle foot orthoses assist users with drop foot, passive and powered prosthetic ankles allow users to walk naturally over ground, wearable robots are assisting users to sit, stand, and walk in manufacturing environments, and lastly, wearable robots are moving into the recreation environment assisting gait and even making it easier to ski.

9.2.1 REHABILITATION

Roboticists have been developing devices to assist gait of stroke survivors and people with spinal cord injuries. Systems include: the Lokomat, AutoAmbulator, Gait Trainer, Haptic Walker, GaitMaster, G-EO System, LokoHelp, LOPES (Lower Extremity Powered Exoskeleton), ARTHur (Ambulation-Assisting Robotic Tool for Human Rehabilitation), POGO (Pneumatically Operated Gait Orthosis), PAM (Pelvic Assist Manipulator), ALEX (Active Leg Exoskeleton, and ALTACRO (Actuated Compliant Robotic Orthosis) (Pennycott et al., 2012; Diaz et al., 2011). These systems allow the user to walk, assisted on a treadmill to perform repeated walking movements. The repetitive task training has shown some success in improving gait after stroke (Pennycott et al., 2012). Many of these systems such as the LOPES are relying on springs to compliantly interact with the human. Principles of energy storage and release and impedance control are used to allow the system to naturally interact with the human.

9.2.2 SPINAL CORD INJURY

Wearable robots are allowing users to get up and walk for a couple of hours a day in the clinic and over ground. Systems include robots from: Ekso Bionics, Parker Hannifin, ReWalk, Rex Bionics, Wandercraft, US Bionics, Technaid, Institute for Human Machine Cognition (IHMC), and ExoAtlet (Esquenazi et al., 2012;

FIGURE 9.2.1   Robots allow spinal cord injured people to walk over ground, Ekso Bionics and ReWalk, courtesy, www.eksobionics.com and www.rewalk.com.

Farris et al., 2011; Strickland, 2012; Aach et al., 2014). These systems use hydraulic or electrical motor servo systems to control the motion of the hip, knee and ankle (Del-Ama et al., 2012), see Fig. 9.2.1. It is important to note that these systems are used only for part of the day; however, because the user stands up, secondary complications are reduced such as pressure sores. Currently most of these systems use predefined gait trajectories from biomechanical literature to drive motor patterns at the hip and knee. Also, most systems use brushless DC motors connected to large gear ratios with harmonic transmissions. Concepts such as energy storage and release using springs to reduce reflected inertia to the user could be used to make these systems more compliant and efficient. Lastly, most systems do not actuate the ankle which is important for human locomotion during push-off and stability and balance during standing.

9.2.3   PASSIVE ANKLE FOOT ORTHOSES

New, carbon fiber, ankle foot orthoses (AFO's) are being developed that allow controlled roll over at the ankle joint while assisting drop foot, see Fig. 9.2.2. These systems tune the stiffness for natural roll over. Manufacturers include Ossur, Otto Bock, Endolite, Freedom Innovations, etc. A group of scientists at the Center for the Intrepid at Brooke Army Medical Center have been investigating the tuned stiffness needed at the ankle (Esposito et al., 2014; Patzkowski et al., 2012). This group is able to tune the stiffness of a strut that links a cuff mounted at the shank to a footplate mounted inside the shoe. These systems make use of the natural dynamics while walking. As the leg rolls over the ankle during early stance, the torsional stiffness at the ankle joint stores energy. This energy is then used to help propel the person forward during push-off.

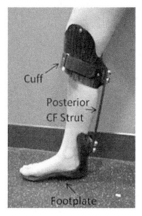

FIGURE 9.2.2 Carbon fiber ankle foot orthoses are designed to assist with drop foot. The angular stiffness of the system is tuned to the user. The carbon fiber brace on the left assists with drop foot, courtesy, www.alimed.com. The brace in the figure on the right is called IDEO for Intrepid Dynamic Exoskeletal Orthosis (Esposito et al., 2014; Patzkowski et al., 2012).

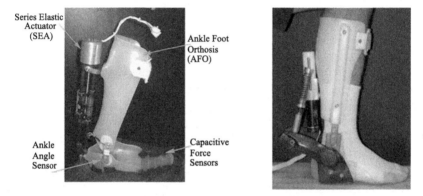

FIGURE 9.2.3 A powered ankle foot orthoses have been designed to assist push-off and lift the toes during the swing phase (Blaya and Herr, 2004; Hollander et al., 2005; Bharadwaj et al., 2004, 2005).

9.2.4 POWERED ANKLE FOOT ORTHOSES

Powered ankle foot orthoses allow for a natural push-off and lift the toe during swing to assist foot-drop. Research into developing powered systems for stroke survivors was started by Herr and Sugar, see Fig. 9.2.3 (Blaya and Herr, 2004; Dollar and Herr, 2008; Hollander et al., 2005; 2006; Ward et al., 2007, 2008, 2011). Typically, these systems store energy as the leg rolls over the ankle for regenerative braking and use this energy plus additional motor energy to achieve a good push-off at the end of the stance phase. Secondly, and importantly, these systems pick the toe up during the swing phase for people with

FIGURE 9.2.4   A carbon fiber foot, Vari-Flex®, used for walking, courtesy of www.ossur.com. On the right, a carbon fiber foot tuned for running, courtesy of www.ossur.com.

drop-foot. By picking the toe up, the foot does not scuff the ground and there is less hip circumduction. These systems have been shown to work in the laboratory but have not been sold commercially as of yet.

9.2.5   PASSIVE PROSTHETIC ANKLES

Passive energy storage and release feet allow for energy storage during early and mid-stance and release energy at push-off. Typically one foot is tuned for walking and a different foot is tuned for running, see Fig. 9.2.4. The new carbon fiber feet allow for energy storage and good moment profiles. However, because there is no hinge joint, there is little plantarflexion at the end of stance and the push-off power is delayed and weak (Ferris et al., 2012; Harper et al., 2014).

9.2.6   POWERED PROSTHETIC ANKLES

Powered ankle systems allow for more natural walking by adding energy into the gait cycle, powering push-off, lifting the toe during swing, and reducing the metabolic cost of walking for amputees (Hollander et al., 2006; Hitt et al., 2010a; Hitt and Sugar, 2010; Hitt et al., 2007, 2009; Hitt et al., 2010b; Holgate et al., 2008; Au et al., 2008), see Fig. 9.2.5. Research into powered systems has been done by Goldfarb, Grimmer, Herr, Rastagaar, Seyfarth, Grimmer, Sugar, Voglewede, Wang, Lefeber, Cherelle, Q. Wang, B. Vanderborght, and others (Versluys et al., 2009; Cherelle et al., 2014). Very good review articles have been written by Grimmer, Seyfarth, Lefeber, Cherelle, Wang and Vanderborght (Grimmer, 2015; Grimmer and Seyfarth, 2014). The goal of the powered ankle systems is to mimic able body gait motion and add energy into the gait cycle at push off. These systems allow users to walk, run, jump, ascend and descend stairs as well as inclines. Researchers are now studying the kine-

FIGURE 9.2.5   The Odyssey ankle is shown on the left and the Bionix ankle is shown on the right, courtesy of www.springactive.com and www.bionix.com.

matics and kinetics while wearing these powered devices (Ferris et al., 2011, 2012; Grabowski and D'Andrea, 2013; Sinitski et al., 2012). These devices have shown positive outcomes with increased push-off and reduced metabolic cost. These ankle systems are using bio-inspired controllers and mechanical designs to achieve their performance. For example, springs can be tuned to the motion or clutched based on gait principles. In the Robotic Tendon, the spring is tuned to the body mass to store and release energy similar to the human's Achilles tendon (Hitt et al., 2007, 2009, 2010b). In Cherelle's ankle, springs are clutched to store additional energy needed for push off (Cherelle et al., 2012, 2014).

9.2.7   WEARABLE ROBOTS FOR MANUFACTURING

Robotic devices are making it easier to walk and carry loads. Devices are being designed to carry heavy tools such as grinders, see Fig. 9.2.6. Hip exoskeletons are being developed to assist gait when carrying luggage in an airport. Full leg exoskeletons are being developed that power the hip, knee, and ankle. Current passive systems use springs for gravity compensation. Constant force mechanisms are designed to reduce the effects of gravity and make carrying a tool easier. Powered hip exoskeletons add assistive torques to make it easier to walk and lift objects. Controllers for these systems are a challenge because the user's intention is difficult to determine. Many researchers are using surface EMG sensors or IMU's to determine the user's motion.

9.2.8   WEARABLE ROBOTS FOR RECREATION

Lastly, systems are being built for the recreation market and everyday use. Collins and Sawicki have developed a passive ankle device that makes it easier to walk (Collins et al., 2015), see Fig. 9.2.7. The spring reduces the muscle

FIGURE 9.2.6 The Lockheed Martin Fortis device is shown on the left. It is a passive exoskeleton that transfers the load of the grinder to the ground, courtesy of http://www.lockheedmartin.com/us/products/exoskeleton/FORTIS.html. A hip exoskeleton makes it easier to carry luggage, courtesy of http://www.cyberdyne.jp/english/.

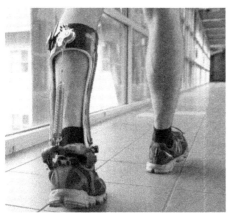 

FIGURE 9.2.7 A passive ankle foot orthosis makes it easier to walk, reducing metabolic cost (Collins et al., 2015). The figure on the right shows an exoskeleton that reduces the load at the knees when skiing, courtesy of www.againer-ski.com.

forces as the tibia rolls over the ankle reducing metabolic cost. Other exoskeletons are being developed to make it easier to go over the bumps on ski slopes, see Fig. 9.2.7.

Wearable robotic systems are assisting, enhancing, and replacing limbs to allow users to move around the world. Systems are being developed for rehabilitation as well as passive and powered orthoses and prostheses. Exoskeleton systems are being developed to assist in manufacturing operations as well as carrying loads. Lastly, systems will move into our everyday lives allowing people to walk easier or walk faster.

REFERENCES

Aach, M., Cruciger, O., Sczesny-Kaiser, M., Höffken, O., Meindl, R.C., Tegenthoff, M., Schwenkreis, P., Sankai, Y., Schildhauer, T.A., 2014. Voluntary driven exoskeleton as a new tool for rehabilitation in chronic spinal cord injury: a pilot study. Spine J. 14, 2847–2853.

Au, S., Berniker, M., Herr, H., 2008. Powered ankle–foot prosthesis to assist level-ground and stair-descent gaits. Neural Netw. 21.

Bharadwaj, K., Hollander, K.W., Mathis, C.A., Sugar, T.G., 2004. Spring over muscle (SOM) actuator for rehabilitation devices. In: 26th Annual International Conference of the IEEE Engineering in Medicine and Biology Society. IEMBS '04.

Bharadwaj, K., Sugar, T.G., Koeneman, J.B., Koeneman, E.J., 2005. Design of a robotic gait trainer using spring over muscle actuators for ankle stroke rehabilitation. ASME J. Biomech. Eng. 127, 1009–1013.

Blaya, J.A., Herr, H., 2004. Adaptive control of a variable-impedance ankle–foot orthosis to assist drop-foot gait. IEEE Trans. Neural Syst. Rehabil. Eng. 12.

Cherelle, P., Matthys, A., Grosu, V., Vanderborght, B., Lefeber, D., 2012. The amp-foot 2.0: Mimicking intact ankle behavior with a powered transtibial prosthesis. In: IEEE RAS & EMBS International Conference on Biomedical Robotics and Biomechatronics (BioRob).

Cherelle, P., Mathijssen, G., Wang, Q., Vanderborght, B., Lefeber, D., 2014. Advances in propulsive bionic feet and their actuation principles. Adv. Mech. Eng. 6, 984046.

Collins, S.H., Wiggin, M.B., Sawicki, G.S., 2015. Reducing the energy cost of human walking using an unpowered exoskeleton. Nature 522.

Del-Ama, A.J., Koutsou, A.D., Moreno, J.C., De-Los-Reyes, A., Gil-Agudo, Á., Pons, J.L., 2012. Review of hybrid exoskeletons to restore gait following spinal cord injury. J. Rehabil. Res. Dev. 49, 497–514.

Diaz, I., Gil, J.J., Sanchez, E., 2011. Lower-limb robotic rehabilitation: literature review and challenges. J. Robot. 2011.

Dollar, A.M., Herr, H., 2008. Lower extremity exoskeletons and active orthoses: challenges and state-of-the-art. IEEE Trans. Robot. 24, 144–158.

Esposito, E.R., Blanck, R.V., Harper, N.G., Hsu, J.R., Wilken, J.M., 2014. How does ankle–foot orthosis stiffness affect gait in patients with lower limb salvage? Clin. Orthop. Relat. Res. 472, 3026–3035.

Esquenazi, A., Talaty, M., Packel, A., Saulino, M., 2012. The ReWalk powered exoskeleton to restore ambulatory function to individuals with thoracic-level motor-complete spinal cord injury. Am. J. Phys. Med. Rehabil. 91, 911–921.

Farris, R.J., Quintero, H., Goldfarb, M., 2011. Preliminary evaluation of a powered lower limb orthosis to aid walking in paraplegic individuals. IEEE Trans. Neural Syst. Rehabil. Eng. 19, 652–659.

Ferris, A.E., Aldridge, J.E., Sturdy, J.T., Wilken, J.M., 2011. Evaluation of the biomimetic properties of a new powered ankle–foot prosthetic system. In: Proceedings of the 2011 Annual Meeting of the American Society of Biomechanics.

Ferris, A.E., Aldridge, J.M., Rábago, C.A., Wilken, J.M., 2012. Evaluation of a powered ankle–foot prosthetic system during walking. Arch. Phys. Med. Rehabil. 93, 1911–1918.

Grabowski, A.M., D'Andrea, S., 2013. Effects of a powered ankle–foot prosthesis on kinetic loading of the unaffected leg during level-ground walking. J. NeuroEng. Rehabil. 10, 1.

Grimmer, M., 2015. Powered Lower Limb Prostheses. University of TU, Darmstadt.

Grimmer, M., Seyfarth, A., 2014. Mimicking human-like leg function in prosthetic limbs. In: Neuro-Robotics. Springer, pp. 105–155.

Harper, N.G., Esposito, E.R., Wilken, J.M., Neptune, R.R., 2014. The influence of ankle–foot orthosis stiffness on walking performance in individuals with lower-limb impairments. Clin. Biomech. 29, 877–884.

Hitt, J., Sugar, T., 2010. Load carriage effects on a robotic transtibial prosthesis. In: 2010 International Conference on Control Automation and Systems (ICCAS), pp. 139–142.

Hitt, J.K., Bellman, R., Holgate, M., Sugar, T.G., Hollander, K.W., 2007. The SPARKy (spring ankle with regenerative kinetics) project: design and analysis of a robotic transtibial prosthesis with regenerative kinetics. In: ASME 2007 International Design Engineering Technical Conferences and Computers and Information in Engineering Conference, pp. 1587–1596.

Hitt, J.K., Holgate, M., Bellman, R., Sugar, T.G., Hollander, K.W., 2009. Robotic transtibial prosthesis with biomechanical energy regeneration. Ind. Robot 36, 441–447.

Hitt, J., Brechue, B., Boehler, A., Ward, J., Hollander, K., Holgate, M., Sugar, T., Audet, D., Kanagaki, G., 2010a. Dismounted soldier biomechanical power regeneration. In: 27th Army Science Conference.

Hitt, J.K., Sugar, T.G., Holgate, M., Bellman, R., 2010b. An active foot–ankle prosthesis with biomechanical energy regeneration. J. Med. Devices 4, 011003.

Holgate, M.A., Hitt, J.K., Bellman, R.D., Sugar, T.G., Hollander, K.W., 2008. The SPARKy (spring ankle with regenerative kinetics) project: choosing a DC motor based actuation method. In: 2nd IEEE RAS & EMBS International Conference on Biomedical Robotics and Biomechatronics, BioRob 2008, pp. 163–168.

Hollander, K.W., Ilg, R., Sugar, T.G., 2005. Design of the robotic tendon. In: Design of Medical Devices.

Hollander, K.W., Ilg, R., Sugar, T.G., Herring, D.E., 2006. An efficient robotic tendon for gait assistance. ASME J. Biomech. Eng. 128, 788–791.

Patzkowski, J.C., Blanck, R.V., Owens, J.G., Wilken, J.M., Kirk, K.L., Wenke, J.C., Hsu, J.R., Consortium, S.T.R., 2012. Comparative effect of orthosis design on functional performance. J. Bone Jt. Surg. 94, 507–515.

Pennycott, A., Wyss, D., Vallery, H., Klamroth-Marganska, V., Riener, R., 2012. Towards more effective robotic gait training for stroke rehabilitation: a review. J. NeuroEng. Rehabil. 9.

Sinitski, E.H., Hansen, A.H., Wilken, J.M., 2012. Biomechanics of the ankle–foot system during stair ambulation: implications for design of advanced ankle–foot prostheses. J. Biomech. 45, 588–594.

Strickland, E., 2012. Good-bye, wheelchair. IEEE Spectr. 49, 30–32.

Versluys, R., Beyl, P., Van Damme, M., Desomer, A., Van Ham, R., Lefeber, D., 2009. Prosthetic feet: state-of-the-art review and the importance of mimicking human ankle–foot biomechanics. Disabil. Rehabil., Assist. Technol. 4, 65–75.

Ward, J.A., Balasubramanian, S., Sugar, T., He, J., 2007. Robotic gait trainer reliability and stroke patient case study. In: IEEE 10th International Conference on Rehabilitation Robotics, ICORR 2007, pp. 554–561.

Ward, J., Boehler, A., Shin, D., Hollander, K., Sugar, T., 2008. Control architectures for a powered ankle foot orthsosis. Int. J. Assist. Robot. Mechatron. 9, 2–13.

Ward, J., Sugar, T., Boehler, A., Standeven, J., Engsberg, J.R., 2011. Stroke survivors' gait adaptations to a powered ankle–foot orthosis. Adv. Robot. 25, 1879–1901.

Chapter 9.3

Related Research Projects and Future Directions

Maziar A. Sharbafi[‡,§], André Seyfarth[§], Koh Hosoda[¶], and Thomas G. Sugar[‖]

[‡]*School of Electrical and Computer Engineering, College of Engineering, University of Tehran, Iran* [§]*Lauflabor Locomotion Laboratory, TU Darmstadt, Germany* [¶]*Department of System Innovation, Graduate School of Engineering Science, Osaka University, Toyonaka, Japan* [‖]*Fulton Schools of Engineering, The Polytechnic School, Arizona State University, AZ, USA*

Different methods in modeling, design, and control of legged systems were presented in this book. In this chapter we explained some applications in daily life and provided a comparison of artificial and biological legged systems' performance, efficiency and robustness. Here, we summarize some recent research projects related to legged locomotion. There are also other prominent associations investing legged locomotion and especially humanoid robots like NASA with the Robonaut2 (http://www.nasa.gov/robonaut2), HONDA with the Asimo robot (http://asimo.honda.com/), and Boston Dynamics with the new ATLAS robot (http://www.bostondynamics.com/robot_Atlas.html).

9.3.1 EUROPEAN PROJECTS

The European Union's Seventh Programme for research (FP7) supports different projects related to application of bioinspired legged locomotion in daily life. In the following we present some of the related projects.

BALANCE: **B**alance **A**ugmentation in **L**ocomotion, through **A**nticipative, **N**atural and **C**ooperative control of **E**xoskeletons.

BALANCE (http://balance-fp7.eu) is funded by FP7 from 2013 to 2017. Research topics include development of exoskeleton (design and manufacturing, adaptive robot control), human motor control (simulation studies, human experiments), gait mechanics, biomechanical sensing and balance assessment technology. Eight partners from different institutes and companies in Europe are involved in this project (Table 9.3.1).

The goal of this project is to realize a robotic exoskeleton to support humans balance in standing and walking especially in challenging situations like occurence of perturbations or for assisting patients. Existing exoskeleton robots can mainly support the body weight or additional loads, or guide impaired legs through a step-like motion. So far, however, they do not support maintaining

TABLE 9.3.1 BALANCE project consortium

	Participant name	Country
1	Tecnalia Research & Innovation	Spain
2	Technische Universität Darmstadt	Germany
3	University of Twente	The Netherlands
4	CEA LIST	France
5	Eidgenössische Technische Hochschule Zürich (ETHZ)	Switzerland
6	Imperial College London,	United Kingdom
7	XSENS 3D Motion Tracking	The Netherlands
8	University Rehabilitation Institute	Slovenia

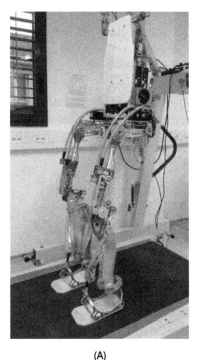
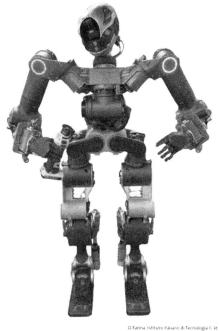

(A)   (B)

FIGURE 9.3.1 (A) EMY exoskeleton developed by CEA List within BALANCE project. (B) WALK-MAN robot, a humanoid robot developed within the EU FP7 project WALK-MAN.

postural balance, especially if the user has remaining control and the robot has to cooperate with the human. This is addressed in the BALANCE project.

The novel exoskeleton developed by CEA List (Fig. 9.3.1A) is designed to support the user and take over the control if needed, e.g., in difficult conditions or in case of malfunction of the user. The project aims at improving the under-

TABLE 9.3.2 BioMot project consortium

	Participant name	Country
1	Agencia del Consejo Superior de Investigaciones Científicas	Spain
2	Vrije Universiteit Brussel	Belgium
3	Università degli Studi di Padova	Italy
4	Ossur hf	Iceland
5	Fundación del Hospital Nacional de Parapléjicos	Spain
6	RIKEN	Japan
7	Universidad Miguel Hernández de Elche	Spain
8	Technaid S.L.	Spain

standing of human postural control, its functionality and robustness in healthy or impaired people, which provides the basis for enabling the exoskeleton to support human postural control. In this project, locomotion subfunction concept (Chapter 2), bioinspired conceptual models and some control concepts like VPP (Chapter 3), stability indices and impedance control (Chapter 4), torque control (Chapter 5) and also Neuromuscular control (Chapter 6) are used to develop novel exoskeleton functions.

BioMot: Smart Wearable Robots with Bioinspired Sensory–Motor Skills

This project (http://biomotproject.eu) founded by FP7 from 2013 to 2016, includes 8 research partners as shown in Table 9.3.2. The main objective of BioMot is improvement of existing wearable robotic exoskeletons exploiting dynamic sensory–motor interactions with the focus on developing cognitive capabilities that can lead to symbiotic gait behavior in the interaction of a human with a wearable robot. Real-time human-like adaptability and flexibility to cope with natural perturbations are a missing part in most advanced wearable robots, due to voluntary control or environmental constraints. In this project a cognitive architecture is proposed for wearable robots exploiting neuronal control and learning mechanisms to enable positive coadaptation and seamless interaction with humans.

They use a neuro-musculoskeletal model for computing neuromuscular activity (EMG) to predict joint moments, prescribe the exoskeleton function and estimate human efforts in real-time to be able to support the body with the exoskeleton when it is necessary. In addition, brain signals (EEG) and kinetic and kinematic sensors are employed for gait detection and decision making, handling transitions or task violations. Finally, local reflex-based joint controllers are designed to allow for automatic adaptation when confronting changes in the interaction. At the physical level, intrinsically compliant actuators are developed

TABLE 9.3.3 SYMBITRON project consortium

	Participant name	Country
1	University of Twente	Netherlands
2	Ecole Polytechnique Fédérale de Lausanne (EPFL)	Switzerland
3	Technical University of Delft	Netherlands
4	Imperial College London	United Kingdom
5	Santa Lucia Foundation	Italy
6	Össur	Iceland

to exploit natural dynamics of movement, orchestrated by the control system for economy and stability. A global learning scheme modulates joint compliance as a function of gait efficiency and semantic signals inferred from user demand.

BioMot's goal is "to improve the efficiency in the management of human–robot interaction in over ground gait exoskeletons by means of mixture of bioinspired control, actuation and learning approaches." Therefore, bioinspired motion control techniques besides efficiency and robustness concepts (Chapter 4), torque control (Chapter 5), neural control (Chapter 6) and bioinspired (compliant) actuation mechanisms (Chapter 8) are the key ingredients in this project "to show how the embodiment of bioinspired and architectural mechanisms can allow a user to conveniently alter the behavior of wearable robots for walking."

SYMBITRON: Symbiotic man–machine interactions in wearable exoskeletons to enhance mobility for paraplegics

SYMBITRON is another FP7 project running from 2013 to 2017 with 6 research partners (see Table 9.3.3). The main goals of Symbitron project (https://www.symbitron.eu) are:

i. To develop a safe, bio-inspired, personalized wearable exoskeleton that enables spinal cord injured (SCI) patients to walk without additional assistance, by complementing their remaining motor function
ii. To develop training environments and training protocols for SCI patients and their clinicians
iii. To provide clinical proof of concept for safety and functionality of the system

CYBERLEGs: The CYBERnetic Lower-Limb Cognitive Ortho-prosthesis

CYBERLEGs is a Collaborative Research project funded by the European Commission under the 7th Framework Programme with five partners from three

TABLE 9.3.4 Cyberlegs project consortium

	Participant name	Country
1	Scuola Superiore Sant' Anna	Italy
2	Université Catholique de Louvain	Belgium
3	Vrije Universiteit Brussel	Belgium
4	Univerza v Ljubljani	Slovenia
5	Fondazione Don Carlo Gnocchi Onlus	Italy

different EU countries (Table 9.3.4). This project (http://www.cyberlegs.eu) started in February 2012.

The aim of the CYBERLEGs project is to develop an artificial cognitive system for trans-femoral amputees' lower-limb functional replacement and assistance. CYBERLEGs wanted to find ways of cognitive control, motivated and validated through the ortho-prosthesis scenario, of a multi-degree-of-freedom system with both lower-limb replacing and assistive capacities. CYBERLEGs is a robotic system constituted of an active cognitive artificial leg for the functional replacement of the amputated limb and a wearable active orthosis for assisting the contralateral sound limb which allows the amputee to have different maneuvers of locomotion (e.g., walk backward/forward, stairs climbing, move from sit-to-stand, and vice versa) with a minimum cognitive and energetic effort.

Main scientific and technological challenges of the CYBERLEGs project are: (i) design and development of an energy-efficient lower-limb ortho-prosthesis with tunable passive compliant joints allowing passive energy transfer from knee to ankle joint (related to Chapters 4, 7 and 8); (ii) modeling based on motor primitives capturing human behavior while executing locomotion-related tasks (Chapter 3); (iii) transfer the primitive-based model into a controller for CYBERLEGs and make the control simple and intuitive (Chapter 4 and 5).

H2R: Integrative Approach for the Emergence of Human-like Locomotion

H2R (http://www.h2rproject.eu) is one of the FP7 projects with 6 partners (Table 9.3.5), started in 2011 to investigate drawbacks of existing walking bipeds regarding stability, energy consumption, and robustness to unknown disturbances in comparison with healthy humans (see Sect. 9.1). Therefore, H2R project attempted to demonstrate human-like gait and posture in a controlled compliant bipedal robot as a result of a combination of the most relevant motor control and cognitive mechanisms found in humans. In that respect, they adopted a three-fold process which can help better understand bioinspired legged locomotion and apply it on robotics: (i) investigating the human behavior in order to formalize the most crucial biomechanical and neuromotor

TABLE 9.3.5 H2R project consortium

	Participant name	Country
1	Consejo Superior de Investigaciones Científicas	Spain
2	University of Kaiserslautern	Germany
3	Vrije Universiteit Brussel	Belgium
4	University Medical school Göttingen	Germany
5	Universitaetsklinikum Freiburg	Germany
6	Technaid S.L.	Spain

TABLE 9.3.6 WALK-MAN project consortium

	Participant name	Country
1	Istituto Italiano di Tecnologia (IIT)	Italy
2	Ecole Polytechnique Fédérale de Lausanne (EPFL)	Switzerland
3	The Università di Pisa	Italy
4	Karlsruhe Institute of Technology (KIT)	Germany
5	Université Catholique de Louvain	Belgium

principles of walking and standing, (ii) testing the formalized biological concepts, by their integration into currently existing robotic platforms and developing a new biped (H2R biped), by iteratively including the components and methods successfully tested, and (iii) introducing a new internationally validated benchmarking scheme to test the human-like properties of robotic bipeds (http://benchmarkinglocomotion.org).

WALK-MAN: Whole-body Adaptive Locomotion and Manipulation

WALK-MAN (https://www.walk-man.eu/) is a 4 year integrated project (IP) funded by the European Commission started on September 2013 with 5 partners (Table 9.3.6) and has the goal to develop a robotic platform (of an anthropomorphic form), which can operate outside the laboratory space in unstructured environments and work spaces as a result of natural and man-made disasters. The robot (shown in Fig. 9.3.1B), demonstrates new skills including:

i. Dexterous, powerful manipulation skills, e.g., turning heavy valves
ii. Robust walking (or crawling) over uneven terrains
iii. Autonomous operation regarding enhanced perception and cognition.

Even if the complete autonomous operation is not required, the ability to be functional with reduced level of teleoperation is very important. The reason is that communication limitations for remote control is expectable due to limited channel bandwidth. The robot is designed to demonstrate human levels of lo-

TABLE 9.3.7 EXO-LEGS project consortium

	Participant name	Country
1	University of Gävle	Sweden
2	Karlsruhe Institute of Technology	Germany
3	Universidad Politécnica de Cartegena	Spain
4	Chas A Blatchford & Sons Limited	United Kingdom
5	Hocoma AG	Switzerland
6	GIGATRONIK Technologies GmbH	Germany
7	MRK Systeme GmbH	Germany
8	Proyecto Control Montaje, S.L.	Spain
9	Mobile Robotics Sweden AB	Sweden
10	Gävle kommun and other Gävleborg partners	Sweden

comotion, balance and manipulation and to be validated in realistic challenge tasks outside the laboratory environment.

EXO-LEGS: Exoskeleton Legs for Elderly Persons

EXO-LEGS (http://www.exo-legs.org/) is a three year project which started in October 2012 with funding under the Ambient Assistive living (AAL) programme. This project was performed as a collaborative work of 10 research partners from 5 different countries (Table 9.3.7), in which they tried to solve the progressive nature of impairments in elderly people which often leads to loss of independence and influences the quality of life. EXO-LEGS targeted developing a range of active lower-limb exoskeletal assistive solutions for improving indoor and outdoor mobility. Different locomotion tasks are considered in this project such as normal walking, standing up, sitting down, stepping over objects, walking on soft and uneven ground and walking up- and down-stairs.

KoroiBot: Improving humanoid walking capabilities by human-inspired mathematical models, optimization and learning

This project (http://www.koroibot.eu/) was funded by EU from 2013 to 2016 in which 7 research institutes (Table 9.3.8) are involved beside an external collaborator (Philipps-Universität Marburg, Germany). The goal of KoroiBot project is to enhance the ability of humanoid robots to walk in a dynamic, versatile and human-like fashion. Moving on difficult situations like rough terrains or in perturbed condition is the main challenge of humanoids locomotion. In this project human walking in different conditions is studied (e.g., on stairs and slopes, on soft and slippery ground or over beams and seesaws), to develop mathematical models. New optimization and learning methods for walking on

TABLE 9.3.8 KoroiBot project consortium

	Participant name	Country
1	Universität Heidelberg Germany	Germany
2	Centre National de la Recherche Scientifique	France
3	Karlsruhe Institute of Technology	Germany
4	Istituto Italiano di Tecnologica	Italy
5	Technische Universiteit Delft	The Netherlands
6	Weizmann Institute	Israel
7	Universität Tübingen	Germany

TABLE 9.3.9 AXO-SUIT project consortium

Participant number	Participant name	Country
1	Aalborg University	Denmark
2	University of Gävle	Sweden
3	University of Limerick	Ireland
4	Welldana A/S	Denmark
5	Bioservo Technologies AB	Sweden
6	MTD Precision Engineering LTD	Ireland
7	Hjälpmedelsteknik Sverige	Sweden
8	COMmeto BVBA	Belgium

two legs can be implemented in practice on real robots or even be applied in medicine, e.g., for designing and controlling intelligent artificial limbs or exoskeletons.

AXO-SUIT

AXO-SUIT (http://www.axo-suit.eu/) is a three year project started on October 2014 and is funded under the Ambient Assisted Living (AAL) Joint Programme. Three universities and five companies from four different European countries work on this project (Table 9.3.9). With their research experience on assistive devices, AXO-SUIT is to comprehensively supplement the strength of elderly persons with feasible exoskeletons in undertaking volunteer work (e.g., maintaining gardens or carrying groceries as well as participating in local social activities). The AXO-SUIT integrates recent advances in assistive technology to study and design exoskeletons and to meet the challenges in helping elderly workers. Novel exoskeletons are developed which consist integrative modules to realize prototypes for upper-, lower- and full-body assistive suits. The exoskeletons could also be extended to be utilizedd for weak or disabled adults, or elder employees with similar their needs.

9.3.2   RESEARCH PROJECTS IN NORTH AMERICA

There are several research foundations and funding agencies in US supporting research related to bioinspired legged locomotion. We have selected few research activities in the following.

9.3.2.1   National Science Foundation (NSF)

NSF provides funding for fundamental research projects in US. Among these projects, several projects are addressing legged locomotion. In the following we present some selected research in this area.

Biped Locomotion Control. This project was performed in University of Michigan from 2000 to 2003. Its objective was to develop a coherent mathematical framework in which asymptotically stabilizing feedback controllers for biped systems may be rigorously analyzed and synthesized. The proposed work was based on a recent breakthrough in locomotion stability analysis by Grizzle, Abba, and Plestan (see HZD controller in Subchapter 4.7). For the important special case of an under actuated biped with a torso and stiff legs (i.e., no knees), these researchers provided the first mathematical proof of the asymptotic stability of a feedback-controlled walking motion. This project systematically explored feedback stabilization and design for a sequence of more realistic biped models. The models have been selected on the basis of their applicability to robotics, biomechanics and prosthetics, and permit largely under-actuated designs.

Passivity Based Control in Bipedal Locomotion. This project (2005–2008) has been conducted by Mark Spong at the University of Illinois at Urbana-Champaign. The goal was to investigate passivity based control in bipedal locomotion. The project explored several extensions of bipedal locomotion in the context of passivity based hybrid nonlinear control and investigated speed regulation, the use of alternate potential functions to increase the basins of attraction of stable limit cycles, the effect of control saturation and underactuation in passivity based control, and the efficiency of passivity based control methods compared to true energy optimal control. In addition, this was tested for control of gait transitions, including gait initiation and stopping. Besides development of new concepts and the design of provably correct control algorithms the project aimed to integrate the theoretical tools of passivity based analysis and control with studies of balance and locomotion in human subjects in order to supplement the descriptive research typical in those studies with more analytical methods. The practical application of this research project is on the design of walking robots that have improved performance capabilities over existing machines. From a broader perspective, the analysis and design tools developed in this project also contribute to a better understanding of human locomotion, which can result in applications in biomechanics and biomedicine, such as the design of improved prosthetic devices, the development of falls prevention programs for the elderly, and rehabilitation techniques.

Unified Model and Robotic Implementation of Bio-Inspired Walking and Running. This project was funded by NSF Dynamical Systems Program from 2011 to 2016 as a collaborative work of Hartmut Geyer at Carnegie Mellon University and Jonathan Hurst at Oregon State University. The research objectives are to develop and implement a biomechanically relevant, unified theory of legged dynamics that spans walking and running, and to demonstrate this theory on a new bipedal robot, called ATRIAS.

To date, a legged machine whose dynamic behavior can approach the performance of human walking and running, including transitions between these two gaits, does not exist. The research results in principled models of gait and gait transitions with human-like leg dynamics, in generalized models for manipulating cyclic hybrid dynamic systems to achieve different goal behaviors, and in the verification and demonstration of this new scientific understanding with a bipedal robot. The research progresses from analyzing simplified gait models that capture the essential dynamics of human locomotion to integrating these modules in a core dynamic model of human gait. The results of this research can provide the opportunity to create control algorithms for powered legged systems that enable functionally versatile behaviors similar to animals and humans. Examples include prosthetic legs and exoskeletons that provide human users with human-like leg dynamics during walking, running, and the transitions between these gaits. Other examples include legged robots that achieve robust and efficient dynamic turning behaviors.

9.3.2.2 Defense Advanced Research Projects Agency (DARPA)

In 2015, DARPA established a new competition called DARPA Robotics Challenge (**DRC**) to address reducing rescue risk in disaster relief by promoting innovation in humanoid robotic technology. The primary technical goal of the DRC was to develop human-supervised ground robots capable of executing complex tasks in dangerous, degraded, human-engineered environments. To achieve its goal, the DRC was advancing the state-of-the-art of supervised autonomy, mounted and dismounted mobility, and platform dexterity, strength, and endurance. Improvements in supervised autonomy, could facilitate control of robots by non-expert supervisors and allow effective operation despite degraded communications (low bandwidth, high latency, intermittent connection).

The DRC program (http://www.theroboticschallenge.org) provides highlights, including the DRC Trials held in December 2013 and the DRC Finals in June 2015. DARPA also supports many other projects in this field. Some of them are presented in the following.

Efficient, Agile and Robust 3D Bipedal Walking and Running The project was funded by the DARPA Maximum Mobility and Manipulation Program from 2011 to 2016. The goal was to develop bipedal robots that navigate natural, uneven terrain with agility, speed and robustness to disturbances. This project was a collaboration between research groups of Hartmut Geyer (Carnegie Mellon University), Jonathan Hurst (Oregon State University), and Jessy Grizzle (University of Michigan). Different types of expertise were combined in this research project to follow a reproducible path from principled models of legged locomotion to robotic implementation, feedback control, and experimental verification.

9.3.2.3   CAREER: Robust Bipedal Locomotion in Real-World Environments

This project runs from 2013 to 2017 and is executed by Katie Byl's Group at University of California, Santa Barbara, who also was one of the finalists in DRC. The objective of this Faculty Early Career Development (CAREER) Program grant is to develop tools for analyzing and optimizing quasiperiodic biped gaits for high-dimensional models of both humans and humanoid devices. The inverted-pendulum dynamics (Chapter 3) that enbles highly maneuverable upright walking under desired control inputs also make it highly susceptible to destabilization. This model becomes more complex by the discontinuities of impulsive footsteps that vary in both width and height, resulting in a "quasiperiodic" gait. In this project, dimension reduction and machine learning techniques are used for control policy evaluation and improvement to quantifiably estimate fall rates, energy consumption, and speed for bipedal walking on stochastic terrain.

This modeling approach provides means of quantifying the reliability for systems with high dimensionality and complexity for which traditional measures of stability cannot be guaranteed. This method can be applied to estimate the risk of falling (e.g. for a stroke survivor) or to design smart lower-limb prostheses.

9.3.2.4   MIT Cheetah Robot

After focusing on body design (leg and tails, (Ananthanarayanan et al., 2012; Briggs et al., 2012)) in the compliant MIT Cheetah robot, Sangbae Kim and his colleagues introduced MIT Cheetah 2 robot with a new actuation technology and without any physical elastic element (Seok et al., 2015). Because of the high bandwidth actuation in Cheetah 2, the ground reaction force can be controlled similar to what is observed in SLIP model without having a real spring in the legs (Ananthanarayanan et al., 2012). This robot is a unique research to study dynamic locomotion capabilities experimentally benefiting from previous research on the MIT Cheetah I robot and optimal actuator design. As a result, efficient running at speeds from 0–6.4 m/s, mild running turns and autonomous jumping over obstacles up to 40 cm in height (80% of leg length) were achieved.

Further information can be found in http://biomimetics.mit.edu/research/dynamic-locomotion-mit-cheetah-2.

9.3.3   RESEARCH IN ASIA

There is an increasing number of research institutes studying legged locomotion in Asia (e.g., in Japan, South Korea, Iran and Singapore). Here, we present some of the institutes and projects.

9.3.3.1   Humanoid Robotics Institute, Waseda University

Prof. Kato is a pioneer in humanioid robotics who started to build bipedal robots around 1970 in Waseda University. The laboratory has a long history and developed a variety of humanoid robots, starting from WABOT series (1970), Humanoid Project (Wabian, 1995). In 2000, Prof. Takanishi started the Humanoid Robotics Institute.

9.3.3.2   JST Laboratory, University of Tokyo

Professors Inoue and Inaba started to build miniature humanoid robots around 1990. These robots did not carry their computers (their "brains") with them (called "remote-brain project"). The project developed many kinds of small humanoids that can achieve a variety of bipedal movements. In the follow-up Humanoid Project, the so-called H-series humanoid robots driven by electric motors were developed. Around 2000, Kenta, Kotaro, Kojiro, Kenzoh, Kenshiro, and Kengoro started to build muscular-skeletal robots. These robots use electric artificial muscle actuators as described in Chapter 8.

9.3.3.3   Honda Robotics

Honda has a long history of developing humanoid robots. They started to build biped robots from 1986 but they kept the development secret until the disclosure of P2, which is the first completely independent bipedal robot released in December 1996. After P2, they continued to develop humanoid robots, and finally they released HONDA ASIMO in 2000 (http://world.honda.com/ASIMO/). Honda claims that ASIMO is the most advanced humanoid robot in the world (http://asimo.honda.com/).

9.3.3.4   Humanoid Robotics Project (HRP)

The Humanoid Robotics Project was launched by Prof. Inoue at the University of Tokyo in 1998. The project was supported by Ministry of International Trade and Industry and Ministry of Economy, Trade and Industry together with NEDO, MSTC, and AIST for 5 years. Their goal was to demonstrate usefulness of the humanoid robots in the real working environment. The first platform was HRP-1 that was designed based on Honda P-3. Afterwards, they developed HRP-2 in 2003. After that project, Kawada Industry and AIST continued to build the next generations of these humanoid robots with HRP-3P and HRP-4C. This project is still leading in the development of life-size humanoid robots.

9.3.3.5   HUBO Project, KAIST

Professor Jun-Ho Oh developed HUBO, a life-size walking humanoid robot. They started research in 2000 and developed the KHR series. They took part in the DARPA Robotics Challenge and won the first prize in 2015.

9.3.3.6   Adaptive Robotics Laboratory, Osaka University (Hosoda Laboratory)

Adaptive Robotics Laboratory (directed by Prof. Hosoda) has been studying humanoid robots since 2000. Study on bipedal walking started in 2003. They developed a series of bipedal robots based on passive dynamics, which are called Air-Leg series, driven by McKibben pneumatic artificial muscles. Utilizing the compliance of the actuator, they developed bipedal walking robots called Pneumat and a jumping monopod Que-Kaku-K that has a similar muscular-skeletal structure as humans. They developed a whole-body pneumatic-driven humanoid robot Pneumat-BS in 2011, described in Chapter 8. In 2011, they started a project, namely "Understanding Human's Adaptive Bipedal Walking by Using a Cadaver Feet / Artificial Muscular-Skeleton Hybrid Robot" together with Prof. Naomichi Ogihara, studying the function of the foot by observing the behavior of the cadaver foot in the context of walking.

9.3.3.7   Surena Bipedal Robot Series

Surena project, founded by different research foundations (e.g., R&D Society of Iranian Industries and mines and Industrial Development and Renovation Organization (IDRO) of Iran) is led by Dr. Yousefi-Koma, the head of the Center for Advanced Systems and Technologies (CAST) in University of Tehran. In this project, a series of bipedal robots was developed with a human-like height and weight. Surena robots do not have the level of mobility and dexterity of advanced robots like ATLAS, ASIMO or participants in DARPA Challenge. However, this Iranian national project has a smooth progress starting from Surena I, a wheeled robot with 8 DOF in 2008, continuing Surena II with 22 DOF in 2010, able to do super slow walking at 0.03 m/s and releasing Surena III with 31 DOF walking 10 times faster and able to walk on stairs and rough terrains. Recently, Surena Mini (50 cm tall) was developed by the same research group. More details about this project can be found in http://surenahumanoid.com.

9.3.4   NOVEL TECHONOLOGIES

9.3.4.1   Soft Exosuit

In **Harvard Biodesign Lab** (directed by Prof. Wash) a new generation of soft wearable robots is under development (http://biodesign.seas.harvard.edu/soft-exosuits). They use innovative textiles to provide a more conformal, unobtrusive, and compliant means to interface to the human body. These robots augment the capabilities of healthy individuals (e.g., improved walking efficiency) in addition to assisting people with muscle weakness or patients who suffer from physical or neurological disorders. Their exoskeletons are expected to provide advantages compared to traditional ones. The wearable parts are designed to be lightweight and remove rigidity resulting in unconstrained (user) joints. These properties minimize the suit's unintentional interference with the body's natural biomechanics and allow for more synergistic interaction with the wearer. The innovative textiles are inspired by human biomechanics and anatomy. These wearable garments provide a means to transmit assistive torques to a wearer's joints without the use of rigid external structures. These systems yield new challenges like the need to design appropriate force transmission systems. A key feature of exosuits is that if the actuated segments are extended, the suit length can increase so that the entire suit is slack. With this, wearing an exosuit feels like wearing a pair of pants and does not restrict the wearer. Regarding actuation, the focus is on cable driven electrical motors. More recently, also pneumatic McKibben actuators are examined.

9.3.4.2 Superflex

Superflex is another new soft exosuit, a full-body suit filled with soft muscle-like actuators that detect body movements and give them a boost. Such an assistance mechanism allows the user to walk normally in physical therapy, or prevents a soldier from fatigue when carrying heavy loads on backpacks. Rich Mahoney with many years of experience in assistive robotics is the founder of Superflex company. The nonprofit research organization SRI International supported Superflex as a spin-off company. The suit was originally part of the DARPA Warrior Web Project aiming at helping soldiers who have to carry heavy loads over extended distances. Similar to the Harvard Exosuit, it pulls at the bootstraps and applies a force at the back of the heel. Both of them can assist movement in passive mode (all motors turned off).

9.3.4.3 JTAR from SpringActive

SpringActive developed JTAR, a joint torque augmentation system based on motors and springs at the hip and at the ankle. A joint torque augmentation robot (JTAR) was developed to aid walking. Two systems were developed, one that powered the hips and one that powered the ankles. The actuation unit is based on a unidirectional, spring based actuator that stores and releases energy. For example, at the ankle, the spring stores energy as the leg rolls over the ankle and the motor pulls on the spring to store additional energy. The spring energy is released in a controlled burst at push-off. Because the system is used to navigate uneven terrain, extra passive degrees-of-freedom were added to allow full ankle motion. Metabolic savings were shown when wearing the device.

9.3.4.4 Bionics at MIT

MIT biomechatronics lab (directed by Prof. Herr) at MIT media lab focuses on seeking to restore function to individuals who have impaired mobility and developing technologies that augment human performance beyond what nature intends. In this lab the scientific discipline of organismal and cellular neuromechanics are combined with the technological discipline of bionic device design in two main aspects: (i) leg prostheses and (ii) muscular system synthesis.

The focus of the leg prostheses research is on adaptation to different gait conditions (e.g., speed). For example, a powered lower limb exoskeleton that emulates the function of a biological ankle during level-ground walking, specifically providing the net positive work required for a range of walking velocities. Walking with the bionic prosthesis resulted in metabolic energy costs, preferred walking velocities and biomechanical patterns that were not significantly different from people without an amputation (Mooney et al., 2014).

9.3.4.5 Quadrupedal Robots of Boston Dynamics

BigDog is a quadrupedal robot designed to navigate rough terrain that walks, runs, climbs and carries heavy loads. Compared to BigDog, the more recent Spot quadruped is 30% lighter. Spot picks its feet off the ground with dainty precision and keeps pace with running humans by adopting an equine canter. With its smaller size, Spot becomes more practical than its big ancestor. It can be used both indoors and outdoors and my help search and rescue, mapping, or accessing disaster zones.

Another system developed at Boston Dynamics is the Orthotic Joint Braces. The recently patented Brace system is composed of a medial brace and a lateral brace securable via cross members, each brace having an upper part, a lower part and a hinge assembly between those two parts. The brace system also includes a force differential actuator subsystem connected to both the medial and lateral braces. With this, the orthotic knee brace can aid in the treatment of people suffering from joint injuries.

9.3.4.6 SRI PROXI Humanoid Robot

To meet both the high-performance and high-efficiency requirements set out by the DARPA Robotics Challenge, the development of the SRI Humanoid (https://www.sri.com/engage/products-solutions/proxi-high-efficiency-humanoid-robot-platform) is following a three-pronged approach:

(i) Design of a novel transmission, which reduces friction and increases efficiency to 97% (60–70% efficiency in current modern commercial transmissions).

(ii) Electric batteries connect directly to low-cost electric motors, enabling throughput of a significant amount of current without overheating.

(iii) The walking gait is dynamically stable by incorporating springs in the legs to store and release energy, and resembles the natural gait of a human as opposed to the standard slow, squatted, methodical gait employed by current humanoids.

In addition, energy from the robot's impact with the ground is stored as mechanical energy in the springs, and does not get converted into another form of energy, resulting in added efficiency.

9.3.4.7 SCHAFT Biped Robot

SCHAFT which was developed at the JSK Robotics Laboratory at the University of Tokyo is owned by Google's Alphabet. Their most recent biped robot is able to traverse uneven terrain and stairs, and carry up to 60 kg.

This humanoid robot has a very different design to Alphabet's other robots made by Boston Dynamics, with a compact two-leg design and central body that can be moved up or down to cope with different tasks. Unlike Alphabet's larger bipedal robots designed either to interact in a human-like fashion with the world – the humanoid Atlas – or to be a robotic packhorse for the US military or dog's plaything,[1] the Schaft robot is designed to be lower cost, lower power, and to be used by civilians, carrying up to 60 kg over uneven terrain and stairs. The robot was successfully demonstrated being able to dealing with real-time foot placement adaptations, standing on a moving pipe and walking on shingle.

9.3.4.8   "Spring–Mass" Technology in the Future of Walking Robots

This study was conducted by Jonathan Hurst at Oregon State University. The goal was to achieve a realistic robotic implementation of human walking dynamics with human-like versatility and performance. The system design is based on the theoretical concept of "spring–mass" walking (BSLIP; see Subchapter 3.6). The work has been supported by the National Science Foundation, the Defense Advanced Research Projects Agency and the Human Frontier Science Program.

The technologies developed at OSU have evolved from intense studies of both human and animal walking and running, to learn how animals achieve a fluidity of motion with a high degree of energy efficiency. Animals combine a sensory input from nerves, vision, muscles and tendons to create locomotion that researchers have translated into a working robotic system.

This robot was pioneered by MABEL (Sreenath et al., 2011) which became the world's fastest bipedal robot, setting a record of 10.9 km/h or 6.8 mph, in 2011. MABEL, or Michigan Anthropomorphic Biped Electric Leg, is part of a research project ("Control Designs for Bipedal Walkers and Runners") funded through the NSF ECCS/EPAS Program. The objective of this research project is to develop systematic, model-based feedback design procedures for a class of bipedal robots that take advantage of compliance in order to enhance locomotion efficiency and robustness when running on smooth terrain and walking on rough terrain.

9.3.4.9   Summary

Bioinspired legged locomotion has become an important research field in both fundamental and applied studies. Main domains of applications are legged robots and assistive devices. Still, the existing systems are much less robust

---

1. See https://www.theguardian.com/technology/2016/mar/01/top-dog-watch-what-happens-when-a-real-canine-meets-a-robo-pooch.

and versatile compared to human and animal locomotion. Also, the hardware designs and locomotion control approaches are often very system-specific and cannot be easily reused in other systems. Taking advantage of biological design and control principles for legged locomotion can help design future legged systems which are able to support humans in daily activities with high level of energy efficiency and adaptability to the user's needs. To further progress in this research field, a tight interaction with adjacent research topics like cognitive science, actuator and sensor technology, material science, and human motor control will be required.

REFERENCES

Ananthanarayanan, A., Azadi, M., Kim, S., 2012. Towards a bio-inspired leg design for high-speed running. Bioinspir. Biomim. 7 (4).

Briggs, R., Lee, J., Haberland, M., Kim, S., 2012. Tails in biomimetic design: analysis, simulation, and experiment. In: 2012 IEEE/RSJ International Conference on Intelligent Robots and Systems (IROS). Vilamoura, Portugal. IEEE.

Mooney, L.M., Rouse, E.J., Herr, H.M., 2014. Autonomous exoskeleton reduces metabolic cost of human walking during load carriage. J. NeuroEng. Rehabil. 11 (1), 1.

Seok, S., Wang, A., Chuah, M.Y.M., Hyun, D.J., Lee, J., Otten, D.M., Kim, S., 2015. Design principles for energy-efficient legged locomotion and implementation on the MIT Cheetah Robot. IEEE/ASME Trans. Mechatron. 20 (3), 1117–1129.

Sreenath, Koushil, Park, Hae-Won, Poulakakis, Ioannis, Grizzle, Jessy W., 2011. A compliant hybrid zero dynamics controller for stable, efficient and fast bipedal walking on MABEL. Int. J. Robot. Res. 30 (9), 1170–1193.

Index

Symbols

α-Motoneurons, 404, 410, 423, 440
  flexor and extensor, 405, 409

A

Accelerations, 17, 32, 141, 169, 174, 200,
      213, 234, 257, 334, 488, 542
Actuators, 1, 25, 89, 103, 135, 172, 188,
      199, 212, 227, 254, 274, 299,
      312, 402, 459, 487, 498, 512,
      525, 546, 581, 591, 613, 625,
      652
  linear, 246, 616
  muscle-like, 613, 654
  stiff, 592, 605
Anchor dynamics, 67, 254
Anchors, 63–72, 255, 502
Angle of attack, 30, 34, 114, 124, 535
Angular momentum, 35, 47, 70, 85, 144,
      187, 257, 271, 286, 325, 354
Animals
  bipeds, 3, 16, 64
  multilegged, 16, 21, 33
  quadrupeds, 16, 33, 527
Ankle exoskeleton, 5, 357, 381
Ankle foot orthoses
  passive, 633
  powered, 634
Ankle prosthesis, 595, 607
Articulated torsos, 72, 552
Artificial muscles, 225, 522, 563, 599,
      617
  actuators, 615

McKibben, 652
pleated pneumatic (PPAM), 493, 600,
      619
pneumatic, 199, 523, 617, 652
Asymmetric spring-loaded inverted
      pendulum (ASLIP), 321
Axial leg function, 109, 582
AXO-SUIT, 647

B

Balance, 3, 14, 30, 46, 281, 354, 359, 424,
      437, 523, 583, 617, 633, 646
  control, 122, 522, 584
Biarticular muscles, 41, 197, 507,
      515–525, 582, 613
Biceps femoris short head (BFsH), 445
Bioinspired legged locomotion, 515, 624,
      640
Bioinspired robot, 507, 527, 556
  BioBiped, 40, 521
Biological muscles, 2, 243, 489, 563
Biomechanics, 2, 140, 158, 165, 198, 323,
      437, 483, 507, 515, 524, 648
Biped locomotion, 135, 333
Biped robots, 136, 332, 339
Bipedal locomotion, 13, 23, 120, 139,
      165, 189, 198, 219, 440, 461,
      465, 648
Bipedal M-SLIP model, 118
Bipedal robots, 66, 165, 167, 281, 292,
      305, 316, 459, 493, 506, 649
  AMBER 2, 294, 318
  and hexapedal, 527

Printed in the United States
By Bookmasters